Student's Solution
and Survival Manual
for
CALCULUS

Sixth Edition
by
Smith, Strauss, and Toda

KARL J. SMITH

Kendall Hunt
publishing company

Cover image © 2014 Shutterstock, Inc.

www.kendallhunt.com
Send all inquiries to:
4050 Westmark Drive
Dubuque, IA 52004-1840

Contents

CHAPTER 1 Functions and Graphs

Chapter Overview This chapter contains the background information you will need to successfully complete this course. Some instructors will cover this material as new material, but many instructors will only treat this material lightly and may even begin the course with Chapter 2. However your instructor teats the material in this chapter, it would be wise for you to make sure that your are familiar with the material of this chapter. Most of the material in this chapter was covered in a *Precalculus* course, or in *College Algebra* and *Trigonometry* courses.

1.1 What is Calculus?, page 11

SURVIVAL HINT: *Your instructor may not "cover" this section, but it would be a good idea to begin this book by reading this section. You have enrolled in a calculus course, probably because someone told you to take it, but can you tell someone what the course is about? This section will answer that question.*

1. Limits, derivatives, and integrals

SURVIVAL HINT: *There are many problems throughout the book that are called "What does this say? The intent is that you write out an answer in your own words. This means, that the answers we show in this manual for this type of problem does not constitute a correct answer. Be sure to write your answer using complete sentences, and when possible, support what you say with examples or by citing sources.*

3. Answers vary; the models are generally continually refined.

5. Answers vary; theoretically, she never reaches the wall, but in reality she must hit the wall because the woman has physical dimensions.

7. Answers vary.

SURVIVAL HINT: *When you see "Answers vary" in this manual, it generally means that you should write a paragraph or two to discuss the given topic. In this problem, discuss the relationship between global warming and modeling, but do this using your general knowledge, since the directions include "do not do research."*

9. $\frac{1}{3}$ 11. π

13.   15.

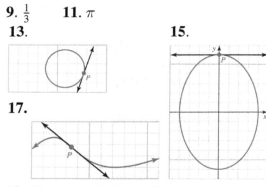

17.

19. Given $s = 0.4w + 10$.
    If $w = 15$, then

$$s = 0.4(15) + 10 = 16$$

The length is 16 inches.

21. Given $s = 0.4w + 10$.
   If $s = 35$, then
$$35 = 0.4w + 10$$
$$25 = 0.4w$$
$$w = 62.5$$
   The weight is 62.5 pounds.

23. Given $s = 0.25w + 5$.
   If $s = 10$, then
$$10 = 0.25w + 5$$
$$5 = 0.25w$$
$$w = 20$$
   The weight is 20 pounds.

25. Given $C = 0.6n + 4$.
   If $n = 15$, then
$$C = 0.6(15) + 4$$
$$= 13$$
   The temperature is 13°C.

27. Given $C = 0.6n + 4$.
   If $C = 40$, then
$$40 = 0.6n + 4$$
$$36 = 0.6n$$
$$n = 60$$
   In one minute you would hear
   $4(60) = 240$ chirps.

29. Given $C = 8.50x + 3,600$
   For $x = 10$:
$$C = 8.50(10) + 3,600$$
$$= 3,685$$
   The cost is $3,685.

31. Given $C = 45x + 405,000$
   For $x = 1$:
$$C = 45(1) + 405,000$$
$$= 405,045$$
   Average for one item is the same,
   namely, $405,045.

33. The revenue for 1 item is $25.

Answers to Problems 35-40 *may vary.*

35. $\lim\limits_{n\to\infty} \dfrac{2n}{n+4}$
   Consider larger and larger values of n:

n	1	10	100	1,000
$\dfrac{2n}{n+4}$	$\dfrac{2(1)}{1+4} = \dfrac{2}{5}$	$\dfrac{2(10)}{10+4} = \dfrac{20}{14}$	$\dfrac{2(100)}{100+4} = \dfrac{200}{104}$	$\dfrac{2(1,000)}{1,000+4} = \dfrac{2,000}{1,004}$

   It looks like the limit is 2.

37. $\lim\limits_{n\to\infty} \dfrac{n+1}{n+2}$
   Consider larger and larger values of n:

n	1	10	100	1,000
$\dfrac{n+1}{n+2}$	$\dfrac{1+1}{1+2} = \dfrac{2}{3}$	$\dfrac{10+1}{10+2} = \dfrac{11}{12}$	$\dfrac{100+1}{100+2} = \dfrac{101}{102}$	$\dfrac{1,000+1}{1,000+2} = \dfrac{1,001}{1,002}$

   It looks like the limit is 1.

39. $\lim\limits_{n\to\infty} \dfrac{3n}{n^2+2}$
   Consider larger and larger values of n:

n	1	10	100
$\dfrac{3n}{n^2+2}$	$\dfrac{3(1)}{1^2+2} = 1$	$\dfrac{3(10)}{10^2+2} \approx 0.29$	$\dfrac{3(100)}{100^2+2} \approx 0.029$

   It looks like the limit is 0.

41. Look at the underlying grid. It looks like the shaded area is about 6 "squares." (Actually they are rectangles with area $1 \times \frac{1}{2} = \frac{1}{2}$). Thus, we estimate that the area of is about 3 square units.

43. Look at the underlying grid. It looks like the shared area is about 13 "squares." (Actually they are rectangles with area $1 \times \frac{1}{2} = \frac{1}{2}$.) Thus, we estimate that the area of is about 6 to 7 square units.

45. Look at the underlying grid. It looks like the shaded area is about 24 "squares." Since four squares account for one unit, we estimate the area to be about 6 square units.

47. a. The width of each rectangle is $\frac{1}{8}$.

$A \approx A_1 + A_2 + \cdots + A_7 + A_8$

$= \frac{1}{8}\left(\frac{1}{8}\right)^2 + \frac{1}{8}\left(\frac{2}{8}\right)^2 + \cdots + \frac{1}{8}\left(\frac{7}{8}\right)^2 + \frac{1}{8}\left(\frac{8}{8}\right)^2$

$= \frac{1}{8}\left(\frac{1 + 4 + \cdots + 49 + 64}{64}\right)$

$= \frac{204}{512} \approx 0.3984375$

The sum is about 0.40.

b. The width of each rectangle is $\frac{1}{16}$.

$A \approx A_1 + A_2 + \cdots + A_{15} + A_{16}$

$= \frac{1}{16}\left(\frac{1}{16}\right)^2 + \frac{1}{16}\left(\frac{2}{16}\right)^2 + \cdots + \frac{1}{16}\left(\frac{15}{16}\right)^2 + \frac{1}{16}\left(\frac{16}{16}\right)^2$

$= \frac{1}{16}\left(\frac{1 + 4 + \cdots + 15^2 + 16^2}{16^2}\right)$

$= \frac{1,496}{16^3} \approx 0.365234375$

The sum is about 0.37.

c. We guess the area is $\frac{1}{3}$.

49. Given $F = 1.8C + 32$

For $F = 70$:

$70 = 1.8C + 32$

$38 = 1.8C$

$C \approx 21$

The temperature is $21°C$.

51. Given $IQ = \frac{M}{C}(100)$

a. For $C = 15$, $M = 18$,

$$IQ = \frac{18}{15}(100)$$

$$= 120$$

b. For $C = 9$, $IQ = 132$

$$132 = \frac{M}{9}(100)$$

$$M = 11.88$$

This is 11.88 years, or approximately 11 years, 11 months.

53. Given $P = \dfrac{F}{A}$

For $A = 2(0.001) = 0.002$, $F = 4$

$$P = \frac{4}{0.002} = 2,000$$

The applied force is 2,000 lb/in.^{2}.

55. Answers vary.

57.-59. These are modeling experiments, and as such, have no specific answers. In doing each of these problems, the students should demonstrate a true modeling process: propose a mathematical model, derive results, test those results, and then go back and revise the model.

1.2 Preliminaries, page 25

1. a. $[-3, 4]$ **b.** $3 \leq x \leq 5$

   c. $-2 \leq x < 1$ **d.** $(2, 7]$

3. a.
```
←+++++●++++○++→
-6 -4 -2  0  2  4  6
```

b.
```
←++++●++++○+++→
-6 -4 -2  0  2  4  6
```

c.
```
←+++++○■○++++→
-6 -4 -2  0  2  4  6
```

d.
```
←++++●■○++++→
-6 -4 -2  0  2  4  6
```

5. a.

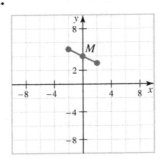

$M = \left(\frac{2-2}{2}, \frac{5+3}{2}\right) = (0, 4)$

$d = \sqrt{(-2-2)^2 + (5-3)^2}$

$\quad = \sqrt{20}$

$\quad = 2\sqrt{5}$

b.

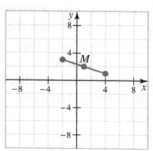

$M = \left(\frac{-2+4}{2}, \frac{3+1}{2}\right) = (1, 2)$

$d = \sqrt{(4+2)^2 + (1-3)^2}$

$\quad = \sqrt{40}$

$\quad = 2\sqrt{10}$

7. $\quad x^2 - x = 0$

$\quad\quad x(x-1) = 0$

$\quad\quad\quad\quad x = 0, 1$

9. $\quad 3x^2 - bx = c$

$\quad 3x^2 - bx - c = 0$

$\quad\quad x = \dfrac{b \pm \sqrt{b^2 - 4(3)(-c)}}{2(3)}$

$\quad\quad\quad = \dfrac{b \pm \sqrt{b^2 + 12c}}{6}$

11. $|3 - 2w| = 7$

$\quad 3 - 2w = 7 \quad$ or $-(3 - 2w) = 7$

$\quad\quad -2w = 4 \quad\quad\quad\quad\quad 2w = 10$

$\quad\quad\quad\quad w = -2 \quad\quad\quad\quad\quad\quad w = 5$

13. No values of x satisfy this equation because an absolute value cannot be negative. We state this simply by using an empty set symbol, $\varnothing$.

15. $\sin x = -\frac{1}{2}$ on $[0, 2\pi)$

This means that the reference angle (in Quadrant I) is $\frac{\pi}{6}$. The sine is negative in Quadrant III and Quadrant IV, so $x = \frac{7\pi}{6}, \frac{11\pi}{6}$.

17. $\left(2\cos x + \sqrt{2}\right)(2\cos x - 1) = 0$ on $[0, 2\pi)$

$2\cos x + \sqrt{2} = 0 \quad$ or $\quad 2\cos x - 1 = 0$

$\cos x = -\dfrac{\sqrt{2}}{2} \quad\quad\quad\quad \cos x = \dfrac{1}{2}$

$x = \dfrac{3\pi}{4}, \dfrac{5\pi}{4} \quad\quad\quad\quad x = \dfrac{\pi}{3}, \dfrac{5\pi}{3}$

19. $3x + 7 < 2$

$\quad 3x < -5$

$\quad\quad x < -\dfrac{5}{3} \quad \left(-\infty, -\dfrac{5}{3}\right)$

21. $-5 < 3x < 0$

$\quad -\dfrac{5}{3} < x < 0 \quad \left(-\dfrac{5}{3}, 0\right)$

23. $\quad 3 \le -y < 8$

$\quad -3 \ge y > -8$

$\quad -8 < y \le -3 \quad (-8, -3]$

25. $\quad t^2 - 2t \le 3$

$\quad t^2 - 2t - 3 \le 0$

$\quad (t+1)(t-3) \le 0$

Plot (closed) critical values of $t = -1$ and $t = 3$, and test each interval to find $[-1, 3]$.

27. Read this problem as a distance function: The distance between x and 8 is less than or equal to 0.001. The interval is $[7.999, 8.001]$.

29. $(x+1)^2 + (y-2)^2 = 9$

31. $x^2 + (y-1.5)^2 = 0.0625$

33.
$$x^2 - 2x + y^2 + 2y + 1 = 0$$
$$(x^2 - 2x + 1) + (y^2 + 2y + 1) = -1 + 1 + 1$$
$$(x - 1)^2 + (y + 1)^2 = 1$$
Circle with center at $(1, -1)$ and $r = 1$.

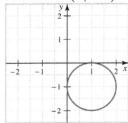

35.
$$x^2 + y^2 + 2x - 10y + 25 = 0$$
$$(x^2 + 2x + 1) + (y^2 - 10y + 25) = 1$$
$$(x + 1)^2 + (y - 5)^2 = 1$$

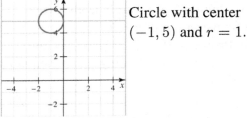

Circle with center $(-1, 5)$ and $r = 1$.

37. $\sin\left(-\dfrac{\pi}{12}\right) = \sin\left(\dfrac{\pi}{4} - \dfrac{\pi}{3}\right)$
$$= \sin\frac{\pi}{4}\cos\frac{\pi}{3} - \cos\frac{\pi}{4}\sin\frac{\pi}{3}$$
$$= \frac{\sqrt{2} - \sqrt{6}}{4} \approx -0.2588$$

39. $\tan\left(\dfrac{\pi}{12}\right) = \tan\left(\dfrac{\pi}{4} - \dfrac{\pi}{6}\right)$
$$= \frac{\tan\frac{\pi}{4} - \tan\frac{\pi}{6}}{1 + \tan\frac{\pi}{4}\tan\frac{\pi}{6}}$$
$$= \frac{1 - \frac{\sqrt{3}}{3}}{1 + \frac{\sqrt{3}}{3}}$$
$$= 2 - \sqrt{3} \approx 0.2679$$

41. If $ax^2 + bx + c = 0$, $a \neq 0$, then
$$x = \frac{-b \pm \sqrt{b^2 - 4ac}}{2a}.$$ This is called the quadratic formula.

43. If $|x| \leq a$, then $-a \leq x \leq a$.

45. a. **b.**

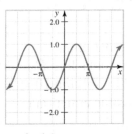 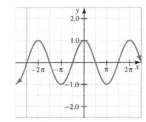

period 2π period 2π

c.

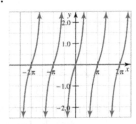

period π

47. a. **b.**

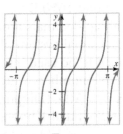 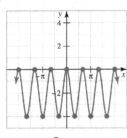

period $\dfrac{\pi}{2}$ period $\dfrac{2\pi}{3}$

49. Let $y - A = B\cos C(x - D)$

31. $u(x) = 2x^2 - 1;\ g(u) = u^4$
33. $u(x) = 5x - 1;\ g(u) = \sqrt{u}$
35. $u(x) = x^2;\ g(u) = \tan u$
37. $u(x) = \sin x;\ g(u) = \sqrt{u}$
39. $P(5, f(5));\ Q(x_0, f(x_0))$

SURVIVAL HINT: *The relationship between coordinates and points on a graph is an essential idea in calculus. Make sure you understand this problem. If you are not sure, you might also want to work Problem 40.*

41. $3x^2 - 5x - 2 = 0$
$(3x + 1)(x - 2) = 0$
$$x = -\frac{1}{3}, 2$$

43. $(x^2 - 10)(x^2 - 12)(x^2 - 20) = 0$
$x = \pm\sqrt{10},\ \pm 2\sqrt{3},\ \pm 2\sqrt{5}$

45. $x^4 - 41x^2 + 400 = 0$
$(x^2 - 16)(x^2 - 25) = 0$
$$x = \pm 4, \pm 5$$

47. $\dfrac{x(x^2 - 3)}{x^2 + 5} = 0$
$x(x^2 - 3) = 0$
$$x = 0,\ \pm\sqrt{3}$$

49. a. $f(2) = -(2)^3 + 6(2) + 15(2)^2$
$= 64$
　b. $f(2) - f(1) = 64 - 20$
$= 44$

51. $S(r) = C(R^2 - r^2)$
$= 1.76 \times 10^5 (1.2^2 \times 10^{-4} - r^2)$
a. $S(0) \approx 25.344$ cm^3/s
b. $S(0.6 \times 10^{-2}) \approx 19.008$ cm^3/s

SURVIVAL HINT: *Spend some time looking over the directory of curves (Table 1.4). These are the curves which will make up the majority of the examples and problems in this text. It will be assumed that you will recognize and be able to graph each of these functions.*

53.

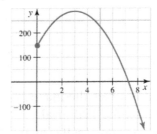

a. $s(0) = -16(0)^2 + 96(0) + 144 = 144$
The height of the cliff is 144 ft.
b. $s(t) = 0$ if
$-16t^2 + 96t + 144 = 0$
$t^2 - 6t - 9 = 0$
$$t = \frac{6 \pm \sqrt{36 - 4(-9)}}{2}$$
$$= 3 \pm 3\sqrt{2}$$
The ball hits the ground when
$t = 3 + 3\sqrt{2} \approx 7.24$ (reject the negative t).
c. The maximum height occurs at the vertex of the parabola
$s = -16t^2 + 96t + 144.$

$$s - 144 = -16(t^2 - 6t)$$
$$s - 144 + (-16)(9) = -16(t^2 - 6t + 9)$$
$$s - 288 = -16(t - 3)^2$$

It takes 3 seconds for the ball to reach its highest point. The maximum height is $s(3) = 288$ ft.

55. a. $D = (-\infty, 0) \cup (0, \infty)$

b. Since n represents the number of trials, n is a positive integer.

c. For the third trial, $n = 3$ so
$$f(3) = 3 + \tfrac{12}{3} = 7 \text{ minutes}$$

d. $f(n) \leq 4$, so
$$3 + \frac{12}{n} \leq 4$$
$$\frac{12}{n} \leq 1$$
$$12 \leq n \qquad \text{\textit{Since} } n > 0$$
The rat first transverses the maze in 4 minutes or less on the 12th trial.

e. $12/n$ gets smaller and smaller as n increases. Thus $12/n \to 0$ as $n \to \infty$ and $f(n)$ gets closer and closer to 3. Note: this mathematical result is not practical since the rat cannot reach this 3-minute barrier no matter how many trials are taken.

57.

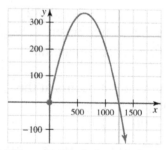

You can find (by trial) the maximum distance is for $\alpha = 45°$. We can also find this explicitly as follows. The distance traveled by the cannonball is $\overline{x}$, where

$$m\overline{x} - \frac{16}{v^2}(1 + m^2)\overline{x}^2 = 0$$

$$\overline{x} = \frac{mv^2}{16(1 + m^2)}$$

Graphing $\overline{x}$ as a function of m, you can see that it is maximized when $m = 1$; that is, when $\tan \alpha = 1$ or when $\alpha = 45°$.

59. Let $x = 11$; then
$$f(11 + 3) = \frac{f(11) - 1}{f(11) + 1} = \frac{10}{12} = \frac{5}{6}$$
$$f(11 + 6) = \frac{f(14) - 1}{f(14) + 1} = \frac{\frac{5}{6} - 1}{\frac{5}{6} + 1} = -\frac{1}{11}$$
$$f(11 + 9) = \frac{f(17) - 1}{f(17) + 1} = \frac{-\frac{1}{11} - 1}{-\frac{1}{11} + 1} = -\frac{6}{5}$$
$$f(11 + 12) = \frac{f(20) - 1}{f(20) + 1} = \frac{-\frac{6}{5} - 1}{-\frac{6}{5} + 1} = 11$$
The pattern repeats in groups of four for any number $N = 11 + 3k$; we have $f(N)$ is

$$\begin{array}{ll} 11 & \text{if } k = 4m \\[4pt] \dfrac{5}{6} & \text{if } k = 4m + 1 \\[8pt] -\dfrac{1}{11} & \text{if } k = 4m + 2 \\[8pt] -\dfrac{6}{5} & \text{if } k = 4m + 3 \end{array}$$

Since $2018 = 11 + 3(669)$, we have $k = 669 = 167(4) + 1$ so that
$$f(2018) = \frac{5}{6}$$

Page 14

1.5 Inverse Functions; Inverse Trigonometric Functions, page 59

> SURVIVAL HINT: *The material of this section may be new to you. Even if your instructor skipped Chapter 1, you might want to take some time to read this section.*

1. Trigonometric functions are not one-to-one. Restrictions make them one-to-one so that the inverses can be defined.

3. $f[g(x)] = 5\left(\dfrac{x-3}{5}\right) + 3 = x$

   $g[f(x)] = \dfrac{(5x+3)-3}{5} = x$

   These are inverse functions.

5. $f[g(x)] = \frac{4}{5}\left(\frac{5}{4}x+3\right) + 4 \neq x$

   These are not inverse functions.

7. $f[g(x)]$ does not exist.

   $g[f(x)] = \sqrt{x^2} = |x| = -x$ *Since $x < 0$*

   These are not inverse functions.

9. $f[g(x)] = \left(-\sqrt{x}\right)^2 = x$

   $g[f(x)] = -\sqrt{x^2} = -|x| = -x$ *Since $x > 0$*

   These are not inverse functions.

11. To find the inverse interchange the domain and range values:

    $$\{(5,4), (3,6), (1,7), (4,2)\}$$

13. Given $y = 2x + 3$; inverse is $x = 2y + 3$

    or: $y = \frac{1}{2}x - \frac{3}{2}$.

15. Given $y = x^2 - 5$, $x \geq 0$; inverse is $x = y^2 - 5$, $y \geq 0$ or $y = \sqrt{x+5}$ (positive value since $y \geq 0$).

17. Given $y = \sqrt{x} + 5$; inverse is $x = \sqrt{y} + 5$

    or: $y = (x-5)^2$.

19. Given $y = \dfrac{2x-6}{3x+3}$; inverse is $x = \dfrac{2y-6}{3y+3}$

    or: $3xy + 3x = 2y - 6$

    $\qquad (3x-2)y = -3x - 6$

    $$y = \frac{3x+6}{2-3x}$$

> SURVIVAL HINT: *Some students know the exact value for the trigonometric functions, and some instructors require that students know the exact values (see Table 1.3) of the trigonometric functions. You can use this knowledge to answer Problem 21-34. You must also pay attention to the quadrant in which the angle is located (from the definition of the inverse trigonometric functions in Table 1.5.*

21. a. $\frac{\pi}{3}$ b. $\frac{2\pi}{3}$ 23. a. $-\frac{\pi}{4}$ b. $\frac{5\pi}{6}$

25. a. $\frac{3\pi}{4}$ b. $-\frac{\pi}{4}$ 27. a. $\frac{\pi}{3}$ b. $-\frac{\pi}{3}$

29. $\frac{\sqrt{3}}{2}$ 31. 3

33. Let $\alpha = \sin^{-1}\frac{1}{5}$ and $\beta = \cos^{-1}\frac{1}{5}$; then $\sin\alpha = \frac{1}{5}$, $\cos\beta = \frac{1}{5}$ and using reference triangles we find $\cos\alpha = \sin\beta = \frac{2\sqrt{6}}{5}$.

   $\cos\left(\sin^{-1}\frac{1}{5} + 2\cos^{-1}\frac{1}{5}\right) = \cos(\alpha + 2\beta)$

   $= \cos\alpha\cos 2\beta - \sin\alpha\sin 2\beta$

   $= \cos\alpha\left(\cos^2\beta - \sin^2\beta\right) - \sin\alpha(2\cos\beta\sin\beta)$

   $= \dfrac{2\sqrt{6}}{5}\left[\dfrac{1}{25} - \dfrac{24}{25}\right] - \dfrac{1}{5}\left[2\cdot\dfrac{1}{5}\cdot\dfrac{2\sqrt{6}}{5}\right]$

   $= \dfrac{2\sqrt{6}}{5}\left[-\dfrac{23}{25} - \dfrac{2}{25}\right] = -\dfrac{2\sqrt{6}}{5}$

35. If $\sin\alpha + \cos\alpha = s$ and $\sin\alpha - \cos\alpha = t$, then (by addition) $2\sin\alpha = s + t$ so that $\sin\alpha = \dfrac{s+t}{2}$. By subtracting, $2\cos\alpha = s - t$ so that $\cos\alpha = \dfrac{s-t}{2}$.

   Thus,

$$\tan \alpha = \frac{\sin \alpha}{\cos \alpha} = \frac{s+t}{s-t}$$

$$\alpha = \tan^{-1}\left(\frac{s+t}{s-t}\right)$$

37.

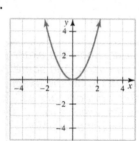

$f(x) = x^2$ for all x does not have an inverse, since the function is not one-to-one.

39.

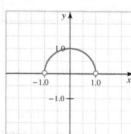

$f(x) = \sqrt{1-x^2}$ does not have an inverse because it is not one-to-one.

41.

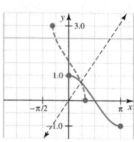

The inverse exists, since it passes the horizontal line test. $f^{-1}(x) = \cos^{-1}x$

43. If $\sin^{-1}x = \theta$, then $\sin \theta = x$ and $\cos \theta = \sqrt{1-x^2}$. Thus,

$$\sin(2\sin^{-1}x) = \sin 2\theta = 2\sin \theta \cos \theta$$
$$= 2x\sqrt{1-x^2}$$

45. If $\cos^{-1}x = \theta$, then $\cos \theta = x$, so (by a reference triangle) $\tan \theta = \dfrac{\sqrt{1-x^2}}{x}$.

47. In any right triangle, the sum of the acute angles is $\frac{\pi}{2}$, so for $|x| < 1$, we know $\sin^{-1}x + \cos^{-1}x = \frac{\pi}{2}$, so

$$\sin(\sin^{-1}x + \cos^{-1}x) = \sin \frac{\pi}{2} = 1$$

49. Consider a reference triangle as shown.

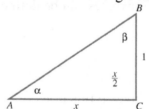

$\cot \alpha = \frac{x}{1}$ so $\cot^{-1}x = \alpha$, $\tan \beta = \frac{x}{1}$ so $\tan^{-1}x = \beta$. Also, since the triangle is a right triangle,

$$\alpha + \beta = \frac{\pi}{2}$$
$$\cot^{-1}x + \tan^{-1}x = \frac{\pi}{2}$$
$$\cot^{-1}x = \frac{\pi}{2} - \tan^{-1}x$$

51. a. 0.9805   **b.** 0.7284   **c.** 0.3964
      d. 2.5657

53. Let the angles θ and α be drawn as shown in the figure.

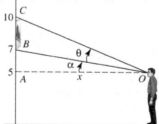

$\tan\alpha = \frac{2}{x}$, so $\alpha = \tan^{-1}\frac{2}{x}$ and
$\tan(\theta + \alpha) = \frac{5}{x}$, so $\theta + \alpha = \tan^{-1}\frac{5}{x}$
Thus, $\theta = \tan^{-1}\frac{5}{x} - \tan^{-1}\frac{2}{x}$.

55. a. $\tan^{-1}1 + \tan^{-1}2 + \tan^{-1}3 \approx 3.141592$
The conjecture is that it is π.
To prove this conjecture, let
$\alpha = \tan^{-1}1$, $\beta = \tan^{-1}2$, and
$\gamma = \tan^{-1}3$; also let

$$\tan^{-1}1 + \tan^{-1}2 + \tan^{-1}3 = A$$

For α, we use a reference triangle with
sides 1, 1, and $\sqrt{2}$ to find
$\sin\alpha = \cos\alpha = \frac{1}{\sqrt{2}}$;
For β, we use a reference triangle with
sides 1, 2, and $\sqrt{5}$ to find
$\sin\beta = \frac{2}{\sqrt{5}}$, $\cos\beta = \frac{1}{\sqrt{5}}$;
For γ, we use a reference triangle with
sides 1, 3, and $\sqrt{10}$ to find
$\sin\gamma = \frac{3}{\sqrt{10}}$, $\cos\gamma = \frac{1}{\sqrt{10}}$;
Then
$$\begin{aligned}
\sin A &= \sin(\alpha + \beta + \gamma)\\
&= \sin[(\alpha + \beta) + \gamma]\\
&= \sin(\alpha + \beta)\cos\gamma + \cos(\alpha + \beta)\sin\gamma\\
&= [\sin\alpha\cos\beta + \sin\beta\cos\alpha]\cos\gamma\\
&\quad + [\cos\alpha\cos\beta - \sin\alpha\sin\beta]\sin\gamma\\
&= \left[\frac{1}{\sqrt{2}}\frac{1}{\sqrt{5}} + \frac{2}{\sqrt{5}}\frac{1}{\sqrt{2}}\right]\frac{1}{\sqrt{10}}\\
&\quad + \left[\frac{1}{\sqrt{2}}\frac{1}{\sqrt{5}} - \frac{1}{\sqrt{2}}\frac{2}{\sqrt{5}}\right]\frac{3}{\sqrt{10}}\\
&= \left[\frac{3}{\sqrt{10}}\right]\frac{1}{\sqrt{10}} + \left[\frac{-1}{\sqrt{10}}\right]\frac{3}{\sqrt{10}}\\
&= 0
\end{aligned}$$

Thus, $A = 0, \pi$, or 2π and only $A = \pi$
is possible since $0 < \alpha + \beta + \gamma < 2\pi$.

b. Let $\alpha = \tan^{-1}1$, $\beta = \tan^{-1}2$, so that

$$\begin{aligned}
\tan(\tan^{-1}1 + \tan^{-1}2) &= \tan(\alpha + \beta)\\
&= \frac{\tan\alpha + \tan\beta}{1 - \tan\alpha\tan\beta}\\
&= \frac{1 + 2}{1 - 1(2)}\\
&= -3
\end{aligned}$$

$$\begin{aligned}
&\tan[(\tan^{-1}1 + \tan^{-1}2) + \tan^{-1}3]\\
&= \frac{\tan(\tan^{-1}1 + \tan^{-1}2) + \tan(\tan^{-1}3)}{1 - [\tan(\tan^{-1}1 + \tan^{-1}2)\tan(\tan^{-1}3)]}\\
&= \frac{-3 + 3}{1 - (-3)(3)}\\
&= \frac{0}{10}\\
&= 0
\end{aligned}$$

We see that

$$\tan^{-1}1 + \tan^{-1}2 + \tan^{-1}3 = 0, \pi, 2\pi$$

Cannot be 0, too small to be 2π, so it
must be π.

57. Since cosecant is positive in Quadrant I
and negative in Quadrant IV, we
conjecture that $A = 0$ and $B = -1$. We
need to prove $\csc^{-1}(-x) = -\csc^{-1}x$ for
$x > 1$. Let $\theta = \csc^{-1}(-x)$, $\csc\theta = -x$ or
$x = -\csc\theta = \csc(-\theta)$. Thus, for $x > 1$

$$\csc^{-1}(-x) = -\csc^{-1}x$$

In particular, $\csc^{-1}(-9.38) \approx -0.1068$
and $\csc^{-1}9.38 \approx 0.1068$, so that

$$\csc^{-1}(-9.38) = -\csc^{-1}9.38 = -0.1068$$

59. Suppose $g_1[f(x)] = x = f[g_1(x)]$ and
$g_2[f(x))] = x = f[g_2(x)]$. Thus,
$x = f[g_1(x)] = f[g_2(x)]$ and

$f^{-1}\{f[g_1(x)]\} = f^{-1}\{f[g_2(x)]\}$, which
implies $g_1(x) = g_2(x)$; that is, the inverse
is unique.

Chapter 1 Review

Studying for a chapter examination is a personal process, one which nobody else can do for
you. Simply take the time to review what you have done.

SURVIVAL HINT: Work all of Chapter 1 problems in the Proficiency Examination (whether they are
assigned or not). Work through all of the problems before looking at the answers, and *then* correct each
of the problems. The answers to all these problems are given in the answer section at the back of the
text. If you worked the problem correctly, move on to the next problem, but if you did not work it correctly
(or you did not know what to do), then look at the solutions below, look back in the chapter to study the
procedure, or ask your instructor.
   Finally, go back over the homework problems you have been assigned. If you worked a problem
correctly, move on to the next problem, but if you missed it on your homework, then you should look back
in the book or talk to your instructor about how to work the problem.
   If you follow these steps, you should be successful with your review of this chapter.

Proficiency Examination, page 62

1. $\mathbb{N} = \{1, 2, 3, \cdots\}$;
$\mathbb{W} = \{0, 1, 2, 3, \cdots\}$;
$\mathbb{Z} = \{\cdots, -3, -2, -1, 0, 1, 2, 3, \cdots\}$;
$\mathbb{Q} = \{\frac{p}{q}$ so that p is an integer and q is
    a nonzero integer$\}$;
$\mathbb{Q}' = \{$nonrepeating or nonterminating
    decimals$\}$;
$\mathbb{R} = \mathbb{Q} \cup \mathbb{Q}'$

2. $|a| = a$ if $a \geq 0$; $|a| = -a$ if $a < 0$

3. $|x + y| \leq |x| + |y|$

4. $d = \sqrt{(x_2 - x_1)^2 + (y_2 - y_1)^2}$

5. $m = \tan \theta$ where θ is the angle of inclination.

6. **a.** $Ax + By + c = 0$
   b. $y = mx + b$
   c. $y - k = m(x - h)$
   d. $y = k$ **e.** $x = h$

7. Lines are parallel if they have the same
slope, and perpendicular if their slopes
are negative reciprocals of one another.

8. A function is a rule that assigns to each
element x of the domain D a unique
element of the range R.

9. $(f \circ g) = f[g(x)]$

10. The graph of a function f consists of
all points (x, y) such that $y = f(x)$ for
x in the domain of f.

11. a.   **b.**

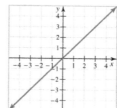   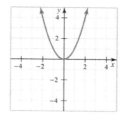

c.

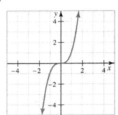

d.

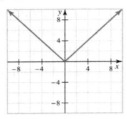

e.

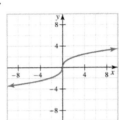

f.

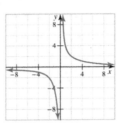

g.

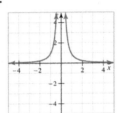

h.

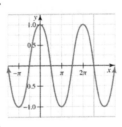

i.

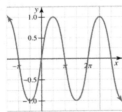

j.

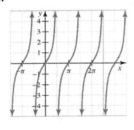

k.

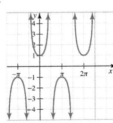

l.

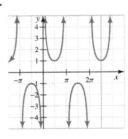

m.

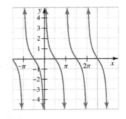

n.

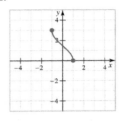

o.

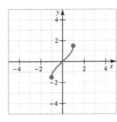

p.

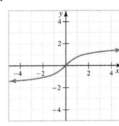

q.

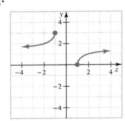

r.

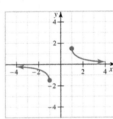

s.

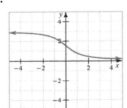

12. $f(x) = a_n x^n + a_{n-1} x^{n-1} + \cdots + a_2 x^2 + a_1 x + a_0$

13. $f(x) = \dfrac{P(x)}{D(x)}$, P and D are polynomials with $D(x) \neq 0$.

14. a. Let f be a function with domain D and range R. Then the function f^{-1} with domain R and range D is the inverse of f if $f^{-1}[f(x)] = x$ for all x in D and $f[f^{-1}(y)] = y$ for all y in R

b. Reflect the graph of $y = f(x)$ in the line $y = x$.

15. The horizontal line test says that a function f has an inverse f^{-1} if and only if no horizontal line meets the graph of f in more than one point (that is, f is a one-to-one function).

16. $\sin(\sin^{-1} x) = x \quad$ for $-1 \le x \le 1$
$\sin^{-1}(\sin y) = y \quad$ for $-\frac{\pi}{2} \le y \le \frac{\pi}{2}$
$\tan(\tan^{-1} x) = x \quad$ for all x
$\tan^{-1}(\tan y) = y \quad$ for $-\frac{\pi}{2} < y < \frac{\pi}{2}$

17. $\cot^{-1} x = \frac{\pi}{2} - \tan^{-1} x$
$\sec^{-1} x = \cos^{-1} \frac{1}{x}$ if $|x| \ge 1$
$\csc^{-1} x = \sin^{-1} \frac{1}{x}$ if $|x| \ge 1$

18. **a.**
$$y - 5 = -\frac{3}{4}\left[x - \left(-\frac{1}{2}\right)\right]$$
$$6x + 8y - 37 = 0$$

b. $m = \dfrac{2 - 5}{7 + 3} = -\dfrac{3}{10}$
$$y - 5 = -\frac{3}{10}[x - (-3)]$$
$$3x + 10y - 41 = 0$$

c. $\dfrac{x}{4} + \dfrac{y}{-\frac{3}{7}} = 1$
$$3x - 28y - 12 = 0$$

d. Writing the given equation in slope-intercept form, $y = -\frac{2}{5}x + \frac{11}{5}$, we see that the slope is $-\frac{2}{5}$. A parallel line must have the same slope. Now use the point-slope form.
$$y - 5 = -\frac{2}{5}\left(x + \frac{1}{2}\right)$$
$$2x + 5y - 24 = 0$$

e. Find the slope of $\overline{PQ}$. A perpendicular line will have a slope which is the

negative reciprocal of the given slope. Find the midpoint of $\overline{PQ}$. Then use the point-slope form for the equation of the line. The slope of $\overline{PQ}$ is $-\frac{3}{4}$. The midpoint of $\overline{PQ}$ is $(1, 4)$.
$$y - 4 = \frac{4}{3}(x - 1)$$
$$4x - 3y + 8 = 0$$

19. $y = -\dfrac{3}{2}x + 6$

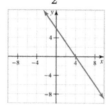

20. $y - 3 = |x + 1|$

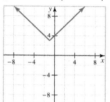

21. $y - 3 = -2(x - 1)^2$

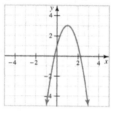

22. $y + 14 = (x - 2)^2$

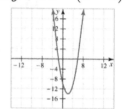

23. $y = 2\cos(x - 1)$

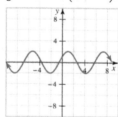

24. $y + 1 = \tan 2\left(x + \dfrac{3}{2}\right)$

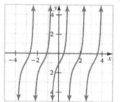

25. $y = \sin^{-1}(2x)$

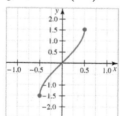

26. $y = \tan^{-1} x^2$

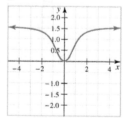

27. If $f(x) = \dfrac{1}{x+1}$, then

$$f\left(\frac{1}{x+1}\right) = \frac{1}{\frac{1}{x+1}+1}, \text{ and}$$

$$f\left(\frac{2x+1}{2x+4}\right) = \frac{1}{\frac{2x+1}{2x+4}+1}$$

We need to find the values of x for which:

$$\frac{1}{\frac{1}{x+1}+1} = \frac{1}{\frac{2x+1}{2x+4}+1}$$

$$\frac{x+1}{1+x+1} = \frac{2x+4}{2x+1+2x+4}$$

$$4x^2 + 9x + 5 = 2x^2 + 8x + 8$$

$$2x^2 + x - 3 = 0$$

$$(2x+3)(x-1) = 0$$

$$x = -\frac{3}{2}, 1$$

28. The root of a quotient equals the quotient of the roots when they have the same domain. The domain of f is $(-\infty, 0] \cup (1, \infty)$. The domain of g is $(1, \infty)$. The two functions are not the same.

29. $f \circ g = \sin\sqrt{1-x^2}$
$g \circ f = \sqrt{1 - \sin^2 x} = |\cos x|$

30. Let the base of the box be x and the height z. Then, $V = x^2 z$, so
sides: xz with a cost of $\$3xz$
bottom: x^2 with a cost of $\$8x^2$
Then

$$96 = 4(3xz) + 8x^2$$

$$24 = 3xz + 2x^2$$

$$z = \frac{24 - 2x^2}{3x}$$

$$V = x^2\left(\frac{24 - 2x^2}{3x}\right) = \frac{2}{3}x(12 - x^2)$$

Supplementary Problems, page 63

1. $m = \dfrac{4}{5}; b = -5; a = \dfrac{25}{4}$

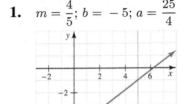

3. $y = -\dfrac{3}{5}x + 3; m = -\dfrac{3}{5}; b = 3; a = 5$

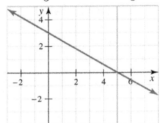

5. $6(1-x) < 2x + 1$
$6 - 6x < 2x + 1$
$-8x < -5$
$x > \dfrac{5}{8}$
$\left(\dfrac{5}{8}, \infty\right)$

7. $\qquad x^2 - x \geq 6$
$x^2 - x - 6 \geq 0$
$(x-3)(x+2) \geq 0$
Critical values: $-2, 3$
$(-\infty, -2] \cup [3, \infty)$

9. $d_1 = \sqrt{(-1+1)^2 + (8-3)^2} = \sqrt{25} = 5$
$d_2 = \sqrt{(11+1)^2 + (8-8)^2} = \sqrt{144} = 12$
$d_3 = \sqrt{(11+1)^2 + (8-3)^2} = \sqrt{169} = 13$

Page 21

$P = 5 + 12 + 13 = 30; A = \frac{1}{2}(5)(12) = 30$

11. $|\overline{AB}| = 8$

$|\overline{BC}| = \sqrt{(8-0)^2 + (5-2)^2} = \sqrt{73}$

$|\overline{CD}| = 2$

$|\overline{DA}| = \sqrt{(8-0)^2 + (0+3)^2} = \sqrt{73}$

$P = 8 + 2\sqrt{73} + 2$

$\quad = 2(5 + \sqrt{73})$

$\quad \approx 27.1$

$A = \frac{2+8}{2}(8) = 40$

13. $(x^2 + y^2 - 5x + 7y) - (x^2 + y^2 + 4y) = 3 - 0$

$\qquad\qquad\qquad\qquad -5x + 3y = 3$

The line is $5x - 3y + 3 = 0$.

15. $r = 4;\ (x-5)^2 + (y-4)^2 = 16$

17. $f(x) = x - \dfrac{3}{x} \qquad\qquad D: x \neq 0$

$f(-1) = -1 - \dfrac{3}{-1} = 2$

$f(1) = 1 - \dfrac{3}{1} = -2$

$f(3) = 3 - \dfrac{3}{3} = 2$

None are zeros of the function.

19. $f(x) = \cos x - \sin x \qquad\qquad D: \mathbb{R}$

$f(-\pi) = \cos(-\pi) - \sin(-\pi) = -1$

$f\left(\dfrac{\pi}{2}\right) = \cos\dfrac{\pi}{2} - \sin\dfrac{\pi}{2} = -1$

$f(\pi) = \cos\pi - \sin\pi = -1$

None are zeros of the function.

21. $\dfrac{f(x+h) - f(x)}{h} = \dfrac{5(x+h) - 5x}{h}$

$\qquad\qquad\qquad = \dfrac{5h}{h}$

$\qquad\qquad\qquad = 5$

23. $\dfrac{f(x+h) - f(x)}{h} = \dfrac{\frac{1}{2(x+h)} - \frac{1}{2x}}{h}$

$\qquad\qquad\qquad = \dfrac{1}{h}\dfrac{2x - 2x - 2h}{2x(x+h)}$

$\qquad\qquad\qquad = \dfrac{-1}{2x(x+h)}$

25. $(f \circ g)(x) = f(x^2)\quad (g \circ f)(x) = g(3x)$

$\qquad\qquad\quad = 3x^2 \qquad\qquad\quad = (3x)^2$

$\qquad\qquad\qquad\qquad\qquad\qquad = 9x^2$

27. $(f \circ g)(u) = f(u+1)$

$\qquad\qquad\quad = (u+1) - 1$

$\qquad\qquad\quad = u$

$\quad (g \circ f)(u) = g(u-1)$

$\qquad\qquad\quad = (u-1) + 1$

$\qquad\qquad\quad = u$

29. $6x^2 + 13x - 10 = 0$

$x = \dfrac{-13 \pm \sqrt{13^2 - 4(6)(-10)}}{2(6)}$

$\quad = \dfrac{-13 \pm \sqrt{409}}{12}$

31. $\qquad\qquad x^4 - 5x^2 + 4 = 0$

$\qquad\qquad (x^2 - 4)(x^2 - 1) = 0$

$(x-2)(x+2)(x-1)(x+1) = 0$

$\qquad\qquad\qquad\qquad x = -2, -1, 1, 2$

33. Since f is not one-to-one, the inverse does not exist.

35. The function f is one-to-one, so the inverse is

$x = y^2 + 1$

$y^2 = x - 1$

$|y| = \sqrt{x-1}$

$y = -\sqrt{x-1}$ for $x > 1$

37.

39.

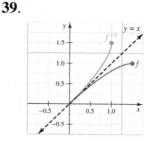

41. $f(x) = 2x^3 - 7$

　　inverse: $x = 2y^3 - 7$

　　　　$2y^3 = x + 7$

　　　　$y = \sqrt[3]{\dfrac{1}{2}(x+7)}$

43. $f(x) = \sqrt{\sin x},\ 0 < x < \dfrac{\pi}{2}$

　　inverse: $x = \sqrt{\sin y},\ 0 < y < \dfrac{\pi}{2}$

　　　　$x^2 = \sin y$

　　　　$y = \sin^{-1} x^2$

45. If $\cos^{-1} x = \theta$, then $\cos\theta = x$ and $\sin\theta = \sqrt{1-x^2}$. Thus,

$$\cos(2\cos^{-1} x) = \cos 2\theta = \cos^2\theta - \sin^2\theta$$
$$= x - (1 - x^2) = 2x^2 - 1$$

47. In any right triangle, the sum of the acute angles is $\frac{\pi}{2}$, so for $|x| < 1$, we know $\sin^{-1} x + \cos^{-1} x = \frac{\pi}{2}$, so

$$\cos(\sin^{-1} x + \cos^{-1} x) = \cos\tfrac{\pi}{2} = 0$$

49. $A = \dfrac{5\pi}{6}$

51.　$\sin 3x = \sin(2x + x)$

　　　　$= \sin 2x \cos x + \cos 2x \sin x$

　　　　$= (2\sin x \cos x)\cos x + (1 - 2\sin^2 x)\sin x$

　　　　$= 2\sin x \cos^2 x + \sin x - 2\sin^3 x$

　　　　$= 2\sin x(1 - \sin^2 x) + \sin x - 2\sin^3 x$

　　　　$= 2\sin x - 2\sin^3 x + \sin x - 2\sin^3 x$

　　　　$= 3\sin x - 4\sin^3 x$

　　$4\sin^3 x = -\sin 3x + 3\sin x$

　　$\sin^3 x = -\dfrac{1}{4}\sin 3x + \dfrac{3}{4}\sin x$

　　Thus, $A = -\frac{1}{4}$ and $B = \frac{3}{4}$.

53. a. function; not one-to-one;
　　　D: $(-\infty, \infty)$; R: $[0, \infty)$

　　b. not a function

　　c. function; one-to-one;
　　　D: $(-\infty, \infty)$; R: $(-\infty, \infty)$

　　d. not a function

　　e. function; not one-to-one;
　　　D: $(-\infty, -2) \cup (-2, 2) \cup (2, \infty)$;
　　　R: $(-\infty, 1) \cup (1, \infty)$

55. The last two equations in Problem 54 lead to

$$\frac{5}{3}(x - 5) + 6 = -\frac{1}{4}(x + 2) + 1$$
$$20x - 100 + 72 = -3x - 6 + 12$$
$$20x - 28 = -3x + 6$$
$$23x = 34$$
$$x = \frac{34}{23}$$

Substitute this value into one of these equations to obtain $y = \frac{3}{23}$. The point is $\left(\frac{34}{23}, \frac{3}{23}\right)$. Repeat these steps for the other line pairs of Problem 54. The medians meet at the point

$$\left(\frac{-2 + 5 + 3}{3}, \frac{1 + 6 - 2}{3}\right) = \left(2, \frac{5}{3}\right)$$

57. The current population is found when $t = 0$:

$$P(0) = \frac{11(0) + 12}{2(0) + 3} = 4$$

(thousand people). In 6 years,

$$P(6) = \frac{11(6) + 12}{2(6) + 3} = \frac{78}{15} = 5.2$$

(thousand). For $P = 5$ we have

$$5 = \frac{11t + 12}{2t + 3}$$
$$10t + 15 = 11t + 12$$
$$t = 3$$

The population will reach 5,000 in 3 years.

59. a. $\sin\left(\cos^{-1}\frac{\sqrt 5}{4}\right) = \sin\theta$ where
　　　$\theta = \cos^{-1}\frac{\sqrt 5}{4}$

$$\sin\theta = \sqrt{1 - \frac{5}{16}} = \frac{\sqrt{11}}{4}$$

b. $\sin(2\tan^{-1}3) = \sin 2\theta$ where
$\theta = \tan^{-1}3$

$$\sin 2\theta = 2\sin\theta\cos\theta$$

$$= 2\left(\frac{3}{\sqrt{10}}\right)\left(\frac{1}{\sqrt{10}}\right)$$

$$= \frac{3}{5}$$

c. $\sin(\cos^{-1}\frac{3}{5} + \sin^{-1}\frac{5}{13}) = \sin(\alpha + \beta)$
where $\alpha = \cos^{-1}\frac{3}{5}$ and $\beta = \sin^{-1}\frac{5}{13}$
$\sin(\alpha + \beta) = \sin\alpha\cos\beta + \cos\alpha\sin\beta$

$$= \sqrt{1 - \frac{9}{25}}\sqrt{1 - \frac{25}{169}} + \left(\frac{3}{5}\right)\left(\frac{5}{13}\right)$$

$$= \frac{63}{65}$$

61. $y = f(x) = \dfrac{x+a}{x-1}$;

The inverse is
$$x = \frac{y+a}{y-1}$$
$$x(y-1) = y+a$$
$$xy - x = y + a$$
$$y = \frac{x+a}{x-1}$$
The domain of f^{-1} is all real x, $x \neq 1$.

63. $y = f(x) = \dfrac{x+1}{x-1}$

The inverse is
$$x = \frac{y+1}{y-1}$$
$$xy - x = y + 1$$
$$y = \frac{x+1}{x-1}$$
The domain of f^{-1} is all real x, $x \neq 1$.

65. a. False; $\tan^{-1}1 = \frac{\pi}{4}$, but $\dfrac{\sin^{-1}1}{\cos^{-1}1}$ is not
defined.

b. False; $\tan^{-1}1 = \frac{\pi}{4}$;
$(\tan 1)^{-1} = \cot 1 \approx 0.642$

c. True; $\cot^{-1}x = \frac{\pi}{2} - \tan^{-1}x$ because
these are complementary angles.

d. True; let $\alpha = \sin^{-1}x$, then $\sin\alpha = x$,
$\cos^2\alpha = 1 - x^2$;
$$\cos(\sin^{-1}\alpha) = \sqrt{1 - x^2}$$

e. True; let $\alpha = \cos^{-1}x$, then $\cos\alpha = x$;
$$\frac{1}{\cos\alpha} = \frac{1}{x} = \sec\alpha \text{ or }$$
$$\alpha = \sec^{-1}\frac{1}{x} = \cos^{-1}x$$

67. a. $s = \dfrac{3d\cos\theta}{\sqrt{7 + 9\cos^2\theta}}$

$$= \frac{3(5)\cos 37°}{\sqrt{7 + 9\cos^2 37°}}$$

$$\approx 3.3562$$

The swimmer is about 3.4 m
underwater.

b.
$$s = \frac{3d\cos\theta}{\sqrt{7 + 9\cos^2\theta}}$$
$$2.5 = \frac{3(5)\cos\theta}{\sqrt{7 + 9\cos^2\theta}}$$
$$2.5\sqrt{7 + 9\cos^2\theta} = 15\cos\theta$$
$$\sqrt{7 + 9\cos^2\theta} = 6\cos\theta$$
$$7 + 9\cos^2\theta = 36\cos^2\theta$$
$$\cos^2\theta = \frac{7}{27}$$
$$\theta \approx 59.39°$$
The angle of incidence is about 60°.

69. a. The domain consists of all $x \neq 300$.
b. x represents a percentage, so
$0 \leq x \leq 100$ in order for $f(x) \geq 0$.
c. If $x = 50$, $f(50) = \dfrac{600(50)}{300 - 50} = 120$
d. With $x = 100$,

$$f(100) = \frac{600(100)}{300 - 100} = 300$$

e. With $f(x) = 150$, $\dfrac{600x}{300 - x} = 150$ or $x = 60$. The percentage of households should be 60%.

71. a. $r = 4$, $V = \frac{4}{3}\pi r^3 = \frac{256}{3}\pi$;
   $S = 4\pi r^2 = 64\pi$

   b. $\ell = 2$, $w = 3$, $h = 5$, so
   $V = 2(3)(5) = 30$;
   $S = 2(2)(3) + (2)(2)(5) + 2(3)(5) = 62$

   c. $r = 2$, $h = 4$; $V = \pi r^2 h = 16\pi$;
   $S = 2\pi r^2 + 2\pi rh = 8\pi + 16\pi = 24\pi$

   d. $r = 3$, $h = 5$;
   $V = \frac{\pi}{3}r^2 h = \frac{5(3^2)}{3}\pi = 15\pi$ For the lateral area cut the cone open along an edge. The flattened surface is a sector of a circle of radius
   $r = \sqrt{9 + 25} = \sqrt{34}$. The full circle of such a radius has area 34π and circumference $2\pi\sqrt{34}$. The lateral surface area is
   $$A = \frac{6\pi\sqrt{34}}{2} = 3\pi\sqrt{34}$$

73. Without loss of generality, let $A(a, 0)$, $B(b, 0)$, and $C(0, c)$. Then, $M_1\left(\frac{a}{2}, \frac{c}{2}\right)$, $M_2\left(\frac{b}{2}, \frac{c}{2}\right)$. The slope is 0, so the line segment $M_1 M_2$ is parallel to $\overline{AB}$. The length of $\overline{AB} = |b - a|$ and the length of $M_1 M_2 = \left|\frac{b}{2} - \frac{a}{2}\right| = \frac{1}{2}|b - a|$.

75. Since the y-intercept is $(0, -5)$, we find $c = -\frac{4}{5}$. Thus, $x^2 + xy - \frac{4}{5}y = 4$. For $(x_0, 0)$, $x_0^2 = 4$, so the intercepts are $(-2, 0)$ and $(2, 0)$.

77.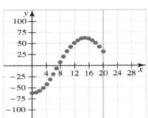

   Answers vary; it looks like, $(h, k) = (7.5, 0)$; $a = 60$; the period is found by solving $30 = \dfrac{2\pi}{b}$ so that $b = \dfrac{\pi}{15}$. A possible equation is $y = 60 \sin \dfrac{\pi}{15}(x - 7.5)$.

79. θ is the angle opposite the 7 ft mural and α is the angle opposite the 5 ft side of the right triangle.
   $$\tan(\alpha + \theta) = \frac{12}{12} = 1$$
   Thus, $\alpha + \theta = \frac{\pi}{4}$; $\alpha = \tan^{-1}\frac{5}{12}$ so that $\theta = \frac{\pi}{4} - \tan^{-1}\frac{5}{12} \approx 0.3906$.

81.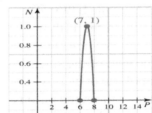

   a. $N = -p^2 + 14p - 48 = 0$ if $p = 6$ or $p = 8$. The operation is profitable for $6 < p < 8$.

   b. The maximum is likely to occur at the vertex or at the midpoint of the interval from 6 to 8. $N(7) = 1$ is the maximum profit.

83. Let x be the number of additional days after today (the eighth day) before the club takes all its glass to the recycling center. Assume that the same quantity of glass is collected daily, namely $2{,}400/8 = 300$ lb. The daily price per pound is $15 - x$ cents. The club's revenue on day x is

$$(2{,}400 + 300x)(0.15 - 0.01x) = 3(8 + x)(15 - x)$$
$$= -3x^2 + 21x + 360$$

The maximum revenue occurs at the vertex; consider the graph:

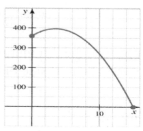

It looks like the maximum revenue is $397 when $x = 3.5$. However, since x must be an integer, we check $x = 3$ with revenue $396 and $x = 4$ with revenue $396. The glass should be taken in on the 11th or 12th day.

85. Let x the length of the side of the square base and h the height of the open box. The area of the top and bottom the base is x^2 ft^2 and the area of the sides is xh ft^2. Since the volume is 250 ft^3, $V = x^2h = 250$, so $h = 250x^{-2}$. The cost of the box (in dollars) is

$$C(x) = \left(\begin{array}{c} \text{TOP} \\ \text{COST} \end{array}\right) + \left(\begin{array}{c} \text{BOTTOM} \\ \text{COST} \end{array}\right) + 4\left(\begin{array}{c} \text{SIDE} \\ \text{COST} \end{array}\right)$$
$$= 2x^2 + 2x^2 + 4(xh)$$
$$= 4x^2 + 4x(250x^{-2})$$
$$= 4x^2 + 1{,}000x^{-1}$$

Answers for the mathematical essays in Problems 87-95 will vary, but each one should be about 500 words in length and include material from several sources.

97. Putnam Examination Problem   This is Putnam Problem 5 in the morning session for 1960. Consider a constant function g, say $g(x) = a$. Then for $f[g(x)] = g[f(x)]$ becomes $f(a) = a$. Since this is true for all real a, f is the identity function; i.e. $f(x) = x$. Remark: the proof does not require the hypothesis that f be a polynomial function.

99. Book report Answers vary.

CHAPTER 2 Limits and Continuity

Chapter Overview
There are three main topics in beginning calculus: limit, derivative, and integral. The definition of derivative (Chapter 3) and the definition of integral (Chapter 5) each use the notion of a limit. Understanding the limiting process is essential to understanding calculus. With the notion of limit, we are able to discuss one of the fundamental properties of functions, that is the idea of a continuous function.

2.1 The Limit of a Function, page 83

1. a. 0   **b.** 2   **c.** 6
3. a. 2   **b.** 7   **c.** 7.5
5. a. 6   **b.** 6   **c.** 6
7. $f(x) = 4x - 5$.

$x \to 5^-$

x	2	3	4	4.5	4.9	4.99
$f(x)$	3	7	11	13	14.6	14.96

$f(x) \to ?$

$\lim\limits_{x \to 5^-} (4x - 5) = 15$

9. $f(x) = \dfrac{x^2 + 2x + 4}{x^3 - 8}$

$x \to 2^-$

x	1	1.9	1.99	1.999
$f(x)$	-1	-10	-100	$-1,000$

$f(x) \to ?$

$2^+ \leftarrow x$

x	2.5	2.01	2.001
$f(x)$	2	100	1,000

$f(x) \to ?$

$\lim\limits_{x \to 2} \dfrac{x^2 + 2x + 4}{x^3 - 8}$ increases without limit.
In other words, we say this limit does not exist.

11. $f(x) = \dfrac{\tan 2x}{\tan 3x}$

$x \to 0^+$

x	1	0.5	0.1	0.01	0.001
$f(x)$	15.33	0.11044	0.65531	0.66656	0.66667

$f(x) \to ?$

$\lim\limits_{x \to 0} \dfrac{\tan 2x}{\tan 3x} = \dfrac{2}{3}$

13. $\lim\limits_{x \to 2} g(x) = 4$ **15.** $\lim\limits_{x \to 0^+} F(x) = 0$
17. $\lim\limits_{x \to 2\pi} t(x) = 1$ **19.** $\lim\limits_{x \to 0^+} x^4 = 0.00$
21. $\lim\limits_{x \to 2^-} (x^2 - 4) = 0$ **23.** $\lim\limits_{x \to 1^+} \dfrac{1}{x - 3} = -0.50$
25. $\lim\limits_{x \to 3} \dfrac{1}{x - 3}$ does not exist
27. a. $\lim\limits_{x \to 0} \dfrac{\cos x}{x}$ does not exist.
   b. $\lim\limits_{x \to \pi} \dfrac{\cos x}{x} = -0.32$
29. a. $\lim\limits_{x \to 0.4} |x| \sin \dfrac{1}{x} = 0.24$
   b. $\lim\limits_{x \to 0} |x| \sin \dfrac{1}{x} = 0.00$
31. $\lim\limits_{x \to 1} \dfrac{\sin \frac{\pi}{x}}{x - 1} = 3.14$
33. $\lim\limits_{x \to 9} \dfrac{\sqrt{x} - 3}{x - 3} = 0.00$

35. $\lim\limits_{x \to 64} \dfrac{\sqrt[3]{x}-8}{\sqrt{x}-4} = -1.00$

37. $\lim\limits_{x \to 9^+} \dfrac{\frac{1}{\sqrt{x}}-\frac{1}{3}}{x-3} = 0.00$

39. $\lim\limits_{x \to 0} \dfrac{1-\frac{1}{x+1}}{x} = 1.00$

41. $\lim\limits_{x \to 0} \dfrac{\sin 3x}{x} = 3.00$

43. $\lim\limits_{x \to 0} \tan \dfrac{1}{x}$ does not exist

45. $\lim\limits_{x \to 3} \dfrac{x^2+3x-10}{x-3}$ does not exist

47. $\lim\limits_{x \to 3^+} \dfrac{\sqrt{x-3}+x}{3-x}$ does not exist

49. **a.**

$$v(t) = \lim_{x \to t} \frac{s(x)-s(t)}{x-t}$$
$$= \lim_{x \to t} \frac{(-16x^2+40x+24)-(-16t^2+40t+24)}{x-t}$$
$$= \lim_{x \to t} (-16x-16t+40)$$
$$= -32t+40$$

b. $v(0) = \lim\limits_{x \to 0}(-32x+40)$
$$= 40 \text{ ft/s}$$

c. $s(t) = 0$ if

$$-16t^2+40t+24 = 0$$
$$2t^2-5t-3 = 0$$
$$(t-3)(2t+1) = 0$$
$$t = 3, -\frac{1}{2}$$

Reject the negative solution, so the time to impact is 3 seconds. The impact velocity is

$$v(3) = -32(3)+40 = -56 \text{ ft/s}$$

d. At the highest point on the trajectory, the ball has stopped moving upward and has not yet started on its downward fall. This occurs at $-32t+40 = 0$ or at $t = 1.25$ seconds.

51. Let $f(x) = \dfrac{x^3-9x^2-45x-91}{x-13}$

x	1	10	12	12.6	12.99
$f(x)$	12	147	199	216.16	227.7

x	14	13.5	13.1	13.01
$f(x)$	259	242.25	231.01	228.3

$$\lim_{x \to 13} \frac{x^3-9x^2-45x-91}{x-13} = 228$$

53. The Cauchy statement (quoted in several sources) is: "When the successive values attributed to a variable approach indefinitely to a fixed value so as to end by differing from it by as little as one wishes, this last is called the limit of all the others. Thus, for example, an irrational number is the limit of diverse fractions which furnish more and more approximate values of it." The δ-ϵ definition was given by H. E. Heine (1821-1881) in his book *Elements* (1872). Heine was a student of Karl Weierstraß (1815-1897) and it is acknowledged that Heine's published definition came from Weierstraß's lectures. This definition, as stated in Carl B. Boyer's *History of Mathematics*, (© 1968 by John Wiley & Sons, Inc., p. 608) is: "If given any ϵ, there is a η_0 such that for $0 \le \eta < \eta_0$ the difference $f(x_0 \pm \eta) - L$ is less in absolute value than ϵ, then L is the limit of $f(x)$ for $x = x_0$."

The use of the Greek letter ϵ in this context was originated by Cauchy and probably first used as an abbreviation for *error.* According to Judith V. Grabiner, *The Origins of Cauchy's Rigorous Calculus* (The MIT Press, Cambridge, 1981, p. 76), "the epsilon in a modern proof may be regarded as an inheritance from the days when inequalities belonged in approximations. The epsilon notation is a reminder that, paradoxically, the development of approximations and estimates of error brought forth many of the techniques necessary for the first exact and rigorous proofs about the concepts of calculus. Eighteenth-century mathematicians were never more exact than when they were being approximate."

55. $|f(t) - L| = |(3t - 1) - 0| < |3t - 1|$
The statement is false. Choose $\epsilon = \frac{1}{3}$ and it is not possible to find a delta.

57. $|f(x) - L| = |(2x - 5) + 3|$
$$= 2|x - 1|$$
$$< 2\delta$$
Choose $\delta = \dfrac{\epsilon}{2}$.

59. $|f(x) - L| = \left| \dfrac{1}{x} - \dfrac{1}{2} \right|$
$$= \dfrac{|x - 2|}{2|x|}$$
Choose $\delta = \dfrac{4\epsilon}{1 + 2\epsilon}$.

2.2 Algebraic Computation of Limits, page 94

SURVIVAL HINT: *This problem set is important because it is testing your ability to perform one of the three fundamental processes of calculus, that of finding limits. The time you can give to working this problems will pay rich dividends in your future work in mathematics,*

1. $\lim\limits_{x \to -1} (x^2 + 3x - 7) = 1 - 3 - 7$
$$= -9$$

3. $\lim\limits_{x \to 3} (x + 5)(2x - 7) = (3 + 5)(2 \cdot 3 - 7)$
$$= -8$$

5. $\lim\limits_{z \to 1} \dfrac{z^2 + z - 3}{z + 1} = \dfrac{1 + 1 - 3}{1 + 1}$
$$= -\dfrac{1}{2}$$

7. $\lim\limits_{x \to \pi/3} \sec x = \sec \dfrac{\pi}{3}$
$$= 2$$

9. $\lim\limits_{x \to 1/3} \dfrac{x \sin \pi x}{1 + \cos \pi x} = \dfrac{\frac{1}{3}\sin \frac{\pi}{3}}{1 + \cos \frac{\pi}{3}}$
$$= \dfrac{\frac{\sqrt{3}}{6}}{\frac{3}{2}}$$
$$= \dfrac{\sqrt{3}}{9}$$

11. $\lim\limits_{u \to -2} \dfrac{4 - u^2}{2 + u} = \lim\limits_{u \to -2} \dfrac{(2 + u)(2 - u)}{(2 + u)}$
$$= \lim\limits_{u \to -2} (2 - u)$$
$$= 4$$

13. $\lim\limits_{x \to 1} \dfrac{\frac{1}{x} - 1}{x - 1} = \lim\limits_{x \to 1} \dfrac{\frac{1 - x}{x}}{x - 1}$
$$= \lim\limits_{x \to 1} \dfrac{-1}{x}$$
$$= -1$$

15. $\lim\limits_{x \to 1} \dfrac{\sqrt{x} - 1}{x - 1} = \lim\limits_{x \to 1} \dfrac{(\sqrt{x} - 1)(\sqrt{x} + 1)}{(x - 1)(\sqrt{x} + 1)}$
$$= \lim\limits_{x \to 1} \dfrac{1}{\sqrt{x} + 1}$$
$$= \dfrac{1}{2}$$

17. $\lim\limits_{x \to 0} \dfrac{\sqrt{x + 1} - 1}{x}$

$$= \lim_{x \to 0} \frac{(\sqrt{x+1}-1)(\sqrt{x+1}+1)}{x(\sqrt{x+1}+1)}$$

$$= \lim_{x \to 0} \frac{x+1-1}{x(\sqrt{x+1}+1)}$$

$$= \lim_{x \to 0} \frac{x}{x(\sqrt{x+1}+1)}$$

$$= \lim_{x \to 0} \frac{1}{\sqrt{x+1}+1}$$

$$= \frac{1}{\sqrt{0+1}+1}$$

$$= \frac{1}{2}$$

19. $\displaystyle \lim_{x \to 0^+} \frac{\sin x}{\sqrt{x}} = \lim_{x \to 0^+} \frac{\sin x}{\sqrt{x}} \cdot \frac{\sqrt{x}}{\sqrt{x}}$

$$= \lim_{x \to 0^+} \left(\frac{\sin x}{x}\right)\sqrt{x}$$

$$= 1 \cdot 0$$

$$= 0$$

21. $\displaystyle \lim_{x \to 0} \frac{\sin 2x}{x} = \lim_{x \to 0} 2\left(\frac{\sin 2x}{2x}\right)$

$$= 2$$

23.

$$\lim_{t \to 0} \frac{\tan 5t}{\tan 2t}$$

$$= \lim_{t \to 0} \left(\frac{\sin 5t}{\cos 5t} \cdot \frac{\cos 2t}{\sin 2t}\right)$$

$$= \left(\lim_{t \to 0} \frac{\sin 5t}{5t}\right)\left(\lim_{t \to 0} \frac{5}{\cos 5t}\right)\left(\lim_{t \to 0} \frac{2t}{\sin 2t}\right)\left(\lim_{t \to 0} \frac{\cos 2t}{2}\right)$$

$$= (1)\left(\frac{5}{1}\right)(1)\left(\frac{1}{2}\right)$$

$$= \frac{5}{2}$$

25. $\displaystyle \lim_{x \to 0} \frac{1-\cos x}{\sin x} = \lim_{x \to 0} \frac{\frac{1-\cos x}{x}}{\frac{\sin x}{x}}$

$$= \frac{\lim_{x \to 0} \frac{1-\cos x}{x}}{\lim_{x \to 0} \frac{\sin x}{x}}$$

$$= \frac{0}{1}$$

$$= 0$$

> **SURVIVAL HINT:** As you work through these limits problems, make sure you repeat the word "limit" each time you simplify the algebra. When you evaluate the limit, then the word "lim" no longer appears.

27. $\displaystyle \lim_{x \to 0} \frac{\sin^2 x}{2x} = \lim_{x \to 0} \frac{1}{2}\left(\frac{\sin x}{x}\right)\sin x$

$$= \frac{1}{2}(1)(0)$$

$$= 0$$

29. $\displaystyle \lim_{x \to 0} \frac{\sec x - 1}{x \sec x} = \lim_{x \to 0} \frac{1-\cos x}{x}$

$$= 0 \quad \textit{From Example 10c.}$$

> **SURVIVAL HINT:** We found this limit using technology in Problem 28, Section 2.1, and got the same answer.

31. The limit of a polynomial can be found by substitution; that is

$$\lim_{x \to a} f(x) = f(a)$$

33. $\displaystyle \lim_{x \to 0} \frac{\sin ax}{x} = \lim_{ax \to 0} a\left(\frac{\sin ax}{ax}\right) = a(1) = a$

35. $\displaystyle \lim_{x \to 1^+} \frac{\sqrt{x-1}+x}{1-2x} = \frac{0+1}{1-2} = -1$

37. $\displaystyle \lim_{x \to 3} |3-x|$

Consider the left- and right-hand limits:

$$\lim_{x \to 3^+} (x-3) = 0$$

$$\lim_{x \to 3^-} (3 - x) = 0$$

Thus, $\lim\limits_{x \to 3} |3 - x| = 0$

39. $\lim\limits_{x \to -2} \dfrac{|x + 2|}{x + 2}$; two possibilities:

$$\lim_{x \to -2^+} \frac{x + 2}{x + 2} = 1$$

$$\lim_{x \to -2^-} \frac{-(x + 2)}{x + 2} = -1$$

The limit does not exist because the left- and right-hand limits are not equal.

41. Consider the left- and right-hand limits.

$$\lim_{s \to 1^+} g(s) = \lim_{s \to 1^+} \frac{s^2 - s}{s - 1}$$
$$= \lim_{s \to 1^+} \frac{s(s - 1)}{s - 1}$$
$$= \lim_{s \to 1^+} s$$
$$= 1$$

$$\lim_{s \to 1^-} g(s) = \lim_{s \to 1^-} \sqrt{1 - x}$$
$$= \sqrt{1 - 1}$$
$$= 0$$

The limit does not exist because the left- and right-hand limits are not equal.

43. $\lim\limits_{x \to 2^+} \dfrac{1}{\sqrt{x - 2}}$

As $x \to 2^+$, the denominator $\sqrt{x - 2}$ is approaching 0, the result of dividing a constant (1 in this case) by a quantity approaching zero, becomes infinite.

45. $\lim\limits_{x \to 3} \dfrac{x^2 + 4x + 3}{x - 3} = \lim\limits_{x \to 3} \dfrac{(x + 3)(x + 1)}{x - 3}$

As $x \to 3$, the numerator is approaching $(6)(4) = 24$ and the denominator $(x - 3)$ is approaching 0, which when divided into a number close to 24, becomes infinite.

47. $\lim\limits_{x \to 1} \csc \pi x$

Does not exist because $\csc \pi x$ increases without bound as $x \to 1$ from the left and decreases without bound as $x \to 1$ from the right.

49. $\lim\limits_{t \to -1} g(t)$, where $g(t) = \begin{cases} 2t + 1 & \text{if } t > -1 \\ 5t^2 & \text{if } t < -1 \end{cases}$

Does not exist because the limit from the left is 5 and the limit from the right is -1.

51. $\lim\limits_{x \to 0} \left(\dfrac{1}{x} - \dfrac{1}{x^2} \right) = \lim\limits_{x \to 0} \left(\dfrac{x - 1}{x^2} \right)$

Since the numerator approaches -1 and the denominator approaches 0, the fraction approaches $-\infty$ as $x \to 0^+$ and $-\infty$ as $x \to 0^-$, so the limit does not exist.

53. $\lim\limits_{t \to 2} g(t) = 4$ since the left- and right-hand limits are both equal to 4.

55. $\lim\limits_{x \to 3} f(x) = 8$ since the left- and right-hand limits both equal to 8.

> **SURVIVAL HINT:** *You may decide to use technology in some of these problems, and if you do, keep in mind that the proper use of technology is essential for limit estimates, and incorrect use often leads to erroneous answers.*

57. $\lim\limits_{h \to 0} \dfrac{\cos h - 1}{h} = \lim\limits_{h \to 0} \left(\dfrac{\cos h - 1}{h} \cdot \dfrac{\cos h + 1}{\cos h + 1} \right)$
$$= \lim_{h \to 0} \frac{\cos^2 h - 1}{h(\cos h + 1)}$$
$$= \lim_{h \to 0} \frac{-\sin^2 h}{h(\cos h + 1)}$$
$$= (-1)\lim_{h \to 0} \left[\left(\frac{\sin h}{h} \right) \left(\frac{\sin h}{\cos h + 1} \right) \right]$$
$$= (-1) \left(\lim_{h \to 0} \frac{\sin h}{h} \right) \left(\lim_{h \to 0} \frac{\sin h}{\cos h + 1} \right)$$

$$= (-1)(1)\left(\frac{0}{2}\right)$$
$$= 0$$

59. To find $\lim\limits_{x \to x_0} \cos x$, let $h = x - x_0$ or

$x = h + x_0$. Note that $h \to 0$ as $x \to x_0$.

$$\begin{aligned}
\lim_{x \to x_0} \cos x &= \lim_{h \to 0} \cos(h + x_0) \\
&= \lim_{h \to 0}[\cos h \cos x_0 - \sin h \sin x_0] \\
&= 1(\cos x_0) - 0(\sin x_0) \\
&= \cos x_0
\end{aligned}$$

2.3 Continuity, page 104

1. Temperature is continuous, so TEMPERATURE $= f$(TIME) would be a continuous function. The domain could be midnight to midnight say, $0 \le t < 24$.

3. The closing price of any stock may be quite different from day to day, so this is not a continuous function. The domain is the dates that the stock market is open on any given year. In fact, all stock price movement is discontinuous because even a change in price of one cent is a jump.

5. The charges (range of the function) consist of rational numbers only (dollars and cents to the nearest cent), so the function CHARGE $= f$(MILEAGE) would be a step function (that is, not continuous). The domain would consist of the mileage from the beginning of the trip to its end.

7. No suspicious points and no points of discontinuity with a polynomial.

9. The denominator factors to $x(x-1)$, so suspicious points would be $x = 0, 1$. There will be a hole discontinuity at $x = 0$ and a pole discontinuity at $x = 1$.

11. $x = 0$ is suspicious and is a point of discontinuity, since the denominator vanishes. Points $x < 0$ are not in the domain since square roots of negative numbers are not defined in the set of real numbers.

13. $t = 0$, $t = -1$ are suspicious points; when the fraction is simplified it becomes
$$\frac{1 - 2t}{t(t + 1)}$$
Both $t = 0$ and $t = -1$ are points of discontinuity.

15. $x = 1$ is a suspicious point; there are no points of discontinuity.

17. The sine and cosine are continuous on the reals, but the tangent is discontinuous at $x = \frac{\pi}{2} + n\pi$, for any integer n. Each of these values will have a pole type discontinuity.

19. $h(x) = \csc x \cot x$
Cosecant is not defined for $x = n\pi$, n an integer, and the cotangent is not defined for the same values, so the function has a discontinuity at multiples of π, as shown by the graph.

21. For continuity, $f(2)$ must equal
$$\begin{aligned}
\lim_{x \to 2} f(x) &= \lim_{x \to 2} \frac{(x - 2)(x + 1)}{x - 2} \\
&= \lim_{x \to 2}(x + 1) \\
&= 3
\end{aligned}$$

23. For continuity, $f(2)$ must equal
$$\begin{aligned}
\lim_{x \to 2} f(x) &= \lim_{x \to 2} \frac{\sin(\pi x)}{x - 2} \\
&= \pi
\end{aligned}$$

(Limit found by table or graphing.)

25. The function is not defined at 2, and since

$$\lim_{x \to 2^-} f(x) = 11 \quad \text{and} \quad \lim_{x \to 2^+} f(x) = 9$$

no value can be assigned to $f(2)$ to "tie together" the two pieces. This is sometimes called an "essential" discontinuity. Only hole-type discontinuities are "removable".

27. a. No suspicious points on $[1, 2]$; continuous
b. Suspicious point $x = 0$. Discontinuous on $[0, 1]$ since the pole $x = 0$ is in the domain. If the interval had been $(0, 1]$, the function would be continuous on the interval.

29. Suspicious point $x = 2$; the limit from the left is 4 while that from the right is 7. The function is discontinuous at $x = 2$.

31. No suspicious points; $y = x$ and $y = \sin x$ are continuous on the reals, so $f(x) = x \sin x$ will be continuous on $(0, \pi)$.

33. $f(x) = \sqrt[3]{x} - x^2 - 2x + 1$ is continuous on $[0, 1]$ and $f(0) = 1$, $f(1) = -1$ so the hypotheses of the intermediate value theorem are met, and we are guaranteed that there is at least one number c on $(0, 1)$ such that $f(c) = 0$.

35. $f(x) = \dfrac{1}{x + 1} - x^2 + x + 1$ is continuous on $[1, 2]$ and $f(1) = \frac{3}{2}$, $f(2) = -\frac{2}{3}$, so the hypotheses of the intermediate value theorem are met, and we are guaranteed that there is at least one number c on $(1, 2)$ such that $f(c) = 0$.

37. $f(x) = \cos x - \sin x - x$ is continuous on the reals and $f(0) = 1$, $f(\frac{\pi}{2}) = -1 - \frac{\pi}{2}$. Since the hypotheses of the intermediate value theorem have been met, we are guaranteed that there exists at least one number c on $(0, \frac{\pi}{2})$ such that $f(c) = 0$.

39. $f(2) = b$; if $x \neq 2$ and $a = 2$, then
$$\frac{2x - 4}{x - 2} = 2, \text{ so } b = 2.$$

41. $F(5) = 8$; from the right, $5a + 3 = 8$, so $a = 1$. From the left, $25 + 5b + 1 = 8$, so $b = -\frac{18}{5}$.

43. $M(0) = 5$; if $x < 0$,

$$\lim_{x \to 0^-} \frac{\sin ax}{x} = a \lim_{x \to 0^-} \frac{\sin ax}{ax} = a = 5$$

Similarly,

$$\lim_{x \to 0^+} M(x) = 0 + b = 5$$

45. $f(2) + 3 = 2a + b + 3$
Also, $f(0) = 0^2 - 4(0) + b + 3$
so

$$2a + b + 3 = b + 3$$
$$2a = 0$$
$$a = 0$$

Furthermore,

$$\lim_{x \to 1^+} (ax + b) = a + b = 3$$

so if $a = 0$, then $b = 3$. Thus, $a = 0$ and $b = 3$.

47. Both f and g are discontinuous at $x = 0$, but

$$f(x) + g(x) = \begin{cases} 4x & \text{if } x \neq 0 \\ 0 & \text{if } x = 0 \end{cases}$$

is a function without a jump at 0, and is therefore continuous at $x = 0$.

49. Answers vary.

$$f(x) = \begin{cases} x & \text{if } x \text{ is rational} \\ 1-x & \text{if } x \text{ is irrational} \end{cases}$$

is continuous only at $x = \frac{1}{2}$.

51. For x minutes after the hour, let $f(x)$ denote the angle from the minute hand to the hour hand (not allowing for multiple revolutions). Clearly, at 1:00 P.M.

$$f(0) < 0$$

and at 1:15 P.M.

$$f(15) > 0$$

Since the sign of $f(x)$ changes between $x = 0$ and $x = 15$ and f is continuous, by the intermediate value theorem there must be a time when $f(x) = 0$; that is, a time when the hands coincide.

53. a. $\displaystyle\lim_{v \to v_w} \frac{cv^k}{v - v_w}$ does not exist because there is a singularity at $v = v_w$. This means that the less variance between the rate at which the fish swim and the rate at which the river flows, the larger the expended energy (because the fish is not making any headway).

b. As $v \to \infty$ the velocity of the fish gets large and the energy expended goes to infinity.

55. $\displaystyle\lim_{\alpha \to 0} \sin\left(\frac{1}{2}\alpha\right) = 0$, so

$$\lim_{\alpha \to 0} \sin(x + \alpha) - \sin x = \lim_{\alpha \to 0}\left[2\sin\frac{\alpha}{2}\cos\left(x + \frac{\alpha}{2}\right)\right]$$
$$= 2(0)\cos x$$
$$= 0$$

so the difference approaches 0 and $\sin x$ is a continuous function of x.

57. a. Carry out the bisection method and verify with a calculator $x \approx 1.25872$.

b. Carry out the bisection method and verify with a calculator $x \approx 0.785398$.

59. $f(x)$ is continuous at $x = c$, it must be continuous from the left and the right (where c is not an endpoint of an interval). Conversely, suppose f is continuous from the right and left at $x = c$. Then,

$$\lim_{x \to c^+} f(x) = f(c) = \lim_{x \to c^-} f(x)$$

This implies continuity at $x = c$.

2.4 Exponential and Logarithmic Functions, page 119

> **SURVIVAL HINT:** *In high school algebra exponential functions are common, but logarithmic functions are sometimes treated as some sort of mystery. Remember, a logarithm is nothing more than an exponent. Find the "What this says" box following the definition on page 111 and make sure you can explain its contents to someone else. The material of this section is essential for your further mathematical life.*

1.

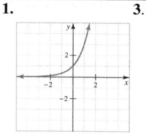

3.

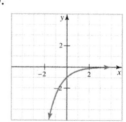

5. $\log_2 4 + \log_3 \dfrac{1}{9} = 2 + (-2) = 0$

7. $5\log_3 9 - 2\log_2 16 = 5(2) - 2(4) = 2$

9. $e^{5\ln 2} = e^{\ln 2^5} = 2^5 = 32$

11. $\ln(\log 10^e) = \ln e = 1$

13. $\log x = 5.1$

$$x = 10^{5.1}$$
$$\approx 125892.5412$$

15. $\ln x^2 = 9$

$\qquad x^2 = e^9$

$\qquad x = \pm\sqrt{e^9}$

$\qquad\quad \approx \pm 90.0171313$

17. $e^{2x} = \ln(4 + e)$

$\qquad 2x = \ln[\ln(4 + e)]$

$\qquad x = \dfrac{1}{2}\ln[\ln(4 + e)]$

$\qquad\quad \approx 0.322197023$

19. $\log_x 16 = 2$

$\qquad x^2 = 16$

$\qquad x = \pm 4$

(Reject the negative value since logarithms of negative numbers are not defined.) Thus, $x = 4$.

21. $\qquad\qquad 3^{x^2 - x} = 9$

$\quad 3^{x^2 - x} = 3^2$

$\qquad\qquad x^2 - x = 2$

$\qquad (x - 2)(x + 1) = 0$

$\qquad\qquad\qquad x = 2, -1$

23. $2^x 5^{x+2} = 25{,}000$

$\qquad 2^x 5^x 5^2 = 2^3 5^5$

$\qquad (2 \cdot 5)^x = 2^3 5^3$

$\qquad\qquad 10^x = 10^3$

$\qquad\qquad\quad x = 3$

25. $\qquad \left(\sqrt[3]{2}\right)^{x+10} = 2^{x^2}$

$\qquad\quad 2^{x/3 + 10/3} = 2^{x^2}$

$\qquad\qquad \dfrac{x}{3} + \dfrac{10}{3} = x^2$

$\qquad 3x^2 - x - 10 = 0$

$\qquad (x - 2)(3x + 5) = 0$

$\qquad\qquad\qquad x = 2, -\dfrac{5}{3}$

27. $e^{2x+3} = 1$

$\qquad e^{2x+3} = e^0$

$\qquad 2x + 3 = 0$

$\qquad\qquad x = -\dfrac{3}{2}$

29. $\log_3 x + \log_3(2x + 1) = 1$

$\qquad \log_3[x(2x + 1)] = 1$

$\qquad\qquad x(2x + 1) = 3$

$\qquad\qquad 2x^2 + x - 3 = 0$

$\qquad (x - 1)(2x + 3) = 0$

$\qquad\qquad\qquad x = 1, -\dfrac{3}{2}$

(Reject the negative value since logarithms of negative numbers are not defined.) Thus, $x = 1$.

31. a. $\lim\limits_{x \to 0} x^2 e^{-x} = 0^2 e^0 = 0$

$\quad$ **b.** $\lim\limits_{x \to 1} x^2 e^{-x} = 1^2 e^{-1} = e^{-1}$

33. a. $\lim\limits_{x \to 0^+} (1 + x)^{1/x} = \lim\limits_{u \to \infty} \left(1 + \dfrac{1}{u}\right)^u$ *Let* $u = \dfrac{1}{x}$ *so* $u \to \infty$ *as* $x = 0^+$

$\qquad\qquad\qquad\qquad\quad = e \qquad$ *Definition of* e.

$\quad$ **b.** $\lim\limits_{x \to 1} (1 + x)^{1/x} = (1 + 1)^{1/1}$

$\qquad\qquad\qquad\qquad = 2$

35. $\log_b 1{,}296 = 4$

$\qquad\qquad b^4 = 1{,}296$

$\qquad\qquad b^4 = 6^4$

$\qquad\qquad b = 6$

Thus, $\left(\dfrac{3}{2} \cdot 6\right)^{3/2} = 9^{3/2}$

$\qquad\qquad\qquad\qquad = 27$

37. Use the graph or solve utility on a graphing calculator to find $x \approx 0.4229976068$.

Thus, to the nearest tenth, $x = 0.4$.

39. logarithmic   **41.** exponential

43. $I(x) = I_0 e^{kx}$

$I(0) = I_0$; $I(2) = 0.05 I_0$, but also
$I(2) = I_0 e^{2k}$ so that

$$0.05 I_0 = I_0 e^{2k}$$
$$e^{2k} = 0.05$$
$$2k = \ln 0.05$$
$$k = -1.497866$$

When does $I(x) = 0.01 I_0$?

$$0.01 I_0 = I_0 e^{kx}$$
$$kx = \ln 0.01$$
$$x \approx 3.074487$$

It is approximately 3 meters below the surface.

45. a. $2P = P e^{et}$

$$2 = e^{rt}$$
$$rt = \ln 2$$
$$t = \frac{\ln 2}{r}$$

It will take about $0.693147/r$ or about $0.69/r$ years to double.

b. $\ln 2 \approx 0.69$, which means that to find the time it takes for money invested into an account paying interest compounded continuously, divided the interest rate into 0.69... hence a "Rule of 69." If you look at Wikipedia, you will see this rule is also known as the "Rule of 70" or the "Rule of 72." In the days before calculators, it was much easier to divide numbers into 72 mentally than into 69, since 72 has so many factors. Today with calculators, we should really think of this as the "Rule of ln 2."

47. $2(8,500) = 8,500 e^{10r}$

$$2 = e^{10r}$$
$$10r = \ln 2$$
$$r \approx 0.0693147181$$

The interest rate is about 6.9% .

49. a. $2P_0 = P_0 \left(2^{60k}\right)$

$$2^1 = 2^{60k}$$
$$1 = 60k$$
$$k = \frac{1}{60}$$

$$P(t) = P_0\left(2^{t/60}\right)$$

Since $P(20) = 1,000$, we have

$$1,000 = P_0\left(2^{20/60}\right)$$
$$P_0 = 1,000 \cdot 2^{-1/3}$$
$$\approx 794$$

b. $5,000 = P_0 \cdot 2^{t/60}$

$$2^{t/60} = \frac{5,000}{P_0}$$
$$2^{t/60} = 5 \cdot 2^{1/3}$$
$$t = 60\log_2\left(5 \cdot 2^{1/3}\right)$$
$$\approx 159.315686$$

The time is about 2 hr and 39 min.

51. a. $p(t) = 100[e^{-0.03t}]\%$ where $t = 40$:
$p(40) \approx 30.12\%$.

b. Failure rate is

$$100\% - p(50) = 100\% - 100\left[e^{-0.03(50)}\right]\%$$
$$\approx 77.69\%$$

c. $p(40) - p(50) = 30.12\% - 22.31\%$
$$= 7.81\%$$

53. a. $m = \dfrac{\$10,000\left(\frac{0.12}{12}\right)}{1 - \left(1 + \frac{0.12}{12}\right)^{-48}} \approx \$263.34$

b. $m = \dfrac{\$210,000(0.80)\left(\frac{0.08}{12}\right)}{1 - \left(1 + \frac{0.08}{12}\right)^{-360}} \approx \$1{,}232.72$

55. a. $b^m b^n = \underbrace{(bbb \cdots \cdot b)}_{m} \underbrace{(bbb \cdots \cdot b)}_{n}$

$= \underbrace{bbb \cdots \cdot b}_{m+n}$

$= b^{m+n}$

b. $\dfrac{b^m}{b^n} = \dfrac{\overbrace{bbb \cdots \cdot b}^{m}}{\underbrace{bbb \cdots \cdot b}_{n}}$

$= \underbrace{bbb \cdots \cdot b}_{m-n}$

$= b^{m-n}$

57. Let $M = \log_b x$ and $N = \log_b y$. Then
$b^M = x$ and $b^N = y$

a. $\quad xy = b^M b^N$

$= b^{M+N}$

$M + N = \log_b(MN)$ *Definition of logarithm*

$\log_b x + \log_b y = \log_b(MN)$ *Substitution*

b. $\quad \dfrac{x}{y} = \dfrac{b^M}{b^N}$

$\dfrac{x}{y} = b^{M-N}$

$M - N = \log_b\left(\dfrac{x}{y}\right)$ *Definition of logarithm*

$\log_b x - \log_b y = \log_b\left(\dfrac{x}{y}\right)$ *Substitution*

59. $b^{x \log_b x} = b^{\log_b x^x} = x^x$

Chapter 2 Review

 Studying for a chapter examination is a personal process, one which nobody else can do for you. Simply take the time to review what you have done.

SURVIVAL HINT: Work all of Chapter 2 problems in the Proficiency Examination (whether they are assigned or not). Work through all of the problems before looking at the answers, and *then* correct each of the problems. The answers to all these problems are given in the answer section at the back of the text. If you worked the problem correctly, move on to the next problem, but if you did not work it correctly (or you did not know what to do), then look at the solutions below, look back in the chapter to study the procedure, or ask your instructor.

Finally, go back over the homework problems you have been assigned. If you worked a problem correctly, move on to the next problem, but if you missed it on your homework, then you should look back in the book or talk to your instructor about how to work the problem.

If you follow these steps, you should be successful with your review of this chapter.

Proficiency Examination, page 122

1. $\lim\limits_{x \to c} f(x) = L$ means that the function values $f(x)$ can be made arbitrarily close to L by choosing x sufficiently close to c.

2. $\lim\limits_{x \to c} f(x) = L$ means that for each $\epsilon > 0$ there exists a number $\delta > 0$ such that $|f(x) - L| < \epsilon$ whenever $0 < |x - c| < \delta$.

3. a. $\lim\limits_{x \to c} k = k$ for any constant k.

b. $\lim\limits_{x \to c}[sf(x)] = s \lim\limits_{x \to c} f(x)$

c. $\lim\limits_{x \to c}[f(x) + g(x)] = \lim\limits_{x \to c} f(x) + \lim\limits_{x \to c} g(x)$

d. $\lim\limits_{x \to c}[f(x) - g(x)] = \lim\limits_{x \to c} f(x) - \lim\limits_{x \to c} g(x)$

e. $\lim\limits_{x \to c}[f(x)g(x)] = \left[\lim\limits_{x \to c} f(x)\right]\left[\lim\limits_{x \to c} g(x)\right]$

f. $\lim\limits_{x \to c} \dfrac{f(x)}{g(x)} = \dfrac{\lim\limits_{x \to c} f(x)}{\lim\limits_{x \to c} g(x)} \qquad \lim\limits_{x \to c} g(x) \neq 0$

g. $\lim\limits_{x \to c}[f(x)]^n = \left[\lim\limits_{x \to c} f(x)\right]^n$

   n is a rational number and the limit on the right exists.

h. If P is a polynomial function, then
$$\lim\limits_{x \to c} P(x) = P(c)$$

i. If Q is a rational function defined by
$$Q(x) = \frac{P(x)}{D(x)}, \text{ then}$$
$$\lim\limits_{x \to c} Q(x) = \frac{P(c)}{D(c)}$$
provided $\lim\limits_{x \to c} D(x) \neq 0$.

j. If T is a trigonometric, exponential, or natural logarithmic function defined at $x = c$, then
$$\lim\limits_{x \to c} T(x) = T(c)$$

4. If $g(x) \leq f(x) \leq h(x)$ for all x on an open interval containing c, and if
$$\lim\limits_{x \to c} g(x) = \lim\limits_{x \to c} h(x) = L$$
then
$$\lim\limits_{x \to c} f(x) = L$$

5. a. $\lim\limits_{x \to 0} \dfrac{\sin x}{x} = 1$ **b.** $\lim\limits_{x \to 0} \dfrac{\cos x - 1}{x} = 0$

6. A function f is continuous at a point $x = c$ if

   (1) $f(c)$ is defined

   (2) $\lim\limits_{x \to c} f(x)$ exists

   (3) $\lim\limits_{x \to c} f(x) = f(c)$

7. If f is a polynomial, rational, power, trigonometric, logarithmic, exponential, or inverse trigonometric function, then f is continuous at any number $x = c$ for which $f(c)$ is defined.

8. If $\lim\limits_{x \to c} g(x) = L$ and f is a function that is continuous at L, then $\lim\limits_{x \to c} f[g(x)] = f(L)$. That is,

$$\lim\limits_{x \to c} f[g(x)] = f(L) = f\left[\lim\limits_{x \to c} g(x)\right]$$

9. If f is a continuous function on the closed interval $[a, b]$ and L is some number strictly between $f(a)$ and $f(b)$, then there exists at least one number c on the open interval (a, b) such that $f(c) = L$.

10. If f is continuous on the closed interval $[a, b]$ and if $f(a)$ and $f(b)$ have opposite algebraic signs (one positive and the other negative), then $f(c) = 0$ for at least one number c on the open interval (a, b).

11. For any real number x, there exist rational numbers r_n such that

$$x = \lim\limits_{n \to \infty} r_n$$

which means that for any number $\epsilon > 0$, there exists a number N such that $|x - r_n| < \epsilon$ whenever $n > N$.

12. a. Let x be a real number, and let r_n be a sequence of rational numbers such that $x = \lim\limits_{n \to \infty} r_n$. Then, the exponential function with base $b > 0$ ($b \neq 1$) is given by

$$b^x = \lim_{n \to \infty} b^{r_n}$$

b. Exponential and logarithmic functions are inverse functions.

13. $e = \lim_{n \to \infty} \left(1 + \dfrac{1}{n}\right)^n$

14. a. If $b > 0$ and $b \neq 1$, the logarithm of x to the base b is the function $y = \log_b x$ that satisfies $b^y = x$; that is

$$y = \log_b x \quad \text{means} \quad b^y = x$$

b. A common logarithm is a logarithm to the base 10, written $\log_{10} x = \log x$.

c. A natural logarithm is a logarithm to the base e, written $\log_e x = \ln x$.

15. a.   **b.**

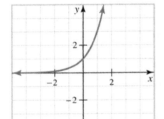 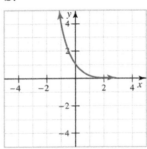

16. a.   **b.**

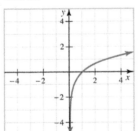 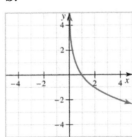

17. a. If $b \neq 1$, then $b^x = b^y$ if and only if $x = y$.

b. If $x > y$ and $b > 1$, then $b^x > b^y$.
If $x > y$ and $0 < b < 1$, then $b^x < b^y$.

c. $b^x b^y = b^{x+y}$

d. $\dfrac{b^x}{b^y} = b^{x-y}$

e. $(b^x)^y = b^{xy}$; $(ab)^x = a^x b^x$; $\left(\dfrac{a}{b}\right)^x = \dfrac{a^x}{b^x}$

18. a. 0   **b.** 1   **c.** x **d.** y **e.** $e^{x \ln b}$

19. $\lim\limits_{x \to 3} \dfrac{x^2 - 4x + 9}{x^2 + x - 8} = \dfrac{(3)^2 - 4(3) + 9}{3^2 + 3 - 8} = \dfrac{3}{2}$

20. $\lim\limits_{x \to 2} \dfrac{x^2 - 5x + 6}{x^2 - 4} = \lim\limits_{x \to 2} \dfrac{(x-2)(x-3)}{(x+2)(x-2)}$

$= \lim\limits_{x \to 2} \dfrac{x - 3}{x + 2}$

$= -\dfrac{1}{4}$

21. $\lim\limits_{x \to 4} \left[\left(\dfrac{\sqrt{x} - 2}{x - 4}\right)\left(\dfrac{\sqrt{x} + 2}{\sqrt{x} + 2}\right)\right] = \lim\limits_{x \to 4} \dfrac{x - 4}{(x - 4)(\sqrt{x} + 2)}$

$= \lim\limits_{x \to 4} \dfrac{1}{\sqrt{x} + 2}$

$= \dfrac{1}{4}$

22. $\lim\limits_{x \to 0} \dfrac{\sin 9x}{\sin 5x} = \lim\limits_{x \to 0} \left[\dfrac{9x}{9x}(\sin 9x) \cdot \dfrac{5x}{5x}\left(\dfrac{1}{\sin 5x}\right)\right]$

$= \lim\limits_{x \to 0} \left[\left(\dfrac{9x}{5x}\right)\left(\dfrac{\sin 9x}{9x}\right)\left(\dfrac{5x}{\sin 5x}\right)\right]$

$= \dfrac{9}{5}(1)(1)$

$= \dfrac{9}{5}$

23. $\lim\limits_{x \to 0} \dfrac{1 - \cos x}{2 \tan x} = \lim\limits_{x \to 0} \dfrac{(1 - \cos x)(1 + \cos x)}{2 \frac{\sin x}{\cos x}(1 + \cos x)}$

$= \lim\limits_{x \to 0} \dfrac{\sin^2 x}{2 \frac{\sin x}{\cos x}(1 + \cos x)}$

$= \lim\limits_{x \to 0} \dfrac{\sin x \cos x}{2(1 + \cos x)}$

$$= \frac{0}{4}$$
$$= 0$$

24. $\displaystyle\lim_{x \to (1/2)^-} \frac{|2x-1|}{2x-1} = \lim_{x \to (\frac{1}{2})^{-1}} \frac{-(2x-1)}{2x-1} = -1$

25. a. exponential   **b.** exponential
    c. logarithmic   **d.** logarithmic

26. We have suspicious points where the denominators are 0, at $t = 0, -1$. There are pole discontinuities at each of these points.

27. Suspicious points $x = -2$ (pole) and $x = 1$ (removable hole) are also points of discontinuity (since the denominator is 0).

28. a. $5{,}000 = 2{,}000 \left(1 + \dfrac{0.08}{4}\right)^t$

$2.5 = 1.02^{4t}$

$4t = \log_{1.02} 2.5$

$t = \dfrac{1}{4}\log_{1.02} 2.5 \approx 11.56779247$

It will take 11 years, 3 quarters (since 11 years 2 quarters is insufficient).

b. $5{,}000 = 2{,}000 \left(1 + \dfrac{0.08}{12}\right)^{12t}$

$2.5 = \left(1 + \dfrac{0.08}{12}\right)^{12t}$

$12t = \log_{(1+0.08/12)} 2.5$

≈ 11.49177065

It will take 11 years, 6 months.

c. $5{,}000 = 2{,}000 e^{0.08t}$

$2.5 = e^{0.08t}$

$0.08t = \ln 2.5$

≈ 11.45363415

It will take 11 years, 166 days.

29. Polynomials are everywhere continuous, so the only problem is at $x = 1$. We need

$\displaystyle\lim_{x \to 1^-}(Ax + 3) = 2$ and $\displaystyle\lim_{x \to 1^+}(x^2 + B) = 2$

Thus,

$A + 3 = 2 \qquad 1 + B = 2$

$A = -1 \qquad\quad B = 1$

30. Let $f(x) = x + \sin x - \dfrac{1}{\sqrt{x+3}}$

This function is continuous on $[0, \infty)$, and

$f(0) = -\dfrac{1}{3} < 0 \qquad f(\pi) = \pi - \dfrac{1}{\sqrt{\pi+3}} > 0$

So by the root location theorem there must be some value c on $(0, \pi)$ where $f(c) = 0$.

Supplementary Problems, page 123

1. $e^{\ln \pi} = \pi$

3. $32^{2/5} + 9^{3/2} = \left(2^5\right)^{2/5} + \left(3^2\right)^{3/2}$
$= 2^2 + 3^3$
$= 31$

5. $(128)^{1/2}(81)^{3/4} = \left(2^7\right)^{1/2}\left(3^4\right)^{3/4}$
$= 3^3 2^3 2^{1/2}$
$= 216\sqrt{2}$

7. $\ln 4.5 \approx 1.50$

9. $5{,}000\left(1 + \frac{0.135}{12}\right)^{12(5)} \approx 9{,}783.23$

11. $\log_2 7 \approx 2.81$

13. $\quad 4^{x-1} = 8$
$2^{2x-2} = 2^3$
$2x - 2 = 3$
$x = \dfrac{5}{2}$

15. $\log_2 2^{x^2} = 4$
$2^4 = 2^{x^2}$
$x^2 = 4$
$x = \pm 2$

17. $\log_2 x + \log_2(x - 15) = 4$
$$\log_2 x(x - 15) = 4$$
$$2^4 = x(x - 15)$$
$$x^2 - 15x - 16 = 0$$
$$(x - 16)(x + 1) = 0$$
$$x = 16, -1$$
Reject $x = -1$ since $\log_2 x$ is not defined at $x = -1$, so the solution is $x = 16$.

19. $\ln(x - 1) + \ln(x + 1) = 2\ln\sqrt{12}$
$$\ln(x - 1) + \ln(x + 1) = \ln 12$$
$$\ln[(x - 1)(x + 1)] = \ln 12$$
$$x^2 - 1 = 12$$
$$x^2 = 13$$
$$x = \pm\sqrt{13}$$
Reject $x = -\sqrt{13}$ since $\ln(x + 1)$ is not defined for negative values. The solution is $x = \sqrt{13}$.

21. $2\ln x - \dfrac{1}{2}\ln 9 = \ln 3(x - 2)$
$$\ln x^2 - \ln 3 = \ln 3(x - 2)$$
$$\ln\frac{x^2}{3} = \ln(3x - 6)$$
$$\frac{x^2}{3} = 3x - 6$$
$$x^2 - 9x + 18 = 0$$
$$(x - 6)(x - 3) = 0$$
$$x = 6, 3$$

23. $\displaystyle\lim_{x \to 2}\frac{3x^2 - 7x + 2}{x - 2} = \lim_{x \to 2}\frac{(x - 2)(3x - 1)}{x - 2}$
$$= \lim_{x \to 2}(3x - 1)$$
$$= 5$$

25. $\displaystyle\lim_{x \to 1}\left(\frac{x^2 - 3x + 2}{x^2 + x - 2}\right)^2 = \lim_{x \to 1}\left[\frac{(x - 1)(x - 2)}{(x - 1)(x + 2)}\right]^2$
$$= \left[\lim_{x \to 1}\frac{x - 2}{x + 2}\right]^2$$
$$= \left[\frac{-1}{3}\right]^2$$
$$= \frac{1}{9}$$

27. $\displaystyle\lim_{x \to 3}\sqrt{\frac{x^2 - 2x - 3}{x - 3}} = \sqrt{\lim_{x \to 3}\frac{(x - 3)(x + 1)}{x - 3}}$
$$= \sqrt{\lim_{x \to 3}(x + 1)}$$
$$= \sqrt{4}$$
$$= 2$$

29. $\displaystyle\lim_{x \to 1}\frac{x^3 - 1}{x^2 - 1} = \lim_{x \to 1}\frac{(x - 1)(x^2 + x + 1)}{(x + 1)(x - 1)}$
$$= \lim_{x \to 1}\frac{x^2 + x + 1}{x + 1}$$
$$= \frac{3}{2}$$

31. $\displaystyle\lim_{x \to 0^-}\frac{|x|}{x} = \lim_{x \to 0^-}\frac{-x}{x} = -1$

33. $\displaystyle\lim_{x \to e}\frac{\ln\sqrt{x}}{x} = \frac{\ln e^{1/2}}{e} = \frac{1}{2}e^{-1}$

35. $\displaystyle\lim_{x \to 0^+}(1 + x)^{4/x} = \left[\lim_{x \to 0^+}(1 + x)^{1/x}\right]^4$
$$= e^4$$

37. $\displaystyle\lim_{x \to 1}\frac{x^5 - 1}{x - 1} = \lim_{x \to 1}(x^4 + x^3 + x^2 + x + 1)$
$$= 5$$

39. $\displaystyle\lim_{x \to 0}\frac{\sin 3x}{\sin 2x} = \lim_{x \to 0}\frac{3}{2}\frac{\frac{\sin 3x}{3x}}{\frac{\sin 2x}{2x}}$
$$= \frac{3}{2} \cdot \frac{1}{1}$$
$$= \frac{3}{2}$$

41. $\lim\limits_{x\to 0}\dfrac{\sin(\cos x)}{\sec x} = \lim\limits_{x\to 0}(\cos x)[\sin(\cos x)]$

$\qquad = (1)\sin 1$

$\qquad = \sin 1$

43. $\lim\limits_{x\to 0}\dfrac{1-\sin x}{\cos^2 x} = \dfrac{1-0}{\cos^2 0}$

$\qquad\qquad = 1$

45. $\lim\limits_{x\to 0}\dfrac{e^{3x}-1}{e^x-1} = \lim\limits_{x\to 0}\left(e^{2x}+e^x+1\right) = 3$

47. $\lim\limits_{\Delta x\to 0}\dfrac{f(x+\Delta x)-f(x)}{\Delta x} = \lim\limits_{\Delta x\to 0}\dfrac{7-7}{\Delta x}$

$\qquad\qquad\qquad\qquad\qquad = 0$

49. $\lim\limits_{\Delta x\to 0}\dfrac{f(x+\Delta x)-f(x)}{\Delta x}$

$\quad = \lim\limits_{\Delta x\to 0}\dfrac{\sqrt{2(x+\Delta x)}-\sqrt{2x}}{\Delta x}$

$\quad = \lim\limits_{\Delta x\to 0}\dfrac{\left(\sqrt{2x+2\Delta x}-\sqrt{2x}\right)\left(\sqrt{2x+2\Delta x}+\sqrt{2x}\right)}{\Delta x\left(\sqrt{2x+2\Delta x}+\sqrt{2x}\right)}$

$\quad = \lim\limits_{\Delta x\to 0}\dfrac{2x+2\Delta x-2x}{\Delta x\left(\sqrt{2x+2\Delta x}+\sqrt{2x}\right)}$

$\quad = \lim\limits_{\Delta x\to 0}\dfrac{2}{\sqrt{2x+2\Delta x}+\sqrt{2x}}$

$\quad = \dfrac{1}{\sqrt{2x}}$

51. $\lim\limits_{\Delta x\to 0}\dfrac{f(x+\Delta x)-f(x)}{\Delta x}$

$\quad = \lim\limits_{\Delta x\to 0}\dfrac{\frac{4}{x+\Delta x}-\frac{4}{x}}{\Delta x}$

$\quad = \lim\limits_{\Delta x\to 0}\dfrac{1}{\Delta x}\left[\dfrac{4x-4x-4\Delta x}{x(x+\Delta x)}\right]$

$\quad = \lim\limits_{\Delta x\to 0}\dfrac{-4}{x(x+\Delta x)}$

$\quad = \dfrac{-4}{x^2}$

53. $\lim\limits_{\Delta x\to 0}\dfrac{f(x+\Delta x)-f(x)}{\Delta x}$

$\quad = \lim\limits_{\Delta x\to 0}\dfrac{\cos(x+\Delta x)-\cos x}{\Delta x}$

$\quad = \lim\limits_{\Delta x\to 0}\dfrac{\cos x\cos\Delta x-\sin x\sin\Delta x-\cos x}{\Delta x}$

$\quad = (-\sin x)\lim\limits_{\Delta x\to 0}\left(\dfrac{\sin\Delta x}{\Delta x}\right)$

$\qquad + (\cos x)\lim\limits_{\Delta x\to 0}\left(\dfrac{\cos\Delta x-1}{\Delta x}\right)$

$\quad = -\sin x(1)+(\cos x)(0)$

$\quad = -\sin x$

55. $\lim\limits_{\Delta x\to 0}\dfrac{\ln(x+\Delta x)-\ln x}{\Delta x}$

$\quad = \lim\limits_{\Delta x\to 0}\dfrac{\ln\left(\frac{x+\Delta x}{x}\right)}{\Delta x}$

$\quad = \lim\limits_{\Delta x\to 0}\dfrac{\ln\left(1+\frac{\Delta x}{x}\right)}{\Delta x}\quad Let\ \dfrac{1}{h}=\dfrac{\Delta x}{x}$

$\quad = \lim\limits_{h\to\infty}\dfrac{h}{x}\ln\left(1+\dfrac{1}{h}\right)$

$\quad = \dfrac{1}{x}\lim\limits_{h\to\infty}\ln\left(1+\dfrac{1}{h}\right)^h$

$\quad = \dfrac{1}{x}\ln\left[\lim\limits_{h\to\infty}\left(1+\dfrac{1}{h}\right)^h\right]$

$\quad = \dfrac{1}{x}\ln e$

$\quad = \dfrac{1}{x}$

57. $\displaystyle\lim_{\Delta x \to 0} \frac{f(x + \Delta x) - f(x)}{\Delta x}$

$\displaystyle= \lim_{\Delta x \to 0} \frac{\tan(x + \Delta x) - \tan x}{\Delta x}$ *Let* $c = x + \Delta x$

$\displaystyle= \lim_{\Delta x \to 0} \frac{1}{\Delta x}\left(\frac{\sin c}{\cos c} - \frac{\sin x}{\cos x}\right)$

$\displaystyle= \lim_{\Delta x \to 0} \frac{1}{\Delta x}\left(\frac{\sin c \cos x - \cos c \sin x}{\cos c \cos x}\right)$

$\displaystyle= \lim_{\Delta x \to 0} \frac{1}{\Delta x}\left[\frac{\sin(c - x)}{\cos c \cos x}\right]$

$\displaystyle= \lim_{\Delta x \to 0} \frac{1}{\Delta x}\left[\frac{\sin(x + \Delta x - x)}{\cos c \cos x}\right]$

$\displaystyle= \lim_{\Delta x \to 0} \left(\frac{\sin \Delta x}{\Delta x}\right)[\sec(x + \Delta x)\sec x]$

$= \sec^2 x$

59. $\displaystyle\lim_{\Delta x \to 0} \frac{\log_2(x + \Delta x) - \log_2 x}{\Delta x}$

$\displaystyle= \lim_{\Delta x \to 0} \frac{\log_2\left(\frac{x + \Delta x}{x}\right)}{\Delta x}$

$\displaystyle= \lim_{\Delta x \to 0} \frac{\log_2\left(1 + \frac{\Delta x}{x}\right)}{\Delta x}$ *Let* $\frac{1}{h} = \frac{\Delta x}{x}$

$\displaystyle= \lim_{h \to \infty} \frac{h}{x}\log_2\left(1 + \frac{1}{h}\right)$

$\displaystyle= \frac{1}{x}\lim_{h \to \infty} \log_2\left(1 + \frac{1}{h}\right)^h$

$\displaystyle= \frac{1}{x}\lim_{h \to \infty} \frac{\ln\left(1 + \frac{1}{h}\right)^h}{\ln 2}$

$\displaystyle= \frac{1}{x\ln 2}\ln\left[\lim_{h \to \infty}\left(1 + \frac{1}{h}\right)^h\right]$

$\displaystyle= \frac{1}{x\ln 2}\ln e$

$\displaystyle= \frac{1}{x\ln 2}$

61. Suspicious point is $x = 8$; not continuous since $f(x)$ is not defined at $x = 8$. It is not possible to redefine f at one point so that f becomes continuous because it is a pole discontinuity.

63. Suspicious point at $x = 36$ is not in the interval. Continuous on $[-5, 5]$.

65. Suspicious point at $x = 2$.
(1) defined;
(2) $\displaystyle\lim_{x \to 2}|x - 2| = 0$
(3) $\displaystyle\lim_{x \to 2}|x - 2| = f(2)$
It is continuous on $[-5, 5]$.

67. a. Continuous on $[0, 5]$.
  b. Not continuous at $x = -2$.
  c. $\displaystyle\lim_{x \to -2^-} g(x) = \lim_{x \to -2^-} \frac{x^2 - 3x - 10}{x + 2}$
  $\displaystyle= \lim_{x \to -2^-} \frac{(x - 5)(x + 2)}{x + 2}$
  $= -7$

Also,

$\displaystyle\lim_{x \to -2^+} g(x) = \lim_{x \to -2^+} (x - 5)$
$= -7$

Thus, $\displaystyle\lim_{x \to 2} g(x) = -7 = g(-2)$ so g is continuous on $[-5, 5]$.

69. a.

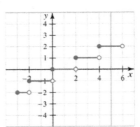

b. $\displaystyle\lim_{x \to 3^-} \left[\!\left[\tfrac{x}{2}\right]\!\right] = 1, \lim_{x \to 3^+} \left[\!\left[\tfrac{x}{2}\right]\!\right] = 1$ so $\displaystyle\lim_{x \to 3} \left[\!\left[x\right]\!\right] = 1$.

c. The limit exists for any number that is not an even integer.

71. $\lim\limits_{x \to 1^-} (Ax + 3) = A + 3;$

$A + 3 = 5$, so $A = 2$.

$\lim\limits_{x \to 1^+} (x^2 + B) = 1 + B;$

$1 + B = 5$, so $B = 4$.

73. The limits exists only if

$$\lim_{x \to 3} (x^3 + cx^2 + 5x + 12) = 0$$

$$3^3 + c(3)^2 + 5(3) + 12 = 0$$

$$9c = -54$$

$$c = -6$$

We have

$$\lim_{x \to 3} \frac{x^3 - 6x^2 + 5x + 12}{x^2 - 7x + 12}$$

$$= \lim_{x \to 3} \frac{(x - 3)(x^2 - 3x - 4)}{(x - 3)(x - 4)}$$

$$= 4$$

75. a.

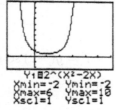

Y1◻2^(X²−2X)
Xmin=-2 Ymin=-2
Xmax=6 Ymax=10
Xscl=1 Yscl=1

b. It crosses the y-axis at $(0, 1)$.

As $x \to \infty$, $y \to \infty$.

As $x \to -\infty$, $y \to \infty$.

c. The smallest value of y is $E = 0.5$.

77. a. $F(5) = 2F(9)$

$$e^{-5k} = 2e^{-9k}$$

$$e^{4k} = 2$$

$$4k = \ln 2$$

$$k = 0.25 \ln 2$$

Thus, $F(7) = e^{-k(7)} \approx 0.2973$; after 7 weeks, approximately 29.7% are burning.

b. $1 - F(10) = 1 - e^{-k(10)}$

$$\approx 0.8232$$

This is about 82.3%.

c. $F(4) - F(5) \approx 0.07955$

This is about 7.96%.

79. If we compare a given sound intensity I and compare it with the threshold intensity I_0, we have

$$D = 10 \log \frac{I}{I_0}$$

$$\frac{D}{10} = \log \frac{I}{I_0}$$

$$\frac{I}{I_0} = 10^{D/10}$$

$$I = I_0 10^{D/10}$$

Thus, for a rock concert at 110 decibels,

$$I_r = I_0 10^{110/10}$$

$$= 10^{-16} \cdot 10^{11}$$

$$= 10^{-5}$$

and for normal conversation,

$$I_n = I_0 10^{50/10} = 10^{-16} \cdot 10^5 = 10^{-11}$$

The difference in loudness between the concert and normal conversation is

$$\frac{I_r}{I_n} = \frac{10^{-5}}{10^{-11}} = 10^6$$

The rock concert is one million times as intense as normal conversation. Finally,

$$D = 10 \log 10^6 = 60$$

times as loud.

81. $f(-1) = (-1)^3 - (-1)^2 + (-1) + 1$
$\qquad = -2$
$\quad f(1) = 1^3 - 1^2 + 1 + 1$
$\qquad = 2$
Since $f(-1) < 0$ and $f(1) > 0$, and since f is continuous, there must be some value x on $(-1, 1)$ so that $f(c) = 0$.

83. The radius to the point of contact $P_0(x_0, y_0)$ is perpendicular to the tangent line. The slope of this radius is $m = \dfrac{y_0}{x_0}$.

The slope of the tangent line is $\dfrac{-x_0}{y_0}$ and the equation of the tangent line is

$$y - y_0 = -\frac{x_0}{y_0}(x - x_0)$$

The desired equation is

$$y_0 y + x x_0 = x_0^2 + y_0^2 = r^2$$

85. $\lim\limits_{x \to 0} f x) = \lim\limits_{x \to 0} \dfrac{1}{x}$ does not exist.

$\lim\limits_{x \to 0} g(x) = \lim\limits_{x \to 0} \dfrac{1}{\sin x}$ does not exist.

$\lim\limits_{x \to 0} \dfrac{f(x)}{g(x)} = \lim\limits_{x \to 0} \dfrac{\sin x}{x} = 1$

87. Answers vary. You should have no trouble finding sources for this question. You should include at least one of the journal sources listed as well as internet sources.

89. $\lim\limits_{x \to c}[f(x) + g(x)] = L_1$ and $\lim\limits_{x \to c} f(x) = L_2$.
Then,

$$\lim\limits_{x \to c} g(x) = \lim\limits_{x \to c} \{[f(x) + g(x)] - f(x)\}$$
$$= L_1 - L_2$$

91. From Figure 2.30 in the text, the area of sector AOB is between that of $\triangle AOB$ and $\triangle BOD$. Thus,

$$\frac{1}{2}(\tan x)(1) \ge \frac{1}{2}x(1)^2 \ge \frac{1}{2}\sin x \cos x$$

Divide by $\frac{1}{2}\sin x > 0$ (in Quadrant I):

$$\frac{1}{\cos x} \ge \frac{x}{\sin x} \ge \cos x$$

Since $1 \ge \cos x$, we have

$$1 \ge \frac{1}{\cos x} \ge \frac{x}{\sin x}$$
$$\lim\limits_{x \to 0} 1 \ge \lim\limits_{x \to 0} \frac{1}{\cos x} \ge \lim\limits_{z \to 0} \frac{x}{\sin x}$$
$$1 \ge \lim\limits_{x \to 0} \frac{1}{\cos x} \ge 1$$

Thus, $\lim\limits_{x \to 0} \cos x = 1$ which implies $\lim\limits_{x \to 0} \cos x = 1$. Since $\cos 0 = 1$, the cosine is continuous at $x = 0$.
$\lim\limits_{x \to c}(\cos x) = \lim\limits_{x \to c}\{\cos[(x - c) + c]\}$
$\qquad = \lim\limits_{x \to c}[\cos(x - c)\cos c - \sin(x - c)\sin c]$
$\qquad = (\cos c)\lim\limits_{x \to c}\cos(x - c) - (\sin c)\lim\limits_{x \to c}\sin(x - c)$
$\qquad = (\cos c)(1) - (\sin c)(0)$
$\qquad = \cos c$

93. Let $H(t)$ be the height of water in the cylinder t minutes after water begins to flow, and let $h(t)$ be the corresponding height of water in the trough. Then $H(t)$ and $h(t)$ are continuous and so is $F(t) = H(t) - h(t)$. Let t_1 be the time needed to drain the cylinder. Then, $h(0) = 0 = H(t_1)$, so

$$F(0) = H(0) - h(0) = H(0) - 0 > 0$$
$$F(t_1) = H(t_1) - h(t_1) = 0 - h(t_1) < 0$$

and it follows that $F(t) = 0$ for some time $\bar{t}$ between $t = 0$ and $t = t_1$.

95. a. The wind chill for $v = 20$ mi/h is

$91.4 + (91.4 - 30)(0.0203)(20) - 0.304\sqrt{20} - 0.474)$
$\approx 3.75°$

and for $v = 50$ mi/h it is

$$1.6(30) - 55 = -7°$$

b. A calculator solution for $T = 30$, and $w = 0$ and the equation

$91.4 + (91.4 - 30)(0.0203v - 0.304\sqrt{v} - 0.474) = 0$
gives two values for v: $v \approx 25.2$ (99.1 is outside $4 < v < 45$).

c. At $v = 4$, the wind chill function is $w(v) = T$ and is continuous for all values of T on $(0, 4)$. As $v \to 4^-$, $w(v) = T$ and as $v \to 4^+$,

$w(v) = 91.4 + (91.4 - T)(0.0203v - 0.304\sqrt{v} - 0.474)$
which equals T if $T \approx 91.4$.

At $v = 45$, $w(v) = 1.6T - 55$, and is continuous for all values of T on $(45, \infty)$. As $v \to 45^+$, $w(v) = 1.6T - 55$, and as $v \to 45^-$,

$w(v) = 91.4 + (91.4 - T)(0.0203v - 0.304\sqrt{v} - 0.474)$
which equals $1.6T - 5$ if $T \approx 868$, a highly unlikely value.

97. Let $\epsilon > 0$ be given and let $\delta = \epsilon/M$. Then, if $|x - c| < \delta$, we have

$$|f(x) - f(c)| < M|x - c| < M\left(\frac{\epsilon}{M}\right) = \epsilon$$

so $\lim\limits_{x \to c} f(x) = f(c)$ and f is continuous at $x = c$.

99. This is Problem A1 of the morning session of the 1956 Putnam Examination. For

$$f(x) = \left[\frac{1}{x} \cdot \frac{a^x - 1}{a - 1}\right]^{1/x}$$

let $x > 0$ and $a > 1$. Then

$$\ln f(x) = -\frac{\ln x}{x} - \frac{\ln(a - 1)}{x} + \frac{\ln(a^x - 1)}{x}$$

as $x \to \infty$, $\dfrac{\ln x}{x} \to 0$ and $\dfrac{\ln(a - 1)}{x} \to 0$,

while

$$\frac{\ln(a^x - 1)}{x} = \frac{\ln(1 - a^{-x})}{x} + \ln a \to \ln a$$

as $x \to \infty$, so

$$\lim_{x \to \infty} f(x) = \ln a$$

If $0 < a < 1$, then

$$\ln f(x) = -\frac{\ln x}{x} - \frac{\ln(1 - a)}{x} + \frac{\ln(1 - a^x)}{x}$$
$$\lim_{x \to \infty} [\ln f(x)] = -0 - 0 + 0$$
$$= 0$$

Since e^x is an continuous function,

$$\lim_{x \to \infty} f(x) = e\left[\lim_{x \to \infty} \ln f(x)\right]$$
$$= \begin{cases} e^{\ln a} = a & \text{if } a > 1 \\ e^0 = 1 & \text{if } 0 < a < 1 \end{cases}$$

CHAPTER 3 Differentiation

Chapter Overview
Calculus is sometimes divided into two parts: *differential calculus* and *integral calculus*.
Differential calculus refers to the derivative and application of the derivative. In this chapter,
we consider various ways of efficiently computing derivatives. We have seen how to apply
the definition of a derivative, but the purpose of this chapter is to provide techniques that,
while based on the definition, provide easy-to-use methods. Your success in this course
depends on your understanding of the material in this chapter.

3.1 An Introduction to the Derivative: Tangent, page 141

SURVIVAL HINT: *The definition of a derivative is given on page 135. Read this, study this, memorize it, and explain it to your best friend! This definition is foundational to this course.*

1. Some describe it as a five-step process:
   (1) Find $f(x)$
   (2) Find $f(x + \Delta x)$
   (3) Find $f(x + \Delta x) - f(x)$
   (4) Find $\dfrac{f(x + \Delta x) - f(x)}{\Delta x}$
   (5) Finally find $\displaystyle\lim_{\Delta x \to 0} \dfrac{f(x + \Delta x) - f(x)}{\Delta x}$

3. **a.** Continuity does not imply differentiability.
   b. Differentiability implies continuity.

5.  |7.

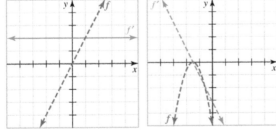

9.

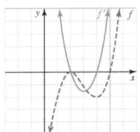

11. **a.** $\dfrac{f(-5 + \Delta x) - f(-5)}{\Delta x} = \dfrac{3 - 3}{\Delta x}$
$= 0$

   b. $f'(-5) = \displaystyle\lim_{\Delta x \to 0} 0 = 0$

13. **a.** $\dfrac{f(1 + \Delta x) - f(1)}{\Delta x} = \dfrac{2 + 2\Delta x - 2}{\Delta x}$
$= 2$

   b. $f'(1) = \displaystyle\lim_{\Delta x \to 0} 2 = 2$

15. **a.** $\dfrac{f(0 + \Delta x) - f(0)}{\Delta x}$
$= \dfrac{2 - (\Delta x)^2 - 2}{\Delta x}$
$= -\Delta x$

   b. $f'(1) = \displaystyle\lim_{\Delta x \to 0} (-\Delta x) = 0$

Page 47

SURVIVAL HINT: *Problems 17-24 are designed to help you understand the definition of derivative.*

17. $f'(x) = \lim\limits_{\Delta x \to 0} \dfrac{f(x + \Delta x) - f(x)}{\Delta x}$

$\quad\quad = \lim\limits_{\Delta x \to 0} \dfrac{5 - 5}{\Delta x}$

$\quad\quad = 0$

Differentiable for all x.

19. $F'(x) = \lim\limits_{\Delta x \to 0} \dfrac{F(x + \Delta x) - F(x)}{\Delta x}$

$\quad\quad = \lim\limits_{\Delta x \to 0} \dfrac{[3(x + \Delta x) - 7] - [3x - 7]}{\Delta x}$

$\quad\quad = \lim\limits_{\Delta x \to 0} \dfrac{3\Delta x}{\Delta x}$

$\quad\quad = \lim\limits_{\Delta x \to 0} 3$

$\quad\quad = 3$

Differentiable for all x.

21. $f'(t) = \lim\limits_{\Delta t \to 0} \dfrac{f(t + \Delta t) - f(t)}{\Delta t}$

$\quad\quad = \lim\limits_{\Delta t \to 0} \dfrac{\left[3(t + \Delta t)^2\right] - \left[3t^2\right]}{\Delta t}$

$\quad\quad = \lim\limits_{\Delta t \to 0} \dfrac{6t\Delta t + 3(\Delta t)^2}{\Delta t}$

$\quad\quad = \lim\limits_{\Delta t \to 0} (6t + 3\Delta t)$

$\quad\quad = 6t$

Differentiable for all t.

23. $h'(x) = \lim\limits_{\Delta x \to 0} \dfrac{h(x + \Delta x) - h(x)}{\Delta x}$

$\quad\quad = \lim\limits_{\Delta x \to 0} \dfrac{1}{\Delta x}\left[\dfrac{1}{2x + 2\Delta x} - \dfrac{1}{2x}\right]$

$\quad\quad = \lim\limits_{\Delta x \to 0} \dfrac{2x - 2x - 2\Delta x}{\Delta x(2x)(2x + 2\Delta x)}$

$\quad\quad = \lim\limits_{\Delta t \to 0} \dfrac{-1}{2x(x + \Delta x)} = \dfrac{-1}{2x^2}$

Differentiable for all $x \neq 0$.

25. $f'(x) = \lim\limits_{\Delta x \to 0} \dfrac{f(x + \Delta x) - f(x)}{\Delta x}$

$\quad\quad = \lim\limits_{\Delta x \to 0} \dfrac{[2(x + \Delta x) + 3] - (2x + 3)}{\Delta x}$

$\quad\quad = \lim\limits_{\Delta x \to 0} \dfrac{2\Delta x}{\Delta x}$

$\quad\quad = 2$

$f'(x) = 2 = m$, so

$$y - 1 = 2(x + 1)$$
$$2x - y + 3 = 0$$

27. $f'(s) = 3s^2$ (See Example 9)

$f'\left(-\frac{1}{2}\right) = 3\left(-\frac{1}{2}\right)^2 = \frac{3}{4} = m$

Also, $f\left(-\frac{1}{2}\right) = \left(-\frac{1}{2}\right)^3 = -\frac{1}{8}$

$$t - \left(-\dfrac{1}{8}\right) = \dfrac{3}{4}\left[s - \left(-\dfrac{1}{2}\right)\right]$$
$$8t + 1 = 6s + 3$$
$$3s - 4t + 1 = 0$$

29. $h'(x) = \lim\limits_{\Delta x \to 0} \dfrac{h(x + \Delta x) - h(x)}{\Delta x}$

$\quad\quad = \lim\limits_{\Delta x \to 0} \dfrac{[(x + \Delta x) - 1] - [x - 1]}{\Delta x}$

$\quad\quad = \lim\limits_{\Delta x \to 0} \dfrac{\Delta x}{\Delta x}$

$\quad\quad = \lim\limits_{\Delta x \to 0} 1$

$\quad\quad = 1$

$h'(0) = 1$

$h(0) = 0 - 1 = -1$

$$y + 1 = 1(x - 0)$$
$$x - y - 1 = 0$$

31. $f'(x) = \lim\limits_{\Delta x \to 0} \dfrac{f(x + \Delta x) - f(x)}{\Delta x}$

$= \lim\limits_{\Delta x \to 0} \dfrac{[3(x + \Delta x) - 5] - [3x - 5]}{\Delta x}$

$= \lim\limits_{\Delta x \to 0} \dfrac{3\Delta x}{\Delta x}$

$= \lim\limits_{\Delta x \to 0} 3$

$= 3$

$f'(3) = 3$, so the normal line has slope $m = -\frac{1}{3}$. Since it passes through $(3, 4)$, the equation is

$$y - 4 = -\frac{1}{3}(x - 3)$$
$$3y - 12 = -x + 3$$
$$x + 3y - 15 = 0$$

33. $f'(t) = \lim\limits_{\Delta x \to 0} \dfrac{f(x + \Delta x) - f(x)}{\Delta x}$

$= \lim\limits_{\Delta x \to 0} \dfrac{1}{\Delta x}\left[\dfrac{1}{x + \Delta x + 3} - \dfrac{1}{x + 3}\right]$

$= \lim\limits_{\Delta x \to 0} \dfrac{x + 3 - x - \Delta x - 3}{\Delta x(x + 3)(x + \Delta x + 3)}$

$= \lim\limits_{\Delta x \to 0} \dfrac{-1}{(x + 3)(x + \Delta x + 3)}$

$= \dfrac{-1}{(x + 3)^2}$

$f'(2) = \dfrac{-1}{(2 + 3)^2} = -\dfrac{1}{25}$, so the normal

line has slope 25.

$f(2) = \dfrac{1}{2 + 3} = \dfrac{1}{5}$, so

$y - \dfrac{1}{5} = 25(x - 2)$

$5y - 1 = 125x - 250$

$125x - 5y - 249 = 0$

35. $\dfrac{dy}{dx} = \lim\limits_{\Delta x \to 0} \dfrac{f(x + \Delta x) - f(x)}{\Delta x}$

$= \lim\limits_{\Delta x \to 0} \dfrac{2(x + \Delta x) - 2x}{\Delta x}$

$= \lim\limits_{\Delta x \to 0} \dfrac{2\Delta x}{\Delta x}$

$= 2$

$\left.\dfrac{dy}{dx}\right|_{x=-1} = 2$

37. $\dfrac{dy}{dx} = \lim\limits_{\Delta x \to 0} \dfrac{f(x + \Delta x) - f(x)}{\Delta x}$

$= \lim\limits_{\Delta x \to 0} \dfrac{\left[1 - (x + \Delta x)^2\right] - (1 - x^2)}{\Delta x}$

$= \lim\limits_{\Delta x \to 0} \dfrac{-2x\Delta x - (\Delta x)^2}{\Delta x}$

$= \lim\limits_{\Delta x \to 0} (-2x - \Delta x)$

$= -2x$

$\left.\dfrac{dy}{dx}\right|_{x=0} = 0$

> **SURVIVAL HINT:** ☠ *Be sure you understand what we have just shown with Example 7 and Theorem 3.2: If a function is differentiable at $x = c$, then it must be continuous at that point. The converse is not true: If a function is continuous at $x = c$, then it may or may not be differentiable at that point. Finally, if a function is discontinuous at $x = c$, then it cannot possibly have a derivative at that point.* ☠

39. a. $f(-2) = 4;\ f(-1.9) = 3.61$

$m_{\text{sec}} = \dfrac{4 - 3.61}{-2 + 1.9} = -3.9$

b. From Example 2, $f'(x) = 2x$

$m_{\text{tan}} = f'(-2) = 2(-2) = -4$

41. $f(x) = x^2 - x$

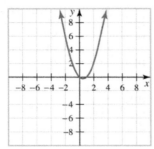

$$f'(x) = \lim_{\Delta x \to 0} \frac{f(x + \Delta x) - f(x)}{\Delta x}$$

$$= \lim_{\Delta x \to 0} \frac{[(x + \Delta x)^2 - (x + \Delta x)] - (x^2 - x)}{\Delta x}$$

$$= \lim_{\Delta x \to 0} \frac{x^2 + 2x\Delta x + (\Delta x)^2 - x - \Delta x - x^2 + x}{\Delta x}$$

$$= \lim_{\Delta x \to 0} (2x + \Delta x - 1)$$

$$= 2x - 1$$

The derivative is zero when $x = \frac{1}{2}$.
The graph has a horizontal tangent at $\left(\frac{1}{2}, -\frac{1}{4}\right)$

43. $f'(x) = \lim_{\Delta x \to 0} \dfrac{f(x + \Delta x) - f(x)}{\Delta x}$

$$f'(2) = \lim_{\Delta x \to 0} \frac{|2 + \Delta x - 2| - |2 - 2|}{\Delta x}$$

$$= \lim_{\Delta x \to 0} \frac{|\Delta x|}{\Delta x}$$

From Example 8, this limit from the left is -1, and from the right it is 1, so the limit, and consequently the derivative, does not exist.

45. If $x < 0$,

$$f'(0) = \lim_{\Delta x \to 0^-} [-2(0) - \Delta x] = 0$$

If $x \geq 0$

$$f'(0) = \lim_{\Delta x \to 0^+} [2(0) + \Delta x] = 0$$

Since these limits are the same, the

derivative exists.

47. Let $\Delta x = 0.1, c = 1, f(x) = (2x - 1)^2$

$$f'(x) \approx \frac{f(x + \Delta x) - f(x)}{\Delta x}$$

$$f'(1) = \frac{f(1.1) - f(1)}{0.1}$$

$$= \frac{1.44 - 1}{0.1}$$

$$= 4.4$$

Let $\Delta x = 0.01$

$$f'(x) \approx \frac{f(x + \Delta x) - f(x)}{\Delta x}$$

$$= \frac{f(1.01) - f(1)}{0.01}$$

$$= \frac{1.0404 - 1}{0.01}$$

$$= 4.04$$

It appears that $f'(1) = 4$.

49. We could proceed as shown in the solution for Problem 47, but here we present data which could be generated using a graphing calculator, spreadsheet, or a computer program.

$$f(x) = \sin x, \ c = \frac{\pi}{3}$$

Δx	$c + \Delta x$	$\dfrac{f(c + \Delta x) - f(c)}{\Delta x}$
0.5	1.5472	0.26739
0.125	1.1722	0.44464
0.03125	1.0784	0.48639
0.00781	1.0550	0.49661
0.00195	1.0491	0.49916
0.00049	1.0477	0.49979

We might guess that the derivative is $f'(\frac{\pi}{3}) = \frac{1}{2} \approx 0.5$
Note: later we will find this to be $\cos \frac{\pi}{3} = \frac{1}{2}$.

51. We could proceed as shown in the solution for Problem 47, but here we present data which could be generated using a graphing calculator, spreadsheet, or a computer program.

$$f(x) = \sqrt{x}, \ c = 4$$

Δx	$c + \Delta x$	$\dfrac{f(c + \Delta x) - f(c)}{\Delta x}$
0.5	4.5	0.24264
0.125	4.125	0.24808
0.03125	4.0313	0.24951
0.00781	4.0078	0.24988
0.00195	4.0020	0.24997
0.00049	4.0005	0.24999

We might guess that the derivative is $f'(4) = \frac{1}{4} \approx 0.25$

53. a. From Example 2, $f'(x) = 2x$
For $g(x) = x^2 - 3$, we have

$$g'(x) = \lim_{\Delta x \to 0} \frac{g(x + \Delta x) - g(x)}{\Delta x}$$

$$= \lim_{\Delta x \to 0} \frac{\left[(x + \Delta x)^2 - 3\right] - [x^2 - 3]}{\Delta x}$$

$$= \lim_{\Delta x \to 0} \frac{x^2 + 2x\Delta x + (\Delta x)^2 - 3 - (x^2 - 3)}{\Delta x}$$

$$= \lim_{\Delta x \to 0} (2x + \Delta x)$$

$$= 2x$$

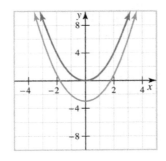

The graph of $y = x^2 - 3$ is the graph of $y = x^2$ lowered by 3 units. Thus, the tangents of the graphs have the same slopes for any value of x. This accounts geometrically for the fact their derivatives are identical.

b. The graph of $y = x^2 + 5$ is also the same as the graph of $y = x^2$ raised by three units. Thus,

$$\frac{d}{dx}(x^2 + 5) = \frac{d}{dx}(x^2) = 2x$$

55. For $x < 0$, the tangent lines have negative slope and for $x > 0$ the tangent lines have positive slope. There is a corner at $x = 0$, and there is no tangent at $x = 0$, and consequently no derivative at that point.

57. a. $\triangle STP \sim \triangle PQR$ because their sides are parallel, so the corresponding sides are proportional: $\dfrac{A}{y_0} = \dfrac{E}{|QR|}$ implies

$|QR| = \dfrac{Ey_0}{A}$. From this Fermat found the coordinates of Q to be

$$x_1 = x_0 + E, \ y_1 = y_0\left(1 + \frac{E}{A}\right)$$

b. $x^3 + y^3 - 2xy = 0$

$$(x_0 + E)^3 + y_0^3\left(1 + \frac{E}{A}\right)^3 - 2y_0(x_0 + E)\left(1 + \frac{E}{A}\right) = 0$$

After several steps find

$$E\left(3x_0^2 + \frac{3y_0^3}{A} - \frac{2x_0y_0}{A} - 2y_0\right)$$
$$+ E^2\left(3x_0 + \frac{3y_0^3}{A^2} - \frac{2y_0}{A}\right)$$
$$+ E^3\left(1 + \frac{y_0^3}{A^3}\right) + 0 = 0$$

c. Divide both sides by E, and *then* let $E = 0$ to find

$$A = -\frac{3y^3 - 2xy}{3x^2 - 2y}$$

d. See Figure 3.14 in textbook. Estimate the slope of the line passing through P and S. We estimate the rise to be about 0.9 and the run to be about 1.3 so we estimate the slope to be $0.9/1.3 \approx 0.7$ to find the equation of the tangent line is

$$y - 0.9 = 0.7(x - 0.5)$$

59. $\dfrac{dy}{dx} = 2Ax; \quad \dfrac{dy}{dx}\bigg|_{x=c} = 2Ac$

The point is $(c, f(c)) = (c, ac^2)$.
The equation of the tangent line is:

$$y - Ac^2 = 2Ac(x - c)$$

The x-intercept occurs when $y = 0$:

$$-Ac^2 = 2Ac(x - c)$$
$$-c = 2(x - c)$$
$$x = \frac{c}{2}$$

The y-intercept occurs when $x = 0$:

$$y - Ac^2 = -2Ac^2$$
$$y = -Ac^2$$

3.2 Techniques of Differentiation, page 152

> SURVIVAL HINT: *This section presents the "meat and potatoes" of differential calculus. To see the value of its content, look at Problems 1-4. Notice how several hours of "work" in Problem Set 3.1 can be condensed to a several minutes or work using the formulas of this section.*

1. (11) $f'(x) = 0$ (12) $f'(x) = 1$
   (13) $f'(x) = 2$ (14) $f'(x) = 4x$
   (15) $f'(x) = -2x$ (16) $f'(x) = -2x$

3. (25) $f'(x) = 2$ (26) $g'(x) = 6x$
   (27) $f'(s) = 3s^2$ (28) $g'(t) = -2t$
   (29) $h(x) = 1$

> SURVIVAL HINT: *Problems 5-20 give you practice in using Theorem 3.5. This theorem is so important, that in a short while, every successful student in this course will have these formulas committed to memory.*

5. a. $f'(x) = 3(4)x^3 - 0$
           $= 12x^3$
  b. $g'(x) = 0 - 1$
           $= -1$

7. a. $f'(x) = 3x^2$
  b. $g'(x) = 1$

SURVIVAL HINT: *Theorem 3.4 (Power rule) is one of the most important formulas in mathematics. From this point on, you will use this formula every day you are working in this book.*

9. $r(t) = t^2 - t^{-2} + 5t^{-4}$
   $r'(t) = 2t + 2t^{-3} + 20t^{-5}$

11. $f(x) = 7x^{-2} + x^{2/3} + C$
    $f'(x) = -14x^{-3} + \frac{2}{3}x^{-1/3}$

13. $f(x) = x + 1 + x^{-1} - 7x^{-2}$
    $f'(x) = 1 - x^{-2} + 14x^{-3}$

15. $f(x) = 2x - 8x^4 + 1 - 4x^3$
    $f'(x) = -32x^3 - 12x^2 + 2$

17. $f'(x) = \dfrac{(x+9)(3) - (3x+5)(1)}{(x+9)^2}$

    $= \dfrac{22}{(x+9)^2}$

19. $g(x) = x^4 + 4x^3 + 4x^2$
    $g'(x) = 4x^3 + 12x^2 + 8x$

21. $f(x) = x^5 - 5x^3 + x + 12$
    $f'(x) = 5x^4 - 15x^2 + 1$
    $f''(x) = 20x^3 - 30x$
    $f'''(x) = 60x^2 - 30$
    $f^{(4)}(x) = 120x$

23. $f(x) = -2x^{-2}$
    $f'(x) = 4x^{-3}$
    $f''(x) = -12x^{-4}$
    $f'''(x) = 48x^{-5}$
    $f^{(4)}(x) = -240x^{-6}$

25. $y = 3x^3 - 7x^2 + 2x - 3$
    $\dfrac{dy}{dx} = 9x^2 - 14x + 2$
    $\dfrac{d^2y}{dx^2} = 18x - 14$

27. $f(-2) = (-2)^2 - 3(-2) - 5$
    $= 5$

Point of tangency is $(-2, 5)$.
$f'(x) = 2x - 3;\ f'(-2) = -7$
$$y - 5 = -7(x + 2)$$
$$7x + y + 9 = 0$$

29. $f(1) = (1^2 + 1)(1 - 1^3)$
    $= 0$
    Point of tangency is $(1, 0)$.
    $f(x) = -x^5 - x^3 + x^2 + 1$
    $f'(x) = -5x^4 - 3x^2 + 2x$
    $f'(1) = -6$
    $$y - 0 = -6(x - 1)$$
    $$6x + y - 6 = 0$$

31. $f(1) = 1$
    Point of tangency is $(1, 1)$.
    $f'(x) = \dfrac{(x+5)(2x) - (x^2+5)(1)}{(x+5)^2}$

    $= \dfrac{x^2 + 10x - 5}{(x+5)^2}$

    $f'(1) = \dfrac{1}{6}$
    $$y - 1 = \frac{1}{6}(x - 1)$$
    $$x - 6y + 5 = 0$$

33. $f'(x) = 6x^2 - 14x + 8$
    $$6x^2 - 14x + 8 = 0$$
    $$2(3x - 4)(x - 1) = 0$$
    $$x = \frac{4}{3},\ 1$$

    $f\left(\frac{4}{3}\right) = -\frac{1}{27}$ for the point $\left(\frac{4}{3}, -\frac{1}{27}\right)$;
    $f(1) = 0$ for the point $(1, 0)$.

35. $f(t) = t^{-2} - t^{-3}$
$\quad f'(t) = -2t^{-3} + 3t^{-4}$

$$-\frac{2}{t^3} + \frac{3}{t^4} = 0$$
$$-2t + 3 = 0$$
$$t = \frac{3}{2}$$

$f\left(\frac{3}{2}\right) = \left(\frac{3}{2}\right)^{-2} - \left(\frac{3}{2}\right)^{-3} = \frac{4}{27}$
The point of tangency is $\left(\frac{3}{2}, \frac{4}{27}\right)$.

37. $h(u) = u^{1/2} + 9u^{-1/2}$
$\quad h'(u) = \frac{1}{2}u^{-1/2} - \frac{9}{2}u^{-3/2}$

$$\frac{1}{2}u^{-1/2} - \frac{9}{2}u^{-3/2} = 0$$
$$u - 9 = 0$$

$h(9) = (9)^{1/2} + 9(9)^{-1/2} = 3 + 3 = 6$
The point of tangency is $(9, 6)$.

39. a. $f'(x) = 4x - 5$
$\quad$ **b.** $f(x) = (2x + 1)(x - 3)$
$\qquad f'(x) = (2x + 1)(1) + 2(x - 3)$
$\qquad\qquad = 4x - 5$

41. The derivative of $y = x^4 - 2x + 1$ is
$y' = 4x^3 - 2$. The given line has a slope 2, so
we want the point(s) of tangency (x_0, y_0) to
satisfy

$$4x_0^3 - 2 = 2$$
$$x_0^3 = 1$$
$$x_0 = 1$$

$y_0 = 1^4 - 2(1) + 1 = 0$. The equation of the
tangent line is

$$y - 0 = 2(x - 1)$$
$$2x - y - 2 = 0$$

43. a. $f(x) = x^4 - 4x^2$
$\qquad f(1) = 1^3 - 4(1)^2 = -3$
$\qquad f'(x) = 4x^3 - 8x;\ f'(1) = -4$
$\qquad$ The equation of the tangent line is

$$y + 3 = -4(x - 1)$$
$$4x + y - 1 = 0$$

$\quad$ **b.** $f'(0) = 0$ so the slope is not defined
$\qquad$ and the normal line is vertical with
$\qquad$ equation $x = 0$ (that is, the y-axis).

45. We are looking for particular points (x_0, y_0) on the graph of $y = 4x^2$ which have a tangent at that point which will pass through the point $(2, 0)$.

$$y' = 8x;\ f(x_0) = 4x_0^2;\ f'(x_0) = 8x_0$$

So we have a point, $(x_0, 4x_0^2)$ and the slope of the line at that point, $8x_0$. We can now write the equation of the line:

$$y - 4x_0^2 = 8x_0(x - x_0)$$

This line must pass through the point $(2, 0)$, so

$$0 - 4x_0^2 = 8x_0(2 - x_0)$$
$$4x_0^2 - 16x_0 = 0$$
$$4x_0(x_0 - 4) = 0$$
$$x_0 = 0, 4$$

Therefore there are two points on the curve at which the tangent line will pass through $(2, 0)$. If $x_0 = 0$, then $y = 4(0)^2 = 0$; point $(0, 0)$; if $x_0 = 4$, then $y = 4(4)^2 = 64$; point $(4, 64)$.

47. $f'(x) = 3x^2 + 2x + 1$
$\quad f''(x) = 6x + 2$
$\quad f'''(x) = 6$

$$y''' + y'' + y' = 3x^2 + 2x + 1 + 6x + 2 + 6$$
$$= 3x^2 + 8x + 9$$

The equation is not satisfied.

49. $f'(x) = 4x + 1$
$f''(x) = 4$
$f'''(x) = 0$

$$y''' + y'' + y' = 0 + 4 + 4x + 1$$
$$= 4x + 5$$

The equation is not satisfied.

51. $P_k(x) = a_k x^k + \cdots$
$P_k'(x) = k a_k x^{k-1} + \cdots$
$P_k''(x) = k(k-1) a_k x^{k-2} + \cdots$
$P_k'''(x) = k(k-1)(k-2) a_k x^{k-3} + \cdots$
$$\vdots$$
$P_k^{(k)}(x) = k! a_k$
The $(k+1)$st derivative is 0.

53. $F(x) = f(x) + g(x)$
$$F'(x) = \lim_{\Delta x \to 0} \frac{F(x + \Delta x) - F(x)}{\Delta x}$$
$$= \lim_{\Delta x \to 0} \frac{[f(x + \Delta x) + g(x + \Delta x)] - [fx) + g(x)]}{\Delta x}$$
$$= \lim_{\Delta x \to 0} \frac{f(x + \Delta x) - f(x)}{\Delta x}$$
$$\quad + \lim_{\Delta x \to 0} \frac{g(x + \Delta x) - g(x)}{\Delta x}$$
$$= f'(x) + g'(x)$$

55. $(fg)' = \frac{1}{2}[(f+g)^2 - f^2 - g^2]'$
$$= \frac{1}{2}[2(f+g)(f+g)' - 2ff' - 2gg']$$
$$= (f+g)(f' + g') - ff' - gg'$$
$$= fg' + f'g$$

57. $r(x) = \dfrac{1}{f(x)}$
By the quotient rule,

$$r'(x) = \frac{f(x)(1)' - (1)f'(x)}{[f(x)]^2}$$
$$= -\frac{f'(x)}{[f(x)]^2}$$

59. a. Write $g = f^3$ where g is f and h is f for the formula in Problem 58. Then
$$g' = f^2 f' + f' f^2 + f^2 f' = 3f^2 f'$$

b. $p(x) = [f(x)]^4$
$$p'(x) = \{f(x)[f(x)]^3\}'$$
$$= f'(x)[f(x)]^3 + [f(x)^3]' f(x)$$
$$= f'(x)[f(x)]^3 + \{3[f(x)]^2 f'(x)]\} f(x)$$
$$= 4[f(x)]^3 f'(x)$$

3.3 Derivatives of Trigonometric, Exponential, and Logarithmic Functions, page 160

1. $f'(x) = \cos x - \sin x$

3. $g'(t) = 2t - \sin t$

5. Write $f(t) = (\sin t)(\sin t)$
$f'(t) = (\sin t)(\cos t) + (\cos t)(\sin t)$
$\quad = 2\sin t \cos t$
$\quad = \sin 2t$

7. $f'(x) = \sqrt{x}(-\sin x) + (\cos x)\left(\frac{1}{2}x^{-1/2}\right)$
$\quad + x(-\csc^2 x) + (\cot x)(1)$
$\quad = -\sqrt{x}\sin x + \frac{1}{2}x^{-1/2}\cos x - x\csc^2 x + \cot x$

9. $p'(x) = x^2(-\sin x) + 2x\cos x$
$\quad = -x^2\sin x + 2x\cos x$

11. $q'(x) = \dfrac{x\cos x - \sin x}{x^2}$

13. $h'(t) = e^t(-\csc t \cot t) + (\csc t)(e^t)$
$\quad = e^t \csc t(1 - \cot t)$

15. $f'(x) = x^2 \left(\dfrac{1}{x} \right) + (\ln x)(2x)$

$\qquad = x + 2x \ln x$

17. $h'(x) = e^x(\sin x - \cos x) + e^x(\cos x + \sin x)$

$\qquad = 2e^x \sin x$

19. $f'(x) = \dfrac{e^x \cos x - \sin x (e^x)}{(e^x)^2}$

$\qquad = \dfrac{e^x(\cos x - \sin x)}{e^{2x}}$

$\qquad = e^{-x}(\cos x - \sin x)$

21. $f'(x) = \dfrac{(1 - 2x)(\sec^2 x) - \tan x(-2)}{(1 - 2x)^2}$

$\qquad = \dfrac{\sec^2 x - 2x \sec^2 x + 2 \tan x}{(1 - 2x)^2}$

23. $f'(t) = \dfrac{(t + 2)(\cos t) - (2 + \sin t)(1)}{(t + 2)^2}$

$\qquad = \dfrac{t \cos t + 2 \cos t - 2 - \sin t}{(t + 2)^2}$

25. $f'(x) = \dfrac{(1 - \cos x)(\cos x) - \sin x (\sin x)}{(1 - \cos x)^2}$

$\qquad = \dfrac{\cos x - (\cos^2 x + \sin^2 x)}{(1 - \cos x)^2}$

$\qquad = \dfrac{\cos x - 1}{(1 - \cos x)^2}$

$\qquad = \dfrac{-1}{1 - \cos x}$

$\qquad = \dfrac{1}{\cos x - 1}$

27. $f'(x) = \dfrac{(2 - \cos x)(\cos x) - (1 + \sin x)(\sin x)}{(2 - \cos x)^2}$

$\qquad = \dfrac{2 \cos x - \cos^2 x - \sin x - \sin^2 x}{(2 - \cos x)^2}$

$\qquad = \dfrac{2 \cos x - \sin x - 1}{(2 - \cos x)^2}$

29. $f'(x) = [(\sin x - \cos x)(\cos x - \sin x)$

$\qquad - (\sin x + \cos x)(\cos x + \sin x)]/(\sin x - \cos x)^2$

$\qquad = \dfrac{-2}{(\sin x - \cos x)^2}$

31. $g(x) = \sec^2 x - \tan^2 x + \cos x$

$\qquad = 1 + \cos x$

$\quad g'(x) = -\sin x$

> **SURVIVAL HINT:** *Now that you have done some derivatives of trigonometric functions, do you feel comfortable with problems like 1-32? If you do not, then perhaps you want to do a bit more practice using Theorems 3.6, 3.7, 3.8, and 3.9.*

---

33. $f(\theta) = \sin \theta$

$\qquad f'(\theta) = \cos \theta$

$\qquad f''(\theta) = -\sin \theta$

35. $f(\theta) = \tan \theta$

$\qquad f'(\theta) = \sec^2 \theta$

$\qquad f''(\theta) = 2 \sec \theta \sec \theta \tan \theta$

$\qquad\qquad = 2 \sec^2 \theta \tan \theta$

37. $f(\theta) = \sec \theta$

$\qquad f'(\theta) = \sec \theta \tan \theta$

$\qquad f''(\theta) = \sec \theta(\sec^2 \theta) + \tan \theta(\sec \theta \tan \theta)$

$\qquad\qquad = \sec^3 \theta + \sec \theta \tan^2 \theta$

39. $f'(x) = \cos x - \sin x$

$\qquad f''(x) = -\sin x - \cos x$

41. $f'(x) = -e^x \sin x + e^x \cos x$

$\qquad f''(x) = -e^x \cos x + (-e^x)\sin x$

$\qquad\qquad + e^x(-\sin x) + \cos x(e^x)$

$\qquad\qquad = -2e^x \sin x$

43. $h'(t) = \sqrt{t} \left(\dfrac{1}{t} \right) + \dfrac{1}{2} t^{-1/2} \ln t$

$\qquad = \dfrac{1}{2} t^{-1/2}(2 + \ln t)$

$$h''(t) = \frac{1}{2}t^{-1/2}\left(\frac{1}{t}\right) - \frac{1}{2}\left(\frac{1}{2}t^{-3/2}\right)(2 + \ln t)$$
$$= \frac{1}{4}t^{-3/2}(2 - 2 - \ln t)$$
$$= -\frac{1}{4}t^{-3/2}\ln t$$

45. $f(\theta) = \tan \theta$
$f(\frac{\pi}{4}) = \tan \frac{\pi}{4} = 1$
$f'(\theta) = \sec^2\theta$
$f'(\frac{\pi}{4}) = \sec^2 \frac{\pi}{4} = 2$
The equation of the tangent line is

$$y - 1 = 2\left(x - \frac{\pi}{4}\right)$$
$$4x - 2y - \pi + 2 = 0$$

47. $f(x) = \sin x$
$f(\frac{\pi}{6}) = \frac{1}{2}$
$f'(x) = \cos x$
$f'(\frac{\pi}{6}) = \frac{\sqrt{3}}{2}$
The equation of the tangent line is

$$y - \frac{1}{2} = \frac{\sqrt{3}}{2}\left(x - \frac{\pi}{6}\right)$$
$$\sqrt{3}x - 2y + \left(1 - \frac{\sqrt{3}\pi}{6}\right) = 0$$

49. $y = x + \sin x$; if $x = 0$, then $y = 0$;
$y' = 1 + \cos x$; if $x = 0$, then $y' = 2$.
The equation of the tangent line is

$$y - 0 = 2(x - 0)$$
$$2x - y = 0$$

51. $y = e^x\cos x$
$y' = e^x(-\sin x) + e^x\cos x$
If $x = 0$, then $y = 1$ and $y' = 1$.
The equation of the tangent line is

$$y - 1 = 1(x - 0)$$
$$x - y + 1 = 0$$

53. $y' = -A \sin x + B \cos x$
$y'' = -A \cos x - B \sin x$

$y'' + 2y' + 3y$
$= (-A\cos x - B\sin x) + 2(-A\sin x + B\cos x)$
$\quad + 3(A\cos x + B\sin x)$
$= (-A + 2B + 3A)\cos x + (-B - 2A + 3B)\sin x$

In order for this to be equal to $2 \sin x$, we see that

$$\begin{cases} -A + 2B + 3A = 0 \\ -B - 2A + 3B = 2 \end{cases}$$

Solving this system, we see that $A = -\frac{1}{2}$ and $B = \frac{1}{2}$.

55. Answers vary; one possibility is
$$f(x) = \begin{cases} x^2\sin\frac{1}{x} & x \neq 0 \\ 0 & x = 0 \end{cases}$$
For $x \neq 0$, the function is differentiable with derivative

$$f'(x) = 2x\sin\frac{1}{x} - \cos\frac{1}{x}$$

For $x = 0$, the squeeze theorem and definition of derivative can be used to show $f'(0) = 0$. This function is not continuous at $x = 0$.

57. $\dfrac{d}{dx}(\cot x) = \dfrac{d}{dx}\left(\dfrac{\cos x}{\sin x}\right)$
$\quad = \dfrac{(\sin x)(-\sin x) - (\cos x)(\cos x)}{\sin^2 x}$
$\quad = \dfrac{-\sin^2 x - \cos^2 x}{\sin^2 x}$

$$= \frac{-1}{\sin^2 x}$$
$$= -\csc^2 x$$

59. $\dfrac{d}{dx}(\csc x) = \dfrac{d}{dx}\left(\dfrac{1}{\sin x}\right)$
$$= \frac{(\sin x)(0) - (1)(\cos x)}{\sin^2 x}$$
$$= -\frac{\cos x}{\sin x}\left(\frac{1}{\sin x}\right)$$
$$= -\cot x \csc x$$

3.4 Rates of Change: Modeling Rectilinear Motion, page 170

> **SURVIVAL HINT:** *The idea of mathematical modeling is central to the way we develop this book. Did you read that subsection which begins on page 164? If you did not, it would be a good idea to do that now. Can you explain what we mean by "mathematical modeling" to someone else?*

1. A mathematical model is a mathematical framework whose results approximate the real world situation. It involves abstractions, predictions, and then interpretations and comparisons with real world events. Comap has an annual contest in mathematical modeling. You can check **www.comap.com** and then click on "Previous Contests" for some good modeling examples.

3. The instantaneous rate of change, or rate of change at a point, is given by $f'(x)$.
$f'(x) = 2x - 3$; $f'(2) = 1$

5. $f'(x) = \dfrac{2}{(x+1)^2}$; $f'(1) = \dfrac{2}{(1+1)^2} = \dfrac{1}{2}$

7. $f'(x) = (x^2 + 2)\left(1 + \tfrac{1}{2}x^{-1/2}\right) + \left(x + x^{1/2}\right)(2x)$
$f'(4) = (18)\left(\tfrac{5}{4}\right) + (6)(8) = \tfrac{141}{2}$

9. $f'(x) = (x+1)\cos x + \sin x$
$f'\left(\tfrac{\pi}{2}\right) = \left(\tfrac{\pi}{2} + 1\right)\cos \tfrac{\pi}{2} + \sin \tfrac{\pi}{2} = 1$

11. $f'(x) = \dfrac{e^x(2x) - x^2 e^x}{(e^x)^2}$
$$= (2x - x^2)e^{-x}$$
$f'(0) = 0$

13. Write $f(x) = \left(x - \dfrac{2}{x}\right)\left(x - \dfrac{2}{x}\right)$
We now use the product rule:
$f'(x) = \left(x - \dfrac{2}{x}\right)\left(1 + \dfrac{2}{x^2}\right)$
$\qquad + \left(x - \dfrac{2}{x}\right)\left(1 + \dfrac{2}{x^2}\right)$
$\qquad = 2\left(\dfrac{x^2 - 2}{x}\right)\left(\dfrac{x^2 + 2}{x^2}\right)$
$\qquad = \dfrac{2(x^4 - 4)}{x^3}$
$f'(1) = \dfrac{2(1^4 - 4)}{1^3} = -6$

15. a. $s'(t) = v(t) = 2t - 2$
b. $s''(t) = a(t) = 2$
c. Object begins at $s(0) = 6$ and ends at $s(2) = 6$; $s'(t) = 0$ when $2t - 2 = 0$ or when $t = 1$. On $[0, 1)$ object retreats to $s(1) = 5$; on $(1, 2]$ object advances. Distance covered:
$$|s(2) - s(1)| + |s(1) - s(0)| = 2$$
d. Because $a(t) > 0$, the object is always accelerating.

17. a. $s'(t) = v(t) = 3t^2 - 18t + 15$
b. $s''(t) = a(t) = 6t - 18$

c. Object begins at $s(0) = 25$ and ends at $s(6) = 7$; $s'(t) = 0$ when

$$3t^2 - 18t + 15 = 0$$
$$3(t - 5)(t - 1) = 0$$
$$t = 1, 5$$

On $[0, 1)$ object retreats to $s(1) = 32$; on $(1, 5)$ object retreats to $s(5) = 0$; and on $(5, 6]$ it advances.
Distance covered:
$$|s(1) - s(0)| + |s(5) - s(1)| + |s(6) - s(5)|$$
$$= |32 - 25| + |0 - 32| + |7 - 0| = 46$$

d. $s''(t) = 0$ when $6t - 18 = 0$ or $t = 3$.
On $[0, 3)$ the object is decelerating and on $(3, 6]$ it is accelerating.

19. Write $s(t) = 2t^{-1} + t^{-2}$

a. $s'(t) = v(t) = -2t^{-2} - 2t^{-3}$

b. $s''(t) = a(t) = 4t^{-3} + 6t^{-4}$

c. Object begins at $s(1) = 3$ and ends at $s(3) = \frac{7}{9}$; $s'(t) = 0$ when

$$-2t^{-2} - 2t^{-3} = 0$$
$$-2t - 2 = 0$$
$$t = -1$$

The number -1 is not in the interval $[1, 3]$. On $[1, 3]$ object retreats.
Distance covered:
$$|s(3) - s(1)| = \left| \frac{7}{9} - 3 \right| = \frac{20}{9}$$

d. $s''(t) = 0$ when

$$4t^{-3} + 6t^{-4} = 0$$
$$4t + 6 = 0$$
$$t = -\frac{3}{2}$$

On $[1, 3]$, $s''(t) > 0$, so the object is accelerating.

21. a. $s'(t) = v(t) = -3 \sin t$

b. $s''(t) = a(t) = -3 \cos t$

c. Object begins at $s(0) = 3$ and ends at $s(2\pi) = 3$; $s'(t) = 0$ when

$$-3 \sin t = 0$$
$$t = 0, \pi, \ 2\pi$$

On $(0, \pi)$ object retreats to $s(\pi) = -3$. On $(\pi, 2\pi)$ object advances to $s(2\pi) = 3$. Distance covered:
$$|s(\pi) - s(0)| + |s(2\pi) - s(\pi)|$$
$$= |-3 - 3| + |3 - (-3)| = 12$$

d. $s''(t) = 0$ when

$$-3 \cos t = 0$$
$$t = \frac{\pi}{2}, \ \frac{3\pi}{2}$$

On $[0, \frac{\pi}{2})$ the object is decelerating, on $\left(\frac{\pi}{2}, \frac{3\pi}{2}\right)$ it is accelerating, and on $\left(\frac{3\pi}{2}, 2\pi\right]$ it is decelerating again.

23. quadratic model   25. exponential model

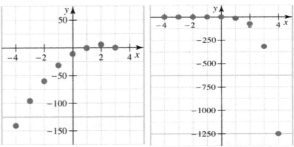

27. quadratic model   **29.** cubic model

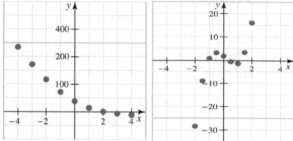

31. logarithmic model

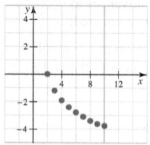

33. a. $f'(x) = -6$

b. The rate of change is negative, scores are declining. The rate of change is constant, the decline will be the same each year.

35. $v(t) = x'(t) = 3t^2 - 18t + 24$
$v(t) = 0$ when

$$3t^2 - 18t + 24 = 0$$
$$3(t - 2)(t - 4) = 0$$
$$t = 2, 4$$

The object is advancing on $[0, 2)$ and $(4, 8]$ and retreating on $(2, 4)$. Thus, the total distance traveled is
$$|x(2) - x(0)| + |x(4) - x(2)| + |x(8) - x(4)|$$
$$= |40 - 20| + |36 - 40| + |148 - 36|$$
$$= 136$$

37. a. $s(t) = 10t + \dfrac{t}{e^t}$

$$v(t) = s'(t) = 10 + \frac{e^t(1) - te^t}{e^{2t}}$$
$$= 10 + e^{-t} - te^{-t}$$
$$v(4) = 10 + e^{-4} - 4e^{-4}$$
$$\approx 9.945 \text{ m/min}$$

b. $v(t) > 0$ so the distance traveled during the 5th minute is

$$s(5) - s(4) \approx 50.0336 - 40.0733$$
$$\approx 9.960 \text{ m}$$

39. $s(t) = -16t^2 + v_0 t + s_0$
$v(t) = s'(t) = -32t + v_0$
Assume Earth's gravity and disregard air resistance and friction.

a. The maximum height is at $t = 2$ when the velocity is 0.

$$-32(2) + v_0 = 0$$
$$v_0 = 64$$

The initial velocity is 64 ft/sec.

b. The rock hits the ground when $s(t) = 0$. This occurs when $t = 7$; so

$$-16(7)^2 + 64(7) + s_0 = 0$$
$$s_0 = 336$$

The cliff is 336 ft high.

c. $v(t) = -32t + 64$ ft/s

d. $v(7) = -32(7) + 64 = -160$
The negative sign indicates downward motion, since upward is positive. The velocity when it hits the ground is 160 ft/s.

SURVIVAL HINT: *You may feel a bit uncomfortable with the entire idea of mathematical modeling. Since it is an interactive process, there is no "quick answer." Take your time as you work through Problems 39-46.*
---

41. Assume earth's gravity and disregard air resistance and friction. The equation guiding the path of the first rock is

$$s_1(t) = -16t^2 + s_1'(0)t + 90$$

and the second rock is

$$s_2(t) = -16(t-1)^2 + s_2'(1)(t-1) + H$$

Time t is 0 when the first rock starts its motion; $s'(0) = 0$ since the rocks are dropped.

The first rock hits the ground at time

$$-16t^2 + 90 = 0$$
$$t^2 = \frac{90}{16}$$
$$t = \frac{3\sqrt{10}}{4} \qquad \textit{Reject negative}$$

The two rocks hit the ground simultaneously, so

$$-16\left(\frac{3\sqrt{10}}{4} - 1\right)^2 + H = 0$$
$$H \approx 30.105336$$

The height is about 30 ft.

43. Assume earth's gravity and disregard air resistance and friction.

$$s(t) = -16t^2 + s_0$$

The pavement, that is ground level, is reached in 3 seconds, so

$$0 = -16(3^2) + s_0$$
$$s_0 = 144$$

The height of the building is 144 ft.

45. On Mars $g = 12$ ft/s^2; so
$$s(t) = -6t^2 + v_0 t + s_0$$
$$v(t) = -12t + v_0$$
Since the rock goes up and back to cliff level in 4 sec, it reaches its maximum height in 2 sec. At that time $v = 0$. So
$$0 = -12(2) + v_0$$
$$v_0 = 24$$
Therefore, the rock passes her on the way down with $v = -24$ ft/s. The equation for the rest of the rock's trip will be:
$$s(t) = -6t^2 + (-24)t + 0$$
$$s(3) = -6(3)^2 - 24(3) = -126$$
The initial velocity is 24 ft/s, and the cliff is 126 ft. high. (The negative means the rock has traveled down 126 ft).

47. a. Because $C(t) = 100t^2 + 400t + 50t \ln t$ is the circulation t years from now, the rate of change of circulation t years from now is $C'(t) = 200t + 50 \ln t + 450$ newspapers per year.

b. The rate of change of circulation 5 years from now is $C'(5) \approx 1{,}530$ newspapers per year.

c. The actual change in the circulation during the sixth year is

$C(6) - C(5)$
$$= \left[100(6^2) + 400(6) + 50 \cdot 6 \cdot \ln 6\right]$$
$$- \left[100(5^2) + 400(5) + 50 \cdot 5 \cdot \ln 5\right]$$
$$\approx 1{,}635 \text{ newspapers}$$

49. $q(t) = 0.05t^2 + 0.1t + 3.4$
  a. $q'(t) = 0.1t + 0.1$
   $q'(1) = 0.2$ ppm/yr
  b. Change in first year:
$$q(1) - q(0) = (0.05 + 0.1 + 3.4) - (3.4)$$
$$= 0.15 \text{ ppm}$$

  c. Change in second year:
$$q(2) - q(1) = (0.2 + 0.2 + 3.4) - (3.55)$$
$$= 0.25 \text{ ppm}$$

51. $P(t) = P_0 + 61t + 3t^2$
   $P'(t) = 61 + 6t$
   $P'(5) = 61 + 30$
   $= 91$ thousand/hr
53. $P(x) = 2x + 4x^{3/2} + 5{,}000$
  a. $P'(x) = 2 + 6x^{1/2}$
   $P'(9) = 2 + 18 = 20$ persons/mo
  b. $\dfrac{P'(9)}{P(9)}(100) = \dfrac{20}{5{,}126}(100)$
   $\approx 0.39\%$ per mo
55. $P = \dfrac{4}{3}\pi N \left(\dfrac{\mu^2}{3kT}\right)$
   $= \dfrac{4\pi\mu^2 N}{9k} T^{-1}$
   $\dfrac{dP}{dT} = -\dfrac{4\pi\mu^2 N}{9k} T^{-2}$
57. a. $s(t) = 7\cos t$
   $v(t) = s'(t) = -7\sin t$
   $a(t) = s''(t) = -7\cos t$

  b. $s(0) = 7$, $s(2\pi) = 7$
   The period (one revolution) is 2π.
  c. The highest point is reached at $t = \pi$
   (downward is positive). $s(\pi) = -7$.
   The amplitude is 7, so the distance
   between the highest point and the lowest
   point is 14.
59. Let x be one edge of the cube. Then
$$V(x) = x^3$$
$$V'(x) = 3x^2 = 0.5S(x)$$

The rate of change is half the surface area.

3.5 The Chain Rule, page 178

> **SURVIVAL HINT:** *Up until this section, the functions we have considered have been carefully selected because the vast majority of functions that exist require a process that requires the "chain rule." Make certain that you understand this process, presented in Theorem 3.10.*

1. The chain rule concerns the
   differentiation of a function of a
   function. If $y = f(u)$ and $u = g(x)$,
   then
$$\frac{dy}{dx} = \frac{dy}{du} \cdot \frac{du}{dx}$$

3. $\dfrac{dy}{dx} = \dfrac{dy}{du}\dfrac{du}{dx}$
   $= \dfrac{d}{du}\left(u^2 + 1\right)\dfrac{d}{dx}(3x - 2)$
   $= 2u(3)$
   $= 6(3x - 2)$

5. $\dfrac{dy}{dx} = \dfrac{dy}{du}\dfrac{du}{dx}$

$\quad = \left(\dfrac{-4}{u^3}\right)(2x)$

$\quad = \dfrac{-8x}{(x^2-9)^3}$

7. $\dfrac{dy}{dx} = \dfrac{dy}{du}\dfrac{du}{dx}$

$\quad = \left[\tan u + u\sec^2 u\right]\left[3 - \dfrac{6}{x^2}\right]$

$\quad$ where $u = 3 - \dfrac{6}{x^2}$

9. a. $g'(u) = 3u^2$

$\quad$ **b.** $u'(x) = 2x$

$\quad$ **c.** $f'(x) = 3(x^2+1)^2(2x)$

$\qquad = 6x(x^2+1)^2$

11. a. $g'(u) = 7u^6$

$\quad$ **b.** $u'(x) = -8 - 24x$

$\quad$ **c.** $f'(x) = 7(5 - 8x - 12x^2)^6(-8 - 24x)$

$\qquad = -7(24x + 8)(12x^2 + 8x - 5)^6$

$\qquad = -56(3x + 1)(12x^2 + 8x - 5)^6$

SURVIVAL HINT: *Problems 13-34 require the chain rule, and the nature of this rule is such that if you are not using it properly you may get an answer that is "just a little" off. So, as you check your work on these problems, and if your answer is not exactly as it should be, you are probably misapplying the chain rule.*

---

13. $f'(x) = 5(5x-2)^4(5)$

$\quad = 25(5x-2)^4$

15. $f'(x) = 4(3x^2 - 2x + 1)^3(6x - 2)$

$\quad = 8(3x - 1)(3x^2 - 2x + 1)^3$

17. $s'(\theta) = [\cos(4\theta + 2)](4)$

$\quad = 4\cos(4\theta + 2)$

19. $f'(x) = e^{-x^2+3x}(-2x + 3)$

21. $y' = e^{\sec x}\sec x \tan x$

23. $f'(t) = (2t + 1)e^{t^2+t+5}$

25. $g'(x) = x(\cos 5x)(5) + \sin 5x$

$\quad = 5x\cos 5x + \sin 5x$

27. $f(x) = (1 - 2x)^{-3}$

$\quad f'(x) = -3(1 - 2x)^{-4}(-2)$

$\quad = 6(1 - 2x)^{-4}$

29. $f'(x)$

$= \dfrac{1}{2}\left(\dfrac{x^2+3}{x^2-5}\right)^{-1/2}\left[\dfrac{(x^2-5)(2x) - (x^2+3)(2x)}{(x^2-5)^2}\right]$

$= \dfrac{1}{2}\left(\dfrac{x^2+3}{x^2-5}\right)^{-1/2}\left[\dfrac{-16x}{(x^2-5)^2}\right]$

31. $g'(x) = \dfrac{12x^3 + 5}{3x^4 + 5x}$

33. $f'(x) = \dfrac{\cos x - \sin x}{\sin x + \cos x}$

35. $f'(x) = x\left(\dfrac{1}{2}\right)(1 - 3x)^{-1/2}(-3) + \sqrt{1 - 3x}$

$\quad = \dfrac{-3x + 2(1 - 3x)}{2(1 - 3x)^{1/2}}$

$\quad = \dfrac{-9x + 2}{2(1 - 3x)^{1/2}}$

$\quad f'(x) = 0$ when $\dfrac{-9x + 2}{2(1 - 3x)^{1/2}} = 0$

$\qquad\qquad\qquad x = \dfrac{2}{9}$

37. $q'(x) = \dfrac{(x+2)^3 2(x-1) - (x-1)^2 3(x+2)^2(1)}{(x+2)^6}$

$\quad = \dfrac{(x-1)(-x+7)}{(x+2)^4}$

$\quad q'(x) = 0$ when $\dfrac{(x-1)(-x+7)}{(x+2)^4} = 0$

$\qquad\qquad (x-1)(-x+7) = 0$

$\qquad\qquad\qquad\qquad x = 1, 7$

39. $T'(x) = x^2 e^{1-3x}(-3) + 2xe^{1-3x}$
$$= -e^{-1-3x}(3x^2 - 2x)$$
$T'(x) = 0$ when $-e^{-1-3x}(3x^2 - 2x) = 0$
$$x(3x - 2) = 0$$
$$x = 0, \frac{2}{3}$$

41. a. The graph indicates that $u \approx 5$ when $x = 2$.
The slope of the tangent line to the curve $u = g(x)$ at $x = 2$ is about 1.

b. The graph indicates that $y \approx 3$ when $u = 5$. The slope of the tangent line is about $\frac{3}{2}$.

c. The slope of $y = f[g(x)]$ is about $(1)\left(\frac{3}{2}\right) = 1.5$.

43. $h(x) = g[f(x)]$, so $h'(x) = g'[f(x)] \cdot f'(x)$

a. From the graphs, $f(-1) \approx \frac{3}{4}$; the slope of g at $\frac{3}{4}$ is $g'\left(\frac{3}{4}\right) \approx -\frac{3}{2}$; the slope of f at -1 is found as follows: estimate the point to be $\left(-1, \frac{3}{4}\right)$ and draw a tangent line at this point. We estimate the point of intersection of the tangent line and the x-axis to be $(4, 0)$, so that from $\left(-1, \frac{3}{4}\right)$ we have a rise of $-\frac{3}{4}$ and a run of 5 for a slope of
$$\frac{-\frac{3}{4}}{5} = -\frac{3}{20}$$
thus, $f'(-1) \approx -\frac{3}{20}$.
Thus,
$$h'(-1) = g'\left(\frac{3}{4}\right) \cdot f'(-1)$$
$$\approx -\frac{3}{2}\left(-\frac{3}{20}\right)$$
$$= \frac{9}{40}$$

b. From the graphs, $f(1) \approx -\frac{1}{2}$; the slope of g at $-\frac{1}{2}$ is $g'\left(-\frac{1}{2}\right) \approx \frac{3}{8}$; the slope of f at 1 is $f'(1) \approx -1$. Thus,
$$h'(1) = g'\left(-\frac{1}{2}\right) \cdot f'(1)$$
$$\approx \frac{3}{8}(-1)$$
$$= -\frac{3}{8}$$

c. From the graphs, $f(3) \approx -\frac{1}{2}$; the slope of g at $-\frac{1}{2}$ is $f'(3) \approx 1$. Thus,
$$h'(3) = g'\left(-\frac{1}{2}\right) \cdot f'(3)$$
$$\approx \frac{3}{8}(1)$$
$$= \frac{3}{8}$$

45. From the chain rule,
$$h'(1) = f'[g(1)] \cdot g'(1)$$
$$= f'(2) \cdot g'(1)$$

We see that we need to find the values of $f'(2)$ and $g'(1)$. Use the given table of values to draw possible graphs for both f and g, along with the appropriate tangent lines:

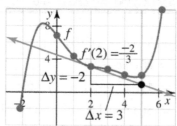

$$f'(2) \approx -\frac{2}{3}$$

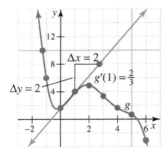

$g'(1) \approx 1$. Thus,

$$h'(1) = f'[g(1)] \cdot g'(1)$$
$$\approx f'(2) \cdot g'(1)$$
$$= -\frac{2}{3} \cdot 1$$
$$= -\frac{2}{3}$$

47. $\dfrac{dD}{dt} = \dfrac{dD}{dp} \cdot \dfrac{dp}{dt}$

$$= 4{,}374\left(-2p^{-3}\right)(0.04t + 0.1)$$
$$= -8{,}748p^{-3}(0.04t + 0.1)$$

When $t = 10$, $p(10) = 9$ and

$$\left.\frac{dD}{dt}\right|_{t=10,\,p=9}$$
$$= -8{,}748(9)^{-3}[0.04(10) + 0.1]$$
$$\approx -6 \text{ lb/wk}$$

Demand is decreasing by 6 pounds per week.

49. a. $I = 20s^{-2}$

$$\frac{dI}{dt} = (20)\left(-2s^{-3}\right)\frac{ds}{dt}$$
$$\frac{ds}{dt} = -2t$$

When $s(t) = 19$, we have
$$28 - t^2 = 19$$
$$t^2 = 9$$
$$t = 3 \qquad \textit{Disregard negative}$$

Thus,
$$\frac{dI}{dt} = (20)\left(-2s^{-3}\right)\frac{ds}{dt}$$
$$= (20)(-2)(19^{-3})(-2 \cdot 3)$$
$$= .0349905$$

Illuminance is increasing by 0.035 lux/s.

b. Since $s(t) = 28 - t^2$, we have
$t = \sqrt{28 - s}$ and the rate of change is

$$\frac{dI}{dt} = \frac{80t}{s^3} = \frac{80\sqrt{28 - s}}{s^3}$$

Using technology to solve

$$\frac{80\sqrt{28 - s}}{s^3} = 1$$

we obtain s ≈ 7.15 m.

51. $T'(t) = 17 \cos\left(\dfrac{\pi t}{10} - \dfrac{5}{6}\right)\left(\dfrac{\pi}{10}\right)$

$$T'(2) = 1.7\pi \cos\left(\frac{\pi}{5} - \frac{5}{6}\right) \approx 5.23$$

Since the derivative is positive, it is getting hotter.

53. $I(\theta) = I_0 \left(\dfrac{\sin \beta}{\beta}\right)^2$

$$I'(\theta) = I_0(2)\left(\frac{\sin \beta}{\beta}\right)\left(\frac{\beta \cos \beta - \sin \beta}{\beta^2}\right)\left(\pi a \cos\frac{\theta}{\lambda}\right)\left(\frac{1}{\lambda}\right)$$
$$= \frac{2a\pi I_0}{\lambda \beta^3} \sin \beta(\beta \cos \beta - \sin \beta)\cos \frac{\theta}{\lambda}$$

55. $g(x) = f[u(x)]$, $g'(x) = f'[u(x)][u'(x)]$

$$g'(-3) = f'[u(-3)][u'(-3)]$$
$$= [f'(5)](2)$$
$$= -6$$
$$g(-3) = f[u(-3)] = f(5) = 3$$
The equation of the tangent line is:
$$y - 3 = -6(x + 3)$$
$$6x + y + 15 = 0$$

57. a. If $-\frac{\pi}{2} < x < \frac{\pi}{2}$, $\cos x > 0$, so
$$F'(x) = (\cos x)^{-1}(-\sin x)$$
$$= -\tan x$$

If $\frac{\pi}{2} < x < \frac{3\pi}{2}$, $\cos x < 0$, so
$$F'(x) = (-\cos x)^{-1}(\sin x)$$
$$= -\tan x$$

b. $F'(x) = \dfrac{\sec x \tan x + \sec^2 x}{\sec x + \tan x}$
$$= \sec x$$

59. $\dfrac{d}{dx}f'[f(x)] = f''[f(x)]f'(x)$
and
$$\dfrac{d}{dx}f[f'(x)] = f'[f'(x)]f''(x)$$

3.6 Implicit Differentiation, page 190

> SURVIVAL HINT: *Before doing the problems in this problem set, review the procedure for implicit differentiation given on page 183,*

1. $x^2 + y^2 = 25$
$$2x + 2y\dfrac{dy}{dx} = 0$$
$$\dfrac{dy}{dx} = -\dfrac{x}{y}$$

3. $xy = 25$
$$xy' + y = 0$$
$$y' = -\dfrac{y}{x}$$

5. $\dfrac{1}{y} + \dfrac{1}{x} = 1$
$$-\dfrac{1}{y^2}\dfrac{dy}{dx} - \dfrac{1}{x^2} = 0$$
$$\dfrac{dy}{dx} = -\dfrac{y^2}{x^2}$$

7. $\cos xy = 1 - x^2$
$$(-\sin xy)\dfrac{d}{dx}(xy) = -2x$$
$$(-\sin xy)\left(x\dfrac{dy}{dx} + y\right) = -2x$$
$$(-x\sin xy)\dfrac{dy}{dx} = y\sin xy - 2x$$
$$\dfrac{dy}{dx} = \dfrac{2x - y\sin xy}{x\sin xy}$$

9. $\ln(xy) = e^{2x}$ Note that $xy \neq 0$.
$$\dfrac{1}{xy}(y + xy') = 2e^{2x}$$
$$\dfrac{1}{x} + \dfrac{y'}{y} = 2e^{2x}$$
$$y' = (2e^{2x} - x^{-1})y$$

11. a. $2x + 3y^2\dfrac{dy}{dx} = 0$
$$\dfrac{dy}{dx} = -\dfrac{2x}{3y^2}$$

b. $x^2 + y^3 = 12$

$$y = (12 - x^2)^{1/3}$$

$$\frac{dy}{dx} = \frac{1}{3}(12 - x^2)^{-2/3}(-2x)$$

$$= \frac{-2x}{3(12 - x^2)^{2/3}}$$

$$= -\frac{2x}{3y^2}$$

13. a. $1 - \frac{1}{y^2}\frac{dy}{dx} = 0$

$$\frac{dy}{dx} = y^2$$

b. $xy + 1 = 5y$

$$(x - 5)y = -1$$

$$y = \frac{-1}{x - 5}$$

$$\frac{dy}{dx} = -\frac{-1}{(x - 5)^2}$$

$$= \frac{1}{(x - 5)^2}$$

$$= y^2$$

SURVIVAL HINT: *The derivatives of the inverse trigonometric formulas are given in Theorem 3.11. You will need this theorem for Problems 15-24.*

---

15. $\frac{dy}{dx} = \frac{1}{\sqrt{1 - (2x + 1)^2}}(2)$

$$= \frac{2}{\sqrt{-4x^2 - 4x}}$$

$$= \frac{1}{\sqrt{-x^2 - x}}$$

17. $\frac{dy}{dx} = \frac{1}{1 + (x^2 + 1)}\frac{x}{\sqrt{x^2 + 1}}$

$$= \frac{x}{(x^2 + 2)\sqrt{x^2 + 1}}$$

19. $\frac{dy}{dx} = \frac{1}{|e^{-x}|\sqrt{e^{-2x} - 1}}(-e^{-x})$

$$= \frac{-1}{\sqrt{e^{-2x} - 1}}$$

21. $\frac{dy}{dx} = \frac{1}{1 + x^{-2}}(-x^{-2})$

$$= \frac{-1}{x^2 + 1}$$

23. $\frac{1}{\sqrt{1 - y^2}}y' + y' = 2xy' + 2y$

$$\left[(1 - y^2)^{-1/2} + 1 - 2x\right]y' = 2y$$

$$\frac{dy}{dx} = \frac{2y}{(1 - y^2)^{-1/2} + 1 - 2x}$$

25. $2x + 2yy' = 0$

$$y' = -\frac{x}{y}$$

$$-\frac{x}{y}\Big|_{(-2,3)} = \frac{2}{3}$$

The equation of the tangent line is

$$y - 3 = \frac{2}{3}(x + 2)$$

$$2x - 3y + 13 = 0$$

27. $\cos(x - y)[1 - y'] = xy' + y$

$$y' = \frac{\cos(x - y) - y}{\cos(x - y) + x}$$

$$y'\big|_{(0,\pi)} = \frac{\cos(-\pi) - \pi}{\cos(-\pi)}$$

$$= \pi + 1$$

The equation of the tangent line is

$$y - \pi = (\pi + 1)(x - 0)$$
$$(\pi + 1)x - y + \pi = 0$$

29. $\tan^{-1} y + x\left(\dfrac{1}{1+y^2}\right)y' = 2x + y'$

$$\left(\dfrac{x}{1+y^2} - 1\right)y' = 2x - \tan^{-1} y$$

$$\left(\dfrac{x - 1 - y^2}{1 + y^2}\right)y' = 2x - \tan^{-1} y$$

$$y' = \dfrac{(2x - \tan^{-1} y)(1 + y^2)}{x - 1 - y^2}$$

At $(0,0)$, $y' = 0$, so the tangent line is horizontal with equation $y = 0$.

31. $2(x^2 + y^2)(2x + 2yy') = 4(2xy + x^2 y')$

At $(1, 1)$:
$$2 + 2y' = 2 + y'$$
$$y' = 0$$

33. $3x^2 + 3y^2 y' - \dfrac{9}{2}[xy' + y] = 0$

$$6x^2 + 6y^2 y' - 9xy' - 9y = 0$$
$$(6y^2 - 9x)y' = 9y - 6x^2$$
$$y' = \dfrac{3(3y - 2x^2)}{3(2y^2 - 3x)}$$
$$= \dfrac{3y - 2x^2}{2y^2 - 3x}$$

At $(2, 1)$:
$$y' = \dfrac{3(1) - 2(2)^2}{2(1)^2 - 3(2)}$$
$$= \dfrac{-5}{-4}$$
$$= \dfrac{5}{4}$$

35. $2x + 2xy' + 2y - 3y'y^2 = 0$

At $(1, -1)$
$$2 + 2y' - 2 - 3y' = 0$$
$$y' = 0$$

The tangent line is horizontal so the normal line is a vertical line with equation $x - 1 = 0$.

37. $7x + 5y^2 = 1$

$$7 + 10yy' = 0$$
$$y' = -\dfrac{7}{10y}$$

For y'' we again differentiate:

$$\dfrac{d}{dx}(y') = \dfrac{d}{dx}\left(-\dfrac{7}{10y}\right)$$
$$y'' = \dfrac{7}{10y^2}y'$$
$$= \left(\dfrac{7}{10y^2}\right)\left(-\dfrac{7}{10y}\right)$$
$$= -\dfrac{49}{100y^3}$$

39. a. Differentiate a variable base, constant exponent; $y' = 2x$.

  b. Differentiate a constant base ($\neq e$), variable exponent; $y' = 2^x \ln 2$.

  c. Differentiate a constant base e, variable exponent; $y' = e^x$.

  d. Differentiate a variable base, constant exponent of e; $y' = ex^{e-1}$.

41. Logarithmic differentiation is advantageous when differentiating a complicated product or quotient. It is a technique for changing the derivative of a product to the derivatives of a sum and the derivative of a quotient to the derivatives of a difference. Along with the product transformation is the changing of the derivative of a power to the differentiation of the base. *Note*: $y = uv/w$ becomes

$$\ln y = \ln u + \ln v - \ln w$$

43. $y = \left(x^{10} + 1\right)^{1/6}\left(x^7 - 3\right)^{4/9}$

$\ln y = \dfrac{1}{6}\ln\left(x^{10} + 1\right) + \dfrac{4}{9}\ln\left(x^7 - 3\right)$

$\dfrac{1}{y}\dfrac{dy}{dx} = \dfrac{1}{6}\left(x^{10} + 1\right)^{-1}(10x^9) + \dfrac{4}{9}\left(x^7 - 3\right)^{-1}(7x^6)$

$\dfrac{dy}{dx} = y\left[\dfrac{5x^9}{3(x^{10} + 1)} + \dfrac{28x^6}{9(x^7 - 3)}\right]$

45. $y = x^x$

$\ln y = x\ln x$

$\dfrac{1}{y}\dfrac{dy}{dx} = xx^{-1} + \ln x$

$\dfrac{dy}{dx} = y(1 + \ln x)$

47. a. $b^2u^2 + a^2v^2 = a^2b^2$

$b^2uu' + a^2v = 0$

$u' = -\dfrac{a^2v}{b^2u}$

b. $b^2u + a^2vv' = 0$

$v' = -\dfrac{b^2u}{a^2v}$

49. $2x - 3xy' - 3y + 4yy' = 0$

$y' = \dfrac{3y - 2x}{4y - 3x}$

The denominator is not defined if $y = \dfrac{3x}{4}$.

Substituting this into the given equation we find:

$x^2 - \dfrac{9x^2}{4} + \dfrac{9x^2}{8} = -2$

$x = \pm 4$

Then $y = \pm 3$. There are two such points: $(4, 3)$, and $(-4, -3)$.

51. $3x^2 + 3y^2\dfrac{dy}{dx} = 2Ax\dfrac{dy}{dx} + 2Ay$

$\left(3y^2 - 2Ax\right)\dfrac{dy}{dx} = 2Ay - 3x^2$

$\dfrac{dy}{dx} = \dfrac{2Ay - 3x^2}{3y^2 - 2Ax}$

At (A, A),

$\dfrac{dy}{dx} = -\dfrac{A^2}{A^2} = -1$

The equation of the tangent line is

$y - A = -(x - A)$
$x + y - 2A = 0$

For the normal, the slope is 1, and the equation of the normal is

$y - A = x - A$
$x - y = 0$

53. $x^2 + y^2 = \left(x^2 + y^2\right)^{1/2} + x$

$2x + 2yy' = \dfrac{1}{2}\left(x^2 + y^2\right)^{-1/2}(2x + 2yy') + 1$

$\left[2y - y\left(x^2 + y^2\right)^{-1/2}\right]y' = x\left(x^2 + y^2\right)^{-1/2} + 1 - 2x$

$y' = \dfrac{x(x^2 + y^2)^{-1/2} + 1 - 2x}{2y - y(x^2 + y^2)^{-1/2}}$

$= \dfrac{x + (1 - 2x)(x^2 + y^2)^{1/2}}{2y(x^2 + y^2)^{1/2} - y}$

There will be a vertical tangent when the denominator is zero. That is,

$y\left[2\left(x^2 + y^2\right)^{1/2} - 1\right] = 0$

$y = 0 \text{ or } (x^2 + y^2)^{1/2} = \dfrac{1}{2}$

For $\boldsymbol{y = 0}$, the original equation is

$$x^2 = \sqrt{x^2} + x$$
$$x^2 = |x| + x$$

If $x \geq 0$, then $x^2 = 2x$, or $x = 0, 2$.
If $x < 0$, then $x^2 = 0$, which is impossible.
At $(0, 0)$, the numerator is also 0, so we
exclude this point. Thus, there is a vertical
tangent at $(2, 0)$.

For $(x^2 + y^2)^{1/2} = \frac{1}{2}$, we substitute into
the original equation to find

$$\frac{1}{4} = \frac{1}{2} + x$$
$$x = -\frac{1}{4}$$

By substitution this gives the points

$$\left(-\frac{1}{4}, \frac{\sqrt{3}}{4} \right) \text{ and } \left(-\frac{1}{4}, -\frac{\sqrt{3}}{4} \right).$$

Hence, the tangent is vertical at $(2, 0)$,

$$\left(-\frac{1}{4}, \frac{\sqrt{3}}{4} \right) \text{ and } \left(-\frac{1}{4}, -\frac{\sqrt{3}}{4} \right).$$

55. $\dfrac{x^2}{a^2} + \dfrac{y^2}{b^2} = 1$

$$\frac{2x}{a^2} + \frac{2yy'}{b^2} = 0$$

$$y' = -\frac{b^2 x}{a^2 y}$$

At (x_0, y_0):

$$m = \frac{-b^2 x_0}{a^2 y_0}$$

so the equation of the tangent line is

$$y - y_0 = -\frac{b^2 x_0}{a^2 y_0}(x - x_0)$$

$$a^2 y y_0 - a^2 y_0^2 + b^2 x_0 x - b^2 x_0^2 = 0$$

$$b^2 x_0 x + a^2 y y_0 = a^2 y_0^2 + b^2 x_0^2$$

$$\frac{b^2 x_0 x}{a^2 b^2} + \frac{a^2 y y_0}{a^2 b^2} = \frac{a^2 y_0^2}{a^2 b^2} + \frac{b^2 x_0^2}{a^2 b^2}$$

$$\frac{x_0 x}{a^2} + \frac{y y_0}{b^2} = \frac{x_0^2}{a^2} + \frac{y_0^2}{b^2}$$

since (x_0, y_0) lies on the curve it satisfies the
equation for the ellipse so that

$$\frac{x_0^2}{a^2} + \frac{y_0^2}{b^2} = 1$$

and therefore

$$\frac{x_0 x}{a^2} + \frac{y_0 y}{b^2} = 1$$

57. Let α be the angle between the tangent to C_1
and the positive x-axis. Let β be the angle
between the tangent to C_2 and the positive
x-axis. Then

$$\theta = \pi - [\alpha + (\pi - \beta)] = \beta - \alpha$$

$$\tan \theta = \tan(\beta - \alpha)$$
$$= \frac{\tan \beta - \tan \alpha}{1 + \tan \alpha \tan \beta}$$
$$= \frac{m_1 - m_2}{1 + m_2 m_1} \quad \text{if } m_2 > m_1$$

Similarly,

$$\tan \theta = \frac{m_1 - m_2}{1 + m_2 m_1} \quad \text{if } m_1 > m_2$$

Thus, $\tan \theta = \dfrac{|m_1 - m_2|}{1 + m_2 m_1}$.

59. Let $y = \tan^{-1} x$; on $\left(-\frac{\pi}{2}, \frac{\pi}{2} \right)$ we have
$\tan y = x$ so

$$\frac{d}{dx}(\tan y) = \frac{d}{dx}(x)$$

$$\sec^2 y \frac{dy}{dx} = 1$$

$$\frac{dy}{dx} = \frac{1}{\sec^2 y}$$

$$= \frac{1}{1 + \tan^2 y}$$

$$= \frac{1}{1 + x^2}$$

Thus, using the chain rule on $\left(-\frac{\pi}{2}, \frac{\pi}{2}\right)$,

$$\frac{d}{dx}(\tan^{-1} u) = \frac{1}{1 + u^2}\frac{du}{dx}$$

To obtain the formula for $y = \sec^{-1} x$, we have
$\sec y = x$, so

$$\frac{d}{dx}(\sec y) = \frac{d}{dx}(x)$$

$$\sec y \tan y \frac{dy}{dx} = 1$$

$$\frac{dy}{dx} = \frac{1}{\sec y \tan y}$$

We know $\sec y = x$ and

$$\tan y = \pm\sqrt{\sec^2 y - 1} = \pm\sqrt{x^2 - 1}$$

so

$$\frac{dy}{dx} = \pm\frac{1}{x\sqrt{x^2 - 1}}$$

The slope of the graph of $y = \sec^{-1} x$ is always positive, so

$$\frac{d}{dx}\left(\sec^{-1}x\right) = \begin{cases} \dfrac{1}{x\sqrt{x^2 - 1}} & \text{if } x > 1 \\[3mm] -\dfrac{1}{x\sqrt{x^2 - 1}} & \text{if } x < -1 \end{cases}$$

We use absolute values to write

$$\frac{d}{dx}\left(\sec^{-1}x\right) = \frac{1}{|x|\sqrt{x^2 - 1}}$$

where $|x| > 1$. Now, we apply the chain rule to obtain

$$\frac{d}{dx}\left(\sec^{-1}u\right) = \frac{1}{|u|\sqrt{u^2 - 1}}\frac{du}{dx}$$

where $|u| > 1$.

3.7 Related Rates and Applications, page 198

1. $2x + 3y - 5 = 0$

$$2\frac{dx}{dt} + 3\frac{dy}{dt} = 0$$

$$2(3) + 3\frac{dy}{dt} = 0$$

$$3\frac{dy}{dt} = -6$$

$$\frac{dy}{dt} = -2$$

3. $y = 2x^2 + 3x + 4$

$$\frac{dy}{dt} = 4x\frac{dx}{dt} + 3\frac{dx}{dt}$$

$$46 = (4 \cdot 5 + 3)\frac{dx}{dt}$$

$$\frac{dx}{dt} = \frac{46}{23} = 2$$

5.
$$5x^2 - y = 100$$
$$10x\frac{dx}{dt} - \frac{dy}{dt} = 0$$
$$10(5)\frac{dx}{dt} - 40 = 0$$
$$50\frac{dx}{dt} = 40$$
$$\frac{dx}{dt} = \frac{4}{5}$$

7.
$$4x^2 - y = 100$$
$$8x\frac{dx}{dt} - \frac{dy}{dt} = 0$$
$$8(8)\frac{dx}{dt} - (-16) = 0$$
$$64\frac{dx}{dt} = -16$$
$$\frac{dx}{dt} = -\frac{16}{64}$$
$$= -\frac{1}{4}$$

9.
$$y = x^{1/2}$$
$$\frac{dy}{dt} = \frac{1}{2}x^{-1/2}\frac{dx}{dt}$$
$$= \frac{1}{2}(144)^{-1/2}(6)$$
$$= \frac{3}{12}$$
$$= \frac{1}{4}$$

11.
$$y = 2x^{1/2} - 9$$
$$\frac{dy}{dt} = x^{-1/2}\frac{dx}{dt}$$
$$5 = (9)^{-1/2}\frac{dx}{dt}$$
$$\frac{dx}{dt} = 15$$

13.
$$y = 5(x + 9)^{1/2}$$
$$\frac{dy}{dt} = \frac{5}{2}(x+9)^{-1/2}\frac{dx}{dt}$$
$$4 = \frac{5}{2}(16+9)^{-1/2}\frac{dx}{dt}$$
$$4 = \frac{1}{2}\frac{dx}{dt}$$
$$\frac{dx}{dt} = 8$$

15. If $x = 5$, then $5y = 10$
$$y = 2$$
$$xy = 10$$
$$y\frac{dx}{dt} + x\frac{dy}{dt} = 0$$
$$2(-2) + 5\frac{dy}{dt} = 0$$
$$\frac{dy}{dt} = \frac{4}{5}$$

17. If $x = \sqrt[3]{500}$, then $500 + y^3 = 1{,}000$
$$y^3 = 500$$
$$y = \sqrt[3]{500}$$
$$x^3 + y^3 = 1{,}000$$
$$3x^2\frac{dx}{dt} + 3y^2\frac{dy}{dt} = 0$$
$$3(500)^{2/3}(20) + 3(500)^{2/3}\frac{dy}{dt} = 0$$
$$\frac{dy}{dt} = -\frac{60(500)^{2/3}}{3(500)^{2/3}}$$
$$= -20$$

19. If $x = 1$, then $5(1)y^2 = 10$
$$y^2 = 2$$
$$y = \sqrt{2}$$

$$5xy^2 = 10$$
$$5y^2 \frac{dx}{dt} + 10xy \frac{dy}{dt} = 0$$
$$5(\sqrt{2})^2(-2) + 10(1)(\sqrt{2})\frac{dy}{dt} = 0$$
$$10\sqrt{2}\frac{dy}{dt} = 20$$
$$\frac{dy}{dt} = \frac{2}{\sqrt{2}} \cdot \frac{\sqrt{2}}{\sqrt{2}}$$
$$= \sqrt{2}$$

21. If $x = 4$, then $4^2 + 4y - y^2 = 20$
$$y^2 - 4y + 4 = 0$$
$$(y-2)^2 = 0$$
$$y = 2$$
$$x^2 + xy - y^2 = 20$$
$$2x\frac{dx}{dt} + y\frac{dx}{dt} + x\frac{dy}{dt} - 2y\frac{dy}{dt} = 0$$
$$(2x+y)\frac{dx}{dt} + (x-2y)\frac{dy}{dt} = 0$$
$$(2 \cdot 4 + 2)\frac{dx}{dt} + (4 - 2 \cdot 2)(5) = 0$$
$$10\frac{dx}{dt} = 0$$
$$\frac{dx}{dt} = 0$$

23. $\dfrac{d}{dt}F(x) = -12\dfrac{dx}{dt} = -12\left(\dfrac{1}{4}\right) = -3$

> **SURVIVAL HINT:** *Notice that since F is a linear function of x, the change in F is a constant, and does not depend upon the value of x. (See Problems 23 and 24.)*

25. $y^2 = 4x$
$$2y\frac{dy}{dt} = 4\frac{dx}{dt}$$

$$-4\frac{dy}{dt} = 4(3)$$
$$\frac{dy}{dt} = -3 \text{ ft/s}$$

27. The area of the ripple is $A = \pi r^2$ in.^{2}
$$\frac{dA}{dt} = 2\pi r \frac{dr}{dt} = 2\pi(8)(3) = 48\pi \text{ in.}^2/\text{s}$$

29. $\dfrac{dQ}{dt} = 2p\dfrac{dp}{dt} + 3\dfrac{dp}{dt}$
$$= 2(30)(2) + 3(2)$$
$$= 126 \text{ units/yr}$$

31. $\dfrac{dN}{dt} = 2p\dfrac{dp}{dt} + 5\dfrac{dp}{dt}$
$$= 40(1.2) + 5(1.2)$$
$$= 54$$
The number of patients is increasing by 54 people/yr.

33. A related rate problem is a problem involving one or more equations in which an unknown rate is computed by relating it to the known rates.

> **SURVIVAL HINT:** *The procedure for solving related rate problems is given on page 194.*

35. $V = \dfrac{4}{3}\pi r^3$
$$\frac{dV}{dt} = 4\pi r^2 \frac{dr}{dt}$$
When $r = 2$ and $\dfrac{dV}{dt} = 3$,
$$3 = 4\pi(2)^2 \frac{dr}{dt}$$
$$\frac{dr}{dt} = \frac{3}{16\pi} \text{ in./s}$$

37. Let x be the length of the shadow and y be the distance of the person from the street light. Given $dy/dt = 5$ ft/s, we wish to find dx/dt. Because of similar triangles,
$$\frac{x}{6} = \frac{x+y}{18}$$
$$y = 2x$$
$$\frac{dy}{dt} = 2\frac{dx}{dt}$$
$$5 = 2\frac{dx}{dt}$$
$$\frac{dx}{dt} = 2.5$$
The shadow is lengthening at 2.5 ft/s.

39. $v_c = 40$ mi/h; $v_t = 30$ mi/h; because the speed is constant, $s_c = 40t$ mi, $s_t = 30t$ mi. The distance is the hypotenuse of a right triangle,
$$D = 10t\sqrt{3^2 + 4^2}$$
$$\frac{dD}{dt} = 50(1)$$
$$= 50 \text{ mi/h}$$

41. Let x be the horizontal distance from the boat to the pier and D the length of the rope; D is the hypotenuse of a right triangle with legs 4 and x, and $dx/dt = 2$:
$$x^2 + 4^2 = D^2$$
$$2x\frac{dx}{dt} = 2D\frac{dD}{dt}$$
$$\frac{dD}{dt} = \frac{x}{D}(2)$$
At the instant when $D = 5$, $x = 3$:

$$\frac{dD}{dt} = \frac{2(3)}{5} = 1.2$$
The rope is unwinding at the rate of 1.2 m/min.

43. Use similar triangles:
$$\frac{s}{10+L} = \frac{6}{L}$$
$$L = \frac{60}{s-6}$$
$$\frac{dL}{dt} = -\frac{60}{(s-6)^2}\frac{ds}{dt}$$
Since $s = 30 - 16t^2$ and $\frac{ds}{dt} = -32t$, when $t = 1$ we know $s = 14$, and $\frac{ds}{dt} = -32$ so
$$\frac{dL}{dt} = -\frac{60}{64}(-32) = 30 \text{ ft/s}$$
The shadow is lengthening at 30 ft/s.

45. Assume the ice is in the shape of a sphere of radius r.
$$V = \frac{4}{3}\pi r^3$$
$$\frac{dV}{dt} = 4\pi r^2\frac{dr}{dt}$$
With $r = 4$ and $\frac{dV}{dt} = -5$,
$$-5 = 4\pi(4)^2\frac{dr}{dt}$$
$$\frac{dr}{dt} = -\frac{5}{64\pi}$$
$$\approx 0.025 \text{ in./min}$$
The radius is decreasing at the rate of about 0.025 in./min.

The surface area is given by

$$S = 4\pi r^2$$
$$\frac{dS}{dt} = 8\pi r \frac{dr}{dt}$$

With $r = 4$ and $\frac{dr}{dt} = \frac{-5}{64\pi}$,

$$\frac{dS}{dt} = 8\pi(4)\left(-\frac{5}{64\pi}\right)$$
$$= -2.5 \text{ in.}^2/\text{min}$$

The surface area is decreasing at the rate of 2.5 in.^{2}/min.

47. The assumptions we need are given: the shape of the tank is a cone and the water is flowing out at a constant rate. Let x be the radius of the top circle of the body of water and y its height. The radius of the top circle is 20, the height of the cone is 40 ft. By similar right triangles,

$$\frac{20}{40} = \frac{x}{y}$$
$$x = \frac{1}{2}y$$

The volume of the body of water is

$$V = \frac{1}{3}\pi r^2 h$$
$$= \frac{1}{3}\pi\left(\frac{1}{2}y\right)^2 (y)$$
$$= \frac{1}{12}\pi y^3$$

Then,

$$\frac{dV}{dt} = \frac{1}{4}\pi y^2 \frac{dy}{dt}$$

When $y = 12$, $dV/dt = 80$, so

$$80 = \frac{1}{4}\pi(12)^2\frac{dy}{dt}$$
$$\frac{dy}{dt} = \frac{80(4)}{\pi(144)}$$
$$\approx 0.71 \text{ ft/min}$$

49. a. $\frac{ds}{dt} = -49 + 49e^{-t/5}$

$$\frac{dp}{dt} = -0.000125e^{-0.000125s}\left(49e^{-t/5} - 49\right)$$

When $t = 2$, $s \approx 2{,}982.77$ so

$$\frac{dp}{dt} \approx 0.00139 \text{ atmospheres/s}$$

b. We find the solution graphically; $s = 0$ when $t \approx 66$ s; at that instant the air pressure is changing at the rate of about 0.006125 atmospheres/s.

51. Let y be the height of the balloon and s the distance between the balloon and the observer. When $y = 400$, $s = 500$, and $dy/dt = 10$ ft/s.

$$s^2 = 300^2 + y^2$$
$$2s\frac{ds}{dt} = 2y\frac{dy}{dt}$$
$$2(500)\frac{ds}{dt} = 2(400)(10)$$
$$\frac{ds}{dt} = 8 \text{ ft/s}$$

53. The velocity of the man walking is

$$dx/dt = -5 \text{ ft/s}$$
$$x = (18 - 6)\cot\theta$$
$$\frac{dx}{dt} = -12\csc^2\theta\frac{d\theta}{dt}$$

If $x = 9$, $\csc\theta = \frac{15}{12} = \frac{5}{4}$ so that

$$-5 = -12\left(\frac{5}{4}\right)^2 \frac{d\theta}{dt}$$

$$\frac{d\theta}{dt} = \frac{4}{15} \text{ rad/s}$$

55. a. Let y be the depth of the water and 10 by $(2 + 2x)$ the dimensions of the horizontal surface of the water. Form a right triangle in the trapezoidal cross-section by drawing a line through an end of the 2 ft base perpendicular to the 5 ft base. Now draw in the height y for the (vertical) water level, and the horizontal dimension x for the water level portion in this triangle.

These two right triangles reveal that

$$\frac{x}{y} = \frac{\frac{5-2}{2}}{2}$$

$$x = \frac{3}{4}y$$

The volume of water in the trough is

$$V = \frac{2 + 2x + 2}{2}(y)(10)$$

$$= 10(2 + x)y$$

$$= 10\left(2 + \frac{3}{4}y\right)y$$

$$= 7.5y^2 + 20y$$

b. $V = 7.5y^2 + 20y$

$$\frac{dV}{dt} = 15y\frac{dy}{dt} + 20\frac{dy}{dt}$$

$$10 = [15(1) + 20]\frac{dy}{dt}$$

$$\frac{dy}{dt} = \frac{2}{7} \approx 0.28 \text{ ft/min} \approx 3.5 \text{ in./min}$$

57. Let y be the depth of the water at the deep end and 25 by x the dimensions of the horizontal rectangular water surface. The volume of the water is (for $y \le 12$)

$$V = \frac{25xy}{2}$$

In a vertical cross-section, the right triangle with legs x and y is similar to the right triangle with legs 60 and 12. Thus,

$$\frac{x}{60} = \frac{y}{12}$$

$$x = 5y$$

By substitution, we have

$$V = \frac{125y^2}{2}$$

Now $y = 5$, $dV/dt = 800$ ft³/min, so we have

$$\frac{dV}{dt} = 125y\frac{dy}{dt}$$

$$800 = 125(5)\frac{dy}{dt}$$

$$\frac{dy}{dt} = \frac{32}{25} = 1.28 \text{ ft/min}$$

59. a. Consider the right triangle with legs 2 and $s(t)$. The acute angle at the vertex with the lighthouse is θ, and $\theta = 6\pi t$.

$$\frac{s(t)}{2} = \tan\theta$$

$$s(t) = 2\tan\theta$$

$$= 2\tan(6\pi t)$$

b. When the point on the cliff is 4 km from the lighthouse

$$s = \sqrt{16 - 4} = 2\sqrt{3}$$

and the corresponding value of θ is

$$\theta = \cos^{-1}\left(\frac{2}{4}\right) = \frac{\pi}{3}$$

Therefore,

$$\frac{ds}{dt} = 2(6\pi)\sec^2\theta = 12\pi(4) = 48\pi$$

This is about 150.8 km/min.

3.8 Linear Approximation and Differentials, page 211

SURVIVAL HINT: *You might be thinking, "Why should I do linear approximations when I can use a computer or a calculator?" If you stop and think about it for a moment, you will realize that almost all the curves we consider in this course are linear if we "ZOOM" into a particular point on that curve. This is an important idea which is used not only in this section, but later in your mathematical work.*

1. $d(2x^3) = 6x^2 dx$

3. $d(2\sqrt{x}) = x^{-1/2} dx$

5. $d(x\cos x) = (\cos x - x\sin x)\,dx$

7. $d\left(\dfrac{\tan 3x}{2x}\right) = \dfrac{x(3\sec^2 3x) - (\tan 3x)(1)}{2x^2}\,dx$

$$= \frac{3x\sec^2 3x - \tan 3x}{2x^2}\,dx$$

9. $d(\ln|\sin x|) = \dfrac{1}{\sin x}\cos x\,dx$

$$= \cot x\,dx$$

11. $d(e^x \ln x) = \left[e^x \dfrac{1}{x} + e^x \ln x\right] dx$

$$= \frac{e^x}{x}(1 + x\ln x)dx$$

13. $d\left(\dfrac{x^2\sec x}{x-3}\right)$

$$= \left[\frac{(x-3)(x^2\sec x\tan x + 2x\sec x)}{(x-3)^2}\right.$$
$$\left. - \frac{(x^2\sec x)(1)}{(x-3)^2}\right]dx$$

15. $d\left(\dfrac{x-5}{\sqrt{x+4}}\right)$

$$= \frac{\sqrt{x+4}(1) - (x-5)(\frac{1}{2})(x+4)^{-1/2}}{x+4}dx$$

$$= \frac{2x+8 - x+5}{2(x+4)^{3/2}}dx$$

$$= \frac{x+13}{2(x+4)^{3/2}}dx$$

17. $dx = \Delta x;\ dy = f'(x)\,dx$

19. Let $f(x) = \sqrt{x} = x^{1/2}$; then $f'(x) = \dfrac{1}{2\sqrt{x}}$

$x_0 = 1.0;\ \Delta x = dx = -0.01$

$$f(x_0 + \Delta x) \approx f(x_0) + f'(x_0)\,dx$$
$$\sqrt{0.99} = \sqrt{1 + (-0.01)}$$
$$\approx \sqrt{1} + \frac{1}{2\sqrt{1}}(-0.01)$$
$$= 0.995$$

By calculator, $\sqrt{0.99} \approx 0.9949874371$

21. Let $f(x) = x^5 - 2x^3 + 3x^2 - 2$; then
$f'(x) = 5x^4 - 6x^2 + 6x$
$x_0 = 3;\ \Delta x = dx = 0.01$
$$f(x_0 + \Delta x) \approx f(x_0) + f'(x_0)\,dx$$

$$f(3.01) \approx f(3) + f'(3)\, dx$$
$$= \left[(3)^5 - 2(3)^3 + 3(3)^2 - 2\right]$$
$$+ \left[5(3)^4 - 6(3)^2 + 6(3)\right](0.01)$$
$$= 217.69$$

By calculator, $f(3.01) \approx 217.7155882$

23. $A(r) = \pi r^2$; $A'(r) = 2\pi r$;

$$\left|\frac{dA}{A}\right| = \left|\frac{2\pi r\, dr}{\pi r^2}\right| = 2\left|\frac{dr}{r}\right| = 0.06$$

The measurement is accurate to within 6%.

25. $V(r) = \frac{4}{3}\pi r^3$; $V'(r) = 4\pi r^2$;

$$\left|\frac{dV}{V}\right| = \left|\frac{3(4\pi r^2)\, dr}{4\,\pi r^3}\right| = 3\left|\frac{dr}{r}\right| = 0.03$$

The measurement is accurate to within 3%.

27. Because the average level of carbon monoxide in the air t years from now will be

$$Q(t) = 0.05t^2 + 0.1t + 3.4$$

parts per million, the change in the carbon monoxide level during the next six months ($\Delta t = 0.5$) will be

$$\Delta Q = Q(0.5) - Q(0)$$
$$\approx Q'(0)(0.5)$$

Since $Q'(t) = 0.1t + 0.1$ and $Q'(0) = 0.1$, it follows that

$$\Delta Q \approx 0.1(0.5) = 0.05 \text{ ppm}$$

29. $Q(L) = 60{,}000L^{1/3}$
$Q'(L) = 20{,}000L^{-2/3}$
$L_0 = 1{,}000,\ \Delta L = -60$

$$\Delta Q \approx Q'(L_0)\Delta L$$
$$= 20{,}000(1{,}000)^{-2/3}(-60)$$
$$= -12{,}000$$

The output will be reduced by 12,000 units.

31. Let x be the length of an edge of the cube. The total cost (in cents) is

$$C(x) = \underbrace{2(4x^2)}_{sides} + \underbrace{3x^2}_{bottom} + \underbrace{4x^2}_{top}$$
$$= 15x^2$$
$$\Delta C \approx C'(x)\Delta x = 30x\Delta x$$

When $x = 20$, $\Delta x = 1$,

$$\Delta C \approx 30(20)(1) = 600 \text{ cents}$$

The actual change is

$$C(21) - C(20) = 615 \text{ cents}$$

The actual increase is 615 cents = \$6.15.

33. Let $C(t)$ be the concentration of the drug.
$$C(t) = \frac{0.12t}{t^2 + t + 1}$$
$$C'(t) = (0.12)\frac{(t^2 + t + 1)(1) - t(2t + 1)}{(t^2 + t + 1)^2}$$
$$= (0.12)\frac{-t^2 + 1}{(t^2 + t + 1)^2}$$

Since 30 minutes is 0.5 hours,
$$C'(0.5) = (0.12)\frac{0.75}{(1.75)^2} \approx 0.02939$$
$$\Delta t = \frac{5}{60} = \frac{1}{12};$$
$$\Delta C = C'(0.5)\Delta t \approx 0.00245$$

The concentration will change by about 0.00245 mg/cc^3.

Page 78

35. Let $S(R)$ be the speed of the blood.

$$S(R) = cR^2$$
$$S'(R) = 2cR$$

We have

$$\frac{\Delta S}{S} \approx \frac{S'(R)\Delta R}{S} = \frac{2cR\Delta R}{cR^2} = 2\frac{\Delta R}{R}$$

A 1% error in R means $\Delta R/R = 0.01$ and the propagated error in S is

$$\frac{\Delta S}{S} \approx 2(0.01) = 0.02$$

The error in S is approximately $\pm 2\%$.

37. $S = 4\pi r^2$; $S' = 8\pi r$

$$\frac{\Delta S}{S} \approx \frac{S'(r)\Delta r}{S} = \frac{8\pi r \Delta r}{4\pi r^2} = 2\frac{\Delta r}{r}$$

A 1% increase in r means $\Delta r/r$ is 0.01, so

$$\frac{\Delta S}{S} \approx 0.02$$

or S increases by 2%. Since $V = \frac{4}{3}\pi r^3$; $V' = 4\pi r^2$ and

$$\frac{\Delta V}{V} \approx \frac{V'(r)\Delta r}{V} = 3\frac{\Delta r}{r}$$

so if $\Delta r/r = 0.01$, $\Delta V/V = 0.03$ and the volume increases by 3%.

39. $\Delta L \approx L'(T))\Delta T$, so

$$\sigma = \frac{L'(T)}{L(T)} \approx \frac{\Delta L}{L(T)\Delta T}$$

If $\sigma = 1.4 \times 10^{-5}$, $L = 75$, and $\Delta T = 40 - (-10) = 50$, then

$$\Delta L \approx \sigma L(T)\Delta T$$
$$= \left(1.4 \times 10^{-5}\right)(75)(50)$$
$$= 0.0525$$

The length will change by about 0.0525 ft or 0.63 in.

41. Let N be the number of alpha particles falling on a unit area of the screen.

$$N(\theta) = \frac{1}{\sin^4\left(\frac{\theta}{2}\right)}$$
$$N'(\theta) = -4\left[\sin^{-5}\left(\frac{\theta}{2}\right)\cos\left(\frac{\theta}{2}\right)\right]\frac{1}{2}$$
$$= -2\sin^{-5}\left(\frac{\theta}{2}\right)\cos\left(\frac{\theta}{2}\right)$$

With $\theta = 1$ and $\Delta\theta = 0.1$,

$$\Delta N \approx N'(\theta)\Delta\theta$$
$$= -2[\sin^{-5}(0.5)\cos(0.5)](0.1)$$
$$\approx -6.93$$

A decrease of about 7 (or 6.93) particles/unit area.

43. a. Since the cost is
$C(q) = 0.1q^3 - 5q^2 + 500q + 200$
The cost of producing the fourth unit is

$$C'(3) = 0.3(3)^2 - 10(3) + 500 = 472.7$$

b. The actual cost of manufacturing the fourth unit is

$$C(4) - C(3)$$
$$= \left[0.1(4)^3 - 5(4)^2 + 500(4) + 200\right]$$
$$\quad - \left[0.1(3)^3 - 5(3)^2 + 500(3) + 200\right]$$
$$= 468.70$$

45. Let $Q(L)$ be the daily output.

$$Q'(L) = 120L^{-2/3}$$

$$Q'(1{,}000) = \frac{120}{1{,}000^{2/3}} = 1.2$$

$\Delta Q \approx 1.2(1) = 1.2$ units

47. Let $x = -\sqrt{2}$; then $x^2 = 2$ or $x^2 - 2 = 0$.
We need to find a zero for $f(x) = x^2 - 2$;
$f'(x) = 2x$;

$$x_{n+1} = x_n - \frac{f(x_n)}{f'(x_n)}$$

$$= x_n - \frac{x_n^2 - 2}{2x_n}$$

$$= \frac{x_n^2 + 2}{2x_n}$$

$x_0 = -1$

$$x_1 = \frac{1^2 + 2}{2(1)} = -1.5$$

$$x_2 = \frac{(-1.5)^2 + 2}{2(-1.5)} \approx -1.416666667$$

$$x_3 = \frac{(-1.416666667)^2 + 2}{2(-1.416666667)} \approx -1.414215686$$

$$x_4 = \frac{(-1.414215686)^2 + 2}{2(-1.414215686)} = -1.414213562$$

The approximation (to four decimal places) is
-1.4142.

49. Let $f(x) = x - e^{-x}$; then $f'(x) = 1 + e^{-x}$

$$x_{n+1} = x_n - \frac{f(x_n)}{f'(x_n)}$$

$$= x_n - \frac{x_n - e^{-x_n}}{1 + e^{-x_n}}$$

$x_0 = 1$

$$x_1 = x_0 - \frac{1 - e^{-1}}{1 + e^{-1}} = 0.5378828427$$

$$x_2 = x_1 - \frac{x_1 - e^{-x_1}}{1 + e^{-x_1}} \approx 0.5669869914$$

$$x_3 = x_2 - \frac{x_2 - e^{-x_2}}{1 + e^{-x_2}} \approx 0.567143286$$

The approximation (to four decimal places) is
0.5671.

51. Solve

$$8 = \frac{\pi}{3} H^2 [3(2) - H]$$

We can use technology where $X = H$ and
$Y = \frac{\pi}{3} X^2 (6 - X) - 8$.

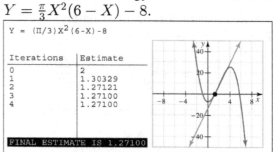

Root $x_1 \approx -1.04159$ not in the domain since
height must be positive. Second root
$x_2 \approx 1.27100$. There is a third root
$x_3 \approx 5.77069$ which is also not in the
domain, since the height cannot be greater
than the radius.

53. a. $f(x) = x^2 - N$
$f'(x) = 2x$

$$x_{n+1} = x_n - \frac{x_n^2 - N}{2x_n}$$

$$= \frac{1}{2x_n}\left(2x_n^2 - x_n^2 + N\right)$$

$$= \frac{1}{2}\left(x_n + \frac{N}{x_n}\right)$$

b. For $\sqrt{1,265}$, $N = 1,265$; Take $x_0 = 35$ since $35^2 = 1,225$.

$$x_1 = \frac{1}{2}\left(35 + \frac{1,265}{35}\right) \approx 35.57143$$

$$x_2 = \frac{1}{2}\left(35.57143 + \frac{1,265}{35.57143}\right)$$
$$\approx 35.56684$$

$$x_3 = \frac{1}{2}\left(35.56684 + \frac{1,265}{35.56684}\right)$$
$$\approx 35.56684$$

55. a. $f(x) = x^{1/2}$; $f'(x) = \frac{1}{2}x^{-1/2}$
With $\Delta x = h$, $f'(1) = \frac{1}{2}$ so

$$f(1 + h) \approx f(1) + f'(1)h = 1 + \frac{h}{2}$$

b. Let $f(x) = \dfrac{1}{x} = x^{-1}$; $f'(x) = -x^{-2}$

$$f(1 + h) \approx f(1) + f'(1)h = 1 - h$$

57. $f(x) = \sqrt{x}$ and $f'(x) = \dfrac{1}{2x^{1/2}}$; $\Delta x = -3$

$$f(97) = f(100 - 3)$$
$$= f(100) + f'(100)(-3)$$
$$= 10 - \frac{3}{2\sqrt{100}}$$
$$= 9.85$$

Calculator check: $\sqrt{97} \approx 9.848857802$

If $\Delta x = 16$,

$$f(97) = f(81 + 16)$$
$$= f(81) + f'(81)(16)$$
$$= 9 + \frac{16}{2\sqrt{81}}$$
$$\approx 9.89$$

59. $f(x) = x^{1/2}$; $f'(x) = \frac{1}{2}x^{-1/2}$;

$$f(0.05) \approx 0.2236; \quad f'(0.05) \approx \frac{1}{0.447} \approx 2.24$$

$$x_1 = 0.05 - \frac{0.2236}{2.24} < 0 \text{ so } f(x_1) \text{ is not}$$

defined. Note that the tangent line is vertical to the curve at $x = 0$. Another value of x_0 would not have made any difference.

Chapter 3 Review

 Studying for a chapter examination is a personal process, one which nobody else can do for you. Simply take the time to review what you have done.

SURVIVAL HINT: Work all of Chapter 3 problems in the Proficiency Examination (whether they are assigned or not). Work through all of the problems before looking at the answers, and *then* correct each of the problems. The answers to all these problems are given in the answer section at the back of the text. If you worked the problem correctly, move on to the next problem, but if you did not work it correctly (or you did not know what to do), then look at the solutions below, look back in the chapter to study the procedure, or ask your instructor.

Finally, go back over the homework problems you have been assigned. If you worked a problem correctly, move on to the next problem, but if you missed it on your homework, then you should look back in the book or talk to your instructor about how to work the problem.

If you follow these steps, you should be successful with your review of this chapter.

Proficiency Examination, page 216

1. a. $m_{\tan} = \lim\limits_{\Delta x \to 0} \dfrac{f(x+\Delta x)-f(x)}{\Delta x}$

b. $m_{\sec} = \dfrac{\Delta y}{\Delta x} = \dfrac{f(x+\Delta x)-f(x)}{\Delta x}$

The slope of the tangent line is the limit (as $\Delta x \to 0$) of the slope of the secant line.

c. A normal line is perpendicular to the tangent line at a point on the graph of a function.

2. a. If $y = f(x)$, then

$$\frac{dy}{dx} = \lim\limits_{\Delta x \to 0} \frac{f(x+\Delta x)-f(x)}{\Delta x}$$

provided this limit exists.

b. Answers should include $f'(x)$, y', and $\dfrac{dy}{dx}$.

3. If a function is differentiable at $x = c$, then it must be continuous at that point. The converse is not true: If a function is continuous at $x = c$, then it may or may not be differentiable at that point. Finally, if a function is discontinuous at $x = c$, then it cannot possibly have a derivative at that point.

4. a. $(cf)' = cf'$

b. $(f+g)' = f' + g'$

c. $(f-g)' = f' - g'$

d. $(af+bg)' = af' + bg'$

e. $d(fg) = fg' + f'g$

f. $\left(\dfrac{f}{g}\right)' = \dfrac{gf'-fg'}{g^2}$ $(g \neq 0)$

g. $d(cf) = c\,df$
$d(f+g) = df + dg$
$d(f-g)' = df - dg$
$d(af+bg) = a\,df + b\,dg$

$d(fg) = f\,dg + g\,df$

$d\left(\dfrac{f}{g}\right) = \dfrac{g\,df - f\,dg}{g^2}$ $(g \neq 0$

5. a. $\dfrac{d}{dx}(k) = 0$

b. $\dfrac{d}{dx}(x^n) = nx^{n-1}$

c. $\dfrac{d}{dx}\sin x = \cos x$ $\quad \dfrac{d}{dx}\cos x = -\sin x$

$\dfrac{d}{dx}\tan x = \sec^2 x$ $\quad \dfrac{d}{dx}\cot x = -\csc^2 x$

$\dfrac{d}{dx}\sec x = \sec x \tan x$

$\dfrac{d}{dx}\csc x = -\csc x \cot x$

d. $\dfrac{d}{dx}e^x = e^x$

e. $\dfrac{d}{dx}\ln x = \dfrac{1}{x}$

f. $\dfrac{d}{dx}\sin^{-1} x = \dfrac{1}{\sqrt{1-x^2}}$

$\dfrac{d}{dx}\cos^{-1} x = \dfrac{-1}{\sqrt{1-x^2}}$

$\dfrac{d}{dx}\tan^{-1} x = \dfrac{1}{1+x^2}$

$\dfrac{d}{dx}\cot^{-1} x = \dfrac{-1}{1+x^2}$

$\dfrac{d}{dx}\sec^{-1} x = \dfrac{1}{|x|\sqrt{x^2-1}}$

$\dfrac{d}{dx}\csc^{-1} x = \dfrac{-1}{|x|\sqrt{x^2-1}}$

6. a. A higher derivative is a derivative of a derivative.

b. y'', y'''; $y^{(4)}$; $\ldots$; $\dfrac{d^2y}{dx^2}$; $\dfrac{d^3y}{dx^3}$; $\dfrac{d^4y}{dx^4}$; $\ldots$

7. a. Rate of change refers to both average and instantaneous rates of change.

b. The average rate of change for a function f is

$$\frac{f(x + \Delta x) - f(x)}{\Delta x}$$

The instantaneous rate of change is

$$\lim_{\Delta x \to 0} \frac{f(x + \Delta x) - f(x)}{\Delta x} = f'(x)$$

c. The relative rate of change of $y = f(x)$ with respect to x is given by the ratio

$$\frac{f'(x)}{f(x)}$$

8. $v(t) = s'(t)$
$a(t) = v'(t) = s''(t)$
The speed is $|v(t))|$.

9. $\dfrac{dy}{dx} = \dfrac{dy}{du}\dfrac{du}{dx}$ or
$[f(u(x))]' = f'[u(x)]u'(x)$

10. Logarithmic differentiation is a procedure in which logarithms are used to trade the task of differentiating products and quotients for that of differentiating sums and differences. It is especially valuable as a means for handling complicated product or quotient functions and power functions where variables appear in both the base and the exponent.

11. Apply all the rules of differentiation, treating y as a function of x and remembering the chain rule.

12. **Step 1.** Draw a figure.
    Step 2. Relate the variables through a
             formula or equation.
    Step 3. Differentiate the equation(s)
             (formulas).
    Step 4. Substitute numerical values and
             solve algebraically for the required
             rate in terms of known rates.

13. **a.** $f(b) \approx f(a) + f'(a)(b - a)$
    b. The propagated error is the difference between $f(x + \Delta x)$ and $f(x)$ and is defined by

$$\Delta f = f(x + \Delta x) - f(x)$$

The relative error is $\Delta f / f$, and the percentage error is $100|\Delta f/f|$.

14. $dx = \Delta x;\ dy = f'(x)\,dx$
    For the sketch, see Figure 3.57.

15. Marginal analysis is the use of the derivative to approximate a change in a function produced by a unit change in the variable. It is especially useful in economics, where the function is cost, revenue, or profit.

16. The Newton-Raphson method approximates a root of a function by locating a point near a root, and then finding where the tangent line at this point intersects the x-axis. That is,

$$x_{x+1} = x_n - \frac{f(x_n)}{f'(x_n)}$$

Repetition of this technique usually closes in on the root.

17. $y = x^3 + x^{3/2} + \cos 2x$
$$\frac{dy}{dx} = 3x^2 + \frac{3}{2}x^{1/2} - 2\sin 2x$$

18. $\dfrac{dy}{dx} = \dfrac{1}{2}\left[\sin(3 - x^2)\right]^{-1/2}\left[\cos(3 - x^2)\right](-2x)$
$$= -x\left[\cos(3 - x^2)\right]\left[\sin(3 - x^2)\right]^{-1/2}$$

19. $x\dfrac{dy}{dx} + y + 3y^2\dfrac{dy}{dx} = 0$
$$\left(x + 3y^2\right)\frac{dy}{dx} = -y$$
$$\frac{dy}{dx} = \frac{-y}{x + 3y^2}$$

20. $y' = x^2 e^{-\sqrt{x}}\left(-\dfrac{1}{2\sqrt{x}}\right) + e^{-\sqrt{x}}(2x)$

$\qquad = -\dfrac{1}{2}e^{3/2}e^{-\sqrt{x}} + 2xe^{-\sqrt{x}}$

$\qquad = \dfrac{1}{2}xe^{-\sqrt{x}}\left(4 - \sqrt{x}\right)$

21. $y' = \dfrac{(\ln 3x)\left(\frac{1}{x}\right) - (\ln 2x)\left(\frac{1}{x}\right)}{(\ln 3x)^2}$

$\qquad = \dfrac{\ln 1.5}{x(\ln 3x)^2}$

22. $y' = \dfrac{3}{\sqrt{1 - (3x+2)^2}}$

23. $y' = \dfrac{1}{1 + (2x)^2}(2)$

$\qquad = \dfrac{2}{1 + 4x^2}$

24. $\ln y = \ln\left[\dfrac{\ln(x^2 - 1)}{\sqrt[3]{x}(1 - 3x)^3}\right]$

$\qquad = \ln[\ln(x^2 - 1)] - \dfrac{1}{3}\ln x - 3\ln(1 - 3x)$

$\dfrac{1}{y}y' = \dfrac{1}{\ln(x^2 - 1)}\dfrac{1}{x^2 - 1}(2x) - \dfrac{1}{3x} - \dfrac{3}{1 - 3x}(-3)$

$\qquad = \dfrac{2x}{(x^2 - 1)\ln(x^2 - 1)} - \dfrac{1}{3x} - \dfrac{9}{3x - 1}$

$y' = y\left[\dfrac{2x}{(x^2 - 1)\ln(x^2 - 1)} - \dfrac{1}{3x} - \dfrac{9}{3x - 1}\right]$

25. $\sin^2 a + \cos^2 a = 1$. So $y = 1$, $\dfrac{dy}{dx} = 0$

26. $y' = x^2(3)(2x - 3)^2(2) + 2x(2x - 3)^3$

$\qquad = 2x(2x - 3)^2(5x - 3)$

$y'' = [2(2x - 3)^2(5x^2 - 3x)]'$

$\qquad = 2(2x - 3)^2(10x - 3)$

$\qquad\quad + 2(2)(2x - 3)(2)(5x^2 - 3x)$

$\qquad = 2(2x - 3)[(2x - 3)(10x - 3)$

$\qquad\quad + 2(2)(5x^2 - 3x)]$

$\qquad = 2(2x - 3)[20x^2 - 36x + 9 + 20x^2 - 12x]$

$\qquad = 2(2x - 3)(40x^2 - 48x + 9)$

27. $f(x) = x - 3x^2$

$f(x + \Delta x) = (x + \Delta x) - 3(x + \Delta x)^2$

$\qquad\qquad = x + \Delta x - 3x^2 - 6x\Delta x - 3\Delta x^2$

$\dfrac{dy}{dx} = \lim\limits_{\Delta x \to 0} \dfrac{f(x + \Delta x) - f(x)}{\Delta x}$

$\qquad = \lim\limits_{\Delta x \to 0} \dfrac{(x + \Delta x - 3x^2 - 6x\Delta x - 3\Delta x^2) - (x - 3x^2)}{\Delta x}$

$\qquad = \lim\limits_{\Delta x \to 0} \dfrac{\Delta x - 6x\Delta x - 3(\Delta x)^2}{\Delta x}$

$\qquad = \lim\limits_{\Delta x \to 0}(1 - 6x - 3\Delta x)$

$\qquad = 1 - 6x$

28. When $x = 1$, $y = 8$, so the point is $(1, 8)$;

$\dfrac{dy}{dx} = (x^2 + 3x - 2)(-3) + (7 - 3x)(2x + 3)$

At $x = 1$, $\dfrac{dy}{dx} = 14$

$\qquad\qquad y - 8 = 14(x - 1)$

$\qquad\qquad 14x - y - 6 = 0$

29. $y = f(1) = \frac{1}{2}$, so the point is $\left(1, \frac{1}{2}\right)$

$\dfrac{dy}{dx} = 2\sin\left(\dfrac{\pi x}{4}\right)\cos\left(\dfrac{\pi x}{4}\right)\left(\dfrac{\pi}{4}\right)$

$\qquad = \dfrac{\pi}{4}\sin\left(\dfrac{\pi}{2}x\right)$

At $x = 1$,

$\dfrac{dy}{dx} = \dfrac{\pi}{4}(1)$

$\qquad = \dfrac{\pi}{4}$

The equation of the tangent line:

$$y - \dfrac{1}{2} = \dfrac{\pi}{4}(x - 1)$$

The equation of the normal line at that same point must have a slope that is the negative reciprocal of the slope of the tangent, $m = -\frac{4}{\pi}$. The equation of the normal:

$$y - \frac{1}{2} = -\frac{4}{\pi}(x - 1)$$

30. $A = \pi r^2$ so $\dfrac{dA}{dt} = 2\pi r \dfrac{dr}{dt}$

$\dfrac{dr}{dt} = 0.5$ when $r = 2$ and

$$\frac{dA}{dt} = 2\pi(2)(0.5)$$
$$= 2\pi \, \text{ft}^2/s$$

Supplementary Problems, page 217

> **SURVIVAL HINT:** *Problems 1-30 this problem set is important because it is testing your ability to perform one of the three fundamental processes of calculus, that of differentiation. The time you can give to working this problems will pay rich dividends in your future work in mathematics,*

1. $y' = 4x^3 + 6x - 7$

3. $y' = \dfrac{1}{2}\left(\dfrac{x^2 - 1}{x^2 - 5}\right)^{-1/2}\left[\dfrac{(x^2 - 5)(2x) - (x^2 - 1)(2x)}{(x^2 - 5)^2}\right]$

$\quad = \dfrac{-4x}{(x^2 - 1)^{1/2}(x^2 - 5)^{3/2}}$

5. $4x - xy' - y + 2y' = 0$

$\quad\quad (x - 2)y' = 4x - y$

$\quad\quad\quad\quad y' = \dfrac{4x - y}{x - 2}$

7. $y' = \sqrt[3]{x}\left[5(x^3 + 1)^4(3x^2)\right] + (x^3 + 1)^5\left(\dfrac{1}{3}\right)x^{-2/3}$

$\quad = \dfrac{1}{3}(x^3 + 1)^4\left[9 \cdot 5x^{7/3} + (x^3 + 1)x^{-2/3}\right]$

$\quad = \dfrac{1}{3}(x^3 + 1)^4\left[x^{-2/3}(45x^3 + x^3 + 1)\right]$

$\quad = \dfrac{(x^3 + 1)^4(46x^3 + 1)}{3x^{2/3}}$

9. $y' = \dfrac{1}{2}[\sin(5x)]^{-1/2}(\cos 5x)(5)$

$\quad = \dfrac{5\cos 5x}{2\sqrt{\sin 5x}}$

11. $y' = (4x + 5)\exp(2x^2 + 5x - 3)$

13. $y' = x(3^{2-x})\ln 3(-1) + 3^{2-x}$

$\quad\quad = 3^{2-x}(1 - x\ln 3)$

15. $\quad e^{xy} + 2 = \ln y - \ln x$

$\quad e^{xy}(xy' + y) = y^{-1}y' - x^{-1}$

$\quad\quad\quad\quad y' = \dfrac{y + xy^2 e^{xy}}{x(1 - xye^{xy})}$

17. $y' = e^{\sin x}(\cos x)$

19.

$$x2^y + y2^x = 3$$
$$x(2^y\ln 2)y' + 2^y + y(2^x\ln 2) + 2^x y' = 0$$
$$(x2^y\ln 2 + 2^x)y' = -2^y - y2^x\ln 2$$
$$y' = -\frac{2^y + y2^x\ln 2}{x(2^y\ln 2) + 2^x}$$

21. $\dfrac{dy}{dx} = e^{-x}\dfrac{1}{2x\sqrt{\ln 2x}} - e^{-x}\sqrt{\ln 2x}$

23. $y' = -[\sin(\sin x)]\cos x$

25. $8x - 32yy' = 0$

$\quad\quad y' = \dfrac{x}{4y}$

27. $y' = e^{1-2x} + x(-2)e^{1-2x}$

$\quad = (1 - 2x)e^{1-2x}$

29. $y' = 2\csc\sqrt{x}\left(-\csc\sqrt{x}\cot\sqrt{x}\right)\left(\dfrac{1}{2\sqrt{x}}\right)$

$\quad = -\dfrac{\csc^2\sqrt{x}\cot\sqrt{x}}{\sqrt{x}}$

31. $\dfrac{dy}{dx} = 5x^4 - 20x^3 + 21x^2$

$\quad \dfrac{d^2y}{dx^2} = 20x^3 - 60x^2 + 42x$

33. $2x + 3y^2y' = 0$

$$y' = -\frac{2x}{3y^2}$$

$$y'' = -\frac{3y^2(2) - (2x)(6yy')}{(3y^2)^2}$$

$$= -\frac{6y^2 - 12xy\left(-\frac{2x}{3y^2}\right)}{9y^4}$$

$$= -\frac{6y^3 + 8x^2}{9y^5}$$

$$= -\frac{2(3y^3 + 4x^2)}{9y^5}$$

35. $2x + (\cos y)y' = 0$

$$y' = -\frac{2x}{\cos y}$$

$$y'' = -2\left[\frac{\cos y - x(-\sin y)y'}{\cos^2 y}\right]$$

$$= \frac{-2}{\cos^2 y}\left[\cos y + x\sin y\left(-\frac{2x}{\cos y}\right)\right]$$

$$= \frac{4x^2\sin y - 2\cos^2 y}{\cos^3 y}$$

37. $y' = 4x^3 - 21x^2 + 2x$

At $P(0, -3)$, $m = y' = 0$.

The equation of the (horizontal) tangent line is

$$y + 3 = 0$$

39. $f(x) = x\cos x$; $f\left(\frac{\pi}{2}\right) = 0$

$f'(x) = -x\sin x + \cos x$; $f'\left(\frac{\pi}{2}\right) = -\frac{\pi}{2}$.

The equation of the tangent line is:

$$y = -\frac{\pi}{2}\left(x - \frac{\pi}{2}\right)$$

$$2\pi x + 4y - \pi^2 = 0$$

41. $2xyy' + x^2y' + y^2 + 2xy = 0$

At $(1, 1)$: $2y' + y' + 1 + 2 = 0$

$$y' = -1$$

The equation of the tangent line is:

$$y - 1 = -(x - 1)$$

$$x + y - 2 = 0$$

43. $e^{xy}(xy' + y) = 1 - y'$

At $x = 1$, $y = 0$, and $y' = \frac{1}{2}$.

The equation of the tangent line is:

$$y = \frac{1}{2}(x - 1)$$

$$x - 2y - 1 = 0$$

45. a. $f'(x) = -4x$

b. The tangent is horizontal when $f'(x) = 0$ or when $x = 0$. When $x = 0$, $y = 4 - 2(0)^2 = 4$. The horizontal line passing through $(0, 4)$ is

$$y - 4 = 0$$

c. the line $8x + 3y = 4$ has slope $m = -\frac{8}{3}$. We want to find x when

$$-4x = -\frac{8}{3}$$

$$x = \frac{2}{3}$$

When $x = \frac{2}{3}$, then $y = 4 - 2\left(\frac{2}{3}\right)^2 = \frac{28}{9}$. The point is $\left(\frac{2}{3}, \frac{28}{9}\right)$.

47. $f'(x) = x^2\left(\cos x^2\right)(2x) + 2x\sin x^2$

$$= 2x^3\cos x^2 + 2x\sin x^2$$

$$f''(x) = 2x^3(-\sin x^2)(2x) + 2(3x^2)(\cos x^2)$$

$$+ 2x(\cos x^2)(2x) + 2\sin x^2$$

$$= -4x^4\left(\sin x^2\right) + 6x^2\left(\cos x^2\right)$$

$$+ 4x^2\left(\cos x^2\right) + 2\sin x^2$$

$$= 2(1 - 2x^4)\left(\sin x^2\right) + 10x^2\left(\cos x^2\right)$$

49. Let $y = \sqrt[3]{\dfrac{x^4 + 1}{x^4 - 2}}$ so $y^3 = \dfrac{x^4 + 1}{x^4 - 2}$

$$3y^2 y' = \frac{(4x^3)(x^4 - 2 - x^4 - 1)}{(x^4 - 2)^2}$$

$$y' = \frac{(4x^3)(-3)}{3y^2(x^4 - 2)^2}$$

$$= \frac{-4x^3}{(x^4 - 2)^{4/3}(x^4 + 1)^{2/3}}$$

51.
$$x^2 + 4xy - y^2 = 8$$
$$2x + 4xy' + 4y - 2yy' = 0$$
$$y' = \frac{x + 2y}{y - 2x}$$
$$y'' = \frac{(y - 2x)(1 + 2y') - (x + 2y)(y' - 2)}{(y - 2x)^2}$$
$$= \frac{5y - 5xy'}{(y - 2x)^2}$$
$$= \frac{5y - 5x\left(\frac{x+2y}{y-2x}\right)}{(y - 2x)^2}$$
$$= \frac{5y^2 - 20xy - 5x^2}{(y - 2x)^3}$$

53. Answers vary; $y = |x - 5|$.

55. $f'(x) = -2x$. Let $P(x_0, y_0)$ be the required point. Since P is on $y = -x^2$, the equation of the tangent at (x_0, y_0) is

$$y + x_0^2 = -2x_0(x - x_0)$$

For the line to pass through $(0, 9)$,

$$9 + x_0^2 = -2x_0(0 - x_0)$$
$$x_0^2 + 9 = 2x_0^2$$
$$x_0 = \pm 3$$

Then, $f(\pm 3) = -9$. Thus, the points are $(3, -9)$ and $(-3, -9)$.

57. Answers vary. For this 𝒬uest write a page or two. You can use Internet sources, but you should check at least one of the sources listed in the problem.

59. Let $t = 0$ for 2009 and $G(t)$ the GDP in billions of dollars.
$G(0) = 125$, $G(2) = 155$, so the slope of the line is

$$m = \frac{155 - 125}{2} = 15 = G'(t)$$

$G(t) = 125 + 15t;$
$G(5) = 125 + 75 = 200$
The percentage rate of change is

$$\frac{100(15)}{200} = 7.5\%$$

per year in 2014.

61. Here we have a function of a function, so the chain rule is required.

$$\frac{dL}{dt} = \frac{dL}{dp} \cdot \frac{dp}{dt}$$
$$= (0.5)\left(\frac{1}{2}\right)(p^2 + p + 58)^{-1/2}(2p + 1)\left[\frac{6}{(t+1)^2}\right]$$
$$= \frac{6(2p + 1)}{4(p^2 + p + 58)^{1/2}(t + 1)^2} \text{ ppm/year}$$

Two years from now, $t = 2$, $p(2) = 18$, and the rate of change is

$$\frac{dL}{dt} = \frac{6[2(18) + 1]}{4(18^2 + 18 + 58)^{1/2}(2 + 1)^2}$$
$$= \frac{37}{120}$$
$$\approx 0.31 \text{ ppm/year}$$

63. $\theta = \theta_M \sin kt$

$$\frac{d\theta}{dt} = k\theta_M \cos kt$$

$$\frac{d^2\theta}{dt^2} = -k^2\theta_M \sin kt$$

$$\frac{d^2\theta}{dt^2} + k^2\theta = -k^2\theta_M \sin kt + k^2(\theta_M \sin kt)$$

$$= 0$$

65. $f'(x) = \sqrt{x^2 + 5}$; $g(x) = x^2 f\left(\frac{x}{x-1}\right)$

$$g'(x) = x^2 f'\left(\frac{x}{x-1}\right)\left[\frac{-1}{(x-1)^2}\right] + 2xf\left(\frac{x}{x-1}\right)$$

$$g'(2) = -2^2 f'(2)(-1) + 2(2)f(2)$$

$$= -4\sqrt{4+5} + 4(-3)$$

$$= -24$$

67. $\dfrac{dy}{dt} = 2$ m/s; $\theta = \tan^{-1}\dfrac{y}{4}$

$$\frac{d\theta}{dt} = \frac{\frac{1}{4}\frac{dy}{dt}}{1 + \frac{y^2}{16}}$$

$$= \frac{4}{16 + y^2}\frac{dy}{dt}$$

When $y = \frac{3}{2}$

$$\frac{d\theta}{dt} = \frac{4(2)}{16 + \frac{9}{4}} = \frac{32}{73} \text{ rad/s}$$

69. $$x^{1/2} + y^{1/2} = C$$

$$\frac{1}{2}x^{-1/2} + \frac{1}{2}y^{-1/2}y' = 0$$

$$y' = -\frac{y^{1/2}}{x^{1/2}}$$

The equation of the tangent line at (x_0, y_0) is

$$y - y_0 = -\frac{y_0^{1/2}}{x_0^{1/2}}(x - x_0)$$

The x-intercept is found by setting $y = 0$:

$$0 - y_0 = -\frac{y_0^{1/2}}{x_0^{1/2}}(x - x_0)$$

$$x = x_0 + \sqrt{y_0 x_0}$$

The y-intercept is found by setting $x = 0$:

$$y - y_0 = -\frac{y_0^{1/2}}{x_0^{1/2}}(0 - x_0)$$

$$y = y_0 + \sqrt{y_0 x_0}$$

The sum of these intercepts is

$$x_0 + \sqrt{y_0 x_0} + y_0 + \sqrt{x_0 y_0} = x + y$$

$$x_0 + 2\sqrt{x_0 y_0} + y_0 = x + y$$

$$\left(\sqrt{x_0} + \sqrt{y_0}\right)^2 = x + y$$

$$C^2 = x + y$$

71. Let $x = \sqrt[5]{23}$; then $x^5 = 23$ or $x^5 - 23 = 0$.
We need to find a zero for $f(x) = x^5 - 23$;
$f'(x) = 5x^4$;

$$x_{n+1} = x_n - \frac{f(x_n)}{f'(x_n)}$$

$$= x_n - \frac{x_n^5 - 23}{5x_n^4}$$

$$= \frac{4x_n^5 + 23}{5x_n^4}$$

$$x_0 = 2$$

$$x_1 = \frac{4(2)^5 + 23}{5(2)^4} = 1.8875$$

$$x_2 = \frac{4(1.8875)^5 + 23}{5(1.8875)^4} \approx 1.872418$$

$$x_3 = \frac{4(1.872418)^5 + 23}{5(1.872418)^4} \approx 1.87217129$$

$$x_4 = \frac{4(1.87217129)^5 + 23}{5(1.87217129)^4} = 1.87217123$$

The approximation (to four decimal places) is
1.87217123.

73. a. Let $P(x)$ be the profit when producing x units at a total cost of $C(x)$ and revenue of $R(x)$.

$$P(x) = R(x) - C(x)$$
$$P'(x) = R'(x) - C'(x)$$

$P'(x) = 0$ if $C'(x) = R'(x)$.

b. $A'(x) = \dfrac{xC'(x) - C(x)}{x^2}$

$$= \frac{1}{x}\left[C'(x) - \frac{C(x)}{x}\right]$$

$A'(x) = 0$ if $C'(x) = A(x)$.

75. Let $f(x) = x^{3/2} + 2x^{1/2}$; $f(16) = 72$.
$f'(x) = \frac{3}{2}x^{1/2} + x^{-1/2}$
$f'(16) = 6.25$; $\Delta x = 0.01$

$$f(16.01) = f(16 + 0.01)$$
$$\approx f(16) + f'(16)(0.01)$$
$$= 72 + 6.26(0.01)$$
$$= 72.0625$$

A calculator approximation is 72.06251.

77. Consider a right triangle with horizontal leg of 600, vertical leg y, and angle of elevation θ.

$$y = 600\tan\theta$$
$$\frac{dy}{dt} = 600\sec^2\theta\,\frac{d\theta}{dt}$$

When $y = 800$, $\sec\theta = \dfrac{1{,}000}{600} = \dfrac{5}{3}$.

$$-20 = 600\left(\frac{5}{3}\right)^2\frac{d\theta}{dt}$$
$$\frac{d\theta}{dt} = \frac{-20(9)}{(600)(25)}$$
$$= -0.012 \text{ rad/min}$$

79. $f'(0) = \lim\limits_{\Delta x \to 0} \dfrac{f(0 + \Delta x) - f(0)}{\Delta x}$

$$= \lim_{\Delta x \to 0} \frac{\Delta x \sin\frac{1}{\Delta x}}{\Delta x}$$

$$= \lim_{\Delta x \to 0} \sin\frac{1}{\Delta x}$$

This limit does not exist so $f'(0)$ is not defined.

81. $V = x^3$; when the volume is 27, $x = 3$,
$\dfrac{dx}{dt} = -1$; the surface area is $S = 6x^2$:

$$\frac{dS}{dt} = 12x\frac{dx}{dt}$$

When $x = 3$ and $\dfrac{dx}{dt} = -1$,

$$\frac{dS}{dt} = 12(3)(-1) = -36\,\text{cm}^2/\text{h}$$

83. $v(0) = 1{,}200$ m/s; $v(2 \times 10^{-3}) = 6{,}000$ m/s;
$v(t) = at + v(0)$
Thus, when $t = 2 \times 10^{-3}$

$$(2 \times 10^{-3})a = 6{,}000 - 1{,}200$$
$$a = \frac{4{,}800}{2 \times 10^{-3}}$$
$$= 2.4 \times 10^6 \text{ m/s}^2$$

85. Let A be the cross-sectional area of the artery. $r_0 = 1.2$;

$$A = \pi r^2$$

$$\frac{dA}{dr} = 2\pi r \frac{dr}{dr}$$

$$= 2\pi(1.2 - 0.3)$$

$$= 1.8\pi \text{ mm}^2/\text{mm}$$

87. a. Let x be the rate in gallons/mi and t the time in hours. The cost of the driver is: $C_d = 26t$, where $t = \dfrac{300}{x}$ hr so the cost of gas is

$$C_g = 4\left(\frac{1}{300}\right)\left(\frac{1,500}{x} + x\right)(300)$$

$$= 4\left(\frac{1,500}{x} + x\right)$$

The total cost is

$$C(x) = 26\left(\frac{300}{x}\right) + 4\left(\frac{1,500}{x} + x\right)$$

$$= 13,800x^{-1} + 4x$$

b. $C'(x) = \dfrac{-13,800}{x^2} + 4$

$C'(55) \approx -0.5620$

A change from 55 to 57 mi/h decreases the cost by about

$$2(-0.5620) \approx -1.12$$

The cost is decreasing by approximately $1.12.

89. Consider a right triangle with the leg along the shore $\overline{PQ}$ with the length labeled x, and the leg from the lighthouse to P labeled 4,000 ft. We know that $dx/dt = 3$ ft/s. Let θ be the acute angle opposite $\overline{PQ}$. Then $x = 4,000 \tan \theta$ and $\theta = \tan^{-1}(x/4,000)$.

$$\frac{d\theta}{dt} = \frac{\frac{1}{4,000}}{1 + \left(\frac{1}{4,000}\right)^2} \frac{dx}{dt}$$

$$= \frac{4,000}{4,000^2 + x^2} \frac{dx}{dt}$$

At $x = 1,000$

$$\frac{d\theta}{dt} = \frac{3(4,000)}{4,000^2 + 1,000^2} = \frac{12}{17,000} \text{ radians/s}$$

91. $s(t) = A\sin 2t$; $\dfrac{ds}{dt} = 2A\cos 2t$;

$$\frac{d^2 s}{dt^2} = -4A\sin 2t$$

Thus,

$$\frac{d^2 s}{dt^2} + 4s = -4A\sin 2t + 4A\sin 2t = 0$$

93. $g[f(x)] = x$

$g'[f(x)][f'(x)] = 1$

Thus,

$$g'[f(x)] = \frac{1}{f'(x)}$$

$$\frac{dg}{dx} = \frac{1}{df/dx}$$

95. For the velocity,

$$\frac{d\theta}{dt} = (3 \text{ rev/s})(2\pi \text{ rad/rev})$$

$$= 6\pi \text{ rad/s}$$

Also,

$$25 = 2^2 + x^2 - 2(2)x\cos\theta$$

so we have

$$0 = 2x\frac{dx}{dt} - 4\left[\frac{dx}{dt}\cos\theta + x(-\sin\theta)\frac{d\theta}{dt}\right]$$

$$\frac{dx}{dt} = \frac{4x\sin\theta}{4\cos\theta - 2x}\frac{d\theta}{dt}$$

$$= \frac{2x\sin\theta}{2\cos\theta - x}\frac{d\theta}{dt}$$

$$= \frac{12\pi x\sin\theta}{2\cos\theta - x}$$

For the acceleration,

$$\frac{d^2x}{dt^2} = \frac{12\pi}{(2\cos\theta - x)^2}\left(\left[\frac{dx}{dt}\sin\theta + x\cos\theta\frac{d\theta}{dt}\right]\times\right.$$

$$(2\cos\theta - x) - x\sin\theta\left[-2\sin\theta\frac{d\theta}{dt} - \frac{dx}{dt}\right]\bigg)$$

$$= \frac{12\pi}{(2\cos\theta - x)^2}\times$$

$$\left(2\sin\theta\cos\theta\frac{dx}{dt} + (2 - x\cos\theta)x\frac{d\theta}{dt}\right)$$

$$= \frac{12\pi}{(2\cos\theta - x)^2}\times$$

$$\left(\frac{2\cdot 12\pi x\sin^2\theta\cos\theta}{2\cos\theta - x} + (2 - x\cos\theta)6\pi x\right)$$

$$= \frac{144\pi^2 x}{(2\cos\theta - x)^3}\times$$

$$\left(2\sin^2\theta\cos\theta + (2\cos\theta - x)\left[1 - \frac{1}{2}x\cos\theta\right]\right)$$

97. This is Putnam Problem 1, morning session in 1946. Suppose that the function

$$f(x) = ax^2 + bx + c$$

where a, b, c are real constants, satisfies the condition $|f(x)| \le 1$. Prove $|f'(x)| \le 4$ for $|x| \le 1$. If $a \ne 0$, the graph of

$$y = ax^2 + bx + c$$

is a parabola which can be assumed without loss of generality to open upward, *i.e.,* $a > 0$. [We discuss the straight line case later.] By symmetry we may assume that b is nonnegative. Then the vertex falls in the left half-plane and it is clear that $\max_{|x|\le 1}|f'(x)|$ occurs when $x = 1$, and this maximum value is $2a + b$. It remains to show that $2a + b \le 4$. Now,

$$f(1) = a + b + c \le 1$$

and $f(0) = c \ge -1$. Thus, $a + b \le 2$. Since a and b are both nonnegative, $a \le 2$ and $2a + b \le 4$.

For the linear case we have $a = 0$, so that

$$f'(x) = b = \frac{f(1) - f(x)}{2}$$

and

$$|f'(x)| = \frac{|f(1) - f(-1)|}{2} \le 1$$

Historical Quest The chemist Mendeleev raised the question as to restrictions on $p_n'(x)$ for $-1 \le x \le 1$ when $|p_n(x)| \le 1$ on $-1 \le x \le 1$, then $|p_n'(x)| \le 1$ on $-1 \le x \le 1$, where p_n is a polynomial of degree n. A.A. Markov answered this question in 1890 by proving that if $|p_n(x)| \le 1$ on $-1 \le x \le 1$, then $|p_n'(x)| \le n^2$ on the same interval. The present problem is thus the special case where $n = 2$. It is known that equality occurs if and only if

$$p_n(x) = \cos(n\cos^{-1}x)$$

i.e. $p_n(x)$ is the polynomial such that $\cos n\theta = p_n(\cos\theta)$. For $n = 2$,

$$\cos 2\theta = 2\cos^2\theta - 1 = 2x^2 - 1$$

The polynomials $p_n(x)$ are called Chebyshev polynomials. See John Todd, *A Survey of Numerical Analysis*, New York, 1962, pp. 138-139. The generalized version appears as Problem 83, in Section 6, Polya and Szego, *Aufgaben und Lehrsätze aus der Analyze*, Vol 2, p. 91 and p. 287.

99. This is Putnam Problem 6, morning session in 1946. Newton's law of motion for a particle of unit mass takes the form

$$F = \text{force} = \frac{dv}{dt}$$

Since we are given that

$$x = at + bt^2 + ct^3$$

it follows that

$$v = \frac{dx}{dt} = a + 2bt + 3ct^2$$

$$\frac{dv}{dt} = 2b + 6ct$$

We now express the force in terms of v:

$$\begin{aligned}
F^2 &= 4b^2 + 24bct + 36c^2t^2 \\
&= 4b^2 + 12c(2bt + 3ct^2) \\
&= 4b^2 + 12c(v - a)
\end{aligned}$$

Hence

$$F = f(v) = \pm\sqrt{4b^2 - 12ac + 12cv}$$

The radical sign is taken to be the sign of $2b + 6ct$ which, if the hypotheses of the problem are satisfied, cannot change for the interval of time under consideration, since then v would take the same value twice but dv/dt would not.

Page 92

CHAPTER 4 Additional Applications of the Derivative

Chapter Overview
This is the first chapter in which you are able to see some of the real power of differential calculus. Optimization (finding the maximum and minimum values) is one of the most important applications to master. As you work your way through this chapter, you will begin to realize the power of mathematics. Some of those applications include Fermat's principle, maximizing profit, minimizing cost, inventory modeling, optimal holding time, concentration of a drug in the bloodstream and optimal angle for vascular branching.

4.1 Extreme Values of a Continuous Function, page 235

In Problems 1-14, the maximum value is denoted by M and the minimum value by m.

1. $f(x) = 5 + 10x - x^2$ on $[-3, 3]$
$f'(x) = 10 - 2x \qquad 10 - 2x = 0$
$$x = 5$$

endpoints	critical numbers
$m = f(-3) = -34$	$x = 5$ is not
$M = f(3) = 26$	in the domain.

3. $f(x) = x^3 - 3x^2$ on $[-1, 3]$
$f'(x) = 3x^2 - 6x \qquad 3x^2 - 6x = 0$
$$3x(x - 2) = 0$$
$$x = 0, 2$$

endpoints	critical numbers
$m = f(-1) = -4$	$M = f(0) = 0$
$f(3) = 0$	$m = f(2) = -4$

5. $f(t) = t^4 - 8t^2$ on $[-3, 3]$
$f'(t) = 4t^3 - 16t$
$$4t^3 - 16t = 0$$
$$4t(t + 2)(t - 2) = 0$$
$$t = 0, -2, 2$$

endpoints	critical numbers
$M = f(-3) = 9$	$f(0) = 0$
$M = f(3) = 9$	$m = f(-2) = -16$
	$m = f(2) = -16$

7. $g(x) = x^5 - x^4$ on $[-1, 1]$
$g'(x) = 5x^4 - 4x^3$
$$5x^4 - 4x^3 = 0$$
$$x^3(5x - 4) = 0$$
$$t = 0, \frac{4}{5}$$

endpoints	critical numbers
$m = f(-1) = -2$	$M = f(0) = 0$
$M = f(1) = 0$	$f\left(\frac{4}{5}\right) \approx -0.0819$

9. $h(t) = te^{-t}$ on $[0, 2]$
$h'(t) = e^{-t}(1 - t)$
$$e^{-t}(1 - t) = 0$$
$$t = 1$$

endpoints	critical numbers
$m = h(0) = 0$	$M = h(1) = e^{-1}$
$h(2) = 2e^{-2}$	

11. $f(x) = |x|$ on $[-1, 1]$
$f'(x)$ is not defined at $x = 0$.

endpoints	critical numbers
$M = f(-1) = 1$	$m = f(0) = 0$
$M = f(1) = 1$	

Page 93

13. $f(u) = \sin^2 u + \cos u$ on $[0, 2]$
$f'(u) = 2\sin u \cos u - \sin u$
$2\sin u \cos u - \sin u = 0$
$\sin u(2\cos u - 1) = 0$
$$u = 0, \frac{\pi}{3}$$

endpoints	critical numbers
$f(0) = 1$	$f(0) = 1$
$m = f(2) \approx 0.41067$	$M = f\left(\frac{\pi}{3}\right) = \frac{5}{4}$

> **SURVIVAL HINT:** *Notice the reasoning in Problems 13 and 14 to determine which one approaches 0 and which one approaches ∞.*

15. Step 1: Find the value of the function at the endpoints of an interval.
Step 2: Find the critical points; that is, points at which the derivative of the function is zero or undefined.
Step 3: Find the value of the function at each critical point.
Step 4: State the absolute extrema.

17. $f(u) = 1 - u^{2/3}$ on $[-1, 1]$
$f(1) = 0; f(-1) = 0;$
$f'(u) = -\frac{2}{3}u^{-1/3}$ which is not defined at $u = 0$. $f(0) = 1$. The maximum value is 1 and the minimum value is 0.

19. $g(x) = 2x^3 - 3x^2 - 36x + 4$ on $[-4, 4]$
$g(-4) = -28; g(4) = -60$
$g'(x) = 6x^2 - 6x - 36$
$6x^2 - 6x - 36 = 0$
$6(x - 3)(x + 2) = 0$
$$x = 3, -2$$
$g(3) = -77; g(-2) = 48$
The maximum value is 48 and the minimum value is -77.

21. $f(x) = \frac{8}{3}x^3 - 5x^2 + 8x - 5$ on $[-4, 4]$

$f(-4) = -\frac{863}{3}; f(4) = \frac{353}{3}$
$f'(x) = 8x^2 - 10x + 8$
$8x^2 - 10x + 8 = 0$
$4x^2 - 5x + 4 = 0$
$$x = \frac{5 \pm \sqrt{25 - 64}}{8}$$
There are no real roots. The maximum value is $\frac{353}{3}$ and the minimum value is $-\frac{863}{3}$.

23. $h(x) = \tan x + \sec x$ on $[0, 2\pi]$
$h(0) = 1$, $h(2\pi) = 1$ but h is not continuous. There is no maximum and no minimum value.

25. $f(x) = e^{-x}\sin x$ on $[0, 2\pi]$
$f(0) = 0; f(2\pi) = 0$
$f'(x) = e^{-x}(\cos x - \sin x)$
$e^{-x}(\cos x - \sin x) = 0$
$$\cos x = \sin x$$
$$x = \frac{\pi}{4}, \frac{5\pi}{4}$$
$$f\left(\frac{\pi}{4}\right) = \frac{\sqrt{2}}{2}e^{-\pi/4}$$
$$f\left(\frac{5\pi}{4}\right) = -\frac{\sqrt{2}}{2}e^{-5\pi/4}$$
The maximum value is $\frac{\sqrt{2}}{2}e^{-\pi/4}$ and the minimum value is $-\frac{\sqrt{2}}{2}e^{-5\pi/4}$.

27. $f(x) = \begin{cases} 9 - 4x & \text{if } x < 1 \\ -x^2 + 6x & \text{if } x \geq 1 \end{cases}$ on $[0, 4]$
$f(0) = 9; f(4) = 8$
$f'(x) = -4$ if $x < 1$;
$f'(x) = -2x + 6$ if $x \geq 1$
$f'(0) = 0$ when $x = 3; f(3) = 9$
$f'(x)$ is not defined at $x = 1; f(1) = 5$
The maximum value is 9 and the minimum value is 5.

29. $f(x) = \dfrac{1}{x(x+1)}$ on $[-0.5, 0)$

$f'(x) = \dfrac{-(2x+1)}{x^2(x+1)^2}$

$f'(x) = 0$ at $x = -\frac{1}{2}$

On $[-0.5, 0)$, $f'(x) < 0$ so f is decreasing on this interval; that is, as $x \to 0^-$ the values of $f \to \infty$.

31. $g(x) = \dfrac{x^2 - 1}{x^2 + 1}$ on $[-1, 1]$; $g(\pm 1) = 0$

$g'(x) = \dfrac{(x^2 + 1)(2x) - (x^2 - 1)(2x)}{(x^2 + 1)^2}$

$ = \dfrac{4x}{(x^2 + 1)^2}$

$g'(x) = 0$ when $x = 0$; $g(0) = -1$
The smallest value is -1.

33. $f(x) = e^x + e^{-x} - x$ on $[0, 2]$

$f(0) = 2$; $f(2) = e^2 + e^{-2} - 2$

$f'(x) = e^x - e^{-x} - 1$

$\dfrac{e^{2x} - e^x - 1}{e^x} = 0$

$e^{2x} - e^x - 1 = 0$

$e^x = \dfrac{1 \pm \sqrt{1 - 4(-1)}}{2}$

The negative root is not in the domain.

$x = \ln\left(\dfrac{1 + \sqrt{5}}{2}\right)$

The smallest value of f is

$f\left(\ln\left(\frac{1+\sqrt{5}}{2}\right)\right) = \sqrt{5} - \ln\left(\frac{1+\sqrt{5}}{2}\right)$.

35. $f(\theta) = \cos^3\theta - 4\cos^2\theta$ on $[-0.1, \pi + 0.1]$

$f(-0.1) \approx -2.98$;

$f(\pi + 0.1) \approx -4.95$;

$f'(\theta) = 3\cos^2\theta(-\sin\theta) - 8\cos\theta(-\sin\theta)$

$ = \cos\theta\sin\theta(-3\cos\theta + 8)$

$f'(\theta) = 0$ when $\theta = \frac{\pi}{2}, 0, \pi$

$f(\frac{\pi}{2}) = 0$; $f(0) = -3$; $f(\pi) = -5$. The maximum value of $f(\theta)$ is 0 and the minimum value is -5.

37. $f(x) = 20\sin(378\pi x)$ on $[-1, 1]$

$f(-1) = 0$; $f(1) = 0$;

$f'(x) = 20(378\pi)\cos(378\pi x)$

$20(378\pi)\cos(378\pi x) = 0$

$378\pi x = \dfrac{n\pi}{2}$, n an odd integer

$x = \dfrac{n}{756}$

for each odd integer with $n \le 756$.

$f\left(\dfrac{n}{756}\right) = \pm 20$; the maximum value is 20 and the minimum value is -20.

SURVIVAL HINT: *It is interesting to note there will be 378 points with a maximum of 20, and 378 points with a minimum of -20. Your graphing calculator will most likely not be able to show this on $[-1, 1]$. A change of the horizontal scale gives the following graph:*

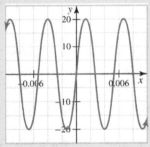

39. $f(w) = \sqrt{w}(w - 5)^{1/3}$ on $[0, 4]$

$f(0) = 0$; $f(4) = -2$

$f'(w) = \dfrac{1}{2}w^{-1/2}(w - 5)^{1/3} + \dfrac{1}{3}w^{1/2}(w - 5)^{-2/3}$

$ = \dfrac{1}{6}w^{-1/2}(w - 5)^{-2/3}[3(w - 5) + 2w]$

$$= \frac{1}{6}w^{-1/2}(w-5)^{-2/3}(5w-15)$$

$$= \frac{5}{6}w^{-1/2}(w-5)^{-2/3}(w-3)$$

$f'(w) = 0$ when $w = 3$, and does not exist when $w = 0$, $w = 5$. The value $w = 5$ is not in the domain and $w = 0$ is in the domain of f (considered above).

$$f(3) = \sqrt{3}(3-5)^{1/3}$$

$$= -\sqrt[6]{108}$$

The maximum value is 0 and the minimum value is $-\sqrt[6]{108}$.

41. $h(x) = \cos^{-1}x \tan^{-1}x$ on $[0, 1]$
$h(0) = h(1) = 0$

$$h'(x) = \frac{\cos^{-1}x}{1+x^2} - \frac{\tan^{-1}x}{\sqrt{1-x^2}}$$

$h'(x) = 0$ when $x \approx 0.602$ (by graphing) and does not exist when $x = 1$.
$h(0.602) \approx 0.501$, so the maximum value is approximately 0.5 and the minimum value is 0.

Answers to Problems 43-48 may vary.

43. a. $f(x) = \begin{cases} x^2 & \text{for } -1 < x < 1 \\ x+2 & \text{for } 1 \le x < 2 \end{cases}$

The minimum is 0, but there is no maximum.

b. $f(x) = \begin{cases} -x^2 & \text{for } -1 < x < 1 \\ -x+2 & \text{for } 1 \le x < 2 \end{cases}$

The maximum is 1, but there is no minimum.

c. $f(x) = \begin{cases} \sin x & \text{for } -\pi < x \le \frac{\pi}{2} \\ 0.5 & \text{for } \frac{\pi}{2} < x < 3 \end{cases}$

The maximum is 1 and the minimum is -1.

d. $f(x) = \begin{cases} \sin x & \text{for } 0 < x < \frac{\pi}{2} \\ 0.5 & \text{for } \frac{\pi}{2} \le x < 3 \end{cases}$

There is no maximum and there is no minimum.

45. a. $f(x) = \sin x$ on $(-\pi, 1)$
The minimum is -1, but there is no maximum.

b. $f(x) = \sin x$ on $(0, 4)$.
The maximum is 1, but there is no minimum.

c. $f(x) = \sin x$ on $(0, 2\pi)$
The maximum is 1 and the minimum is -1.

d. $f(x) = \sin x$ on $(-1, 1)$.
There is no maximum and no minimum.

47. On $[-1, 1]$, let $f(x) = \begin{cases} x^{-2} & x \ne 0 \\ 0 & x = 0 \end{cases}$

This function has no maximum.

49. $v(t) = s'(t) = 3t^2 - 12t - 15$
$v(0) = -15;\ v(4) = -15$
$v'(t) = 6t - 12 = 0$ when $t = 2$.
$v(2) = -27$
The maximum value for the velocity is -15.

51. Let x and y be the numbers we are seeking on $[0, 8]$. Then $x + y = 8$ and $P = x^2(8-x)^2$.
$$P'(x) = x^2(2)(8-x)(-1) + 2x(8-x)^2$$
$$= 2x(8-x)(8-2x)$$
$P'(x) = 0$ when $x = 0, 8$, and 4.
$P(0) = P(8) = 0;\ P(4) = 256$.
The largest product occurs when $x = y = 4$.

53. $P = xy^3 = x(80-3x)^3$ on $[0, \frac{80}{3}]$.
$P(0) = P(\frac{80}{3}) = 0$
$$P'(x) = x(3)(80-3x)^2(-3) + (80-3x)^3$$
$$= (80-3x)^2(80-3x-9x)$$
$$= (80-3x)^2(80-12x)$$
$P'(x) = 0$ when $x = \frac{80}{3}$ and $x = \frac{20}{3}$
$$P\left(\frac{20}{3}\right) = \frac{20}{3}\left[80 - 3\left(\frac{20}{3}\right)\right]^2$$
$$= 1{,}200^2$$

The largest product occurs when $x = \frac{20}{3}$ and $y = 60$.

55. Let x and y be the sides of a rectangle. Then the perimeter is $P = 2x + 2y$ or

$$y = \frac{P - 2x}{x}.$$

$$A(x) = x\left(\frac{P - 2x}{2}\right) = \frac{1}{2}(xP - 2x^2)$$

$$A'(x) = \frac{1}{2}(P - 4x)$$

$A'(x) = 0$ when $x = \dfrac{P}{4}$

$$A(0) = A\left(\frac{P}{2}\right) = 0;\ A\left(\frac{P}{4}\right) = \frac{P^2}{16}$$

The largest area occurs when $x = \dfrac{P}{4}$.

$$y = \frac{P - 2\left(\frac{P}{4}\right)}{2} = \frac{P}{4}.\ \text{Thus, } x = y.$$

57. a.
$$x \geq x^2$$
$$x - x^2 \geq 0$$
$$x(1 - x) \geq 0$$

Solution is in $[0, 1]$ since it is impossible for $x < 0$ and $1 - x < 0$ simultaneously because $1 - x < 0$ is the same as $1 < x$.

Let $P(x) = x - x^2$;
$$P'(x) = 1 - 2x$$
$P'(x) = 0$ when $x = \frac{1}{2}$.
$$P(0) = P(1) = 0;\ P\left(\tfrac{1}{2}\right) = \tfrac{1}{4}$$

The greatest difference occurs at $x = \frac{1}{2}$.

b.
$$x \geq x^3$$
$$x - x^3 \geq 0$$
$$x(1 - x^2) \geq 0$$
$$x(1 - x)(1 + x) \geq 0$$

Solution for nonnegative x is in $[0, 1]$.

Let $P(x) = x - x^3$;
$$P'(x) = 1 - 3x^2$$

$P'(x) = 0$ when $x = \dfrac{1}{\sqrt{3}}$ (disregard negative value).

$$P(0) = P(1) = 0;\ P\left(\frac{1}{\sqrt{3}}\right) = \frac{2}{3\sqrt{3}};$$

The greatest difference occurs at
$$x = \frac{1}{\sqrt{3}}.$$

c.
$$x \geq x^n$$
$$x - x^n \geq 0$$
$$x(1 - x^{n-1}) \geq 0$$

Solution for nonnegative x is in $[0, 1]$.

Let $P(x) = x - x^n$;
$$P'(x) = 1 - nx^{n-1}$$

$P'(x) = 0$ when $x = \left(\dfrac{1}{n}\right)^{1/(n-1)}$

$$P(0) = P(1) = 0$$
$$P\left[\left(\tfrac{1}{n}\right)^{1/(n-1)}\right] = n^{-1/(n-1)} - n^{-n/(n-1)} > 0$$

The greatest difference occurs at
$$x = \left(\frac{1}{n}\right)^{1/(n-1)}$$

59. Since $f(x)$ has a relative minimum at $x = c$,

$$f(c) \leq f(c + \Delta x)$$
$$f(c) - f(c + \Delta x) \leq 0$$
$$f(c + \Delta x) - f(c) \geq 0$$

If $\Delta x > 0$, then

$$\frac{f(c + \Delta x) - f(c)}{\Delta x} \geq 0$$
$$\lim_{\Delta x \to 0} \frac{f(c + \Delta x) - f(c)}{\Delta x} \geq 0$$
$$f'(x) \geq 0$$

If $\Delta x < 0$, then

23. $g(t) = (t^3 + t)^2$

$g'(t) = 2(t^3 + t)(3t^2 + 1)$

$\quad = 2t(t^2 + 1)(3t^2 + 1)$

$g''(t) = 2(15t^4 + 12t^2 + 1)$

critical point:

$\quad (0, 0)$, relative minimum;

increasing on $(0, \infty)$;

decreasing on $(-\infty, 0)$;

concave up on $(-\infty, \infty)$

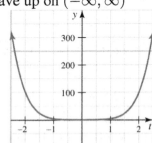

25. $f(x) = \dfrac{x}{x^2 + 1}$

$f'(x) = \dfrac{x^2 + 1 - x(2x)}{(x^2 + 1)^2}$

$\quad = \dfrac{1 - x^2}{(x^2 + 1)^2}$

$f''(x) = \dfrac{(x^2 + 1)^2(-2x) - (1 - x^2)(2)(x^2 + 1)(2x)}{(x^2 + 1)^4}$

$\quad = \dfrac{2x(x^2 - 3)}{(x^2 + 1)^3}$

critical points:

$(-1, -\frac{1}{2})$, relative minimum;

$(1, \frac{1}{2})$, relative maximum;

points of inflection: $(0, 0)$, $(\sqrt{3}, \frac{\sqrt{3}}{4})$,

$(-\sqrt{3}, -\frac{\sqrt{3}}{4})$;

increasing on $(-1, 1)$;

decreasing on $(-\infty, -1) \cup (1, \infty)$;

concave up on $(-\sqrt{3}, 0) \cup (\sqrt{3}, \infty)$;

concave down on $(-\infty, -\sqrt{3}) \cup (0, \sqrt{3})$

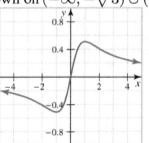

27. $f(t) = t^2 e^{-3t}$

$f'(t) = t^2(-3)e^{-3t} + 2te^{-3t}$

$\quad = (2 - 3t)te^{-3t}$

$f''(t) = (9t^2 - 12t + 2)e^{-3t}$

critical points:

$\quad (0, 0)$, relative minimum;

$\quad (0.67, 0.06)$, relative maximum;

$\quad (0.195, 0.021)$, point of inflection;

$\quad (1.138, 0.043)$, point of inflection;

decreasing on $(-\infty, 0) \cup (0.67, \infty)$;

increasing on $(0, 0.67)$;

concave up on $(-\infty, 0.195) \cup (1.138, \infty)$;

concave down on $(0.195, 1.138)$

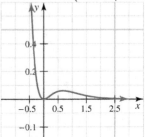

29. $f(x) = (\ln x)^2$

$f'(x) = 2(\ln x)x^{-1}$

$f''(x) = 2\left[(\ln x)(-x^{-2}) + x^{-2}\right]$

$\quad = 2(x^{-2})(1 - \ln x)$

critical points

$\quad (1, 0)$, relative minimum;

$(e, 1)$, point of inflection;
decreasing on $(0, 1)$,
increasing on $(1, \infty)$;
concave up on $(0, e)$;
concave down on (e, ∞)

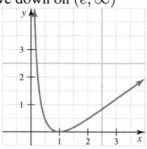

31. $t(\theta) = \theta + \cos 2\theta$ for $0 \leq \theta \leq \pi$
$t'(\theta) = 1 - 2\sin 2\theta$
$t''(\theta) = -4\cos 2\theta$
critical points:
   $\left(\frac{\pi}{12}, 1.13\right)$, relative maximum;
   $\left(\frac{5\pi}{12}, 0.44\right)$, relative minimum;
   inflection points $\left(\frac{\pi}{4}, 0.79\right)$, $\left(\frac{3\pi}{4}, 2.36\right)$;
increasing on $\left(0, \frac{\pi}{12}\right) \cup \left(\frac{5\pi}{12}, \pi\right)$;
decreasing on $\left(\frac{\pi}{12}, \frac{5\pi}{12}\right)$;
concave up on $\left(\frac{\pi}{4}, \frac{3\pi}{4}\right)$;
concave down on $\left(0, \frac{\pi}{4}\right) \cup \left(\frac{3\pi}{4}, \pi\right)$

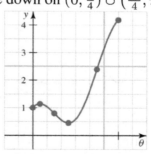

33. $f(x) = x^3 + \sin x$ on $\left[-\frac{\pi}{2}, \frac{\pi}{2}\right]$
$f'(x) = 3x^2 + \cos x$
$f''(x) = 6x - \sin x$

There are no critical points; the function is
increasing for all x. Inflection point $(0,0)$,
concave down $\left(-\frac{\pi}{2}, 0\right)$; concave up $\left(0, \frac{\pi}{2}\right)$

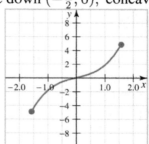

35. $f'(x) = \dfrac{(3 - 2x)\left(-2xe^{-x^2}\right) - e^{-x^2}(-2)}{(3 - 2x)^2}$
$= \dfrac{2e^{-x^2}(2x - 1)(x - 1)}{(2x - 3)^2}$
At $x = \frac{1}{2}$ (check left and right sides),
relative maximum; at $x = 1$, relative
minimum

37. $f'(x) = \frac{1}{3}(x^3 - 48x)^{-2/3}(3x^2 - 48)$
At $x = 4$ (check left and right sides),
relative minimum

39. $f(x) = \dfrac{x^2 - x + 5}{x + 4}$
$f'(x) = \dfrac{x^2 + 8x - 9}{(x + 4)^2}$
$f''(x) = \dfrac{50}{(x + 4)^3}$
$f''(1) = \frac{2}{5} > 0$, relative minimum;
$f''(-9) = -\frac{2}{5} < 0$, relative maximum

41. $f(x) = \sin x + \frac{1}{2}\cos 2x$
$f'(x) = \cos x - \sin 2x$
$f''(x) = -2\cos 2x - \sin x$
$f''\left(\frac{\pi}{2}\right) = 1 > 0$, relative minimum;
$f''\left(\frac{\pi}{6}\right) = -\frac{3}{2} < 0$, relative maximum

Problems 42–45, answers may vary.

43.   **45.**

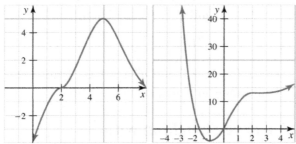

47. $v'(T) = v_0 \left(\dfrac{1}{2}\right)\left(1 + \dfrac{1}{273}T\right)^{-1/2}\left(\dfrac{1}{273}\right)$

For $T > 0$ this is always positive, so the function is monotonic increasing.

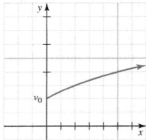

49. a. At 10:00 A.M.:
$$-(2)^3 + 6(2)^2 + 13(2) = 42$$
At 12:15 P.M.:
$$-\tfrac{1}{3}(2)^3 + \tfrac{1}{2}(2)^2 + 25(2) = 49\tfrac{1}{3}$$

b. $N(x) = -x^3 + 6x^2 + 13x - \dfrac{1}{3}(4-x)^3$
$$+\,\dfrac{1}{2}(4-x)^2 + 25(4-x)$$

c. $N'(x) = -2x^2 + 5x$
$N'(x) = 0$ when $x = 2.5$, so the optimum time for the break is 10:30 A.M. It is a maximum because $N''(2.5) < 0$.

51. $D(x) = \dfrac{9}{4}x^4 - 7\ell x^3 + 5\ell^2 x^2$ on $[0, \ell]$

$D'(x) = 9x^3 - 21\ell x^2 + 10\ell^2 x$
$$= x\left(9x^2 - 21\ell x + 10\ell^2\right)$$
$$= x(3x - 5\ell)(3x - 2\ell)$$
$D'(x) = 0$ when $x = 0, \tfrac{2}{3}\ell$
($\tfrac{5}{3}\ell$ is not in the domain)
$D''(x) = 27x^2 - 42\ell x + 10\ell^2$
$D''\left(\tfrac{2}{3}\ell\right) < 0$, so $\tfrac{2}{3}\ell$ is a maximum;
$D\left(\tfrac{2}{3}\ell\right) = \tfrac{16}{27}\ell^4$
Check the endpoints:
$$D(0) = 0, \quad D(\ell) = \dfrac{\ell^4}{4}$$
Maximum deflection at $x = \dfrac{2\ell}{3}$.

53. Let $f(x) = \dfrac{1}{\sin x} - \dfrac{1}{x}$
To show $f(x) > 0$ for $0 < x \le \tfrac{\pi}{2}$;
Note $x > \sin x$ or
$$\dfrac{1}{x} < \dfrac{1}{\sin x}$$
so that
$$\dfrac{1}{\sin x} - \dfrac{1}{x} > 0$$
To show that f is strictly increasing, find
$$f'(x) = \dfrac{-\cos x}{\sin^2 x} + \dfrac{1}{x^2}$$
$$= \dfrac{-x^2\cos x + \sin^2 x}{x^2 \sin^2 x}$$
$$> 0$$
because $-x^2\cos x + \sin^2 x = 0$ has solution $x = 0$ (found using technology), which is not in the domain $(0, \tfrac{\pi}{2}]$.

55. $y = Ax^2 + Bx + C$

We assume $A \neq 0$ because if it were, the equation would not be quadratic.

$y' = 2Ax + B; y'' = 2A$

The graph is concave up if $y'' > 0$, or when $A > 0$; it is concave down if $y'' < 0$ or when $A < 0$.

57. $f(x) = Ax^3 + Bx^2 + C$

$f'(x) = 3Ax^2 + 2Bx$

$f''(x) = 6Ax + 2B$

It is given that $(2, 11)$ is an extremum, so

$$f'(2) = 12A + 4B f'(x) = 0$$

when $B = -3A$. It is also given that $(1, 5)$ is a point of inflection, so

$$f''(1) = 6A + 2B f''(x) = 0$$

when $B = -3A$. Since the points $(2, 11)$ and $(1, 5)$ are on the curve, these components must satisfy its equation. Thus,

$$\begin{cases} f(2) = 8A + 4B + C \\ f(1) = A + B + C \end{cases}$$

and by substitution we have

$$\begin{cases} 8A - 12A + C = 11 \\ A - 3A + C = 5 \end{cases}$$

Solving these equations simultaneously, we find $A = -3$, $B = 9$, and $C = -1$.

Thus, $f(x) = -3x^3 + 9x^2 - 1$.

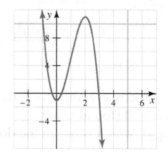

59. Let f and g be two functions concave up on $[a, b]$. Then, for all c on the interval, $f''(c) > 0$ and $g''(c) > 0$. Then,

$$(f + g)''(c) = f''(c) + g''(c) > 0$$

since $f''(c) > 0$ and $g''(c) > 0$. Thus, the sum function $f + g$ is concave up.

4.4 Curve Sketching with Asymptotes: Limits Involving Infinity, page 271

1. In general, to sketch a curve, check extent, find intercepts, look for symmetry, set $f'(x) = 0$, rough out intervals of curve increase and decrease, set $f''(x) = 0$, rough out intervals of curve concavity, look for horizontal and vertical asymptotes. Details are in Table 4.1 on page 270.

3. Concavity indicates the shape of the curve as cupped up or cupped down; a point of inflection indicates where the curve changes concavity.

5. $\displaystyle\lim_{x \to \infty} \frac{2,000}{x + 1} = 0$

7. $\displaystyle\lim_{x \to \infty} \frac{3x + 5}{x - 2} = \lim_{x \to \infty} \frac{3 + \frac{5}{x}}{1 - \frac{2}{x}}$

$$= \frac{3 + 0}{1 - 0}$$

$$= 3$$

9. $\displaystyle\lim_{t \to \infty} \frac{9t^5 + 50t^2 + 800}{t^5 - 1,000} = \lim_{t \to \infty} \frac{9 + \frac{50}{t^3} + \frac{800}{t^5}}{1 - \frac{1,000}{t^5}}$

$$= 9$$

11. $\lim\limits_{x\to\infty} \dfrac{x}{\sqrt{x^2+1{,}000}} = \lim\limits_{x\to\infty} \dfrac{1}{\sqrt{1+\frac{1{,}000}{x^2}}}$

$= \dfrac{1}{\sqrt{1+0}}$

$= 1$

13. $\lim\limits_{x\to\infty} \dfrac{x^{5.916}+1}{x^{\sqrt{35}}} = 0$

$\sqrt{35} > 5.916$, so the denominator overpowers the numerator.

15. $\lim\limits_{x\to 1^-} \dfrac{x-1}{|x^2-1|} = \lim\limits_{x\to 1^-} \dfrac{x-1}{|x+1||x-1|}$

$= \lim\limits_{x\to 1^-} \dfrac{-1}{x+1}$

$= -\dfrac{1}{2}$

17. $\lim\limits_{x\to 0^+} \dfrac{x^2-x+1}{x-\sin x} = \infty$

Since $\sin x < x$ for $x > 0$, the denominator will approach 0 through positive values. Meanwhile the numerator is approaching 1.

19. $\lim\limits_{x\to\infty} \left(x\sin\dfrac{1}{x}\right) = \lim\limits_{u\to 0^+} \dfrac{\sin u}{u}$ *where* $u = \frac{1}{x}$

$= 1$

21. $\lim\limits_{x\to 0^+} \dfrac{\ln\sqrt[3]{x}}{\sin x} = \dfrac{1}{3}\lim\limits_{x\to 0^+} \dfrac{\ln x}{\sin x}$

$= -\infty$

23. $\lim\limits_{x\to -\infty} e^x \sin x = \lim\limits_{x\to -\infty} \dfrac{\sin x}{e^{-x}}$

$= 0$

$\sin x$ is bounded and $e^{-x} \to \infty$ as $x \to -\infty$

25. $f'(x) = \dfrac{26}{(7-x)^2}$; $f''(x) = \dfrac{52}{(7-x)^3}$

asymptotes: $x = 7$, $y = -3$;
graph rising on $(-\infty, 7) \cup (7, \infty)$;
concave up on $(-\infty, 7)$;
concave down on $(7, \infty)$;

no critical points; no points of inflection

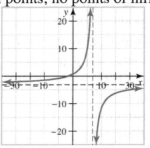

27. $f'(x) = \dfrac{-6}{(x-3)^2}$; $f''(x) = \dfrac{12}{(x-3)^3}$

asymptotes: $x = 3$, $y = 6$;
graph falling on $(-\infty, 3) \cup (3, \infty)$;
concave up on $(3, \infty)$;
concave down on $(-\infty, 3)$;
no critical points; no points of inflection;

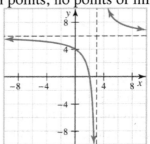

29. $f'(x) = \dfrac{-27x^2}{(x^3-8)^2}$; $f''(x) = \dfrac{108x(x^3+4)}{(x^3-8)^3}$

asymptotes: $x = 2$, $y = 1$;
graph falling on $(-\infty, 2) \cup (2, \infty)$;
concave up on $(-\sqrt[3]{4}, 0) \cup (2, \infty)$;

concave down on $(-\infty, -\sqrt[3]{4}) \cup (0, 2)$;
critical point is $(0, -\frac{1}{8})$;
points of inflection $(0, -\frac{1}{8}), (-\sqrt[3]{4}, \frac{1}{4})$;

31. $g'(t) = 2t(t^2 + 1)(3t^2 + 1)$
$g''(t) = 2(15t^4 + 12t^2 + 1)$
no asymptotes;
graph rising on $(0, \infty)$;
graph falling on $(-\infty, 0)$;
concave up on $(-\infty, \infty)$; critical point is
$(0, 0)$, which is a relative minimum; no points
of inflection;

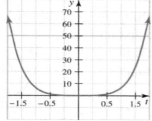

33. $f'(x) = 4x(x - 3)(x + 3)$
$f''(x) = 12(x^2 - 3)$
no asymptotes;
graph rising on $(-3, 0) \cup (3, \infty)$;
graph falling on $(-\infty, -3) \cup (0, 3)$;
concave up on $(-\infty, -\sqrt{3}) \cup (\sqrt{3}, \infty)$;
concave down on $(-\sqrt{3}, \sqrt{3})$;
critical points are
   $(-3, 0)$, relative minimum;
   $(0, 81)$, relative maximum;
   $(3, 0)$, relative minimum;

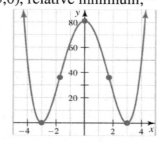

points of inflection are $(-\sqrt{3}, 36), (\sqrt{3}, 36)$;

35. $f'(x) = \dfrac{4(x - 1)}{3x^{2/3}}$; $f''(x) = \dfrac{4(x + 2)}{9x^{5/3}}$
no asymptotes;
graph rising on $(1, \infty)$;
graph falling on $(-\infty, 1)$;
concave up on $(-\infty, -2) \cup (0, \infty)$;
concave down on $(-2, 0)$;
critical points are
   $(1, -3)$, relative minimum;
   $(0, 0)$, vertical tangent;
$(0, 0$ and $(-2, 6\sqrt[3]{2})$ are points of inflection;

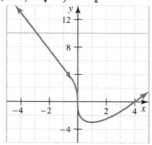

37. $f'(x) = \dfrac{2x}{1 + x^4}$; $f''(x) = \dfrac{-2(3x^4 - 1)}{(1 + x^4)^2}$
$f'(x) = 0$ when $x = 0$;
$\lim\limits_{x \to \infty} \tan^{-1} x^2 = \frac{\pi}{2}$
The second-order critical number is at $x^4 = \frac{1}{3}$
or when $x = \pm \sqrt[4]{\frac{1}{3}} \approx \pm 0.76$.
horizontal asymptote $y = \frac{\pi}{2}$
graph falling on $(-\infty, 0)$
graph rising on $(0, \infty)$
concave down on $(-\infty, -0.76) \cup (0.76, \infty)$
concave up on $(-0.76, 0.76)$
critical point is
   $(0, 0)$, relative minimum
point of inflection $(\pm 0.76, 0.11)$

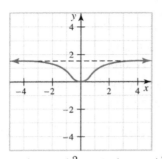

39. $g'(x) = -\dfrac{8(x+4)^2 + 27(x-1)^2}{(x-1)^2(x+4)^2}$

$g''(x) = \dfrac{10(x+1)(7x^2 - 4x + 97)}{(x-1)(x+4)^3}$

asymptotes: $x = -4$, $x = 1$, $y = 0$;
graph falling on
$(-\infty, -4) \cup (-4, 1) \cup (1, \infty)$;
concave up on $(-4, -1) \cup (1, \infty)$;
concave down on $(-\infty, -4) \cup (-1, 1)$;
no critical points;
$\quad$ point of inflection is $(-1, 5)$;

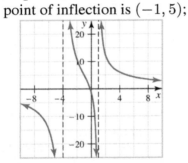

41. $t'(\theta) = \cos\theta + \sin\theta$
critical numbers at $\theta = \frac{3\pi}{4}, \frac{7\pi}{4}$
$t''(\theta) = -\sin\theta + \cos\theta$
critical numbers at $\theta = \frac{\pi}{4}, \frac{5\pi}{4}$
no asymptotes;
graph rising on $(0, \frac{3\pi}{4}) \cup (\frac{7\pi}{4}, 2\pi)$;
graph falling on $(\frac{3\pi}{4}, \frac{7\pi}{4})$;
concave up on $(0, \frac{\pi}{4}) \cup (\frac{5\pi}{4}, 2\pi)$;
concave down on $(\frac{\pi}{4}, \frac{5\pi}{4})$;

critical points are
$\quad \left(\frac{3\pi}{4}, \sqrt{2}\right)$, relative maximum;
$\quad \left(\frac{7\pi}{4}, -\sqrt{2}\right)$, relative minimum;
points of inflection: $\left(\frac{\pi}{4}, 0\right)$, $\left(\frac{5\pi}{4}, 0\right)$

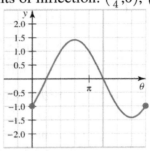

43. $f'(x) = 2\cos x(\sin x - 1)$
critical number at $x = \frac{\pi}{2}$
$f''(x) = -4\sin^2 x + 2\sin x + 2$
$\qquad = -2(\sin x - 1)(2\sin x + 1)$
critical number at $x = \frac{\pi}{2}$
no asymptotes;
graph rising on $(\frac{\pi}{2}, \pi)$;
graph falling on $(0, \frac{\pi}{2})$;
concave up on $(0, \pi)$;
critical points are
$\quad (\frac{\pi}{2}, 0)$, relative minimum;
$\quad (0, 1)$ and $(\pi, 1)$, maxima;
no points of inflection

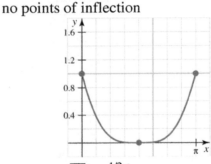

45. $\quad v = \sqrt{gr}\,\tan^{1/2}\theta$

$$v' = \frac{\sqrt{gr}}{2}\left(\frac{\sec^2\theta}{\sqrt{\tan\theta}}\right)$$

This is not defined at $\theta = 0$ and $\frac{\pi}{2}$

$$v'' = \frac{\sqrt{gr}}{2}\left(\frac{\sqrt{\tan\theta}(2)\sec^2\theta\tan\theta}{\tan\theta}\right)$$
$$-\frac{\sqrt{gr}}{2}\left(\frac{\sec^2\theta(\frac{1}{2})\tan^{-1/2}\theta\sec^2\theta}{\tan\theta}\right)$$
$$=\frac{\sqrt{gr}}{4}\left[\frac{\sec^2\theta(4\tan^2\theta-\sec^2\theta)}{\tan\theta}\right]$$

In terms of sines and cosines,
$$v''=\frac{\sqrt{gr}}{4}\left[\frac{4\sin^2\theta-1}{\sin^{3/2}\theta\cos^{5/2}\theta}\right]$$
$$v''=0 \text{ when } 4\sin^2-1=0$$
$$\sin\theta=\pm\frac{1}{2}$$

On $[0,\frac{\pi}{2}]$ the solution is $\theta=\frac{\pi}{6}$.
There is an asymptote at $x=\frac{\pi}{2}$
Point of inflection at $\left(0.52, 0.76\sqrt{gr}\right)$.

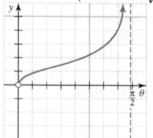

47. Answers may vary.

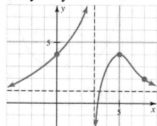

49. $P'(t)=e^{-0.04t}(1)+e^{-0.04t}(t+1)(-0.04)$
$=0.04e^{-0.04t}(24-t)$
$P'(t)=0$ when $t=24$.
The population is the largest after 24 minutes.
In the long run,
$$\lim_{t\to\infty}\left[5+e^{-0.04t}(t+1)\right]=5$$
The population approaches 5,000. For the inflection point, we find
$$P''(t)=0.04e^{-0.04t}(-1.96+0.04t)$$
$P''(0)=0$ when $t=49$ and $P''(t)>0$ for $t>49$. After 49 minutes, the rate at which the population decreases per minute starts to increase.

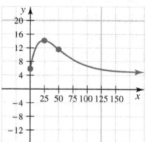

51. In order to have a vertical asymptote, it is necessary that $3-5b=0$, or $b=3/5$. Thus,
$$\lim_{x\to\infty}\frac{ax+5}{3-\frac{3x}{5}}=\lim_{x\to\infty}\frac{5ax+25}{15-3x}=-\frac{5a}{3}$$
Since this limit is to be equal to -3, we see $a=9/5$.

53. False; $f(x)=x^6$ is concave up and $g(x)=-x^2$ is concave down so the hypotheses are satisfied. Let
$$F(x)=f(x)g(x)=-x^8$$
$F''(x)<0$ and F is concave down.

55. False; let $f(x) = -\cos x$ on $[0, \frac{\pi}{4}]$ and

$$g(x) = [f(x)]^2 = \cos^2 x$$

$f(x) < 0$, and $f''(x) = \cos x > 0$, but g is concave down on the interval.

57. a. $\lim\limits_{x \to c^+} f(x) = -\infty$

Given $M < 0$, there exists a $\delta > 0$ such that $f(x) < M$ for all x with $0 < x - c < \delta$.

b. $\lim\limits_{x \to c^-} f(x) = \infty$

Given $M > 0$, there exists a $\delta > 0$ such that $f(x) > M$ for all x with $0 < c - x < \delta$.

59. Let $M > 0$ and $\epsilon > 0$ be given. Since $\lim\limits_{x \to c} g(x) = A$, there exists a δ_1 such that

$$|g(x) - A| < \epsilon \text{ when } 0 < |x - c| < \delta_1$$

that is,

$$-\epsilon < g(x) - A < \epsilon$$

so $A - \epsilon < g(x)$. Since $\lim\limits_{x \to c} f(x) = \infty$, there exists a $\delta_2 > 0$ so that

$$f(x) > \frac{M}{A - \epsilon}$$

when $0 < |x - c| < \delta_2$. Let $\delta = \min(\delta_1, \delta_2)$. Then, if $0 < |x - c| < \delta$,

$$f(x)g(x) > \left(\frac{M}{A - \epsilon}\right)(A - \epsilon) = M$$

so $\lim\limits_{x \to c} f(x)g(x) = \infty$ as required.

4.5 l'Hôpital's Rule, page 280

1. a. The limit is not an indeterminate

form. In fact,

$$\lim_{x \to \pi} \frac{1 - \cos x}{x} = \frac{1 - (-1)}{\pi} = \frac{2}{\pi}$$

b. The limit is not an indeterminate form.

$$\lim_{x \to \frac{\pi}{2}} \frac{\sin x}{x} = \frac{1}{\frac{\pi}{2}} = \frac{2}{\pi}$$

3. $\lim\limits_{x \to 1} \dfrac{x^3 - 1}{x^2 - 1} = \lim\limits_{x \to 1} \dfrac{3x^2}{2x} = \dfrac{3}{2}$

5. $\lim\limits_{x \to 1} \dfrac{x^{10} - 1}{x - 1} = \lim\limits_{x \to 1} \dfrac{10x^9}{1} = 10$

7. $\lim\limits_{x \to 0^+} \dfrac{1 - \cos^2 x}{\sin^3 x} = \lim\limits_{x \to 0^+} \dfrac{-2 \cos x(-\sin x)}{3 \sin^2 x \cos x}$

$$= \frac{2}{3} \lim_{x^+ \to 0} \frac{1}{\sin x}$$

$$= \infty$$

This limit is not defined.

9. $\lim\limits_{x \to \pi} \dfrac{\cos \frac{x}{2}}{\pi - x} = \lim\limits_{x \to \pi} \dfrac{-\frac{1}{2}\sin \frac{x}{2}}{-1} = \dfrac{1}{2}$

11. $\lim\limits_{x \to 0} \dfrac{1 - \cos x}{x^2} = \lim\limits_{x \to 0} \dfrac{\sin x}{2x}$

$$= \frac{1}{2} \lim_{x \to 0} \cos x$$

$$= \frac{1}{2}$$

13. $\lim\limits_{x \to 0} \dfrac{\tan 3x}{\sin 5x} = \lim\limits_{x \to 0} \dfrac{3 \sec^2 3x}{5 \cos 5x} = \dfrac{3}{5}$

15. $\lim\limits_{x \to 0} \dfrac{x + \sin^3 x}{x^2 + 2x} = \lim\limits_{x \to 0} \dfrac{1 + 3 \sin^2 x \cos x}{2x + 2}$

$$= \frac{1}{2}$$

17. $\lim\limits_{x \to 0} \dfrac{\sin 3x \sin 2x}{x \sin 4x} = \lim\limits_{x \to 0} \dfrac{6\left(\frac{\sin 3x}{3x}\right)\left(\frac{\sin 2x}{2x}\right)}{4\left(\frac{\sin 4x}{4x}\right)}$

$$= \frac{6(1)(1)}{4} = \frac{3}{2}$$

51. $E(\lambda) = \sqrt{\left(\dfrac{hc}{\lambda}\right)^2 + m_0^2 c^4}$

$E'(\lambda) = -\dfrac{h^2}{\lambda}\left|\dfrac{c}{\lambda}\right|\left(c^2\lambda^2 m_0^2 + h^2\right)^{-1/2}$

Since $E'(\lambda) \neq 0$, there are no critical values. As $\lambda \to \infty$, $E(\lambda) \to m_0 c^2$.

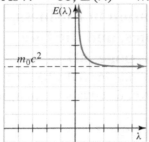

53. The amount of material is

$$M = 2(\pi r^2) + 22\pi r h$$

and the volume is $V = \pi r^2 h = 355$.
Solving for h we obtain $h = (355/\pi)r^{-2}$
so that

$$M(r) = 2\pi r^2 + 2\pi r\left(\dfrac{355}{\pi}r^{-2}\right)$$

$$= 2\pi r^2 + 710 r^{-1}$$

$M'(r) = 4\pi r - 710 r^{-2}$.
$M' = 0$ when $r \approx 3.84$ and $h \approx 7.67$ cm.
$M'' > 0$ for $r > 0$, so this is a minimum.

The actual dimensions of a Coke can are $h \approx 12$ cm (larger than necessary) and $r \approx 3.5$ cm (a bit smaller than necessary). The actual dimensions of a can of cola differ from the optimal dimensions for historical and marketing reasons.

55. Refer to Figure 4.67 in the text.
Note: $\cos\theta = \dfrac{x}{L}$ and

$$\cos(\pi - 2\theta) = \dfrac{20 - x}{x}$$

then $\cos 2\theta = (x - 20)/x$; where
$0 \le x \le 15$

$$\dfrac{x - 20}{x} = \cos 2\theta$$

$$= 2\cos^2\theta - 1$$

$$= 2\left(\dfrac{x}{L}\right)^2 - 1$$

$$L^2(x - 20) = 2x^3 - xL^2$$
$$L^2(x - 20) + xL^2 = 2x^3$$

$$L^2 = \dfrac{2x^3}{2x - 20}$$

$$L = \dfrac{x^{3/2}}{(x - 10)^{1/2}}$$

$$L' = \dfrac{(x - 10)^{1/2}\left(\frac{3}{2}\right)x^{1/2} - x^{3/2}\left(\frac{1}{2}\right)(x - 10)^{-1/2}}{x - 10}$$

$$= \dfrac{3(x - 10)x^{1/2} - x^{3/2}}{2(x - 10)^{3/2}}$$

$$= \dfrac{x^{1/2}(2x - 30)}{2(x - 10)^{3/2}}$$

$L' = 0$ when $x = 0$, 15 (undefined if $x = 10$)

$$L(15) = \sqrt{\dfrac{15^2}{5}} = 15\sqrt{3} \approx 26 \text{ cm}$$

By the first derivative test, we see this is a minimum.

57. $V'(T) = -6.42\left(10^{-5}\right) + 17.02\left(10^{-6}\right)T$
$\qquad\qquad - 20.37\left(10^{-8}\right)T^2$

$V'(T) = 0$ when

$$T = \dfrac{0.85\left(10^{-5}\right) \pm \sqrt{0.724\left(10^{-10}\right) - 0.1308\left(10^{-10}\right)}}{2.037\left(10^{-7}\right)}$$

$$\approx 4, \ 80$$

Reject 80 °C since the mathematical model is only valid around the freezing temperature. $T = 4°$ is a minimum since V' is falling on $(0, 4)$ and rising on $(4, 80)$.

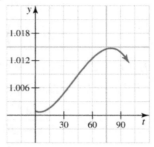

Liquid water and solid ice can coexist at 0°C. The water below the surface is denser since the trapped temperature is a little higher than 0. As the temperature above the surface drops, the change in temperature is passed to the next higher level of water and transforms to ice. The minimum value occurs when $T \approx 4°$.

59. a.

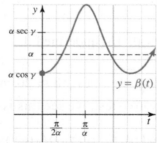

b. $\beta(t) = \dfrac{\alpha \cos \gamma}{1 - \sin^2\gamma \, \sin^2 \alpha t}$

$\beta'(t) = \dfrac{2\alpha^2 \cos \gamma \, \cos(\alpha t) \sin^2\gamma \, \sin(\alpha t)}{[1 - \sin^2\gamma \, \sin^2(\alpha t)]^2}$

$\quad = \dfrac{\alpha^2 \sin^2\gamma \, \cos \gamma \, \sin(2\alpha t)}{[1 - \sin^2\gamma \, \sin^2(\alpha t)]^2}$

$\beta'(t) = 0$ when $2\alpha t = n\pi$, $n = 0, 1, 2, \cdots$

or $t = \dfrac{n\pi}{2\alpha}$. Lowest when $t = 0$;

highest when $t = \frac{\pi}{2\alpha}$

$\beta_{min} = \beta(0) = \alpha \cos \gamma$

$\beta_{max} = \beta(\dfrac{\pi}{2\alpha}) = \dfrac{\alpha \cos \gamma}{1 - \sin^2\gamma} = \alpha \sec \gamma$

4.7 Optimization in Business, Economics, and the Life Sciences, page 307

1. The profit is maximized when the marginal revenue equals the marginal cost.

$C'(x) = \frac{1}{8}(2x) + 5$

$R(x) = xp = \frac{1}{2}(75x - x^2)$

$R'(x) = \frac{1}{2}(75 - 2x)$

$$R'(x) = C'(x)$$

$$\frac{1}{2}(75 - 2x) = \frac{1}{4}x + 5$$

$$150 - 4x = x + 20$$

$$5x = 130$$

$$x = 26$$

3. The profit is maximized when the marginal revenue equals the marginal cost.

$C'(x) = 50$

$R(x) = xp = 200x - 0.04x^2$

$R'(x) = 200 - 0.08x$

$$R'(x) = C'(x)$$

$$200 - 0.08x = 50$$

$$0.08x = 150$$

$$x = 1,875$$

5. The profit is maximized when the marginal revenue equals the marginal cost.

$C'(x) = 6x$

$$R(x) = xp = 200x - 2x^2$$
$$R'(x) = 200 - 4x$$
$$R'(x) = C'(x)$$
$$200 - 4x = 6x$$
$$10x = 200$$
$$x = 20$$

7. $\overline{P}(x) = \dfrac{x^3 - 8x^2 + 2x + 50}{x}$

$\quad = x^2 - 8x + 2 + 50x^{-1}$

$\overline{P}'(x) = 2x - 8 - 50x^{-2}$

9. $\overline{P}(x) = \dfrac{5x - 0.05x^2}{x}$

$\quad = 5 - 0.05x$

$\overline{P}'(x) = -0.05$

11. $\overline{P}(x) = \dfrac{x^3 - 50x^2 + 5x + 200}{x}$

$\quad = x^2 - 50x + 5 + 200x^{-1}$

$\overline{P}'(x) = 2x - 50 - 200x^{-2}$

13. Average cost is

$A(x) = \dfrac{C(x)}{x} = 3x + 1 + 48x^{-1}$

$C'(x) = 6x + 1.$ $A(x)$ is minimized when

$$6x + 1 = 3x + 1 + 48x^{-1}$$
$$3x^2 - 48 = 0$$
$$(x - 4)(x + 4) = 0$$
$$x = \pm 4$$

We disregard the negative value. We find

$$A(4) = 3(4) + 1 + \frac{48}{4}$$
$$= 25$$

The minimum average cost is $25.

15. a. Total cost, $C(x)$, is the average cost, $A(x)$, times the number of items produced, x.

$$C(x) = x\left(5 + \frac{x}{50}\right) = 5x + \frac{x^2}{50}$$
$$R(x) = x\left(\frac{380 - x}{20}\right)$$
$$P(x) = R(x) - C(x)$$
$$= \frac{380x - x^2}{20} - \frac{250x + x^2}{50}$$
$$= \frac{-7x^2 + 1{,}400x}{100}$$
$$= -0.07x^2 + 14x$$

b. The maximum profit occurs when

$$R'(x) = C'(x)$$
$$19 - \frac{x}{10} = 5 + \frac{x}{25}$$
$$14 = \frac{7}{50}x$$
$$x = 100$$

This gives a price per item,

$$p = \frac{380 - x}{20} = \$14/\text{item}$$

The maximum profit is

$$P(x) = \frac{(-7x + 1{,}400)x}{100}$$
$$= \frac{(-700 + 1{,}400)100}{100}$$
$$= \$700$$

17. The profit is $P(x) = 1{,}000e^{-0.1x}(x - 50).$
$P'(x) = 1{,}000\left[e^{-0.01x} + e^{-0.1x}(-0.1)(x - 50)\right]$
$\quad = 1{,}000e^{-0.1x}(6 - 0.1x)$
$P'(x) = 0$ when $x = 60.$ That is, profit is maximized when $x = 60.$

19. a. $P(t) = 50e^{0.02t}$; $P'(t) = e^{0.02t}$

$P'(10) = e^{0.2} \approx 1.22$ million people/yr

b. The percentage rate is

$$\frac{100e^{0.02t}}{50e^{0.02t}} = 2$$

that is, 2%.

21. We want to find the largest value of the population rate $R(t) = p'(t)$.

$p(t) = 160(1 + 8e^{-0.01t})^{-1}$

$p'(t) = 160(1 + 8e^{-0.01t})^{-2}(0.01)(8e^{-0.01t})$

$\qquad = 12.8e^{-0.01t}(1 + 8e^{-0.01t})^{-2}$

Using the product rule, we find that

$R'(t) = p''(t)$

$\quad = 12.8(-0.01)e^{-0.01t}(1 + 8e^{-0.01t})^{-2}$

$\quad + (-2)(12.8e^{-0.01t})(1 + 8e^{-0.01t})^{-3}(8)(-0.01)e^{-0.01t}$

$\quad = 12.8(-0.01)e^{-0.01t}(1 + 8e^{-0.01t})^{-3}$

$\qquad \cdot [1 - 8e^{-0.01t}]$

$R'(t) = 0$ when

$$1 - 8e^{-0.01t} = 0$$
$$e^{0.01t} = 8$$
$$t = 100 \ln 8$$
$$\approx 208$$

The population will be growing most rapidly 208 years from now.

23. Let x denote the number of cases of connectors in each shipment, and $C(x)$ the corresponding (variable) cost. Then,

$C(x) = \text{STORAGE COST} + \text{ORDERING COST}$

$\qquad = \dfrac{4.5x}{2} + 20\left(\dfrac{18{,}000}{x}\right)$

$\qquad = 2.25x + 360{,}000x^{-1}$

on $(0, 18{,}000]$

$\qquad C'(x) = 2.25 - 360{,}000x^{-2}$

$C'(x) = 0$ when $x = 400$. Since this is the only critical point in the interval, and since

$C''(x) = 720{,}000x^3 > 0$ when $x = 400$, C is minimized when $x = 400$. The number of shipments should be $18{,}000/400 = 45$ times per year.

25. a. Average revenue per unit is total revenue divided by number of units:

$$A(x) = -2x + 68 - \frac{128}{x}$$

The marginal revenue is $R'(x) = -4x + 68$. They are equal when:

$$-2x + 68 - \frac{128}{x} = -4x + 68$$
$$2x = \frac{128}{x}$$
$$x^2 = 64$$
$$x = \pm 8 \,(\text{reject } x = -8)$$

b. $A'(x) = -2 + \dfrac{128}{x^2}$

$A'(8) = 0;\ A'(8^-) > 0,\ A'(8^+) < 0$

So there is a relative maximum at $(8, 36)$. The function is increasing on $[0, 8)$ and decreasing on $(8, \infty)$.

c.

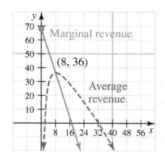

27. $P(t) = 300e^{\sqrt{3t}}e^{-0.08t} = 300e^{\sqrt{3t}-0.08t}$

$P'(t) = 300\exp\left(\sqrt{3t} - 0.08t\right)\left[\tfrac{1}{2}(3t)^{-1/2}(3) - 0.08\right]$

$P'(t) = 0$ if $t = 117.19$

Since the optimum solution is over 100 years, you should will the book to your heirs so they can sell it in 117.19 years. (Some things—like rare books ... take time!)

29. $P(x) = R(x) - C(x)$

Let x represent the increase in price, $40 + x$ is the cost per board, $45 - 3x$ is the number of boards sold.

$P(x) = (40 + x)(45 - 3x) - 29(45 - 3x)$
$\quad = -3x^2 + 12x + 495$

$P'(x) = -6x + 12$ which is 0 when $x = 2$. Therefore, raise the price \$2, sell the boards at a price of \$42, sell 39 per month, and have a maximum profit of \$507.

31. Let x be the number of people above the 100 level ($0 \le x \le 50$). Then, the number of travelers is $100 + x$. The fare per traveler is $2{,}000 - 10x$. The revenue is

$$R(x) = (100 + x)(2{,}000 - 10x)$$

and the profit is

$$P(x) = (100 + x)(200 - 10x)$$
$$\quad - 125{,}000 - 500(x + 100)$$
$$\quad = -10x^2 + 500x + 25{,}000$$
$$P'(x) = -20x + 500$$

$P'(x) = 0$ when $x = 25$. Thus, lower the fare by $10(25) = \$250$.

33. We want to maximize the yield, $Y(x)$. Let x be the additional number of trees to be planted.

NUMBER OF TREES $= 60 + x$
AVERAGE YIELD $= 400 - 4x$

$Y(x) = (60 + x)(400 - 4x)$
$\quad = -4x^2 + 160x + 24{,}000$

$Y'(x) = -8x + 160 = 0$ when $x = 20$. Plant 80 total trees, have an average yield per tree of 320 oranges, and a maximum total crop of 25,600 oranges.

35. Let x denote the number of additional grapevines to be planted and $N(x)$ the corresponding yield. Since there are 50 grapevines to begin with and x additional grapevines are planted, the total number is $50 + x$. For each additional grapevine, the average yield is decreasing by 2 lb, so that the yield is $150 - 2x$ lb. Thus, on $[0, 20]$

$N(x) = $ (LBS OF GRAPES PER VINE)(NO. OF GRAPEVINES)
$\quad = (150 - 2x)(50 + x)$
$\quad = 2\left(3{,}750 + 25x - x^2\right)$

$N'(x) = 2(25 - 2x) = 0$ when $x = 12.5$.
$N(0) = 7{,}500$; $N(12) = 7{,}812$;
$N(13) = 7{,}812$; $N(20) = 7{,}700$.
The greatest possible yield is 7,812 pounds of grapes, which is generated by planting 12 or 13 (choose 12 for practical reasons) additional vines; that is, 62 vines are planted.

37. a. Let x be the number of days and $R(x)$ the corresponding revenue. Over the period of 80 days, 24,000 pounds of glass have been collected at a rate of 300 pounds per day, and so for each day over 80, an additional 300 pounds will be collected. Thus, the total number of pounds collected and sold is $24{,}000 + 300x$. Currently the recycling center pays 1¢ per pound. For each additional day, it reduces the price it pays by 1¢ per 100 pounds; that is, by 0.01¢/lb. Hence, after x additional days, the price per pound will be $1 - 0.01x$ cents. On $[0, 100]$,

$$R(x) = (\text{NO. OF LBS OF GLASS})(\text{PRICE/LB})$$
$$= (24{,}000 + 300x)(1 - 0.01x)$$
$$= 24{,}000 + 60x - 3x^2$$

$R'(x) = 60 - 6x = 0$ when $x = 10$.
$R(0) = 24{,}000$; $R(100) = 0$;
$R(10) = 24{,}300$

The most profitable time to conclude the project is 10 days from now.

b. Assume R is continuous over $[0, 10]$ and that the glass is coming in at a constant rate during the time period.

39. a. $R(x) = xp(x) = \dfrac{bx - x^2}{a}$ on $[0, b]$.

$R'(x) = \dfrac{1}{a}(b - 2x) = 0$ if $x = \dfrac{b}{2}$.

R is increasing on $\left(0, \frac{b}{2}\right)$ and decreasing on $\left(\frac{b}{2}, b\right)$.

b.

41. $t = \dfrac{s}{v - v_1}$

$$E = cv^k t = cs\left(\dfrac{v^k}{v - v_1}\right)$$

$$\dfrac{dE}{dv} = cs\left[\dfrac{(v - v_1)kv^{v-1} - v^k}{(v - v_1)^2}\right]$$

$\dfrac{dE}{dv} = 0$ when

$$\dfrac{csv^{k-1}}{(v - v_1)^2}(kv - kv_1 - v) = 0$$

$$kv - kv_1 - v = 0$$

$$v = \dfrac{kv_1}{k - 1}$$

43. a. $P'(x) = Asx^{s-1}e^{-sx/r}$
$$+ Ax^s\left(-\dfrac{s}{r}\right)e^{-sx/r}$$
$$= \dfrac{As}{r}x^{s-1}e^{-sx/r}(r - x)$$
$P'(x) = 0$ when $x = r$
$$P''(x) = \dfrac{As}{r^2}x^{s-2}e^{-sx/r}(sx^2$$
$$- 2rsx + r^2 s - r^2)$$
$P''(r) = -Ase^{-s}r^{s-2} < 0$
so $x = r$ is a maximum.

b. $P''(x) = 0$ when

$$x = \dfrac{r}{\sqrt{x}}(\sqrt{s} \pm 1)$$
$$= \dfrac{r}{s}(s \pm \sqrt{s})$$

These are inflection points, so if $s > 1$, there are two inflection points.

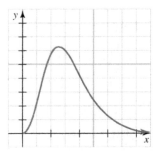

The production rate $P'(x)$ is increasing for $0 < x < \dfrac{r}{s}(s - \sqrt{s})$ and for $x > \dfrac{r}{s}(s + \sqrt{s})$. The rate is decreasing for

$$\frac{r}{s}(s - \sqrt{s}) < x < \frac{r}{s}(s + \sqrt{s})$$

c. $s = 0$ means $P = A$, a horizontal line. If $0 < s < 1$ one of the inflection points is negative, so only one appears in $x \geq 0$.

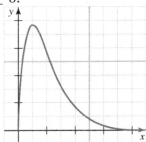

The production rate $P'(x)$ is decreasing for $0 < x < \dfrac{r}{s}(s + \sqrt{s})$ and increasing for $x > \dfrac{r}{s}(s + \sqrt{s})$.

45. $E' = \dfrac{1}{v}[2a(v - b)] - \dfrac{1}{v^2}[a(v - b)^2 + c]$

$= \dfrac{1}{v^2}(av^2 - ab^2 - c)$

$E' = 0$ when

$$v^2 = \frac{ab^2 + c}{a}$$

When $a = 0.04$, $b = 36$, and $c = 9$, then

$$v^2 = \frac{(0.04)(36)^2 + 9}{0.04} = 1{,}521$$

Thus, $v = 39$ units of length per unit of time.

47. a.

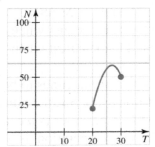

$N'(t) = -1.7T + 45.45 = 0$ when $T \approx 26.74$. The largest survival percentage is 60.56% when $T \approx 26.74$ and the smallest survival percentage is 22% for $t = 20$.

b. $S'(T) = -(-0.06T + 1.67)(-0.03T^2 + 1.67T - 13.67)^{-2}$

$S'(T) = 0$ when $T = \dfrac{167}{6}$

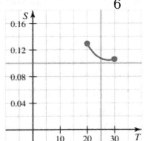

c.

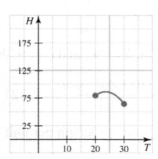

$H'(T) = -1.06T + 25$
$H'(T) = 0$ when $T \approx 23.58$
The largest hatching occurs when
$T \approx 23.58$ and the smallest when
$T = 30$.

49. a. Let x be number of machines to be set up. The set-up cost is Sx, the number of units produced per hour is nx, the number of hours of operation is $pQ/(nx)$, and the total cost is

$$C = Sx + \frac{pQ}{nx}$$
$$C' = S - \frac{pQ}{nx^2}$$

$C' = 0$ when $x^2 = \dfrac{pQ}{nS}$ or $x = \sqrt{\dfrac{pQ}{nS}}$.

b. The set-up cost equals the operating cost if

$$x = \sqrt{\frac{pQ}{nS}}$$

51. The epidemic is spreading most rapidly when the rate of change $R(t$ is a maximum, that is, when
$R'(t) = f''(t) = 0$.
$$f'(t) = A(-1)\big(1 + Ce^{-kt}\big)^{-2}\big(-Cke^{-kt}\big)$$
$$= kACe^{-kt}\big(1 + Ce^{-kt}\big)^{-2}$$

$$f''(t) = kAC\big(1 + Ce^{-kt}\big)^{-4}$$
$$\cdot \Big[\big(1 + Ce^{-kt}\big)^2\big(-ke^{-kt}\big)$$
$$- \big(e^{-kt}\big)(2)\big(1 + Ce^{-kt}\big)$$
$$\cdot \big(-Cke^{-kt}\big)\Big]$$
$$= \frac{k^2 ACe^{-kt}(Ce^{-kt} - 1)}{(1 + Ce^{-kt})^3}$$

$f''(t) = 0$ when $Ce^{-kt} = 1$ or $t = k^{-1}\ln C$.
Substituting in the original equation leads to
$f(k^{-1}\ln C) = A\big(1 + Ce^{-\ln C}\big)^{-1} = \frac{1}{2}A$
This means half of the total number of susceptible residents. $R'(t) < 0$ when
$t > (\ln C)k^{-1}$ (the graph of the curve is decreasing) and $R'(t) > 0$ when
$t < (\ln C)k^{-1}$ (the graph of the curve is increasing).

53. $\text{COST} = \text{STORAGE COST} + \text{ORDERING COST}$
$$C(x) = \frac{tx}{2} + \frac{QS}{x}$$
$$C'(x) = \frac{t}{2} - \frac{QS}{x^2} = 0 \text{ if}$$
$$x^2 = \frac{2QS}{t}$$
$$x = \sqrt{\frac{2QS}{t}}$$

We now substitute this value into the cost equations:
$$\text{STORAGE COST} = \frac{tx}{2}$$
$$= \frac{t}{2}\sqrt{\frac{2QS}{t}}$$
$$= \frac{1}{\sqrt{2}}\sqrt{QSt}$$

$$\text{ORDERING COST} = \frac{QS}{x}$$

$$= QS\sqrt{\frac{t}{2QS}}$$

$$= \frac{1}{\sqrt{2}}\sqrt{QSt}$$

Thus, the total cost is minimized when the storage cost is equal to the ordering cost.

55. The revenue is $R(x) = px$ and its derivative

$$\frac{dR}{dx} = p + x\frac{dp}{dx}$$

$$= p\left[1 + \frac{x}{p}\frac{dp}{dx}\right]$$

$$= \frac{xp}{x}\left[1 + \frac{x}{p}\frac{dp}{dx}\right]$$

$$= \frac{R(x)}{x}\left[1 + \frac{1}{E(x)}\right]$$

57. $C'(t) = \dfrac{k}{b-a}\left(-ae^{-at} + be^{-bt}\right), a < b$

$C'(t) = 0$ when $t = t_c = \dfrac{1}{b-a}\ln\left(\dfrac{b}{a}\right)$

$C''(t) = \dfrac{k}{b-a}\left(a^2e^{-at} - b^2e^{-bt}\right)$

Since $ae^{-at_c} = be^{-bt_c}$ and $a < b$, we have

$$a^2e^{-at_c} = abe^{-bt_c} < b^2e^{-bt_c}$$

so

$$C''(t_c) = \frac{k}{b-a}\left[a^2e^{-at_c} - b^2e^{-bt_c}\right] < 0$$

and it follows that a relative maximum occurs where $t = t_c$.

59. $S'(\theta) = 1.5s^2(\csc^2\theta - \sqrt{3}\csc\theta\cot\theta)$

$$= \frac{1.5s^2}{\sin^2\theta}\left(1 - \sqrt{3}\cos\theta\right)$$

$S'(\theta) = 0$ when $\theta \approx 0.9553$; this is about $55°$.

Chapter 4 Review

 Studying for a chapter examination is a personal process, one which nobody else can do for you. Simply take the time to review what you have done.

SURVIVAL HINT: Work all of Chapter 4 problems in the Proficiency Examination (whether they are assigned or not). Work through all of the problems before looking at the answers, and *then* correct each of the problems. The answers to all these problems are given in the answer section at the back of the text. If you worked the problem correctly, move on to the next problem, but if you did not work it correctly (or you did not know what to do), then look at the solutions below, look back in the chapter to study the procedure, or ask your instructor.

Finally, go back over the homework problems you have been assigned. If you worked a problem correctly, move on to the next problem, but if you missed it on your homework, then you should look back in the book or talk to your instructor about how to work the problem.

If you follow these steps, you should be successful with your review of this chapter.

Proficiency Examination, page 313

1. A relative extremum is an extremum only in the neighborhood of the point of interest. An absolute extremum is largest (or smallest) for all values in the domain.

2. A continuous function f on a closed interval $[a, b]$ has an absolute maximum and an absolute minimum.

3. A critical number of a function is a value of the independent variable at which the derivative of the function is zero or is not defined. A critical point $(c, f(c))$ is the point on $y = f(x)$ that corresponds to the critical number c.

4. Find critical numbers, evaluate the function at these values and at the endpoints of the closed interval. Finally, determine the absolute extrema by selecting the largest and smallest of the evaluated functional values.

5. Mean value theorem: If f is continuous on the closed interval $[a, b]$ and differentiable on the open interval (a, b), then there exists at least one number c such that

$$\frac{f(b) - f(a)}{b - a} = f'(c)$$

for $a < c < b$. Rolle's theorem is a special case where

$$f(a) = f(b)$$

so $f'(c) = 0$.

6. Suppose f is a continuous function on the closed interval $[a, b]$ and is differentiable on the open interval (a, b),

with $f'(x) = 0$ for all x on (a, b). Then the function f is constant on $[a, b]$.

7. Use the second-derivative test first: Given $f(x)$ and c such that $f'(c) = 0$. If $f''(x) > 0$ there is a relative minimum at $x = c$ (concave up). If $f''(x) < 0$, there is a relative maximum at $x = c$ (concave down). If the this test fails, *i.e.* $f''(c) = 0$, $f''(c)$ is difficult to evaluate, or $f''(c)$ does not exist, then use the first derivative test: Given $f(x)$, find c such that $f'(c) = 0$ or $f'(c)$ is not defined (that is, c is a critical number). If $f'(x) < 0$ for $x < c$ and $f'(x) > 0$ for $x > c$ (what we have been calling the ↓ ↑ pattern), there is a relative minimum at $x = c$. If $f'(c) > 0$ for $x < c$ and $f'(c) < 0$ for $x > c$ (the ↑ ↓ pattern), there is a relative minimum at $x = c$.

8. For a plane curve, an asymptote is a line which has the property that the distance from a point P on the curve to the line approaches zero as the distance from P to the origin increases without bound and P is on a suitable portion of the curve.

9. Suppose the function f is continuous at the point $P(c, f(c))$. Then, the graph of f has a cusp at P if $\lim_{x \to c^-} f'(x)$ and $\lim_{x \to c^+} f'(x)$ have opposite signs. The graph has a vertical tangent at P if $\lim_{x \to c^-} f'(x)$ and $\lim_{x \to c^+} f'(x)$ are either both positive or both negative.

10. Let f and g be functions that are differentiable on an open interval containing c (except possibly at c itself).

If

$$\lim_{x \to c} \frac{f(x)}{g(x)}$$

produces an indeterminate form $\frac{0}{0}$ or $\frac{\infty}{\infty}$, and

$$\lim_{x \to c} \frac{f'(x)}{g'(x)}$$

exists, then

$$\lim_{x \to c} \frac{f(x)}{g(x)} = \lim_{x \to c} \frac{f'(x)}{g'(x)}$$

11. $\lim_{x \to \infty} f(x) = L$, given an $\epsilon > 0$, there exists a number N_1 such that

$$|f(x) - L| < \epsilon$$

whenever $x > N_1$ for x in the domain of f. $\lim_{x \to c} f(x) = \infty$ if for any number $N > 0$, it is possible to find a number $\delta > 0$ such that $f(x) > N$ whenever $0 < |x - c| < \delta$.

12. Find the domain and range of a function, locate intercepts, if any. Investigate symmetry, asymptotes, and find extrema and/or points of inflection, if any. Determine where the graph is rising, where it is falling, and determine the concavity. (See Table 4.1).

13. An optimization problem involves finding the largest, or smallest value of a function. Procedure:
Step 1 Draw a figure (if appropriate) and label all quantities relevant to the problem.

Step 2 Find a formula for the quantity to be maximized or minimized.
Step 3 Use conditions in the problem to eliminate variables in order to express the quantity to be maximized or minimized in terms of a single variable.
Step 4 Find the practical domain for the variables in Step 3; that is, the interval of possible values determined from the physical restrictions in the problem.
Step 5 If possible, use the methods of calculus to obtain the required optimum value.

14. Light travels between two points in such a way as to minimize the time of transit.

15. If a beam of light strikes the boundary between two media with angle of incidence α and is refracted through an angle β, then

$$\frac{\sin \alpha}{\sin \beta} = \frac{v_1}{v_2}$$

where v_1 and v_2 are the rates at which light travels through the first and second media, respectively. The constant ratio

$$n = \frac{\sin \alpha}{\sin \beta}$$

is called the relative index of refraction of the two media.

16. Profit is maximized when marginal revenue equals marginal cost. Average cost is minimized at the level of production where the marginal cost equals the average cost.

17. The resistance to the flow of blood in an artery is directly proportional to the

artery's length and inversely proportional to the fourth power of its radius.

18. $\lim\limits_{x\to\frac{\pi}{2}} \dfrac{\sin 2x}{\cos x} = \lim\limits_{x\to\frac{\pi}{2}} \dfrac{2\cos 2x}{-\sin x} = \dfrac{-2}{-1} = 2$

19. $\lim\limits_{x\to 1} \dfrac{1-\sqrt{x}}{x-1} = \lim\limits_{x\to 1} \dfrac{-\frac{1}{2}x^{-1/2}}{1} = -\dfrac{1}{2}$

20. $\lim\limits_{x\to\infty} \left(\dfrac{1}{x} - \dfrac{1}{\sqrt{x}}\right) = \lim\limits_{x\to\infty} \dfrac{1-\sqrt{x}}{x}$

$\qquad = \lim\limits_{x\to\infty} \dfrac{-\frac{1}{2}x^{-1/2}}{1}$

$\qquad = 0$

21. $\lim\limits_{x\to\infty} \left(1 + \dfrac{2}{x}\right)^{3x}$ *Let* $\dfrac{1}{u} = \dfrac{2}{x}$.

$\qquad = \lim\limits_{x\to\infty} \left(1 + \dfrac{1}{u}\right)^{3(2u)}$

$\qquad = \lim\limits_{x\to\infty} \left[\left(1 + \dfrac{1}{u}\right)^{u}\right]^{6}$

$\qquad = e^6$

22. $f(x) = x^3 + 3x^2 - 9x + 2$

$\quad f'(x) = 3x^2 + 6x - 9$

$\qquad\quad = 3(x+3)(x-1)$

$\quad f'(x) = 0$ when $x = -3, 1$

$\quad f''(x) = 6x + 6$

$\quad f''(x) = 0$ when $x = -1$

Relative maximum at $(-3, 29)$; relative minimum at $(1, -3)$; inflection point at $(-1, 13)$

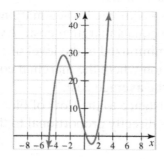

23. $f(x) = x^{1/3}(27 - x)$

$\quad f'(x) = 9x^{-2/3} - \frac{4}{3}x^{1/3} = 0$ when

$$\dfrac{9}{x^{2/3}} = \dfrac{4x^{1/3}}{3}$$

$$x = \dfrac{27}{4}$$

Not defined at $x = 0$.

$f''(x) = -6x^{-5/3} - \frac{4}{9}x^{-2/3} = 0$ when $x = -\frac{27}{2}$.

$f''\left(-\frac{27}{2}^{-}\right) > 0$; $f''\left(-\frac{2}{7}^{+}\right) < 0$ as well as $f''(0^-) > 0$, $f''(0^+) < 0$ so there are inflection points at approximately $\left(-\frac{27}{2}, -96.43\right)$ and $(0, 0)$. Relative maximum at $\left(\frac{27}{4}, 38.27\right)$.

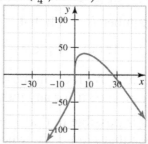

24. $f(x) = \dfrac{x^2 - 1}{x^2 - 4}$

Checking for horizontal asymptotes:

$\lim\limits_{x\to\infty} f(x) = 1$; $\lim\limits_{x\to-\infty} f(x) = 1$

So $y = 1$ is a horizontal asymptote.

Factoring the denominator we see that there are two candidates for vertical asymptotes; $x = 2$ and $x = -2$. Testing these:

$\lim\limits_{x\to-2^-} f(x) = \infty$; $\lim\limits_{x\to-2^+} f(x) = -\infty$

$\lim\limits_{x\to 2^-} f(x) = -\infty$; $\lim\limits_{x\to 2^+} f(x) = \infty$

So there are two vertical asymptotes.

$$f'(x) = \frac{(x^2 - 4)2x - (x^2 - 1)2x}{(x^2 - 4)^2}$$

$$= \frac{-6x}{(x^2 - 4)^2}$$

This is equal to 0 when $x = 0$, and it is easy to see that the slope is positive to the left of 0, and negative to the right of 0. So there is a relative maximum at $\left(0, \frac{1}{4}\right)$. This should be sufficient information to sketch the graph:

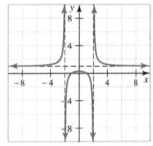

25. $f(x) = (x^2 - 3)e^{-x}$
relative minimum at $(-1, -2e)$;
relative maximum at $(3, 6e^{-3})$;
inflection points at approximately $(-0.24, -3.73)$ and $(4.24, 0.22)$;
concave up on $(-\infty, -0.24) \cup (4.24, \infty)$ and concave down on $(-0.24, 4.24)$;

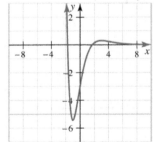

26. $f(x) = x + \tan^{-1}x$
$f'(x) = 1 + \left(1 + x^2\right)^{-1} > 0$; the curve is rising for all x.

$f''(x) = -2x(1 + x^2)^{-2} = 0$ when $x = 0$; $(0, 0)$ is a point of infection. The graph is concave up for $x < 0$ and down for $x > 0$.

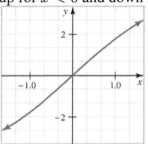

27. $f(x) = \sin^2 x - 2\cos x$ on $[0, 2\pi]$
relative maximum at $(\pi, 2)$;
inflection points at $\left(\frac{\pi}{3}, -\frac{1}{4}\right)$ and $\left(\frac{5\pi}{3}, -\frac{1}{4}\right)$;
concave up on $\left(0, \frac{\pi}{3}\right) \cup \left(\frac{5\pi}{3}, 2\pi\right)$ and concave down on $\left(\frac{\pi}{3}, \frac{5\pi}{3}\right)$;

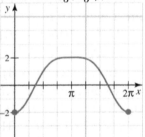

28. This is a continuous function on a closed interval, so the extreme value theorem guarantees an absolute maximum and an absolute minimum. The candidates are the critical points and the endpoints.

$$f'(x) = 4x^3 - 10x^4 = 2x^3(2 - 5x)$$

which is 0 at $x = 0, \frac{2}{5}$. Testing our candidates: $f(0) = 5$;
$f\left(\frac{2}{5}\right) = \frac{15{,}641}{3{,}125} \approx 5.00512$, $f(1) = 4$. The absolute maximum is at $(0.4, 5.005)$ and the absolute minimum at $(1, 4)$.

29. We are asked to minimize the amount of material, $S = x^2 + 4xh$, with the restriction that $V = x^2 h = 2$. We can use the volume formula to find $h = 2/x^2$ to express S as a function of x.

$$S = x^2 + 4x\left(\frac{2}{x^2}\right) = x^2 + \frac{8}{x}$$

$$S' = 2x - \frac{8}{x^2}$$

$$S' = 0 \text{ when } x^3 = 4, \ x = \sqrt[3]{4};$$

$$h = \frac{2}{x^2} = \frac{2}{\sqrt[3]{4^2}}$$

The dimensions of the box are approximately 1.587 by 1.587 by 0.794 ft, or to the nearest inch: 19 in. × 19 in. × 10 in.

30. $R(N) = -3N^4 + 50N^3 - 261N^2 + 540N$
$R'(N) = -12N^3 + 150N^2 - 522N + 540$
$\qquad = -6(N-2)(N-3)(2n-15)$
$R'(N) = 0$ when $N = 2, 3, 7.5$
$R(0) = 0$, $R(9) = 486$, $R(2) = 388$,
$R(3) = 378$, $R(7.6) = 970.3125$
We can't hire 7.7 people, so we look at $N = 7$ and $N = 8$: $R(7) = 938$ and $R(8) = 928$, so we should hire 7 people.

Supplementary Problems, page 314

1.   **3.**

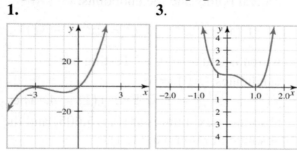

5.   **7.**

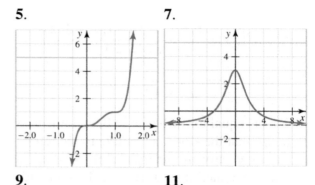

9.   **11.**

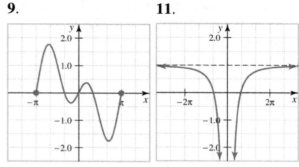

13.   **15.**

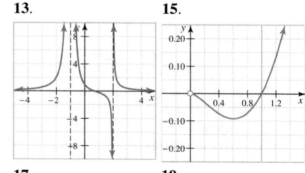

17.   **19.**

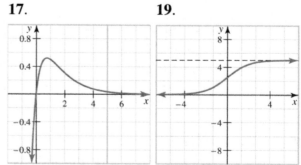

21. C   **23.** B

25. $f(x) = x^4 - 8x^2 + 12$ on $[-1, 2]$
$f'(x) = 4x^3 - 16 = 4x(x + 2)(x - 2)$
critical numbers $x = 0, -2$ (not in
interval), 2; $f(-1) = 5$; $f(2) = -4$;
$f(0) = 12$;
maximum $f(0) = 12$;
minimum $f(2) = -4$

27. $f(x) = 2x - \sin^{-1}x$ on $[0, 1]$
$$f'(x) = 2 - \frac{1}{\sqrt{1 - x^2}}$$
critical numbers $x = \pm\dfrac{\sqrt{3}}{2}$; negative is
not in the domain. $f(1) \approx 0.429$;
$f(0) = 0$, minimum;
$$f\left(\frac{\sqrt{3}}{2}\right) = \sqrt{3} - \frac{\pi}{3} \approx 0.68$$
maximum

29. Since $\dfrac{x\sin^2 x}{x^2 + 1} \geq 0$ as $x \to \infty$ and since
$$\lim_{x\to\infty} \frac{x\sin^2 x}{x^2 + 1} \leq \lim_{x\to\infty}\frac{x}{x^2 + 1}$$
$$= \lim_{x\to\infty}\frac{\frac{1}{x}}{1 + \frac{1}{x}}$$
$$= 0$$
we see $\lim\limits_{x\to\infty}\dfrac{x\sin^2 x}{x^2 + 1} = 0$ by the squeeze
theorem.

31. $\lim\limits_{x\to\infty}\left(\sqrt{x^2 - x} - x\right)$
$$= \lim_{x\to\infty}\frac{\left(\sqrt{x^2 - x} - x\right)\left(\sqrt{x^2 - x} + x\right)}{\sqrt{x^2 - x} + x}$$
$$= \lim_{x\to\infty}\frac{x^2 - x - x^2}{\sqrt{x^2 - x} + x}$$

$$= \lim_{x\to\infty}\frac{-1}{\sqrt{1 - \frac{1}{x}} + 1}$$
$$= -\frac{1}{2}$$

33. $\lim\limits_{x\to 0}\dfrac{x\sin x}{x + \sin^3 x} = \lim\limits_{x\to 0}\dfrac{x\cos x + \sin x}{1 + 3\sin^2 x\cos x} = 0$

35. $\lim\limits_{x\to 0}\dfrac{x\sin^2 x}{x^2 - \sin^2 x}$
$$= \lim_{x\to 0}\frac{x(2\sin x\cos x) + \sin^2 x}{2x - 2\sin x\cos x}$$
$$= \lim_{x\to 0}\frac{x\sin 2x + \sin^2 x}{2x - \sin 2x}$$
$$= \lim_{x\to 0}\frac{\sin 2x + 2x\cos 2x + 2\sin x\cos x}{2 - 2\cos 2x}$$
$$= \lim_{x\to 0}\frac{2\sin 2x + 2x\cos 2x}{2 - 2\cos 2x}$$
$$= \lim_{x\to 0}\frac{\sin 2x + x\cos 2x}{1 - \cos 2x}$$
$$= \lim_{x\to 0}\frac{2\cos 2x + \cos 2x - 2x\sin x}{2\sin 2x}$$
$$= \lim_{x\to 0}\frac{3\cos 2x - 2x\sin 2x}{2\sin 2x}$$
$$= \lim_{x\to 0}\left(\frac{3}{2}\cot 2x - 1\right)$$
This limit does not exist.

37. $\lim\limits_{x\to 0}\dfrac{\sin^2 x}{\sin x^2} = \lim\limits_{x\to 0}\dfrac{2\sin x\cos x}{2x\cos x^2}$
$$= \left[\lim_{x\to 0}\frac{\sin x}{x}\right]\left[\lim_{x\to 0}\frac{\cos x}{\cos x^2}\right]$$
$$= (1)\left(\frac{1}{1}\right)$$
$$= 1$$

39. $\lim\limits_{x\to\left(\frac{\pi}{2}\right)^-}\dfrac{\sec^2 x}{\sec^2 3x} = \lim\limits_{x\to\left(\frac{\pi}{2}\right)^-}\dfrac{\cos^2 3x}{\cos^2 x}$

$$= \left[\lim_{x \to \left(\frac{\pi}{2}\right)^-} \frac{\cos 3x}{\cos x}\right]^2$$

$$= \left[\lim_{x \to \left(\frac{\pi}{2}\right)^-} \frac{-3\sin 3x}{-\sin x}\right]^2$$

$$= \left[\frac{-3(-1)}{-1}\right]^2$$

$$= 9$$

41. $\displaystyle\lim_{x \to \left(\frac{\pi}{2}\right)^-} (\sec x - \tan x)$

$$= \lim_{x \to \left(\frac{\pi}{2}\right)^-} \frac{1 - \sin x}{\cos x}$$

$$= \lim_{x \to \left(\frac{\pi}{2}\right)^-} \frac{-\cos x}{-\sin x}$$

$$= 0$$

43. $\displaystyle\lim_{x \to 1} \frac{(x-1)\sin(x-1)}{1 - \cos(x-1)}$ *Let* $u = x - 1$

$$= \lim_{u \to 0} \frac{u \sin u}{1 - \cos u}$$

$$= \lim_{u \to 0} \frac{u \cos u + \sin u}{\sin u}$$

$$= \lim_{u \to 0} \frac{-u \sin u + 2\cos u}{\cos u} = 2$$

45. $\displaystyle\lim_{x \to \infty} \left(\sqrt{x^2 + 4} - \sqrt{x^2 - 4}\right)$

$$= \lim_{x \to \infty} \frac{\left(\sqrt{x^2+4} - \sqrt{x^2-4}\right)\left(\sqrt{x^2+4} + \sqrt{x^2-4}\right)}{\left(\sqrt{x^2+4} + \sqrt{x^2-4}\right)}$$

$$= \lim_{x \to \infty} \frac{x^2 + 4 - x^2 + 4}{\sqrt{x^2+4} + \sqrt{x^2-4}}$$

$$= 0$$

47. $\displaystyle\lim_{x \to 1^-} \left(\frac{1}{1-x}\right)^x = \infty$

49. $\displaystyle\lim_{x \to 0} \frac{5^x - 1}{x} = \lim_{x \to 0} \frac{5^x \ln 5}{1} = \ln 5$

51. Let $L = \displaystyle\lim_{x \to \infty} \left(\frac{1}{x}\right)^x$, then

$$\ln L = \lim_{x \to \infty} x\ln\left(\frac{1}{x}\right)$$

$$= \lim_{x \to \infty} (-x)\ln x$$

$$= -\infty$$

$$L = e^{-\infty} = 0$$

53. $\displaystyle\lim_{x \to \infty} \frac{e^x \cos x - 1}{x} = \lim_{x \to \infty} \frac{-e^x \sin x + e^x \cos x}{1}$

$$= \lim_{x \to \infty} e^x(-\sin x + \cos x)$$

Limit diverges by oscillation.

55. The function is f and the derivative is g. Note: the original article in the *Mathematics Teacher* had the graph presented upside-down.

57. $y' = 3x^2 + 2bx + c = 0$ when

$$x = \frac{-b \pm \sqrt{b^2 - 3c}}{3}$$

If $b^2 - 3c > 0$, there are both a relative maximum and a relative minimum. The distance between the extrema changes as b varies. If $b^2 - 3c = 0$ there is a horizontal tangent but no relative extremum. If $b^2 - 3c < 0$ the are no horizontal tangents.

59.

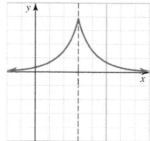

The derivative does not exist at $x = 1$.

61. $f(x) = Ax^3 + Bx^2 + Cx + D$
$f'(x) = 3Ax^2 + 2Bx + C$

$f''(x) = 6Ax + 2B$

Because $x = \pm 1$ are critical points we need

$$\begin{cases} f'(-1) = 3A - 2B + C = 0 \\ f''(1) = 3A + 2B + C = 0 \end{cases}$$

These equations imply $B = 0$ and $C = -3A$. Thus, $f(x) = Ax^2 - 3Ax + D$ so

$$\begin{cases} f(-1) = -A + 3A + D = 1 \\ f(1) = A - 3A + D = -1 \end{cases}$$

These equations imply $D = 0$, $A = \frac{1}{2}$, and $C = -\frac{3}{2}$. Thus,

$$f(x) = \frac{1}{2}x^3 - \frac{3}{2}x$$

63. Let x be the number of \$20 increases. The number of units rented is $200 - 5x$ $(0 \leq x \leq 40)$ and the amount of each rent payment is $600 + 20x$. The profit is

$$P(x) = (200 - 5x)(600 + 20x) - (200 - 5x)(80)$$
$$= (200 - 5x)(520 + 20x)$$
$$P'(x) = (200 - 5x)(20) + (-5)(520 + 20x)$$
$$= 200(7 - x)$$

$P'(x) = 0$ if $x = 7$;
$P(0) = 104{,}000; P(40) = 0$;
$P(7) = 108{,}900$
The maximum profit of \$108,900 is reached when 165 units are rented at \$740 each.

65. YIELD $= f(x) = (200 - 5x)(30 + x)$
$$= -5x^2 + 50x + 6{,}000$$
$f'(x) = -10x + 50 = 0$ when $x = 5$

$f''(x) < 0$, thus $x = 5$ leads to a maximum yield of $f(5) = 6{,}125$ for 35 trees per acre.

67. Let Q be the point on the shore straight across from the oil rig and P the point on the shore where the pipe starts on land. With $\overline{PQ} = x$, the distance along the bank is $8 - x$. The distance across the water is given by the Pythagorean theorem to be $\sqrt{9 + x^2}$. The cost C is

$C = $ COST IN THE WATER $+$ COST ON THE SHORE
$$= (1.5)\sqrt{9 + x^2} + (1)(8 - x)$$
$$C'(x) = \frac{(1.5)(2x)}{2\sqrt{9 + x^2}} - 1$$

$C'(x) = 0$ when

$$\frac{(1.5)(2x)}{2\sqrt{9 + x^2}} - 1 = 0$$
$$3x = 2\sqrt{9 + x^2}$$
$$2.25x^2 = 9 + x^2$$
$$x^2 = \frac{36}{5}$$
$$x = \pm \frac{6}{5}\sqrt{5}$$

(reject negative value)
$C(0) = 12.5$; $C(8) \approx 12.81$;
$C\left(\frac{6}{5}\sqrt{5}\right) \approx 11.35$ which is the lowest cost. Thus, the pipe should be laid so that it hits the shore $\frac{6}{5}\sqrt{5}$ mi $\approx 14{,}167$ ft from the 3 mile point on the shoreline. This means that 28,072 ft of pipe laid on the shore gives the minimum cost.

69. *Here is the solution to the first part (where the answer is given):* Let x be the length and y be the width; then $xy = 600$. We wish to minimize the cost

$$C = 10x + 5x + 5y + 5y$$
$$= 15x + 10y$$
$$= 15x + 10\left(\frac{600}{x}\right)$$
$$= 15x + 6{,}000x^{-1}$$
$$C' = 15 - 6{,}000x^{-2}$$

$C' = 0$ when $x = 20$. The dimensions are 20 by 30. This is a minimum because $C''(20) > 0$.

Here is the solution to the extended part: Consider the cost function $C = ax + by$ for some numbers a and b with C to be minimized. In this problem, $a = 10 + 15$ dollars/ft, whereas left and right sides yield $b = 5 + 5 = 10$ per ft. The area is $xy = C$, the given constant area that will later be set equal to the minimum cost. Solve for y in terms of x (and constants a and b) to obtain

$$C = ax + by$$
$$= ax + \frac{bC}{x}$$

Show this has a minimum over the set of positive reals; differentiate, equate to zero, and solve for x and y. Substitute into C and equate the resulting minimum cost to the given area C to obtain the necessary and sufficient condition: $C = 4ab$. For the given example, $600 = 4 \times 15 \times 10$.

71. Let x be the number of units of telephones sold. The price per telephone is
$$p(x) = 150 - x$$
The total production cost is
$$C(x) = 2{,}500 + 30x$$
The profit is

$$P(x) = x(150 - x) - 2{,}500 - 30x$$
$$= -\left(x^2 - 120x + 2{,}500\right)$$
$$P'(x) = -2x + 120 = 0 \text{ when } x = 60$$

The price per unit is
$$p(60) = 150 - 60 = 90$$
and the maximum profit is
$$P(60) = -\left[60^2 - 120(60) + 2{,}500\right]$$
$$= 1{,}100$$
This is a maximum because $P''(x) < 0$.

73. If $f(x)$ is a polynomial of degree n, then $f''(x)$ is a polynomial of degree $(n - 2)$ and will have at most $(n - 2)$ real roots. Thus, there can be at most $(n - 2)$ points of inflection.

75. Let $P_0(x_0, y_0)$ be the point of contact between the line $y = mx + b$ and the circle $x^2 + y^2 = 1$. The slope of the line is

$$\frac{dy}{dx} = -\frac{x}{y}$$

At P_0, $m = -\dfrac{x_0}{y_0}$ so the equation of the tangent line is

$$y - y_0 = -\frac{x_0}{y_0}(x - x_0)$$
$$y_0 y - y_0^2 + x_0 x - x_0^2 = 0$$
$$y_0 y + x_0 x = x_0^2 + y_0^2 = 1$$

Intercepts of this line are $\left(\dfrac{1}{x_0}, 0\right)$ and $\left(0, \dfrac{1}{y_0}\right)$, so we wish to minimize

$$f(x_0) = \frac{1}{x_0} + \frac{1}{y_0} = \frac{1}{x_0} + \frac{1}{\sqrt{1 - x_0^2}}$$
$$f'(x_0) = -\frac{1}{x_0^2} + \frac{x_0}{\left(1 - x_0^2\right)^{3/2}}$$
$$f'(x_0) = 0 \text{ when } x_0^3 = \left(1 - x_0^2\right)^{3/2}$$

or $x_0 = \dfrac{1}{\sqrt{2}}$, $y_0 = \dfrac{1}{\sqrt{2}}$

The minimal line is $x_0 + y_0 = \sqrt{2}$.

77. $\dfrac{f(x) - f(a)}{x - a} = f'(d)$ for d in (a, x).

Then, since $f'(d) = c$,

$$f(x) = f'(d)(x - a) + f(a)$$
$$= c(x - a) + f(a)$$

79. $f'(x) = \dfrac{a^{-1}}{1 + a^{-2}x^2} - \dfrac{b^{-1}}{1 + b^{-2}x^2}$

$\qquad = \dfrac{a}{a^2 + x^2} - \dfrac{b}{b^2 + x^2}$

$f'(x) = 0$ when

$$a(b^2 + x^2) = b(a^2 + x^2)$$
$$ab^2 + ax^2 = ba^2 + bx^2$$
$$(a - b)x^2 = ab(a - b)$$
$$x = \pm\sqrt{ab}$$

$$f''(x) = \dfrac{-2ax}{(a^2 + x^2)^2} + \dfrac{2bx}{(b^2 + x^2)^2}$$

$$f''(\sqrt{ab}) = \dfrac{-2a^{3/2}b^{1/2}}{(a^2 + ab)^2} + \dfrac{2a^{1/2}b^{3/2}}{(b^2 + ab)^2}$$

$$\qquad = \dfrac{2\sqrt{ab}}{(a + b)^2}(-a^{-1} + b^{-1})$$

$$\qquad > 0$$

so $x = \sqrt{ab}$ leads to a relative minimum. Similarly, $x = -\sqrt{ab}$ leads to a relative maximum.

81. By l'Hôpital's rule, $\displaystyle\lim_{x\to\infty} \dfrac{x^n}{x^m} = \infty$ for $n > m$. Thus, in any given polynomial of degree n, for x sufficiently large and M given, each of the terms of order less than n are less than $|x^n|(a_n + M)$. M can be

made as close to 0 as we want, so the bound becomes $|a_n x_n|$.

83. $f(x) = x^3 - x^2 - x + 1$ on $\left[-\frac{3}{2}, 2\right]$

$f'(x) = 3x^2 - 2x - 1$

$f''(x) = 6x - 2$

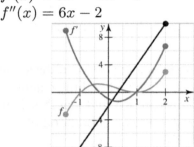

a. $f'(x) > 0$ on $\left(-\frac{3}{2}, -\frac{1}{3}\right) \cup (1, 2)$

b. $f'(x) < 0$ on $\left(-\frac{1}{3}, 1\right)$

c. $f''(x) > 0$ on $\left(\frac{1}{3}, 2\right)$

d. $f''(x) < 0$ on $\left(-\frac{3}{2}, \frac{1}{3}\right)$

e. $f'(x) = 0$ at $x = -\frac{1}{3}, 1$

f. f' exists everywhere

g. $f''(x) = 0$ at $x = \frac{1}{3}$

85. $f(x) = x^3 - x + \dfrac{1}{x} + 1$ on $[-2, 2]$

$f'(x) = 3x^2 - 1 - x^{-2}$

$f''(x) = 6x + 2x^{-3}$

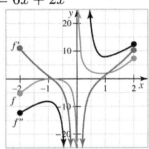

a. $f'(x) > 0$ on $(-2, -0.876) \cup (0.876, 2)$

b. $f'(x) < 0$ on $(-0.876, 0) \cup (0, 0.876)$

c. $f''(x) > 0$ on $(0, 2)$

d. $f''(x) < 0$ on $(-2, 0)$

e. $f'(x) = 0$ at $x \approx \pm 0.876$

f. $f'(x)$ does not exist at $x = 0$

g. $f''(x) \neq 0$

87. a. ∞ **b.** ∞ **c.** $-\infty$ **d.** $-\infty$ **e.** ∞ **f.** $-\infty$

89. With $4x^2 + y^2 = 1$, $y = \sqrt{1 - 4x^2}$ and

$$d = \frac{1}{x} - \sqrt{1 - 4x^2}$$

is the vertical distance between the shore and the road. Using a graphing calculator or computer software we find that this distance is minimized in the neighborhood of $x = 0.460355$.

91.   Detail of graph:

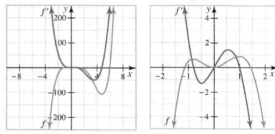

Estimate the relative minimum at $(0, 0)$ and $(4.8, -107.9)$, and relative maxima at $(1, 1)$ and $(-0.8, 0.7)$.

93. $\lim\limits_{\epsilon \to 0} \dfrac{a^\epsilon - 1}{\epsilon} = \lim\limits_{\epsilon \to 0} \dfrac{a^\epsilon \ln a}{1} = \ln a$

95. This is Putnam Problem 1 in the morning session of 1941. Make the substitution $x = a(1 - y)$. Then the polynomial becomes

$$a^6 y^2 \left(y^4 - 3y^3 + \tfrac{5}{2}y^2 - \tfrac{1}{2} \right)$$

Since $a^6 y^2$ is surely positive, it suffices to prove that

$$g(y) = y^4 - 3y^3 + \tfrac{5}{2}y^2 - \tfrac{1}{2} < 0$$

for $0 < y < 1$. Since

$$g'(y) = 4y^3 - 9y^2 + 5y = y(y - 1)(4y - 5)$$

the critical numbers for g are $0, 1, \frac{5}{4}$. Between consecutive critical numbers g does not change sign. Therefore, since $g(0) = -\frac{1}{2}$ and $g(1) = 0$, we have

$$-\frac{1}{2} < g(y) < 0$$

for $0 < y < 1$.

97. This is Problem A2 of the morning session of 1985 Putnam examination.

$$\frac{A(R) + A(S)}{A(T)} = \frac{ay + bz}{\frac{1}{2}hx}$$

where T is the triangle and the height of the triangle is $h = a + b + c$. By similar triangles,

$$\frac{x}{h} = \frac{y}{b + c} = \frac{z}{c}$$

which implies

$$y = \frac{(b + c)x}{h} \quad \text{and} \quad z = \frac{cx}{h}$$

so

$$\frac{A(R) + A(S)}{A(T)} = \frac{a\left[\frac{(b+c)x}{h}\right] + b\left[\frac{cx}{h}\right]}{\frac{1}{2}hx}$$
$$= \frac{2}{h^2}(ab + ac + bc)$$

We need to maximize $ab + ac + bc$ subject to $a + b + c = h$. To do this, we fix a so that $b + c = h - a$ and

$$ab + ac + bc = a(h - a) + bc$$

Then bc is maximized when $b = c$ and $2ab + b^2$ is maximized (subject to

$a + 2b = h$) when $a = b = c = h/3$. Thus, the maximum ratio is 2/3, which is independent of T.

99. This is Putnam Problem 1 of the morning session of 1961. In the first quadrant the given equation is equivalent to

$$\frac{1}{y}\ln y = \frac{1}{x}\ln x$$

Consider the function given by $f(t) = t^{-1}\ln t$ for $t > 0$. Since

$$f'(t) = (1 - \ln t)t^{-2}$$

it is clear that f is strictly increasing for $t < e$, is strictly decreasing for $t > e$ and achieves its maximum value for e^{-1} for $t = e$.

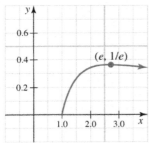

Moreover, $f(t) \to -\infty$ as $t \to 0$ and $f(t) \to 0$ as $t \to \infty$. It follows for α in $(0, e^{-1})$ the equation $f(t) = \alpha$ has two

solutions, one in $(1, e)$, the other in (e, ∞).

For α near 0, the lower solution is just above 1 and the upper solution is large. As α increases to e^{-1}, the lower solution increases to e and the upper solution decreases to e. Therefore, the set of points satisfying the displayed equation consists of the line $y = x$ and a curve M lying in the quadrant $x > 1, y > 1$ and asymptotic to the lines $x = 1$ and $y = 1$, as shown in the figure. M is evidently symmetric in the line $y = x$ and crosses that line at (e, e).

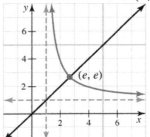

To establish the smoothness of the curve requires material not yet discussed in the book. For reference see "The Real Function Defined by $x^r = y^s$" in *American Mathematical Monthly*, Vol 23 (1916), pp. 233-237. R. Robinson Rowe called the curve "mutuabola" in *Journal of Recreational Mathematics*, Vol. 3 (1970) pp. 176-178.

CHAPTER 5 Integration

Chapter Overview

In this chapter, we begin our study of the third great idea of calculus, namely the idea of integration. As we see in the first section of this chapter, this part of calculus, called *integral calculus*, reversed the operation of differentiation. After defining the process of integration we examine several techniques of integration. Additional techniques are discussed in the next chapter.

5.1 Antidifferentiation, page 334

1. $\displaystyle\int 2\,dx = 2x + C$

SURVIVAL HINT: *Don't forget the "+ C" every time you evaluate an indefinite integral.*

3. $\displaystyle\int (2x + 3)\,dx = x^2 + 3x + C$

5. $\displaystyle\int (4t^3 + 3t^2)\,dt = t^4 + t^3 + C$

7. $\displaystyle\int \frac{dx}{2x} = \frac{1}{2}\ln|x| + C$

9. $\displaystyle\int (6u^2 - 3\cos u)\,du = 2u^3 - 3\sin u + C$

11. $\displaystyle\int \sec^2\theta\,d\theta = \tan\theta + C$

13. $\displaystyle\int 2\sin\theta\,d\theta = -2\cos\theta + C$

15. $\displaystyle\int \frac{5}{\sqrt{1 - y^2}}\,dy = 5\sin^{-1}y + C$

17. $\displaystyle\int x(x + \sqrt{x})\,dx = \int (x^2 + x^{3/2})\,dx$

$$= \frac{1}{3}x^3 + \frac{2}{5}x^{5/2} + C$$

19. $\displaystyle\int (u^{3/2} - u^{1/2} + u^{-10})\,du$

$$= \frac{u^{5/2}}{\frac{5}{2}} - \frac{u^{3/2}}{\frac{3}{2}} + \frac{u^{-9}}{-9} + C$$

$$= \frac{2}{5}u^{5/2} - \frac{2}{3}u^{3/2} - \frac{1}{9}u^{-9} + C$$

21. $\displaystyle\int \left(\frac{1}{t^2} - \frac{1}{t^3} + \frac{1}{t^4}\right)\,dt$

$$= -t^{-1} + \frac{1}{2}t^{-2} - \frac{1}{3}t^{-3} + C$$

23. $\displaystyle\int (2x^2 + 5)^2\,dx = \int (4x^4 + 20x^2 + 25)\,dx$

$$= \frac{4}{5}x^5 + \frac{20}{3}x^2 + 25x + C$$

25. $\displaystyle\int \left(\frac{x^2 + 3x - 1}{x^4}\right)\,dx$

$$= \int (x^{-2} + 3x^{-3} - x^{-4})\,dx$$

$$= -x^{-1} - \frac{3}{2}x^{-2} + \frac{1}{3}x^{-3} + C$$

27. $\displaystyle\int \frac{x^2 + x - 2}{x^2}\,dx$

$$= \int (1 + x^{-1} - 2x^{-2})\,dx$$

$$= x + \ln|x| + 2x^{-1} + C$$

29. $\displaystyle\int \frac{\sqrt{1-x^2}-1}{\sqrt{1-x^2}}\,dx$

$\displaystyle = \int\left[1-\frac{1}{\sqrt{1-x^2}}\right]dx$

$\displaystyle = x - \sin^{-1}x + C$

31. $F(x) = \displaystyle\int \left(x^2 + 3x\right)dx$

$\displaystyle = \frac{1}{3}x^3 + \frac{3}{2}x^2 + C$

$F(0) = \frac{1}{3}(0) + \frac{3}{2}(0) + C = 0$

so $C = 0$;

$F(x) = \frac{1}{3}x^3 + \frac{3}{2}x^2$

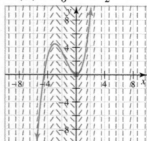

33. $F = \displaystyle\int \left(\sqrt{x}+3\right)^2 dx$

$\displaystyle = \int \left(x + 6x^{1/2} + 9\right)dx$

$\displaystyle = \frac{1}{2}x^2 + 4x^{3/2} + 9x + C$

$F(4) = \dfrac{1}{2}(4)^2 + 4(4)^{3/2} + 9(4) + C$

$= 36$

so $C = -40$;

$F(x) = \frac{1}{2}x^2 + 4x^{3/2} + 9x - 40$

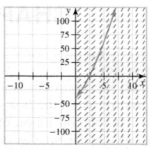

35. $F(x) = \displaystyle\int \frac{x+1}{x^2}dx = \int \left(x^{-1} + x^{-2}\right)dx$

$\displaystyle = \ln|x| - x^{-1} + C$

$F(1) = 0 - 1 + C = -2$, so $C = -1$

$F(x) = \ln|x| - x^{-1} - 1$

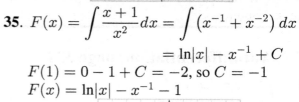

37. $F(x) = \displaystyle\int \left(x + e^x\right)dx$

$\displaystyle = \frac{1}{2}x^2 + e^x + C$

$F(0) = 1 + C = 2$, so $C = 1$

$F(x) = \frac{1}{2}x^2 + e^x + 1$

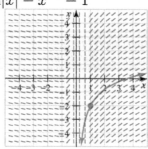

39. $F(x) = \int \left(x^{-1/2} - 4 \right) dx$

$$= 2x^{1/2} - 4x + C$$
$$F(1) = 2 - 4 + C = 0, \text{ so } C = 2$$
$$F(x) = 2\sqrt{x} - 4x + 2$$

a.

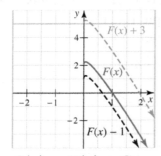

b. $G(x) = F(x) + C_0$

$$G'(x) = F'(x) = \frac{1}{\sqrt{x}} - 4 = 0$$

At $x = \frac{1}{16}$,
$$G\left(\tfrac{1}{16}\right) = 2\left(\tfrac{1}{4}\right) - 4\left(\tfrac{1}{16}\right) + 2 + C_0 = 0$$
so $C_0 = -\frac{9}{4}$

41. $C(x) = \int C'(x)\, dx$

$$= \int \left(6x^2 - 2x + 5 \right) dx$$

$$= 2x^3 - x^2 + 5x + C$$
$$C(1) = 2(1)^3 - (1)^2 + 5(1) + C = 500,$$
so $C = 494$.

$$C(x) = 2x^3 - x^2 + 5x + 494$$
$$C(5) = 250 - 25 + 25 + 494$$
$$= \$744$$

43. Let $P(t)$ be the population in the town at time t in months.
$$P'(t) = 4 + 5t^{2/3}$$

$$P(t) = \int \left(4 + 5t^{2/3} \right) dt$$

$$= 4t + 3t^{5/3} + C$$
$$P(0) = 10{,}000, \text{ so } C = 10{,}000$$

$$P(t) = 4t + 3t^{5/3} + 10{,}000$$

The population in 8 months will be

$$P(8) = 4(8) + 3(8)^{5/3} + 10{,}000 = 10{,}128$$

45. $a(t) = k$,
$$v(t) = \int a(t)\, dt = \int k\, dt = kt + C$$
$$v(0) = 0, \text{ so } C = 0,$$

$$s(t) = \int v(t)\, dt = \int kt\, dt = \frac{kt^2}{2} + C$$
We know that $s(0) = C$ and
$s(6) = 18k + C$ and are given

$$s(6) - s(0) = 18k = 360$$

Thus, $k = 20$ ft/s^2.

47. With $a(t) = k$, $v(t) = kt + C_1$, $C_1 = 0$
because the plane starts from rest ($v_0 = 0$)
and the distance is

$$s(t) = \frac{kt^2}{2} + C_2$$

For convenience, measure the distance
from the point where the plane begins
moving so $s(0) = 0$ and $C_2 = 0$. Let t_1
be the time required for liftoff. Since

$$s(t_1) = \frac{kt_1^2}{2} = 900 \quad \text{and} \quad v(t_1) = kt_1 = 88$$

we find (by dividing these equations)

$$\frac{kt_1^2}{kt_1} = \frac{1,800}{88}$$

$$t_1 \approx 20.4545 \text{ seconds}$$

From the velocity equation,

$$k = \frac{88}{t_1} = \frac{88}{20.4545} \approx 4.3$$

The acceleration of the plane at lift off is approximately 4.3 ft/s^2.

49. With $a(t) = k$, $v(t) = kt + C_1$; the initial speed of the car is $v_0 = 88$, so $v(0) = C_1 = 88$ and the distance is

$$s(t) = \frac{kt^2}{2} + 88t + C_2$$

For convenience, measure the distance from the point where the car begins moving so $s(0) = 0$ so $C_2 = 0$. Let t_1 be the time required for stopping.

$$s(t_1) = \frac{kt_1^2}{2} + 88t_1 = 121$$

and $v(t_1) = 0$ or $t_1 = -\dfrac{88}{k}$. Substituting into the distance formula leads to

$$\frac{88^2 k}{2k^2} - \frac{88^2}{k} - 121 = 0$$

$$k = -\frac{88^2}{242}$$

$$= -32 \text{ ft/s}^2$$

51. $C(x) = \displaystyle\int \left(0.1e^x + 21x^{1/2} \right) dx$

$$= 0.1e^x + 21\left(\frac{2}{3}\right)x^{3/2} + C_0$$

$C(1) = 0.1e^1 + 14(1)^{2/3} + C_0 = 100$, so
$C_0 \approx 85.73$.

$$C(x) = 0.1e^x + 14x^{3/2} + 85.73$$
$$C(4) \approx 203.19$$

It costs \$203.19 to produce 4 units.

53. As in Example 10, $A(x)$ is an antiderivative of $\sqrt{x}$.

$$A(x) = \int \sqrt{x}\, dx$$
$$= \frac{2}{3}x^{3/2} + C$$

Clearly $A(1) = 0$, so $C = -\frac{2}{3}$.
$$A(4) = \frac{2}{3}(4)^{3/2} + C = \frac{16}{3} - \frac{2}{3}$$
$$= \frac{14}{3}$$

55. As in Example 10, $A(x)$ is an antiderivative of $\dfrac{x+1}{x}$.

$$A(x) = \int \frac{x+1}{x}\, dx$$
$$= \int \left(1 + x^{-1}\right) dx$$
$$= x + \ln|x| + C$$
Clearly $A(1) = 0$, so $C = -1$.
$$A(2) = 2 + \ln 2 + C$$
$$= 2 + \ln 2 - 1$$
$$= 1 + \ln 2$$

57. As in Example 10, $A(x)$ is an antiderivative of $\left(1 - x^2\right)^{-1/2}$.

$$A(x) = \int \left(1 - x^2\right)^{-1/2} dx$$
$$= \sin^{-1}x + C$$
Clearly $A(0) = 0$, so $C = 0$.

$$A\left(\frac{1}{2}\right) = \sin^{-1}\frac{1}{2} + C$$

$$= \sin^{-1}\frac{1}{2}$$

$$= \frac{\pi}{6}$$

59. $\displaystyle\int [af(x) + bg(x) + ch(x)]\,dx$

$$= \int \left\{ [af(x) + bg(x)] + ch(x) \right\} dx$$

$$= \int [af(x) + bg(x)]\,dx + \int ch(x)\,dx$$

$$= \int af(x)\,dx + \int bg(x)\,dx + \int ch(x)\,dx$$

$$= a\int f(x)\,dx + b\int g(x)\,dx + c\int h(x)\,dx$$

5.2 Area as the Limit of a Sum, page 342

1. $\displaystyle\sum_{k=1}^{6} 1 = 6(1) = 6$

SURVIVAL HINT: *You will want to be familiar with summation notation and the properties listed in Theorem 5.4.*

3. $\displaystyle\sum_{k=4}^{10} 3 = (10 - 3)3 = 21$

5. $\displaystyle\sum_{k=1}^{10}(k + 1) = \sum_{k=1}^{10}k + \sum_{k=1}^{10}1$

$$= \frac{(10)(11)}{2} + 10$$

$$= 65$$

7. $\displaystyle\sum_{k=1}^{5} k^3 = \frac{5^2 6^2}{4} = 225$

9. $\displaystyle\sum_{k=1}^{100}(2k - 3) = 2\sum_{k=1}^{100}k - 3\sum_{k=1}^{100}1$

$$= 2\left[\frac{100(101)}{2}\right] - 300$$

$$= 9{,}800$$

SURVIVAL HINT: *Bookmark the summation formulas on page 340. You will not only use them in this section, but will use them later in this book and in your future mathematics courses.*

11. $\displaystyle\lim_{n\to\infty}\sum_{k=1}^{n}\frac{k}{n^2} = \lim_{n\to\infty}\frac{1}{n^2}\sum_{k=1}^{n}k$

$$= \lim_{n\to\infty}\frac{1}{n^2}\left[\frac{n(n+1)}{2}\right]$$

$$= \lim_{n\to\infty}\left[(1)\left(\frac{1}{2}\right)\left(1 + \frac{1}{n}\right)\right]$$

$$= \frac{1}{2}$$

13. $\displaystyle\lim_{n\to\infty}\sum_{k=1}^{n}\frac{1}{n} = \lim_{n\to\infty}\frac{1}{n}(n)$

$$= 1$$

15. $\displaystyle\lim_{n\to\infty}\sum_{k=1}^{n}\frac{k}{n^3} = \lim_{n\to\infty}\frac{1}{n^3}\sum_{k=1}^{n}k$

$$= \lim_{n\to\infty}\frac{1}{n^3}\left[\frac{n(n+1)}{2}\right]$$

$$= \lim_{n\to\infty}\left[(1)\left(\frac{1}{2}\right)\left(1 + \frac{1}{n}\right)\left(\frac{1}{n}\right)\right]$$

$$= 0$$

17. $\displaystyle\lim_{n\to\infty}\sum_{k=1}^{n}\left(1 + \frac{k}{n}\right)\left(\frac{2}{n}\right)$

$$= \lim_{n\to\infty}\left\{\frac{2}{n}(n) + \frac{2}{n^2}\left[\frac{n(n+1)}{2}\right]\right\}$$

$$= \lim_{n \to \infty} \left[2 + \left(1 + \frac{1}{n} \right) \right] = 3$$

19. a. $n = 4, \Delta x = 0.25$ **b.** $n = 8, \Delta x = 0.125$

$f(a + k\Delta x) = k + 1$

$f(a + k\Delta x) = \dfrac{k}{2} + 1$

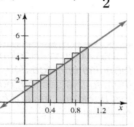

$S = 3.5$ $S = 3.25$

21. a. $n = 4, \Delta x = 0.25$ **b.** $n = 6, \Delta x = \dfrac{1}{6}$

$f(a + k\Delta x) = \left(1 + \dfrac{k}{4} \right)^2$ $f(a + k\Delta x) = \left(1 + \dfrac{k}{6} \right)^2$

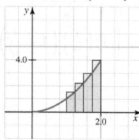 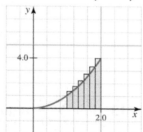

$S \approx 2.719$ $S \approx 2.588$

23. $n = 4, \Delta x = \dfrac{\pi}{8}$ **25.** $n = 4, \Delta x = 0.25$

$f(a + k\Delta x)$

$= \cos\left(-\dfrac{\pi}{2} + \dfrac{k\pi}{8} \right)$

$f(a + k\Delta x) = \left(1 + \dfrac{k}{4} \right)^{-2}$

$S \approx 1.183$ $S \approx 0.415$

27. $n = 4, \Delta x = 0.75$

$f(a + k\Delta x) = \sqrt{1 + \dfrac{3k}{4}}$

$S \approx 5.030$

29. $f(x) = 4x^3 + 2x$ on $[0, 2]$

$$\Delta x = \frac{b - a}{n} = \frac{2}{n}$$

$$f\left(\frac{2k}{n} \right) = \frac{32k^3}{n^3} + \frac{4k}{n}$$

$$A = \lim_{n \to \infty} \sum_{k=1}^{n} f(0 + k\Delta x)\Delta x$$

$$= \lim_{n \to \infty} \sum_{k=1}^{n} f\left(\frac{2k}{n} \right) \Delta x$$

$$= \lim_{n \to \infty} \sum_{k=1}^{n} \left(\frac{32k^3}{n^3} + \frac{4k}{n} \right) \left(\frac{2}{n} \right)$$

$$= \lim_{n \to \infty} \left[\frac{32}{n^3} \sum_{k=1}^{n} k^3 + \frac{4}{n} \sum_{k=1}^{n} k \right] \left(\frac{2}{n} \right)$$

$$= \lim_{n \to \infty} \left[\frac{64}{n^4} \frac{n^2(n+1)^2}{4} + \frac{8n(n+1)}{2n^2} \right]$$

$$= 16 + 4$$

$$= 20$$

31. $y = 6x^2 + 2x + 4$ on $[0, 3]$

$$\Delta x = \frac{b - a}{n} = \frac{3}{n}$$

$$f\left(\frac{3k}{n}\right) = 6\left(\frac{3k}{n}\right)^2 + 2\left(\frac{3k}{n}\right) + 4$$

$$A = \lim_{n\to\infty} \sum_{k=1}^{n} f\left(\frac{3k}{n}\right)\Delta x$$

$$= \lim_{n\to\infty} \sum_{k=1}^{n}\left[6\left(\frac{3k}{n}\right)^2 + 2\left(\frac{3k}{n}\right) + 4\right]\left(\frac{3}{n}\right)$$

$$= \lim_{n\to\infty} \sum_{k=1}^{n}\left(\frac{54k^2}{n^2} + \frac{6k}{n} + 4\right)\left(\frac{3}{n}\right)$$

$$= \lim_{n\to\infty}\left[\frac{162}{n^3}\sum_{k=1}^{n}k^2 + \frac{18}{n^2}\sum_{k=1}^{n}k + \frac{12}{n}\sum_{k=1}^{n}1\right]$$

$$= \lim_{n\to\infty}\left[\frac{162}{n^3}\frac{n(n+1)(2n+1)}{6} + \frac{18}{n^2}\frac{n(n+1)}{2} + \frac{12n}{n}\right]$$

$$= \frac{162}{3} + 9 + 12$$

$$= 75$$

33. $y = 3x^2 + 2x + 1$ on $[0, 1]$

$$\Delta x = \frac{b-a}{n} = \frac{1}{n}$$

$$f\left(\frac{k}{n}\right) = \frac{3k^2}{n^2} + \frac{2k}{n} + 1$$

$$A = \lim_{n\to\infty} \sum_{k=1}^{n} f\left(\frac{k}{n}\right)\Delta x$$

$$= \lim_{n\to\infty} \sum_{k=1}^{n}\left(\frac{3k^2}{n^2} + \frac{2k}{n} + 1\right)\left(\frac{1}{n}\right)$$

$$= \lim_{n\to\infty}\left[\frac{3}{n^3}\sum_{k=1}^{n}k^2 + \frac{2}{n}\cdot\frac{1}{n}\sum_{k=1}^{n}k + \frac{1}{n}\sum_{k=1}^{n}1\right]$$

$$= \lim_{n\to\infty}\left[\frac{3}{n^3}\frac{n(n+1)(2n+1)}{6} + \frac{2}{n^2}\frac{n(n+1)}{2} + \frac{n}{n}\right]$$

$$= 1 + 1 + 1$$

$$= 3$$

35. The statement is true. We are dealing with a rectangle of height C and base $b - a$.

37. The statement is true. Consider the trapezoid of base $(b - a)$ and parallel sides of lengths a^2 and b^2. The area is

$$A = \tfrac{1}{2}(b - a)(b^2 + a^2)$$

The area under the parabola is less than the area of the trapezoid.

39. The statement is false. Let $f(x) = x^2$ on $[0, 1]$, then $[f(x)]^2 = x^4$. Now,

$$[f(x)]^2 \le f(x)$$

and the area under $[f(x)]^2$ is less than that under $f(x)$.

41. $f(x) = x^3$ on $[0, 1]$; $\Delta x = \dfrac{1}{n}$;

$$f(a + kx) = f\left(\frac{k}{n}\right) = \frac{k^3}{n^3}.$$

$$A = \lim_{n\to\infty}\frac{1}{n^4}\sum_{k=1}^{n}k^3$$

$$= \lim_{n\to\infty}\frac{1}{n^4}\frac{n^2(n+1)^2}{4}$$

$$= \frac{1}{4}$$

43. $f(x) = -f(x) = -\dfrac{h}{b}x + h$ where h is the height and b the base of the right triangle;

$$\Delta x = \frac{b}{n};\; f\left(\frac{bk}{n}\right) = -\frac{h}{b}\left(\frac{bk}{n}\right) + h$$

$$A = \lim_{n\to\infty}\sum_{k=1}^{n}\left[-\frac{h}{b}\left(\frac{bk}{n}\right) + h\right]\left(\frac{b}{n}\right)$$

$$= \lim_{n\to\infty}\left[-\frac{bh}{n^2}\frac{n(n+1)}{2} + \left(\frac{bh}{n}\right)n\right]$$

$$= -\frac{bh}{2} + bh$$

$$= \frac{1}{2}bh$$

45. The area seems to be 2 square units.

n	Sum of rectangular areas
4	2.5
8	2.25
1,024	2.00

47. The area seems to be 1 square unit.

n	Sum of rectangular areas
4	1.18
8	1.09
1,024	1.00

49. The area seems to be 3.4 square units.

n	Sum of rectangular areas
4	4.30
8	3.85
1,024	3.41

51. The area seems to be 1 square unit.

n	Sum of rectangular areas
4	1.16
8	1.09
1,024	1.00

53. a.
$$\frac{1}{2}\sum_{k=1}^{n}\left[k^2 - \left(k^2 - 2k + 1\right)\right] + \frac{1}{2}\sum_{k=1}^{n}1$$
$$= \frac{1}{2}\sum_{k=1}^{n}(2k)$$
$$= \frac{1}{2}[2(1 + 2 + \cdots + n)]$$
$$= \sum_{k=1}^{n}k$$

b. $\sum_{k=1}^{n}\left[k^2 - (k-1)^2\right]$
$$= \left(1^2 - 0^2\right) + \left(2^2 - 1^2\right) + \cdots$$
$$+ \left[n^2 - (n-1)^2\right]$$
$$= n^2$$

c. $\sum_{k=1}^{n}k = \frac{1}{2}\sum_{k=1}^{n}[k^2 - (k^2 - 2k + 1)] + \frac{1}{2}n$

From part a.

$$= \frac{1}{2}\left(n^2 + n\right) \quad \textit{From part b.}$$
$$= \frac{n(n+1)}{2}$$

55. $\lim_{n \to \infty} \sum_{k=1}^{n}\left(\frac{k-1}{n}\right)^2\left(\frac{1}{n}\right)$

$$= \lim_{n \to \infty}\frac{1}{n^3}\sum_{k=1}^{n}\left(k^2 - 2k + 1\right)$$
$$= \lim_{n \to \infty}\frac{1}{n^3}\left[\frac{n(n+1)(2n+1)}{6} - 2\frac{n(n+1)}{2} + n\right]$$
$$= \frac{2}{6} - 0 + 0$$
$$= \frac{1}{3}$$

The result is the same as that shown in Example 1.

57. $\sum_{k=1}^{n}c = \sum_{k=1}^{n}ck^0$
$$= c\left(1^0 + 2^0 + \cdots + n^0\right)$$
$$= nc$$

59. $\sum_{k=1}^{n}ca_k = ca_1 + ca_2 + \cdots + ca_n$
$$= c(a_1 + a_2 + \cdots + a_n)$$
$$= c\sum_{k=1}^{n}a_k$$

5.3 Riemann Sums and Definite Integral, page 354

1. $f(x) = 2x + 1$; $a = 0$; $\Delta x = \frac{1}{4}$

$$f(a + k\Delta x) = f\left(\frac{k}{4}\right) = \frac{k}{2} + 1$$

$$\int_0^1 (2x + 1)\, dx \approx \sum_{k=1}^4 \left(\frac{k}{2} + 1\right)\left(\frac{1}{4}\right)$$

$$= 2.25$$

SURVIVAL HINT: *Distinguish between the indefinite integral (a function) and the definite integral (a number). The indefinite integral will have a "$+ C$" but the definite integral will be a number, as seen in Problem 1.*

---

3. $f(x) = x^2;\ a = 1;\ \Delta x = \frac{1}{2}$

$$f(a + k\Delta x) = f\left(1 + \frac{k}{2}\right) = \left(1 + \frac{k}{2}\right)^2$$

$$\int_1^3 x^2\, dx \approx \sum_{k=1}^4 \left(1 + \frac{k}{2}\right)^2 \left(\frac{1}{2}\right)$$

$$= 10.75$$

5. $f(x) = 1 - 3x;\ a = 0;\ \Delta x = \frac{1}{4}$

$$f(a + k\Delta x) = f\left(\frac{k}{4}\right) = 1 - \frac{3k}{4}$$

$$\int_0^1 (1 - 3x)\, dx \approx \sum_{k=1}^4 \left(1 - \frac{3k}{4}\right)\left(\frac{1}{4}\right)$$

$$= -0.875$$

7. $f(x) = \cos x;\ a = -\frac{\pi}{2};\ \Delta x = \frac{\pi}{8}$

$$f(a + k\Delta x) = f\left(-\frac{\pi}{2} + \frac{k\pi}{8}\right)$$

$$= \cos\left(-\frac{\pi}{2} + \frac{k\pi}{8}\right)$$

$$\int_{-\frac{\pi}{2}}^0 \cos x\, dx \approx \sum_{k=1}^4 \cos\left(-\frac{\pi}{2} + \frac{k\pi}{8}\right)\left(\frac{\pi}{8}\right)$$

$$\approx 1.183$$

9. $f(x) = e^x;\ a = 0;\ \Delta x = \frac{1}{4}$

$$f(a + k\Delta x) = f\left(\frac{k}{4}\right) = e^{k/4}$$

$$\int_0^1 e^x\, dx \approx \sum_{k=1}^4 \left(e^{k/4}\right)\left(\frac{1}{4}\right)$$

$$\approx 1.942$$

11. $f(x) = 2x + 1;\ a = 0;\ \Delta x = \frac{1}{4}$

$$f(a + (k-1)\Delta x) = f\left(\frac{k}{4} - \frac{1}{4}\right)$$

$$= \frac{k}{2} + \frac{1}{2}$$

$$\int_0^1 (2x + 1)\, dx \approx \sum_{k=1}^4 \left(\frac{k}{2} + \frac{1}{2}\right)\left(\frac{1}{4}\right)$$

$$= 1.75$$

13. $f(x) = x^2;\ a = 1;\ \Delta x = \frac{1}{2}$

$$f(a + (k-1)\Delta x) = f\left(1 + \frac{k-1}{2}\right)$$

$$= \left(\frac{k+1}{2}\right)^2$$

$$\int_1^3 x^2\, dx \approx \sum_{k=1}^4 \left(\frac{k+1}{2}\right)^2 \left(\frac{1}{2}\right)$$

$$= 6.75$$

15. $f(x) = 1 - 3x;\ a = 0;\ \Delta x = \frac{1}{4}$

$$f(a + (k-1)\Delta x) = f\left(\frac{k-1}{4}\right)$$

$$= 1 - 3\left(\frac{k-1}{4}\right) = \frac{7 - 3k}{4}$$

$$\int_0^1 (1 - 3x)\, dx \approx \sum_{k=1}^4 \left(\frac{3 - 3k}{4}\right)\left(\frac{1}{4}\right)$$

$$= -0.125$$

17. $f(x) = \cos x;\ a = -\frac{\pi}{2};\ \Delta x = \frac{\pi}{8}$

$$f(a + (k-1)\Delta x)$$

$$= f\left(-\frac{\pi}{2} + \frac{(k-1)\pi}{8}\right)$$
$$= \cos\left(\frac{(k-5)\pi}{8}\right)$$

$$\int_{-\frac{\pi}{2}}^{0} \cos x \, dx \approx \sum_{k=1}^{4} \cos\left(\frac{(k-5)\pi}{8}\right)\left(\frac{\pi}{8}\right)$$
$$\approx 0.791$$

19. $f(x) = e^x$; $a = 0$; $\Delta x = \frac{1}{4}$
$$f(a + (k-1)\Delta x) = f\left(\frac{k-1}{4}\right)$$
$$= e^{(k-1)/4}$$

$$\int_{0}^{1} e^x \, dx \approx \sum_{k=1}^{4} \left(e^{(k-1)/4}\right)\left(\frac{1}{4}\right)$$
$$\approx 1.512$$

21. $v(t) = 3t + 1$; $a = 1$; $\Delta t = \frac{3}{4}$
$$v(a + k\Delta t) = v\left(1 + \frac{3k}{4}\right)$$
$$= 3\left(1 + \frac{3k}{4}\right) + 1$$
$$S_4 = \sum_{k=1}^{4}\left(4 + \frac{9k}{4}\right)\left(\frac{3}{4}\right)$$
$$= \frac{231}{8}$$
$$= 28.875$$

23. $v(t) = \sin t$; $a = 0$; $\Delta t = \frac{\pi}{4}$
$$v(a + k\Delta t) = v\left(\frac{k\pi}{4}\right)$$
$$= \sin\frac{k\pi}{4}$$
$$S_4 = \sum_{k=1}^{4}\sin\left(\frac{k\pi}{4}\right)\left(\frac{\pi}{4}\right)$$

$$= \frac{\pi}{4}\left[\sqrt{2} + 1\right]$$
$$\approx 1.896$$

25. $v(t) = e^{-t}$; $a = 0$; $\Delta t = \frac{1}{4}$
$$v(a + k\Delta t) = v\left(\frac{k}{4}\right)$$
$$= e^{-k/4}$$
$$S_4 = \sum_{k=1}^{4} e^{-k/4}\left(\frac{1}{4}\right)$$
$$\approx 0.556$$

27. $v(t) = 3t + 1$; $a = 1$; $\Delta t = \frac{3}{4}$
$$v(a + (k-1)\Delta t) = v\left(1 + \frac{3(k-1)}{4}\right)$$
$$= 3\left(\frac{3k+1}{4}\right) + 1$$
$$= \frac{9k+7}{4}$$
$$S_4 = \sum_{k=1}^{4}\left[\frac{9k+7}{4}\right]\left(\frac{3}{4}\right)$$
$$= \frac{177}{8}$$
$$= 22.125$$

29. $v(t) = \sin t$; $a = 0$; $\Delta t = \frac{\pi}{4}$
$$v(a + (k-1)\Delta t) = v\left[\frac{(k-1)\pi}{4}\right]$$
$$= \sin\frac{k\pi - \pi}{4}$$
$$S_4 = \sum_{k=1}^{4}\sin\left(\frac{k\pi - \pi}{4}\right)\left(\frac{\pi}{4}\right)$$
$$= \frac{\pi}{4}\left[\sqrt{2} + 1\right]$$
$$\approx 1.896$$

31. $v(t) = e^{-t}$; $a = 0$; $\Delta t = \frac{1}{4}$

$$v(a + (k-1)\Delta t) = v\left(\frac{k-1}{4}\right)$$
$$= e^{(1-k)/4}$$

$$S_4 = \sum_{k=1}^{4} e^{(1-k)/4}\left(\frac{1}{4}\right)$$
$$\approx 0.714$$

33. $\displaystyle\int_0^{-1} x^2\, dx = -\int_{-1}^0 x^2\, dx = -\frac{1}{3}$

35. $\displaystyle\int_{-1}^2 \left(2x^2 - 3x\right) dx = 2\int_{-1}^2 x^2\, dx - 3\int_{-1}^2 x\, dx$
$$= 2(3) - 3\left(\frac{3}{2}\right)$$
$$= \frac{3}{2}$$

37. $\displaystyle\int_{-1}^0 x\, dx = \int_{-1}^2 x\, dx + \int_2^0 x\, dx$
$$= \int_{-1}^2 x\, dx - \int_0^2 x\, dx$$
$$= \frac{3}{2} - 2$$
$$= -\frac{1}{2}$$

39. On $[0, 1]$, $x^3 \le x$, so

$$\int_0^1 x^3\, dx \le \int_0^1 x\, dx$$

Now $\int_0^1 x\, dx$ is the same as the area of a right triangle with base 1 and height 1, so

$$\int_0^1 x^3\, dx \le \int_0^1 x\, dx \le (1)\left(\frac{1}{2}\right) = \frac{1}{2}$$

41. On $[0, 0.7]$, $x^2 \le \frac{1}{2}$, so

$$\int_0^{0.7} x^2\, dx \le \int_0^{0.7} \frac{1}{2}\, dx$$

Now $\int_0^{0.7} \frac{1}{2}\, dx$ is the same as the area of a rectangle with length 1 and height $\frac{1}{2}$, so

$$\int_0^{0.7} x^2\, dx \le \int_0^{0.7} \frac{1}{2}\, dx$$
$$= (0.7)\left(\frac{1}{2}\right) < \frac{1}{2}$$

43. Let $\displaystyle F = \int_{-2}^4 f(x)\, dx$ and $\displaystyle G = \int_{-2}^4 g(x)\, dx$. Then

$$\int_{-2}^4 [5f(x) + 2g(x)]\, dx = 5F + 2G = 7$$

and

$$\int_{-2}^4 [3f(x) + g(x)]\, dx = 3F + G = 4$$

Subtracting the first from twice the second leads to $F = 1$ and $G = 1$.

45. By the subdivision and opposite properties,

$$\int_{-1}^2 f(x)\, dx = \int_{-1}^1 f(x)\, dx + \int_1^3 f(x)\, dx + \int_3^2 f(x)\, dx$$
$$= \int_{-1}^1 f(x)\, dx + \int_1^3 f(x)\, dx - \int_2^3 f(x)\, dx$$
$$= 3 + 5 - (-2)$$
$$= 10$$

Page 157

47.

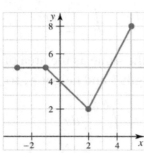

For continuity, suspicious points are $x = -1$ and $x = 2$. By considering the left- and right-hand limits at these points, it is easy to show f is continuous on $[-3, 5]$.

$$\int_{-3}^{5} f(x)\, dx$$

$$= \int_{-3}^{-1} 5\, dx + \int_{-1}^{2} (4 - x)\, dx + \int_{2}^{5} (2x - 2)\, dx$$

$$= 10 + \frac{21}{2} + 15 = \frac{71}{2}$$

49. $\displaystyle \int_{a}^{b} f(x)\, dx = \int_{a}^{c} f(x)\, dx + \int_{c}^{b} f(x)\, dx$

$$= \int_{a}^{c} f(x)\, dx + \left[\int_{c}^{d} f(x)\, dx + \int_{d}^{b} f(x)\, dx \right]$$

51. $f(x) = x^2$ and $\Delta x = \dfrac{b - a}{n}$; $x_k = a + \dfrac{b - a}{n} k$

$$S_n = \sum_{k=1}^{n} \left[a + \frac{b - a}{n} k \right]^2 \left(\frac{b - a}{n} \right)$$

$$= \sum_{k=1}^{n} \frac{b - a}{n} \left[a^2 + \frac{2(b - a)}{n} ak + \frac{(b - a)^2}{n^2} k^2 \right]$$

$$= \frac{b - a}{n} \left[a^2 \sum_{k=1}^{n} 1 + \frac{2(b - a)}{n} a \sum_{k=1}^{n} k \right.$$
$$\left. + \frac{(b - a)^2}{n^2} \sum_{k=1}^{n} k^2 \right]$$

$$= \frac{b - a}{n} \left[a^2 n + \frac{2(b - a)}{n} \frac{an(n + 1)}{2} \right.$$
$$\left. + \frac{(b - a)^2}{n^2} \frac{n(n + 1)(2n + 1)}{6} \right]$$

$$\int_{a}^{b} x^2\, dx = \lim_{n \to \infty} S_n$$

$$= \left[a^2 + ab - a^2 + \frac{(b - a)^2}{3} \right] (b - a)$$

$$= \left(\frac{b - a}{3} \right) (3a^2 + 3ab - 3a^2 + b^2 - 2ab + a^2)$$

$$= \frac{1}{3} (b - a)(b^2 + ab + a^2)$$

$$= \frac{1}{3} (b^3 - a^3)$$

53. $f(x) = 4 - 5x$

k	1	2	3	4	5
x_k^*	-0.5	0.8	1	1.3	1.8
$f(x_k^*)$	6.5	0	-1	-2.5	-5
Δx_k	0.8	1.1	0.4	0.4	0.3

$$R_5 = \sum_{k=1}^{5} f(x_k) \Delta x_k$$

$$= 6.5(0.8) + 0 + (-1)(0.4)$$
$$\quad + (-2.5)(0.4) + (-5)(0.3)$$

$$= 2.3$$

$$\|P\| = 1.1$$

55. Assume that $f(x) \leq g(x)$ on $[a, b]$.

$$\int_{a}^{b} f(x)\, dx = \lim_{n \to \infty} \sum_{k=1}^{n} f(x_k^*) \Delta x_k$$

$$\leq \lim_{n \to \infty} \sum_{k=1}^{n} g(x_k^*) \Delta x_k$$

$$= \int_{a}^{b} g(x)\, dx$$

57. It is false; for example, consider the function $y = x$ on $[-1, 1]$. The value of the integral

$$\int_{-1}^{1} x\, dx = A_1 - A_2$$

where A_1 is the area of the triangle to the

right of the origin and A_2 is the area of the triangle to the left of the origin. That is,

$$A_1 = \frac{1}{2}(1)(1) = \frac{1}{2} \quad \text{and} \quad A_2 = \frac{1}{2}(1)(1) = \frac{1}{2}$$

Thus,

$$\int_{-1}^{1} x\,dx = \frac{1}{2} - \frac{1}{2} = 0$$

However, the area enclosed by the two triangles is

$$\frac{1}{2} + \frac{1}{2} = 1$$

59. Since $m \le f(x) \le M$ on $[a, b]$, we have

$$\int_{a}^{b} m\,dx \le \int_{a}^{b} f(x)\,dx \le \int_{a}^{b} M\,dx$$

so

$$m(b-a) \le \int_{a}^{b} f(x)\,dx \le M(b-a)$$

5.4 The Fundamental Theorems of Calculus, page 362

1. $\displaystyle \int_{-10}^{10} 7\,dx = 7x\big|_{-10}^{10}$

$$= 7[10 - (-10)]$$
$$= 140$$

3. $\displaystyle \int_{-3}^{5} (2x + a)\,dx$

$$= \left(x^2 + ax\right)\big|_{-3}^{5}$$
$$= (5)^2 + 5a - \left[(-3)^2 + (-3)a\right]$$
$$= 16 + 8a$$

5. $\displaystyle \int_{-1}^{2} ax^3\,dx = \frac{1}{4}ax^4\Big|_{-1}^{2}$

$$= \frac{1}{4}a[16 - 1]$$
$$= \frac{15}{4}a$$

7. $\displaystyle \int_{1}^{2} \frac{c}{x^3}\,dx = \frac{cx^{-2}}{-2}\Big|_{1}^{2}$

$$= -\frac{1}{2}c\left[\frac{1}{4} - 1\right]$$
$$= \frac{3}{8}c$$

9. $\displaystyle \int_{0}^{9} \sqrt{x}\,dx = \frac{2}{3}x^{3/2}\Big|_{0}^{9}$

$$= \frac{2}{3}(27 - 0)$$
$$= 18$$

11. $\displaystyle \int_{0}^{1} (5u^7 + \pi^2)\,du = \left(\frac{5}{8}u^8 + \pi^2 u\right)\Big|_{0}^{1}$

$$= \frac{5}{8} + \pi^2$$

13. $\displaystyle \int_{1}^{2} (2x)^\pi\,dx = 2^\pi \frac{x^{\pi+1}}{\pi + 1}\Big|_{1}^{2}$

$$= \frac{2^\pi 2^{\pi+1}}{\pi + 1} - \frac{2^\pi \cdot 1}{\pi + 1}$$
$$= \frac{2^\pi(2^{\pi+1} - 1)}{\pi + 1}$$

15. $\displaystyle \int_{0}^{4} \sqrt{x}(x + 1)\,dx = \int_{0}^{4} \left(x^{3/2} + x^{1/2}\right)\,dx$

$$= \left(\frac{2}{5}x^{5/2} + \frac{2}{3}x^{3/2}\right)\Big|_{0}^{4}$$
$$= \frac{272}{15}$$

17. $\displaystyle\int_1^2 \frac{x^3+1}{x^2}\,dx = \int_1^2 \left(x+x^{-2}\right) dx$

$$= \left(\frac{1}{2}x^2 - x^{-1}\right)\Big|_1^2$$

$$= 2$$

19. $\displaystyle\int_1^{\sqrt{3}} \frac{6a}{1+x^2}\,dx = 6a\tan^{-1}x\Big|_1^{\sqrt{3}}$

$$= 6a\left(\frac{\pi}{3} - \frac{\pi}{4}\right)$$

$$= \frac{a\pi}{2}$$

21. $\displaystyle\int_0^1 \frac{x^2-4}{x-2}\,dx = \int_0^1 \frac{(x-2)(x+2)}{x-2}\,dx$

$$= \int_0^1 (x+2)\,dx$$

$$= \left(\frac{x^2}{2} + 2x\right)\Big|_0^1 = \frac{5}{2}$$

23. $\displaystyle\int_{-1}^2 (x+|x|)\,dx$

$$= \int_{-1}^0 (x-x)\,dx + \int_0^2 (x+x)\,dx$$

$$= \frac{2x^2}{2}\Big|_0^2$$

$$= 4$$

25. $\displaystyle\int_{-2}^3 (\sin^2 x + \cos^2 x)\,dx = \int_{-2}^3 dx$

$$= x\big|_{-2}^3$$

$$= 5$$

27. $f(x) = x^2 + 1$ is continuous on $[-1,1]$ and $f(x) \geq 0$ on the interval, so we have

$$\int_{-1}^1 (x^2+1)\,dx = \left(\frac{x^3}{3} + x\right)\Big|_{-1}^1 = \frac{8}{3}$$

29. $f(x) = \sec^2 x$ is continuous on $[0, \frac{\pi}{4}]$ and $f(x) \geq 0$ on the interval, so we have

$$\int_0^{\pi/4} \sec^2 x\,dx = \tan x\big|_0^{\pi/4}$$

$$= \tan\frac{\pi}{4} - \tan 0$$

$$= 1$$

31. $f(t) = e^t - t$ is continuous on $[0,1]$ and $f(t) \geq 0$ on the interval, so we have

$$\int_0^1 (e^t - t)\,dt = \left(e^t - \frac{1}{2}t^2\right)\Big|_0^1$$

$$= e - \frac{3}{2}$$

33. $f(x) = \dfrac{x^2 - 2x + 3}{x}$ is continuous on $[1,2]$ and $f(x) \geq 0$ on the interval, so we have

$$\int_1^2 (x - 2 + 3x^{-1})\,dx$$

$$= \left(\frac{x^2}{2} - 2x + 3\ln|x|\right)\Big|_1^2$$

$$= 3\ln 2 - \frac{1}{2}$$

35. $F'(x) = \dfrac{x^2 - 1}{\sqrt{x+1}}$

$$= (x-1)\sqrt{x-1}$$

SURVIVAL HINT: *Do you see how the solution to Problem 35 relates to the fundamental theorem of calculus? Problems 35-40 all use the fundamental theorem of calculus.*

37. $F'(t) = \dfrac{\sin t}{t}$ **39.** $F'(x) = \dfrac{-1}{\sqrt{1 + 3x^2}}$

41. $\dfrac{d}{du}\left(\dfrac{u}{2} + \dfrac{\sin 2au}{4a} + C\right)$

$= \dfrac{1}{2} + \dfrac{2a}{4a}\cos 2au$

$= \dfrac{1}{2} + \dfrac{1}{2}\cos 2au$

$= \dfrac{1}{2} + \dfrac{1}{2}(2\cos^2 au - 1))$

$= \cos^2 au$

43. $\dfrac{d}{du}\left(\dfrac{1}{\sqrt{a^2 - u^2}} + C\right)$

$= -\dfrac{1}{2}(a^2 - u^2)^{-3/2}(-2u)$

$= \dfrac{u}{(a^2 - u^2)^{3/2}}$

45. $\dfrac{d}{du}\left(-\sqrt{a^2 - u^2} + C\right)$

$= -\dfrac{1}{2}(a^2 - u^2)^{-1/2}(-2u)$

$= \dfrac{u}{\sqrt{a^2 - u^2}}$

47. If $f(x) \geq 0$ on $[a, b]$, the integral

$$\int_a^b f(x)\, dx$$

is the same as the area under $y = f(x)$ and above the x-axis on $[a, b]$. If $f(x) \leq 0$ on $[a, b]$, the integral is the opposite of the area above $y = f(x)$ and below the x-axis on $[a, b]$. If $f(x) < 0$ over part of the

interval of integration, the integral and the area are not the same. Thus, area is an integral, but an integral is not necessarily an area.

49. $\displaystyle\int_0^a x^{1/2}\, dx = 8\int_0^c x^{1/2}\, dx$

$\dfrac{2}{3}x^{3/2}\Big|_0^a = 8\left(\dfrac{2}{3}x^{3/2}\right)\Big|_0^c$

$\dfrac{2}{3}a^{3/2} = \dfrac{16}{3}c^{3/2}$

$\left(\dfrac{2}{3}a^{3/2}\right)^{2/3} = \left(\dfrac{16}{3}c^{3/2}\right)^{2/3}$

$a = 4c$

$c = \dfrac{a}{4}$

51. $\displaystyle\int_0^2 f(x)\, dx = \int_0^1 f(x)\, dx + \int_1^2 f(x)\, dx$

$= \displaystyle\int_0^1 x^3\, dx + \int_1^2 x^4\, dx$

$= \dfrac{1}{4}x^4\Big|_0^1 + \dfrac{1}{5}x^5\Big|_1^2 = 6.45$

53. a. It is a relative minimum because the derivative (function f) shows the $\downarrow\ \uparrow$ pattern. Also, $g(a) < 0$, g' is rising.

b. Obviously $f'' = g'''$ is concave down at $x = 0.5$ and concave up at $x = 1.25$, so there must be an inflection point for f somewhere between those points; this inflection points looks like $x = 1$. We estimate $g'''(1) = f''(1) = 0$.

c. The function f crosses the x-axis 3 times, at $x = a$, $x = b$, and $x = c$. Now $g(a)$ is a relative minimum, and at $x = c$, $g(c)$ is also a relative minimum (function f has the $\downarrow\ \uparrow$ pattern). Thus, g has a relative maximum at $x = b$ (f

has the ↑ ↓ pattern), and we estimate this to be at $x \approx 0.75$.

d.

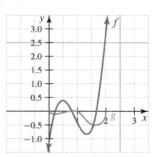

55. a. For our purpose with these counter examples, we let all the constants of integration be zero.

$$\int x\sqrt{x}\,dx = \int x^{3/2}\,dx = \frac{2}{5}x^{5/2}$$

$$\int x\,dx = \frac{1}{2}x^2; \quad \int \sqrt{x}\,dx = \frac{2}{3}x^{3/2}$$

Since $\frac{2}{5}x^{5/2} \neq \left(\frac{1}{2}x^2\right)\left(\frac{2}{3}x^{3/2}\right)$ the result follows.

b. $\int \frac{\sqrt{x}}{x}\,dx = \int x^{-1/2}\,dx = 2\sqrt{x}$

Since $2\sqrt{x} \neq \dfrac{\frac{2}{3}x^{3/2}}{\frac{1}{2}x^2}$, the result follows.

57. $\dfrac{d}{dx}\displaystyle\int_0^{x^4}(2t-3)\,dt$

$$= \frac{d}{du}\left[\int_0^u (2t-3)\,dt\right]\frac{du}{dx}$$

$$= (2u-3)(4x^3)$$

$$= (2x^4-3)(4x^3)$$

$$= 8x^7 - 12x^3$$

Check: $\displaystyle\int_0^{x^4}(2t-3)\,dt = \left(t^2-3t\right)\Big|_0^{x^4}$

$$= x^8 - 3x^4$$

Thus, $\dfrac{d}{dx}(x^8-3x^4) = 8x^7 - 12x^3$.

59. $\displaystyle\int_u^v f(t)\,dt = \int_u^a f(t)\,dt + \int_a^v f(t)\,dt$

$$= -\int_a^u f(t)\,dt + \int_a^v f(t)\,dt$$

Differentiate both sides with respect to x:

$$\frac{d}{dx}\left[\int_u^v f(t)\,dt\right]$$

$$= \frac{d}{dx}\left[-\int_a^u f(t)\,dt\right] + \frac{d}{dx}\left[\int_a^v f(t)\,dt\right]$$

$$= -f(u)\frac{du}{dx} + f(v)\frac{du}{dx}$$

Thus, $F'(x) = f(v)\dfrac{dv}{dx} - f(u)\dfrac{du}{dx}$

5.5 Integration by Substitution, page 369

1. a. $\displaystyle\int_0^4 (2t+4)\,dt = \left(t^2+4t\right)\Big|_0^4 = 32$

b. $\displaystyle\int_0^4 (2t+4)^{-1/2}\,dt$ | $u = 2t+4; \, du = 2\,dt$
If $t = 0$, $u = 4$;
if $t = 4$, $u = 12$

$$= \int_4^{12} u^{-1/2}\left(\frac{1}{2}du\right)$$

$$= u^{1/2}\Big|_4^{12}$$

$$= 2\sqrt{3} - 2$$

3. a. $\displaystyle\int_0^\pi \cos t\,dt = \sin t\Big|_0^\pi = 0$

b. $\displaystyle\int_0^{\sqrt{\pi}} t\cos t^2\,dt$ $\boxed{\begin{array}{l} u = t^2;\ du = 2t\,dt \\ \text{If } t = 0,\ u = 0; \\ \text{if } t = \sqrt{\pi},\ u = \pi \end{array}}$

$$= \frac{1}{2}\int_0^{\pi}\cos u\,du$$

$$= \frac{1}{2}\sin u\,\Big|_0^{\pi}$$

$$= 0$$

5. a. $\displaystyle\int_0^{16}\sqrt[4]{x}\,dx = \frac{4}{5}x^{5/4}\,\Big|_0^{16} = \frac{128}{5}$

b. $\displaystyle\int_{-16}^{0}\sqrt[4]{-x}\,dx$ $\boxed{\begin{array}{l} u = -x;\ du = -dx \\ \text{If } x = -16,\ u = 16; \\ \text{if } x = 0,\ u = 0 \end{array}}$

$$= -\int_{16}^{0} u^{1/4}\,du$$

$$= -\frac{4}{5}u^{5/4}\,\Big|_{16}^{0}$$

$$= \frac{128}{5}$$

7. a. $\displaystyle\int x^2\sqrt{2x^3}\,dx = \sqrt{2}\int x^{7/2}\,dx$

$$= \frac{2}{9}\sqrt{2}\,x^{9/2} + C$$

b. $\displaystyle\int x^2\sqrt{2x^3-5}\,dx$ $\boxed{\begin{array}{l} u = 2x^3 - 5; \\ du = 6x^2\,dx \end{array}}$

$$= \frac{1}{6}\int u^{1/2}\,du$$

$$= \frac{1}{9}\left(2x^3 - 5\right)^{3/2} + C$$

> **SURVIVAL HINT:** *Don't forget the "+ C" every time you evaluate an indefinite integral.*

9. $u = 2x + 3;\ du = 2\,dx$

$$\int (2x+3)^4\,dx = \frac{1}{2}\int u^4\,du$$

$$= \frac{1}{2}\cdot\frac{1}{5}u^5 + C$$

$$= \frac{1}{10}(2x+3)^5 + C$$

11. For the second part of the integral, let
$$u = 3x, \quad du = 3\,dx.$$

$$\int (x^2 - \cos 3x)\,dx = \frac{1}{3}x^3 - \frac{1}{3}\int\cos u\,du$$

$$= \frac{1}{3}x^3 - \frac{1}{3}\sin 3x + C$$

13. $u = 4 - x;\ du = -dx$

$$\int\sin(4-x)\,dx = -\int\sin u\,du$$

$$= \cos(4-x) + C$$

15. $u = t^{3/2} + 5;\ du = \frac{3}{2}t^{1/2}dt$

$$\int\sqrt{t}(t^{3/2}+5)^3\,dt = \frac{2}{3}\int u^3\,du$$

$$= \frac{1}{6}\left(t^{3/2}+5\right)^4 + C$$

17. $u = 3 + x^2;\ du = 2x\,dx$

$$\int x\sin(3+x^2)\,dx$$

$$= \frac{1}{2}\int\sin u\,du$$

$$= -\frac{1}{2}\cos\left(3+x^2\right) + C$$

19. $u = 2x^2 + 3;\ du = 4x\,dx$

$$\int\frac{x\,dx}{2x^2+3} = \frac{1}{4}\int\frac{du}{u}$$

$$= \frac{1}{4}\ln|u| + C$$

$$= \frac{1}{4}\ln\left(2x^2+3\right) + C$$

21. $u = 2x^2 + 1;\ du = 4x\,dx$

$$\int x\sqrt{2x^2+1}\,dx = \frac{1}{4}\int u^{1/2}\,du$$

$$= \frac{1}{4}\left(\frac{2}{3}\right)u^{3/2} + C$$

$$= \frac{1}{6}\left(2x^2+1\right)^{3/2} + C$$

23. $u = x\sqrt{x} = x^{3/2};\ du = \frac{3}{2}x^{1/2}dx$

$$\int \sqrt{x}e^{x\sqrt{x}}dx = \frac{2}{3}\int e^u\,du$$

$$= \frac{2}{3}e^{x^{3/2}} + C$$

25. $u = x^2 + 4;\ du = 2x\,dx$

$$\int x(x^2+4)^{1/2}\,dx = \frac{1}{2}\int u^{1/2}\,du$$

$$= \frac{1}{2}\cdot\frac{2}{3}u^{3/2} + C$$

$$= \frac{1}{3}\left(x^2+4\right)^{3/2} + C$$

27. $u = \ln x;\ du = \frac{1}{x}\,dx$

$$\int \frac{\ln x}{x}\,dx = \int u\,du$$

$$= \frac{1}{2}u^2 + C$$

$$= \frac{1}{2}(\ln x)^2 + C$$

29. $u = x^{1/2} + 7;\ du = \frac{1}{2}x^{-1/2}dx$

$$\int \frac{dx}{\sqrt{x}(\sqrt{x}+7)} = 2\int u^{-1}du$$

$$= 2\ln|u| + C$$

$$= 2\ln\left|\sqrt{x}+7\right| + C$$

31. $u = e^t + 1;\ du = e^t\,dt$

$$\int \frac{e^t dt}{e^t+1} = \int u^{-1}\,du$$

$$= \ln|u| + C$$

$$= \ln\left(e^t+1\right) + C$$

33. Let $u = 2x^3 + 1;\ du = 6x^2\,dx$

If $x = 0,\ u = 1;$ if $x = 1,\ u = 3.$

$$\int_0^1 \frac{5x^2\,dx}{2x^3+1} = \frac{5}{6}\int_0^1 \frac{6x^2\,dx}{2x^3+1}$$

$$= \frac{5}{6}\int_1^3 \frac{du}{u}$$

$$= \frac{5}{6}(\ln 3 - \ln 1)$$

$$= \frac{5}{6}\ln 3$$

35. $\displaystyle\int_{-\ln 2}^{\ln 2} \frac{1}{2}\left(e^x - e^{-x}\right)dx$

$$= \frac{1}{2}\int_{-\ln 2}^{\ln 2} e^x\,dx + \frac{1}{2}\int_{-\ln 2}^{\ln 2} e^{-x}(-dx)$$

$$= \frac{1}{2}\left(e^{\ln 2} - e^{-\ln 2} + e^{-\ln 2} - e^{\ln 2}\right)$$

$$= 0$$

37. Let $u = \frac{1}{x};\ du = -\frac{dx}{x^2}$

If $x = 1,\ u = 1;$ if $x = 2,\ u = \frac{1}{2}.$

$$\int_1^2 \frac{e^{1/x}\,dx}{x^2} = -\int_1^2 e^{1/x}\left(-\frac{dx}{x^2}\right)$$

$$= -\int_1^{1/2} e^u\,du$$

$$= -e^u\big|_1^{1/2}$$

$$= e - e^{1/2}$$

39. $u = \cos 2x,\ du = -2\sin 2x\,dx$

If $x = 0,\ u = 1;$ if $x = \frac{\pi}{6},\ u = \frac{1}{2}$

$$\int_0^{\frac{\pi}{6}} \tan 2x\,dx = -\frac{1}{2}\int_0^{1/2} \frac{du}{u}$$

$$= -\frac{1}{2}\ln|u|\Big|_1^{1/2}$$

$$= \frac{1}{2}\ln 2$$

41. Let $u = 1 + e^x$, $du = e^x\,dx$

If $x = 0$, $u = 2$; if $x = 1$, $u = 1 + e$

$$\int_0^1 \frac{e^x\,dx}{1 + e^x} = \int_2^{1+e} \frac{du}{u}$$

$$= \ln u\big|_2^{1+e}$$

$$= \ln(1 + e) - \ln 2$$

43. a. We take 1 Frdor as the variable so the note from the students reads, "Because of illness I cannot lecture between Easter and Michaelmas."

b. The Dirichlet function is defined as a function f so that $f(x)$ equals a determined constant c (usually 1) when the variable x takes a rational value, and equals another constant d (usually 0) when this variable is irrational. This famous function is one which is discontinuous everywhere.

45. $f(t) = \dfrac{1}{t^2}\sqrt{5 - \dfrac{1}{t}}$ is continuous and

positive on $\left[\frac{1}{5}, 1\right]$. If $u = 5 - t^{-1}$,

$du = t^{-2}\,dt$. If $t = \frac{1}{5}$, $u = 0$; if $t = 1$,

$u = 4$.

$$\int_{1/5}^1 \frac{1}{t^2}\sqrt{5 - \frac{1}{t}}\,dt = \int_0^4 u^{1/2}\,du$$

$$= \frac{2}{3}u^{3/2}\Big|_0^4$$

$$= \frac{16}{3}$$

47. $f(x) = \dfrac{x}{\sqrt{x^2 + 1}}$ is continuous and

positive on $[1, 3]$. Let $u = x^2 + 1$, $du = 2x\,dx$.

If $x = 1$, $u = 2$; if $x = 3$, $u = 10$.

$$\int_1^3 \frac{x}{\sqrt{x^2 + 1}}\,dx = \int_2^{10} \frac{1}{2}\frac{1}{\sqrt{u}}\,du$$

$$= \frac{1}{2}\int_2^{10} u^{-1/2}\,du$$

$$= \frac{1}{2}\cdot\frac{2}{1}u^{1/2}\Big|_2^{10}$$

$$= \sqrt{10} - \sqrt{2}$$

> **SURVIVAL HINT:** *Problem 48 restates the results of two previous problems. Problem 58 in Section 5.3 defines an odd functions and Problem 40 in Section 5.2 defined an even function. These terms are worth remembering.*

49. $\displaystyle\int_{-\pi}^{\pi} \sin x\,dx = 0$ since $\sin x$ is odd.

51. $\displaystyle\int_{-3}^{3} x\sqrt{x^4 + 1}\,dx = 0$ since $x\sqrt{x^4 + 1}$ is odd.

53. a. $f(x) = 7x^{1,001} + 14x^{99}$ is odd, so the integral is 0, and so it is true.

b. $\displaystyle\int_0^{\pi} \left(\sin^2 x - \cos^2 x\right)dx$

$$= -\int_0^{\pi} \cos 2x\,dx$$

$$= -\frac{\sin 2x}{2}\Big|_0^{\pi}$$

$$= 0$$

So the statement is true.

c. $\displaystyle\int_{-\pi/2}^{\pi/2} \cos x\,dx = \sin x\big|_{-\pi/2}^{\pi/2} = 2$

$\displaystyle\int_{-\pi}^{0} \sin x\,dx = -\cos x\big|_{-\pi}^0 = -2$

The given statement is false.

55. $\dfrac{dy}{dx} = \dfrac{2x}{1 - 3x^2}$

Let $u = 1 - 3x^2$, $du = -6x\,dx$

$$F(x) = \int \frac{2x}{1 - 3x^2}\,dx$$

$$= -\frac{1}{3}\int \frac{du}{u}$$

$$= -\frac{1}{3}\ln|u| + C$$

$$= -\frac{1}{3}\ln\left|1 - 3x^2\right| + C$$

$F(x) = -\frac{1}{3}\ln|1| + C$ implies $C = 5$.

$$F(x) = -\frac{1}{3}\ln\left|1 - 3x^2\right| + 5$$

57. Water flows into the tank at the rate of

$$R(t) = V'(t) = t\left(3t^2 + 1\right)^{-1/2} \text{ ft}^3/\text{s}$$

The volume at time t is

$$V(t) = \frac{1}{6}\int \left(3t^2 + 1\right)^{-1/2}(6t\,dt)$$

$$= \frac{1}{3}\sqrt{3t^2 + 1} + C$$

The tank is empty to start, so
$V(0) = 0 = \frac{1}{3} + C$, so $C = -\frac{1}{3}$. Thus,

$$V(t) = \frac{1}{3}\left(\sqrt{3t^2 + 1} - 1\right)$$

$$V(4) = \frac{1}{3}\left(\sqrt{49} - 1\right) = 2 \text{ ft}^3$$

The amount of water at 4 seconds is 2 ft^3.
The height h is given by the equation

$$100h = 2$$

$$h = \frac{1}{50} \text{ ft or } 0.24 \text{ in.}$$

The depth at that time is about $\frac{1}{4}$ in.

59. a. Let $u = 36 + 16t - t^2$,

$$du = (16 - 2t)\,dt$$

$$L(t) = \int \frac{0.24 - 0.03t}{\sqrt{36 + 16t - t^2}}\,dt$$

$$= \int \frac{0.015(16 - 2t)\,dt}{\sqrt{36 + 16t - t^2}}$$

$$= 0.015\int \frac{du}{\sqrt{u}}$$

$$= 0.015(2)u^{1/2} + C$$

$$= 0.03\sqrt{36 + 16t - t^2} + C$$

If $t = 0$, then $4 = 0.18 + C$ or
$C = 3.82$.

$$L(t) = 0.03\sqrt{36 + 16t - t^2} + 3.82$$

b.

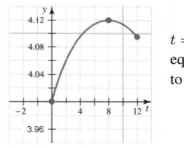

$t = 0$ is
equivalent
to 7:00 A.M.

The highest level occurs when $L'(t) = 0$
or when

$$16 - 2t = 0 \text{ (at 3 P.M.).}$$

$$t = 8$$

The highest level is 4.12 ppm.

c. ·

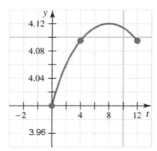

It is the same at 11 A.M. ($t = 4$) as at 7 P.M. ($t = 12$)

5.6 Introduction to Differential Equations, page 382

1. $x^2 + y^2 = 7$

$$2x + 2y\frac{dy}{dx} = 0$$

$$\frac{dy}{dx} = -\frac{x}{y}$$

3. $xy = C$

$$x\frac{dy}{dx} + y = 0$$

$$\frac{dy}{dx} = -\frac{y}{x}$$

5. $\dfrac{dy}{dx} = A\cos(Ax + B)$

$$\frac{d^2y}{dx^2} = -A^2\sin(Ax + B)$$

Thus,

$$\frac{d^2y}{dx^2} + A^2y$$

$$= -A^2\sin(Ax + B) + A^2\sin(Ax + B)$$

$$= 0$$

7. $y = 2e^{-x} + 3e^{2x}$

$$y' = -2e^{-x} + 6e^{2x}$$

$$y'' = 2e^{-x} + 12e^{2x}$$

$$y'' - y' - 2y$$
$$= \left(2e^{-x} + 12e^{2x}\right) - \left(-2e^{-x} + 6e^{2x}\right)$$
$$\qquad - 2\left(2e^{-x} + 3e^{2x}\right)$$
$$= 0$$

9. $\dfrac{dy}{dx} = -\dfrac{x}{y}$

$$y\,dy = -x\,dx$$

$$\int y\,dx = -\int x\,dx$$

$$\frac{y^2}{2} = -\frac{x^2}{2} + C_1$$

$$x^2 + y^2 = C$$

Passes through $(2, 2)$, so $4 + 4 = C$;

$$x^2 + y^2 = 8$$

11. $\dfrac{dy}{dx} = y^2 + 1$

$$\int \frac{dy}{y^2 + 1} = x + C$$

$$\tan^{-1}y = x + C$$

$$y = \tan(x + C)$$

> **SURVIVAL HINT:** *Note:*
> $$y = \tan(x + C) \neq \tan x + C,$$
> *so be careful here; in other words, we know that the constant must be supplied immediately after the last integration.*

Since the curve passes through $(\pi, 1)$,

$$\frac{\pi}{4} = \pi + C$$

$$-\frac{3\pi}{4} = C$$

Thus, $y = \tan\left(x - \frac{3\pi}{4}\right)$.

13. $\dfrac{dy}{dx} = \sqrt{\dfrac{x}{y}}$

$$y^{1/2}\,dy = x^{1/2}\,dx$$

$$\int_o y^{1/2}\,dy = \int x^{1/2}\,dx$$

$$\frac{2}{3}y^{3/2} = \frac{2}{3}x^{3/2} + C_1$$

$$x^{3/2} - y^{3/2} = C$$

Since the curve passes through $(4,1)$,

$$8 - 1 = C$$
$$C = 7$$

Thus, $x^{3/2} - y^{3/2} = 7$.

15. **17.**

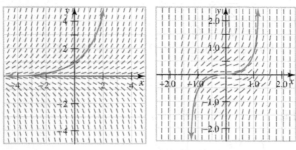

19.

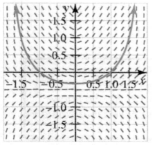

21. $\quad \dfrac{dy}{dx} = 3xy$

$$\int y^{-1}\,dy = \int 3x\,dx$$

$$\ln|y| = \frac{3}{2}x^2 + C$$

$$y = \exp\left(\frac{3}{2}x^2 + C\right)$$

$$= Be^{(3/2)x^2}$$

for a constant B.

23. $\quad \dfrac{dy}{dx} = \dfrac{x}{y}\sqrt{1-x^2}$

$$\int y\,dy = \int x\sqrt{1-x^2}\,dx$$

$$\int y\,dy = \int\left(-\frac{1}{2}\right)u^{1/2}\,du \quad \boxed{\begin{array}{l} u = 1 - x^2 \\ du = -2x\,dx \end{array}}$$

$$\frac{y^2}{2} = -\frac{1}{2}\frac{(1-x^2)^{3/2}}{\frac{3}{2}} + C_1$$

$$\frac{(1-x^2)^{3/2}}{3} + \frac{y^2}{2} = C_1$$

$$2(1-x^2)^{3/2} + 3y^2 = C$$

25. $\quad \dfrac{dy}{dx} = \dfrac{\sin x}{\cos y}$

$$\cos y\,dy = \sin x\,dx$$

$$\int \cos y\,dy = \int \sin x\,dx$$

$$\sin y = -\cos x + C$$

$$\cos x + \sin y = C$$

27. $\quad xy\dfrac{dy}{dx} = \dfrac{\ln x}{\sqrt{1-y^2}}$

$$\int y\sqrt{1-y^2}\,dy = \int \frac{\ln x}{x}\,dx$$

$$\boxed{\begin{array}{l} u = 1 - y^2; \ v = \ln x \\ du = -2y\,dy; \ dv = \frac{1}{x}\,dx \end{array}}$$

$$\int \sqrt{u}\left(-\frac{1}{2}\,du\right) = \int v\,dv$$

$$-\frac{1}{2}\frac{u^{3/2}}{\frac{3}{2}} = \frac{v^2}{2} + C$$

$$-\frac{1}{3}(1-y^2)^{3/2} = \frac{1}{2}(\ln x)^2 + C$$

29. $x\,dy + y\,dx = 0$

$$d(xy) = 0$$
$$xy = C$$

31.
$$y\,dx = x\,dy$$
$$\frac{x\,dy - y\,dx}{x^2} = 0$$
$$\frac{d}{dx}\left(\frac{y}{x}\right) = 0$$
$$\frac{y}{x} = C$$
$$y = Cx$$

33. Family of curves: $2x - 3y = C$; differentiating with respect to x leads to the slope of the tangent lines

$$\frac{dy}{dx} = \frac{2}{3}$$

For the orthogonal trajectories, the slope is the negative reciprocal, or

$$\frac{dY}{dX} = -\frac{3}{2}$$

Integrating leads to the orthogonal trajectories: $2Y + 3X = K$.

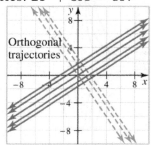

35. Family of curves $y = x^3 + C$; differentiating with respect to x leads to the slope of the tangent lines
$$\frac{dy}{dx} = 3x^2$$
For the orthogonal trajectories, the slope is the negative reciprocal, or
$$\frac{dY}{dX} = -\frac{X^{-2}}{3}$$

Integrating leads to $Y = \frac{1}{3}X^{-1} + K$.

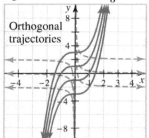

Note the differences in scale between the x- and y-axis. This makes the orthogonal curves "look like" they do not meet at right angles.

37. Family of curves $xy^2 = C$; differentiating with respect to x leads to the slope of the tangent lines

$$\frac{dy}{dx} = -\frac{y}{2x}$$

For the orthogonal trajectories, the slope is the negative reciprocal, or

$$\frac{dY}{dX} = \frac{2X}{Y}$$

Integrating leads to $Y^2 - 2X^2 = K$ or
$$X^2 - \frac{Y^2}{2} = K.$$

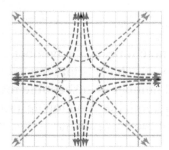

39. $x^2 + y^2 = r^2$
$$2x + 2yy' = 0$$
$$y' = -\frac{x}{y}$$

The slope of the orthogonal trajectory is the negative reciprocal:
$$\frac{dY}{dX} = \frac{Y}{X}$$
$$\int Y^{-1}\,dY = \int X^{-1}\,dX$$
$$\ln|Y| = \ln|X| + K$$
$$\left|\frac{Y}{X}\right| = e^K = C$$
$$Y = CX$$

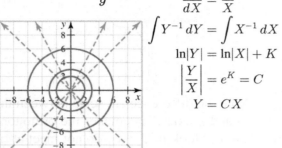

41. Let Q denote the number of bacteria. Then, dQ/dt is the rate of change of Q, and since this rate of change is proportional to Q, it follows that

$$\frac{dQ}{dt} = kQ$$

where k is a positive constant of proportionality.

43. Let T be temperature, t be time, T_m be the temperature of the surrounding medium, and c be the constant of proportionality. Then:

$$\frac{dT}{dt} = c(T - T_m)$$

45. Let t denote time, Q the number of residents who have been infected, and P the total number of susceptible residents. The differentiable equation describing the spread of the epidemic is

$$\frac{dQ}{dt} = kQ(P - Q)$$

where k is the positive constant of proportionality.

47. Family of curves: $x^2 - y^2 = C$ (dashed blue shows two positive values for C); differentiating with respect to x leads to the slope of the tangent lines

$$\frac{dy}{dx} = \frac{x}{y}$$

For the orthogonal trajectories, the slope is the negative reciprocal of

$$\frac{dY}{dX} = -\frac{Y}{X}$$

Integrating leads to $Y = C/X$ or $XY = C$ (for positive C), which is a family of hyperbolas (red curves)

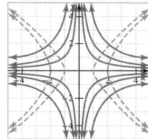

49. a.   **b.**

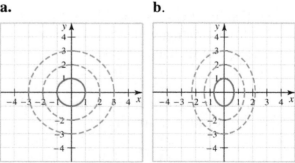

c.   **d.**

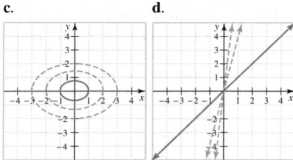

e.   **f.**

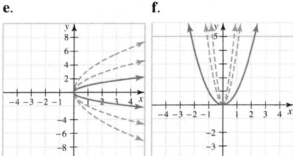

a and *d* are orthogonal trajectories;
b and *e* are orthogonal trajectories;
c and *f* are orthogonal trajectories.

51. $0.28Q_0 = Q_0 e^{kt}$

$kt = \ln 0.28$

$t = \dfrac{\ln 0.28}{k}$ where $k = \dfrac{\ln \frac{1}{2}}{5{,}730}$

≈ 10523

It is about 10,523 years.

53. Let V be the volume in ft^3 and h the height at time t. Then $V = 9\pi h$,

$$\frac{dV}{dt} = 9\pi \frac{dh}{dt}$$

The hole has area

$$A_0 = \pi \left(\frac{2.0}{12}\right)^2 = \frac{1}{36}\pi \text{ ft}^2$$

By Torricelli's law

$$\frac{dV}{dt} = -4.8 \left(\frac{1}{36}\pi\right)\sqrt{h}$$

Thus,

$$9\pi \frac{dh}{dt} = -4.8 \left(\frac{1}{36}\pi\right)\sqrt{h}$$

$$\frac{dh}{dt} = -\frac{2}{135}\sqrt{h}$$

$$\int h^{-1/2} dh = -\int \frac{2}{135} dt$$

$$2\sqrt{h} = -\frac{2}{135}t + C$$

When $t = 0$, $h = 5$, so

$$2\sqrt{5} = C$$

and

$$2\sqrt{h} = -\tfrac{2}{135}t + 2\sqrt{5}$$

When $h = 0$,

$$t = 135\sqrt{5} \approx 302 \text{ sec} \approx 5 \text{ min}$$

55. a. $v^2 = \dfrac{2gR^2}{s} + v_0^2 - 2gR$

$$= \frac{2(32)(3{,}956)^2(5{,}280)^2}{(3{,}956)(5{,}280) + 200}$$

$$+ 150^2 - 2(32)(3{,}956)(5{,}280)$$

$$\approx 9700.1226$$

$$v \approx 98.489200$$

The velocity is about 98.5 ft/s.

b. The velocity at the maximum height is 0:

$$0 = \frac{2gR^2}{s} + v_0^2 - 2gR$$

$$\frac{R}{s} = 1 - \frac{v_0^2}{2gR}$$

$$\approx 0.9999831689$$

$$s \approx 3{,}956.067$$

We now find

$$h \approx 0.067 \text{ mi} \approx 352 \text{ ft}$$

57. Differentiating $V = 9\pi h^3$ leads to

$$\frac{dV}{dt} = 9\pi\left(3h^2\frac{dh}{dt}\right) = -4.8A_0\sqrt{h}$$

from Torricelli's law. Since the area of the hole (in ft^2) is $A_0 = \pi\left(\dfrac{1}{12}\right)^2 = \dfrac{\pi}{144}$.

$$27\pi h^2\frac{dh}{dt} = (-4.8)\left(\frac{\pi}{144}\right)\sqrt{h}$$

$$\int h^{3/2}dh = -\int \frac{4.8\pi}{144(27\pi)}dt$$

$$\frac{2}{5}h^{5/2} = -\frac{1}{810}t + C$$

If $t = 0$, then $h = 4$, so that $C = \frac{64}{5}$. The height is zero when

$$0 = -\frac{1}{810}t + \frac{64}{5}$$
$$t = 10{,}368$$

10,368 sec is about 173 min or 2 hr and 53 min.

59. On the surface of the planet, $s = R$ and $a = -g$. Since $F = ma$

$$F = ma = \frac{mk}{R^2}$$

so $k = -gR^2$. In general, at a distance of s,

$$F = ma = \frac{mk}{s^2}$$

$$a = \frac{k}{s^2} = -\frac{gR^2}{s^2}$$

5.7 The Mean Value Theorem for Integrals; Average Value, page 389

1. $f(x)$ is continuous on $[1, 2]$, so the MVT guarantees the existence of c such that:

$$\int_1^2 4x^3\, dx = f(c)(2-1)$$
$$15 = f(c)$$
$$15 = 4c^3$$
$$c^3 = \frac{15}{4}$$
$$c = \frac{\sqrt[3]{30}}{2} \approx 1.55$$

1.55 is in the interval $[1, 2]$.

3. $f(x)$ is continuous on $[0, 2]$, so the MVT guarantees the existence of c such that:

$$\int_0^2 \left(x^2 + 4x + 1\right) dx = f(c)(2-0)$$
$$\left(\frac{1}{3}x^3 + 2x^2 + x\right)\Big|_0^2 = (c^2 + 4c - 1)(2)$$
$$\frac{38}{6} = c^2 + 4c + 1$$
$$3c^2 + 12c - 16 = 0$$
$$c \approx 1.055, -5.055$$

1.055 is in the interval $[0, 2]$.

5. $f(x)$ is continuous on $[1, 5]$, so the MVT guarantees the existence of c such that:

$$\int_1^5 15x^{-2}\, dx = f(c)(5-1)$$
$$-15x^{-1}\Big|_1^5 = 15c^{-2}(4)$$
$$12 = 60c^{-2}$$

$$c^2 = 5$$
$$c = \pm\sqrt{5}$$

$\sqrt{5} \approx 2.24$ is in the interval $[1, 5]$.

7. The mean value theorem does not apply because $f(x) = \csc x$ is discontinuous at $x = 0$ or $\left[-\frac{\pi}{3}, \frac{\pi}{3}\right]$.

9. $f(x)$ is continuous on $\left[-\frac{1}{2}, \frac{1}{2}\right]$, so the MVT guarantees the existence of c such that:

$$\int_{-1/2}^{1/2} e^{2x}\, dx = f(c)\left(\frac{1}{2} + \frac{1}{2}\right)$$

$$\frac{1}{2} e^{2x}\Big|_{-1/2}^{1/2} = e^{2c}$$

$$\frac{1}{2}\left(e - e^{-1}\right) = e^{2c}$$

$$c \approx 0.0807$$

0.0807 is in the interval $\left[-\frac{1}{2}, \frac{1}{2}\right]$.

11. $f(x)$ is continuous on $[0, 1]$, so the MVT guarantees the existence of c such that:

$$\int_0^1 \frac{x}{1+x}\, dx = f(c)(1-0)$$

$$\int_0^1\left(1 - \frac{1}{1+x}\right)dx = \frac{c}{1+c}$$

$$x - \ln(x+1)\big|_0^1 = \frac{c}{1+c}$$

$$1 - \ln 2 = \frac{c}{1+c}$$

$$c = \frac{1 - \ln 2}{\ln 2} \approx 0.4427$$

The value is in the interval $[0, 1]$.

13. The mean value theorem does not apply because $f(x) = \cot x$ is not continuous at $x = 0$ which is in the interval $[-2, 2]$.

15. $A = \displaystyle\int_0^{10} \frac{x}{2}\, dx$

$$= \frac{x^2}{4}\bigg|_0^{10}$$

$$= 25$$

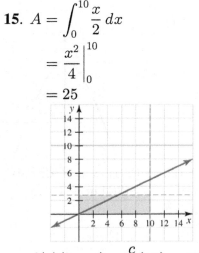

$$f(c)(b-a) = \frac{c}{2}(10) = 25$$

so $c = 5$ and $f(5) = 2.5$.

17. $A = \displaystyle\int_0^2 \left(x^2 + 2x + 3\right) dx$

$$= \left(\frac{1}{3}x^3 + x^2 + 3x\right)\bigg|_0^2$$

$$= \frac{38}{3}$$

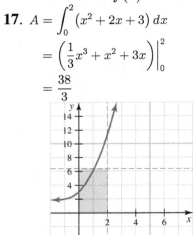

$$f(c)(b-a) = (c^2 + 2c + 3)(2) = \frac{38}{3}$$

so $3c^2 + 6c - 10 = 0$

$$c = \frac{-6 \pm \sqrt{36 - 4(3)(-10)}}{2(6)}$$

$$\approx 1.08,\ -3.08\ (\text{reject})$$

$$f(1.08) \approx 6.33.$$

ant рахунок

19. $A = \displaystyle\int_{-1}^{1.5} \cos x \, dx$

$= \sin x \big|_{-1}^{1.5}$

$= \sin 1.5 - \sin(-1)$

≈ 1.83897

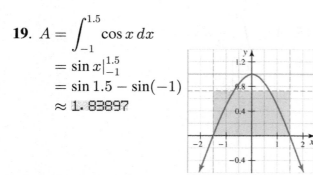

$f(c)(b-a) = (\cos c)(2.5) \approx 1.83897$ so

$c \approx 0.744264$ or -0.744264 in $[-1, 1.5]$.

$f(c) \approx 0.735586.$

21. $\dfrac{1}{2-(-1)} \displaystyle\int_{-1}^{2} (x^2 - x + 1)\, dx$

$= \dfrac{1}{3}\left(\dfrac{1}{3}x^3 - \dfrac{1}{2}x^2 + x\right)\Big|_{-1}^{2}$

$= \dfrac{3}{2}$

23. $\dfrac{1}{1-(-2)} \displaystyle\int_{-2}^{1} (x^3 - 3x^2)\, dx$

$= \dfrac{1}{3}\left(\dfrac{1}{4}x^4 - x^3\right)\Big|_{-2}^{1}$

$= -\dfrac{17}{4}$

25. $\dfrac{1}{1-(-1)} \displaystyle\int_{-1}^{1} (e^x - e^{-x})\, dx$

$= \dfrac{1}{2}(e^x + e^{-x})\Big|_{-1}^{1}$

$= 0$

27. $\dfrac{1}{1-0} \displaystyle\int_{0}^{1} \dfrac{x}{2x+3}\, dx$ *Use long division.*

$= \displaystyle\int_{0}^{1}\left[\dfrac{1}{2} - \dfrac{3}{2}\cdot\dfrac{1}{2x+3}\right] dx$

$= \left(\dfrac{1}{2}x - \dfrac{3}{4}\ln|2x+3|\right)\Big|_{0}^{1}$

$= \dfrac{1}{2} - \dfrac{3}{4}\ln\dfrac{5}{3}$

29. $\dfrac{1}{\frac{\pi}{4}-0} \displaystyle\int_{0}^{\pi/4} \sin x \, dx$

$= -\dfrac{4}{\pi}\cos x \Big|_{0}^{\pi/4}$

$= -\dfrac{4}{\pi}\left(\cos\dfrac{\pi}{4} - 1\right)$

$= -\dfrac{1}{\pi}\left(2\sqrt{2} - 4\right) = \dfrac{1}{\pi}\left(4 - 2\sqrt{2}\right)$

31. $\dfrac{1}{4-0} \displaystyle\int_{0}^{4} \sqrt{4-x}\, dx$

$= \dfrac{1}{4}\left(-\dfrac{2}{3}\right)(4-x)^{3/2}\Big|_{0}^{4}$

$= \dfrac{4}{3}$

33. $\dfrac{1}{0-(-7)} \displaystyle\int_{-7}^{0} (1-x)^{1/3}\, dx$

$= -\dfrac{1}{7}\cdot\dfrac{3}{4}(1-x)^{4/3}\Big|_{-7}^{0}$

$= \dfrac{45}{28}$

35. $\dfrac{1}{1-0} \displaystyle\int_{0}^{1} (2x-3)^3\, dx$

$= \dfrac{1}{2}\dfrac{(2x-3)^4}{4}\Big|_{0}^{1}$

$= -10$

37. $\dfrac{1}{1-0} \displaystyle\int_{0}^{1} x(2x^2 + 7)^{1/2}\, dx$

$$= \frac{1}{4} \frac{2(2x^2+7)^{3/2}}{3} \Big|_0^1$$

$$= \frac{1}{6}\left(27 - 7^{3/2}\right)$$

39. $\dfrac{1}{3-0} \displaystyle\int_0^3 \dfrac{x}{\sqrt{x^2+1}}\, dx$

$$= \frac{1}{6} \int_0^3 \left(x^2+1\right)^{-1/2}(2x\,dx)$$

$$= \frac{2}{6}\left(x^2+1\right)^{1/2}\Big|_0^3$$

$$= \frac{1}{3}\left(\sqrt{10}-1\right)$$

41. $\dfrac{1}{3-(-3)} \displaystyle\int_{-3}^3 \sqrt{9-x^2}\, dx$

$$= \frac{1}{6}(\text{AREA OF SEMICIRCLE WITH } r = 3)$$

$$= \frac{1}{6}\left(\frac{9\pi}{2}\right)$$

$$= \frac{3\pi}{4}$$

43. $\dfrac{1}{50-10} \displaystyle\int_{10}^{50} 10\left(e^{0.03x}-1\right) dx$

$$= \frac{1}{4} \int_{10}^{50} \left(e^{0.03x}-1\right) dx$$

$$= \frac{1}{4}\left(\frac{e^{0.03x}}{\frac{3}{100}} - x\right)\Big|_{10}^{50}$$

$$= \frac{25}{3}e^{1.5} - \frac{25}{3}e^{0.3} - 10$$

$$= 16.0985855230$$

The average price is \$16.10.

45. $\dfrac{1}{100-0} \displaystyle\int_0^{100} 50e^{0.03x}\, dx$

$$= \frac{1}{2} \int_0^{100} e^{0.03x}\, dx$$

$$= \frac{1}{2} \frac{e^{0.03}}{\frac{3}{100}} \Big|_0^{100}$$

$$= \frac{50}{3}\left(e^3 - 1\right)$$

$$\approx 318.092282053$$

The average value is approximately \$318.

47. $\dfrac{1}{24-0} \displaystyle\int_0^{24} (6.44t - 0.23t^2 + 30)\,dt$

$$= \frac{1}{24}\left(3.22t^2 - \frac{23}{300}t^3 + 30t\right)\Big|_0^{24}$$

$$= 63.12$$

The average daily temperature is about 63°F.

49. $\dfrac{ds}{dt} = v = -gt + v_0$

We calculate the average velocity:

$$\frac{1}{t_1-t_0} \int_{t_0}^{t_1} (-gt + v_0)\,dt$$

$$= \frac{1}{t_1-t_0}\left(-\frac{1}{2}gt^2 + v_0 t\right)\Big|_{t_0}^{t_1}$$

$$= -\frac{g}{2}(t_1-t_0) + v_0$$

51. We calculate the average temperature:

$$\frac{1}{15-12} \int_{12}^{15} (-0.1t^2 + t + 50)\,dt$$

$$= \frac{1}{3}\left[-\frac{1}{30}t^3 + \frac{1}{2}t^2 + 50t\right]\Big|_{12}^{15}$$

$$= \frac{226}{5}$$

The average temperature is $45.2°F$.

53. Using the result from Problem 52, we use technology to solve

$te^{-0.01t^2} \approx 1.434480245$

$t \approx 1.46563, 15.31941$

This is measured in years. Converting to years and months, we find that in the first three years the carbon dioxide level is reached in 18 months.

55. Using the result from Problem 54, we solve

$$\frac{2,000}{1 + 0.3e^{-0.276t}} \approx 1987.24$$

$$t \approx 13.93$$

The average population is reached at just under 14 minutes.

57.   The area under the curve in Problem 56 is

$$\int_0^x f(t)\,dt = A(x) = \tan x \text{ on } \left[0, \frac{\pi}{2}\right)$$

so

$$\frac{d}{dx}\int_0^x f(t)\,dt = (\tan x)'$$

Thus, $f(x) = \sec^2 x$.

59. $A = 0$, $B = k$, $C = 0$, $g(\lambda) = kte^{-\lambda t}$

5.8 Numerical Integration; The Trapezoidal Rule and Simpson's Rule, page 398

1. Approximate the area under a curve (evaluate an integral) by taking the sum of areas of trapezoids whose upper line segments join two consecutive points on an arc of the curve.

3. $\Delta x = \dfrac{b-a}{n} \qquad f(x) = \sqrt{x+3}$

    $\quad = \dfrac{6-1}{1} \qquad f(1) = 2$

    $\quad = 5 \qquad\qquad f(6) = 3$

    a. $A_1 = \Delta x[f(x_0)]$

    $\quad = 5(2)$

    $\quad = 10$

    b. $T_1 = \dfrac{\Delta x}{2}[f(x_0) + f(x_1)]$

    $\quad = \dfrac{5}{2}[2+3]$

    $\quad = \dfrac{25}{2}$

    c. $\Delta x = \dfrac{b-a}{n} \qquad f\left(\dfrac{7}{2}\right) = \sqrt{\dfrac{7}{2}+3}$

    $\quad = \dfrac{6-1}{2} \qquad f(2) = 3$

    $\quad = \dfrac{5}{2}$

    $S_2 = \dfrac{\Delta x}{3}[f(x_0) + 4f(x_1) + f(x_2)]$

    $\quad = \dfrac{5}{6}\left[2 + 4\cdot\sqrt{\dfrac{13}{2}} + 3\right]$

    $\quad \approx 12.6650325227$

5. $\Delta x = \dfrac{b-a}{n} \qquad f(x) = \sqrt{x+3}$

    $\quad = \dfrac{6-1}{3} \qquad f(1) = 2$

    $\quad = \dfrac{5}{3} \qquad\quad f\left(\dfrac{8}{3}\right) = \sqrt{\dfrac{8}{3}+3}$

    $\qquad\qquad\qquad f\left(\dfrac{13}{3}\right) = \sqrt{\dfrac{13}{3}+3}$

    $\qquad\qquad\qquad f(2) = 3$

    a. $A_3 = \Delta x[f(x_0) + f(x_1) + f(x_2)]$

    $\quad = \dfrac{5}{3}\left[2 + \sqrt{\dfrac{17}{3}} + \sqrt{\dfrac{22}{3}}\right]$

    $\quad \approx 11.8141482406$

b. $T_3 = \dfrac{\Delta x}{2}[f(x_0) + 2f(x_1)$
$\qquad + 2f(x_2) + f(x_3)]$

$\qquad = \dfrac{5}{6}\left[2 + 2\sqrt{\dfrac{17}{3}} + 2\sqrt{\dfrac{22}{3}} + 3\right]$

$\qquad \approx 12.647481574$

c. $\Delta x = \dfrac{b-a}{n}$ $\qquad f\left(\dfrac{11}{6}\right) = \sqrt{\dfrac{11}{6}} + 3$

$\qquad = \dfrac{6-1}{6}$ $\qquad f\left(\dfrac{16}{6}\right) = \sqrt{\dfrac{16}{6}} + 3$

$\qquad = \dfrac{5}{6}$ $\qquad f\left(\dfrac{21}{6}\right) = \sqrt{\dfrac{21}{6}} + 3$

$\qquad\qquad\qquad f\left(\dfrac{26}{6}\right) = \sqrt{\dfrac{26}{6}} + 3$

$\qquad\qquad\qquad f\left(\dfrac{31}{6}\right) = \sqrt{\dfrac{31}{6}} + 3$

$\qquad\qquad\qquad f(6) = 3$

$S_6 = \dfrac{\Delta x}{3}[f(x_0) + 4f(x_1) + 2f(x_2)$
$\qquad + 4f(x_3) + 2f(x_4) + 4f(x_5) + f(x_6)]$

$\qquad = \dfrac{5}{18}\left[2 + 4 \cdot \sqrt{\dfrac{29}{6}} + 2 \cdot \sqrt{\dfrac{34}{6}} + 4 \cdot \sqrt{\dfrac{39}{6}}\right.$

$\qquad\qquad \left. + 2 \cdot \sqrt{\dfrac{44}{6}} + 4 \cdot \sqrt{\dfrac{49}{6}} + 3\right]$

$\qquad \approx 12.6666406540$

7. $\Delta x = \dfrac{b-a}{n}$

$\qquad = \dfrac{4-2}{1}$

$\qquad = 2$

a. $A_1 = \Delta x[f(x_0)]$

$\qquad = 2f(2)$

$\qquad = 2\left[\dfrac{2}{(1+4)^2}\right]$

$\qquad = 0.16$

b. $T_1 = \dfrac{\Delta x}{2}[f(x_0) + f(x_1)]$

$\qquad = 1[f(2) + f(4)]$

$\qquad = 1\left[\dfrac{2}{(1+4)^2} + \dfrac{4}{(1+2\cdot4)^2}\right]$

$\qquad \approx 0.129382716049$

c. $\Delta x = \dfrac{b-a}{n}$

$\qquad = \dfrac{4-2}{2}$

$\qquad = 1$

$S_2 = \dfrac{\Delta x}{3}[f(x_0) + 4f(x_1) + f(x_2)]$

$\qquad = \dfrac{1}{3}[f(2) + 4f(3) + f(4)]$

$\qquad = \dfrac{1}{3}\left[\dfrac{2}{(1+4)^2} + \dfrac{4\cdot3}{(1+2\cdot3)^2} + \dfrac{4}{(1+8)^2}\right]$

$\qquad \approx 0.124760225078$

9. $\Delta x = \dfrac{b-a}{n}$

$\qquad = \dfrac{4-2}{3}$

$\qquad = \dfrac{2}{3}$

a. $A_3 = \Delta x[f(x_0) + f(x_1) + f(x_2)]$

$\qquad = \dfrac{2}{3}\left[f(2) + f\left(\dfrac{8}{3}\right) + f\left(\dfrac{10}{3}\right)\right]$

$\qquad = \dfrac{2}{3}\left[\dfrac{2}{(1+4)^2} + \dfrac{\frac{8}{3}}{\left(1+\frac{8}{3}\right)^2} + \dfrac{\frac{10}{3}}{\left(1+\frac{10}{3}\right)^2}\right]$

$\qquad \approx 0.135461846338$

b. $T_3 = \dfrac{\Delta x}{2}[f(x_0) + 2f(x_1) + 2f(x_2) + f(x_3)]$

$\qquad = \dfrac{1}{3}\left[f(2) + 2f\left(\dfrac{8}{3}\right) + 2f\left(\dfrac{10}{3}\right) + f(4)\right]$

$\qquad = \dfrac{1}{3}\left[\dfrac{2}{(1+4)^2} + \dfrac{2\cdot\frac{8}{3}}{(1+\frac{16}{3})^2} + \dfrac{2\cdot\frac{10}{3}}{(1+\frac{20}{3})^2}\right.$

$\qquad\qquad \left. + \dfrac{4}{(1+8)^2}\right]$

$\qquad \approx 0.125256085021$

c.　$\Delta x = \dfrac{b-a}{n}$

$= \dfrac{4-2}{6}$

$= \dfrac{1}{3}$

$S_6 = \dfrac{\Delta x}{3}[f(x_0) + 4f(x_1) + 2f(x_2) + 4f(x_3)$

$+ 2f(x_4) + 4f(x_5) + f(x_6)]$

$= \dfrac{1}{9}\left[f(2) + 4f\left(\dfrac{7}{3}\right) + 2f\left(\dfrac{8}{3}\right) + 4f(3)\right.$

$\left. + 2f\left(\dfrac{10}{3}\right) + 4f\left(\dfrac{11}{3}\right) + f(4)\right]$

$= \dfrac{1}{9}\left[\dfrac{2}{(1+4)^2} + \dfrac{4\cdot\frac{7}{3}}{(1+2\cdot\frac{7}{3})^2} + \dfrac{2\cdot\frac{8}{3}}{(1+2\cdot\frac{8}{3})^2}\right.$

$\left. + \dfrac{4\cdot 3}{(1+2\cdot 3)^2} + \dfrac{4\cdot\frac{11}{3}}{(1+2\cdot\frac{11}{3})^2} + \dfrac{4}{(1+2\cdot 4)^2}\right]$

≈ 0.1247247850410

11.　$\Delta x = \dfrac{2-1}{2} = \dfrac{1}{2}$

Trapezoidal rule: 2.375
Simpson's rule: 2.333
Check with exact value:

$$\int_1^2 x^2\, dx = \left.\dfrac{x^3}{3}\right|_1^2 = \dfrac{7}{3}$$

13.　$\Delta x = \dfrac{4-0}{4} = 1$

Trapezoidal rule: 5.146
Simpson's rule: 5.252
Check with exact value:

$$\int_0^4 \sqrt{x}\, dx = \left.\dfrac{2}{3}x^{3/2}\right|_0^4 = \dfrac{16}{3}$$

15.　$\Delta x = \dfrac{4-0}{6} = \dfrac{2}{3}$

　a.　Trapezoidal rule: 0.783

　b.　Simpson's rule: 0.785

17.　$\Delta x = \dfrac{4-2}{4} = \dfrac{1}{2}$

　a.　Trapezoidal rule: 2.038

　b.　Simpson's rule: 2.049

19.　$\Delta x = \dfrac{2-0}{6} = \dfrac{1}{3}$

　a.　Trapezoidal rule: 0.584

　b.　Simpson's rule: 0.594

21.　For the trapezoidal rule,

$$|E_n| \le \dfrac{(b-a)^3}{12n^2}M$$

where M is the maximum value of $|f''(x)|$ on $[a, b]$.

$$f(x) = \dfrac{1}{x^2+1}$$

$$f'(x) = \dfrac{-2x}{(x^2+1)^2}$$

$$f''(x) = \dfrac{6x^2-2}{(x^2+1)^3}$$

The maximum of $|f''(x)|$ on $[0, 1]$ is 2 (at $x = 0$), so we need

$$\dfrac{1}{12n^2}(2) \le 0.05$$

$$\dfrac{1}{n^2} \le 0.30$$

$$n^2 \ge \dfrac{10}{3}$$

$$n \ge 2$$

$$A = \dfrac{1}{2}[f(0) + 2f(0.5) + f(1)]\left(\dfrac{1}{2}\right)$$

$$= \dfrac{1}{4}[1 + 2(0.8) + 0.5]$$

$$= 0.775$$

The exact answer is between $0.775 - 0.05$ and $0.775 + 0.05$.

23. $f(x) = \cos 2x$

$f'(x) = -2\sin 2x$

$f''(x) = -4\cos 2x$

$f'''(x) = 8\sin 2x$

$f^{(4)}(x) = 16\cos 2x$

The maximum of $\left|f^{(4)}(x)\right|$ on $[0, 1]$ is 16, so we need

$$\frac{1^5}{180n^4}(16) < 0.0005$$

$$n^4 > 178$$

$$n > 3.65$$

We pick $n = 4$ (n must be even)

$$A = \frac{1}{3}[f(0) + 4f(0.25) + 2f(0.5)$$

$$+ 4f(0.75) + f(1)]\left(\frac{1}{4}\right)$$

$$= \frac{1}{12}\Big[1 + 4(0.878) + 2(0.54)$$

$$+ 4(0.071) - 0.416\Big]$$

$$\approx 0.455$$

The exact answer is between

$0.455 - 0.0005$ and $0.455 + 0.0005$.

25. $f(x) = x(4 - x)^{1/2}$

$$f'(x) = \frac{8 - 3x}{2(4 - x)^{1/2}}$$

$$f''(x) = \frac{3x - 16}{4(4 - x)^{3/2}}$$

The maximum of $\left|f''(x)\right|$ on $[0, 2]$ occurs at the right endpoint since f is an increasing function.

$\left|f''(2)\right| \approx 0.88388$.

$$\frac{2^3}{12n^2}(0.88388) \leq 0.01$$

$$n^2 \geq 58.92533$$

$$n \geq 7.68$$

We pick $n = 8$ terms. The trapezoidal approximation gives $A \approx 3.25$. The exact answer is between $3.25 - 0.01$ and $3.25 + 0.01$.

27. $f(x) = \tan^{-1}x$

$$f'(x) = \frac{1}{x^2 + 1}$$

$$f''(x) = \frac{-2x}{\left(x^2 + 1\right)^2}$$

$$f'''(x) = \frac{2(3x^2 - 1)}{\left(x^2 + 1\right)^3}$$

$$f^{(4)}(x) = \frac{-24x(x^2 - 1)}{\left(x^2 + 1\right)^4}$$

The maximum of $\left|f^{(4)}(x)\right|$ on $[0, 1]$ is approximately 4.669 so we need

$$\frac{(1 - 0)^5}{180n^4}(4.669) < 0.01$$

$$n^4 > 2.5939$$

$$n > 1.27$$

We pick $n = 2$ (n must be even). Using Simpson's rule to find $A \approx 0.440$, the exact answer is between $0.440 - 0.01$ and $0.440 + 0.01$.

29. $f(x) = x^{-1}$

$f'(x) = -x^{-2}$

$f''(x) = 2x^{-3}$

$f'''(x) = -6x^{-4}$

$f^{(4)}(x) = 24x^{-5}$

a. $\dfrac{2^3(2)}{12n^2} \leq 0.00005$

$$n \approx 163.3$$

Pick $n = 164$.

b. $\dfrac{2^5(24)}{180n^4} \le 0.00005$

$n \approx 17.09$

Pick $n = 18$.

31. $f(x) = x^{-1/2}$

$f'(x) = -\dfrac{1}{2}x^{-3/2}$

$f''(x) = \dfrac{3}{4}x^{-5/2}$

$f'''(x) = -\dfrac{15}{8}x^{-7/2}$

$f^{(4)}(x) = \dfrac{105}{16}x^{-9/2}$

a. $\dfrac{3^3}{12n^2}\left(\dfrac{3}{4}\right) \le 0.00005$

$n \approx 183.71$

Pick $n = 184$.

b. $\dfrac{3^5}{180n^4}\left(\dfrac{105}{16}\right) \le 0.00005$

$n \approx 20.52$

Pick $n = 22$.

33. $f(x) = e^{-2x}$

$f'(x) = -2e^{-2x}$

$f''(x) = 4e^{-2x}$

$f'''(x) = -8e^{-2x}$

$f^{(4)}(x) = 16e^{-2x}$

a. $\dfrac{1^3(4)}{12n^2} \le 0.00005$

$n^2 \approx 6,666.67$

$n \approx 81.65$

Pick $n = 82$.

b. $\dfrac{1^5(16)}{180n^4} \le 0.00005$

$n^4 \approx 1,777.78$

$n \approx 6.49$

Pick $n = 8$.

35. $f(x) = (1 - x^2)^{1/2}$; the second derivative is unbounded on [0, 1], so the number of intervals needed to guarantee a certain accuracy cannot be predicted. If $n = 8$,

$$T_8 \approx (0.5)(12.347)(0.125) \approx 0.772$$

Thus, $\pi \approx 4(0.772) \approx 3.09$; to one decimal place this is 3.1.

37. $f(x) = x^{-1}$

$f'(x) = -x^{-2}$

$f''(x) = 2x^{-3}$

This is a decreasing function, so the maximum occurs at the left endpoint: $f''(1) = 2$.

$$\dfrac{1(2)}{12n^2} \le 0.0000005$$

$n > 577.35$; pick $n = 578$.

39. $\Delta x = 5$, given.

$\text{Area} \approx \dfrac{5}{2}\Big[0 + 2(10) + 2(11) + 2(13) + 2(16)$
$\quad + 2(18) + 2(23) + 2(25) + 2(27)$
$\quad + 2(25) + 10\Big]$
$= 865$

Since it is 3 ft deep, $865(3) = 2,595$ ft^3

$= \dfrac{2,595}{27}$ yd^3

≈ 96.1 yd^3

About 100 yd^3 of fill is needed.

41. The distance traveled is

$$s = \int_0^{60} v(t)\, dt$$

since $v(t) \geq 0$ for $0 \leq t \leq 60$. By the trapezoidal rule (using $\Delta t = 1/12$ hr):

$$s \approx \frac{1}{2}\Big[54 + 2(57) + 2(50) + \cdots$$
$$+ 2(42) + 2(48) + 53\Big]\left(\frac{1}{12}\right)$$
$$\approx 50.38 \text{ miles}$$

43. $\displaystyle\int_0^5 f(x)\,dx \approx \frac{1}{3}\Big[10 + 4(9.75) + 2(10) + 4(10.75)$
$$+ 2(12) + 4(13.75) + 2(16)$$
$$+ 4(18.75) + 2(22) + 4(25.75) + 30\Big](0.5)$$
$$= 79.1\overline{6}$$

45. Rectangular approximation; right endpoints

	10	20	40	80
I	1.9835235	1.9958860	1.9989718	1.9997430
E_n	0.0164765	0.0041140	0.0010282	0.0002570
$E_n \cdot n$	0.164765	0.08228	0.041128	0.02056
$E_n \cdot n^2$	1.64765	1.6456	1.64512	1.645
$E_n \cdot n^3$	16.4765	32.912	65.8041	131.598
$E_n \cdot n^4$	164.765	658.24	$2,632.165$	$10,527.85$

The order of convergence is n^2. In numerical analysis, it is shown that trapezoidal approximations generally converge about twice as fast as rectangular approximations.

47. Simpson's rule

	10	20	40	80
I	2.0001095	2.0000068	2.0000004	2.0000000
E_n	0.0001095	0.0000068	0.0000004	0.0000000
$E_n \cdot n$	0.001095	0.000136	0.000017	0.0000000
$E_n \cdot n^2$	0.01095	0.000271	0.00068	0.0000000
$E_n \cdot n^3$	0.1095	0.0543	0.0271	0.0000000
$E_n \cdot n^4$	1.095	1.086	1.083	0.0000000

The last approximation has no error, but the others seem to indicate that Simpson's approximation has order of convergence n^4.

49. $\displaystyle\int_0^\pi \left(9x - x^3\right) dx = \left[\frac{9}{2}x^2 - \frac{1}{4}x^4\right]\Big|_0^\pi$
$$= \frac{1}{4}\left(18\pi^2 - \pi^4\right)$$
$$\approx 20.06094705$$

	10	20	40	80
Simpsons	20.06094705	20.06094705	20.06094705	20.06094705

For Simpson's rule the same values occur. The error term involves $f^{(4)}(x)$ which is 0 for cubics; that is, the Simpson error is 0.

51. Answers vary, but the calculated value should be about 314 cm^2.

53. Newton-Cotes
$$\frac{3}{8}\left(\tan^{-1}0 + 3\tan^{-1}1 + 3\tan^{-1}2 + \tan^{-3}x\right)(1)$$
$$\approx 2.597507406$$

Type of estimate	Estimate
Left endpoint	2.084150122
Trapezoid	2.552542287
Simpson	2.599731411
Newton-Cotes	2.597507406

The exact answer (by computer) is 2.5958, so the Newton-Cotes method is the most accurate.

55. $a = -1$, $b = 2$, $b - a = 3$, $\dfrac{b-a}{6} = \dfrac{1}{2}$, and $\dfrac{a+b}{2} = \dfrac{1}{2}$

$$\int_{-1}^2 \left(x^3 - 3x + 4\right) dx$$
$$= \Big\{2^3 - 3(2) + 4$$
$$+ 4\big[(0.5)^3 - 3(0.5) + 4\big] + (-1)^3$$
$$- 3(-1) + 4\Big\}(0.5)$$
$$= \frac{45}{4}$$

$$= \frac{1}{2^{10}} \int \sin^9 2x (2\cos 2x) dx$$

$$= \frac{1}{10 \cdot 2^{10}} (\sin 2x)^{10} + C$$

37.   **39.**

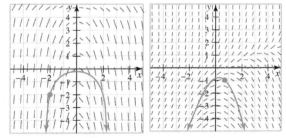

41.

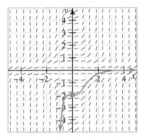

43. The function f is continuous and nonnegative on $[1, 4]$

$$\int_1^4 x^{-1} dx = \ln|x|\big|_1^4 = \ln 4$$

45. The function f is continuous and positive on $[0, 2]$.

$$\int_0^2 e^{4x} dx = \frac{1}{4} e^{4x} \bigg|_0^2 = \frac{1}{4}(e^8 - 1)$$

47. $f'(t) = \int (\sin 4t - \cos 2t)\, dt$

$$= -\frac{1}{4}\cos 4t - \frac{1}{2}\sin 2t + C_1$$

Since $f'(\frac{\pi}{2}) = 1$, $C_1 = \frac{5}{4}$, so

$$f'(t) = -\frac{1}{4}\cos t - \frac{1}{2}\sin 2t + \frac{5}{4}$$

$$f(t) = \int \left(-\frac{1}{4}\cos 4t - \frac{1}{2}\sin 2t + \frac{5}{4}\right) dt$$

$$= -\frac{1}{16}\sin 4t + \frac{1}{4}\cos 2t + \frac{5}{4}t + C_2$$

Since $f(\frac{\pi}{2}) = 1$, $C_2 = \frac{5}{4} - \frac{5}{8}\pi$

$$f(t) = -\frac{1}{16}\sin 4t + \frac{1}{4}\cos 2t$$
$$+ \frac{5}{4}t + \frac{5}{4} - \frac{5}{9}\pi$$

49. $\dfrac{dy}{dx} = (1 - y)^2$

$$\int (1 - y)^{-2} dy = \int dx$$

$$-[-(1 - y)^{-1}] = x + C$$

$$1 - y = \frac{1}{x + C}$$

$$y = 1 - \frac{1}{x + C}$$

51. $\dfrac{dy}{dx} = \left(\dfrac{\cos y}{\sin x}\right)^2$

$$\int \sec^2 y\, dy = \int \csc^2 x\, dx$$

$$\tan y = -\cot x + C$$

$$\tan y + \cot x = C$$

53. $\dfrac{dy}{dx} = y(x^2 + 1)$

$$\int y^{-1} dy = \int (x^2 + 1)\, dx$$

$$\ln|y| = \frac{1}{3}x^3 + x + C_1$$

$$y = Ce^{\left(x + \frac{1}{3}x^3\right)}$$

55. $\dfrac{dy}{dx} = \dfrac{\cos^2 y}{\cot x}$

$$\int \sec^2 y \, dy = \int \tan x \, dx$$

$$\tan y = -\ln|\cos x| + C$$

$$y = \tan^{-1}(C - \ln|\cos x|)$$

57. $\dfrac{dy}{dx} = (y-4)^2$

$$\int (y-4)^{-2} \, dy = \int dx$$

$$-(y-4)^{-1} = x + C_1$$

$$x + (y-4)^{-1} = C$$

$$y = 4 + \frac{1}{C-x}$$

59. $\dfrac{4}{\pi} \displaystyle\int_0^{\pi/4} \dfrac{\sin x}{\cos^2 x} \, dx = \dfrac{4}{\pi} \displaystyle\int_0^{\pi/4} \tan x \sec x \, dx$

$$= \frac{4}{\pi} \sec x \Big|_0^{\pi/4}$$

$$= \frac{4}{\pi}\left(\sqrt{2} - 1\right)$$

61. Exact value:

$$\int_0^{\pi} \sin x \, dx = -\cos x \Big|_0^{\pi} = 2$$

$$\Delta x = \frac{\pi - 0}{6} = \frac{\pi}{6}$$

Trapezoidal rule:

$$A \approx \frac{1}{2}\Big[1(0) + 2(0.5) + 2(0.8660)$$

$$+ 2(1) + 2(0.8660) + 2(0.5) + 1(0)\Big]\left(\frac{\pi}{6}\right)$$

$$= \frac{\pi}{12}(7.4640)$$

$$\approx 1.9541$$

63. $\Delta x = \dfrac{1-0}{8} = \dfrac{1}{8}$

$$A \approx \frac{1}{2}\Big[1(1.000) + 2(0.999) + 2(0.992)$$

$$+ 2(0.975) + 2(0.943) + 2(0.897)$$

$$+ 2(0.839) + 2(0.774) + 1(0.707)\Big]\left(\frac{1}{8}\right)$$

$$= \frac{1}{16}(14.5425)$$

$$\approx 0.9089$$

65. By calculator, the maximum value of $f''(x)$ is about 2, so let $M = 2$.

$$n^2 > 33.3$$

$$n > 5.77$$

Choose $n = 6$. $A \approx 0.216$

67. Answers vary (answers rounded to the nearest unit). The better approximation is in boldface

n	$G(n)$	$\pi(n)$	Trapezoidal $(n=4)$
100	**29**	25	31
1,000	**177**	168	320
10,000	1,245	**1,229**	2,832

Gauss' approximation is closer than the trapezoidal approximation.

69. Here is the result from the trapezoidal rule with $n = 6$: 1. 01405563943
To check, we can use a calculator to find a more accurate approximation:

$$\frac{1}{1-0}\int \frac{\cos x}{1 - \frac{x^2}{2}} \, dx \approx 1.01297$$

71. a. $a = \dfrac{dv}{dt} = \dfrac{dv}{ds}\dfrac{ds}{dt} = -4s$

$$v\frac{dv}{ds} = -4s$$

$$\int v \, dv = \int (-4s) \, ds$$

$$\frac{v^2}{2} = -2s^2 + C$$

Substituting $v = 0$ and $s = 5$ makes the constant $C = 50$ so that

$$v^2 + 4s^2 = 100.$$

b. $v^2 + 4(3^2) = 100$ implies $v = \pm 8$
m/s. Since we are asked when it first
reaches this value, the answer is
$v = -8$.

73. $R(x) = \displaystyle\int_0^x \left(1{,}575 - 5t^2\right) dt$

$= \left(1{,}575t - \dfrac{5}{3}t^3\right)\Big|_0^x$

$= 1{,}575x - \dfrac{5}{3}x^3$

$R(5) = 1{,}575(5) - \dfrac{5}{3}(5)^3$

$\approx 7{,}666.67$

The revenue for the five years is
$7,666.67.

75. $\dfrac{dh}{dt} = 1 + \dfrac{1}{(t+1)^2}$

$\displaystyle\int dh = \int \left[1 + \dfrac{1}{(t+1)^2}\right] dt$

$h = 1 - \dfrac{1}{t+1} + C$

Since the tree was 5 ft tall after 2 years,
$C = \frac{10}{3}$ and $h(0) = \frac{7}{3}$. Thus the tree was
2.33 ft tall when it was transplanted.

77. $\dfrac{dy}{dx} = x\sqrt{x^2 + 5}$

$\displaystyle\int dy = \dfrac{1}{2}\int \left(x^2 + 5\right)^{1/2}(2x\, dx)$

$y = \dfrac{1}{3}\left(x^2 + 5\right)^{3/2} + C$

Since the curve passes through $(2, 10)$

$10 = \dfrac{1}{3}\left(2^2 + 5\right)^{3/2} + C$

$1 = C$

Thus,

$y = \dfrac{1}{3}\left(x^2 + 5\right)^{3/2} + 1$

79. $\dfrac{dQ}{dt} = 0.1t + 0.2$

$Q(t) = 0.05t^2 + 0.2t + C$

Since the current level of carbon monoxide
is 3.4 ppm, $C = 3.4$. In 3 years,
$Q(3) = 0.05(9) + 0.2(3) + 3.4 = 4.45$ ppm

81. $P(x) = \displaystyle\int \left(10 + 2\sqrt{x}\right) dx$

$= 10x + \dfrac{4}{3}x^{3/2} + C$

$P(9) - P(0) = 90 + 36 + C - C$

$= 126$ people

83. Because the price of turkey t months after
the beginning of the year is
$P(t) = 0.06t^2 - 0.2t + 1.2$
dollars per pound, the average price during
the first six months is

$\dfrac{1}{6-0}\displaystyle\int_0^6 \left(0.6t^2 - 0.2t + 1.2\right) dt$

$= \left[\dfrac{1}{6}\left(0.02t^3 - 0.1t^2 + 1.2t\right)\right]\Big|_0^6$

≈ 1.32

The average price was $1.32 per pound.

85. $\displaystyle\int dP = \int \left[-(\ln 2)2^{5-t}\right] dt$

$= -32(\ln 2)\displaystyle\int 2^{-t}\, dt$

$= 32\left(2^{-t}\right) + C$

Since $P = 1$ when $t = 0$:

$1 = 32 + C$

$C = -31$

$P(t) = 31(2^{-t}) - 31$

The colony dies out when $P(t) = 0$ or when $t = 0.0458$ hr (about 2 min, 45 seconds).

87. a. Because

$$-|f(x)| \le f(x) \le |f(x)|$$

we have

$$-\int_a^b |f(x)|\, dx \le \int_a^b f(x)dx \le \int_a^b |f(x)|\, dx$$

Dominance property of integrals

$$\left| \int_a^b f(x)\, dx \right| \le \int_a^b |f(x)|\, dx$$

b. $\left| \int_1^4 \dfrac{\sin x}{x} dx \right| \le \int_1^4 \left| \dfrac{\sin x}{x} \right| dx \le \int_1^4 \dfrac{1}{x} dx$

since $|\sin x| \le 1$. Finally,

$$\int_1^4 \frac{1}{x}\, dx = \ln 4 < \frac{3}{2}$$

89. a. $Q(t) = Q_0 e^{kt}$ and $\dfrac{1}{2} = e^{5.25k}$ or

$$k = \frac{\ln \frac{1}{2}}{5.25} = \frac{-\ln 2}{5.25}$$

$$Q(5) = Q_0 e^{-5\ln 2/5.25} \approx 0.5168 Q_0$$

The percentage remaining after 5 years is

$$\frac{100Q(5)}{Q_0} \approx 51.7\%$$

b. $Q(t_1) = Q_0 e^{-t_1 \ln 2/5.25}$ and also
$Q(t_1) = (1 - 0.9)Q_0$. Thus,

$$0.1Q_0 = Q_0 e^{-t_1 \ln 2/5.25}$$

$$-\frac{t_1 \ln 2}{5.25} = \ln 0.1$$

$$t_1 = \frac{-5.25 \ln 0.1}{\ln 2}$$

$$\approx 17.44$$

The time for 90% to disintegrate is about $17\frac{1}{2}$ years.

91. $Q_0 = 10$ million; $Q(1) = 1.02 Q_0$;

$$1.02 = e^k$$
$$k = \ln 1.02$$
$$Q(t) = Q_0 e^{(\ln 1.02)t}$$
$$Q(10) = Q_0 e^{10(\ln 1.02)} \approx 12.19 \text{ million}$$

For the population to double,

$$2 = e^{t_1 \ln 1.02}$$

$$t_1 = \frac{\ln 2}{\ln 1.02}$$

This is approximately 35 years.

93. Let R_x and R_y be the radii of planets X and Y, respectively. With $g_x = (8/9)g_y$, and $v_{ex} = 6$, we have

$$\frac{v_{ex}^2}{v_{ey}^2} = \frac{2g_x R_x}{2g_y R_y} = \frac{2}{9}$$

which leads to $v_{ey} = 9\sqrt{2}$ ft/s.

95. $Q(t) = Q_0 e^{kt}$

$$0.50 = e^{46.5k}$$

$$k = \frac{\ln 0.5}{46.5}$$

$$Q(30) - Q(35) = 100e^{k(30)} - 100e^{k(35)}$$
$$= 4.59246357$$

The isotope loses

$$100\left(\frac{4.5924635}{100e^{30(\ln 0.5)/46.5}} \right) \approx 7.182\%$$

of its original volume over the 5 hour period. For any other 5 hour period, beginning at time t_1, the percentage lost is

$$100\left[\frac{100\left(e^{k_1} - e^{k(t_1 - 5)}\right)}{100e^{kt_1}} \right] = 100\left(1 - e^{kt}\right)$$

$$\approx 7.182\%$$

The percentage lost is always the same.

97. Let $X = \dfrac{dx}{dt}$ and $Y = \dfrac{dy}{dt}$; then
$2X + 5Y = t$ and $X + 3Y = 7\cos t$. Solve this system to find
$X = -35\cos t + 3t$ and $Y = 14\cos t - t$
Thus,

$$x = \int (3t - 35\cos t)\, dt$$

$$= \frac{3}{2}t^2 - 35\sin t + C_2$$

and

$$y = \int (14\cos t - t)\, dt$$

$$= 14\sin t - \frac{1}{2}t^2 + C_1$$

99. This is Putnam Problem 1 from the morning session in 1958. If

$$a_0 + a_1 x + a_2 x^2 + \cdots + a_n x^n = 0$$

then

$$\int_0^1 f(x)\, dx = \frac{a_0}{1} + \frac{a_1}{2} + \cdots + \frac{a_n}{n+1} = 0$$

Hence, by the mean value theorem for integrals, there exists a number c between 0 and 1 such that

$$f(c) = \int_0^1 f(x)\, dx = 0$$

Remark: this problem appears in G. H. Hardy, *A Course in Pure Mathematics, 7th ed.* Cambridge University Press, 1938, p. 243. It is stated that the problem appeared in the *Cambridge Mathematical Tripos* for 1929.

Cumulative Review
Chapters 1-5, page 418

1. Formally, the limit statement
$$\lim_{x \to c} f(x) = L$$
means that for each $\epsilon > 0$, there corresponds a number $\delta > 0$ with the property that
$$|f(x) - L| < \epsilon$$
whenever $0 < |x - c| < \delta$. The notation
$$\lim_{x \to c} f(x) = L$$
is read "the limit of $f(x)$ as x approaches c is L" and means that the functional values $f(x)$ can be made arbitrarily close to L by choosing x sufficiently close to c (but not equal to c).

3. If f is defined on the closed interval $[a, b]$ we say f is integrable on $[a, b]$ if
$$I = \lim_{\|P\| \to 0} \sum_{k=1}^{n} f(x_k^*)\Delta x_k$$
exists. This limit is called the definite integral of f from a to b. The definite integral is denoted by
$$I = \int_a^b f(x)\, dx$$

5. $\displaystyle \lim_{x \to 2} \frac{3x^2 - 5x - 2}{3x^2 - 7x + 2} = \lim_{x \to 2} \frac{(3x + 1)(x - 2)}{(3x - 1)(x - 2)}$

$\displaystyle = \lim_{x \to 2} \frac{3x + 1}{3x - 1}$

$\displaystyle = \frac{7}{5}$

7. $\displaystyle \lim_{x \to \infty} \left(\sqrt{x^2 + x} - x \right)$

$\displaystyle = \lim_{x \to \infty} \frac{\left(\sqrt{x^2 + x} - x \right)\left(\sqrt{x^2 + x} + x \right)}{\sqrt{x^2 + x} + x}$

$\displaystyle = \lim_{x \to \infty} \frac{x}{\sqrt{x^2 + x} + x}$

$$= \lim_{x \to \infty} \frac{1}{\sqrt{1 + \frac{1}{x}} + 1} = \frac{1}{2}$$

9. $\displaystyle\lim_{x \to 0} \frac{x \sin x}{x + \sin^2 x} = \lim_{x \to 0} \frac{x}{\frac{x}{\sin x} + \sin x}$

$$= 0$$

11. $\quad L = \displaystyle\lim_{x \to \infty} (1 + x)^{2/x}$

$$\ln L = \lim_{x \to \infty} \frac{2 \ln(1 + x)}{x}$$

$$= \lim_{x \to \infty} \frac{2}{1 + x}$$

$$= 0$$

Thus, $L = e^0 = 1$.

13. $L = \displaystyle\lim_{x \to 0^+} x^{\sin x}$

$$= \lim_{x \to 0^+} (\sin x) \ln x$$

$$= \lim_{x \to 0^+} \frac{\ln x}{\csc x}$$

$$= \lim_{x \to 0^+} \frac{\frac{1}{x}}{-\csc x \cot x}$$

$$= \lim_{x \to 0^+} \frac{-\sin^2 x}{x \cos x}$$

$$= \lim_{x \to 0^+} \left(\frac{-\sin x}{x}\right)\left(\frac{\sin x}{\cos x}\right)$$

$$= -1 \cdot 0$$

$$= 0$$

Thus, $L = e^0 = 1$.

15. $y = (x^2 + 1)^3 (3x - 4)^2$

$\quad y' = (x^2 + 1)^3 (2)(3x - 4)(3)$

$\qquad + 3(x^2 + 1)^2 (2x)(3x - 4)^2$

$\quad = 6(x^2 + 1)^2 (3x - 4)(x^2 + 1 + 3x^2 - 4x)$

$\quad = 6(x^2 + 1)^3 (3x - 4)(2x - 1)^2$

17. $\qquad x^2 + 3xy + y^2 = 9$

$\quad 2x + 3xy' + 3y + 2yy' = 0$

$\qquad (3x + 2y)y' = -(2x + 3y)$

$$y' = -\frac{2x + 3y}{3x + 2y}$$

19. $y = (\sin x + \cos x)^3$

$\quad y' = 3(\sin x + \cos x)^2 (\cos x - \sin x)$

21. $y = e^{-x} \log_5 3x = e^{-x} \dfrac{\ln 3x}{\ln 5}$

$$y' = \frac{-e^{-x} \ln 3x + e^{-x}\left(\frac{1}{x}\right)}{\ln 5}$$

$$= \frac{1 - x \ln 3x}{e^x x \ln 5}$$

23. $\displaystyle\int_4^9 d\theta = 9 - 4 = 5$

25. $\displaystyle\int_0^1 \frac{x \, dx}{\sqrt{9 + x^2}} = \frac{1}{2} \int_0^1 (9 + x^2)^{-1/2} (2x \, dx)$

$$= 2\left(\frac{1}{2}\right)(9 + x^2)^{1/2} \Big|_0^1$$

$$= \sqrt{10} - 3$$

27. $\displaystyle\int \frac{e^x \, dx}{e^x + 2} = \ln|e^x + 2| + C$

$$= \ln(e^x + 2) + C$$

29. $\displaystyle\int_0^4 \frac{dx}{\sqrt{1 + x^3}}$

$\quad \approx \dfrac{1}{3}\Big[1(1) + 4(0.8783) + 2(0.5447) + 4(0.3333)$

$\qquad + 2(0.2238) + 4(0.1621) + 1(0.124)\Big]\dfrac{2}{3}$

$\quad \approx \dfrac{1}{3}(8.1558)\left(\dfrac{2}{3}\right) \approx 1.812$

31. $y = \dfrac{3x - 4}{3x^2 + x - 5}$

$\quad y' = \dfrac{(3x^2 + x - 5)(3) - (3x - 4)(6x + 1)}{(3x^2 + x - 5)^2}$

$\quad = -\dfrac{9x^2 - 24x + 11}{(3x^2 + x - 5)^2}$

When $x = 1$, $y = 1$ and $y' = 4$

The equation of the tangent line is

$$y - 1 = 4(x - 1)$$

33. $y = (x^3 - 3x^2 + 3)^3$

$y' = 3(x^3 - 3x^2 + 3)^2(3x^2 - 6x)$

$\quad = 9x(x-2)(x^3 - 3x^2 + 3)^2$

When $x = -1$, $y = -1$ and $y' = 27$ so

$\quad y + 1 = 27(x + 1)$

35. $(1 + x^2)\, dy = (x+1)y\, dx$

$\displaystyle \int y^{-1}\, dy = \int \frac{x+1}{1+x^2}\, dx$

$\displaystyle \ln|y| = \tan^{-1}x + \frac{1}{2}\ln(x^2 + 1) + C$

37. $\displaystyle \frac{dy}{dx} = e^y \sin x$

$\displaystyle \int e^{-y}\, dy = \int \sin x\, dx$

$-e^{-y} = -\cos x - C$

$e^{-y} = \cos x + C$

Since $y = 5$ when $x = 0$, $C = e^{-5} - 1$;

Thus,

$\quad e^{-y} = e^{-5} - 1 + \cos x$

$\quad y = -\ln\left|\cos x + e^{-5} - 1\right|$

39. Let $f(t) = 3t^2 + 1$

$\displaystyle \frac{dy}{dt} = \frac{dy}{dx} \cdot \frac{dx}{dt}$

$\quad = (6x)(6t)$

$\quad = (6x)\left[6(3x^2 + 1)\right]$

$\quad = 36x(3x^2 + 1)$

41. $f(x) = ax^2 + bx + c$

$f(0) = c = 0$, so $c = 0$.

$f(5) = 25a + 5b = 0$ so $5a + b = 0$

$f'(x) = 2ax + b$

$f'(2) = 4a + b = 1$

Solve the system

$\begin{cases} 5a + b = 0 \\ 4a + b = 1 \end{cases}$

to find $a = -1$, $b = 5$. Thus,

$\quad f(x) = -x^2 + 5x$

42.

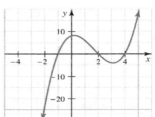

45. $f'(x) = x^2 - 4x + 3$

$\quad = (x-1)(x-3)$

$f'(x) = 0$ when $x = 1, 3$

$f(0) = f(3) = -10$; $f(1) = -\frac{26}{3}$; $f(6) = 8$

The maximum value is 8.

47. $V = \frac{1}{3}\pi r^2 h$; $\dfrac{dV}{dh} = \dfrac{1}{3}\pi r^2$

At $h = 10$, $r = 2$, and $\Delta h = 0.01$

$\quad \Delta V = V'(10)\Delta h$

$\quad = \frac{1}{3}\pi 2^2 (0.01)^1$

$\quad \approx 0.0418879$

A calculator approximation is 0.0418879.

49. a. $\displaystyle F(7) = \int_0^7 f(t)\, dt$

$\quad = (\text{AREA OF QUARTER CIRCLE})$

$\qquad\quad - (\text{AREA OF TRIANGLE})$

$\quad = \frac{1}{4}\pi(3)^2 - \frac{1}{2}\cdot 4 \cdot 4$

$\quad = \frac{9}{4}\pi - 8$

b. $F(x)$ has a relative maximum at $x = 3$ because $F'(x) = f(x)$ changes from positive to negative at $x = 3$.

c. $F(7) = \frac{9}{4}\pi - 8$

$F'(7) = f(7) = -4$

$\quad y - \left(\frac{9}{4}\pi - 8\right) = -4(x - 7)$

$\quad 4x + y - 20 - \frac{9}{4}\pi = 0$

d. $F''(x) = f'(x)$ changes from increasing to decreasing at $x = 0$, and from decreasing to increasing at $x = 7$. Thus, the graph of F has points of inflection at $x = 0$ and $x = 7$.

51. Consider a triangle with legs 5,000 and y, where y is the altitude of the rocket. Use θ for the angle of elevation, opposite the y leg.

$$y = 5{,}000 \tan \theta$$

$$\frac{dy}{dt} = 5{,}000 \sec^2\theta \frac{d\theta}{dt}$$

$$\frac{d\theta}{dt} = \frac{1}{5{,}000}\cos^2\theta \frac{dy}{dt}$$

Given $dy/dt = 850$; if $y = 4{,}000$, $\cos^2\theta = 25/41$:

$$\frac{d\theta}{dt} = \frac{1}{5{,}000}\left(\frac{25}{41}\right)(850)$$

$$= \frac{17}{164}$$

The angle of elevation is changing at $\frac{17}{164}$ rad/s.

53. $v_c = 60$ mi/h; $v_t = 45$ mi/h

Because the speed is constant, $s_c = 60t$ mi, $s_t = 30t$ mi. The distance, D, is the hypotenuse of a right triangle:

$$D = \sqrt{(60t)^2 + (45t)^2}$$

$$= 75t$$

$$\frac{dD}{dt} = 75$$

The distance between the vehicles is changing at 75 mi/h at exactly 45 minutes after they leave the intersection.

55. a. $m = \dfrac{2-0}{0-\frac{\pi}{2}} = -\dfrac{4}{\pi}$

$$y - 2 = -\frac{4}{\pi}(x - 0)$$

$$y = -\frac{4}{\pi}x + 2$$

b. $f'(x) = -2\sin x$

$$f'\left(\frac{\pi}{2}\right) = -2\sin\frac{\pi}{2} = -2$$

$$y - 0 = -2\left(x - \frac{\pi}{2}\right)$$

$$y = -2x + \pi$$

c. $f'(x) = -2\sin x = -\dfrac{4}{\pi}$

$$\sin x = \frac{2}{\pi}$$

$$x = \sin^{-1}\frac{2}{\pi}$$

Mean value theorem

57. a. Let $Q(t)$ be the amount of radioactive substance present at time t. Then,

$$\frac{dQ}{dt} = kQ^2$$

$$\int Q^{-2}\,dQ = \int k\,dt$$

$$-Q^{-1} = kt + C$$

Since $Q_0 = 100$, $C = -\frac{1}{100}$ and

$$-Q^{-1} = kt - \frac{1}{100}$$

If $t = 1$, $Q = 80$, and $k = -\frac{1}{400}$, so that

$$Q^{-1} = \frac{1}{400}t + \frac{1}{100}$$

In 6 days,

$$Q^{-1} = \frac{1}{400}(6) + \frac{1}{100}$$

$$Q = 40$$

that is, 40 g.

b. $\dfrac{1}{10} = \dfrac{1}{400}t + \dfrac{1}{100}$

$$t = 36 \text{ days}$$

59. Answers vary.

CHAPTER 6
Additional Applications of the Integral

Chapter Overview
The goal of this chapter is to consider applications of integration such as volume, arc length, surface area, work, hydrostatic force, centroids of planar regions, future and present value, net earnings, consumer's and producer's surplus, survival and renewal, and the flow of blood through an artery. With a list such as this, is it any wonder that we see that integration is a wonderful and useful operation to master!

6.1 Area Between Curves, page 428

1. Since $f(x) > g(x)$ on $[a, b]$, we see
$$A = \int_a^b [f(x) - g(x)]dx$$

3. Since $g(x) > f(x)$ on $[a, b]$, we see
$$A = \int_a^b [g(x) - f(x)]dx$$

5. Since $g(x) > f(x)$ on $[a, b]$, we see
$$A = \int_a^b [g(x) - f(x)]dx$$

7. Since $f(x) > g(x)$ on $[a, b]$, we see
$$A = \int_a^b [f(x) - g(x)]dx$$

9. On $[a, c]$, $g(x) > f(x)$ and on $[c, b]$, $f(x) > g(x)$, so
$$A = \int_a^c [g(x) - f(x)]dx + \int_c^b [f(x) - g(x)]dx$$

11. On $[a, c]$, $g(x) > f(x)$ and on $[c, b]$, $f(x) > g(x)$, so
$$A = \int_a^c [g(x) - f(x)]dx + \int_c^b [f(x) - g(x)]dx$$

SURVIVAL HINT: *Remember, that to find the area between two curves, you must either be given the boundaries, or if not, you must find the points where the curves intersect.*

13.
$$-x^2 + 6x - 5 = \frac{3}{2}x - \frac{3}{2}$$
$$2x^2 - 9x + 7 = 0$$
$$(2x - 7)(x - 1) = 0$$
$$x = \frac{7}{2}, 1$$

Use vertical strips.

$$\int_1^{7/2}\left[(-x^2 + 6x - 5) - \left(\frac{3}{2}x - \frac{3}{2}\right)\right]dx$$

$$= \int_1^{7/2}\left(-x^2 + \frac{9}{2}x - \frac{7}{2}\right)dx$$

$$= \left(-\frac{1}{3}x^3 + \frac{9}{4}x^2 - \frac{7}{2}x\right)\Big|_1^{7/2}$$

$$= \frac{125}{48}$$

15. $\sin 2x = 0$
$$x = 0, \frac{\pi}{2}, \pi$$
Use vertical strips.

$$\int_0^{\pi/2} \sin 2x \, dx + \int_{\pi/2}^{\pi} (-\sin 2x) \, dx$$

$$= \left(-\frac{1}{2}\cos 2x\right)\Big|_0^{\pi/2} + \left(\frac{1}{2}\cos 2x\right)\Big|_{\pi/2}^{\pi}$$

$$= 1 + 1$$

$$= 2$$

17. $y^2 - 5y = 0$

$y(y - 5) = 0$

$\qquad y = 0, 5$

Use horizontal strips.

$$\int_0^5 [0 - (y^2 - 5y)] \, dy = \int_0^5 (5y - y^2) \, dy$$

$$= \left(\frac{5y^2}{2} - \frac{y^3}{3}\right)\Big|_0^5$$

$$= \frac{125}{6}$$

19. The curves $y = x^2$
and $y = x$ intersect
at $(0, 0)$ and $(1, 1)$.

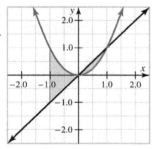

$$A = \int_{-1}^0 (x^2 - x) \, dx + \int_0^1 (x - x^2) \, dx$$

$$= \left(\frac{1}{3}x^3 - \frac{1}{2}x^2\right)\Big|_{-1}^0 + \left(\frac{1}{2}x^2 - \frac{1}{3}x^3\right)\Big|_0^1$$

$$= 1$$

21. The curves $y = 4x^3$
and $y = 0$ intersect
at $(0, 0)$; not in
interval.

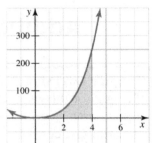

$$A = \int_1^4 4x^3 \, dx = x^4\Big|_1^4 = 255$$

23. The curves $y = x^{-1}$
and $y = 0$ do not
intersect.

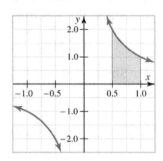

$$A = \int_{0.5}^1 x^{-1} \, dx = \ln x\Big|_{0.5}^1 = 0 - \ln\frac{1}{2} = \ln 2$$

25. The curves intersect
at $(0, 0)$ and $(1, 1)$.

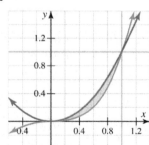

$$A = \int_0^1 (x^2 - x^3) \, dx = \left(\frac{1}{3}x^3 - \frac{1}{4}x^4\right)\Big|_0^1 = \frac{1}{12}$$

27. The curves
$$y = x^2 - 1$$
and $y = 0$ intersect
at $(-1, 0)$ and $(1, 0)$.

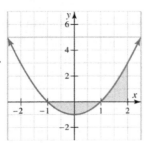

$$A = \int_{-1}^{1} (1 - x^2)\, dx + \int_{1}^{2} (x^2 - 1)\, dx$$

$$= \left(-\frac{1}{3}x^3 + x \right) \Big|_{-1}^{1} + \left(\frac{1}{3}x^3 - x \right) \Big|_{1}^{2} = \frac{8}{3}$$

29. The curves intersect
at $(0, 0)$, $(-3, 54)$,
and $(3, 54)$.

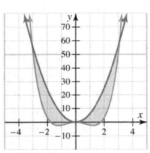

$$A = 2\int_{0}^{3} (9x^2 - x^4)\, dx \qquad \textit{By symmetry}$$

$$= 2\left(\frac{9}{3}x^3 - \frac{1}{5}x^5 \right) \Big|_{0}^{3} = \frac{324}{5}$$

31. The curves intersect
at $(1, 1)$, and
$(-2, -2)$.

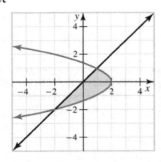

$$A = \int_{-2}^{1} (2 - y^2 - y)\, dy$$

$$= \left(2y - \frac{1}{3}y^3 - \frac{1}{2}y^2 \right) \Big|_{-2}^{1} = \frac{9}{2}$$

33. $2x^3 + x^2 - x - 1 = x^3 + 2x^2 + 5x - 1$
$$x^3 - x^2 - 6x = 0$$
$$x(x - 3)(x + 2) = 0$$
$$x = 0, -2, 3$$

The curves intersect
at $(0, -1)$,
$(-2, -11)$, and
$(3, 59)$.

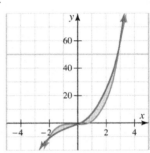

$$A = \int_{-2}^{0} (x^3 - x^2 - 6x)\, dx + \int_{0}^{3} (-x^3 + x^2 + 6x)\, dx$$

$$= \left(\frac{1}{4}x^4 - \frac{1}{3}x^3 - 3x^2 \right) \Big|_{-2}^{0} + \left(-\frac{1}{4}x^4 + \frac{1}{3}x^3 + 3x^2 \right) \Big|_{0}^{3}$$

$$= \frac{253}{12}$$

35. The curves $y = \sin x$
and $y = \sin 2x$
intersect when $x = 0$,
$\pi/3$, and π.

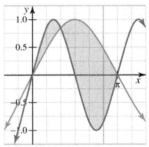

$$A = \int_{0}^{\pi/3} (\sin 2x - \sin x)\, dx + \int_{\pi/3}^{\pi} (\sin x - \sin 2x)\, dx$$

$$= \left(-\frac{1}{2}\cos 2x + \cos x\right)\Big|_0^{\pi/3} + \left(-\cos x + \frac{1}{2}\cos 2x\right)\Big|_0^{\pi.3}$$

$$= \frac{5}{2}$$

37. The curves $y = |4x - 1|$ and $y = x^2 - 5$ do not intersect on $[0, 4]$, but the absolute value function causes the equation of the line to change at $x = 1/4$.

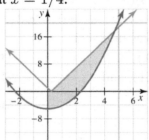

$$A = \int_0^{1/4} (-4x + 1 - x^2 + 5)\, dx$$

$$+ \int_{1/4}^4 (4x - 1 - x^2 + 5)\, dx$$

$$= \left(-2x^2 + x - \frac{1}{3}x^3 + 5x\right)\Big|_0^{1/4}$$

$$+ \left(2x^2 - x - \frac{1}{3}x^3 + 5x\right)\Big|\Big|_{1/4}^4$$

$$= \frac{323}{12}$$

39.

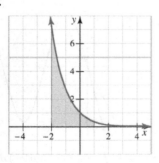

$$A = \int_{-2}^1 e^{-x}\, dx$$

$$= -e^{-x}\big|_{-2}^1 = -e^{-1} + e^2 = e^2 - e^{-1}$$

41. The curves intersect at $(0, 2)$, $(0, -2)$, and $(0, 3)$.

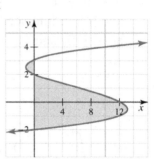

$$A = \int_{-2}^2 (y^3 - 3y^2 - 4y + 12)\, dy + \int_2^3 (-y^3 + 3y^2 + 4y - 12)\, dy$$

$$= \left(\frac{1}{4}y^4 - y^3 - 2y^2 + 12y\right)\Big|_{-2}^2 + \left(-\frac{1}{4}y^4 + y^3 + 2y^2 - 12y\right)\Big|_2^3$$

$$= \frac{131}{4}$$

43.

$$\frac{1}{\sqrt{1 - x^2}} = \frac{2}{x+1}$$

$$(x + 1)^2 = 4(1 - x^2)$$

$$x^2 + 2x + 1 = 4 - 4x^2$$

$$5x^2 + 2x - 3 = 0$$

$$(5x - 3)(x + 1) = 0$$

$$x = \frac{3}{5}, -1$$

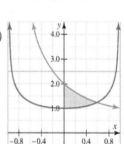

In the domain, the curves intersect at $x = \frac{3}{5}$.

$$\int_0^{3/5} \left(\frac{2}{x + 1} - \frac{1}{\sqrt{1 - x^2}}\right) dx$$

$$= \left(2\ln(x + 1) - \sin^{-1} x\right)\Big|_0^{3/5}$$

$$= 2\ln 1.6 - \sin^{-1} 0.6$$

> **SURVIVAL HINT:** *Notice that in Problems 19-43, you were asked to sketch the regions bounded by the given curves, but even when you are not asked to do so, it is a good idea to sketch the curves. This will help you find the correct limits of integration.*

45.

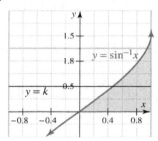

We want

$$\int_0^k (1 - \sin y)\, dy = \frac{1}{2} \int_0^{\pi/2} (1 - \sin y)\, dy$$

$$(y + \cos y)|_0^k = \frac{1}{2}(y + \cos y)\Big|_0^{\pi/2}$$

$$(k + \cos k) - (0 + 1) = \frac{1}{2}\left[\left(\frac{\pi}{2} + 0\right) - (0 + 1)\right]$$

$$k + \cos k = \frac{\pi}{4} + \frac{1}{2}$$

$$k \approx 0.34 \qquad \textit{By calculator}$$

47.

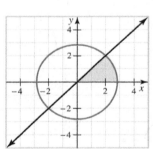

This is one-eighth of a circle with radius $\sqrt{8}$.

$$A = \frac{1}{8}\pi(\sqrt{8})^2 = \pi$$

49. $\displaystyle\int_0^{2,000} \left[100(x + 25)^{-1/2} + 50\right] dx$

$$= \left[\frac{100(x + 25)^{1/2}}{\frac{1}{2}} + 50x\right]\Bigg|_0^{2,000}$$

$$= 200(45) + 50(2,000) - 0$$

$$= 108,000$$

The total cost is \$108,000.

51. $R(t) = \displaystyle\int 500e^{0.01t}\, dt$

$$= \frac{500}{0.01}e^{0.01t} + C$$

$$R(0) = 50,000 + C$$

$$0 = 50,000 + C$$

$$C = -50,000$$

and $C(t) = \displaystyle\int (50 - 0.1t)\, dt$

$$= 50t - 0.05t^2 + C$$

$$C(0) = C$$

$$0 = C$$

Thus, the profit, P, is

$$P(t) = R(t) - C(t)$$
$$= 50,000e^{0.01t} - 50,000 - 50t + 0.05t^2$$

Since t is the number of years, we find the total profit for the first year:

$$\int_0^{12} \left(50,000e^{0.01t} - 50,000 - 50t + 0.05t^2\right) dt$$

$$= \left(50,000e^{0.01t} + \frac{1}{60}t^3 - 25t^2 - 50,000t\right)\Bigg|_0^{12}$$

$$= 33,913.0578970$$

The total profit for the first year is \$33,913.06.

53. $\displaystyle\int_0^{10} 200e^{0.05t}\, dt = 4,000e^{0.05t}\big|_0^{10}$

$$= 4,000(e^{1/2} - 1)$$

The area is about 2,595. This area represents the accumulated units in the culture.

55. The graphs intersect when

$$2e^{-0.1t} = 1$$

$$-0.1t = \ln \frac{1}{2}$$

$$t = -10 \ln 0.5$$

55.

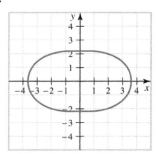

57. Because $r = f(\theta)$ and $x = r\cos\theta$, $y = r\sin\theta$, we have $x = f(\theta)\cos\theta$ and $y = f(\theta)\sin\theta$. Using the chain rule, we find that

$$\frac{dy}{d\theta} = \left(\frac{dy}{dx}\right)\left(\frac{dx}{d\theta}\right)$$

$$\frac{dy}{dx} = \frac{\frac{dy}{d\theta}}{\frac{dx}{d\theta}}$$

Because $x = f(\theta)\cos\theta$ and $y = f(\theta)\sin\theta$, it follows that

$$\frac{dx}{d\theta} = f(\theta)\frac{d}{d\theta}(\cos\theta) + \cos\theta\frac{df}{d\theta}$$
$$= -f(\theta)\sin\theta + f'(\theta)\cos\theta$$
$$\frac{dy}{d\theta} = f(\theta)\frac{d}{d\theta}(\sin\theta) + \sin\theta\frac{df}{d\theta}$$
$$= f(\theta)\cos\theta + f'(\theta)\sin\theta$$

and the slope of the tangent line is given by:

$$m = \frac{dy}{dx} = \frac{\frac{dy}{d\theta}}{\frac{dx}{d\theta}} = \frac{f(\theta)\cos\theta + f'(\theta)\sin\theta}{-f(\theta)\sin\theta + f'(\theta)\cos\theta}$$

59. a. $\tan\alpha = \dfrac{a\cos\theta}{-a\sin\theta} = -\cot\theta$

  b. $\dfrac{dr}{d\theta} = 2\sin\theta$; $\tan\alpha = \dfrac{1-\cos\theta}{\sin\theta}$

  c. $\dfrac{dr}{d\theta} = 6e^{3\theta}$; $\tan\alpha = \dfrac{2e^{3\theta}}{6e^{3\theta}} = \dfrac{1}{3}$

6.4 Arc Length and Surface Area, page 466

1. $\sqrt{1+[f'(x)]^2} = \sqrt{1+3^2} = \sqrt{10}$

$$s = \int_{-1}^{2} \sqrt{10}\,dx = 3\sqrt{10}$$

3. $\sqrt{1+[f'(x)]^2} = \sqrt{1+(-2)^2} = \sqrt{5}$

$$s = \int_{1}^{3} \sqrt{5}\,dx = 2\sqrt{5}$$

5. $\sqrt{1+[f'(x)]^2} = \sqrt{1+x}$

$$s = \int_{0}^{4} \sqrt{1+x}\,dx = \frac{10\sqrt{5}}{3} - \frac{2}{3}$$

7. $\sqrt{1+[f'(x)]^2} = \sqrt{1+x^2(2+x^2)}$
$$= 1+x^2$$

$$s = \int_{0}^{3} \left(1+x^2\right)dx = 12$$

9. $\sqrt{1+[f'(x)]^2} = \sqrt{1+\left(\dfrac{5}{12}x^4 - \dfrac{3}{5}x^{-4}\right)^2}$

$$= \sqrt{\left(\frac{5}{12}x^4\right)^2 + \frac{1}{2} + \left(\frac{3}{5}x^{-4}\right)^2}$$

$$= \sqrt{\left(\frac{5}{12}x^4 + \frac{3}{5}x^{-4}\right)^2}$$

$$= \frac{5}{12}x^4 + \frac{3}{5}x^{-4}$$

$$s = \int_{1}^{2} \left(\frac{5}{12}x^4 + \frac{3}{5}x^{-4}\right)dx$$

$$= \left(\frac{1}{12}x^5 - \frac{1}{5}x^{-3}\right)\Big|_{1}^{2}$$

$$= \frac{331}{120}$$

11. $\sqrt{1+[f'(x)]^2} = \sqrt{1+\left(\dfrac{1}{4}x - \dfrac{1}{x}\right)^2}$

$$= \sqrt{\frac{1}{16}x^2 + \frac{1}{2} + x^{-2}}$$

$$= \sqrt{\left(\frac{1}{4}x + \frac{1}{x}\right)^2}$$

$$= \frac{1}{4}x + x^{-1}$$

$$s = \int_1^2 \left(\frac{1}{4}x + x^{-1} \right) dx$$

$$= \left(\frac{1}{8}x^2 + \ln x \right) \Big|_1^2$$

$$= \frac{3}{8} + \ln 2$$

13. $f'(x) = \frac{1}{2}(e^{2x} - 1)^{-1/2} e^{2x}(2) - e^{-x}(e^{2x} - 1)^{-1/2} e^x$

$$= (e^{2x} - 1)(e^{2x} - 1)^{-1/2}$$

$$= (e^{2x} - 1)^{1/2}$$

$$ds = \sqrt{1 + e^{2x} - 1}\, dx = e^x dx$$

$$s = \int_0^{\ln 2} e^x \, dx$$

$$= e^x \Big|_0^{\ln 2}$$

$$= 1$$

15. $\sqrt{1 + [f'(x)]^2} = \sqrt{1 + \left(3x^2 - \frac{1}{12}x^{-2} \right)^2}$

$$= \sqrt{1 + \left(\frac{36x^4 - 1}{12x^2} \right)^2}$$

$$= \sqrt{\left(\frac{36x^4 + 1}{12x^2} \right)^2}$$

$$= 3x^2 + \frac{1}{12}x^{-2}$$

$$s = \int_1^2 \left(3x^2 + \frac{1}{12}x^{-2} \right) dx$$

$$= \left(x^3 - \frac{1}{12}x^{-1} \right) \Big|_1^2$$

$$= \frac{169}{24}$$

17. $\sqrt{1 + [f'(x)]^2} = \sqrt{1 + \left(\frac{3}{2}x^{1/2} - \frac{1}{6}x^{-1/2} \right)^2}$

$$= \sqrt{1 + \frac{(9x - 1)^2}{36x}}$$

$$= \sqrt{\frac{(9x + 1)^2}{36x}}$$

$$= \frac{3}{2}x^{1/2} + \frac{1}{6}x^{-1/2}$$

$$s = \int_1^9 \left(\frac{3}{2}x^{1/2} + \frac{1}{6}x^{-1/2} \right) dx$$

$$= \left(x^{3/2} + \frac{1}{3}x^{1/2} \right) \Big|_1^9$$

$$= \frac{80}{3}$$

19. $9x^2 = 4y^3$

$$y = \left(\frac{9x^2}{4} \right)^{1/3}$$

$$= \left(\frac{3}{2}x \right)^{2/3}$$

Since $\sqrt{1 + [f'(x)]^2}$ is not very easy to integrate, solve for $x = g(y)$ and use a dy integral.

$$x = \sqrt{\frac{4y^3}{9}} = \frac{2y^{3/2}}{3}, \text{ so } g'(y) = y^{1/2}$$

$$s = \int_0^3 \sqrt{1 + [g'(y)]^2}$$

$$= \int_0^3 \sqrt{1 + y} \, dy$$

$$= \frac{2}{3}(1 + y)^{3/2} \Big|_0^3$$

$$= \frac{14}{3}$$

21. Let $f(x) = \int_1^x \sqrt{t^2 - 1} \, dt$ so that

$$f'(x) = \sqrt{x^2 - 1} \text{ and } [f'(x)]^2 = x^2 - 1.$$

$$s = \int_1^2 \sqrt{1 + [f'(x)]^2}\, dx$$

$$= \int_1^2 \sqrt{1 + x^2 - 1}\, dx$$

$$= \int_1^2 x\, dx$$

$$= \frac{x^2}{2}\Big|_1^2$$

$$= \frac{3}{2}$$

> **SURVIVAL HINT:** *Compare the formulas for arc length and surface area:*
>
> *ARC LENGTH:* $s = \int_a^b \sqrt{1 + [f'(x)]^2}\, dx$
>
> *SURFACE AREA:*
>
> $S = 2\pi \int_a^b f(x)\sqrt{1 + [f'(x)]^2}\, dx$
>
> *We emphasize this similarity by signify the former by lower case s and the latter by a capital S.*

23. $S = 2\pi \int_0^2 (2x + 1)\sqrt{1 + (2)^2}\, dx$

$$= 2\pi\sqrt{5}(x^2 + x)\Big|_0^2$$

$$= 12\pi\sqrt{5}$$

25. $S = 2\pi \int_2^6 \sqrt{x}\left(1 + \frac{1}{4}x^{-1}\right)^{1/2} dx$

$$= 2\pi \int_2^6 \left(x + \frac{1}{4}\right)^{1/2} dx$$

$$= \frac{4}{3}\pi\left(x + \frac{1}{4}\right)^{3/2}\Big|_2^6$$

$$= \frac{49\pi}{3}$$

27. $dy = (x^2 - \frac{1}{4}x^{-2})dx$

$$ds = \sqrt{1 + \left(x^2 - \frac{1}{4}x^{-2}\right)^2}\, dx$$

$$= \sqrt{1 + (x^2)^2 - \frac{1}{2} + \left(\frac{1}{4}x^{-2}\right)^2}\, dx$$

$$= \sqrt{\left(x^2 + \frac{1}{4}x^{-2}\right)^2}\, dx$$

$$= \left(x^2 + \frac{1}{4}x^{-2}\right) dx$$

Also, $dS = 2\pi y\, ds$

$$= 2\pi\left(\frac{1}{3}x^3 + \frac{1}{4}x^{-1}\right)\left(x^2 + \frac{1}{4}x^{-2}\right) dx$$

$$= 2\pi\left(\frac{1}{3}x^5 + \frac{1}{12}x + \frac{1}{4}x + \frac{1}{16}x^{-3}\right) dx$$

$$S = 2\pi \int_1^2 \left(\frac{1}{3}x^5 + \frac{1}{3}x + \frac{1}{16}x^{-3}\right) dx$$

$$= \frac{515\pi}{64}$$

29. The circle $r = \cos\theta$ is traced out for $0 \le \theta < \pi$.

$$s = \int_0^\pi \sqrt{(\cos\theta)^2 + (-\sin\theta)^2}\, d\theta$$

$$= \int_0^\pi d\theta$$

$$= \pi$$

31. $s = \int_0^{\pi/2} \sqrt{(e^{3\theta})^2 + (3e^{3\theta})^2}\, d\theta$

$$= \int_0^{\pi/2} \sqrt{10}\, e^{3\theta}\, d\theta$$

$$= \frac{\sqrt{10}}{3}\left(e^{3\pi/2} - 1\right)$$

33. $s = \int_0^1 \sqrt{(\theta^2)^2 + (2\theta)^2}\, d\theta$

$= \int_0^1 \theta\sqrt{\theta^2 + 4}\, d\theta$

$= \frac{1}{3}\left(\theta^2 + 4\right)^{3/2}\Big|_0^1 = \frac{5\sqrt{5} - 8}{3}$

35. The curve $r = \cos^2\frac{\theta}{2}$ is traced out for $0 \le \theta < 2\pi$. We find the arc length by using symmetry.

$s = 2\int_0^\pi \sqrt{\left(\cos^2\frac{\theta}{2}\right)^2 + \left[\left(\cos^2\frac{\theta}{2}\right)'\right]^2}\, d\theta$

$= 2\int_o^\pi \sqrt{\cos^4\frac{\theta}{2} + \left[2\cos\frac{\theta}{2}\left(-\sin\frac{\theta}{2}\right)\left(\frac{1}{2}\right)\right]^2}\, d\theta$

$= 2\int_0^\pi \cos\frac{\theta}{2}\sqrt{\cos^2\frac{\theta}{2} + \sin^2\frac{\theta}{2}}\, d\theta$

$= 2\int_0^\pi \cos\frac{\theta}{2}\, d\theta$

$= 4\sin\frac{\theta}{2}\Big|_0^\pi = 4$

37. $S = \int_0^{\pi/3} 2\pi(5)\sin\theta\sqrt{(5)^2 + [(5)']}\, d\theta$

$= \int_0^{\pi/3} 2\pi(5)^2\sin\theta\, dx$

$= 50\pi(-\cos\theta)|_0^{\pi/3} = 25\pi$

39. $S = \int_{\pi/4}^{\pi/3} 2\pi\csc\theta\sin\theta\sqrt{(\csc\theta)^2 + [(\csc\theta)']}\, d\theta$

$= \int_{\pi/4}^{\pi/3} 2\pi\sqrt{\csc^2\theta + (-\csc\theta\cot\theta)^2}\, d\theta$

$= \int_{\pi/4}^{\pi/3} 2\pi\csc\theta\sqrt{1 + \cot^2\theta}\, d\theta$

$= \int_{\pi/4}^{\pi/3} 2\pi\csc^2\theta\, d\theta$

$= 2\pi(-\cot\theta)|_{\pi/4}^{\pi/3} = \frac{2\pi}{3}\left(3 - \sqrt{3}\right)$

41.

NO. OF TERMS	ARC LENGTH ESTIMATE
2	3.724192
4	3.790091
8	3.812529
16	3.818275
32	3.819717
64	3.820077
128	3.820168
256	3.82019

It appears the arc length is approximately 3.82.

43. $f'(x) = \frac{1}{3}(3x^2) + \frac{1}{4}(-x^{-2})$

$= x^2 - \frac{1}{4x^2}$

$1 + [f'(x)]^2 = 1 + x^4 - \frac{1}{2} + \frac{1}{16x^4}$

$= \left(x^2 + \frac{1}{4x^2}\right)^2$

a. About the x-axis:

$S = 2\pi\int_1^3 \left[\frac{1}{3}x^3 + (4x)^{-1}\right]\left(x^2 + \frac{1}{4x^2}\right)\, dx$

$= 2\pi\left[\frac{1}{18}x^6 + \frac{1}{6}x^2 - \frac{1}{32}x^{-2}\right]\Big|_1^3$

$= \frac{1,505\pi}{18}$

b. About the y-axis:

$S = 2\pi\int_1^3 x\sqrt{1 + [f'(x)]^2}\, dx$

$= 2\pi\int_1^3 x\left(x^2 + \frac{1}{4x^2}\right)\, dx$

$= 2\pi\left[\frac{1}{4}x^4 + \frac{1}{4}\ln x\right]_1^3$

$= 2\pi\left(20 + \frac{1}{4}\ln 3\right)$

Numerical approximation is 127.4.

45. $dy = \frac{1}{3}dx$; $ds = \sqrt{1 + \left(\frac{1}{3}\right)^2}dx = \frac{1}{3}\sqrt{10}\,dx$

$dS = 2\pi x\,ds$

$\quad = 2\pi x\left(\frac{1}{3}\sqrt{10}\right)dx$

$S = \frac{2}{3}\sqrt{10}\pi\int_0^3 x\,dx$

$\quad = 3\pi\sqrt{10} \approx 29.80$

47. $dy = \left(\frac{1}{2}x^{-1/2} - \frac{1}{2}x^{1/2}\right)dx$

$ds = \sqrt{1 + \left(\frac{1}{2}x^{-1/2} - \frac{1}{2}x^{1/2}\right)^2}\,dx$

$\quad = \left(\frac{1}{2}x^{-1/2} + \frac{1}{2}x^{1/2}\right)dx$

$dS = 2\pi x\,ds$

$\quad = 2\pi x\left(\frac{1}{2}x^{-1/2} + \frac{1}{2}x^{1/2}\right)dx$

$S = 2\pi\int_1^3 x\left(\frac{1}{2}x^{-1/2} + \frac{1}{2}x^{1/2}\right)dx$

$\quad = \pi\int_1^3 \left(x^{1/2} + x^{3/2}\right)dx$

$\quad = \pi\left(\frac{2}{3}x^{3/2} + \frac{2}{5}x^{5/2}\right)\Big|_1^3$

$\quad = \pi\left(\frac{28\sqrt{3}}{5} - \frac{16}{15}\right) \approx 27.12$

49. $f(x) = \left(1 - x^{2/3}\right)^{3/2}$

$f'(x) = \frac{3}{2}\left(1 - x^{2/3}\right)^{1/2}\left(-\frac{2}{3}x^{-1/3}\right)$

$\quad = -x^{-1/3}\left(1 - x^{2/3}\right)^{1/2}$

$ds = \sqrt{1 + x^{-2/3}(1 - x^{2/3})}\,dx$

$\quad = x^{-1/3}dx$

$s = 4\int_0^1 x^{-1/3}\,dx = 6$

Note: from the hint about 4 times arc length in first quadrant.

51. $L = \int_0^3 \sqrt{1 + (y')^2}\,dx$

x_k	$f'(x_k)$	$\sqrt{1 + [f'(x)]^2}$
0	3.7	3.8328
0.3	3.9	4.0262
0.6	4.1	4.2202
0.9	4.1	4.2202
1.2	4.2	4.3174
1.5	4.4	4.5122
1.8	4.6	4.7074
2.1	4.9	5.0010
2.4	5.2	5.2953
2.7	5.5	5.5902
3.0	6.0	6.0828

$L \approx \frac{1}{2}\{3.8328 + 2(4.0262) + \cdots$

$\qquad + 2(5.5902) + 6.0828](0.3)$

≈ 14.0543

53. $y = \frac{r}{h}x$; $y' = \frac{r}{h}$; $ds = \sqrt{1 + \frac{r^2}{h^2}}\,dx$

$dS = 2\pi y\,ds$

$\quad = 2\pi\left(\frac{r}{h}x\right)\frac{\sqrt{h^2 + r^2}}{h}\,dx$

$S = 2\pi\left(\frac{r}{h^2}\right)\sqrt{h^2 + r^2}\int_0^h x\,dx$

$\quad = \pi r\sqrt{h^2 + r^2}$

55. By the MVT there exists an x_k^* such that

$f'(x_k^*) = \frac{f(x_{k-1}) - f(x_k)}{x_{k-1} - x_k} = \frac{\Delta y_k}{\Delta x_k}$

Thus,

$\Delta s = \sqrt{(\Delta x_k)^2 + (\Delta y_k)^2}$

$\quad = \sqrt{1 + \left(\frac{\Delta y_k}{\Delta x_k}\right)^2}\,\Delta x_k$

$\quad = \sqrt{1 + [f'(x_k^*)]^2}\,\Delta x_k$

57. Since $8n(n-1)CD = \frac{1}{2}$, we have

$$y = Cx^{2n} + Dx^{2(1-n)}$$

$$y' = 2nCx^{2n-1} + 2(1-n)Dx^{1-2n}$$

$$1 + (y')^2 = 1 + \left[2nCx^{2n-1} + 2(1-n)Dx^{1-2n}\right]^2$$

$$= 1 + 4n^2C^2x^{2(2n-1)} - 8n(n-1)CD$$

$$\qquad + 4(1-n)^2D^2x^{2(1-2n)}$$

$$= 4n^2C^2x^{2(2n-1)} + \frac{1}{2}$$

$$\qquad + 4(1-n)^2D^2x^{2(1-2n)}$$

$$= \left[2nCx^{2n-1} + 2(n-1)Dx^{1-2n}\right]^2$$

Thus, the arc length is

$$L = \int_a^b \left[2nCx^{2n-1} + 2(n-1)Dx^{1-2n}\right] dx$$

$$= \left[Cx^{2n} - Dx^{2(1-n)}\right]_a^b$$

$$= C\left[b^{2n} - a^{2n}\right] - D\left[b^{2(1-n)} - a^{2(1-n)}\right]$$

59. We are given

$$L(x) = \int_0^x \sqrt{1 + [f'(x)]^2}\, dt$$

$$= \ln(\sec x + \tan x)$$

for all x in $[0,1]$. Differentiating both sides of this equation, we obtain

$$\sqrt{1 + [f'(x)]^2} = \frac{\sec x \tan x + \sec^2 x}{\sec x + \tan x} = \sec x$$

$$1 + [f'(x)]^2 = \sec^2 x$$

$$[f'(x)]^2 = \sec^2 x - 1 = \tan^2 x$$

$$f'(x) = \pm \tan x$$

$$f(x) = \mp \ln|\cos x| + C$$

Since $f(0) = 0$ (the curve $y = f(x)$ passes through the origin), $C = 0$ so

$$f(x) = \pm \ln|\cos x|$$

6.5 Physical Applications: Work, Liquid Force, and Centroids, page 480

1. The work W done on a object moving a distance d in a straight line against a constant force F is $W = Fd$. If the force varies with x for $a \le x \le b$, the work is given by the integral

$$W = \int_a^b F(x)\, dx$$

3. Density ρ is a ratio of mass per unit length, or per unit area, or per unit volume. If the density varies in terms of a variable, find the element of mass and then sum up (integrate). Let f and g be continuous and satisfy $f(x) \ge g(x)$ on the interval $[a,b]$, and consider a thin plate (lamina) of uniform density ρ that covers the region R between the graphs of $y = f(x)$ and $y = g(x)$ on the interval $[a,b]$. Then the centroid of F is the point $(\overline{x}, \overline{y})$ such that

$$\overline{x} = \frac{M_y}{m} = \frac{\rho \int_a^b x[f(x) - g(x)]\, dx}{\rho \int_a^b [f(x) - g(x)]\, dx}$$

and

$$\overline{y} = \frac{M_x}{m} = \frac{\frac{1}{2}\rho \int_a^b \left\{[f(x)]^2 - [g(x)]^2\right\} dx}{\rho \int_a^b [f(x) - g(x)]\, dx}$$

5. $W = 850(15) = 12{,}750$ ft-lb

7. The difference in F is 65 lb, the distance 100 ft for each. The additional work for the full bucket:

$$W = 65(100) = 6{,}500 \text{ ft-lb}$$

9. $F(x) = kx$;
$F(0.75) = 5$; $0.75k = 5$ or $k = \frac{20}{3}$.

$$W = \frac{20}{3} \int_0^1 x \, dx = \frac{10}{3}$$

The necessary amount of work is $3\frac{1}{3}$ ft-lb.

11. The cable weighs $20/50$ lb/ft, so
$F = \frac{20}{50} \frac{\text{lbs}}{\text{ft}} = \frac{2}{5}$ lbs/ft. Let $(50 - x)$ be the length of cable hanging over the cliff. The work of the chain is

$$W = \frac{2}{5} \int_0^{50} (50 - x) \, dx = 500$$

The work done by the ball is
$W = 30(50) = 1{,}500$. Thus, the total work is

$$500 + 1{,}500 = 2{,}000 \text{ ft-lb}$$

13. $W = \displaystyle\int_0^\pi \sin x \, dx - \int_\pi^{2\pi} \sin x \, dx$

$$= 2 \int_0^\pi \sin x \, dx$$

$$= 4 \text{ ergs}$$

15. $\Delta F = 64x(2y)\Delta x$ where $\dfrac{y}{1} = \dfrac{3 - x}{3}$.

$$F = 64 \int_0^3 2x \left(\frac{3 - x}{3} \right) dx$$

$$= \frac{64(2)}{3} \left(\frac{3x^2}{2} - \frac{x^3}{3} \right) \Big|_0^3$$

$$= 192 \text{ lb}$$

17. $\Delta F = 57x(2y)\Delta x$ where $y = \sqrt{4 - x^2}$.
Note: density of SAE20 oil is 57 lb/ft^3.

$$F = 57 \int_0^2 2x\sqrt{4 - x^2} \, dx$$

$$= 57 \left[-\frac{2}{3} \left(4 - x^2 \right)^{3/2} \right] \Big|_0^2$$

$$= 304 \text{ lb}$$

19. $\Delta F = 51.2(x - 1)(2y)\Delta x$ where
$y = 1 + u$ and $\dfrac{u}{1} = \dfrac{2 - x}{2}$.

$$F = 51.2 \int_1^2 2(x - 1) \left[1 + \left(\frac{2 - x}{2} \right) \right] dx$$

$$= 51.2 \left(\frac{5x^2}{2} - 4x - \frac{x^3}{3} \right) \Big|_1^2$$

$$\approx 59.7 \text{ lb}$$

21. $\Delta F = 64.5(x + \frac{1}{24})(2y)\Delta x$ where
$y = \sqrt{\left(\frac{1}{24} \right)^2 - x^2} \, dx$.

$$F = 64.5 \int_0^{1/24} 2 \left(x + \frac{1}{24} \right) \sqrt{\left(\frac{1}{24} \right)^2 - x^2} \, dx$$

23. $\Delta F = 57x(y)\Delta x$ where $y = 5$.

$$F = 57 \int_2^5 5x \, dx$$

25. $\Delta F = 51.2|-3 - x|(2y)\,\Delta x$ where
$y = \sqrt{4 - x^2}$.

$$F = 51.2 \int_{-2}^0 2(x + 3)\sqrt{4 - x^2} \, dx$$

27. The curves intersect where $x^2 - 9 = 0$ or when $x = -3, 3$.

$$m = 2 \int_0^3 \left[0 - (x^2 - 9) \right] dx = 36$$

$M_y = 0$ (by symmetry)

$$M_x = 2\int_0^3 \frac{1}{2}\left[0^2 - (x^2 - 9)^2\right]dx$$

$$= \int_0^3 \left(18x^2 - x^4 - 81\right)dx$$

$$= -\frac{648}{5}$$

$$(\overline{x}, \overline{y}) = \left(0, -\frac{\frac{648}{5}}{36}\right) = \left(0, -\frac{18}{5}\right)$$

29. $m = \displaystyle\int_1^2 x^{-1}\,dx = \ln 2$

$$M_y = \int_1^2 (x)x^{-1}\,dx = 1$$

$$M_x = \frac{1}{2}\int_1^2 \left(x^{-1}\right)^2 dx = \frac{1}{4}$$

$$(\overline{x}, \overline{y}) = \left(\frac{1}{\ln 2}, \frac{\frac{1}{4}}{\ln 2}\right) = \left(\frac{1}{\ln 2}, \frac{1}{4\ln 2}\right)$$

31. $m = \displaystyle\int_1^4 x^{-1/2}\,dx = 2$

$$M_y = \int_1^4 x\left(x^{-1/2}\right)dx = \frac{14}{3}$$

$$M_x = \int_1^4 \frac{1}{2}\left(x^{-1/2}\right)^2 dx = \ln 2$$

$$(\overline{x}, \overline{y}) = \left(\frac{\frac{14}{3}}{2}, \frac{\ln 2}{2}\right) = \left(\frac{7}{3}, \frac{1}{2}\ln 2\right)$$

> **SURVIVAL HINT:** *The volume theorem of Pappus (Theorem 6.2) is a useful theorem that can simplify the calculation of a volume. It is easy to think of using it when specifically requested (as in Problems 33-36. The trick, it to think of using this problem when not specifically requested to use it.*

33. $A = \frac{1}{2}(5)(5) = \frac{25}{2}$

The centroid of a triangle is located where the three medians meet, at the point $(\overline{x}, \overline{y})$ located $\frac{2}{3}$ of the distance from each vertex to the midpoint of the opposite side. The midpoint of the side with vertices $(-3, 0)$ and $(2, 0)$ is $\left(-\frac{1}{2}, 0\right)$, so

$$(\overline{x}, \overline{y}) = (0, 5) + \frac{2}{3}\left(-\frac{1}{2} - 0, 0 - 5\right) = \left(-\frac{1}{3}, \frac{5}{3}\right)$$

The distance from $(\overline{x}, \overline{y})$ to the line $y = -1$ is $s_1 = \frac{5}{3} + 1 = \frac{8}{3}$, so the volume of the solid formed by revolving the triangle about $y = -1$ is

$$V = 2\pi s_1 A = 2\pi\left(\frac{8}{3}\right)\left(\frac{25}{2}\right) = \frac{200\pi}{3}$$

35. $A = \frac{1}{2}m(2)^2 = 2\pi$

$$\overline{x} = \frac{1}{2\pi}\int_{-2}^2 \frac{1}{2}\left[\sqrt{4 - y^2}\right]^2 dy = \frac{8}{3\pi}$$

$$V = (2\pi)\left[2\pi\left(\frac{8}{3\pi} + 2\right)\right] = \frac{32\pi}{3} + 8\pi^2$$

37. **a.** When the tank is full, its volume is $\frac{1}{3}\pi(3)^2(6) = 18\pi$. Let h be the height of water and r the corresponding radius when the tank is half full. Then,

$$\frac{h}{6} = \frac{r}{3} \qquad \text{and} \qquad \frac{1}{3}\pi r^2 h = 9\pi$$

so $h = \sqrt[3]{108}$. To compute the work W required to pump all the water over the top of the tank, we proceed as in Example 3. Since the radius x of the disk of water satisfies

$$\frac{x}{3} = \frac{y}{6}$$

so $x = \frac{y}{2}$ and the distance moved by the disk is $6 - y$, we have

$$\Delta W = \underbrace{62.4\pi x^2 \Delta y}_{\text{weight}} \underbrace{(6 - y)}_{\text{distance}}$$

$$= 62.4\pi \left(\frac{1}{2}y\right)^2 (6 - y)\Delta y$$

$$W = 62.4\pi \int_0^{\sqrt[3]{108}} \left(\frac{1}{2}y\right)^2 (6 - y)\, dy$$

$$\approx 4{,}284$$

The amount of work is approximately 4,284 ft-lb.

b. Think of all the weight concentrated at the center of gravity. Thus,

$$W = 62.4\pi \int_0^{\sqrt[3]{108}} \left(\frac{1}{2}y^2\right) y\, dy \approx 12{,}603 \text{ ft-lb}$$

39. $W = 62.4\pi \int_4^{10} y\left(100 - y^2\right) dy$

$$= 62.4\pi \int_4^{10} \left(100y - y^3\right) dy$$

$$\approx 345{,}800 \text{ ft-lb}$$

41. $W = 40 \int_0^2 \pi(3)^2 (12 - y)dx$

$$= 360\pi \int_0^2 (12 - y)\, dx$$

$$= 7{,}920\pi \text{ ft-lb}$$

43. The horizontal force is 9 ft below the surface, so the force on this face is

$$F_1 = 62.4(1)^2(9) = 561.6 \text{ lb}$$

The force on each of the four vertical faces is

$$F_2 = 62.4 \int_9^{10} x(1)\, dx = 592.8 \text{ lb}$$

Thus, the total force on the five exposed sides is

$$F = F_1 + 4F_2 = 2{,}932.8 \text{ lb}$$

45. On the earth's surface, $F = -800$ lb and $x = 4{,}000$ mi. Thus,

$$-800 = -\frac{1}{4{,}000^2}k$$

$$k = 800(4{,}000)^2$$

$$W = 800(4{,}000)^2 \int_{4{,}000}^{4{,}200} x^{-2}\, dx$$

$$= \frac{3{,}200{,}000}{21}$$

The amount of work is 152,381 mi-lb.

47. $F = -\frac{k}{x^2}$, so $12 = \frac{k}{5^2}$ and $k = 300$.

a. $W = -\int_{10}^8 300x^{-2}\, dx = 7.5$

The amount of work is 7.5 ergs.

b. $W = \lim_{s \to \infty} \int_s^8 \left(-300x^{-2}\right) dx$

$$= \lim_{s \to \infty} \left[\frac{300}{8} - \frac{300}{s}\right]$$

$$\approx 37.5$$

The amount of work is 37.5 ergs.

49. The force on the bottom of the flat shallow end is

$$F_1 = 62.4(3)[(4)(12)] = 8{,}985.6 \text{ lb}$$

To find the force on the inclined plane part of the bottom, put the x-axis along one edge of the bottom as shown:

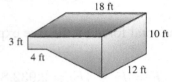

Note that the length of the incline is

$$\sqrt{14^2 + 7^2} = 7\sqrt{5}$$

If $h(x)$ is the depth of the water above point x on the bottom edge, then by similar triangles,

$$\frac{h(x) - 3}{7} = \frac{x}{7\sqrt{5}}$$

$$h(x) = \frac{1}{\sqrt{5}}x + 3$$

Thus, the fluid force on the inclined plane part of the bottom is

$$F_2 = 62.4 \int_0^{7\sqrt{5}} \left(\frac{1}{\sqrt{5}}x + 3\right)(12)\, dx$$

$$\approx 76{,}183.7$$

$F = F_1 + F_2 \approx 8{,}985.6 + 76{,}183.7 = 85{,}169.3$

The total force on the bottom is about 85,169 lb.

51. $I_x = \int_0^2 y^2 [(4 - y^2)\, dy] \qquad I_y = \int_0^4 x^2 [\sqrt{x}\, dx]$

$$= \int_0^2 (4y^2 - y^4)\, dy \qquad = \int_0^4 x^{5/2}\, dx$$

$$= \frac{64}{15} \qquad\qquad\qquad = \frac{256}{7}$$

53. $I_y = \int_0^4 x^2 [y\, dx]$

$$= \int_0^4 (4x^2 - x^3)\, dx$$

$$= \frac{64}{3}$$

Since $A = \frac{1}{2}(4)(4) = 8$ and $I_y = \rho^2 A$ we see

$$\rho = \frac{2\sqrt{2}}{\sqrt{3}} = \frac{2}{3}\sqrt{6}$$

55. $m = A = \int_0^r x\, dx = \frac{1}{2}r^2$

$$\bar{y} = \frac{\frac{1}{2}\int_0^r x^2\, dx}{m} = \frac{\frac{1}{2}\frac{r^3}{3}}{\frac{1}{2}r^2} = \frac{r}{3}$$

Distance traveled by centroid is $2\pi\bar{y} = 2\pi\left(\frac{r}{3}\right)$. Thus, using Pappus' theorem we have:

$$V = \left(2\pi\frac{r}{3}\right)\left(\frac{1}{2}r^2\right) = \frac{\pi r^3}{3}$$

57. The volume theorem of Pappus states that the volume represents the distance traveled by the centroid, multiplied by the area of the square. The area of the square is L^2 and the distance traveled is $2\pi\left(s + \frac{L}{2}\right)$, so

$$V = 2\pi L^2\left(s + \frac{L}{2}\right) = \pi\left(2sL^2 + L^3\right)$$

59. Assume the region is under a curve $y = f(x) > 0$ over $[0, a_2]$. Then,

$$\overline{x} = \frac{\int_0^{a_2} xy \, dx}{\int_0^{a_2} y \, dx}$$

$$= \frac{\int_0^{a_1} xy \, dx + \int_{a_1}^{a_2} xy \, dx}{\int_0^{a_1} y \, dx + \int_{a_1}^{a_2} y \, dx}$$

$$= \frac{1}{A_1 + A_2}\left[A_1 \frac{\int_0^{a_1} xy \, dx}{A_1} + A_2 \frac{\int_{a_1}^{a_2} xy \, dx}{A_2} \right]$$

$$= \frac{1}{A_1 + A_2}\left[A_1 \frac{\int_0^{a_1} xy \, dx}{\int_0^{a_1} y \, dx} + A_2 \frac{\int_{a_1}^{a_2} xy \, dx}{\int_{a_1}^{a_2} y \, dx} \right]$$

$$= \frac{A_1 \overline{x_1} + A_2 \overline{x_2}}{A_1 + A_2}$$

Similarly, $\overline{y} = \dfrac{A_1 \overline{y_1} + A_2 \overline{y_2}}{A_1 + A_2}$.

Any region can be subdivided into sub-regions whose sum and/or differences are under a curve $y = f(x)$ over some interval $[0, a_2]$.

6.6 Applications to Business, Economics, and Life Sciences, page 492

1. For a quantity $Q(x)$, the net change, or *cumulative change* from $x = a$ to $x = b$ is given by the definite integral
$$\int_a^b Q'(x) \, dx$$

3. If q_0 units of a commodity are sold at a price of p_0 dollars per unit, and if $p = D(q)$ is the consumer's demand function for the commodity, then
$$\begin{bmatrix} \text{CONSUMER'S} \\ \text{SURPLUS} \end{bmatrix}$$
$$= \begin{bmatrix} \text{TOTAL AMOUNT CONSUM-} \\ \text{ERS ARE WILLING TO} \\ \text{SPEND FOR } q_0 \text{ UNITS} \end{bmatrix} - \begin{bmatrix} \text{ACTUAL CONSUMER} \\ \text{EXPENDITURE FOR} \\ q_0 \text{ UNITS} \end{bmatrix}$$

$$= \int_0^{q_0} D(q) \, dq - p_0 q_0$$

If q_0 units of a commodity are sold at a price of p_0 dollars per unit, and if $p = S(q)$ is the producer's supply function for the commodity, then

$$\begin{bmatrix} \text{PRODUCER'S} \\ \text{SURPLUS} \end{bmatrix}$$
$$= \begin{bmatrix} \text{ACTUAL CONSUMER} \\ \text{EXPENDITURE} \\ \text{FOR } q_0 \text{ UNITS} \end{bmatrix} - \begin{bmatrix} \text{TOTAL AMT PRODUCERS} \\ \text{RECEIVE WHEN} \\ q_0 \text{ UNITS ARE SUPPLIED} \end{bmatrix}$$
$$= p_0 q_0 - \int_0^{q_0} S(q) \, dq$$

5. a. $q_0 = 5$, $p_0 = 150 - (6)(5) = 120$;

$$CS = \int_0^{q_0} D(q) \, dq - p_0 q_0$$
$$= \int_0^5 (150 - 6q) \, dq - (120)(5)$$
$$= 75$$

b. $q_0 = 12$, $p_0 = 150 - (6)(12) = 78$;

$$CS = \int_0^5 (150 - 6q) \, dq - (78)(12)$$
$$= 432$$

7. a. $q_0 = 1$, $p_0 = 2.5 - (1.5)(1) = 1$;

$$CS = \int_0^{q_0} D(q) \, dq - p_0 q_0$$
$$= \int_0^1 (2.5 - 1.5q) \, dq - 1$$
$$= 0.75$$

b. $q_0 = 0$, $p_0 = 2.5 - 0 = 2.5$;

$$CS = \int_0^0 (2.5 - 1)\, dq - 0$$
$$= 0$$

9. The consumer's demand function is

$$D(q) = 150 - 2q - 3q^2$$

dollars per unit. For the market price of 6 units, $p_0 = 150 - 12 - 108 = 30$.

$$CS = \int_0^6 \left(150 - 2q - 3q^2\right) dq - (30)(6)$$
$$= 468$$

Thus, the consumer's surplus is $468.00.

The consumer's surplus is the area of the region under the demand curve from $q = 0$ to $q = 6$ from which the actual spending is subtracted.

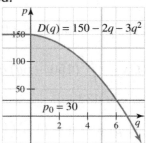

11. The consumer's demand function is

$$D(q) = 75e^{-0.04q}$$

dollars per unit. For the market price of 3 units, $p_0 = 75e^{-0.04(3)} \approx 66.52$.

$$CS = \int_0^3 75e^{-0.04q}\, dq - (3)(66.52)$$
$$= 12.47$$

Thus, the consumer's surplus is $12.47.

The consumer's surplus is the area of the region under the demand curve from $q = 0$ to $q = 3$ from which the actual spending is subtracted.

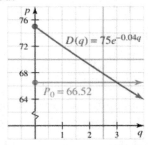

13. $S(q) = 0.5q + 15$, $p_0 = S(5) = 17.5$.

$$PS = 5(17.5) - \int_0^5 (0.5q + 15)\, dq$$
$$= 6.25$$

The producer's surplus for $q_0 = 5$ is $6.25.

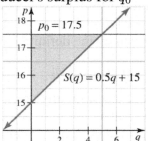

15. $S(q) = 17 + 11e^{0.01q}$, $p_0 = S(7) \approx 28.8$.

$$PS = 7(28.8) - \int_0^7 \left(17 + 11e^{0.01q}\right) dq$$
$$\approx 2.82$$

Page 226

The producer's surplus for $q_0 = 7$ is $2.82.

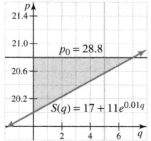

17. Equilibrium occurs when $D(q) = S(q)$.

$$27 - q^2 = \frac{1}{4}q^2 + \frac{1}{2}q + 5$$
$$5q^2 + 2q - 88 = 0$$
$$(q - 4)(5q + 22) = 0$$
$$q = 4 \qquad \text{(Disregard negative)}$$

$p = D(4) = 11.$
Consumer's surplus is:

$$\int_0^4 (27 - q^2)dq - 4(11) = \$42.67$$

19. Equilibrium occurs when $D(q) = S(q)$.

$$25 - q^2 = 5q^2 + 1$$
$$6q^2 = 24$$
$$q = 2 \qquad \text{(Disregard negative)}$$

$p = D(2) = 21.$
Consumer's surplus is:
$$\int_0^2 (25 - q^2)dq - 2(21) = \$5.33$$

21. a. The campaign generates revenue at the rate of

$$R'(t) = 5,000e^{-0.2t}$$

dollars/wk, and accumulates expenses

at the rate of $676/wk. The campaign will be profitable as long as $R(t)$ is greater than 676; that is, until

$$5,000e^{-0.2t} = 676$$
$$-0.2t = \ln \frac{676}{5,000}$$
$$t = -5\ln \frac{676}{5,000} \approx 10$$

The time is about 10 weeks.

b. For $0 \le t \le 10$, the difference $R'(t) - 676$ is the rate of change with respect to time of the net earnings generated by the campaign. Hence, the net earnings during the 10-week period is the definite integral

$$\int_0^{10} [R'(t) - 676]\, dt = \int_0^{10} [5,000e^{-0.2t} - 676]\, dt$$
$$\approx 14,857$$

The change in net earnings is about $14,857.

c. In geometric terms, the net earnings in part **b** is the area between the curve $y = R'(t)$ and the horizontal line $y = 676$ from $t = 0$ to $t = 10$.

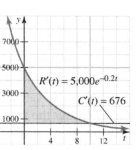

23. a. $\text{FV} = \int_0^3 \$3,000e^{0.08(3-t)}\, dt$
$= \$10,171.84$

b. $\text{PV} = \int_0^3 \$3,000e^{-0.08t}\, dt$
$= \$8,001.46$

25. a. The machine will be used until

$$R'(x) = C'(x)$$
$$6{,}025 - 10x^2 = 4{,}000 + 15x^2$$
$$25x^2 = 2{,}025$$
$$x^2 = 81$$
$$x = \pm 9 \qquad \textit{Reject negative value.}$$

The machine will be sold in 9 years. (The negative value is not in the domain.)

b. The difference $R'(x) - C'(x)$ represents the rate of change of the net earnings generated by the machine at time x. Using integration to "add up" the net earnings over the period of profitability ($0 \leq x \leq 9$), we obtain

$$\int_0^9 [R'(x) - C'(x)]\, dx$$
$$= \int_0^9 [(6{,}025 - 10x^2) - (4{,}000 + 15x^2)]\, dx$$
$$= \int_0^9 (2{,}025 - 25x^2)\, dx$$
$$= 12{,}150$$

In geometric terms, the net earnings is represented by the area of the region between the curves $y = R'(x)$ and $y = C'(x)$ from $x = 0$ to $x = 9$. We show the result below using one common software program, called *Converge*, but there are many available products that you might want to use from time to time. Notice that this software approximates the result by computing Riemann sums; the final result for 64 rectangles is 12,291.64.

More rectangles should give better approximations.

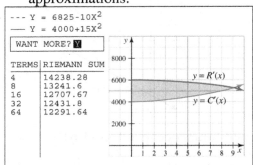

27. a. Integrating, we find $P_1(x) = 100x + \frac{1}{3}x^3$ and $P_2(x) = 190x + x^2$, since $P_1(0) = P_2(0) = 0$. Solving $P_1(x) = P_2(x)$, we obtain $x = 18$. The second plan is more profitable for the first 18 years.

b. Excess profit is found:

$$\int_0^{18} \left[(190x + x^2) - \left(100x + \frac{1}{3}x^3 \right) \right] dx = 7{,}776$$

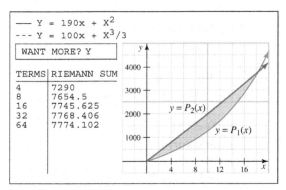

In geometric terms, the excess profit generated by the second plan is the area of the region between the curves $y = P_2(x)$ and $y = P_1(x)$ from $x = 0$ to $x = 18$.

29. a. $P(x) = xp(x) - C(x)$
$$= (124 - 2x)x - (2x^3 - 59x^2 - 4x + 76)$$
$$= -2x^3 + 57x^2 + 120x - 76$$
$$P'(x) = -6x^2 + 114x + 120$$
$$= -6(x - 20)(x + 1)$$
$P'(x) = 0$ when $x = 20$ (disregard negative root);
$P''(x) = -12x + 114 < 0$ when $x = 20$, so $x = 20$ is a maximum.

b. Consumer's surplus
$$\int_0^{20} (124 - 2x)dx - (20)(84) = 400$$

31. a. The cost function is
$$C(q) = \int \left(\frac{3}{4}q^2 + 5 \right) dq$$
$$= \frac{1}{4}q^3 + 5q + K$$
$K = 0$ since there is no overhead. The revenue function is
$$R(q) = qp(q) = \tfrac{1}{4}q(10 - q)^2$$
The marginal revenue function is
$$R'(q) = \frac{1}{4}\left(100 - 40q + 3q^2\right)$$
$$= \frac{1}{4}(10 - q)(10 - 3q)$$

b. The profit function is
$$P(q) = R(q) - C(q)$$
$$= \frac{1}{4}(100q - 20q^2 + q^3) - \frac{1}{4}q^3 - 5q$$
$$= 20q - 5q^2$$
$$P'(q) = 20 - 10q$$
$P'(q) = 0$ when $q = 2$.

c. The consumer's surplus is
$$\frac{1}{4}\int_0^2 (100 - 20q + q^2)dq - (2)(16) \approx 8.67$$
The consumer's surplus is $8.67.

33. $R(x) = q(20 - 4q^2) - (q^2 + 6q)$
$$= -4q^3 - q^2 + 14q$$
$$R'(x) = -12q^2 - 2q + 14$$
$$= -2(q - 1)(6q + 7)$$
$R'(x) = 0$ when $q = 1$ (disregard negative).
$$\int_0^1 \left(20 - 4q^2\right) dq - (1)(16) = \frac{8}{3}$$
The consumer's surplus is $2.67.

35. At 9:00 A.M., $t = 0$; at 10:00 A.M., $t = 1$, and at noon, $t = 3$; we have
$$Q'(t) = -4(t + 2)^3 + 54(t + 2)^2$$
$$Q(t) = \int -4(t + 2)^3 + 54(t + 2)^2 dt$$
$$= -(t + 2)^4 + 18(t + 2)^3 + Q_0$$
$$Q(3) = -5^4 + 18(5)^3 + Q_0$$
$$Q(1) = -3^4 + 18(3)^3 + Q_0$$
Thus,
$$Q(3) - Q(1) = 5^3(18 - 5) - 3^3(18 - 3)$$
$$= 1{,}220$$
The number of people entering the fair during the prescribed period is 1,220 people.

37. Let $P(t)$ denote the price of the oil after t months. Then, $0 \le t \le 24$ (in months) and
$$P'(t) = 400(108 + 0.03t)$$
$$P(t) = 400 \int_0^{24} (108 + 0.03t)\, dt$$
$$= 1{,}040{,}256$$

The total future revenue is $1,040,256.

39. Let $D(t)$ denote the demand for the product. Since the current demand is 5,000 and the demand increases exponentially,

$$D(t) = 5{,}000e^{0.02t}$$

units/yr. Let $R(t)$ denote the total revenue t years from now. Then, the rate of change of revenue is

$$\frac{dR}{dt} = 400\left(5{,}000e^{0.02t}\right)$$

The increase in revenue over the next two years is

$$R(2) - R(0) = \int_0^2 \$2{,}000{,}000e^{0.02t}\, dt$$

$$\approx \$4{,}081{,}077$$

41. a. The first plan generates profit at the rate of
$$P_1'(t) = 100 + t^2$$
dollars/yr and the second generates profit at the rate of
$$P_2'(t) = 220 + 2t$$
dollars/yr. The second plan will be more profitable until
$$P_1'(t) = P_2'(t)$$
that is, until
$$100 + t^2 = 220 + 2t$$
$$(t - 12)(t + 10) = 0$$
$$t = 12, -10 \qquad \textit{Disregard negative}$$
The second plan will be more profitable for 12 years.

b. For $0 \le t \le 12$, the rate at which the profit generated by the second plan exceeds that of the first plan is

$$P_2'(t) - P_1'(t)$$

Hence, the net excess profit,

$$E(t) = P_2(t) - P_1(t)$$

generated by the second plan over the twelve-year period is given by

$$E(12) - E(0) = \int_0^{12} \left[P_2'(t) - P_1'(t)\right] dt$$

$$= \int_0^{12} \left[220 + 2t - \left(100 + t^2\right)\right] dt$$

$$= 1{,}008$$

This means that the excess profit is $100,800.

c. In geometric terms, the net excess profit generated by the second plan is the area of the region between the curves $y = P_2'(t)$ and $y = P_1'(t)$ from $t = 0$ to $t = 12$.

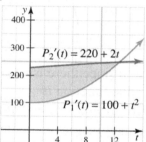

43. a. The first plan generates profit at the rate

$$P_2'(t) = 60e^{0.12t}$$

thousand dollars/yr and the second generates profit at the rate of

$$P_2'(t) = 160e^{0.08t}$$

thousand dollars/yr. The second plan will be more profitable until

$$P_1'(t) = P_2'(t)$$

that is, until

$$60e^{0.12t} = 160e^{0.08t}$$

$$e^{0.04t} = \frac{8}{3}$$

$$0.04t = \ln \frac{8}{3}$$

$$t \approx 24.52$$

The second plan will be more profitable for about 24.5 years.

b. For $0 \le t \le 24.52$, the rate at which the profit generated by the second plan exceeds that of the first plan is

$$P_2'(t) - P_1'(t)$$

Hence, the net excess profit,

$$E(t) = P_2(t) - P_1(t)$$

generated by the second plan over the time period is given by

$$E(24.52) - E(0) = \int_0^{24.52} [P_2'(t) - P_1'(t)]\, dt$$

$$= \int_0^{24.52} \left[160e^{0.08t} - 60e^{0.12t}\right] dt$$

$$\approx 3{,}240.74$$

This means that the excess profit is about $3.2 million.

c. In geometric terms, the net excess profit generated by the second plan is the area of the region between the curves $y = P_2'(t)$ and $y = P_1'(t)$ from $t = 0$ to $t = 24.52$.

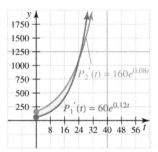

45. a. The machine generates revenue at the rate of $R'(t) = 6{,}025 - 8t^2$ dollars/yr, and costs accumulate at the rate of

$$C'(t) = 4{,}681 + 13t^2$$

dollars/yr. The use of the machine will be profitable as long as the rate at which revenue is generated is greater than the rate at which costs accumulate; that is, until

$$R'(t) = C'(t)$$
$$6{,}025 - 8t^2 = 4{,}681 + 13t^2$$
$$21t^2 = 1{,}344$$
$$t^2 = 64$$
$$t = \pm 8 \qquad \textit{Disregard negative}$$

Since the negative value is not in the domain, we see that the machine will be profitable for 8 years.

b. The difference $R'(t) - C'(t)$ represents the rate of change of the net earnings generated by the machine. Hence, the net earnings over the next eight years is given by the definite integral

$$\int_0^8 [R'(t) - C'(t)]\, dt$$

$$= \int_0^8 \left[(6{,}025 - 8t^2) - (4{,}681 + 13t^2) \right] dt$$

$$= \int_0^8 (1{,}344 - 21t^2)\, dt$$

$$= 7{,}168$$

Thus, the net earnings are $7,168.

c. In geometric terms, the net earnings in part **b** is the area of the region between the curves $y = R'(t)$ and $y = C'(t)$ from $t = 0$ to $t = 8$.

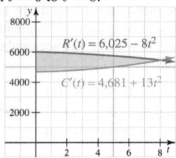

47. a. $Q_1'(t) = 60 - 2(t - 1)^2$
$Q_2'(t) = 50 - 5t$
The excess production is given by

$$\int_0^4 \left[60 - 2(t - 1)^2 - 50 + 5t \right] dt = \frac{184}{3}$$

b. The excess production is the difference between the areas under the production curves.

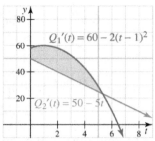

49. The density at a point r miles from the center is

$$D(r) = 25{,}000 e^{-0.05r^2}$$

people/mi². Using the fact that the area of an annulus of radii $A > B$ is
$$\pi(A^2 - B^2) = \pi(A + B)(A - B)$$
we have that for a thin annular ring of radii $r + dr > r$ the area is

$$\pi(2r + dr)dr \approx 2\pi r\, dr$$

Thus,
$$\left(25{,}000 e^{-0.05r^2}\right)(2\pi r\, dr) = 50{,}000\pi r e^{-0.05r^2}\, dr$$
people live in that ring. The total number, N, of people living between one and two miles from the center is

$$N = \int_1^2 50{,}000\pi r e^{-0.054^2}\, dr$$

Let $u = -0.05r^2$; $du = -0.1r\, dr$; if $r = 1$, $u = -0.05$ and if $r = 2$, $u = -0.2$:

$$N = \int_{-0.05}^{-0.2} 50{,}000\pi r e^{u} \frac{du}{-0.1r} \approx 208{,}128$$

There would be approximately 208,128 people living in the ring.

51. The rate of flow through a ring of area $2\pi r\, dr$ (See Problem 49 solution) is
$$2\pi r(8 - 800r^2)dr$$
The rate though the artery is

$$16\pi \int_0^{0.1} \left(r - 100r^3\right) dr = 0.04\pi$$

This is about 0.126 cm^3/s.

53. Let $P(x)$ denote the population of the town x months from now. Then,

$$\frac{dP}{dx} = 5 + 3x^{2/3}$$

and the number by which the population will increase during the next eight months is

$$P(8) - P(0) = \int_0^8 \left(5 + 3x^{2/3}\right) dx \approx 97.6$$

The population will increase by about 98 people.

55. Let $D(t)$ denote the demand for oil after t years. Then,
$$D'(t) = D_0 e^{0.1t} \quad \text{and} \quad D_0 = 30$$
and
$$D(t) = 30\int_0^{10} e^{0.1t}\, dt = 300(e - 1)$$
This is approximately 515.48 billion barrels.

57. Since $f(t) = e^{-t/10}$ is the fraction of members after t months, and since there are 8,000 charter members, the number of charter members still active at the end of 10 months is

$$8,000 f(10) = 8,000e^{-1}$$

Now divide the interval $0 \le t \le 10$ into n equal subintervals of length Δt years and let t_{j-1} denote the beginning of the jth

subinterval. During the jth subinterval, $200\Delta t$ members join, and at the end of 10 months, $(10 - t_{j-1})$ months later, the number of these retaining memberships is

$$200 f(10 - t_{j-1}) = 200e^{-(10-t_{j-1})/10}\Delta t$$

Hence, the number of new residents still active 10 months from now is approximately

$$\lim_{n\to\infty} \sum_{j=1}^n 200 e^{-(10-t_{j-1})/10}\Delta t$$
$$= 200 \int_0^{10} 200e^{-(10-t)/10} dt$$

Thus, the total number, N, of active members 10 months from now is

$$N = 8,000e^{-1} + 200 \int_0^{10} e^{-(10-t)/10}\, dt$$
$$= 8,000e^{-1} + 200e^{-1}\int_0^{10} e^{-(10-t)/10}\, dt$$
$$\approx 4,207$$

There should be approximately 4,207 active members.

59. a. $f(t_k)e^{-rt_{k-1}}$

b. $\sum_{k=1}^n f(t_k)e^{-rt_{k-1}}\Delta_n t$ where $\Delta t_n = \frac{N}{n}$

c. $PV = \lim_{n\to\infty}\sum_{k=1}^n f(t_k)e^{-rt_{k-1}}\Delta_n t$
$$= \int_0^N f(t)e^{-rt}$$

Chapter 6 Review

 Studying for a chapter examination is a personal process, one which nobody else can do for you. Simply take the time to review what you have done.

> SURVIVAL HINT: Work all of Chapter 6 problems in the Proficiency Examination (whether they are assigned or not). Work through all of the problems before looking at the answers, and *then* correct each of the problems. The answers to all these problems are given in the answer section at the back of the text. If you worked the problem correctly, move on to the next problem, but if you did not work it correctly (or you did not know what to do), then look at the solutions below, look back in the chapter to study the procedure, or ask your instructor.
>
> Finally, go back over the homework problems you have been assigned. If you worked a problem correctly, move on to the next problem, but if you missed it on your homework, then you should look back in the book or talk to your instructor about how to work the problem.
>
> If you follow these steps, you should be successful with your review of this chapter.

Proficiency Examination, page 497

1. If f and g are continuous and satisfy $f(x) \geq g(x)$ on the closed interval $[a, b]$, then the area between the two curves $y = f(x)$ and $y = g(x)$ is given by
$$A = \int_a^b [f(x) - g(x)]dx$$

2. Take cross sections perpendicular to a convenient line — say the x-axis. If $A(x)$ is the area of the cross-section at x and the base region extends from $x = a$ to $x = b$, the volume is given by
$$V = \int_a^b A(x)\, dx$$

3. Use disks or washers when the approximating strip is perpendicular to the axis of revolution. Use shells when the strip is parallel to the axis.

4. $x = r\cos\theta;\ y = r\sin\theta$
$r = \sqrt{x^2 + y^2};\ \theta = \tan^{-1}\left|\frac{y}{x}\right|$

5. Answers vary; see Table 6.2.

6. **a.** rose curve   **b.** cardioid
   (one circular leaf)

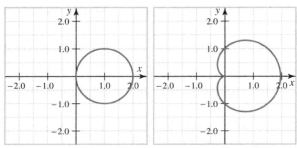

c. spiral   **d.** lemniscate

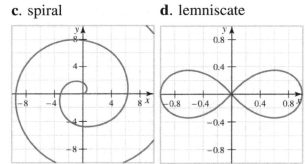

7. **Step 1.** Find the simultaneous solution of the given system of equations.
Step 2. Determine whether the pole lies on the two graphs.

Step 3. Graph the curves to look for other points of intersection. Some may require an alternate representation for (r, θ); namely, $(-r, \theta + \pi)$.

8. Let $r = f(\theta)$ define a polar curve, where f is continuous and $f(\theta) \geq 0$ on the closed interval $0 \leq \alpha \leq \theta \leq \beta \leq 2\pi$. Then the region bounded by the curve $r = f(\theta)$ and the rays $\theta = \alpha$ and $\theta = \beta$ has area

$$A = \frac{1}{2} \int_\alpha^\beta r^2 \, d\theta = \frac{1}{2} \int_\alpha^\beta [f(\theta)]^2 \, d\theta$$

9. Let f be a function whose derivative f' is continuous on the interval $[a, b]$. Then the arc length, s, of the graph of $y = f(x)$ between $x = a$ to $x = b$ is given by the integral

$$s = \int_a^b \sqrt{1 + [f'(x)]^2} \, dx$$

Similarly, for the graph of $x = g(y)$, where g' is continuous on the interval $[c, d]$, the arc length from $y = c$ to $y = d$ is

$$s = \int_c^d \sqrt{1 + [g'(y)]^2} \, dy$$

10. Suppose f' is continuous on the interval $[a, b]$. Then the surface generated by revolving about the x-axis the arc of the curve $y = f(x)$ on $[a, b]$ has surface area

$$S = 2\pi \int_a^b f(x)\sqrt{1 + [f'(x)]^2} \, dx$$

11. The work done by the variable force $F(x)$ as an object moves along the x-axis from $x = a$ to $x = b$ is given by

$$W = \int_a^b F(x) \, dx$$

12. Suppose a flat surface (a plate) is submerged vertically in a fluid of weight density ρ (lb/ft^3, for example) and that the submerged portion of the plate extends from $x = a$ to $x = b$ on a vertical axis. Then the total force, F, exerted by the fluid is given by

$$F = \int_a^b \rho h(x) L(x) \, dx$$

where $h(x)$ is the depth at x and $L(x)$ is the corresponding length of a typical horizontal approximating strip.

13. Let f and g be continuous and satisfy $f(x) \geq g(x)$ on the interval $[a, b]$, and consider a thin plate (lamina) of uniform density ρ that covers the region R between the graphs of $y = f(x)$ and $y = g(x)$ on the interval $[a, b]$. Then the mass of R is:

$$m = \rho \int_a^b [f(x) - g(x)] \, dx$$

The centroid of R is the point $(\overline{x}, \overline{y})$ such that

$$\overline{x} = \frac{M_y}{m} = \frac{\int_a^b x[f(x) - g(x)] \, dx}{\int_a^b [f(x) - g(x)] \, dx}$$

and

$$\overline{y} = \frac{M_x}{m} = \frac{\frac{1}{2} \int_a^b \{[f(x)]^2 - [g(x)]^2\} \, dx}{\int_a^b [f(x) - g(x)] \, dx}$$

14. The solid generated by revolving a region R about a line outside its boundary (but in the same plane) has volume $V = As$, where A is the area of R and s is the distance traveled by the centroid of R.

15. Let $f(t)$ be the amount of money deposited at time t over the time period $[0, T]$ in an account that earns interest at the annual rate r compounded continuously. Then the *future value* of the income flow over the time period is given by

$$\text{FV} = \int_0^T f(t)e^{r(T-t)}\, dt$$

and the *present value* of the same income flow over the time period is

$$\text{PV} = \int_0^T f(t)e^{-rt}\, dt$$

16. Suppose f' is the rate of change of a quantity f and you are required to find the net change $f(b) - f(a)$ in $f(x)$ as x varies from $x = a$ to $x = b$. But since f is an antiderivative of f', the fundamental theorem of calculus tells us that the net change is given by the definite integral

$$f(b) - f(a) = \int_a^b f'(x)\, dx$$

17. If q_0 units of a commodity are sold at a price of p_0 dollars per unit, and if $p = D(q)$ is the consumers' demand function and $p = S(q)$ is the supply function for the commodity, then

$$\begin{bmatrix}\text{CONSUMER'S} \\ \text{SURPLUS}\end{bmatrix}$$

$$= \begin{bmatrix}\text{TOTAL AMOUNT} \\ \text{CONSUMERS ARE} \\ \text{WILLING TO SPEND} \\ \text{FOR } q_0 \text{ UNITS}\end{bmatrix} - \begin{bmatrix}\text{ACTUAL CONSUMER} \\ \text{EXPENDITURE FOR} \\ q_0 \text{ UNITS}\end{bmatrix}$$

$$= \int_0^{q_0} D(q)\, dq - p_0 q_0$$

$$\begin{bmatrix}\text{PRODUCER'S} \\ \text{SURPLUS}\end{bmatrix}$$

$$= \begin{bmatrix}\text{ACTUAL CONSUMER} \\ \text{EXPENDITURE} \\ \text{FOR } q_0 \text{ UNITS}\end{bmatrix} - \begin{bmatrix}\text{TOTAL AMOUNT} \\ \text{PRODUCER WILLING TO} \\ \text{ACCEPT WHEN } q_0 \\ \text{UNITS ARE SUPPLIED}\end{bmatrix}$$

$$= p_0 q_0 - \int_0^{q_0} S(q)\, dq$$

18. A survival function gives the fraction of individuals in a population that can be expected to remain in the group and a renewal function gives the rate at which new members can be expected to join the group.

19. $\text{RATE OF FLOW} = \int_0^R 2\pi k\left(R^2 r - r^3\right) dr$

$$= \frac{\pi k R^4}{2} \text{ in volume/time}$$

20. The definite integrals could represent the following:

A. Disks revolved about the x-axis.
B. Disks revolved about the y-axis.
C. Slices taken perpendicular to the x-axis.
D. Slices taken perpendicular to the y-axis.
E. Mass of a lamina with density π.
F. Washers taken along the x-axis.
G. Washers taken along the y-axis.

a. All but E are formulas for volumes of solids.
b. A, B, F, G   **c.** F, G
d. C, D   **e.** A, F
f. B. G

21. We graph the given curve and shade the area.

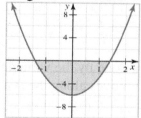

By symmetry,

$$A = 2 \int_0^{\sqrt{2}} \left[0 - \left(3x^2 - 6 \right) \right] dx$$

$$= 2 \int_0^{\sqrt{2}} \left(6 - 3x^2 \right) dx$$

$$= 8\sqrt{2}$$

22. $x = y^3 = \left(x^2 \right)^3 = x^6$, so $x = 0, 1$.

$$A = \int_0^1 \left(x^{1/3} - x^2 \right) dx$$

$$= \frac{5}{12}$$

23. $V = 2 \int_0^2 2\sqrt{4 - x^2} 4 \sqrt{4 - x^2} \, dx$

$$= 16 \int_0^2 \left(4 - x^2 \right) dx$$

$$= \frac{256}{3}$$

24. a. $A = 2 \int_0^2 \left[4 - (x-2)^2 \right] dx$

$$= 2 \int_0^2 \left(4x - x^2 \right) dx$$

$$= \frac{32}{3}$$

b. $V = 2\pi \int_0^2 \left[R^2 - r^2 \right] dx$

$$= 2\pi \int_0^2 \left[4^2 - (x-2)^4 \right] dx$$

$$= 2\pi \int_0^2 \left(-x^4 + 8x^3 - 24x^2 + 32x \right) dx$$

$$= \frac{256\pi}{5}$$

c. $V = \int_0^4 2\pi r h \, dx$

$$= 2\pi \int_0^4 x \left[4 - (x-2)^2 \right] dx$$

$$= 2\pi \int_0^4 \left(4x^2 - x^3 \right) dx$$

$$= \frac{128\pi}{3}$$

25. Solving simultaneously, or by symmetry, we see that the intersection is at $\left(\sqrt{2}\, a, \frac{\pi}{4} \right)$.

$$A = \frac{1}{2} \int_0^4 (2a \sin\theta)^2 d\theta + \frac{1}{2} \int_{\pi/4}^{\pi/2} (2a\cos\theta)^2 d\theta$$

$$= 2a^2 \int_0^{\pi/4} \left(\frac{1 - \cos 2\theta}{2} \right) d\theta + 2a^2 \int_{\pi/4}^{\pi/2} \left(\frac{1 + \cos 2\theta}{2} \right) d\theta$$

$$= a^2 \left(\frac{\pi}{2} - 1 \right)$$

26. $\dfrac{dy}{dt} = -\dfrac{3}{2} x^{1/2}$; $ds = \sqrt{1 + \dfrac{9}{4} x}\, dx$

$$s = \int_0^1 \sqrt{1 + \frac{9}{4} x}\, dx$$

$$= \frac{1}{27} \left(13^{3/2} - 8 \right)$$

27. $s = 2\pi \int_0^1 \sqrt{x} \sqrt{1 + \left(\dfrac{1}{2\sqrt{x}} \right)^2}\, dx$

$$= 2\pi \int_0^1 \sqrt{x} \sqrt{\frac{4x+1}{4x}}\, dx$$

$$= \pi \int_0^1 \sqrt{4x+1}\, dx$$

$$= \frac{\pi}{4} \int_0^1 (4x+1)^{1/2} (4\, dx)$$

$$= \frac{\pi}{6} \left(5\sqrt{5} - 1 \right)$$

28. $m = \int_0^1 \left[(x - x^3) - (x^2 - x) \right] dx$

$= \int_0^1 (2x - x^2 - x^3) \, dx$

$= \dfrac{5}{12}$

$M_y = \int_0^1 x(2x - x^2 - x^3) \, dx$

$= \int_0^1 (2x^2 - x^3 - x^4) \, dx$

$= \dfrac{13}{60}$

$M_x = \dfrac{1}{2} \int_0^1 \left[(x - x^3)^2 - (x^2 - x)^2 \right] dx$

$= \dfrac{1}{2} \int_0^1 (x^6 - 3x^4 + 2x^3) \, dx$

$= \dfrac{3}{140}$

$\overline{x} = \dfrac{M_y}{M} \qquad \overline{y} = \dfrac{M_x}{M}$

$= \dfrac{\frac{13}{60}}{\frac{5}{12}} \qquad = \dfrac{\frac{3}{140}}{\frac{5}{12}}$

$= \dfrac{13}{25} \qquad = \dfrac{9}{175}$

The centroid is $\left(\frac{13}{25}, \frac{9}{175} \right)$.

29.

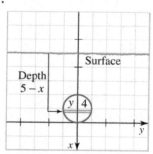

$\Delta F = 64(5 - x)(2y)\Delta x$ where
$y = \sqrt{1 - (x + 1)^2}$.

$F = 128 \int_0^2 (5 - x) \left(\sqrt{1 - (x + 1)^2} \right) dx$

30. This is the future value (FV) of money transferred into an account.

$\text{FV} = \int_0^5 1{,}200 e^{0.08(5-t)} dt$

$= 1{,}200 e^{0.4} \int_0^5 e^{-0.08t} \, dt$

$= 16{,}000 \left(e^{2/5} - 1 \right)$

The amount in the account will be $7,377.37.

Supplementary Problems, page 498

1. circle

3. line

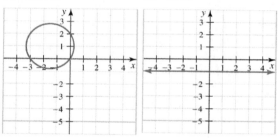

5. circle

7. parabola

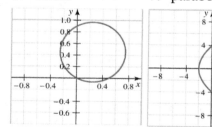

9. The pole is a point of intersection

$r_1 = r_2$

$2 \cos \theta = 1 + \cos \theta$

$\cos \theta = 1$

$\theta = 0$

The points of intersection are
$P_1(0, 0), \; P_2(2, 0)$.

11.

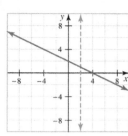

Recognize that these equations have simple rectangular forms:
$$\begin{cases} x + 2y = 4 \\ x = 2 \end{cases}$$
Intersection:
rectangular: $(2, 1)$
polar: $(2.24, 0.464)$

13. $\displaystyle\int_{-1/2}^{1} \left(2y + 2 - 4y^2\right) dy = \frac{9}{4}$

15. $\displaystyle\int_{0}^{1} \left(y^{2/3} - y^2\right) dy = \frac{4}{15}$

17. $A = \dfrac{8}{2} \displaystyle\int_{0}^{\pi/4} \left(16 \cos^2 2\theta\right) d\theta = 8\pi$

19. $A = \dfrac{2}{2} \displaystyle\int_{0}^{\pi/4} \left(\cos 2\theta\right) d\theta = \dfrac{1}{2}$

21. The curves intersect when
$$2a \sin\theta = a$$
$$\theta = \frac{\pi}{6}, \frac{5\pi}{6}$$
Using symmetry, the area is
$$A = 2 \int_{\pi/6}^{\pi/2} \frac{1}{2} \left[(2a \sin\theta)^2 - a^2\right] d\theta$$
$$= a^2 \left(\frac{\sqrt{3}}{2} + \frac{\pi}{3}\right)$$

23. $2\pi \displaystyle\int_{0}^{3/\sqrt{2}} x\left[(9 - x^2) - x^2\right] dx = \dfrac{81\pi}{4}$

25. $\pi \displaystyle\int_{\pi/4}^{\pi/3} \left(\sqrt{\cos x}\right)^2 dx = \dfrac{\pi}{2}\left(\sqrt{3} - \sqrt{2}\right)$

27. $\pi \displaystyle\int_{0}^{1} \left[(4 - x)^2 - (4 - 2x)^2\right] dx$
$$= \pi \int_{0}^{1} \left(8x - 3x^2\right) dx$$
$$= 3\pi$$

29. $2\pi \displaystyle\int_{0}^{1} y\left[\sqrt{y} - y^2\right] dy$
$$= 2\pi \int_{0}^{1} \left(y^{3/2} - y^3\right) dy$$
$$= \frac{3}{10}\pi$$

31. Use disks:
$$V = \pi \int_{0}^{\pi} \left(\sqrt{2\sin x}\right)^2 dx$$
$$= 2\pi \int_{0}^{\pi} \sin x \, dx$$
$$= 4\pi$$

33. $y = \tan x, \; y' = \sec^2 x,$
$$ds = \sqrt{1 + \sec^4 x} \, dx$$
$$S = 2\pi \int_{0}^{1} \tan x \sqrt{1 + \sec^4 x} \, dx$$
$$\approx 8.6322 \quad \textit{Use technology}$$

35. $\dfrac{1}{2} \displaystyle\int_{0}^{1} \pi \left[\dfrac{\sqrt{x} - x}{2}\right]^2 dx$
$$= \frac{1}{2} \int_{0}^{1} \frac{\pi}{4}\left(x^2 - 2x^{3/2} + x\right) dx$$
$$= \frac{\pi}{240}$$

37. $-100 = k(-2 - 0)$
$$k = 50$$
$$F(x) = 50x$$
$$W = 50 \int_{0}^{-3} x \, dx = 225 \text{ in.-lb}$$

39. $V_1 = \displaystyle\int_{-1}^{1} \left(1 - x^2\right)^2 dx = \dfrac{16}{15}$
$$V_2 = \int_{0}^{1} \left(2\sqrt{y}\right)^2 dy = 2$$
S_2 has the greater volume.

41. The cone is formed by revolving $y = \dfrac{h}{r}x$

about the y-axis. Since $\dfrac{dy}{dx} = \dfrac{h}{r}$, we can

find the surface area with

$ds = \sqrt{1 + \dfrac{h^2}{r^2}}\, dx$ and $dx = \dfrac{1}{r}\sqrt{r^2 + h^2}$.

$$S = 2\pi \int_0^r \dfrac{x}{r} \sqrt{1 + \left(\dfrac{h}{r}\right)^2}\, dx$$

$$= 2\pi \int_0^r \dfrac{x}{r}\sqrt{r^2 + h^2}\, dx$$

$$= \dfrac{2\pi}{r} \cdot \dfrac{x^2}{2}\sqrt{r^2 + h^2}\Bigg|_0^r$$

$$= \pi r \sqrt{r^2 + h^2}$$

43. The side consists of a 30×3 rectangle and a right triangle with legs 30×5; for the rectangle, the depth is y, and $dF = 62.4(30y)\,dy$.

$$F = 1{,}872 \int_0^3 y\, dy = 8{,}424$$

For the triangle,

$$\Delta F = 62.4(y+3)L(y)\Delta y \text{ where } \dfrac{L(y)}{30} = \dfrac{5-y}{5}$$

$$F = \int_0^5 62.4(y+3)(6)(5-y)\, dy = 21{,}840$$

The total force on the side is

$$8{,}424 + 21{,}840 = 30{,}264 \text{ ft-lb}$$

45. Half the vertical cross section of the tank is fitted on a Cartesian coordinate plane with the vertex of the cone at the origin and the 12 ft leg along the y-axis.

$$\Delta W = \left(22\pi x^2 \Delta y\right)D(y)$$

where $\dfrac{x}{3} = \dfrac{y}{12}$ and $D(y) = 14 - y$

$$W = \int_0^6 22\pi \left(\dfrac{y}{4}\right)^2 (14 - y)\, dy$$

$$= \dfrac{1{,}881\pi}{2} \text{ ft-lb}$$

47. $W = \displaystyle\int_0^1 62.4\pi\left(y^{1/3}\right)^2(1-y)\, dy$

$$= 62.4\pi \int_0^1 \left(y^{2/3} - y^{5/3}\right)\, dy$$

$$= 14.04\pi \text{ ft-lb}$$

49. $F = -30x$;

$$P = \int_0^s (30x)\, dx = 15s^2$$

When $15s^2 = 20$ ft-lb, $s = \dfrac{2}{\sqrt{3}}$ ft beyond the spring's equilibrium position. The spring should be stretched about 14 in.

51. Since 1 kg = 1,000 g, $m = r^{-2}$ kg; $m = \dfrac{1}{r^2}$;

$F = r^{-2}$;

$$W = \int_{6{,}400}^{10{,}000} r^{-2}\, dr = 5.625 \times 10^{-5} \text{ joules}$$

53. The curves intersect at $(2, 1)$; assume the density is 1.

$$A = \int_0^1 [(5-3y) - 2y]\, dy = \dfrac{5}{2}$$

$$M_x = \int_0^1 y[(5-3y) - (2y)]\, dy = \dfrac{5}{6}$$

$$M_y = \dfrac{1}{2}\int_0^1 \left[(5-3y)^2 - (2y)^2\right]\, dy$$

$$= \dfrac{5}{2}\int_0^1 (y^2 - 6y + 5)\, dy$$

$$= \dfrac{35}{6}$$

$$(\bar{x}, \bar{y}) = \left(\frac{\frac{35}{6}}{\frac{5}{2}}, \frac{\frac{5}{6}}{\frac{5}{2}} \right) = \left(\frac{7}{3}, \frac{1}{3} \right)$$

55. First rotate about the x-axis. Use disks with vertical strips and double the portion to the right of the y-axis. The element of volume is $dV = \pi y^2\, dx = \pi a^{-2} b^2 (a^2 - x^2)\, dx$.

$$V_x = 2a^{-2} b^2 \pi \int_0^a \left(a^2 - x^2 \right) dx = \frac{4}{3} \pi a b^2$$

Now rotate about the y-axis. Use disks with horizontal strips and double the portion above the x-axis. The element of volume is $dV = \pi x^2\, dx = \pi b^{-2} a^2 (b^2 - y^2)\, dy$.

$$V_y = 2b^{-2} a^2 \pi \int_0^b \left(b^2 - y^2 \right) dy = \frac{4}{3} \pi a^2 b$$

57.
$$V = \int_a^r \pi y^2\, dx$$
$$= \int_a^r \pi b^2 \left(\frac{x^2}{a^2} - 1 \right) dx$$
$$= \int_a^r \pi a^{-2} b^2 \left(x^2 - a^2 \right) dx$$
$$= \pi a^{-2} b^2 \left(\frac{1}{3} x^3 - a^2 x \right) \Big|_a^b$$
$$= \pi b^2 \left(\frac{r^3}{3a^2} - r + \frac{2}{3} a \right)$$
$$= \frac{\pi b^2}{3a^2} \left(r^3 - 3a^2 r + 2a^3 \right)$$

59. This is a present value problem, so

$$PV = \int_0^{10} 1{,}000 e^{-0.07t}\, dt$$
$$= -\frac{1{,}000}{0.07} \left(e^{-0.7} - 1 \right)$$

This is approximately $7{,}191.64.

61. Divide the interval $0 \le r \le 5$ into n equal subintervals of length Δr, and let r_{j-1} denote the beginning of the jth subinterval. This divides the circular disk of radius 5 into n concentric circles. If Δr is small, the area of the jth ring is $2\pi r_j \Delta r$ where $2\pi r_j$ is the circumference of the circle of radius r_j that forms the outer boundary of the ring and Δr is the width of the ring. Then, since $D(r) = 5{,}000 e^{-0.02r^2}$ is the population density (people/mi^2) r miles from the center, it follows that the number of people in the jth ring is

$$D(r_j)(\text{AREA OF } j\text{TH RING}) = 5{,}000 e^{-0.02 r_j^2} (2\pi r_j \Delta r)$$

Hence, if N is the total number of people within 5 miles of the center of the city,

$$N = \lim_{n \to \infty} \sum_{j=1}^n 5{,}000 e^{-0.02 r_j^2} (2\pi r_j \Delta r)$$
$$= \int_0^5 5{,}000 (2\pi) r e^{-0.02 r^2}\, dr$$
$$\approx 309{,}030$$

63. a. The demand function is

$$D(q) = 50 - 3q - q^2$$

dollars/unit. To find the number of units bought when the price is 32, we solve

$$32 = 50 - 3q - q^2$$
$$(q + 6)(q - 3) = 0$$
$$q = -6, 3$$

The negative value is not in the domain, so there are 3 units bought.

b. The amount that consumers are willing to spend to obtain 3 units of the commodity is

$$\int_0^3 (50 - 3q - q^2)\, dq = 127.5$$

When the market price is $32/unit, 3 units will be bought and the consumer's surplus is found:

$$\int_0^3 (50 - 3q - q^2)\, dq - 32(3) = 31.5$$

The consumer's surplus is $31.50.

c.

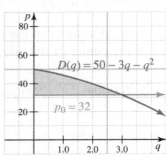

65. $W = \int_0^{100} (200 - 0.5x)\, dx = 17{,}500$ ft-lb

67. Half the vertical side is a right triangle with base $\frac{3}{2}$ and height $\frac{3}{2}\sqrt{3}$. The vertical leg of this right triangle is along the y-axis and the lower end of the side is at the origin. A rectangular element of area is at a height y with area $dA = 2x\, dy$. By similar triangles, $\frac{3}{2}y = \frac{3\sqrt{3}}{2}x$.

$$F = \int_0^{3\sqrt{3}/2} 40\left(\frac{2}{\sqrt{3}}y\right)\left(\frac{3\sqrt{3}}{2} - y\right) dy$$

$$= \frac{80}{\sqrt{3}} \int_0^{3\sqrt{3}/2} \left(\frac{3\sqrt{3}}{2}y - y^2\right) dy$$

$= 135$ lb

69. $W_1 = \int_0^6 (mt + b)\, dt = 18m + 6b$

$W_2 = \int_0^{12} (mt + b)\, dt = 72m + 12b$

Since $W_1 = 13$ and $W_2 = 44$, so we solve the system

$$\begin{cases} 18m + 6b = 13 \\ 72m + 12b = 44 \end{cases}$$

to find $(m, b) = \left(\frac{1}{2}, \frac{2}{3}\right)$. Thus,

$$F(t) = \frac{1}{2}t + \frac{2}{3}$$

71. Use a Cartesian coordinate system with origin at the center of the semicircle. For the hemisphere, $dV = \pi x^2 dy = \pi(4 - y^2)dy$, and the depth is $10 + y$.

$$W_h = \int_0^2 24\pi (4 - y^2)(10 + y)\, dy$$

$$= 24\pi \int_0^2 (-y^3 - 10y^2 + 4y + 40)\, dy$$

$$= 1{,}376\pi$$

For the cylinder, $dV = 4\pi\, dy$; the depth is $10 - y$;

$$W_c = \int_0^{10} (10 - y)(96\pi\, dy) = 4{,}800\pi$$

The total work is

$$1{,}376\pi + 4{,}800\pi = 6{,}176\pi$$

73. Let the cut-out rectangle have dimensions x by y with its centroid at $(\overline{x}_c, \overline{y}_c)$ where the origin is at one corner of the square. The original square has area 256 cm^2 and its centroid is at $(8, 8)$. The area of the

figure remaining after the cutting is
$256 - 156 = 100$.

$$100(4.88) + 156\overline{x}_c = 8(256)$$
$$\overline{x}_c = 10$$

This leaves $16 - 10 = 6$ cm for $\frac{1}{2}x$ or
$x = 12$.

$$y = 156x^{-1} = 13 \text{ so } \overline{y}_c = 6.5 \text{ cm}$$

from its (top or bottom) edge with $\overline{y}$ the y-value of the centroid of the remaining area.

$$100\overline{y} + 156\left(\frac{13}{2}\right) = 8(256)$$
$$\overline{y} = 10.34$$

Thus, the desired distances are 4.88 cm
(from the left; given), $16 - 4.88 = 11.12$
cm from the right, 10.34 cm from the top,
and $16 - 10.34 = 5.66$ cm from the
bottom.

75. a. The vertical cross section of the cone
is an isosceles triangle with height 1 ft
and base 2 ft. A horizontal element of
volume is a cylinder of radius x and
height dy. The volume is
$dV = \pi x^2 \, dy$. This element is y ft
from the base of the cone. From
similar right triangles (half of the
isosceles triangle), $x = 1 - y$. Thus,
$dV = \pi(1 - y)^2 dy$, the distance
through which the element was lifted
is y.

$$W = \int_0^1 140\pi y (1 - y)^2 \, dy$$
$$= 140\pi \int_0^1 \left(y - 2y^2 + y^3\right) dy$$
$$= \frac{35\pi}{3} \text{ ft-lb}$$

b. A horizontal element is a cylinder of
radius 4 in. or $1/3$ ft and height dy.
The volume is $dV = (1/9)\pi \, dy$. This
element is y ft from the base of the
cone, a distance through which the
element was lifted.

$$W = \int_0^1 \frac{140}{9}\pi y \, dy = \frac{70}{9}\pi \text{ ft-lb}$$

77. a. The height of the plate is

$$h = \sqrt{3^2 - \left(\frac{5}{2}\right)^2} = \frac{\sqrt{11}}{2}$$

A thin horizontal strip x ft below the
surface has length $L(x)$, where

$$\frac{L(x)}{5} = \frac{x - 4}{\sqrt{\frac{11}{4}}}$$
$$L(x) = \frac{10}{\sqrt{11}}(x - 4)$$

The total force on the plate is

$$F = \int_4^{4+\sqrt{11}/2} 62.4x \left[\frac{10}{\sqrt{11}}(x - 4)\right] dx$$
$$= 1{,}320.70$$

The force is approximately 1,321 lb.

b. In general,

$$F = \int_D^{D+h} 62.4x \left[B\frac{x - D}{h}\right] dx$$

where $h = \sqrt{A^2 - B^2/4}$.

79. The generating line is below the x-axis on
the interval $[1, 2]$. The surface area is
actually given by the integral

$$S = \int_1^3 2\pi|x - 2|\sqrt{2}\,dx = 2\pi\sqrt{2}$$

81. At the Earth's surface, the force is the object's weight and $r = 4{,}000$ mi. If the object is s ft below the surface then the distance from the center of the earth (in miles) is $t = 4{,}000 - \dfrac{s}{5{,}280}$. Thus, $P = -k/(4{,}000)^2$ and

$$W = -(4{,}000)^2 P \int_t^{4{,}000} r^{-2}\,dx$$

$$= (4{,}000)^2 P \left(\frac{1}{4{,}000} - \frac{1}{t} \right) \text{ mi-lb}$$

83. $V = \displaystyle\int_{-1}^9 \pi\{[f(x) + 1]^2\}\,dx$

$\approx \dfrac{\pi}{2}[(5^2) + 2(4^2) + 2(5^2) + 2(7^2)$

$\qquad + 2(6^2) + 2(7)^2 + 2(8^2) + 2(7^2)$

$\qquad + 2(5^2) + 2(4^2) + (5^2)](1)$

$= 354\pi$

85. a. For $V = 500$, we find $h \approx -0.644391$

   b. For $V = 800$, $h \approx 0.865202$

87. Consider a right triangle formed by a vertical cross section with height 480 ft and base 750 ft. Pick a horizontal slab of height dy and base x^2 located at y ft from the top. By similar triangles,

$$\frac{x}{y} = \frac{750}{480} \quad \text{or} \quad x = \frac{750y}{480}$$

$$W = \int_0^{480} 160 \left(\frac{750}{480} \right)^2 y^2(480 - y)\,dy$$

$$\approx 1.728 \times 10^{12} \text{ ft-lb}$$

89. Use shells;

$$V = 2\pi \int_0^2 x(9 - x^2)^{-1/2}\,dx = 2\pi\left(3 - \sqrt{5}\right)$$

91. $A = \displaystyle\int_0^{\pi/4} \left(\sec^2 x - \tan^2 x \right)\,dx$

$= \displaystyle\int_0^{\pi/4} dx$

$= \dfrac{\pi}{4}$

93. $V \approx \dfrac{\sqrt{3}}{4} \displaystyle\int_0^1 e^{2x}\,dx = \dfrac{\sqrt{3}}{8}\left(e^2 - 1\right)$

95. This is Putnam Examination Problem 1 of the morning session in 1939. The arc in the first quadrant is represented by the equation $y = x^{3/2}$, and its slope is $\frac{3}{2}x^{1/2}$. The point $P(x_0, y_0)$ where the tangent makes an angle of $\pi/4$ is determined from the relation

$$\frac{3}{2}x_0^{1/2} = 1$$

$$x_0 = \frac{4}{9}$$

The desired length is therefore

$$\int_0^{4/9} \sqrt{1 + \frac{9x}{4}}\,dx = \frac{8}{27}\left(2\sqrt{2} - 1\right)$$

97. This is Putnam Examination Problem 1 of the morning session in 1993. Let (b, c) denote the second intersection point. Find c so that

$$\int_0^b [c - (2x - 3x^3)]\,dx = 0$$

This leads to $cb - b^2 + \frac{3}{4}b^4 = 0$. After substituting $c = 2b - 3b^3$ and solving, we find that $b = \frac{2}{3}$. This implies that $c = \frac{4}{9}$.

99. **Book report** Answers vary.

CHAPTER 7 Methods of Integration

Chapter Overview
By now you should understand the concept of integration as well as have a grasp on hwy it is important. It is this chapter, however, that gives you the confidence and ability to actually carry the integration of most common functions and applications. The most important technique, substitution, and this method becomes the foundation for much that is done in this chapter. It is worthwhile to repeat a sentence found on the first page of this chapter: "Learning to integrate is like learning to play a musical instrument: At first, it may seem impossibly complicated, but if you preserve, after a while music starts to happen."

7.1 Review of Substitution and Integration by Table, page 516

1. $\int \dfrac{2x+5}{\sqrt{x^2+5x}}dx \quad \boxed{u=x^2+5x}$

$= \int (x^2+5x)^{-1/2}[(2x+5)dx]$

$= 2(x^2+5x)^{1/2}+C$

3. $\int \dfrac{dx}{x\ln x} \quad \boxed{u=\ln x}$

$= \int \dfrac{1}{\ln x}\dfrac{dx}{x}$

$= \ln|u|+C$

$= \ln|\ln x|+C$

5. $\int \dfrac{x\,dx}{4+x^4} \quad \boxed{u=x^2}$

$= \dfrac{1}{2}\int \dfrac{2x\,dx}{4+x^4}$

$= \left(\dfrac{1}{2}\right)^2 \tan^{-1}\dfrac{1}{2}x^2+C$

$= \dfrac{1}{4}\tan^{-1}\dfrac{x^2}{2}+C$

7. $\int (1+\cot x)^4\csc^2 x\,dx \quad \boxed{u=1+\cot x}$

$= \int (-u^4)\,du$

$= -\dfrac{1}{5}(1+\cot x)^5+C$

9. $\int \dfrac{x^3-x}{(x^4-2x^2+3)^2}\,dx \quad \boxed{u=x^4-2x^2+3}$

$= \dfrac{1}{4}\int (x^4-2x^2+3)^{-2}[(4x^3-4x)]dx$

$= \dfrac{1}{4}\int u^{-2}du$

$= -\dfrac{1}{4}(x^4-2x^2+3)^{-1}+C$

11. $\int \dfrac{2x+1}{x^2+x+1}\,dx \quad \boxed{u=x^2+x+1}$

$= \int (x^2+x+1)^{-1}[(2x+1)\,dx]$

$= \ln(x^2+x+1)+C$

13. $\int \dfrac{dx}{x^2\sqrt{x^2-a^2}} \quad \boxed{\text{Formula 103}}$

$= \dfrac{\sqrt{x^2-a^2}}{a^2 x}+C$

15. $\displaystyle\int x \ln x\, dx$ ☐ Formula 199

$\displaystyle = \frac{1}{2}x^2\left(\ln x - \frac{1}{2}\right) + C$

17. $\displaystyle\int xe^{ax}\, dx$ ☐ Formula 184

$\displaystyle = a^{-1}e^{ax}\left(x - a^{-1}\right) + C$

19. $\displaystyle\int \frac{x^2\, dx}{\sqrt{x^2+1}}$ ☐ Formula 91

$\displaystyle = \frac{1}{2}x\sqrt{x^2+1} - \frac{1}{2}\ln\left(x + \sqrt{x^2+1}\right) + C$

21. $\displaystyle\int \frac{x\, dx}{\sqrt{4x^2+1}}$ ☐ Formula 90; $u = 2x$

$\displaystyle = \frac{1}{4}\sqrt{4x^2+1} + C$

23. $\displaystyle\int e^{-4x}\sin 5x\, dx$ ☐ Formula 192

$\displaystyle = (16+25)^{-1}e^{-4x}(-4\sin 5x - 5\cos 5x) + C$

$\displaystyle = \frac{-4\sin 5x - 5\cos 5x}{41e^{4x}} + C$

25. $\displaystyle\int (1+bx)^{-1}\, dx$ ☐ Formula 34

$\displaystyle = b^{-1}\ln|1 + bx| + C$

27. $\displaystyle\int x(1+x)^3\, dx$ ☐ Formula 31

$\displaystyle = \frac{(x+1)^5}{5} - \frac{(x+1)^4}{4} + C$

$\displaystyle = \frac{1}{20}(x+1)^4(4x-1) + C$

29. $\displaystyle\int xe^{4x}\, dx$ ☐ Formula 184

$\displaystyle = \frac{1}{4}e^{4x}\left(x - \frac{1}{4}\right) + C$

31. $\displaystyle\int \frac{dx}{1 + e^{2x}}$ ☐ Formula 189

$\displaystyle = x - \frac{1}{2}\ln(1 + e^{2x}) + C$

33. $\displaystyle\int \sec^3\left(\frac{x}{2}\right) dx$ ☐ Formula 160

$\displaystyle = \sec\frac{x}{2}\tan\frac{x}{2} + \ln\left|\sec\frac{x}{2} + \tan\frac{x}{2}\right| + C$

35. $\displaystyle\int \sin^4 x\, dx$ ☐ Formula 139

$\displaystyle = \frac{3x}{8} - \frac{\sin 2x}{4} + \frac{\sin 4x}{32} + C$

37. $\displaystyle\int \sqrt{9 - x^2}\, dx$ ☐ Formula 117

$\displaystyle = \frac{x}{2}\sqrt{9 - x^2} + \frac{9}{2}\sin^{-1}\frac{x}{3} + C$

39. $\displaystyle\int \cos^2 x\, dx = \int \frac{1 + \cos 2x}{2}\, dx$

$\displaystyle = \frac{1}{2}x + \frac{1}{2}\frac{\sin 2x}{2} + C$

$\displaystyle = \frac{1}{2}x + \frac{1}{4}\sin 2x + C$

41. Let $u = \cos x,\ du = -\sin x\, dx$;

$\displaystyle\int \sin^3 x \cos^4 x\, dx = \int \sin^2 x \cos^4 x(\sin x\, dx)$

$\displaystyle = \int (1 - \cos^2 x)\cos^4 x(\sin x\, dx)$

$\displaystyle = \int \cos^4 x(\sin x\, dx) - \int \cos^6 x(\sin x\, dx)$

$\displaystyle = -\int u^4\, du + \int u^6\, du$

$\displaystyle = -\frac{u^5}{5} + \frac{u^7}{7} + C$

$\displaystyle = -\frac{1}{5}\cos^5 x + \frac{1}{7}\cos^7 x + C$

43. $\displaystyle\int \sin^2 x \cos^2 x\, dx$

$\displaystyle = \frac{1}{4}\int \sin^2 2x\, dx$

$$= \frac{1}{4}\left(\frac{1}{2}x - \frac{1}{8}\sin 4x\right) + C$$

$$= \frac{x}{8} - \frac{1}{32}\sin 4x + C$$

45. $\displaystyle\int \frac{dx}{x^{1/2} + x^{1/4}}$ $\boxed{\text{Let } u = x^{1/4};\ du = \frac{1}{4}x^{-3/4}dx}$

$$= \int \frac{4u^3 du}{u^2 + u}$$

$$= 4\int \frac{u^2 du}{u + 1}$$ $\boxed{\text{Formula 36}}$

$$= 4\left[\frac{(u+1)^2}{2} - \frac{2(u+1)}{1} + \ln|u+1|\right] + C_1$$

$$= 4\left[\frac{1}{2}u^2 + u + \frac{1}{2} - 2u - 2 + \ln|u+1|\right] + C_1$$

$$= 4\left[\frac{1}{2}u^2 - u + \ln|u+1| + C\right]$$

$$= 4\left[\frac{x^{1/2}}{2} - x^{1/4} + \ln\left(x^{1/4} + 1\right)\right] + C$$

47. $\displaystyle\int \frac{18\tan^2 t \sec^2 t}{(2 + \tan^3 t)^2}\,dt$ $\boxed{\begin{array}{l}\text{Let } u = 2 + \tan^3 t; \\ du = 3\tan^2 t \sec^2 t\,dt\end{array}}$

$$= \int \frac{6\,du}{u^2}$$

$$= -6u^{-1} + C$$

$$= -6\left(2 + \tan^3 t\right)^{-1} + C$$

49. $\displaystyle V = \pi\int_0^9 \left(\frac{x^{3/2}}{\sqrt{x^2 + 9}}\right)^2 dx$

$$= \pi\int_0^9 \frac{x^3}{x^2 + 9}\,dx$$

$$= \pi\left[\int_0^9 x\,dx - \int_0^9 \frac{9x}{x^2 + 9}\,dx\right]$$

$\boxed{\begin{array}{l}\text{Let } u = x^2 + 9; du = 2x\,dx \\ \text{If } x = 0, \text{ then } u = 9 \text{ and if } x = 9, \text{ then } u = 90.\end{array}}$

$$= \pi\frac{x^2}{2}\Big|_0^9 - \pi\int_9^{90} \frac{9x\left(\frac{du}{2x}\right)}{u}$$

$$= \frac{81}{2}\pi - \frac{9}{2}\pi\ln u\Big|_9^{90}$$

$$= \frac{81}{2}\pi - \frac{9}{2}\pi(\ln 90 - \ln 9)$$

$$= \frac{81}{2}\pi - \frac{9}{2}\pi\ln 10$$

$$= \frac{9\pi}{2}(9 - \ln 10)$$

51. $\displaystyle V = 2\pi\int_1^4 x\left[\frac{1}{\sqrt{x}}\left(1 + \sqrt{x}\right)^{1/3}\right]dx$

$$= 2\pi\int_1^4 \sqrt{x}\left(1 + \sqrt{x}\right)^{1/3}dx$$

$\boxed{\begin{array}{l}\text{Let } u = 1 + \sqrt{x},\ du = \dfrac{1}{2\sqrt{x}}\,dx \\ \text{If } x = 1,\ u = 2 \text{ and if } x = 4, \text{ then } u = 3.\end{array}}$

$$= 4\pi\int_2^3 (u - 1)^2 u^{1/3}du$$

$$= \frac{3}{35}\pi u^{4/3}\left(14u^2 - 40u + 35\right)\Big|_2^3$$

$$= \frac{\pi}{35}\left(369\sqrt[3]{3} - 66\sqrt[3]{2}\right)$$

53. $xy' = \sqrt{(\ln x)^2 - x^2}$

$$y' = \frac{1}{x}\sqrt{(\ln x)^2 - x^2}$$

$$= \sqrt{\frac{(\ln x)^2 - x^2}{x^2}}$$

Now, we find the arc length

$$s = \int_{1/4}^{1/2} \sqrt{1 + \frac{(\ln x)^2}{x^2} - 1}\,dx$$

$$= -\int_{1/4}^{1/2} \frac{\ln x}{x}\,dx$$ $\boxed{\begin{array}{l}\text{Let } u = \ln x; \\ du = \dfrac{1}{x}dx\end{array}}$

$$= -\ln^2 x\Big|_{1/4}^{1/2} = \frac{3}{2}\ln^2 2$$

55. $ds = \sqrt{1 + 4x^2}\, dx; \; dS = 2\pi y \, ds$

$S = 2\pi \displaystyle\int_0^1 x^2 \sqrt{1 + 4x^2}\, dx$

$= 2\pi \displaystyle\int_0^2 \frac{u^2}{4} \sqrt{1 + u^2} \left(\frac{du}{2}\right) \boxed{\begin{array}{l} \text{Formula 87;} \\ u = 2x \end{array}}$

$= \dfrac{\pi}{4} \left[\dfrac{u(u^2 + 1)^{3/2}}{4} - \dfrac{u(u^2 + 1)^{1/2}}{8} \right.$

$\left. \qquad\quad - \dfrac{1}{8} \ln\left(u + \left(u^2 + 1^2\right)^{1/2} \right) \right]_0^2$

$= \dfrac{9\sqrt{5}\,\pi}{16} - \dfrac{\pi \ln\left(\sqrt{5} + 2\right)}{32}$

$= \dfrac{\pi}{32} \left[18\sqrt{5} - \ln\left(\sqrt{5} + 2\right) \right]$

57. $\displaystyle\int \csc x \, dx = -\int \dfrac{-\csc x (\csc x + \cot x)}{\csc x + \cot x}\, dx$

$\qquad\qquad = -\ln|\csc x + \cot x| + C$

59. Let $u = \pi - x$, $du = -dx$. Then

$\sin x = \sin u$

$\displaystyle\int_0^\pi x f(\sin x)\, dx$

$= \displaystyle\int_\pi^0 (\pi - u) f(\sin u)(-du)$

$= \displaystyle\int_0^\pi (\pi - u) f(\sin u)\, du$

$= \displaystyle\int_0^\pi \pi f(\sin u)\, du - \int_0^\pi u f(\sin u)\, du$

The value of the integral is independent of the variable used so we can write:

$= \pi \displaystyle\int_0^\pi f(\sin x)\, dx - \int_0^\pi x f(\sin x)\, dx$

Add the second integral to both sides

and divide by 2 to obtain the desired result.

7.2 Integration by Parts, page 521

> **SURVIVAL HINT:** *The formula for the integration by parts:*
>
> $$\int u \, dv = uv - \int v \, du$$
>
> *is one of the most important formulas in this book. Make sure you that thoroughly understand how to use it.*

1. $\displaystyle\int x e^{-2x}\, dx \quad \boxed{u = x; \; dv = e^{-2x}\, dx}$

$= -\dfrac{1}{2} x e^{-2x} + \dfrac{1}{2} \displaystyle\int e^{-2x}\, dx$

$= -\dfrac{1}{2} x e^{-2x} - \dfrac{1}{4} e^{-2x} + C$

$= -\dfrac{1}{4} e^{-2x}(2x + 1) + C$

3. $\displaystyle\int x \ln x \, dx \quad \boxed{u = \ln x; \; dv = x \, dx}$

$= \dfrac{1}{2} x^2 \ln x - \dfrac{1}{2} \displaystyle\int x \, dx$

$= \dfrac{1}{2} x^2 \ln x - \dfrac{1}{4} x^2 + C$

$= \dfrac{x^2}{4}(2 \ln x - 1) + C$

5. $\displaystyle\int \sin^{-1} x \, dx \quad \boxed{u = \sin^{-1} x; \; dv = dx}$

$= x \sin^{-1} x - \displaystyle\int \dfrac{x \, dx}{\sqrt{1 - x^2}}$

$= x \sin^{-1} x + \sqrt{1 - x^2} + C$

7. $I = \displaystyle\int e^{-3x}\cos 4x\, dx$ $\boxed{u = e^{-3x};\ dv = \cos 4x\, dx}$

$\quad = \dfrac{1}{4}e^{-3x}\sin 4x + \dfrac{3}{4}\displaystyle\int e^{-3x}\sin 4x\, dx$

$\hspace{4cm}\boxed{u = e^{-3x};\ dv = \sin 4x\, dx}$

$\quad = \dfrac{1}{4}e^{-3x}\sin 4x - \dfrac{3}{16}\left(e^{-3x}\cos 4x\right)$

$\qquad\quad - \dfrac{9}{16}\displaystyle\int e^{-3x}\cos 4x\, dx$

$\dfrac{25}{16}I = \dfrac{1}{4}e^{-3x}\sin 4x - \dfrac{3}{16}\left(e^{-3x}\cos 4x\right) + C_1$

$\quad I = \dfrac{4}{25}e^{-3x}\sin 4x - \dfrac{3}{25}e^{-3x}\cos 4x + C$

9. $\displaystyle\int x^2\ln x\, dx\ \boxed{u = \ln x;\ dv = x^2\, dx}$

$\quad = \dfrac{1}{3}x^3\ln x - \dfrac{1}{3}\displaystyle\int x^2\, dx$

$\quad = \dfrac{1}{3}x^3\ln x - \dfrac{1}{9}x^3 + C$

$\quad = \dfrac{x^3}{9}\left(3\ln x - 1\right) + C$

11.

$\displaystyle\int \sin(\ln x)\, dx\ \boxed{u = \sin(\ln x);\ dv = dx}$

$\quad = x\sin(\ln x) - x\displaystyle\int \cos(\ln x)\, dx$

$\hspace{3cm}\boxed{u = \cos(\ln x);\ dv = dx}$

$\quad = x\sin(\ln x) - x\cos(\ln x)\, dx - \displaystyle\int \sin(\ln x)\, dx$

$2\displaystyle\int \sin(\ln x)\, dx = x[\sin(\ln x) - \cos(\ln x)] + C_1$

$\displaystyle\int \sin(\ln x)\, dx = \dfrac{x}{2}[\sin(\ln x) - \cos(\ln x)] + C$

13. $\displaystyle\int \ln(x^2 + 1)\, dx\ \boxed{u = \ln(x^2 + 1);\ dv = dx}$

$\quad = x\ln(x^2 + 1) - \displaystyle\int \dfrac{2x^2}{x^2 + 1}\, dx$

$\quad = x\ln(x^2 + 1) - 2\displaystyle\int\left(1 - \dfrac{1}{x^2 + 1}\right)dx$

$\quad = x\ln(x^2 + 1) - 2x + 2\tan^{-1}x + C$

15. $\displaystyle\int \dfrac{xe^{-x}}{(x-1)^2}\, dx$ $\boxed{\begin{array}{l} u = xe^{-x};\ dv = \frac{dx}{(x-1)^2} \\ du = (1-x)e^{-x}dx;\ v = \frac{-1}{x-1} \end{array}}$

$\quad = \dfrac{-xe^{-x}}{x - 1} - \displaystyle\int \dfrac{-1}{x - 1}(1 - x)e^{-x}dx$

$\quad = \dfrac{-xe^{-x}}{x - 1} + e^{-x} + C$

$\quad = e^{-x}\left(\dfrac{x}{1 - x} + 1\right) + C$

$\quad = e^{-x}\left(\dfrac{1}{1 - x}\right) + C$

17. $\displaystyle\int_1^4 \sqrt{x}\ln x\, dx\ \boxed{u = \ln x;\ dv = \sqrt{x}\, dx}$

$\quad = \dfrac{2}{3}x^{3/2}\ln x\,\Big|_1^4 - \dfrac{2}{3}\displaystyle\int_1^4 x^{1/2}\, dx$

$\quad = \dfrac{16}{3}\ln 4 - \dfrac{4}{9}x^{3/2}\,\Big|_1^4$

$\quad = \dfrac{32}{3}\ln 2 - \dfrac{28}{9}$

19. $\displaystyle\int_1^e (\ln x)^2\, dx\ \boxed{u = \ln^2 x;\ dv = dx}$

$\quad = x(\ln x)^2\big|_1^e - 2\displaystyle\int_1^e \ln x\, dx\ \boxed{u = \ln x;\ dv = dx}$

$\quad = e(\ln e)^2 - 2\left[x\ln x\big|_1^e - \displaystyle\int_1^e dx\right]$

$\quad = e - 2[e\ln e - 1\ln 1 - x\big|_1^e] = e - 2$

21. $I = \displaystyle\int_0^\pi e^{2x}\cos 2x\, dx\ \boxed{u = e^{2x};\ dv = \cos 2x\, dx}$

$\quad = \dfrac{1}{2}e^{2x}\sin 2x\,\Big|_0^\pi - \displaystyle\int_0^\pi e^{2x}\sin 2x\, dx$

$\hspace{3cm}\boxed{u = e^{2x};\ dv = \sin 2x\, dx}$

$$= \frac{1}{2}e^{2\pi}(0) - \frac{1}{2}(1)(0)$$

$$- \left[-\frac{1}{2}e^{2x}\cos 2x \Big|_0^\pi + \int_0^\pi e^{2x}\cos 2x \, dx \right]$$

$$= \frac{1}{2}e^{2\pi}(1) - \frac{1}{2}(1)(1) - \int_0^\pi e^{2x}\cos 2x \, dx$$

$$2I = \frac{1}{2}e^{2\pi} - \frac{1}{2}$$

$$I = \frac{1}{4}\left(e^{2\pi} - 1\right)$$

23. Integration by parts is the application of the formula

$$\int u \, dv = uv - \int v \, du$$

The u factor is a part of the integrand that is differentiated and dv is the part that is integrated. Generally, pick dv as complicated as possible yet still integrable, so that the integral on the right is easier to integrate than the original integral.

25. $\int [\sin 2x \ln(\cos x)] \, dx$

$$= \int 2\sin x \cos x \ln(\cos x) \, dx \quad \boxed{\begin{array}{l} t = \cos x; \\ dt = -\sin x \, dx \end{array}}$$

$$= -2\int t \ln t \, dt \quad \boxed{u = \ln t; \ dv = t \, dt}$$

$$= -2\left[\frac{1}{2}t^2\ln t - \frac{1}{2}\int t \, dt \right]$$

$$= -t^2\ln t + \frac{1}{2}t^2 + C$$

$$= \frac{1}{2}\cos^2 x - (\cos^2 x)\ln(\cos x) + C$$

$$= \frac{1}{2}\cos^2 x[1 - 2\ln(\cos x)] + C$$

27. $\int [\sin x \ln(2 + \cos x)] \, dx \quad \boxed{\begin{array}{l} t = 2 + \cos x; \\ dt = -\sin x \, dx \end{array}}$

$$= -\int \ln t \, dt \quad \boxed{u = \ln t; \ dv = dt}$$

$$= -t \ln t + t + C$$

$$= -(2 + \cos x)\ln(2 + \cos x) + (2 + \cos x) + C$$

29. a. $\int \frac{x^3}{x^2 - 1} \, dx \quad \boxed{u = x^2; \ dv = \frac{x \, dx}{x^2 - 1}}$

$$= \frac{1}{2}x^2\ln|x^2 - 1| - \int x \ln|x^2 - 1| \, dx$$

$$= \frac{1}{2}x^2\ln|x^2 - 1| - \frac{1}{2}\int \ln|x^2 - 1|(2x \, dx)$$

$$= \frac{1}{2}x^2\ln|x^2 - 1| - \frac{1}{2}(x^2 - 1)\ln|x^2 - 1|$$

$$\quad + \frac{1}{2}(x^2 - 1) + C_1$$

$$= \frac{1}{2}(\ln|x^2 - 1| + x^2) + C$$

b. $\int \frac{x^3}{x^2 - 1} \, dx = \int [x + x(x^2 - 1)^{-1}] \, dx$

$$= \frac{1}{2}x^2 + \frac{1}{2}\ln|x^2 - 1| + C$$

$$= \frac{1}{2}(\ln|x^2 - 1| + x^2) + C$$

31. $I = \int \cos^2 x \, dx \quad \boxed{u = \cos x; \ dv = \cos x \, dx}$

$$= \sin x \cos x + \int \sin^2 x \, dx$$

$$= \sin x \cos x + \int (1 - \cos^2 x) \, dx$$

$$= \sin x \cos x + x - \int \cos^2 x \, dx$$

$$2I = \sin x \cos x + x + C$$

$$I = \frac{1}{2}(x + \sin x \cos x) + C$$

$$= \frac{x}{2} + \frac{1}{4}\sin 2x + C$$

33. $\displaystyle\int x\cos^2 x\,dx$ $\boxed{\begin{array}{l}u=x;\ dv=\cos^2 x\,dx\\ du=dx;\ v=\frac{x}{2}+\frac{\sin 2x}{4}\\ \text{(From Problem 31)}\end{array}}$

$$= x\left(\frac{x}{2}+\frac{\sin 2x}{4}\right)-\int\left(\frac{x}{2}+\frac{\sin 2x}{4}\right)dx$$

$$= \frac{1}{2}x^2+\frac{1}{4}x\sin 2x-\frac{1}{4}x^2+\frac{1}{8}\cos 2x+C$$

$$= \frac{1}{4}x^2+\frac{1}{4}x\sin 2x+\frac{1}{8}\cos 2x+C$$

35. $\displaystyle\int x^n\ln x\,dx$ $\boxed{u=\ln x;\ dv=x^n\,dx}$

$$= \frac{x^{n+1}}{n+1}\ln x-\frac{1}{n+1}\int x^n dx$$

$$= \frac{x^{n+1}}{n+1}\ln x-\frac{x^{n+1}}{(n+1)^2}+C$$

$$= \frac{x^{n+1}}{n+1}\left(\ln x-\frac{1}{n+1}\right)+C$$

37. Let $Q(t)$ be the number of units produced in the first t hours.

$$Q(3)=\int_0^3 Q'(t)\,dt$$

$$= \int_0^3 100te^{-0.5t}dt \quad \boxed{\begin{array}{l}u=t;\ dv=e^{-0.5t}\,dt\\ du=dt;\ v=-2e^{-0.5t}\end{array}}$$

$$= 100\left[-2te^{-0.5t}\big|_0^3+2\int_0^3 e^{-0.5t}dt\right]$$

$$= 100\left[-2te^{-0.5t}+2(-2e^{-0.5t})\right]\big|_0^3$$

$$= -200e^{-0.5t}(t+2)\big|_0^3$$

$$= -200e^{-0.5(3)}(3+2)+400$$

$$= \text{176.8698398}$$

The production is about 177 units.

39. Let $Q(t)$ be the amount of money raised in t weeks.

$$\frac{dQ}{dt}=2{,}000te^{-0.2t}$$

$$Q(t)=\int_0^5 2{,}000te^{-0.2t}\,dt \quad \boxed{u=t;\ dv=e^{-0.2t}\,dt}$$

$$= 2{,}000(-5te^{-0.2t})\big|_0^5+10{,}000\int_0^5 e^{-0.2t}\,dt$$

$$= \left[-10{,}000te^{-0.2t}-50{,}000e^{-0.2t}\right]\big|_0^5$$

$$= -100{,}000e^{-1}+50{,}000$$

$$\approx \$13{,}212$$

41. $\displaystyle V=2\pi\int_0^2 xe^{-x}\,dx$ $\boxed{u=x;\ dv=e^{-x}\,dx}$

$$= 2\pi\left(-xe^{-x}\big|_0^2+\int_0^2 e^{-x}\,dx\right)$$

$$= 2\pi(-xe^{-x}-e^{-x})\big|_0^2$$

$$= 2\pi\left(1-3e^{-2}\right)$$

43. The curves intersect when $x=0$. Assume the density is 1.

$$A=\int_0^1(e^x-e^{-x})\,dx$$

$$= e+e^{-1}-2$$

$$M_x=\frac{1}{2}\int_0^1(e^x-e^{-x})(e^x+e^{-x})\,dx$$

$$= \frac{1}{2}\int_0^1\left(e^{2x}-e^{-2x}\right)dx$$

$$= \frac{1}{4}\left(e^2+e^{-2}-2\right)$$

$$M_y=\int_0^1 x(e^x-e^{-x})\,dx \quad \boxed{u=x;\ dv=(e^x-e^{-x})dx}$$

$$= x(e^x+e^{-x})\big|_0^1-\int_0^1(e^x+e^{-x})dx$$

$$= [x(e^x+e^{-x})-(e^x-e^{-x})]\big|_0^1$$

$$= \frac{2}{e}$$

$$(\bar{x},\bar{y})=\left(\frac{\frac{2}{e}}{e+e^{-1}-2},\ \frac{\frac{1}{4}(e^2+e^{-2}-2)}{e+e^{-1}-2}\right)$$

$$\approx (0.68,1.27)$$

45.
$$\frac{dy}{dx} = \sqrt{xy}\ln x$$
$$y^{-1/2}dy = x^{1/2}\ln x\, dx$$
$$\int y^{-1/2}dy = \int x^{1/2}\ln x\, dx \quad \boxed{u = \ln x;\ dv = x^{1/2}\, dx}$$
$$2y^{1/2} = \frac{2}{3}x^{3/2}\ln x - \frac{2}{3}\int x^{1/2}dx$$
$$2y^{1/2} = \frac{2}{3}x^{3/2}\ln x - \frac{4}{9}x^{3/2} + C$$
$$2y^{1/2} = \frac{2}{3}x^{3/2}\left(\ln x - \frac{2}{3}\right) + C$$

47.
$$\frac{dy}{dx} = \frac{-xy}{\sec x}$$
$$-y^{-1}dy = x\cos x\, dx$$
$$-\int y^{-1}dy = \int x\cos x\, dx$$
$$\boxed{\text{By parts or by Formula 123}}$$
$$-\ln|y| = \cos x + x\sin x + C$$
At $(0,1), 0 = 1 + C$ so $C = -1$.
$$y = \exp(1 - \cos x - x\sin x)$$

49. $\displaystyle\int_0^\pi [f(x) + f''(x)]\sin x\, dx$

$$= \int_0^\pi f(x)\sin x\, dx + \int_0^\pi f''(x)\sin x\, dx$$
$$\boxed{\begin{array}{l} u = f(x);\ dv = \sin x\, dx \\ u = \sin x;\ dv = f''(x)\, dx \end{array}}$$
$$= -\cos x\, f'(x)\big|_0^\pi + \int_0^\pi f'(x)\cos x\, dx$$
$$+ \sin x\, f'(x)\big|_0^\pi - \int_0^\pi f'(x)\cos x\, dx$$
$$= -\cos\pi\, f(\pi) + \cos 0\, f(0)$$
This is zero (given), so we see
$$-\cos\pi\, f(\pi) + \cos 0\, f(0) = 0$$
$$f(\pi) = -f(0)$$
$$= -3$$
since it is also given that $f(0) = 3$.

51. $v(t) = -r\ln\dfrac{w - kt}{w} - gt$

$$s(t) = \int v(t)\, dt$$
$$= \frac{r(w - kt)}{k}\ln\left(\frac{w - kt}{w}\right) - \frac{1}{2}gt^2 + rt$$

Since $s(0) = 0$ (the rocket starts at ground level), we find (by substituting the given information),

$$s(120) = \frac{8{,}000[30{,}000 - 200(120)]}{200}\ln\left(\frac{30{,}000 - 200(120)}{30{,}000}\right)$$
$$- \frac{1}{2}(32)(120)^2 + 8{,}000(120)$$
$$\approx 343{,}335 \text{ ft}$$
$$\approx 65 \text{ miles}$$

53. Average displacement is:
$$\frac{1}{\frac{\pi}{5}}\int_0^{\pi/5} 2.3e^{-0.25t}\cos 5t\, dt$$
$$= \frac{5}{\pi}(2.3)\left[\frac{e^{-0.25t}(-0.25\cos 5t + 5\sin 5t)}{(-0.25)^2 + (5)^2}\right]\Bigg|_0^{\pi/5}$$
$$\approx 0.07$$

55. $\displaystyle\int\frac{\sin x\, dx}{\sqrt{\cos 2x}} = \int\frac{\sin x\, dx}{\sqrt{2\cos^2 x - 1}} \quad \boxed{\text{Let } u = \cos x}$

$$= -\int\frac{du}{\sqrt{2u^2 - 1}}$$
$$= -\frac{1}{\sqrt{2}}\int\frac{du}{\sqrt{u^2 - \frac{1}{2}}} \quad \boxed{\text{Formula 98}}$$
$$= -\frac{1}{\sqrt{2}}\ln\left|\cos x + \sqrt{\cos^2 x - \frac{1}{2}}\right| + C$$

57. $\displaystyle\int x^n e^x\, dx \quad \boxed{u = x^n;\ dv = e^x\, dx}$

$$= x^n e^x - n\int x^{n-1}e^x\, dx$$

59. $I = \displaystyle\int_0^{\pi/2}\sin^n x\, dx$, n is even

$$\boxed{u = (\sin x)^{n-1};\ dv = \sin x\, dx}$$

$$= \left(-\sin^{n-1}x\cos x\right)\Big|_0^{\pi/2}$$

$$+ \int_0^{\pi/2} (n-1)\sin^{n-2}x\cos^2x\,dx$$

$$= 0 + (n-1)\int_0^{\pi/2} \left[\sin^{n-2}x\left(1-\sin^2x\right)\right]dx$$

$$= (n-1)\int_0^{\pi/2}\sin^{n-2}x\,dx - (n-1)I$$

$$nI = (n-1)\int_0^{\pi/2}\sin^{n-2}x\,dx$$

$$I = \frac{n-1}{n}\int_0^{\pi/2}\sin^{n-2}x\,dx$$

This recursive formula will be used repeatedly with values of n that decrease by 2.

$$I = 0 + \frac{n-1}{n}\int_0^{\pi/2}\sin^{n-3}x\,dx$$

$$= \frac{n-1}{n}\left[-\frac{\sin^{n-2}x\cos x}{n-3}\right]\Bigg|_0^{\pi/2}$$

$$+ \left(\frac{n-1}{n}\right)\left(\frac{n-3}{n-2}\right)\int_o^{\pi/2}\sin^{n-4}x\,dx$$

$$= \cdots$$

$$= \frac{(n-1)(n-3)(n-5)\cdots(5)(3)}{n(n-2)\cdots(4)}\int_0^{\pi/2}\sin^2x\,dx$$

$$= \frac{(n-1)(n-3)(n-5)\cdots(5)(3)}{n(n-2)\cdots(2)}\int_0^{\pi/2}(1+\cos 2x)\,dx$$

$$= \frac{1\cdot 3\cdot 5\cdots(n-3)(n-1)}{2\cdot 4\cdot 6\cdots(n-2)n}\cdot\frac{\pi}{2}$$

For $\displaystyle\int_0^{\pi/2}\cos^n dx$, proceed as shown above.

7.3 Trigonometric Methods, page 530

SURVIVAL HINT: *If you saved your trigonometry book, now would be a good time to get it off the shelf and dust it off. If you did not, or if your memory of trigonometry goes back to an algebra or precalculus course in high school, you might want to find a source book for trigonometry.*

1. Convert $\sin^2 x = \frac{1}{2}(1 - \cos 2x)$ and $\cos^2 x = \frac{1}{2}(1 + \cos 2x)$ and substitute into the given integral.

3. Substitute $u = a\tan\theta$ and then integrate the resulting integral in trigonometric form. It may help to draw a reference triangle.

5. $\displaystyle\int\cos^3 x\,dx = \int\cos^2 x(\cos x\,dx)$ $\boxed{u = \sin x}$

$$= \int\left(1 - u^2\right)du$$

$$= u - \frac{u^3}{3} + C$$

$$= \sin x - \frac{1}{3}\sin^3 x + C$$

7. $\displaystyle\int\sin^2 x\cos^3 x\,dx = \int\sin^2 x\cos^2 x(\cos x\,dx)$

$$= \int\left(\sin^2 x - \sin^4 x\right)(\cos x)\,dx\quad\boxed{u = \sin x}$$

$$= \int\left(u^2 - u^4\right)du$$

$$= \frac{1}{3}u^3 - \frac{1}{5}u^5 + C$$

$$= \frac{1}{3}\sin^3 x - \frac{1}{5}\sin^5 x + C$$

9. $\displaystyle\int\sqrt{\cos t}\sin t\,dt\quad\boxed{u = \sin t}$

$$= \int u^{1/2}(-du)$$

$$= -\frac{2}{3}u^{3/2} + C$$

$$= -\frac{2}{3}(\cos t)^{3/2} + C$$

11. $\displaystyle\int e^{\cos x}\sin x\,dx\quad\boxed{u = \sin x}$

$$= \int e^u(-du)$$

$$= -e^{\cos x} + C$$

13. $\displaystyle\int \sin^2 x \cos^2 x \, dx$

$$= \int \left[\frac{1 - \cos 2x}{2}\right]\left[\frac{1 + \cos 2x}{2}\right] dx$$

$$= \frac{1}{4}\int (1 - \cos^2 2x) \, dx$$

$$= \frac{1}{4}x - \frac{1}{4}\int \left[\frac{1 + \cos 4x}{2}\right] dx$$

$$= \frac{1}{4}x - \frac{1}{8}x - \frac{1}{8}\left[\frac{\sin 4x}{4}\right] + C$$

$$= \frac{1}{8}x - \frac{1}{32}\sin 4x + C$$

15. $\displaystyle\int \tan 2\theta \, d\theta = -\frac{1}{2}\ln|\cos 2\theta| + C$

17. $\displaystyle\int \tan^3 x \sec^4 x \, dx = \int \tan^3 x \sec^2 x(\sec^2 x \, dx)$

$$= \int \tan^3 x(\tan^2 x + 1)(\sec^2 x \, dx)$$

$\boxed{u = \tan x}$ $\quad = \displaystyle\int (u^5 + u^3)du$

$$= \frac{1}{6}u^6 + \frac{1}{4}u^4 + C$$

$$= \frac{1}{6}\tan^6 x + \frac{1}{4}\tan^4 x + C$$

19. $\displaystyle\int (\tan^2 x + \sec^2 x) \, dx = \int (2\sec^2 x - 1) \, dx$

$$= 2\tan x - x + C$$

21. $\displaystyle\int \tan^2 u \sec u \, du = \int (\sec^2 u - 1)\sec u \, du$

$$= \int (\sec^3 u - \sec u) \, du$$

$$= \frac{\sec u \tan u}{2} + \frac{1}{2}\ln|\sec u + \tan u| \quad \boxed{\text{Formula 160}}$$

$$- \ln|\sec u + \tan u| \quad \boxed{\text{Formula 158}}$$

$$= \frac{1}{2}\sec u \tan u - \frac{1}{2}\ln|\sec u + \tan u| + C$$

23. $\displaystyle\int \sqrt[3]{\tan x}\, \sec^2 x \, dx = \int u^{1/3}du \boxed{u = \tan x}$

$$= \frac{3}{4}(\tan x)^{4/3} + C$$

25. $\displaystyle\int x \sin x^2 \cos x^2 \, dx \quad \boxed{u = x^2}$

$$= \frac{1}{2}\int \sin u(\cos u \, du)$$

$$= \frac{1}{4}\sin^2 x^2 + C$$

27. $\displaystyle\int \tan^4 t \sec t \, dt = \int (\sec^2 t - 1)^2 \sec t \, dt$

$$= \int (\sec^5 t - 2\sec^3 t + \sec t)dt$$

$$= \frac{\sec^3 t \tan t}{4} + \frac{3}{4}\int \sec^3 t \, dt$$

$$- 2\int \sec^3 t \, dt + \int \sec t \, dt \quad \boxed{\text{Formula 161}}$$

$$= \frac{1}{4}\sec^3 t \tan t - \frac{5}{4}\int \sec^3 t \, dt + \ln|\sec t + \tan t|$$

$$= \frac{1}{4}\sec^3 t \tan t - \frac{5}{8}\sec t \tan t + \frac{3}{8}\ln|\sec t + \tan t| + C$$

29. $\displaystyle\int \csc^3 x \cot x \, dx$

$$= -\int \csc^2 x(-\csc x \cot x \, dx)$$

$$= -\frac{1}{3}\csc^3 x + C$$

31. $\displaystyle\int \csc^2 x \cos x \, dx = \int \frac{\cos x \, dx}{\sin^2 x} \quad \boxed{u = \sin x}$

$$= \int \frac{du}{u^2}$$

$$= -(\sin x)^{-1} + C$$

$$= -\csc x + C$$

33. $\displaystyle\int \sqrt{9 - 4t^2}\, dt \quad \boxed{t = \frac{3}{2}\sin\theta}$

$$= \int \sqrt{9 - 9\sin^2\theta}\left(\frac{3}{2}\cos\theta\, d\theta\right)$$

$$= \frac{9}{2}\int \cos^2\theta\, d\theta$$

$$= \frac{9}{2}\left[\frac{\theta}{2} + \frac{\sin\theta\cos\theta}{2}\right] + C$$

$$= \frac{9}{4}\sin^{-1}\frac{2t}{3} + \frac{3}{2}t\sqrt{9 - 4t^2} + C$$

35. $\displaystyle\int \frac{x + 1}{\sqrt{4 + x^2}}\, dx$ $\boxed{\text{Formula } 89}$

$$= \frac{1}{2}\int \frac{2x\, dx}{\sqrt{4 + x^2}} + \int \frac{dx}{\sqrt{4 + x^2}}$$

$$= \sqrt{4 + x^2} + \ln\left|x + \sqrt{4 + x^2}\right| + C$$

37. $\displaystyle\int \frac{dx}{\sqrt{x^2 - 7}}$ $\boxed{\text{Formula } 98}$

$$= \ln\left|x + \sqrt{x^2 - 7}\right| + C$$

39. $\displaystyle\int \frac{dx}{\sqrt{5 - x^2}} = \sin^{-1}\frac{x}{\sqrt{5}} + C$

41. $\displaystyle\int \frac{dx}{x^2\sqrt{4 - x^2}}$ $\boxed{x = 2\sin\theta}$

$$= \int \frac{2\cos\theta\, d\theta}{(4\sin^2\theta)\sqrt{4 - 4\sin^2\theta}}$$

$$= \frac{1}{4}\int \csc^2\theta\, d\theta$$

$$= -\frac{1}{4}\cot\theta + C$$

$$= -\frac{\sqrt{4 - x^2}}{4x} + C$$

43. $\displaystyle\int \frac{\sqrt{x^2 - 4}}{x}\, dx$ $\boxed{\begin{array}{l}\text{Let } x = 2\sec\theta; \\ dx = 2\sec\theta\tan\theta\, d\theta \\ \text{or Formula } 109\end{array}}$

$$= \int \frac{\sqrt{4\sec^2\theta - 4}}{2\sec\theta}(2\sec\theta\tan\theta\, d\theta)$$

$$= \int 2\tan^2\theta\, d\theta$$

$$= 2\int \left(\sec^2\theta - 1\right) d\theta$$

$$= 2\tan\theta - 2\theta + C$$

$$= \sqrt{x^2 - 4} - 2\sec^{-1}\frac{x}{2} + C$$

45. $\displaystyle\int \frac{dx}{9 - (x + 1)^2}$ $\boxed{\begin{array}{l}x + 1 = 3\sin\theta \\ \text{or Formula } 52\end{array}}$

$$= \int \frac{3\cos\theta\, d\theta}{9 - 9\sin^2\theta}$$

$$= \frac{1}{3}\int \frac{d\theta}{\cos\theta}$$

$$= \frac{1}{3}\int \sec\theta\, d\theta$$

$$= \frac{1}{3}\ln|\sec\theta + \tan\theta| + C$$

$$= \frac{1}{3}\ln\left|\frac{3}{\sqrt{9 - (x+1)^2}} + \frac{x + 1}{\sqrt{9 - (x+1)^2}}\right| + C$$

$$= \frac{1}{3}\ln\left|\frac{x + 4}{\sqrt{9 - (x+1)^2}}\right| + C$$

47. $\displaystyle\int \frac{dx}{\sqrt{x^2 - 2x + 6}}$

$$= \int \frac{dx}{\sqrt{(x - 1)^2 + 5}}$$ $\boxed{x - 1 = \sqrt{5}\tan\theta}$

$$= \int \frac{\sqrt{5}\sec^2\theta\, d\theta}{\sqrt{5\tan^2\theta + 5}}$$

$$= \int \sec\theta\, d\theta$$

$$= \ln|\sec\theta + \tan\theta| + C_1$$

This can be written $\ln\left|\sqrt{x^2 - 2x + 6} + x - 1\right| + C$.

49. $\displaystyle\int \frac{\sin^3 u\, du}{\cos^5 u}$

$$= \int \tan^3 u\sec^2 u\, du$$ $\boxed{w = \tan u}$

Page 255

$$= \frac{1}{4}\tan^4 u + C$$

51. $A = \dfrac{1}{\pi - 0}\displaystyle\int_0^\pi \sin^2 x\, dx$

$$= \frac{1}{\pi}\left[\frac{x}{2} - \frac{\sin x \cos x}{2}\right]_0^\pi$$

$$= \frac{1}{\pi}\left[\frac{\pi}{2}\right]$$

$$= \frac{1}{2}$$

53. $V = 2\pi\displaystyle\int_0^\pi x\sin^2 x\, dx = \dfrac{\pi^3}{2} \approx 15.5031$

55. $\displaystyle\int \sin 3x \sin 5x\, dx$

$$= \int \frac{1}{2}[\cos(-2x) - \cos 8x]\, dx$$

$$= -\frac{1}{4}\sin(-2x) - \frac{1}{16}\sin 8x + C$$

$$= \frac{1}{4}\sin 2x - \frac{1}{16}\sin 8x + C$$

57. $\displaystyle\int \sin^2 3x \cos 4x\, dx$

$$= \int \frac{1}{2}(1 - \cos 6x)\cos 4x\, dx$$

$$= \int \left\{\frac{1}{2}\cos 4x - \frac{1}{4}[\cos 2x + \cos 10x]\right\} dx$$

$$= \frac{1}{8}\sin 4x - \frac{1}{8}\sin 2x - \frac{1}{40}\sin 10x + C$$

59. $\qquad f'' = -\dfrac{1}{2}(\tan x)f'$

$$\int \frac{f''}{f'}\, dx = \int \left[-\frac{1}{2}(\tan x)\, dx\right]$$

$$\ln f' = \frac{1}{2}\ln|\cos x| + C_1$$

$$f' = C\sqrt{\cos x}$$

$$f'(0) = 1 = C, \text{ so } f' = \sqrt{\cos x}$$

$$s = \int_0^{\pi/2} \sqrt{1 + \left(\sqrt{\cos x}\right)^2}\, dx$$

$$= \int_0^{\pi/2} \sqrt{2\cos^2 \frac{x}{2}}\, dx = 2$$

7.4 Methods of Partial Fractions, page 539

1. $\quad \dfrac{1}{x(x-3)} = \dfrac{A_1}{x} + \dfrac{A_2}{x-3}$

$$1 = A_1(x-3) + A_2 x$$

If $x = 0$, then $A_1 = -\frac{1}{3}$;

if $x = 3$, then $A_2 = \frac{1}{3}$. Thus,

$$\frac{1}{x(x-3)} = \frac{-1}{3x} + \frac{1}{3(x-3)}$$

3. $\dfrac{3x^2 + 2x - 1}{x(x+1)} = 3 - \dfrac{x+1}{x(x+1)}$

$$= 3 - \frac{1}{x}$$

5. $\dfrac{4}{2x^2 + x} = \dfrac{A_1}{x} + \dfrac{A_2}{2x+1}$

$$4 = A_1(2x+1) + A_2 x$$

If $x = 0$, then $A_1 = 4$;

if $x = -\frac{1}{2}$, then $A_2 = -8$. Thus,

$$\frac{4}{2x^2 + x} = \frac{4}{x} + \frac{-8}{2x+1}$$

7. $\dfrac{4x^3 + 4x^2 + x - 1}{x^2(x+1)^2}$

$$= \frac{A_1}{x} + \frac{A_2}{x^2} + \frac{A_3}{x+1} + \frac{A_4}{(x+1)^2}$$

$$4x^3 + 4x^2 + x - 1$$

$$= A_1 x(x^2 + 2x + 1) + A_2(x^2 + 2x + 1)$$

$$+ A_3(x^3 + x^2) + A_4 x^2$$

If $x = 0$, then $A_2 = -1$;
if $x = -1$, then $A_4 = -2$;
equating the coefficients of x^3 as well
as x:
$$\begin{cases} 4 = A_1 + A_3 \\ 1 = A_1 + 2A_2 \end{cases}$$
so $A_1 = 3$ and $A_3 = 1$. Thus,
$$\frac{4x^3 + 4x^2 + x - 1}{x^2(x+1)^2} = \frac{3}{x} + \frac{-1}{x^2} + \frac{1}{x+1} + \frac{-2}{(x+1)^2}$$

9. $\dfrac{x^3 + 3x^2 + 3x - 4}{x^2(x+3)^2}$

$$= \frac{A_1}{x^2} + \frac{A_2}{x} + \frac{A_3}{(x+3)^2} + \frac{A_4}{x+3}$$

$x^3 + 3x^2 + 3x - 4$
$= A_1(x+3)^2 + A_2 x(x+3)^2$
$\quad + A_3 x^2 + A_4 x^2(x+3)$
$= (A_2 + A_4)x^3 + (A_1 + 6A_2 + A_3 + 3A_4)x^2$
$\quad + (6A_1 + 9A_2)x + 9A_1$

If $x = 0$, then $A_1 = -\frac{4}{9}$;
if $x = -3$, then $A_3 = -\frac{13}{9}$;
Equating the coefficients of x^3, x^2, and x:
$$\begin{cases} A_2 + A_4 = 1 \\ A_1 + 6A_2 + A_3 + 3A_4 = 3 \\ 6A_1 + 9A_2 = 3 \end{cases}$$
Solving this system, we find $A_2 = \frac{17}{27}$
and $A_4 = \frac{10}{27}$. Thus,
$$\frac{x^3 + 3x^2 + 3x - 4}{x^2(x+3)^2}$$
$$= \frac{-\frac{4}{9}}{x^2} + \frac{\frac{17}{27}}{x} + \frac{-\frac{13}{9}}{(x+3)^2} + \frac{\frac{10}{27}}{x+3}$$
$$= \frac{-4}{9x^2} + \frac{17}{27x} - \frac{13}{9(x+3)^2} + \frac{10}{27(x+3)}$$

11. $\dfrac{1}{1-x^4} = \dfrac{1}{(1-x)(1+x)(1+x^2)}$

$$= \frac{A_1}{1-x} + \frac{A_2}{1+x} + \frac{A_3 x + B_1}{1+x^2}$$

$1 = A_1(1+x)(1+x^2) + A_2(1-x)(1+x^2)$
$\quad + A_3 x(1 - x^2) + B_1(1 - x^2)$
$= A_1(x^3 + x^2 + x + 1) + A_2(-x^3 + x^2 - x + 1)$
$\quad + A_3(x - x^3) + B_1 - B_1 x^2$

Equating the coefficients of x^3, x^2, x, and constants:
$$\begin{cases} A_1 - A_2 - A_3 = 0 \\ A_1 + A_2 - B_1 = 0 \\ A_1 - A_2 + A_3 = 0 \\ A_1 + A_2 + B_1 = 1 \end{cases}$$
Solving this system, we find $A_1 = \frac{1}{4}$,
$A_2 = \frac{1}{4}$, $A_3 = 0$, and $B_1 = \frac{1}{2}$. Thus,
$$\frac{1}{1-x^4} = \frac{\frac{1}{4}}{1-x} + \frac{\frac{1}{4}}{1+x} + \frac{(0)x + \frac{1}{2}}{1+x^2}$$
$$= -\frac{1}{4(x-1)} + \frac{1}{4(x+1)} + \frac{1}{2(x^2+1)}$$

13. $\dfrac{x^2 + x - 1}{x(x+1)(2x-1)} = \dfrac{A_1}{x} + \dfrac{A_2}{x+1} + \dfrac{A_3}{2x-1}$ If

$x^2 + x - 1 = A_1(x+1)(2x-1)$
$\qquad\qquad + A_2 x(2x - 1)$
$\qquad\qquad + A_3 x(x+1)$

$x = 0$, then $A_1(1)(-1) = -1$, so $A_1 = 1$;
if $x = -1$, then $A_2(-1)(-3) = -1$, $A_2 = -\frac{1}{3}$;
if $x = \frac{1}{2}$, then $A_3(\frac{1}{2})(\frac{3}{2}) = -\frac{1}{4}$, so $A_3 = -\frac{1}{3}$.
Thus,
$$\frac{x^2 + x - 1}{x(x+1)(2x-1)} = \frac{1}{x} + \frac{-\frac{1}{3}}{x+1} + \frac{-\frac{1}{3}}{2x-1}$$
$$= \frac{1}{x} - \frac{1}{3(x+1)} - \frac{1}{3(2x-1)}$$

15. $\dfrac{2x^3 + 9x - 1}{x^2(x^2 - 1)} = \dfrac{A_1}{x} + \dfrac{A_2}{x^2} + \dfrac{A_3}{x+1} + \dfrac{A_4}{x-1}$

Partial fraction decomposition gives
$A_1 = -9$, $A_2 = 1$, $A_3 = 6$, $A_4 = 5$.
$$\int \frac{2x^3 + 9x - 1}{x^2(x^2 - 1)}\, dx$$
$$= -9\int \frac{dx}{x} + \int x^{-2}\,dx + 6\int \frac{dx}{x+1} + 5\int \frac{dx}{x-1}$$
$$= -9\ln|x| - x^{-1} + 6\ln|x+1| + 5\ln|x-1| + C$$

17. $\dfrac{x^2+1}{x^2+x-2} = 1 + \dfrac{-x+3}{x^2+x-2}$

$= 1 + \dfrac{-x+3}{(x-1)(x+2)}$

$= 1 + \dfrac{A_1}{x-1} + \dfrac{A_2}{x+2}$

Partial fraction decomposition gives
$A_1 = \frac{2}{3}, A_2 = -\frac{5}{3}$,

$\displaystyle\int \dfrac{x^2+1}{x^2+x-2}\,dx$

$= x - \dfrac{5}{3}\ln|x+2| + \dfrac{2}{3}\ln|x-1| + C$

19. Use the results of Problem 11.

$\displaystyle\int \dfrac{x^4+1}{x^4-1}\,dx = \int\left[1 + \dfrac{2}{x^4-1}\right]dx$

$= \displaystyle\int dx + 2\left[\dfrac{1}{4}\int\dfrac{dx}{x-1} - \dfrac{1}{4}\int\dfrac{dx}{x+1} - \dfrac{1}{2}\int\dfrac{dx}{x^2+1}\right]$

$= x + \dfrac{1}{2}\ln|x-1| - \dfrac{1}{2}\ln|x+1| - \tan^{-1}x + C$

21. $\displaystyle\int \dfrac{x\,dx}{(x+1)^2} = \int \dfrac{dx}{x+1} - \int\dfrac{dx}{(x+1)^2}$

$= \ln|x+1| + (x+1)^{-1} + C$

23. $\dfrac{1}{x(x+1)(x-2)} = \dfrac{A_1}{x} + \dfrac{A_2}{x+1} + \dfrac{A_3}{x-2}$

Partial fraction decomposition gives
$A_1 = -\frac{1}{2}, A_2 = \frac{1}{3}, A_3 = \frac{1}{6}$.

$\displaystyle\int \dfrac{dx}{x(x+1)(x-2)}$

$= -\dfrac{1}{2}\int\dfrac{dx}{x} + \dfrac{1}{3}\int\dfrac{dx}{x+1} + \dfrac{1}{6}\int\dfrac{dx}{x-2}$

$= -\dfrac{1}{2}\ln|x| + \dfrac{1}{3}\ln|x+1| + \dfrac{1}{6}\ln|x-2| + C$

25. $\dfrac{x}{(x+1)(x+2)^2} = \dfrac{A_1}{x+1} + \dfrac{A_2}{(x+2)^2} + \dfrac{A_3}{x+2}$

Partial fraction decomposition gives
$A_1 = -1, A_2 = 2, A_3 = 1$.

$\displaystyle\int \dfrac{x\,dx}{(x+1)(x+2)^2}$

$= -\displaystyle\int\dfrac{dx}{x+1} + 2\int\dfrac{dx}{(x+2)^2} + \int\dfrac{dx}{x+2}$

$= -\ln|x+1| - \dfrac{2}{x+2} + \ln|x+2| + C$

27. $\dfrac{5x+7}{x^2+2x-3} = \dfrac{5x+7}{(x-1)(x+3)}$

$= \dfrac{A_1}{x-1} + \dfrac{A_2}{x+3}$

Partial fraction decomposition gives
$A_1 = 3, A_2 = 2$.

$\displaystyle\int \dfrac{5x+7}{x^2+2x-3}\,dx = 3\int\dfrac{dx}{x-1} + 2\int\dfrac{dx}{x+3}$

$= 3\ln|x-1| + 2\ln|x+3| + C$

29. $\displaystyle\int \dfrac{3x^2-2x+4}{x^3-x^2+4x-4}\,dx \quad \boxed{u = x^3 - x^2 + 4x - 4}$

$= \ln|x^3 - x^2 + 4x - 4| + C$

31. Answers vary; expand the integrands as a sum of partial fractions, then integrate each term separately. See the procedure box on page 535.

33. $\displaystyle\int \dfrac{e^x\,dx}{2e^{2x} - 5e^x - 3} \quad \boxed{u = e^x;\ du = e^x\,dx}$

$= \displaystyle\int \dfrac{du}{2u^2 - 5u - 3}$

$= \displaystyle\int \dfrac{du}{(2u+1)(u-3)}$

$= \displaystyle\int \dfrac{-\frac{2}{7}}{2u+1}\,du + \int\dfrac{\frac{1}{7}}{u-3} \qquad \textit{Use partial fractions}$

$= -\dfrac{1}{7}\int\dfrac{2\,du}{2u+1} + \dfrac{1}{7}\int\dfrac{du}{u-3}$

$= -\dfrac{1}{7}\ln|2u+1| + \dfrac{1}{7}\ln|u-3| + C$

$= \dfrac{1}{7}\ln|e^x - 3| - \dfrac{1}{7}\ln|2e^x + 1| + C$

35. $\displaystyle\int \frac{\sin x\, dx}{(1+\cos x)^2} = -\int (1+\cos x)^{-2}(-\sin x\, dx)$

$\displaystyle = \frac{1}{1+\cos x} + C$

37. $\displaystyle\int \frac{\sec^2 x\, dx}{\tan x + 4}$ $\boxed{u = \tan x + 4;\ du = \sec^2 x\, dx}$

$\displaystyle = \int \frac{du}{u}$

$\displaystyle = \ln|u| + C$

$\displaystyle = \ln|\tan x + 4| + C$

39. $\displaystyle\int \frac{dx}{x^{2/3} - x^{1/2}}$ $\boxed{x = u^6;\ dx = 6u^5\, du}$

$\displaystyle = \int \frac{6u^5\, du}{u^4 - u^3}$

$\displaystyle = \int \frac{6u^2\, du}{u - 1}$

$\displaystyle = 6\int \left(u + 1 + \frac{1}{u-1}\right) du$

$\displaystyle = 6\left(\frac{u^2}{2} + u + \ln|u - 1|\right) + C$

$\displaystyle = 3x^{1/3} + 6x^{1/6} + 6\ln\left|x^{1/6} - 1\right| + C$

41. $\displaystyle\int \frac{dx}{3\cos x + 4\sin x}$ $\boxed{\text{Weierstrass substitution}}$

$\displaystyle = \int \frac{\frac{2\, du}{1+u^2}}{3\left(\frac{1-u^2}{1+u^2}\right) + 4\left(\frac{2u}{1+u^2}\right)}$

$\displaystyle = \int \frac{-2\, du}{3u^2 - 8u - 3}$

$\displaystyle = \int \frac{-2}{(u-3)(3u+1)}\, du$ *Use partial fractions*

$\displaystyle = \int \frac{-\frac{1}{5}}{u - 3}\, du + \int \frac{\frac{3}{5}}{3u + 1}\, du$

$\displaystyle = -\frac{1}{5}\ln|u - 3| + \frac{1}{5}\ln|3u + 1| + C$

$\displaystyle = -\frac{1}{5}\ln\left|\tan\frac{x}{2} - 3\right| + \frac{1}{5}\ln\left|3\tan\frac{x}{2} + 1\right| + C$

43. $\displaystyle\int \frac{\sin x - \cos x}{\sin x + \cos x}\, dx$

$\boxed{t = \sin x + \cos x;\ dt = (\cos x - \sin x)\, dx}$

$\displaystyle = -\int \frac{dt}{t}$

$\displaystyle = -\ln|t| + C$

$\displaystyle = -\ln|\sin x + \cos x| + C$

45. $\displaystyle\int \frac{dx}{\sec x - \tan x} = \int \frac{\cos x\, dx}{1 - \sin x}$

$\displaystyle = -\ln(1 - \sin x) + C$

47. $\displaystyle\int \frac{dx}{4\sin x - 3\cos x - 5}$ $\boxed{\text{Weierstrass substitution}}$

$\displaystyle = \int \frac{\frac{2\, du}{1+u^2}}{4\left(\frac{2u}{1+u^2}\right) - 3\left(\frac{1-u^2}{1+u^2}\right) - 5}$

$\displaystyle = \int \frac{2\, du}{8u - 3(1 - u^2) - 5(1 + u^2)}$

$\displaystyle = \int \frac{2\, du}{-2u^2 + 8u - 8}$

$\displaystyle = -\int \frac{du}{u^2 - 4u + 4}$

$\displaystyle = -\int \frac{du}{(u - 2)^2}$

$\displaystyle = \frac{1}{u - 2} + C$

$\displaystyle = \frac{1}{\tan\frac{x}{2} - 2} + C$

49. $\displaystyle\int \frac{dx}{x(3 - \ln x)(1 - \ln x)}$ $\boxed{t = \ln x;\ dt = x^{-1}\, dx}$

$\displaystyle = \int \frac{dt}{(3 - t)(1 - t)}$ *Use partial fractions*

$\displaystyle = \int \frac{\frac{1}{2}}{t - 3}\, dt + \int \frac{-\frac{1}{2}}{t - 1}\, dt$

$\displaystyle = \frac{1}{2}\ln|t - 3| - \frac{1}{2}\ln|t - 1| + C$

$\displaystyle = \frac{1}{2}\ln|\ln x - 3| - \frac{1}{2}\ln|\ln x - 1| + C$

51. $A = \displaystyle\int_{4/3}^{7/4} \frac{dx}{x^2 - 5x + 6}$

$= \displaystyle\int_{4/3}^{7/4} \frac{dx}{(x-3)(x-2)}\, dx$

$= -\displaystyle\int_{4/3}^{7/4} \frac{dx}{x-2} + \int_{4/3}^{7/4} \frac{1}{x-3}\, dx$

$= \left[-\ln|x-2| + \ln|x-3|\right]_{4/3}^{7/4}$

$= \ln 2$

53. We use technology to evaluate these definite integrals.

a. $V = \pi \displaystyle\int_0^3 \left(\frac{1}{\sqrt{x^2+4x+3}}\right)^2 dx \approx 1.0888$

b. $V = 2\pi \displaystyle\int_0^3 \frac{x\, dx}{\sqrt{x^2+4x+3}} \approx 7.6402$

55. Substitution:

$\displaystyle\int \frac{x\, dx}{x^2-9} \quad \boxed{u = x^2 - 9}$

$= \dfrac{1}{2}\displaystyle\int \frac{du}{u}$

$= \dfrac{1}{2}\ln|x^2 - 9| + C$

Partial fractions:

$\displaystyle\int \frac{x\, dx}{x^2-9} = \int \frac{dx}{2(x-3)} + \int \frac{dx}{2(x+3)}$

$= \dfrac{1}{2}\ln|x-3| + \dfrac{1}{2}\ln|x+3| + C$

$= \dfrac{1}{2}\ln|x^2 - 9| + C$

Trigonometric substitution:

$\displaystyle\int \frac{x\, dx}{x^2-9} \quad \boxed{x = 3\sin\theta}$

$= \displaystyle\int \frac{3\sin\theta(3\cos\theta\, d\theta)}{9\sin^2\theta - 9}$

$= -\displaystyle\int \frac{\sin\theta}{\cos\theta}\, d\theta$

$= -\displaystyle\int \tan\theta\, d\theta$

$= \ln|\cos\theta| + C$

$= \ln\sqrt{9 - x^2} + C$

$= \dfrac{1}{2}\ln|x^2 - 9| + C$

57. Partial fraction decomposition gives

$\displaystyle\int \frac{dx}{x(ax+b)} = \frac{1}{b}\int \left(\frac{1}{x} - \frac{a}{ax+b}\right) dx$

$= \dfrac{1}{b}\left[\ln|x| - \ln|ax+b|\right] + C$

$= \dfrac{1}{b}\ln\left|\dfrac{x}{ax+b}\right| + C$

59. $\displaystyle\int \sec x\, dx = \int \frac{dx}{\cos x}$ $\boxed{\text{Weierstrass substitution}}$

$= \displaystyle\int \frac{\frac{2\, du}{1+u^2}}{\frac{1-u^2}{1+u^2}}\, du$

$= \displaystyle\int \frac{2\, du}{1 - u^2}$

$= \displaystyle\int \left(\frac{1}{1+u} + \frac{1}{1-u}\right) du$

$= \ln|1+u| - \ln|1-u| + C$

$= \ln\left|\dfrac{1+u}{1-u}\right| + C$

$= \ln\left|\dfrac{1+\tan\frac{x}{2}}{1-\tan\frac{x}{2}}\right| + C$

$= \ln\left|\dfrac{\cos\frac{x}{2} + \sin\frac{x}{2}}{\cos\frac{x}{2} - \sin\frac{x}{2}}\right| + C$

$= \ln\left|\dfrac{(\cos\frac{x}{2} + \sin\frac{x}{2})^2}{\cos^2\frac{x}{2} - \sin^2\frac{x}{2}}\right| + C$

$$= \ln\left|\frac{1 + 2\sin\frac{x}{2}\cos\frac{x}{2}}{\cos x}\right| + C$$

$$= \ln\left|\frac{1 + \sin x}{\cos x}\right| + C$$

$$= \ln|\sec x + \tan x| + C$$

7.5 Summary of Integration Techniques, page 543

SURVIVAL HINT: *This problem set is important because it is testing your ability to perform one of the three fundamental processes of calculus, that of integration. Deciding what to do is just as important as actually doing it. The time you can give to working this problems will pay rich dividends in your future work in mathematics,*

1. $\displaystyle\int \frac{2x-1}{(x-x^2)^3}dx$ $\boxed{u = x - x^2; \ du = (1-2x)dx}$

$$= -\int u^{-3}du$$

$$= \frac{1}{2}u^{-2} + C$$

$$= \frac{1}{2(x-x^2)^2} + C$$

$$= \frac{1}{2x^2(x-1)^2} + C$$

3. $\displaystyle\int (x\sec 2x^2)\,dx$ $\boxed{u = 2x^2; \ du = 4x\,dx}$

$$= \frac{1}{4}\int \sec u\,du$$

$$= \frac{1}{4}\ln|\sec u + \tan u| + C$$

$$= \frac{1}{4}\ln\left|\sec 2x^2 + \tan 2x^2\right| + C$$

5. $\displaystyle\int (e^x\cot e^x)\,dx$ $\boxed{u = e^x; \ du = e^x\,dx}$

$$= \int \cot u\,du$$

$$= \ln|\sin e^x| + C$$

7. $\displaystyle\int \frac{\tan(\ln x)\,dx}{x}$ $\boxed{u = \ln x; \ du = x^{-1}\,dx}$

$$= \int \tan u\,du$$

$$= -\ln|\cos u| + C$$

$$= -\ln|\cos(\ln x)| + C$$

9. $\displaystyle\int \frac{e^{2t}dt}{1 + e^{4t}}$ $\boxed{u = e^{2t}; \ du = 2e^{2t}\,dt}$

$$= \frac{1}{2}\int \frac{du}{1 + u^2}$$

$$= \frac{1}{2}\tan^{-1}u + C$$

$$= \frac{1}{2}\tan^{-1}e^{2t} + C$$

11. $\displaystyle\int \frac{x^2 + x + 1}{x^2 + 9}dx$

$$= \int\left(1 + \frac{x-8}{x^2+9}\right)dx$$

$$= \int dx + \int \frac{x\,dx}{x^2+9} - 8\int \frac{dx}{x^2+9}$$

$$= x + \frac{1}{2}\ln(x^2+9) - \frac{8}{3}\tan^{-1}\frac{x}{3} + C$$

13. $\displaystyle\int \frac{1+e^x}{1-e^x}dx$

$$= \int\left(1 + \frac{2e^x}{1-e^x}\right)dx \quad \boxed{u = 1 - e^x}$$

$$= x + 2\int \frac{-du}{u}$$

$$= x - 2\ln|1 - e^x| + C$$

15. $\displaystyle\int \frac{dx}{1+e^{2x}}$ $\boxed{u = e^{-2x} + 1;\ du = -2e^{-2x}\,dx}$

$$= \int \frac{e^{-2x}\,dx}{e^{-2x}+1}$$

$$= -\frac{1}{2}\int \frac{du}{u}$$

$$= -\frac{1}{2}\ln|u| + C$$

$$= -\frac{1}{2}\ln\left|e^{-2x}+1\right| + C$$

$$= x - \frac{1}{2}\ln\left(e^{2x}+1\right) + C$$

17. $\displaystyle\int \frac{dx}{x^2+2x+2} = \int \frac{dx}{(x+1)^2+1}$

$$= \tan^{-1}(x+1) + C$$

19. $\displaystyle\int e^{-x}\cos x\,dx$ $\boxed{\text{Parts or Formula 193}}$

$$= \frac{e^{-x}(-\cos x + \sin x)}{2} + C$$

$$= \frac{1}{2}e^{-x}(\sin x - \cos x) + C$$

21. $\displaystyle\int \sin^3 x\,dx = \int \sin^2 x \sin x\,dx$

$$= -\int (1 - \cos^2 x)(-\sin x)\,dx$$

$$= -\cos x + \frac{1}{3}\cos^3 x + C$$

23. $\displaystyle\int \sin^3 x \cos^2 x\,dx$

$$= -\int (1 - \cos^2 x)\cos^2 x(-\sin x\,dx)$$

$$= \int (\cos^4 x - \cos^2 x)(-\sin x\,dx)$$

$$= \frac{1}{5}\cos^5 x - \frac{1}{3}\cos^3 x + C$$

25. $\displaystyle\int \sin^2 x \cos^4 x\,dx$

$$= \frac{1}{8}\int (1 - \cos 2x)(1 + \cos 2x)^2\,dx$$

$$= \frac{1}{8}\int (1 - \cos^2 2x)(1 + \cos 2x)\,dx$$

$$= \frac{1}{8}\int \sin^2 2x(1 + \cos 2x)\,dx$$

$$= \frac{1}{8}\int \sin^2 2x\,dx + \frac{1}{8}\int \sin^2 2x(\cos 2x)\,dx$$

$$= \frac{1}{16}\left(x - \frac{1}{4}\sin 4x\right) + \frac{1}{16}\cdot\frac{1}{3}\sin^3 2x + C$$

$$= \frac{1}{16}x - \frac{1}{64}\sin 4x + \frac{1}{48}\sin^3 2x + C$$

27. $\displaystyle\int \tan^5 x \sec^4 x\,dx$

$$= \int \tan^5 x(\tan^2 x + 1)\sec^2 x\,dx$$

$$= \int \tan^7 x \sec^2 x\,dx + \int \tan^5 x \sec^2 x\,dx$$

$$= \frac{1}{8}\tan^8 x + \frac{1}{6}\tan^6 x + C$$

29. $\displaystyle\int \frac{\sqrt{1-x^2}}{x}\,dx$ $\boxed{x = \sin\theta \text{ or Formula 121}}$

$$= \sqrt{1-x^2} - \ln\left|\frac{1 + \sqrt{1-x^2}}{x}\right| + C$$

31. $\displaystyle\int \frac{\cos x\,dx}{\sqrt{1+\sin^2 x}}$ $\boxed{u = \sin x;\ du = \cos x\,dx}$

$$= \int \frac{du}{\sqrt{1+u^2}}$$ $\boxed{u = \tan\theta \text{ or Formula 89}}$

$$= \ln\left(u + \sqrt{1+u^2}\right) + C$$

$$= \ln\left(\sin x + \sqrt{1+\sin^2 x}\right) + C$$

33. $\displaystyle\int \sin^5 x\,dx = -\int (\sin^4 x)(-\sin x\,dx)$

$$= -\int (1 - \cos^2 x)^2(-\sin x\,dx)$$

$$= -\int \left(\cos^4 x - 2\cos^2 x + 1 \right)(-\sin x \, dx)$$

$$= -\frac{\cos^5 x}{5} + \frac{2\cos^3 x}{3} - \cos x + C$$

35. $\displaystyle\int_0^2 \sqrt{4 - x^2} \, dx$ $\boxed{u = 2\sin\theta \text{ or Formula } 117}$

$$= \frac{x\sqrt{4 - x^2}}{2} + \frac{4}{2}\sin^{-1}\frac{x}{2} \Bigg|_0^2$$

$$= \pi$$

37. $\displaystyle\int_0^{\ln 2} e^t \sqrt{1 + e^{2t}} \, dt$ $\boxed{u = e^t}$

$$= \int_1^2 \sqrt{1 + u^2} \, du$$ $\boxed{u = \tan\theta \text{ or Formula } 85}$

$$= \left[\frac{u\sqrt{1 + u^2}}{2} + \frac{1}{2}\ln\left| u + \sqrt{1 + u^2} \right| \right]_1^2$$

$$= \frac{1}{2}\ln\left(\sqrt{5} + 2 \right) - \frac{1}{2}\ln(\sqrt{2} + 1) + \sqrt{5} - \frac{\sqrt{2}}{2}$$

39. $\displaystyle\int_1^2 \frac{dx}{x^4 \sqrt{x^2 + 3}}$ $\boxed{\begin{array}{l} x = \sqrt{3}\tan\theta; \\ dx = \sqrt{3}\sec^2\theta \, d\theta \end{array}}$

$$= \int_{x=1}^{x=2} \frac{\sqrt{3}\sec^2\theta \, d\theta}{9\tan^4\theta \sqrt{3}\sec\theta}$$

$$= \frac{1}{9}\int_{x=1}^{x=2} \frac{(1 - \sin^2\theta)(\cos\theta \, d\theta)}{\sin^4\theta}$$

$$= \frac{1}{9}\left(-\frac{1}{3}\csc^3\theta + \csc\theta \right)\Bigg|_{x=1}^{x=2}$$

$$= \left[\frac{1}{9}\left(\frac{\sqrt{3 + x^2}}{x} \right)\left(-\frac{1}{3}\cdot\frac{3 + x^2}{x^2} + 1 \right) \right]_1^2$$

$$= \frac{1}{27}\left(\frac{5}{8}\sqrt{7} + 2 \right)$$

41. $\displaystyle\int_{-2}^{2\sqrt{3}} x^3 \sqrt{x^2 + 4} \, dx$ $\boxed{x = 2\tan\theta \text{ or Formula } 88}$

$$= \left[\frac{(x^2 + 4)^{5/2}}{5} - \frac{4(x^2 + 4)^{3/2}}{3} \right]\Bigg|_{-2}^{2\sqrt{3}}$$

$$= \frac{1{,}792 - 64\sqrt{2}}{15}$$

43. $\displaystyle\int \frac{e^x \, dx}{\sqrt{1 + e^{2x}}}$ $\boxed{u = e^x; \ du = e^x \, dx}$

$$= \int \frac{du}{\sqrt{1 + u^2}}$$ $\boxed{\text{Formula } 89}$

$$= \ln\left(\sqrt{1 + e^{2x}} + e^x \right) + C$$

45. $\displaystyle\int \frac{x^2 + 4x + 3}{x^3 + x^2 + x} \, dx$

$$= \int \frac{x^2 + 4x + 3}{x(x^2 + x + 1)} \, dx \quad \textit{Use partial fractions}$$

$$= \int \frac{3}{x} dx + \int \frac{-2x + 1}{x^2 + x + 1} dx$$

$$= 3\ln|x| + \int \frac{-2x - 1}{x^2 + x + 1} dx + 2\int \frac{dx}{\left(x + \frac{1}{2}\right)^2 + \frac{3}{4}}$$

$$= 3\ln|x| - \ln|x^2 + x - 1|$$
$$+ 2\left(\frac{2}{\sqrt{3}} \right)\left(\tan^{-1}\frac{2\left(x + \frac{1}{2}\right)}{\sqrt{3}} \right) + C$$

$$= 3\ln|x| - \ln|x^2 + x + 1|$$
$$+ \frac{4}{\sqrt{3}}\tan^{-1}\frac{\sqrt{3}}{3}(2x + 1) + C$$

47. $\displaystyle\int \frac{3x + 5}{x^2 + 2x + 1} \, dx$

$$= 3\int \frac{x + 1}{(x + 1)^2} dx + 2\int \frac{dx}{(x + 1)^2}$$

$$= 3\ln|x + 1| - 2(x + 1)^{-1} + C$$

49. $\displaystyle\int \frac{5x^2 + 18x + 34}{(x - 7)(x + 2)^2} \, dx \quad \textit{Use partial fractions}$

$$= \int \frac{5}{x - 7} dx + \int \frac{0}{x + 2} dx + \int \frac{-2}{(x + 2)^2} dx$$

$$= 5\ln|x - 7| + 2(x + 2)^{-1} + C$$

51. Answers vary; see page 540-543:
   Step 1: simplify
   Step 2: use basic formulas (check table)
   Step 3: substitute
   Step 4: classify; parts, trig powers,
              Weierstrass substitution, trig
              substitutions, or partial fractions
   Step 5: try again

53. Since $y' = 2x$, the arc length is given by

$$s = \int_{-1}^{1} \sqrt{1 + (2x)^2}\, dx \quad \boxed{2x = \tan\theta;\ 2\,dx = \sec^2\theta\, d\theta}$$

$$= \int_{x=-1}^{x=1} \sqrt{1 + \tan^2\theta} \left(\frac{1}{2}\sec^2\theta\right) d\theta$$

$$= \frac{1}{2}\int_{x=-1}^{x=1} \sec^3\theta\, d\theta \quad \boxed{\text{Formula 160}}$$

$$= \frac{1}{2}\left[\frac{\sec\theta\tan\theta}{2} + \frac{1}{2}\ln|\sec\theta + \tan\theta|\right]_{x=-1}^{x=1}$$

$$= \frac{1}{4}\left[\sqrt{1+4x^2}(2x) + \ln\left(\sqrt{1+4x^2} + 2x\right)\right]_{-1}^{1}$$

$$\approx 2.9579$$

55. $V = \pi \int_0^{\pi/2} \cos^2 x\, dx$

$$= \pi\left[\frac{x}{2} + \frac{1}{4}\sin 2x\right]\Big|_0^{\pi/2}$$

$$= \frac{\pi^2}{4}$$

57. $I = \int e^{ax}\sin bx\, dx \quad \boxed{u = e^{ax};\ dv = \sin bx\, dx}$

$$= -\frac{1}{b}e^{ax}\cos bx + \frac{a}{b}\int e^{ax}\cos bx\, dx$$

$$\boxed{u = e^{ax};\ dv = \cos bx\, dx}$$

$$= -\frac{1}{b}e^{ax}\cos bx$$

$$+ \frac{a}{b}\left(\frac{1}{b}e^{ax}\sin bx - \frac{a}{b}\int e^{ax}\sin bx\, dx\right)$$

$$= -\frac{1}{b}e^{ax}\cos bx + \frac{a}{b}\left(\frac{1}{b}e^{ax}\sin bx - \frac{a}{b}I\right)$$

$$= -\frac{1}{b}e^{ax}\cos bx + \frac{a}{b^2}e^{ax}\sin bx - \frac{a^2}{b^2}I$$

$$I + \frac{a^2}{b^2}I = -\frac{1}{b}e^{ax}\cos bx + \frac{a}{b^2}e^{ax}\sin bx + C_1$$

$$\left(\frac{b^2 + a^2}{b^2}\right)I = \frac{(a\sin bx - b\cos bx)e^{ax}}{b^2} + C_1$$

$$I = \frac{(a\sin bx - b\cos bx)e^{ax}}{a^2 + b^2} + C$$

59. $\int x^m(\ln x)^n dx \quad \boxed{\begin{array}{l} u = (\ln x)^n;\ du = n(\ln x)^{n-1}\frac{dx}{x} \\ dv = x^m dx;\ v = \frac{x^{m+1}}{m+1} \end{array}}$

$$= \frac{x^{m+1}(\ln x)^n}{m+1} - \frac{n}{m+1}\int x^m(\ln x)^{n-1}dx$$

For example,

$$\int x^2(\ln x)^3\, dx = \frac{x^3(\ln x)^3}{3} - \frac{3}{3}\int x^2(\ln x)^2 dx$$

$$= \frac{x^3}{3}(\ln x)^3 - \left[\frac{x^3}{3}(\ln x)^2 - \frac{2}{3}\int x^2\ln x\, dx\right]$$

$$= \frac{x^3}{3}(\ln x)^3 - \frac{x^3}{3}(\ln x)^2 + \frac{2x^3}{9}(\ln x) - \frac{2x^3}{27} + C$$

7.6 First-Order Differential Equations, page 553

> **SURVIVAL HINT:** *This section provides an introduction to differential equations. It is considered again in the last chapter, and at most school is offered as a first course following calculus. As such, understanding the ideas of this section will give you a head start on many topics yet to be considered.*

1. $\dfrac{dy}{dx} = x^2$

$$dy = x^2 dx$$

$$\int dy = \int x^2 dx$$

$$y = \frac{x^3}{3} + C$$

3. $12\dfrac{dy}{dx} = \sqrt{5x+1}$

$$dy = \frac{1}{12}(5x+1)^{1/2}dx$$

$$y = \int \frac{1}{12}(5x+1)^{1/2}dx \quad \boxed{u = 5x+1;\ du = 5\,dx}$$

$$= \int \frac{1}{60}u^{1/2}du$$

$$= \frac{1}{60} \cdot \frac{2}{3}u^{3/2} + C$$

$$= \frac{1}{90}(5x+1)^{3/2} + C$$

5. $\quad \dfrac{dP}{dt} = 0.02P$

$$\int P^{-1}dP = 0.02\int dt$$

$$\ln|P| = 0.02t + C_1$$

$$P = e^{0.02t + C_1}$$

$$= Ce^{0.02t}$$

7. $\quad P(x) = \dfrac{3}{x};\ Q(x) = x;$

$$I(x) = e^{\int \frac{3}{x}dx} = e^{3\ln x} = e^{\ln x^3} = x^3$$

$$y = \frac{1}{x^3}\left(\int x(x^3)dx + C\right)$$

$$= \frac{1}{x^3}\left(\frac{x^5}{5} + C\right)$$

$$= \frac{1}{5}x^2 + Cx^{-3}$$

9. Divide both sides by x^4: $\dfrac{dy}{dx} + \dfrac{2}{x}y = 5x^{-4}$

$$P(x) = \frac{2}{x};\ Q(x) = 5x^{-4};$$

$$I(x) = e^{\int \frac{2}{x}dx} = e^{2\ln x} = e^{\ln x^2} = x^2$$

$$y = \frac{1}{x^2}\left(\int x^2(5x^{-4})dx + C\right)$$

$$= \frac{1}{x^2}\left[-5x^{-1} + C\right]$$

$$= -5x^{-3} + Cx^{-2}$$

11. Divide both sides by x: $\dfrac{dy}{dx} + \dfrac{2}{x}y = e^{x^3}$

$$P(x) = \frac{2}{x};\ Q(x) = e^{x^3};$$

$$I(x) = e^{\int \frac{2}{x}dx} = e^{2\ln x} = e^{\ln x^2} = x^2$$

$$y = \frac{1}{x^2}\left(\int x^2\left(e^{x^3}\right)dx + C\right)$$

$$= \frac{1}{x^2}\left[\frac{1}{3}e^{x^3} + C\right]$$

$$= \frac{1}{3}x^{-2}e^{x^3} + Cx^{-2}$$

13. $P(x) = \dfrac{1}{x};\ Q(x) = \tan^{-1}x;$

$$I(x) = e^{\int \frac{1}{x}dx} = e^{\ln x} = x$$

$$y = \frac{1}{x}\left(\int x\tan^{-1}x\,dx + C\right) \quad \boxed{\text{Formula 181}}$$

$$= \left(\frac{1}{2}x + \frac{1}{2x}\right)\tan^{-1}x - \frac{1}{2} + \frac{C}{x}$$

15. $P(x) = \tan x;\ Q(x) = \sin x;$

$$I(x) = e^{\int \tan x\,dx} = e^{-\ln|\cos x|} = \frac{1}{\cos x}$$

$$y = \cos x\left(\int \sin x\,\frac{1}{\cos x}dx + C\right)$$

$$= \cos x(-\ln|\cos x| + C)$$

$$= -\cos x\ln|\cos x| + C\cos x$$

17. $\quad \dfrac{dy}{dx} = x^2y^{-2}$

$$y^2dy = x^2dx$$

$$\int y^2dy = \int x^2dx$$

$$\frac{y^3}{3} = \frac{x^3}{3} + C$$

$$y^3 = x^3 + C$$

For the particular case, substitute
$y = 5$ when $x = 0$ to find C:

$$y^3 = x^3 + C$$

$$5^3 = 0^3 + C$$

$C = 125$

The particular solution is:

$y^3 = x^3 + 125$

$y = \left(x^3 + 125\right)^{1/3}$

19. $x\dfrac{dy}{dx} - y\sqrt{x} = 0$

$x\dfrac{dy}{dx} = x^{1/2}y$

$y^{-1}dy = x^{-1/2}dx$

$\displaystyle\int \dfrac{dy}{y} = \int x^{-1/2}dx$

$\ln|y| = 2x^{1/2} + C$

For the particular case, substitute:

$y = 1$ when $x = 1$ to find C:

$\ln|1| = 2(1)^{1/2} + C$

$0 = 2 + C$

$C = -2$

The particular solution is:

$\ln|y| = 2\sqrt{x} - 2$

$y = e^{2\sqrt{x}-2}$

21. $P(x) = \dfrac{x}{1+x}$; $Q(x) = x(1+x)$;

$I(x) = e^{\int \frac{x}{1+x}\,dx} = e^{x-\ln(x+1)} = \dfrac{e^x}{x+1}$

$y = \dfrac{x+1}{e^x}\left(\displaystyle\int x(1+x)\cdot\dfrac{e^x}{1+x}\,dx + C\right)$

$\boxed{\text{Formula } 184}$

$= \dfrac{x+1}{e^x}\left[e^x(x-1) + C\right]$

$= x^2 - 1 + C\left(\dfrac{x+1}{e^x}\right)$

Since $y = -1$ when $x = 0$, $C = 0$; thus,

$y = x^2 - 1$ for $x > -1$.

23. $P(x) = \dfrac{2x}{1+x^2}$; $Q(x) = \sin x$;

$I(x) = e^{\int \frac{2x}{1+x^2}\,dx} = e^{\ln(x^2+1)} = x^2 + 1$

$y = \dfrac{1}{x^2+1}\left(\displaystyle\int (x^2+1)\sin x\,dx + C\right)$

$= \dfrac{1}{x^2+1}\left(\displaystyle\int x^2\sin x\,dx + \int \sin x\,dx + C\right)$

$\boxed{\text{Formula } 133}$

$= \dfrac{1}{x^2+1}\left[2x\sin x + \left(1-x^2\right)\cos x + C\right]$

Since $y = 1$ when $x = 0$, we find $C = 0$; thus,

$y = \dfrac{1}{x^2+1}\left[2x\sin x + \left(1-x^2\right)\cos x\right]$

25. Divide both sides by x: $\dfrac{dy}{dx} - \dfrac{2}{x}y = 2x^2$

$P(x) = -\dfrac{2}{x}$; $Q(x) = 2x^2$;

$I(x) = e^{\int -\frac{2}{x}\,dx} = e^{-2\ln x} = x^{-2}$

$y = \dfrac{1}{x^{-2}}\left(\displaystyle\int (2x^2)(x^{-2})\,dx + C\right)$

$= x^2(2x + C)$

$= 2x^3 + Cx^2$

Since $y = 0$ when $x = 3$,

$0 = 2(27) + 9C$

$C = -6$

Thus, $y = 2x^3 - 6x^2$.

27. $y^2 = 4kx$

$2yy' = 4k$

$y' = \dfrac{2k}{y}$

$= \dfrac{2\frac{y^2}{4x}}{y}$

$= \dfrac{y}{2x}$

The slope of the orthogonal trajectory is the negative reciprocal:

$$\frac{dY}{dX} = -\frac{2X}{Y}$$

$$\int Y\,dY = \int -2X\,dX$$

$$\frac{1}{2}Y^2 = -X^2 + K$$

$$2X^2 + Y^2 = C$$

29. $x^2 + y^2 = r^2$

$$2x + 2yy' = 0$$

$$y' = -\frac{x}{y}$$

The slope of the orthogonal trajectory is the negative reciprocal:

$$\frac{dY}{dX} = \frac{Y}{X}$$

$$\int Y^{-1}\,dY = \int X^{-1}\,dX$$

$$\ln Y = \ln X + K$$

$$Y = CX$$

31. Assume the growth rate for the period 2009 to 2010 remained at 1.80%.

$$Q(t) = 14.26e^{0.018t}$$

$$Q(10) = 14.26e^{0.018(10)} \approx 17.0723195$$

The GDP in the year 2020 will be about $17.0723 trillion.

33.   Assume that the marriage rate remains constant at 0.71%.

$$Q(t) = 2{,}230e^{0.0071t}$$

$$Q(5) = 2{,}230e^{0.0071(5)} \approx 2310.586955$$

The predicted number of marriages in 2010 is 2,310,000.

35. a. Let $Q(t)$ be the amount of salt in the solution at time t (minutes). Then

$$\frac{dQ}{dt} = \underbrace{(1)(2)}_{inflow} - \underbrace{\frac{Q}{30}(2)}_{outflow}$$

$$\frac{dQ}{dt} + \frac{Q}{15} = 2$$

The integrating factor is $e^{\int 1/15\,dt} = e^{t/15}$.

$$Q(t) = e^{-t/15}\left[\int 2e^{t/15}\,dt + C\right]$$

$$= e^{-t/15}\left[2\frac{e^{t/15}}{\frac{1}{15}} + C\right]$$

$$= 30 + Ce^{-t/15}$$

$$Q(0) = 30 + Ce^0$$

$$10 = 30 + C \qquad \textit{Given } Q(0) = 10.$$

$$C = -20$$

Thus, $Q(t) = 30 - 20e^{-t/15}$.

b. When $Q = 15$, we have

$$15 = 30 - 20e^{-t/15}$$

$$-\frac{t}{15} = \ln\frac{15}{20}$$

$$t = -15\ln\frac{3}{4}$$

$$\approx 4.31523$$

This is about 4 minutes, 19 seconds.

37. $\dfrac{db}{dt} = \alpha - \beta b$

$$\frac{db}{dt} + \beta b = \alpha$$

The integrating factor is $e^{\int \beta\, dt} = e^{\beta t}$.

$$b = e^{-\beta t}\left[\int \alpha e^{\beta t} dt + C\right]$$

$$= e^{-\beta t}\left[\frac{\alpha e^{\beta t}}{\beta} + C\right]$$

$$= \frac{\alpha}{\beta} + Ce^{-\beta t}$$

When $t = 0$, $b(0) = 0$ implies $C = -\dfrac{\alpha}{\beta}$ so we have

$$b(t) = \frac{\alpha}{\beta}\left(1 - e^{-\beta t}\right)$$

In the "long run" (as $t \to \infty$), the concentration will be

$$\lim_{t \to \infty} \frac{\alpha}{\beta}\left(1 - e^{-\beta t}\right) = \frac{\alpha}{\beta}$$

The "half-way point" is reached when

$$\frac{1}{2} \cdot \frac{\alpha}{\beta} = \frac{\alpha}{\beta}\left(1 - e^{-\beta t}\right)$$

$$e^{-\beta t} = \frac{1}{2}$$

$$-\beta t = \ln\frac{1}{2}$$

$$t = \frac{\ln\frac{1}{2}}{-\beta}$$

$$= \frac{\ln 2}{\beta}$$

39. With the uninhibited growth model where t is the time after 2000 in years and $Q(t)$ is the number of Hispanics in year t, and $Q_0 = 15.5$.

$$Q(10) = 31.1 = 15.5e^{10k}$$

$$e^{10k} = \frac{31.1}{15.5}$$

$$k = \frac{1}{10}\ln\left(\frac{31.1}{15.5}\right)$$

$$\approx 0.069636779526$$

We would expect the Hispanic population to be

$$Q(20) = 15.5e^{20k}$$

$$\approx 62.4006451613$$

The population is about 62.4 million.

41.
$$\frac{dM}{dt} = r\left(\frac{k}{r} - M\right)$$

$$\frac{dM}{\frac{k}{r} - M} = r\, dt$$

$$\int \frac{dM}{\frac{k}{r} - M} = \int r\, dt$$

$$-\ln\left|\frac{k}{r} - M\right| = rt + K$$

$$\frac{k}{r} - M = e^{-rt}e^{-K}$$

Since, $M(0) = 0$, we see $e^{-K} = \frac{k}{r}$.
Thus, $M(t) = \frac{k}{r}(1 - e^{-rt})$.

43.
$$V = 5h$$

$$\frac{dV}{dt} = 5\frac{dh}{dt}$$

$$-4.8(0.07)\sqrt{h} = 5\frac{dh}{dt} \quad \textit{By Toricelli's law}$$

$$-0.0672\, dt = \frac{1}{\sqrt{h}}\, dh$$

$$\int -0.0672 \, dt = \int \frac{1}{\sqrt{h}} \, dh$$

$$2h^{1/2} + C = -0.0672t$$

When $t = 0$, $h = 4$, so $4 + C = 0$ or
$C = -4$.

$$2h^{1/2} - 4 = -0.0672t$$

When $h = 0$, $t \approx 59.5$. It will take
about one minute to drain.

45. a. We have $y^{(4)} = -k$ with
$y(0) = y(L) = 0$.

$$y^{(4)} = -k$$

$$y''' = -kx + C_1$$

$$y'' = -\frac{k}{2}x^2 + C_1 x + C_2$$

$$y' = -\frac{k}{6}x^3 + \frac{C_1}{2}x^2 + C_2 x + C_3$$

$$y = -\frac{k}{24}x^4 + \frac{C_1}{6}x^3 + \frac{C_2}{2}x^2 + C_3 x + C_4$$

Since $y(0) = 0$, and $y''(x) = y''(L) = 0$, we
have $C_4 = 0$ and $C_2 = 0$, and the conditions
$y(L) = 0$ and $y''(L) = 0$ tell us that

$$-\frac{kL^4}{24} + \frac{c_1 L^3}{6} + C_3 L = 0$$

$$-\frac{kL^2}{2} + C_1 L = 0$$

$$C_1 = \frac{kL}{2}$$

Thus, $C_3 = -\frac{kL^3}{24}$ so that

$$y = -\frac{k}{24}\left(x^4 - 2Lx^3 + L^3 x\right)$$

b. Maximum deflection occurs where $y' = 0$.
Solve

$$y' = -\frac{k}{24}\left[4x^3 - 6Lx^2 + L^3\right] = 0$$

to obtain $x = L/2$; reject $(1 \pm \sqrt{3})L/2$

because these points lie outside the interval
$[0, L]$. This is the maximum deflection
since y is minimized at $x = L/2$; note

$$y''\left(\frac{L}{2}\right) = \frac{kL^2}{8} > 0$$

The maximum deflection is

$$y_m = y\left(\frac{L}{2}\right)$$

$$= -\frac{k}{24}\left[\frac{L^4}{16} - 2K\left(\frac{L^3}{8}\right) + L^3\left(\frac{L}{2}\right)\right]$$

$$\approx -0.0130kL^4$$

c. For a cantilevered beam, we have

$$y = -\frac{k}{24}x^4 + \frac{C_1}{6}x^3 + \frac{C_2}{2}x^2 + C_3 x + C_4$$

with boundary conditions $y(0) = y(L) = 0$
and $y''(0) = y'(L) = 0$. The conditions
$y(0) = 0$ and $y''(0) = 0$ imply
$C_2 = C_4 = 0$, and the other two conditions
yield

$$C_1 = \frac{3}{8}kL \text{ and } C_3 = -\frac{1}{48}kL^3$$

Thus,

$$y = -\frac{k}{48}\left(2x^4 - 3Lx^3 + L^3 x\right)$$

To find where maximum deflection occurs,
we find

$$y' = -\frac{k}{48}\left(8x^3 - 9Lx^2 + L^3\right)$$

$y' = 0$ when $x = L$, $\left(\dfrac{1 \pm \sqrt{33}}{16}\right)L$.

Checking, we find that the maximum deflection occurs at

$$x = \left(\frac{1 \pm \sqrt{33}}{16}\right) L \approx 0.4215L$$

and the maximum deflection is

$$y_m = y(0.4215L) \approx -0.0054kL^4$$

The maximum deflection in the cantilevered case is less than that in part **b**.

47. Let $S_1(t)$ and $S_2(t)$ be the amounts of salt in the first and second tanks, respectively, at time t.

a. $\dfrac{dS_1}{dt} = 2(1) - \dfrac{S_1}{100}(1)$

$\dfrac{dS_1}{dt} + \dfrac{S_1}{100} = 2$

$I(t) = e^{\int (1/100)dt} = e^{t/100}$

$$S_1(t) = e^{-t/100}\left[\int 2e^{t/100}dt + C\right]$$

$$= 200 + Ce^{-t/100}$$

Since $S_1(0) = 0$ (only pure water at $t = 0$), we obtain

$S_1(0) = 0 = 200 + C$

so that $C = -200$ and

$S_1(t) = 200\left(1 - e^{-t/100}\right)$

b. $\dfrac{dS_2}{dt} = \dfrac{S_1}{100}(1) - \dfrac{S_2}{100}(1)$

$\dfrac{dS_2}{dt} + \dfrac{S_2}{100} = \dfrac{1}{100}\left[200(1 - e^{-t/100}\right]$

from part **a**. The integrating factor is

$I(t) = e^{\int (1/100)\, dt} = e^{t/100}$

$$S_2(t) = e^{-t/100}\left[\int e^{t/100}\left(2 - 2e^{-t/100}\right)dt + C_1\right]$$

$$= 200 - 2te^{-t/100} + C_1 e^{-t/100}$$

Since $S_2(0) = 0$, we obtain $0 = 200 + C_1$ or $C_1 = -200$, and

$$S_2(t) = 200 - 2te^{-t/100} - 200e^{-t/100}$$

c. The excess is

$S(t) = S_1 - S_2 = 2te^{-t/100}$

which is maximized when

$S'(t) = (-0.02t + 2)e^{-t/100} = 0$

or $t = 100$ minutes. The maximum excess is

$$S(100) \approx 73.58 \text{ lbs}$$

49. $\dfrac{dP}{dt} = kP(B - \ln P)$

Let $P = e^u$,

$$\frac{dP}{dt} = e^u \frac{du}{dt}$$

$$e^u \frac{du}{dt} = ke^u(B - u)$$

$$\int \frac{du}{B - u} = \int k\, dt$$

$$-\ln|B - u| = kt + C_1$$

$$B - u = Ce^{-kt}$$

$$u = B - Ce^{-kt}$$

Thus,

$$P(t) = e^{B - Ce^{-kt}} = e^B e^{-Ce^{-kt}}$$

Applying the initial conditions,

$$P_0 = P(0) = e^{B-C}$$

$$B - C = \ln P_0$$

$$P_\infty = \lim_{t \to \infty} e^{B - Ce^{-kt}} = e^B.$$

Thus, $B = \ln P_\infty$ and

$$C = \ln P_\infty - \ln P_0 = \ln \tfrac{P_\infty}{P_0}$$

so

$$P(t) = P_\infty \exp\left[-\left(\ln \frac{P_\infty}{P_0}\right)e^{-kt}\right]$$

51. $\dfrac{dy}{dx} = \dfrac{1+y}{xy + e^y(1+y)}$

$\dfrac{dx}{dy} = \dfrac{xy + e^y(1+y)}{1+y}$

$\quad = \dfrac{y}{1+y}x + e^y$

This is a first order linear differential equation in x. The integrating factor is

$\quad I(t) = e^{\int [-y/(1+y)]dy} = (y+1)e^{-y}$

so

$x = \dfrac{1}{(y+1)e^{-y}}\left[\int (y+1)e^{-y}e^y\, dy + C\right]$

$\quad = \dfrac{e^y}{y+1}\left[\dfrac{1}{2}(y+1)^2\right] + C$

$\quad = \dfrac{1}{2}(y+1)e^y + \dfrac{Ce^y}{y+1}$

53. The governing differential equation is

$$5\frac{dI}{dt} + 10I = E(t)$$

$$\frac{dI}{dt} + 2I = \frac{1}{5}E(t)$$

The integrating factor is $e^{\int 2\, dt} = e^{2t}$. Then,

$$I(t) = e^{-2t}\left[\int \frac{1}{5}e^{2t}E(t)\, dt + C\right]$$

a. If $E = 15$, then

$$I(t) = e^{-2t}\left[\int \frac{15}{5}e^{2t}\, dt + C\right]$$

$$= \frac{3}{2} + Ce^{-2t}$$

Since $I(0) = 0$, $C = -\frac{3}{2}$, so

$$I(t) = \frac{3}{2}\left(1 - e^{-2t}\right)$$

b. If $E = 5e^{-2t}\sin t$, then

$$I(t) = e^{-2t}\left[\int \frac{1}{5}e^{2t}\left(5e^{-2t}\sin t\right)dt + C\right]$$

$$= e^{-2t}[-\cos t + C]$$

55. Let $Q(t)$ be the amount of pollutant in the lake at time t. Then

$$\frac{dQ}{dt} = (0.006)(350) - \frac{Q}{6{,}000}(350)$$

where units are in millions of ft^3. Thus,

$$\frac{dQ}{dt} + 0.0583Q = 0.21$$

The integrating factor is

$$I(t) = e^{\int 0.0583\, dt} = e^{0.0583t}$$

so

$$Q(t) = e^{-0.0583t}\left[\int e^{0.0583t}(0.21)\, dt + C\right]$$

$$= 3.602 + Ce^{-0.0583t}$$

Since the lake initially contains $Q(0) \approx 13.2$ million cubic feet of pollutant, we find $C \approx 9.598$, so

$$Q(t) = 3.602 + 9.598e^{-0.0583t}$$

The lake will contain 0.15% pollutant when

$$0.0015(6{,}000) = 3.602 + 9.598e^{-0.0583t}$$

Solving this equation, we obtain $t \approx 9.872$ days.

57. $A_2 = \displaystyle\int_0^y x\,dy$ and $A_1 = \displaystyle\int_0^x y\,dx$

$$2A_2 = A_1$$

$$2\int_0^y x\,dy = \int_0^x y\,dx$$

$$2x\frac{dy}{dx} = y$$

$$\frac{dy}{dx} = \frac{y}{2x}$$

$$\int y^{-1}dy = \int \frac{1}{2}x^{-1}\,dx$$

$$\ln|y| = \frac{1}{2}\ln|x| + C$$

$$y = Bx^{1/2}$$

59. a. $\qquad m\dfrac{dv}{dt} = -mg - kv$

$$\frac{dv}{dt} + \frac{k}{m}v = -g$$

The integrating factor is

$$I(t) = e^{\int k/m\,dt} = e^{kt/m}$$

so

$$v(t) = e^{-kt/m}\left[\int e^{kt/m}(-g)\,dt + C\right]$$

$$= -\frac{mg}{k} + Ce^{-kt/m}$$

Since $v(0) = v_0$, we have

$$v_0 = \frac{-mg}{k} + C, \text{ so } C = v_0 + \frac{mg}{k}$$

and

$$v(t) = \frac{-mg}{k} + \left(\frac{mg}{k} + v_0\right)e^{-kt/m}$$

Integrating and using $s(0) = 0$ (the object begins at ground level), we obtain

$$s(t) = \int v(t)\,dt$$

$$= \frac{-mg}{k}t + \left(\frac{mg}{k} + v_0\right)\left(-\frac{m}{k}\right)e^{-kt/m}$$
$$+ \left(\frac{mg}{k} + v_0\right)\left(\frac{m}{k}\right)$$

$$= \frac{-mg}{k}t + \frac{m}{k}\left(\frac{mg}{k} + v_0\right)\left(1 - e^{-kt/m}\right)$$

b. The object reaches its maximum height when $v(t) = 0$.

$$\frac{-mg}{k} + \left(\frac{mg}{k} + v_0\right)e^{-kt/m} = 0$$

$$e^{-kt/m} = \frac{mg}{mg + v_0 k}$$

$$t_{\max} = \frac{m}{k}\ln\left(1 + \frac{kv_0}{mg}\right)$$

The maximum height is

$$s_{\max} = s(t_{\max})$$

$$= \frac{-m^2 g}{k^2}\ln\left(1 + \frac{kv_0}{mg}\right)$$

$$+ \left(\frac{mg}{k} + v_0\right)\left(\frac{-m}{k}\right)\left(\frac{mg}{mg + v_0 k}\right)$$

$$+ \left(\frac{mg}{k} + v_0\right)\left(\frac{m}{k}\right)$$

$$= \frac{mv_0}{k} - \frac{m^2 g}{k^2}\ln\left(1 + \frac{kv_0}{mg}\right)$$

c. With $k = 0.75$, $v_0 = 150$, and $m = \frac{20}{32} = 0.625$ slugs, we find that the maximum height is

$$s_{\max} = \frac{0.625(150)}{0.75} - \frac{(0.625)^2(32)}{(0.75)^2}\ln\left[1 + \frac{0.75(150)}{0.625(32)}\right]$$

$$\approx 82.98 \text{ ft}$$

To the nearest foot, the maximum height is 83 ft. The object hits the ground when $s(t) = 0$:

$$0 = \left[-\frac{(0.625)(32)}{(0.75)}\right]t$$
$$+ \left(\frac{0.625}{0.75}\right)\left[\frac{0.625(32)}{0.75} + 150\right]\left(1 - e^{-0.75t/0.625}\right)$$

We use technology to find $t \approx 5.5$ seconds.

d. With $k = 0$, we have
$$v = -gt + v_0$$
$$s = -\frac{gt^2}{2} + v_0 t$$
The objects hits the ground when $s(t) = 0$:
$$\frac{-32t^2}{2} + 150t = 0$$
$$-16t^2 + 150t = 0$$
$$-16t\left(t - \frac{150}{16}\right) = 0$$
$$t = 0, \frac{150}{16}$$
The object will hit the ground in 9.375 seconds, so it will take longer.

7.7 Improper Integrals, page 566

1. An improper integral is a definite integral whose interval of integration is unbounded or whose integrand is unbounded on the interval of integration.

3. $\displaystyle\int_1^\infty \frac{dx}{x^3} = \lim_{N\to\infty} \int_1^N \frac{dx}{x^3}$
$$= \lim_{N\to\infty} \left(-\frac{1}{2x^2}\right)\Big|_1^N$$
$$= \lim_{N\to\infty} \left(-\frac{1}{2N^2} + \frac{1}{2}\right)$$
$$= \frac{1}{2}$$

5. $\displaystyle\int_1^\infty \frac{dx}{x^{0.99}} = \lim_{N\to\infty} \int_1^N x^{-0.99}\, dx$
$$= 100 \lim_{N\to\infty} x^{0.01}\Big|_1^N$$
$$= 100 \lim_{N\to\infty} \left(N^{0.01} - 1\right)$$
$$= \infty; \text{ diverges}$$

7. $\displaystyle\int_1^\infty \frac{dx}{x^{1.01}} = \lim_{N\to\infty} \int_1^N \frac{dx}{x^{1.01}}$
$$= -100 \lim_{N\to\infty} \left(\frac{1}{x^{0.1}}\right)\Big|_1^N$$
$$= -100 \lim_{N\to\infty} \left(\frac{1}{N^{0.1}} - 1\right) = 100$$

> **SURVIVAL HINT:** *Compare the exponents on Problems 3-8. Note how those exponents greater than 1 converge and those with exponents less than 1 diverge. Do you remember what happens when the exponent is 1?*

9. $\displaystyle\int_3^\infty \frac{dx}{2x - 1} = \frac{1}{2}\lim_{N\to\infty} \int_3^N \frac{2\, dx}{2x - 1}$
$$= \frac{1}{2}\lim_{N\to\infty} \ln|2x - 1|\Big|_3^N$$
$$= \infty; \text{ diverges}$$

11. $\displaystyle\int_3^\infty \frac{dx}{(2x - 1)^2} = \lim_{N\to\infty} \int_3^\infty \frac{dx}{(2x - 1)^2}$
$$= -\frac{1}{2}\lim_{N\to\infty} (2x - 1)^{-1}\Big|_3^N$$
$$= \lim_{N\to\infty} \left[-\frac{1}{4N - 2} + \frac{1}{10}\right]$$
$$= \frac{1}{10}$$

SURVIVAL HINT: *Compare the exponents on Problems 9-11. Note how those exponents greater than 1 converge and those with exponents less than 1 diverge. Do you remember what happens when the exponent is 1?*

13. $\displaystyle\int_1^\infty \frac{x^2\, dx}{(x^3+2)^2} = \frac{1}{3}\lim_{N\to\infty}\int_1^N \frac{3x^2\, dx}{(x^3+2)^2}$

$\boxed{u = x^3 + 2}$

$\displaystyle = \frac{1}{3}\lim_{N\to\infty}\left(-\frac{1}{x^3+2}\right)\Big|_1^N$

$\displaystyle = \frac{1}{3}\left(\frac{1}{3}\right)$

$\displaystyle = \frac{1}{9}$

15. $\displaystyle\int_1^\infty \frac{x^2\, dx}{\sqrt{x^3+2}}$

$\displaystyle = \frac{1}{3}\lim_{N\to\infty}\int_1^N (x^3+2)^{-1/2}(3x^2\, dx)$

$\boxed{u = x^3 + 2}$

$\displaystyle = \frac{2}{3}\lim_{N\to\infty}\sqrt{x^3+2}\,\Big|_1^N$

$= \infty;\ \text{diverges}$

17. $\displaystyle\int_1^\infty \frac{e^{-\sqrt{x}}}{\sqrt{x}}\, dx = \lim_{N\to\infty}\int_1^N \frac{e^{-\sqrt{x}}}{\sqrt{x}}\, dx$

$\boxed{u = -\sqrt{x}}$

$\displaystyle = -2\lim_{N\to\infty} e^{-\sqrt{x}}\,\Big|_1^N$

$\displaystyle = -2\lim_{N\to\infty}\left(\frac{1}{e^{\sqrt{N}}} - \frac{1}{e}\right)$

$\displaystyle = \frac{2}{e}$

19. $\displaystyle\int_0^\infty 5xe^{10-x}\, dx$

$\displaystyle = 5\lim_{N\to\infty}\int_0^N xe^{10-x}\, dx$

$\boxed{u = x;\ dv = -e^{-10-x}\, dx}$

$\displaystyle = \lim_{N\to\infty}\left[-5(x+1)e^{10-x}\right]_0^N$

$\displaystyle = \lim_{N\to\infty}\left[-5(N+1)e^{10-N} + 5e^{10}\right]$

$= 5e^{10}$

21. $\displaystyle\int_2^\infty \frac{dx}{x\ln x} = \lim_{N\to\infty}\int_x^N (\ln x)^{-1}\frac{dx}{x}$

$\displaystyle = \lim_{N\to\infty}\ln(\ln x)\,\Big|_2^N$

$= \infty;\ \text{diverges}$

23. $\displaystyle\int_{-\infty}^0 \frac{2x\, dx}{x^2+1} = \lim_{t\to-\infty}\int_t^0 \frac{2x\, dx}{x^2+1}$

$\displaystyle = \lim_{t\to-\infty}\ln(x^2+1)\,\Big|_t^0$

$= -\infty;\ \text{diverges}$

25. $\displaystyle\int_{-\infty}^0 \frac{dx}{\sqrt{2-x}}$

$\displaystyle = \lim_{t\to-\infty}\int_t^0 (2-x)^{-1/2}\, dx$

$\displaystyle = \lim_{t\to-\infty}\left(-2\sqrt{2-x}\right)\Big|_t^0$

$= \infty;\ \text{diverges}$

27. $\displaystyle\int_{-\infty}^\infty xe^{-|x|}\, dx$

$\displaystyle = \int_{-\infty}^0 xe^{-|x|}\, dx + \int_0^\infty xe^{-|x|}\, dx$

$\displaystyle = \lim_{t\to-\infty}\int_t^0 xe^{-|x|}\, dx + \lim_{N\to\infty}\int_0^N xe^{-x}\, dx$

$= -1 + 1$

$= 0$

29. $\displaystyle\int_0^1 \frac{dx}{x^{1/5}} = \lim_{t\to0^+} \int_t^1 x^{-1/5}\, dx$

$\displaystyle\qquad = \frac{5}{4}\lim_{t\to0^+} x^{4/5}\Big|_t^1$

$\displaystyle\qquad = \frac{5}{4}$

31. $\displaystyle\int_0^1 \frac{dx}{(1-x)^{1/2}} = \lim_{N\to1^-} \int_0^N (1-x)^{-1/2}\, dx$

$\displaystyle\qquad = -2\lim_{N\to1^-} \sqrt{1-x}\Big|_0^N$

$\displaystyle\qquad = \lim_{N\to1^-}\left[-2\left(\sqrt{1-N}-1\right)\right]$

$\displaystyle\qquad = 2$

33. $\displaystyle\int_{-1}^1 \frac{e^x}{\sqrt[3]{1-e^x}}\, dx$

$\displaystyle\quad = \int_{-1}^0 \frac{e^x\, dx}{(1-e^x)^{1/3}}\, dx + \int_0^1 \frac{e^x\, dx}{(1-e^x)^{1/3}}\, dx$

$\displaystyle\quad = \lim_{N\to0^-} \int_{-1}^N \frac{e^x\, dx}{(1-e^x)^{1/3}}\, dx + \lim_{N\to0^+} \int_N^1 \frac{e^x\, dx}{(1-e^x)^{1/3}}\, dx$

$\displaystyle\quad = \lim_{N\to0^-}\left[\frac{3}{2}(-1)(1-e^x)^{2/3}\right]_{-1}^N$

$\displaystyle\qquad + \lim_{N\to0^+}\left[\frac{3}{2}(-1)(1-e^x)^{2/3}\right]_N^1$

$\displaystyle\quad = \frac{3}{2}\left[(1-e^{-1})^{2/3} - (1-e)^{2/3}\right]$

35. $\displaystyle\int_0^1 \ln x\, dx = \lim_{t\to0^+} \int_t^1 \ln x\, dx$ $\boxed{\text{parts or Formula 196}}$

$\displaystyle\qquad = \lim_{t\to0^+} (x\ln x - x)\Big|_t^1$

$\displaystyle\qquad = -1 - \lim_{t\to0^+}(t\ln t - t))$ $\boxed{\text{l'Hôpital's rule}}$

$\displaystyle\qquad = -1$

37. $\displaystyle\int_e^\infty \frac{dx}{x(\ln x)^2} = \lim_{N\to\infty} \int_e^N \frac{dx}{x(\ln x)^2}$

$\displaystyle\qquad = \lim_{N\to\infty}\left[-\frac{1}{\ln x}\right]_e^N$

$\displaystyle\qquad = \lim_{N\to\infty}\left(-\frac{1}{\ln N}+1\right)$

$\displaystyle\qquad = 1$

39. $\displaystyle\int_0^1 e^{-\frac{1}{2}\ln x}\, dx = \lim_{t\to0^+} \int_t^1 x^{-1/2}\, dx$

$\displaystyle\qquad = \lim_{t\to0^+} 2\sqrt{x}\Big|_t^1$

$\displaystyle\qquad = 2$

41. The integrand becomes discontinuous where the denominator $1 - \tan x$ is 0; that is, at $x = \frac{\pi}{4}$.

$\displaystyle\int_0^{\pi/3} \frac{\sec^2 x\, dx}{1-\tan x}$

$\displaystyle= \lim_{t\to(\frac{\pi}{4})^-} \int_0^t \frac{\sec^2 x\, dx}{1-\tan x} + \lim_{t\to(\frac{\pi}{4})^+} \int_t^{\pi/3} \frac{\sec^2 x\, dx}{1-\tan x}$

provided both limits exist. However,

$\displaystyle\lim_{t\to(\frac{\pi}{4})^-} \int_0^t \frac{\sec^2 x\, dx}{1-\tan x}$

$\boxed{u = 1-\tan x;\ du = -\sec^2 x\, dx}$

$\displaystyle= \lim_{t\to(\frac{\pi}{4})^-} (-1)\ln|1-\tan x|\,\Big|_0^t$

$\displaystyle= \infty$

Since the limit does not exist, the improper integral diverges.

43. $A = \displaystyle\int_6^\infty \frac{2\,dx}{(x-4)^3}$

$= \displaystyle\lim_{N\to\infty} \left(-\frac{1}{2}\right)(x-4)^{-2}\bigg|_6^N$

$= \displaystyle\lim_{N\to\infty}\left[-(N-4)^{-2}+(6-4)^{-2}\right]$

$= \dfrac{1}{4}$

45. $A = \displaystyle\lim_{T\to\infty}\int_0^T 200e^{-0.002t}\,dt$

$= -100{,}000\displaystyle\lim_{T\to\infty} e^{-0.002t}\bigg|_0^T$

$= 100{,}000$

There will be 100,000 millirads.

47. $I = \displaystyle\int_2^\infty (\ln x)^{-p}\frac{dx}{x}$

$= \displaystyle\lim_{N\to\infty}\int_2^N (\ln x)^{-p}\frac{dx}{x}$

$= \displaystyle\lim_{N\to\infty}\frac{(\ln x)^{-p+1}}{1-p}\bigg|_2^N$

$= \displaystyle\lim_{N\to\infty}\frac{1}{(1-p)(\ln x)^{-1+p}}\bigg|_2^N$

$= (p-1)^{-1}(\ln 2)^{1-p}$ if $p>1$

and diverges if $p<1$.

If $p=1$,

$I = \displaystyle\int_2^\infty \frac{dx}{x\ln x}$

$= \displaystyle\lim_{N\to\infty}\ln(\ln x)\bigg|_2^N$

$= \infty$

Thus, I converges if $p>1$, and diverges if $p\le 1$.

49. $I = \displaystyle\int_0^{1/2}\frac{dx}{x(\ln x)^p}\,dx$

$= \displaystyle\lim_{t\to 0^+}\int_t^{1/2}(\ln x)^{-p}\frac{dx}{x}$

$= \displaystyle\lim_{t\to 0^+}\frac{(\ln x)^{-p+1}}{1-p}\bigg|_t^{1/2}$

$= (1-p)^{-1}(-\ln 2)^{1-p}$ if $p>1$,

and diverges if $p<1$.

If $p=1$,

$I = \displaystyle\lim_{t\to 0^+}\int_t^{1/2}\frac{dx}{x\ln x}$

$= \displaystyle\lim_{t\to 0^+}\ln|\ln x|\bigg|_t^{1/2}$

$= \infty$

Thus, I converges if $p>1$, and diverges if $p\le 1$.

51. Although the answer is correct, the student did not work the problem correctly because l'Hôpital's rule was applied to a quotient that has not become indeterminate.

53. $\mathcal{L}\{af+bg\} = \displaystyle\int_0^\infty e^{-st}(af+bg)\,dt$

$= a\displaystyle\int_0^\infty e^{-st}f\,dt + b\displaystyle\int_0^\infty e^{-st}g\,dt$

$= a\mathcal{L}\{f\} + b\mathcal{L}\{g\}$

55. $F(s) = \mathcal{L}\{f(t)\} = \displaystyle\int_0^\infty e^{-st}f(t)\,dt$

$\mathcal{L}\{f(t)e^{at}\} = \displaystyle\int_0^\infty e^{-st}e^{at}f(t)\,dt$

$= \displaystyle\int_0^\infty e^{-(s-a)t}f(t)\,dt$

Let $s_1 = s-a$; then

$$\mathcal{L}\{f(t)e^{at}\} = \int_0^\infty e^{-s_1 t} f(t)\, dt$$
$$= F(s_1)$$
$$= F(s - a)$$

57. a. $\quad F'(s) = \dfrac{d}{ds} F(s)$

$$= \frac{d}{ds} \int_0^\infty e^{-st} f(t)\, dt$$

$$= \int_0^\infty \frac{d}{ds} \left[e^{-st} f(t) \right] dt$$

$$= \int_0^\infty \left[-t e^{-st} \right] f(t)\, dt$$

$$= -\int_0^\infty \left[t e^{-st} \right] f(t)\, dt$$

$$= -\mathcal{L}\{t f(t)\}$$

b. Since $\mathcal{L}\{\cos 2t\} = \dfrac{s}{s^2 + 4}$, we have

$$\mathcal{L}\{t \cos 2t\} = -\frac{d}{ds}\left(\frac{s}{s^2 + 4}\right)$$

$$= -\left[\frac{(s^2 + 4) - s(2s)}{(s^2 + 4)^2} \right]$$

$$= \frac{s^2 - 4}{(s^2 + 4)^2}$$

59. To find $\mathcal{L}\{t^2 f(t)\}$, apply the result of Problem 57, to $g(t) = t f(t)$. Since
$$G(s) = \mathcal{L}\{t f(t)\} = -F'(s)$$
we have
$$\mathcal{L}\{t^2 f(t)\} = \mathcal{L}\{t\, g(t)\}$$
$$= -G'(s)$$
$$= -[-F'(s)]'$$
$$= F''(s)$$

In general,

$$\mathcal{L}\{t^n f(t)\} = (-1)^n \frac{d^n}{ds^n} \mathcal{L}\{f(t)\}$$
$$= (-1)^n F^{(n)}(s)$$

7.8 Hyperbolic and Inverse Hyperbolic Functions, page 574

1. 3.6269   **3.** -0.7616 **5.** 0.0000

7. 1.1995   **9.** 0.7500   **11.** 2.2924

13. $y' = (\cosh 3x)(3)$
$$= 3 \cosh 3x$$

15. $y' = -4x \sinh\left(1 - 2x^2\right)$

17. $y' = \left(\cosh x^{-1}\right)\left(-x^{-2}\right)$
$$= -x^{-2} \cosh x^{-1}$$

19. $y' = \dfrac{3x^2}{\sqrt{1 + x^6}}$

21. $y' = \dfrac{\sec^2 x}{\sqrt{1 + \tan^2 x}}$
$$= \frac{\sec^2 x}{|\sec x|}$$
$$= |\sec x|$$

23. $y' = \dfrac{\cos x}{1 - \sin^2 x}$
$$= \frac{\cos x}{\cos^2 x}$$
$$= \sec x$$

25. $y' = \dfrac{x\left(\frac{1}{\sqrt{1 + x^2}} - \sinh^{-1} x\right)}{x^2}$
$$= \frac{x - \sqrt{1 + x^2}\, \sinh^{-1} x}{x^2 \sqrt{1 + x^2}}$$

27. $y' = \cosh^{-1} x + \dfrac{x}{\sqrt{x^2 - 1}} - \dfrac{1}{2}\left(x^2 - 1\right)^{-1/2}(2x)$
$$= \cosh^{-1} x$$

29. $\displaystyle\int x\cosh(1-x^2)\,dx$

$\displaystyle= -\frac{1}{2}\int \cosh(1-x^2)\,(-2x\,dx)$

$\displaystyle= -\frac{1}{2}\sinh(1-x^2)+C$

31. $\displaystyle\int \frac{\sinh\frac{1}{x}\,dx}{x^2} = -\int \sinh(x^{-1})(-x^{-2}\,dx)$

$\displaystyle= -\cosh x^{-1}+C$

33. $\displaystyle\int \frac{dt}{\sqrt{9t^2-16}} = \frac{1}{3}\int \frac{\left(\frac{3\,dt}{4}\right)}{\sqrt{\left(\frac{3t}{4}\right)^2-1}}$

$\displaystyle= \frac{1}{3}\cosh^{-1}\frac{3t}{4}+C$

35. $\displaystyle\int \frac{\cos x\,dx}{\sqrt{1+\sin^2 x}} = \sinh^{-1}(\sin x)+C$

37. $\displaystyle\int \frac{x^2\,dx}{1-x^6} = \frac{1}{3}\int \frac{3x^2\,dx}{1-(x^3)^2}$

$\displaystyle= \frac{1}{3}\tanh^{-1}x^3+C$

39. $\displaystyle\int_2^3 \frac{dx}{1-x^2} = \coth^{-1}x\Big|_2^3$

$\displaystyle= \coth^{-1}3 - \coth^{-2}2$

41. $\displaystyle\int_1^2 \frac{e^x\,dx}{\sqrt{e^{2x}-1}} = \cosh^{-1}e^x\Big|_1^2$

$\displaystyle= \cosh^{-1}e^2 - \cosh^{-1}e$

43. $\displaystyle\int_0^1 x\,\mathrm{sech}^2 x^2\,dx = \frac{1}{2}\int_0^1 \mathrm{sech}^2 x^2(2x\,dx)$

$\displaystyle= \frac{1}{2}\tanh x^2\Big|_0^1 = \frac{1}{2}\tanh 1$

45. a. $\displaystyle\tanh(x+y) = \frac{\sinh(x+y)}{\cosh(x+y)}$

$\displaystyle= \frac{\sinh x\cosh y + \cosh x\sinh y}{\cosh x\cosh y + \sinh x\sinh y}$

$\displaystyle= \frac{\tanh x + \tanh y}{1+\tanh x\tanh y}$

b. $\sinh 2x = \sinh x\cosh x + \cosh x\sinh x$

$= 2\sinh x\cosh x$

c. $\cosh 2x = \cosh x\cosh x + \sinh x\sinh x$

$= \cosh^2 x + \sinh^2 x$

47. $y' = \mathrm{sech}^2 x \geq 0$, so the curve is rising for all x.
$y'' = -2\,\mathrm{sech}^2 x\tanh x$
$y'' = 0$ when $x=0$; the curve is concave up on $(-\infty,0)$, and concave down on $(0,\infty)$.

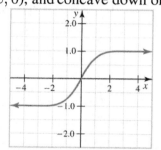

49. $\displaystyle(\cosh x + \sinh x)^n = \left[\frac{1}{2}(e^x+e^{-x}+e^x-e^{-x})\right]^n$

$= (e^x)^n$

$= e^{nx}$

$\cosh x + \sinh x = e^x$ *Take nth root of both sides.*
Since this is true for all x, we can replace x by nx and we have

$\cosh nx + \sinh nx = e^{nx}$
$\cosh nx + \sinh nx = (\cosh x + \sinh x)^n$

51. a. $y' = ac\sinh cx + bc\cosh cx$
$y'' = c^2(a\cosh cx + b\sinh cx) = c^2 y$
Thus, $y'' - c^2 y = 0$.

b. Let $c=2$; then $y = a\cosh 2x + b\sinh 2x$ is a solution of $y'' - 4y = 0$. Since $y'(0) = 2$, $2 = 2b$ or $b=1$. Since $y=1$ when $x=0$, $a=1$ and
$y = \cosh 2x + \sinh 2x$

53. $s = \displaystyle\int_{-a}^{a} \sqrt{1 + (y')^2}\, dx$

$ = 2 \displaystyle\int_{0}^{a} \sqrt{1 + \sinh^2 \frac{x}{a}}\, dx$

$ = 2a \displaystyle\int_{0}^{a} \cosh \frac{x}{a} \left(\frac{1}{a}\, dx \right)$

$ = 2a \sinh \dfrac{x}{a} \Big|_{0}^{a}$

$ = 2a(\sinh 1 - \sinh 0)$

$ = (e - e^{-1})a$

55. $y = \cosh x;$

$ds = \sqrt{1 + \sinh^2 x}\, dx = \cosh x\, dx$

$\qquad S = 2 \displaystyle\int_{0}^{1} 2\pi \cosh^2 x\, dx$

$\qquad = 2\pi \displaystyle\int_{0}^{1} (\cosh 2x + 1)\, dx$

$\qquad = 2\pi \left(\dfrac{1}{2} \sinh 2x + x \right) \Big|_{0}^{1}$

$\qquad = \pi \sinh 2 + 2\pi$

$\qquad = \dfrac{\pi}{2}\left(e^2 - e^{-2} + 4\right)$

57. $\quad y = \cosh u = \dfrac{1}{2}(e^u + e^{-u})$

$\dfrac{dy}{dx} = \dfrac{1}{2}(e^u - e^{-u}) \dfrac{du}{dx}$

$\phantom{\dfrac{dy}{dx}} = \sinh u \dfrac{du}{dx}$

$y = \tanh u = \dfrac{\sinh u}{\cosh u}$

$\dfrac{dy}{dx} = \dfrac{\cosh u \left(\cosh u \frac{du}{dx} \right) - \sinh u \left(\sinh u \frac{du}{dx} \right)}{\cosh^2 u}$

$\phantom{\dfrac{dy}{dx}} = \dfrac{\cosh^2 u - \sinh^2 u}{\cosh^2 u} \dfrac{du}{dx}$

$\phantom{\dfrac{dy}{dx}} = \dfrac{1}{\cosh^2 u} \dfrac{du}{dx}$

$\phantom{\dfrac{dy}{dx}} = \text{sech}^2 u \dfrac{du}{dx}$

$y = \text{sech}\, u = (\cosh u)^{-1}$

$\dfrac{dy}{dx} = -(\cosh u)^{-2} \sinh u \dfrac{du}{dx}$

$\phantom{\dfrac{dy}{dx}} = -\dfrac{\sinh u}{\cosh u} \dfrac{1}{\cosh u} \dfrac{du}{dx}$

$\phantom{\dfrac{dy}{dx}} = -\tanh u\, \text{sech}\, u \dfrac{du}{dx}$

59. a. $y = \cosh^{-1} x$, so $x = \cosh y = \dfrac{e^y + e^{-y}}{2}$

$\qquad x = \dfrac{e^y + e^{-y}}{2}$

$\qquad 2x = e^y + e^{-y}$

$\qquad e^{2y} - 2x e^y + 1 = 0$

$\qquad e^y = \dfrac{2x \pm \sqrt{4x^2 - 4}}{2}$

Reject negative choice since it corresponds to the second branch of solution $y = \cosh^{-1} x$. Thus, for $x \geq 1$,

$\qquad y = \ln\left(x + \sqrt{x^2 - 1} \right)$

b. $\qquad\qquad x = \tanh y$

$\qquad\qquad x = \dfrac{e^y - e^{-y}}{e^y + e^{-y}}$

$\qquad x e^y + x e^{-y} = e^y - e^{-y}$

$\qquad (x - 1)e^y = (-x - 1)e^{-y}$

$\qquad\qquad e^{2y} = \dfrac{1 + x}{1 - x}$

$\qquad\qquad 2y = \ln \dfrac{1 + x}{1 - x}$

Thus, for $|x| < 1$,

$\qquad\qquad \tanh^{-1} x = \dfrac{1}{2} \ln \dfrac{1 + x}{1 - x}$

Chapter 7 Review

Studying for a chapter examination is a personal process, one which nobody else can do for you. Simply take the time to review what you have done.

SURVIVAL HINT: Work all of Chapter 7 problems in the Proficiency Examination (whether they are assigned or not). Work through all of the problems before looking at the answers, and *then* correct each of the problems. The answers to all these problems are given in the answer section at the back of the text. If you worked the problem correctly, move on to the next problem, but if you did not work it correctly (or you did not know what to do), then look at the solutions below, look back in the chapter to study the procedure, or ask your instructor.

Finally, go back over the homework problems you have been assigned. If you worked a problem correctly, move on to the next problem, but if you missed it on your homework, then you should look back in the book or talk to your instructor about how to work the problem.

If you follow these steps, you should be successful with your review of this chapter.

Proficiency Examination, page 576

1. Let u replace a more complicated expression in the variable of integration, say x. Obtain all forms of x in terms of u. Substitute, integrate, return the form of answers from u back to x. In the case of a definite integral transform the limits of x into limits for u and evaluate using the fundamental theorem.

2. **a.** $\displaystyle\int u\, dv = uv - \int v\, du$

   b. A reduction formula for integration expresses an integral involving a power of a particular function in terms of an integral involving a lower power of the same function. By using the formula repeatedly, the given integral can be reduced to one that is more manageable.

3. **a.** A trigonometric substitution may be handy when the integrand contains one of the following forms:

$$\sqrt{x^2 + a^2}$$
$$\sqrt{x^2 - a^2}$$
$$\sqrt{a^2 - x^2}$$

   b. The Weierstrass substitutions are:
$$u = \tan\frac{x}{2},\ \sin x = \frac{2u}{1+u^2},\ \cos x = \frac{1-u^2}{1+u^2}$$
   and $dx = \dfrac{2\, du}{1+u^2}$

4. The method of partial fractions may be handy when integrating a rational function.

5. See Procedure Box in Section 7.5, on page 541.
   Step 1: simplify
   Step 2: use basic formulas (check table)
   Step 3: substitute
   Step 4: classify; parts, trig powers, Weierstrass substitution, trig substitutions, or partial fractions
   Step 5: try again

6. Equations that can be expressed in the form

$$\frac{dy}{dx} + P(x)y = Q(x)$$

are called first-order linear differential equations. The general solution is given by

$$y = \frac{1}{I(x)}\left[\int Q(x)I(x)\,dx + C\right]$$

where $I(x) = e^{\int P(x)\,dx}$.

7. An improper integral is one in which a limit of integration is infinite and/or at least one value in the interval of integration leads to an undefined integrand.

8. $\sinh x = \frac{1}{2}(e^x - e^{-x})$

$\cosh x = \frac{1}{2}(e^x + e^{-x})$

$\tanh x = \dfrac{e^x - e^{-x}}{e^x + e^{-x}}$

9. Let u be a differentiable function of x. Then:

$\dfrac{d}{dx}(\sinh u) = \cosh u \dfrac{du}{dx}$

$\dfrac{d}{dx}(\cosh u) = \sinh u \dfrac{du}{dx}$

$\dfrac{d}{dx}(\tanh u) = \operatorname{sech}^2 u \dfrac{du}{dx}$

$\dfrac{d}{dx}(\coth u) = -\operatorname{csch}^2 u \dfrac{du}{dx}$

$\dfrac{d}{dx}(\operatorname{sech} u) = -\operatorname{sech} u \tanh u \dfrac{du}{dx}$

$\dfrac{d}{dx}(\operatorname{csch} u) = -\operatorname{csch} u \cosh u \dfrac{du}{dx}$

$\displaystyle\int \sinh x\,dx = \cosh x + C$

$\displaystyle\int \cosh x\,dx = \sinh x + C$

$\displaystyle\int \operatorname{sech}^2 x\,dx = \tanh x + C$

$\displaystyle\int \operatorname{csch}^2 x\,dx = \coth x + C$

$\displaystyle\int \operatorname{sech} x \tanh x\,dx = -\operatorname{sech} x + C$

$\displaystyle\int \operatorname{csch} x \coth x\,dx = -\operatorname{csch} x + C$

10. $\sinh^{-1}x = \ln\left(x + \sqrt{x^2 + 1}\right)$, all x

$\cosh^{-1}x = \ln\left(x + \sqrt{x^2 - 1}\right)$, $x \geq 1$

$\tanh^{-1}x = \dfrac{1}{2}\ln\dfrac{1+x}{1-x}$, $|x| < 1$

$\operatorname{sech}^{-1}x = \ln\left(\dfrac{1 + \sqrt{1 - x^2}}{x}\right)$, $0 < x \leq 1$

$\operatorname{csch}^{-1}x = \ln\left(\dfrac{1}{x} + \dfrac{\sqrt{1 + x^2}}{|x|}\right)$, $x \neq 0$

$\coth^{-1}x = \dfrac{1}{2}\ln\dfrac{x+1}{x-1}$, $|x| > 1$

11. $\dfrac{d}{dx}\left(\sinh^{-1}u\right) = \dfrac{1}{\sqrt{1 + u^2}}\dfrac{du}{dx}$

$\dfrac{d}{dx}\left(\cosh^{-1}u\right) = \dfrac{1}{\sqrt{u^2 - 1}}\dfrac{du}{dx}$

$\dfrac{d}{dx}\left(\tanh^{-1}u\right) = \dfrac{1}{1 - u^2}\dfrac{du}{dx}$

$\dfrac{d}{dx}\left(\operatorname{sech}^{-1}u\right) = \dfrac{-1}{u\sqrt{1 - u^2}}\dfrac{du}{dx}$

$\dfrac{d}{dx}\left(\operatorname{csch}^{-1}u\right) = \dfrac{-1}{|u|\sqrt{1 + u^2}}\dfrac{du}{dx}$

$\dfrac{d}{dx}\left(\coth^{-1}u\right) = \dfrac{1}{1 - u^2}\dfrac{du}{dx}$

$\displaystyle\int \dfrac{du}{\sqrt{1 + u^2}} = \sinh^{-1}u + C$

$\displaystyle\int \dfrac{du}{\sqrt{u^2 - 1}} = \cosh^{-1}u + C$

$\displaystyle\int \dfrac{du}{1 - u^2} = \tanh^{-1}u + C$

$$\int \frac{du}{u\sqrt{1-u^2}} = \operatorname{sech}^{-1}|u| + C$$

$$\int \frac{du}{u\sqrt{1+u^2}} = -\operatorname{csch}^{-1}|u| + C$$

$$\int \frac{du}{1-u^2} = \coth^{-1} u + C$$

12. a. $\tanh^{-1} 0.5 = \dfrac{1}{2}\ln\dfrac{1+0.5}{1-0.5} = \dfrac{1}{2}\ln 3$

b. $\sinh(\ln 3) = \dfrac{e^{\ln 3} - e^{-\ln 3}}{2} = \dfrac{4}{3}$

c. $\coth^{-1} 2 = \dfrac{1}{2}\ln\dfrac{2+1}{2-1} = \dfrac{1}{2}\ln 3$

13. $\displaystyle\int \frac{2x+3}{\sqrt{x^2+1}}\,dx$

$\displaystyle = \int \frac{2x\,dx}{\sqrt{x^2+1}} + \int \frac{3\,dx}{\sqrt{x^2+1}}$ \boxed{\text{Formula 89}}

$= 2\sqrt{x^2+1} + 3\sinh^{-1} x + C$

14. $\displaystyle\int x\sin 2x\,dx$ \boxed{u = x;\ dv = \sin 2x\,dx}

$\displaystyle = -\frac{x}{2}\cos 2x - \int\left(-\frac{1}{2}\cos 2x\,dx\right)$

$\displaystyle = -\frac{x}{2}\cos 2x + \frac{1}{4}\sin 2x + C$

15. $\displaystyle\int \sinh(1-2x)\,dx$

$\displaystyle = -\frac{1}{2}\int \sinh(1-2x)(-2\,dx)$

$\displaystyle = -\frac{1}{2}\cosh(1-2x) + C$

16. $\displaystyle\int \frac{dx}{\sqrt{4-x^2}} = \sin^{-1}\frac{x}{2} + C$

17. $\dfrac{x^2}{(x^2+1)(x-1)} = \dfrac{A_1 x + B_1}{x^2+1} + \dfrac{A_2}{x-1}$

Partial fraction decomposition gives
$A_1 = \frac{1}{2},\ A_2 = \frac{1}{2},\ B_1 = \frac{1}{2}.$

$\displaystyle\int \frac{x^2\,dx}{(x^2+1)(x-1)}$

$\displaystyle = \frac{1}{2}\int \frac{x+1}{x^2+1}\,dx + \frac{1}{2}\int \frac{dx}{x-1}$

$\displaystyle = \frac{1}{4}\int \frac{2x\,dx}{x^2+1} + \frac{1}{2}\int \frac{dx}{x^2+1} + \frac{1}{2}\int \frac{dx}{x-1}$

$\displaystyle = \frac{1}{4}\ln(x^2+1) + \frac{1}{2}\tan^{-1} x + \frac{1}{2}\ln|x-1| + C$

18. $\displaystyle\int \frac{x^3\,dx}{x^2-1} = \int\left(x + \frac{x}{x^2-1}\right)dx$

$\displaystyle = \int x\,dx + \frac{1}{2}\int \frac{2x\,dx}{x^2-1}$

$\displaystyle = \frac{x^2}{2} + \frac{1}{2}\ln|x^2-1| + C$

19. $\displaystyle\int_1^2 x\ln x^3\,dx = 3\int_1^2 x\ln x\,dx$

\boxed{\text{Parts or Formula 199}}

$\displaystyle = 3\left(\frac{x^2}{2}\ln x - \frac{x^2}{4}\right)\Big|_1^2$

$\displaystyle = 6\ln 2 - \frac{9}{4}$

20. $\dfrac{1}{(x-1)^2(x+2)} = \dfrac{A_1}{(x-1)^2} + \dfrac{A_2}{x-1} + \dfrac{A_3}{x+2}$

Partial fraction decomposition gives
$A_1 = \frac{1}{3},\ A_2 = -\frac{1}{9},\ A_3 = \frac{1}{9}$

$\displaystyle\int_2^3 \frac{dx}{(x-1)^2(x+2)}$

$\displaystyle = \frac{1}{9}\int_2^3\left[\frac{3}{(x-1)^2} - \frac{1}{x-1} + \frac{1}{x+2}\right]dx$

$\displaystyle = \frac{1}{9}\left[-\frac{3}{x-1} - \ln(x-1) + \ln(x+2)\right]_2^3$

$\displaystyle = \frac{1}{9}\left(\ln\frac{5}{8} + \frac{3}{2}\right)$

21. $\displaystyle\int_3^4 \frac{dx}{2x - x^2} = \int_3^4 \left(\frac{\frac{1}{2}}{x} + \frac{\frac{1}{2}}{2 - x} \right) dx$

$\displaystyle = \left[\frac{1}{2}\ln x - \frac{1}{2}\ln|2 - x| \right]\Big|_3^4$

$\displaystyle = \frac{1}{2}\ln\frac{2}{3}$

22. $\displaystyle\int_0^{\frac{\pi}{4}} \sec^3 x \tan x \, dx$

$\displaystyle = \int_0^{\frac{\pi}{4}} \left(\sec^2 x \right)\left(\sec x \tan x \right) dx$

$\displaystyle = \frac{\sec^3 x}{3}\Big|_0^{\frac{\pi}{4}}$

$\displaystyle = \frac{2\sqrt{2} - 1}{3}$

23. $\displaystyle\int_0^{\infty} x e^{-2x}\, dx$

$\displaystyle = \lim_{N\to\infty}\int_0^N x e^{-2x}\, dx \quad \boxed{\text{Parts or Formula 184}}$

$\displaystyle = \lim_{N\to\infty}\left[-\frac{x}{2}e^{-2N} - \frac{1}{4}e^{-2x} \right]\Big|_0^N$

$\displaystyle = \lim_{N\to\infty}\left[-\frac{1}{4}e^{-2N}(2N + 1) + \frac{1}{4} \right]$

$\displaystyle = \frac{1}{4}$

24. $\displaystyle\int_0^{\frac{\pi}{4}} \frac{\sec^2 x \, dx}{\sqrt{\tan x}} = \int_0^1 \frac{du}{\sqrt{u}} \quad \boxed{u = \tan x}$

$\displaystyle = \lim_{t\to 0^+}\int_t^1 \frac{du}{\sqrt{u}}$

$\displaystyle = \lim_{t\to 0^+} 2\sqrt{u}\,\Big|_t^1$

$\displaystyle = 2$

25. $\displaystyle\frac{2x + 3}{x^2(x - 2)} = \frac{A_1}{x^2} + \frac{A_2}{x} + \frac{A_3}{x - 2}$

Partial fraction decomposition gives
$A_1 = -\frac{3}{2},\ A_2 = -\frac{7}{4};\ A_3 = \frac{7}{4}$

$\displaystyle\int_0^1 \frac{2x + 3}{x^2(x - 2)}\, dx$

$\displaystyle = \lim_{t\to 0^+}\int_t^1 \left(\frac{-\frac{3}{2}}{x^2} + \frac{-\frac{7}{4}}{x} + \frac{\frac{7}{4}}{x - 2} \right) dx$

$\displaystyle = \lim_{t\to 0^+}\left[\frac{3}{2x} + \frac{7}{4}\ln\left| \frac{x - 2}{x} \right| \right]\Big|_t^1$

$\displaystyle = \lim_{t\to 0^+}\left[\frac{3}{2} + \frac{7}{4}\ln 1 - \frac{3}{2t} - \frac{7}{4}\ln\left| \frac{t - 2}{t} \right| \right]$

$= -\infty$

The integral diverges.

26. $\displaystyle\int_0^{\infty} e^{-x}\sin x \, dx$

$\displaystyle = \lim_{N\to\infty}\int_0^N e^{-x}\sin x \, dx \quad \boxed{\text{Formula 192}}$

$\displaystyle = \lim_{N\to\infty}\left[\frac{e^{-x}(-\sin x - \cos x)}{2} \right]\Big|_0^N$

$\displaystyle = \lim_{N\to\infty}\left[\frac{e^{-N}(-\sin N - \cos N)}{2} + \frac{1}{2} \right]$

$\displaystyle = \frac{1}{2}$

27. $\displaystyle y' = \frac{1}{2\sqrt{\tanh^{-1}2x}}\left[\frac{2}{1 - 4x^2} \right]$

$\displaystyle = \frac{1}{\sqrt{\tanh^{-1}2x}}\left[\frac{1}{1 - 4x^2} \right]$

28. $\displaystyle V = 2\pi\int_0^2 x\left(\frac{1}{\sqrt{9 - x^2}} \right) dx$

$\displaystyle = -\pi\int_0^2 \frac{-2x\, dx}{\sqrt{9 - x^2}}$

$\displaystyle = -2\pi\sqrt{9 - x^2}\,\Big|_0^2$

$\displaystyle = 2\pi\left(3 - \sqrt{5} \right)$

29. $\dfrac{dy}{dx} + \left(\dfrac{x}{x+1}\right)y = e^{-x}$

$P(x) = \dfrac{x}{x+1};\ Q(x) = e^{-x}$

$I(x) = e^{\int x/(x+1)\,dx} = \dfrac{e^x}{x+1}$

$y = \dfrac{x+1}{e^x}\left[\int \dfrac{e^x}{x+1}(e^{-x}) + C\right]$

$= \left(\dfrac{x+1}{e^x}\right)[\ln|x+1| + C]$

Since $y = 1$ when $x = 0$, $C = 1$; thus,

$$y = \dfrac{x+1}{e^x}(\ln|x+1| + 1)$$

30. Let $S(t)$ be the amount of salt in the solution at time t.

$$\dfrac{dS}{dt} = (1.3)(5) - \left(\dfrac{S}{200+2t}\right)(3)$$

$\dfrac{dS}{dt} + \dfrac{3S}{200+2t} = 6.5$

$I(t) = e^{\int 3/(200+2t)\,dt} = (t+100)^{3/2}$

$S(t) = (t+100)^{-3/2}\int 6.5(t+100)^{3/2}\,dt + C$

$= (t+100)^{-3/2}\left[2.6(t+100)^{5/2} + C\right]$

$= 2.6(t+100) + C(t+100)^{-3/2}$

Since $S(0) = 200(2) = 400$, we find $C = 140{,}000$ so

$$S(t) = 2.6(t+100) + 140{,}000(t+100)^{-3/2}$$

After one hour, $S(60) \approx 485.2$ lb

Supplementary Problems, page 577

1. $y' = \dfrac{1}{1-(x^{-1})^2}(-x^{-2})$

$= (1-x^2)^{-1}$

3. $y = \dfrac{\sinh x}{e^x}$

$= \dfrac{1}{2}e^{-x}(e^x - e^{-x})$

$= \dfrac{1}{2}(1 - e^{-2x})$

$y' = \dfrac{1}{2}(2e^{-2x}) = e^{-2x}$

5. $y' = x\cosh x + \sinh x$
$+ (e^x - e^{-x})\cosh(e^x + e^{-x})$

7. $\displaystyle\int \dfrac{x^2\,dx}{\sqrt{4-x^2}}$ [Formula 112]

$= -\dfrac{x}{2}\sqrt{4-x^2} + 2\sin^{-1}\dfrac{x}{2} + C$

9. $\dfrac{3x-2}{x^3-2x^2} = \dfrac{3x-2}{x^2(x-2)} = \dfrac{A_1}{x} + \dfrac{A_2}{x^2} + \dfrac{A_3}{x-2}$

Partial fraction decomposition gives
$A_1 = -1$, $A_2 = 1$, $A_3 = 1$.

$\displaystyle\int \dfrac{3x-2}{x^3-2x^2}\,dx$

$= -\int \dfrac{dx}{x} + \int x^{-2}\,dx + \int \dfrac{dx}{x-2}$

$= -\ln|x| - x^{-1} + \ln|x-2| + C$

11. $\displaystyle\int \dfrac{dx}{\sqrt{x}(1+\sqrt[4]{x})}$ $\boxed{x = u^4;\ dx = 4u^3\,du}$

$= \displaystyle\int \dfrac{4u^3\,du}{u^2(1+u)}$

$= 4\displaystyle\int \dfrac{u\,du}{1+u}$

$= 4\displaystyle\int du - 4\int \dfrac{du}{1+u}$

$= 4u - 4\ln|1+u| + C$

$= 4x^{1/4} - 4\ln|1+x^{1/4}| + C$

13. $\displaystyle\int x^2\tan^{-1}x\,dx$ $\boxed{u = \tan^{-1}x;\ dv = x^2\,dx}$

$= \dfrac{x^3}{3}\tan^{-1}x - \dfrac{1}{3}\int \dfrac{x^3}{1+x^2}\,dx$

$$= \frac{x^3}{3}\tan^{-1}x - \frac{1}{3}\int\left[x - \frac{x}{1+x^2}\right]dx$$

$$= \frac{x^3}{3}\tan^{-1}x - \frac{x^2}{6} + \frac{1}{6}\ln(1+x^2) + C$$

Note: you can also use Formula 182.

15. $\displaystyle\int e^x\sqrt{4 - e^{2x}}\,dx$ $\boxed{t = e^x;\ dt = e^x\,dx}$

$$= \int \sqrt{4 - t^2}\,dt \quad \boxed{t = 2\sin\theta \text{ or Formula 117}}$$

$$= \frac{t\sqrt{4 - t^2}}{2} + 2\sin^{-1}\frac{t}{2} + C$$

$$= \frac{1}{2}e^x\sqrt{4 - e^{2x}} + 2\sin^{-1}\frac{e^x}{2} + C$$

17. $\displaystyle\int \frac{\sqrt{1 + \frac{1}{x^2}}}{x^5}\,dx$

$$= \int \frac{1}{x^2}\sqrt{1 + \frac{1}{x^2}}\,\frac{dx}{x^3} \quad \boxed{u = 1 + x^{-2};\ du = -2x^{-3}dx}$$

$$= \int (u - 1)\sqrt{u}\left(\frac{du}{-2}\right)$$

$$= -\frac{1}{2}\int \left[u^{3/2} - u^{1/2}\right]du$$

$$= -\frac{1}{5}\left(1 + x^{-2}\right)^{5/2} + \frac{1}{3}(1 + x^{-2})^{3/2} + C$$

19. $\displaystyle\int \sqrt{1 + \sin x}\,dx = \int \frac{\sqrt{1 - \sin^2 x}\,dx}{\sqrt{1 - \sin x}}$

$$= \int \frac{\cos x\,dx}{\sqrt{1 - \sin x}} \quad \boxed{u = \sin x}$$

$$= \int \frac{du}{\sqrt{1 - u}}$$

$$= -2\sqrt{1 - u} + C$$

$$= -2\sqrt{1 - \sin x} + C$$

21. $\displaystyle I = \int \sin(\ln x)\,dx$ $\boxed{u = \sin(\ln x);\ dv = dx}$

$$= x\sin(\ln x) - \int \cos(\ln x)\,dx$$

$$\boxed{u = \cos(\ln x);\ dv = dx}$$

$$= x\sin(\ln x) - \left[x\cos(\ln x) + \int \sin(\ln x)\,dx\right]$$

$$= x\sin(\ln x) - x\cos(\ln x) - I + C_1$$

$$2I = x\sin(\ln x) - x\cos(\ln x) + C_1$$

$$I = \frac{x}{2}\left[\sin(\ln x) - \cos(\ln x)\right] + C$$

23. $\displaystyle\int \frac{\sinh x\,dx}{2 + \cosh x}$ $\boxed{u = 2 + \cosh x}$

$$= \int \frac{du}{u}$$

$$= \ln(2 + \cosh x) + C$$

When using computer software, you might obtain the equivalent form

$$\ln(e^{2x} + 4e^x + 1) - x + C$$

25. $\displaystyle\int x^2\cot^{-1}x\,dx$ $\boxed{u = \cot^{-1}x;\ dv = x^2\,dx}$

$$= \frac{1}{3}x^3\cot^{-1}x + \frac{1}{3}\int \frac{x^3\,dx}{1 + x^2}$$

$$= \frac{1}{3}x^3\cot^{-1}x + \frac{1}{3}\int \left(x - \frac{x}{1 + x^2}\right)dx$$

$$= \frac{x^3}{3}\cot^{-1}x + \frac{x^2}{6} - \frac{1}{6}\ln(1 + x^2) + C$$

27. $\displaystyle\int \frac{x^2 + 2}{x^3 + 6x + 1}\,dx$ $\boxed{u = x^3 + 6x + 1}$

$$= \frac{1}{3}\int \frac{du}{u}$$

$$= \frac{1}{3}\ln\left|x^3 + 6x + 1\right| + C$$

29. $\displaystyle\int \cos\sqrt{x + 2}\,dx$ $\boxed{t = \sqrt{x + 2};\ dx = 2t\,dt}$

$$= 2\int t\cos t\,dt \quad \boxed{\text{Parts or Formula 123}}$$

$$= 2|\cos t + t\sin t| + C$$

$$= 2\cos\sqrt{x + 2} + 2\sqrt{x + 2}\sin\sqrt{x + 2} + C$$

31. $\displaystyle\int \frac{x^3 + 2x}{x^4 + 4x^2 + 3}\, dx$ $\boxed{u = x^4 + 4x^2 + 3}$

$\displaystyle = \frac{1}{4}\int \frac{du}{u}$

$\displaystyle = \frac{1}{4}\ln\left(x^4 + 4x^2 + 3\right) + C$

33. $\displaystyle\int \frac{\sqrt{5 - x^2}\, dx}{x}$ $\boxed{u = \sqrt{5}\sin\theta \text{ or Formula 121}}$

$\displaystyle = \sqrt{5 - x^2} - \sqrt{5}\ln\left|\frac{\sqrt{5} + \sqrt{5 - x^2}}{x}\right| + C$

If you use computer software you might obtain the following equivalent form

$\displaystyle\sqrt{5}\ln\left|\frac{\sqrt{5 - x^2} - \sqrt{5}}{x}\right| + \sqrt{5 - x^2} + C$

35. $\displaystyle\int x^3(x^2 + 4)^{-1/2}dx$ $\boxed{u = x^2;\ dv = \frac{x\,dx}{\sqrt{x^2+4}}}$

$\displaystyle = x^2\sqrt{x^2 + 4} - \int \sqrt{x^2 + 4}\,(2x\,dx)$

$\displaystyle = x^2\sqrt{x^2 + 4} - \frac{2}{3}(x^2 + 4)^{3/2} + C$

$\displaystyle = \frac{1}{3}\sqrt{x^2 + 4}\,(x^2 - 8) + C$

37. $\displaystyle\int \sec^3 x \tan x\, dx$

$\displaystyle = \int \sec^2 x(\sec x \tan x\, dx)$ $\boxed{u = \sec x}$

$\displaystyle = \frac{\sec^3 x}{3} + C$

39. $\displaystyle\int \frac{x^{1/3}\, dx}{x^{1/2} + x^{2/3}}$ $\boxed{u^6 = x;\ 6u^5\, du = dx}$

$\displaystyle = \int \frac{u^2(6u^5\, du)}{u^3 + u^4}$

$\displaystyle = 6\int \frac{u^4}{1 + u}\,du$

$\displaystyle = 6\int \left[u^3 - u^2 + u - 1 + \frac{1}{u + 1}\right]du$

$\displaystyle = 6\left[\frac{u^4}{4} - \frac{u^3}{3} + \frac{u^2}{2} - u + \ln|u + 1| + C\right]$

$\displaystyle = \frac{3}{2}x^{2/3} - 2x^{1/2} + 3x^{1/3} - 6x^{1/6}$

$\displaystyle\qquad + 6\ln\left|x^{1/6} + 1\right| + C$

41. $\displaystyle\int \frac{\cos x + \sin x}{1 + \cos x - \sin x}\, dx$ $\boxed{u = 1 + \cos x - \sin x}$

$\displaystyle = \int \frac{-du}{u}$

$\displaystyle = -\ln|1 + \cos x - \sin x| + C$

43. $\displaystyle\int \frac{e^x\, dx}{e^{x/3} - e^{x/2}}$ $\boxed{e^x = u^6;\ e^x\, dx = 6u^5\, du}$

$\displaystyle = \int \frac{6u^5\, du}{u^2 - u^3}$

$\displaystyle = 6\int \frac{u^3\, du}{1 - u}$

$\displaystyle = -6\int \left(u^2 + u + 1 + \frac{1}{u - 1}\right)du$

$\displaystyle = -6\left[\frac{u^3}{3} + \frac{u^2}{2} + u + \ln|u - 1|\right] + C$

$\displaystyle = -2e^{x/2} - 3e^{x/3} - 6e^{x/6}$

$\displaystyle\qquad - 6\ln\left|e^{x/6} - 1\right| + C$

45. $\displaystyle\int \sec^4 x\, dx = \int \sec^2 x\left(\tan^2 x + 1\right)dx$

$\displaystyle = \frac{1}{3}\tan^3 x + \tan x + C$

47. $\displaystyle\frac{x}{(x + 1)(x + 2)(x + 3)}$

$\displaystyle = \frac{A_1}{x + 1} + \frac{A_2}{x + 2} + \frac{A_3}{x + 3}$

Partial fraction decomposition gives
$A_1 = -\frac{1}{2},\ A_2 = 2,\ A_3 = -\frac{3}{2}.$

$$\int \frac{x\,dx}{(x+1)(x+2)(x+3)}$$
$$= -\frac{1}{2}\int\frac{dx}{x+1} + 2\int\frac{dx}{x+2} - \frac{3}{2}\int\frac{dx}{x+3}$$
$$= -\frac{1}{2}\ln|x+1| + 2\ln|x+2|$$
$$\qquad - \frac{3}{2}\ln|x+3| + C$$

49.
$$\frac{dy}{dx} = \frac{1-y}{1+x}$$
$$\frac{dy}{dx} + \frac{y}{1+x} = \frac{1}{1+x}$$
$$P(x) = \frac{1}{1+x};\; Q(x) = \frac{1}{1+x}$$
$$I(x) = e^{\int 1/(1+x)\,dx} = e^{\ln|x+1|} = x+1$$
$$y = \frac{1}{1+x}\left[\int\frac{x+1}{x+1}dx + C\right]$$
$$= \frac{1}{x+1}[x+C]$$

51. $xy' - yx^{1/2} = 0$
$$y' = yx^{-1/2}$$
$$y^{-1}y' = x^{-1/2}$$
$$\ln|y| = 2x^{1/2} + C$$
$$y = ke^{2\sqrt{x}}$$

53. $(\cos^3 x)y' = \sin^3 x \cos y$
$$\int\frac{1}{\cos y}dy = \int\frac{\sin^3 x}{\cos^3 x}dx$$
$$\int \sec y\,dy = \int(\sec^2 x - 1)\tan x\,dx$$
$$\ln|\sec y + \tan y| = \frac{1}{2}\tan^2 x + \ln|\cos x| + C$$

55. $\int_1^\infty e^{1-x}dx = \lim_{N\to\infty}\int_1^N e^{1-x}\,dx$
$$= -\lim_{N\to\infty} e^{1-x}\Big|_1^N$$
$$= 1$$

57. $\int_0^{\pi/2}\frac{\cos x\,dx}{\sqrt{\sin x}} = \lim_{t\to 0^+}\int_t^{\pi/2}\frac{\cos x\,dx}{\sqrt{\sin x}}$
$$= \lim_{t\to 0^+} 2\sqrt{\sin x}\Big|_t^{\pi/2}$$
$$= 2 - 0$$
$$= 2$$

59. $\int_{-1}^1\frac{dx}{(2x+1)^{1/3}}$
$$= \lim_{N\to(-\frac{1}{2})^-}\int_{-1}^N\frac{dx}{(2x+1)^{1/3}} + \lim_{N\to(-\frac{1}{2})^+}\int_N^1\frac{dx}{(2x+1)^{1/3}}$$
$$= \lim_{N\to(-\frac{1}{2})^-}\frac{3}{4}(2x+1)^{2/3}\Big|_{-1}^N + \lim_{N\to(\frac{1}{2})^+}\frac{3}{4}(2x+1)^{2/3}\Big|_N^1$$
$$= \lim_{N\to(-\frac{1}{2})^-}\left[\frac{3}{4}(2N+1)^{2/3} - \frac{3}{4}\right]$$
$$\quad + \lim_{N\to(-\frac{1}{2})^+}\left[-\frac{3}{4}(2N-1)^{2/3} + \frac{3^{5/3}}{4}\right]$$
$$= -\frac{3}{4} + \frac{3^{5/3}}{4} = \frac{3}{4}(3^{2/3} - 1)$$

61. $\int_0^1 xf''(3x)\,dx \quad \boxed{t = 3x;\; dt = 3\,dx}$
$$= \frac{1}{9}\int_0^3 tf''(t)\,dt \quad \boxed{u = t;\; dv = f''(t)\,dt}$$
$$= \frac{1}{9}\left[tf'(t)\big|_0^3 - \int_0^3 f'(t)\,dt\right]$$
$$= \frac{1}{9}[tf'(t) - f(t)]\Big|_0^3$$
$$= \frac{1}{9}\Big\{[3(-2) - 4] - (0 - 1)\Big\}$$
$$= -1$$

63.
$$\frac{dT}{dt} = k(T - T_0)$$

$$\int \frac{dT}{T - T_0} = \int k\, dt$$

$$\ln|T - T_0| = kt + C$$

$$T - T_0 = e^{kt+C}$$

$$T = T_0 + Be^{kt} \text{ for a constant } B$$

65. a. $\Gamma(s) = \displaystyle\int_0^\infty e^{-t}t^{s-1}\, dt$

$$= \int_0^1 e^{-t}t^{s-1}\, dt + \int_1^\infty e^{-t}t^{s-1}\, dt$$

The integral on the right converges since

$$\int_1^\infty e^{-t}t^{s-1}\, dt < \int_1^\infty t^{-2}\, dt = 1$$

For the integral on the left, let $u = t^{-1}$:

$$\int_a^1 e^{-t}t^{s-1}\, dt = \lim_{N\to\infty} \int_1^N e^{1-u}u^{-s-1}\, du$$

$$< \lim_{N\to\infty} \int_1^N u^{-s-1}\, du = \frac{1}{s}$$

Therefore, we see $\Gamma(s)$ converges.

b. $\Gamma(s) = \displaystyle\lim_{N\to\infty}\left[-\frac{t^{s-1}}{e^t}\Big|_0^N + (s-1)\int_0^N e^{-t}t^{s-2}\, dt\right]$

The first fraction vanishes as seen by using l'Hôpital's rule. The integral can be rewritten as

$$\Gamma(s) = (s-1)\Gamma(s-1)$$

which can be restated

$$\Gamma(s+1) = s\Gamma(s)$$

c. $\Gamma(n+1) = n\Gamma(n)$

$$= n(n-1)\Gamma(n-1)$$

$$= n(n-1)\cdots(3)(2)\Gamma(1)$$

$$= n!\lim_{N\to\infty}\int_0^N e^{-t}\, dt$$

$$= n!$$

67. $m = \displaystyle\int_0^{\pi/3} \sec^2 x\, dx$

$$= \tan x\big|_0^{\pi/3}$$

$$= \sqrt{3}$$

$$M_y = \int_0^{\pi/3} x\sec^2 x\, dx$$

$$\boxed{u = x;\ dv = \sec^2 x\, dx}$$

$$= [x\tan x]_0^{\pi/3} - \int_0^{\pi/3} \frac{\sin x\, dx}{\cos x}$$

$$= \frac{1}{3}\pi\sqrt{3} - \ln 2$$

$$M_x = \int_0^{\pi/3} \frac{1}{2}\sec^4 x\, dx$$

$$= \frac{1}{2}\int_0^{\pi/3}(1 + \tan^2 x)\sec^2 x\, dx$$

$$= \frac{1}{2}\left(\tan x + \frac{1}{3}\tan^3 x\right)\bigg|_0^{\pi/3}$$

$$= \sqrt{3}$$

$$(\bar{x}, \bar{y}) = \left(\frac{1}{3}\pi - \frac{\ln 2}{\sqrt{3}}, 1\right)$$

69. $V = 2\pi \displaystyle\int_0^{\cosh^{-1} 2} x(2 - \cosh x)\, dx$

$$= 2\pi x^2\big|_0^{\cosh^{-1} 2} - 2\pi\int_0^{\cosh^{-1} 2} x\cosh x\, dx$$

$$\boxed{u = x;\ dv = \cosh\, dx}$$

$$= \left[2\pi x^2 - 2\pi x\sinh x\right]\big|_0^{\cosh^{-1} 2}$$

$$+ 2\pi\int_0^{\cosh^{-1} 2} \sinh x\, dx$$

$$\approx 2.85$$

71. $y = \frac{4}{5}x^{5/4};\ y' = x^{1/4};\ ds = \sqrt{1 + x^{1/2}}\, dx$

$$s = \int_0^1 \sqrt{1 + x^{1/2}} \, dx \qquad \boxed{u = 1 + \sqrt{x}}$$

$$= \int_1^2 2\sqrt{u}(u - 1) \, du$$

$$= \int_1^2 \left(2u^{3/2} - 2u^{1/2}\right) du$$

$$= \left[\frac{4}{5}u^{5/2} - \frac{4}{3}u^{3/2}\right]_1^2$$

$$= \frac{8}{15}\left(\sqrt{2} + 1\right)$$

73. $V = \int_0^4 2\pi x \left(\frac{1}{1 + x^4}\right) dx$

$$= \pi \int_0^4 \frac{2x \, dx}{1 + (x^2)^2} \, dx$$

$$= \pi \tan^{-1} x^2 \big|_0^4$$

$$= \pi \tan^{-1} 16$$

75. $dy = \left(e^x - \frac{1}{4}e^{-x}\right) dx$

$$ds = \sqrt{1 + \left(e^x - \frac{1}{4}e^{-x}\right)^2} \, dx$$

$$= \left(e^x + \frac{1}{4}e^{-x}\right) dx$$

$$S = 2\pi \int_0^1 \left(e^x + \frac{1}{4}e^{-x}\right)^2 dx$$

$$= 2\pi \left[\frac{1}{2}e^{2x} + \frac{1}{2}x - \frac{1}{32}e^{-2x}\right]\Big|_0^1$$

$$= \pi \left(e^2 - \frac{1}{16}e^{-2} + \frac{1}{16}\right)$$

77. $A_1 = \int_1^a \frac{dx}{x} = \ln|a|$

$$A_2 = \int_k^{ka} \frac{dx}{x}$$

$$= \ln|x| \big|_k^{ka}$$

$$= \ln|ka| - \ln|k|$$

$$= \ln\left|\frac{ka}{k}\right|$$

$$= \ln|a|$$

Therefore, we see $A_1 = A_2$.

79. $dx = \left(\frac{2ay}{b^2} - \frac{b^2}{8ay}\right) dy$

$$ds = \sqrt{1 + \left(\frac{2ay}{b^2} - \frac{b^2}{8ay}\right)^2} \, dy$$

$$= \left(\frac{2ay}{b^2} + \frac{b^2}{8ay}\right) dy$$

$$s = \int_b^{2b} \left(\frac{2ay}{b^2} + \frac{b^2}{8ay}\right) dy$$

$$= \left[\frac{ay^2}{b^2} + \frac{b^2}{8a}\ln|y|\right]_b^{2b}$$

$$= 3|a| + \frac{b^2}{8|a|}\ln 2$$

81. $\int \frac{\sin^n x \, dx}{\cos^{m-1} x} \qquad \boxed{u = \sin^{n-1} x; \ dv = \frac{\sin x}{\cos^m x} dx}$

$$= \frac{\sin^{n-1} x}{(m-1)\cos^{m-1} x} + \frac{n-1}{1-m} \int \frac{\sin^{n-2} x \, dx}{\cos^{m-2} x}$$

$$= \frac{\sin^{n-1} x}{(m-1)\cos^{m-1} x} - \frac{n-1}{m-1} \int \frac{\sin^{n-2} x \, dx}{\cos^{m-2} x}$$

83. $I = \displaystyle\int x^n \left(x^2 + a^2\right)^{-1/2} dx$

$\boxed{u = x^{n-1};\; dv = x(x^2 + a^2)^{-1/2} dx}$

$= x^{n-1}(x^2 + a^2)^{1/2}$

$\quad - (n-1) \displaystyle\int x^{n-2}(x^2 + a^2)^{1/2}\, dx$

$= x^{n-1}(x^2 + a^2)^{1/2}$

$\quad - (n-1) \displaystyle\int x^{n-2}(x^2 + a^2)(x^2 + a^2)^{-1/2}\, dx$

$= x^{n-1}(x^2 + a^2)^{1/2}$

$\quad - (n-1) \displaystyle\int x^n \left(x^2 + a^2\right)^{-1/2}\, dx$

$\quad - a^2(n-1) \displaystyle\int x^{n-2}(a^2 + x^2)^{-1/2} dx$

$= x^{n-1}(x^2 + a^2)^{1/2} - (n-1)I$

$\quad - a^2(n-1) \displaystyle\int x^{n-2}(a^2 + x^2)^{-1/2} dx$

$nI = x^{n-1}(x^2 + a^2)^{1/2}$

$\quad - a^2(n-1) \displaystyle\int x^{n-2}(a^2 + x^2)^{-1/2} dx$

$I = \dfrac{x^{n-1}(x^2 + a^2)^{1/2}}{n}$

$\quad - \dfrac{a^2(n-1)}{n} \displaystyle\int x^{n-2}(a^2 + x^2)^{-1/2} dx$

85. $I = \displaystyle\int_1^\infty \left[\dfrac{2Ax^3}{x^4 + 1} - \dfrac{1}{x+1}\right] dx$

$= \lim_{t\to\infty} \left[\dfrac{A}{2}\ln(x^4 + 1) - \ln|x+1|\right]\Big|_1^t$

$= \lim_{t\to\infty} \left[\ln\dfrac{(t^4 + 1)^{A/2}}{t+1} - \ln 2^{A/2-1}\right]$

The limit will be ∞ if $\dfrac{A}{2}(4) > 1$ and

$-\infty$ if $\dfrac{A}{2}(4) < 1$, so we must have

$A = \dfrac{1}{2}$. Then

$I = \lim_{t\to\infty} \left[\ln\dfrac{(t^4 + 1)^{1/4}}{t+1} - \ln 2^{1/4-1}\right]$

$= \dfrac{3}{4}\ln 2$

87. $I = \displaystyle\int_0^{\pi/2} \sin^4 x\, dx$ $\boxed{\text{Formula 139}}$

$= \left[\dfrac{3x}{8} - \dfrac{\sin 2x}{4} + \dfrac{\sin 4x}{32}\right]_0^{\pi/2}$

$= \dfrac{3\pi}{16}$

89.

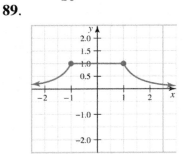

$I = \displaystyle\int_{-\infty}^\infty f(x)\, dx$

$= \lim_{t\to-\infty} \displaystyle\int_t^{-1} e^{x+1}\, dx + \displaystyle\int_{-1}^1 dx$

$\quad + \lim_{N\to\infty} \displaystyle\int_1^N x^{-2}\, dx$

$= 1 - \lim_{t\to-\infty} e^{t+1} + 2 + \lim_{N\to\infty} \dfrac{-1}{N} + 1$

$= 4$

91. $A = \displaystyle\int_0^\infty \left[\dfrac{1}{2}\left(e^{-2\theta}\right)^2 - \dfrac{1}{2}\left(e^{-5\theta}\right)^2\right] d\theta$

$= \lim_{N\to\infty} \displaystyle\int_0^N \dfrac{1}{2}\left(e^{-4\theta} - e^{-10\theta}\right) d\theta$

$= \lim_{N\to\infty} \left[\dfrac{1}{2}\left(-\dfrac{1}{4}\right)e^{-4\theta} - \dfrac{1}{2}\left(-\dfrac{1}{10}\right)e^{-10\theta}\right]\Big|_0^N$

$= \lim_{N\to\infty} \left[\left(-\dfrac{1}{8}e^{-4N} + \dfrac{1}{20}e^{-10N}\right) - \left(-\dfrac{1}{8} + \dfrac{1}{20}\right)\right]$

$$= 0 - \left(-\frac{1}{8} + \frac{1}{20}\right)$$

$$= \frac{3}{40}$$

93. a. Answers vary.

$$m = \frac{W}{g}$$

$$\underbrace{\frac{W}{g} \cdot \frac{dv}{dt}}_{total\ force} = \underbrace{W}_{weight} - \underbrace{B}_{buoying} - \underbrace{kv}_{drag\ force}$$

$$\frac{dv}{dt} + \left(\frac{kg}{W}\right)v = g - \frac{Bg}{W}$$

$$I = e^{\int(kg/W)dt} = e^{(kg/W)t}$$

$$v(t) = e^{-(kg/W)t}\left[\int\left(g - \frac{Bg}{W}\right)e^{(kg/W)t}dt + C_1\right]$$

$$= e^{-(kg/W)t}\left(g - \frac{Bg}{W}\right)\left[\left(\frac{W}{kg}\right)e^{(kg/W)t} + C_2\right]$$

$v(0) = 0$ implies $C_2 = -\frac{W}{k}\left(1 - \frac{B}{W}\right)$ so

$$v(t) = \left(\frac{W - B}{k}\right)\left(1 - e^{(-kg/W)t}\right)$$

b. $s(t) = \int v(t)\,dt$

$$= \frac{W - B}{k}\left[t + \frac{W}{kg}e^{-(kg/W)t} + C_2\right]$$

$s(0) = 0$ implies $C_2 = -\frac{W}{kg}\left(\frac{W - B}{k}\right)$

$$s(t) = \frac{W - B}{k}\left[t + \frac{W}{kg}e^{-(kg/W)t} - \frac{W}{kg}\right]$$

c. With $W = 1,125$; $B = 1,100$; $k = 0.64$; $g = 9.8$.

$$\frac{kg}{W} = \frac{0.64(9.8)}{1,125} \approx 0.0056$$

and

$$\frac{W - B}{k} = \frac{1,125 - 1,100}{0.64} \approx 39.06$$

Therefore, we have

$$v(t) \approx 39.06\left(1 - e^{-0.0056t}\right)$$

$$s(t) \approx 39.06\left[t + \frac{1,125}{(0.64)(9.8)}e^{-0.0056t} - \frac{1,125}{(0.64)(9.8)}\right]$$

$$= 39.06t + 7,006.14e^{-0.0056t} - 7,006.14$$

Solve $v(t) = 10$ for the time when the container breaks:

$$39.06\left(1 - e^{-0.0056t}\right) = 10$$

$$t \approx 52.81\ \sec$$

Then the critical depth is

$s_{max} = s(52.81)$

$$\approx 39.06(52.81) + 7,006.14\left(e^{-0.0056(52.81)} - 1\right)$$

$$\approx 269.07\ \text{meters}$$

95. Let $M(t)$ be the amount of old currency in circulation at time t, in millions. Then,

$$\frac{dM}{dt} = (0)(12) - \frac{M}{5,000 + (12 - 10)t}(10)$$

$$\int \frac{dM}{M} = \int \frac{-5\,dt}{2,500 + t}$$

$$\ln|M| = -5\ln|2,500 + t| + C_1$$

$$M = C(2,500 + t)^{-5}$$

Since $M(0) = 5,000$, we have

$$5{,}000 = C(2{,}500 + t)^{-5}$$
$$C = 5{,}000(2{,}500)^5$$

so that

$$M(t) = \frac{5{,}000(2{,}500)^5}{(2{,}500 + t)^{-5}}$$

There are 95% new style bills in circulation when $M(t) = 5\%$. Solving,

$$0.05(5{,}000) = \frac{5{,}000(2{,}500)^5}{(2{,}500 + t)^{-5}}$$

we obtain $t = 2{,}051$ days (about 5.6 years).

97. This is Putnam examination Problem 1 in the morning session of 1968. The standard approach, from elementary calculus, applies. By division, rewrite the integrand as a polynomial plus a rational function with numerator of degree less than 2. The solution follows easily.

99. This is Putnam examination Problem B5 in the afternoon session of 1985. Let

$$I(x) = \int_0^\infty t^{-1/2} e^{-at - at^{-1}} \, dt$$

Let $u = t^{-1}$, $du = -\dfrac{1}{t^2} \, dt$. Then,

$$I'(x) = -\int_\infty^0 u^{3/2} e^{-au^{-1} - xu} \left(\frac{-du}{u^2} \right)$$
$$= -\int_0^\infty u^{-1/2} e^{-au^{-1} - xu} \, du$$

Now let $w = \left(\dfrac{x}{a} \right) u$, $dw = \dfrac{x}{a} \, du$.

$$I'(x) = -\int_0^\infty \left(\frac{aw}{x} \right)^{-1/2} e^{-x/w - aw} \left(\frac{a}{x} \right) dw$$
$$= -\left(\frac{a}{x} \right)^{1/2} \int_0^\infty w^{-1/2} e^{-x/w - aw} \, dw$$
$$= -\left(\frac{a}{x} \right)^{1/2} I(x)$$

Solve this differential equation to obtain:

$$\int \frac{I'(x)}{I(x)} = \int \left[-\left(\frac{a}{x} \right)^{1/2} \right] dx$$
$$I(x) = -2\sqrt{ax} + C$$

$$I(x) = k e^{-2\sqrt{ax}}$$

$$I(0) = k e^0 = k = \int_0^\infty t^{-1/2} e^{-a/t - 0} \, dt = \sqrt{\frac{\pi}{a}}$$

Thus,

$$I(a) = \sqrt{\frac{\pi}{a}} e^{-2\sqrt{a^2}} = \sqrt{\frac{\pi}{a}} e^{-2a}$$

In particular,

$$I(1985) = \int_0^\infty t^{-1/2} e^{-1985(t + t^{-1})} \, dt$$
$$= \sqrt{\frac{\pi}{1985}} e^{-2(1985)}$$

CHAPTER 8 Infinite Series

Chapter Overview
The first seven chapters of this book make up what is known as single variable differential and integral calculus. We have studied limits, derivatives, and integrals, as well as a variety of techniques dealing with these topics. We are all familiar with a sequence of numbers; for example, $1, 2, 3, 4, \cdots$ or $2, 4, 6, 8, \cdots$ or $1, 1, 2, 3, 5, 8, \cdots$. One of my favorite books is *A Handbook of Integer Sequences* by N.J.A. Sloane (New York: Academic Press 1973). This is a book, much like a dictionary, which lists over 3,000 known sequences. Now, if we take any infinite sequence and find the sum its terms, we have a series. This chapter provides an introduction into infinite series which, as we will see, will a become a significant tool in our future mathematical work.

8.1 Sequences and Their Limits, page 594

1. The limit of a sequence is that unique number that the elements of a sequence approaches as more and more numbers in the sequence are considered.

3. $0, 2, 0, 2, 0$

5. $1, \frac{1}{2}, \frac{1}{3}, \frac{1}{4}, \frac{1}{5}$

7. $\frac{4}{3}, \frac{7}{4}, \frac{10}{5}, \frac{13}{6}, \frac{16}{7}$

9. $256, 16, 4, 2, \sqrt{2}$

11. $1, 3, 13, 183, 33673$

13. $\lim\limits_{n\to\infty} \dfrac{5n}{n+7} = \lim\limits_{n\to\infty} \dfrac{5}{1+\frac{7}{n}} = 5$

15. $\lim\limits_{n\to\infty} \dfrac{4-7n}{8+n} = \lim\limits_{n\to\infty} \dfrac{\frac{4}{n}-7}{\frac{8}{n}+1} = -7$

17. $\lim\limits_{n\to\infty} \dfrac{100n+7{,}000}{n^2-n-1} = \lim\limits_{n\to\infty} \dfrac{\frac{100}{n}+\frac{7{,}000}{n^2}}{1-\frac{1}{n}-\frac{1}{n^2}} = 0$

19. $\lim\limits_{n\to\infty} \dfrac{n^3-6n^2+85}{2n^3-5n+170}$
$= \lim\limits_{n\to\infty} \dfrac{1-\frac{6}{n}+\frac{85}{n^3}}{2-\frac{5}{n^2}+\frac{170}{n^3}}$
$= \dfrac{1}{2}$

21. $\lim\limits_{n\to\infty} \dfrac{8n-500\sqrt{n}}{2n+800\sqrt{n}} = \lim\limits_{n\to\infty} \dfrac{8-\frac{500}{\sqrt{n}}}{2+\frac{800}{\sqrt{n}}} = 4$

23. $\lim\limits_{n\to\infty} \dfrac{\ln n}{n^2} = \lim\limits_{n\to\infty} \dfrac{\frac{1}{n}}{2n}$ *l'Hôpital's rule*
$= \lim\limits_{n\to\infty} \dfrac{1}{2n^2}$
$= 0$

25. Let $L = \lim\limits_{n\to\infty} n^{3/n}$. Then,
$\ln L = \ln \lim\limits_{n\to\infty} n^{3/n}$
$= \lim\limits_{n\to\infty} \ln n^{3/n}$
$= \lim\limits_{n\to\infty} \dfrac{3\ln n}{n}$
$= \lim\limits_{n\to\infty} \dfrac{3\left(\frac{1}{n}\right)}{1}$ *l'Hôpital's rule*
$= 0$
Then $L = e^0 = 1$.

27. Let $L = \lim\limits_{n\to\infty} (n+4)^{1/n}$
$\ln L = \ln \lim\limits_{n\to\infty} (n+4)^{1/n}$

$$= \lim_{n\to\infty} \frac{\ln(n+4)}{n}$$

$$= \lim_{n\to\infty} \frac{\frac{1}{n+4}}{1} \qquad \textit{l'Hôpital's rule}$$

$$= 0$$

Then $L = e^0 = 1$.

29. Let $L = \lim_{n\to\infty} (\ln n)^{1/n}$

$$\ln L = \ln \lim_{n\to\infty} (\ln n)^{1/n}$$

$$= \lim_{n\to\infty} \ln(\ln n)^{1/n}$$

$$= \lim_{n\to\infty} \frac{\ln(\ln n)}{n}$$

$$= \lim_{n\to\infty} \frac{\frac{1}{\ln n}\left(\frac{1}{n}\right)}{1} \qquad \textit{l'Hôpital's rule}$$

$$= 0$$

Then, $L = e^0 = 1$.

31. $\lim_{n\to\infty} \left(\sqrt{n^2+n} - n\right)$

$$= \lim_{n\to\infty} \frac{\left(\sqrt{n^2+n} - n\right)\left(\sqrt{n^2+n} + n\right)}{\sqrt{n^2+n} + n}$$

$$= \lim_{n\to\infty} \frac{n^2+n-n^2}{\sqrt{n^2+n} + n}$$

$$= \lim_{n\to\infty} \frac{1}{\sqrt{1+\frac{1}{n}} + 1} = \frac{1}{2}$$

33. Let $L = \lim_{n\to\infty} \sqrt[n]{n}$

$$\ln L = \ln \lim_{n\to\infty} n^{1/n}$$

$$= \lim_{n\to\infty} \frac{\ln n}{n}$$

$$= \lim_{n\to\infty} \frac{\frac{1}{n}}{1} \qquad \textit{l'Hôpital's rule}$$

$$= 0$$

Then, $L = e^0 = 1$.

35. Let $L = \lim_{n\to\infty} (an+b)^{1/n}$

$$\ln L = \ln \lim_{n\to\infty} (an+b)^{1/n}$$

$$= \lim_{n\to\infty} \frac{\ln(an+b)}{n}$$

$$= \lim_{n\to\infty} \frac{\frac{a}{an+b}}{1} \qquad \textit{l'Hôpital's rule}$$

$$= 0$$

Then, $L = e^0 = 1$.

37. The elements of $\{a_n\} = \left\{\ln\left(\frac{n+1}{n}\right)\right\}$

lie on the curve $f(x) = \ln(x+1) - \ln x$.

$$f'(x) = (x+1)^{-1} - x^{-1} < 0$$

Thus, $f(x)$ and $\{a_n\}$ are both decreasing.
$M = 0$ is a lower bound of the sequence
(the elements are positive since
$\ln(n+1) > \ln n$, so $\{a_n\}$ converges.

39. The elements of $\{a_n\} = \left\{\frac{4n+5}{n}\right\}$ lie on

the curve $f(x) = 4 + 5x^{-1}$.

$$f'(x) = -5x^{-2} < 0$$

Thus, $f(x)$ and $\{a_n\}$ are both decreasing.
$M = 4$ is a lower bound of the sequence, so
$\{a_n\}$ converges.

41. The elements of $\{a_n\} = \left\{\sqrt[n]{n}\right\}$ lie on the
curve

$$f(x) = \sqrt[x]{x}$$

$$= \exp\left[\ln x^{1/x}\right]$$

$$= \exp\left[x^{-1}\ln x\right]$$

$$f'(x) = \exp\left[\frac{\ln x}{x}\right]\left(x^{-2} - x^{-2}\ln x\right)$$

$$< 0$$

(when $e < x$). *Note*: if you use technology for the derivative you may obtain the following equivalent form:

$$f'(x) = x^{(1-2x)/x}(1 - \ln x)$$

Thus, $f(x)$ and $\{a_n\}$ are both decreasing. $M = 0$ is a lower bound of the sequence since $n^{1/n} > 0$ for all n, so $\{a_n\}$ converges.

43. The elements $\{a_n\} = \{\cos n\pi\}$ alternate between -1 and 1, so the sequence diverges by oscillation.

45. The sequence $\{a_n\} = \{\sqrt{n}\}$ diverges because $\lim\limits_{n\to\infty} a_n = \infty$.

47. a. The sequence $\{a_n\} = \left\{\dfrac{\sin n}{n}\right\} \to 0$ because $-1 \le \sin n \le 1$ and

$$\lim_{n\to\infty} \frac{1}{n} = 0 \qquad \textit{by the squeeze theorem}$$

so $\lim\limits_{n\to\infty} a_n = 0$.

b. The sequence $\{b_n\} = \left\{n\sin\dfrac{1}{n}\right\} \to 1$ because if we let $k = \dfrac{1}{n}$, then $k \to 0$ as $n \to \infty$, we have

$$L = \lim_{k\to 0} \frac{\sin k}{k}$$
$$= 1$$

c. If we let $m = \frac{1}{n}$ in part **a** we see that $\{a_m\} = \left\{m\sin\dfrac{1}{m}\right\}$ which is the same as the sequence in part **b** except as $n \to \infty$, $m \to 0$.

49. $\displaystyle\lim_{n\to\infty} \sum_{k=1}^{n} \frac{n}{n^2 + k^2} = \lim_{n\to\infty} \sum_{k=1}^{n} \frac{1}{1 + \left(\frac{k}{n}\right)^2}\left(\frac{1}{n}\right)$

$$= \int_0^1 \frac{dx}{1 + x^2}$$
$$= \tan^{-1} x \Big|_0^1$$
$$= \frac{\pi}{4}$$

The series converges.

51. $\dfrac{n}{n + 1} = 1 - \dfrac{1}{n + 1}$ which is monotone increasing for all $n > 0$.

$$\left|\frac{n}{n + 1} - 1\right| < 0.01$$
$$\left|\frac{n - n - 1}{n + 1}\right| < 0.01$$
$$n + 1 > 100$$
$$n > 99$$

Choose $N = 99$.

53. $\dfrac{n^2 + 1}{n^3} = \dfrac{1}{n} + \dfrac{1}{n^3}$ and each of these fractions decreases as n increases.

$$\left|\frac{n^2 + 1}{n^3} - 0\right| < 0.001$$
$$\frac{n^2 + 1}{n^3} < 0.001$$

Since $n > 0$. $\quad n^2 + 1 < 0.001n^3$

$$n^3 - 1{,}000n^2 - 1{,}000 > 0$$

Since

$$n^3 - 1{,}000n^2 - 1{,}000 < 0 \text{ when } n = 1{,}000$$

and

$$n^3 - 1{,}000n^2 - 1{,}000 > 0 \text{ when } n = 1{,}001$$

we choose $N = 1{,}000$.

55. Let A be the least upper bound of the nondecreasing sequence $\{a_n\}$. Since $\epsilon > 0$, it follows that $A - \epsilon < A$ and

$$A - \epsilon < a_N < A$$

for some integer N. If $n > N$, then $a_n \geq a_N$ since the sequence is nondecreasing, and

$$A - \epsilon < a_N \leq a_n \leq A$$

57. Recall the binomial theorem

$$(a + b)^n = \sum_{k=0}^{n} \binom{n}{k} a^{n-k} b^k$$

Thus,

$$\left(1 + \frac{1}{n}\right)^n = \frac{n!}{n!} + \frac{n!}{(n-1)!} \frac{1}{n} + \frac{n!}{(n-2)!2!} \frac{1}{n^2} + \cdots + \frac{1}{n^n}$$

$$= 1 + 1 + \frac{1}{2}\left(1 - \frac{1}{n}\right) + \frac{1}{3!}\left(\frac{n-1}{n}\right)\left(\frac{n-2}{n}\right)$$

$$+ \cdots + \frac{1}{n!}\left(\frac{n-1}{n}\right)\left(\frac{n-2}{n}\right) \cdots \left(\frac{n-(n-1)}{n}\right)$$

$$= 1 + 1 + \frac{1}{2}\left(1 - \frac{1}{n}\right) + \frac{1}{3!}\left(1 - \frac{1}{n}\right)\left(1 - \frac{2}{n}\right)$$

$$+ \cdots + \frac{1}{n!}\left(1 - \frac{1}{n}\right)\left(1 - \frac{2}{n}\right) \cdots \left(1 - \frac{n-1}{n}\right)$$

59. Since $1 - \dfrac{k}{n+1} > 1 - \dfrac{k}{n}$ for all k, we have

$$\left(1 - \frac{1}{n+1}\right)\left(1 - \frac{2}{n+1}\right) \cdots \left(1 - \frac{k-1}{n+1}\right)$$

$$> \left(1 - \frac{1}{n}\right)\left(1 - \frac{2}{n}\right) \cdots \left(1 - \frac{k-1}{n}\right)$$

From Problem 57,

$$a_{n+1} > a_n \text{ where } a_n = \left(1 + \frac{1}{n}\right)^n$$

Thus, the sequence $\{a_n\}$ is increasing and since it is bounded above by 3 (Problem 58), the series converges by the BMCT.

8.2 Introduction to Infinite Series: Geometric Series, page 603

1. A sequence is a succession of elements. A series is a sum of the elements of the sequence.

> **SURVIVAL HINT:** *Your understanding of these first two problems in this problem set are crucial to your understanding of this chapter. For that reason, we include the answers to both.*

2. A geometric sequence is a sequence whose successive terms have a common ratio. A geometric series is the sum of the elements of a geometric sequence. The sum of n terms of a geometric series is

$$S_n = \frac{a(r^n - 1)}{r - 1}$$

The sum of an infinite geometric series is

$$S = \frac{a}{1 - r} \qquad \text{if } |r| < 1$$

3. $\displaystyle\sum_{k=0}^{\infty} \left(\frac{4}{5}\right)^k = \frac{1}{1 - \frac{4}{5}} = 5$

5. $\displaystyle\sum_{k=0}^{\infty} \frac{2}{3^k} = \frac{2}{1 - \frac{1}{3}} = 3$

7. $\displaystyle\sum_{k=1}^{\infty} \left(\frac{3}{2}\right)^k$ Geometric series with $r = \frac{3}{2} > 1$, so it diverges.

9. $\displaystyle\sum_{k=0}^{\infty} \frac{(-2)^k}{5^{2k+1}} = \frac{1}{5} - \frac{2}{5^3} + \frac{2^2}{5^5} - \frac{2^3}{5^7} + \cdots$

$$= \frac{1}{5}\left[1 - \frac{2}{5^2} + \left(\frac{2}{5^2}\right)^2 - \left(\frac{2}{5^2}\right)^3 + \cdots\right]$$

$$= \frac{1}{5}\left[\frac{1}{1-\left(-\frac{2}{5^2}\right)}\right]$$

$$= \frac{5}{27}$$

11. $\displaystyle\sum_{k=1}^{\infty} e^{-0.2k} = e^{-0.2}\left(1 - e^{-0.2}\right)^{-1} = \frac{1}{e^{1/5}-1}$

13. $\displaystyle\sum_{k=2}^{\infty} \frac{(-2)^{k-1}}{3^{k+1}} = -\frac{2}{27}\left(\frac{1}{1+\frac{2}{3}}\right) = -\frac{2}{45}$

15. $\displaystyle\frac{1}{2} - \frac{1}{2^2} + \frac{1}{2^3} - \frac{1}{2^4} + \cdots = \frac{1}{2}\left(\frac{1}{1+\frac{1}{2}}\right) = \frac{1}{3}$

17. $\displaystyle\frac{1}{4} + \left(\frac{1}{4}\right)^4 + \left(\frac{1}{4}\right)^7 + \left(\frac{1}{4}\right)^{10} + \cdots$

$$= \frac{1}{4}\left(\frac{1}{1-\frac{1}{64}}\right) = \frac{16}{63}$$

19. $2 + \sqrt{2} + 1 + \dfrac{1}{\sqrt{2}} + \cdots$

$$= 2\left[1 + \frac{1}{\sqrt{2}} + \left(\frac{1}{\sqrt{2}}\right)^2 + \left(\frac{1}{\sqrt{2}}\right)^3 + \cdots\right]$$

$$= 2\left(\frac{1}{1-\frac{1}{\sqrt{2}}}\right) = 2\left(2+\sqrt{2}\right)$$

21. $\left(1+\sqrt{2}\right) + 1 + \left(-1+\sqrt{2}\right) + \left(3-2\sqrt{2}\right) + \cdots$

$$= \left(1+\sqrt{2}\right)\left[1 + \left(\frac{1}{1+\sqrt{2}}\right) + \left(\frac{1}{1+\sqrt{2}}\right)^2 + \cdots\right]$$

$$= \left(1+\sqrt{2}\right)\left[\frac{1}{1-\frac{1}{1+\sqrt{2}}}\right]$$

$$= \frac{(1+\sqrt{2})^2}{\sqrt{2}}$$

$$= \frac{1}{2}\left(4+3\sqrt{2}\right)$$

23. $S_n = \displaystyle\sum_{k=1}^{n}\left[\frac{1}{k^{0.1}} - \frac{1}{(k+1)^{0.1}}\right]$

$$= \left(1 - \frac{1}{2^{0.1}}\right) + \left(\frac{1}{2^{0.1}} - \frac{1}{3^{0.1}}\right)$$

$$+ \cdots + \left[\frac{1}{n^{0.1}} - \frac{1}{(n+1)^{0.1}}\right]$$

$$= 1 - \frac{1}{(n+1)^{0.1}}$$

$$S = \lim_{n\to\infty} S_n$$

$$= \sum_{k=1}^{\infty}\left[\frac{1}{k^{0.1}} - \frac{1}{(k+1)^{0.1}}\right]$$

$$= \lim_{n\to\infty}\left[1 - \frac{1}{(n+1)^{0.1}}\right]$$

$$= 1$$

The series converges to 1.

SURVIVAL HINT: *Pay attention to the starting value for each summation.*

25. $S_n = \displaystyle\sum_{k=0}^{n} \frac{1}{(k+1)(k+2)}$

$$= \sum_{k=0}^{n}\left[\frac{1}{k+1} - \frac{1}{k+2}\right] \quad \textit{Partial fractions}$$

$$= \left(1 - \frac{1}{2}\right) + \left(\frac{1}{2} - \frac{1}{3}\right) +$$

$$\cdots + \left(\frac{1}{n+1} - \frac{1}{n+2}\right)$$

$$= 1 - \frac{1}{n+2}$$

$$S = \lim_{n\to\infty} S_n$$

$$= \lim_{n\to\infty}\left(1 - \frac{1}{n+2}\right)$$

$$= 1$$

The series converges to 1.

27. $S_n = \sum_{k=1}^{n} \ln\left(1 + \frac{1}{k}\right)$

$= \sum_{k=1}^{n} [\ln(k+1) - \ln k]$

$= (\ln 2 - \ln 1) + (\ln 3 - \ln 2)$
$\quad + (\ln 4 - \ln 3) + \cdots + [\ln(n+1) - \ln n]$

$= \ln(n+1) - \ln 1$

$= \ln(n+1)$

$S = \lim_{n \to \infty} S_n$

$= \lim_{n \to \infty} \ln(n+1)$

The series diverges.

29. $S_n = \sum_{k=1}^{n} \frac{2k+1}{k^2(k+1)^2}$

$= \sum_{k=1}^{n} \left[\frac{1}{k^2} - \frac{1}{(k+1)^2}\right]$ *Partial fractions*

$= \left(1 - \frac{1}{4}\right) + \left(\frac{1}{4} - \frac{1}{9}\right) + \cdots$

$\quad + \left(\frac{1}{n^2} - \frac{1}{(n+1)^2}\right)$

$= 1 - \frac{1}{(n+1)^2}$

$S = \lim_{n \to \infty} S_n$

$= \lim_{n \to \infty} \left[1 - \frac{1}{(n+1)^2}\right]$

$= 1$

The series converges to 1.

31. $0.\overline{01} = 0.0101010\cdots$

$= \frac{1}{100} + \frac{1}{10,000} + \frac{1}{1,000,000} + \cdots$

This is a geometric series with $r = \frac{1}{100}$.

$S = \frac{\frac{1}{100}}{1 - \frac{1}{100}}$

$= \frac{\frac{1}{100}}{\frac{99}{100}}$

$= \frac{1}{99}$

33. $1.\overline{405} = 1 + 0.405\left[1 + \frac{1}{1,000} + \left(\frac{1}{1,000}\right)^2 + \cdots\right]$

This is a geometric series with $r = \frac{1}{1,000}$.

$S = 1 + \frac{\frac{405}{1,000}}{1 - \frac{1}{1,000}}$

$= 1 + \frac{405}{999}$

$= \frac{52}{37}$

35. a. $\frac{k-1}{2^{k+1}} = \frac{Ak}{2^k} - \frac{B(k+1)}{2^{k+1}}$

$\qquad = \frac{2Ak - B(k+1)}{2^{k+1}}$

We see that

$\quad k = (2A - B)k \quad \text{and} \quad -1 = -B$

$2A - B = 1 \qquad\qquad\qquad B = 1$

This implies that $A = 1$.

b. $S_n = \sum_{k=1}^{n} \frac{k-1}{2^{k+1}}$

$= \sum_{k=1}^{n} \left(\frac{k}{2^k} - \frac{k+1}{2^{k+1}}\right)$

$= \left[\frac{1}{2} - \frac{2}{2^2}\right] + \left[\frac{2}{2^2} - \frac{3}{2^3}\right]$

$\quad + \left[\frac{3}{2^3} - \frac{4}{2^4}\right] + \cdots + \left[\frac{n}{2^n} - \frac{n+1}{2^{n+1}}\right]$

$$= \frac{1}{2} - \frac{n+1}{2^{n+1}}$$

$$S = \lim_{n \to \infty} S_n$$

$$= \lim_{n \to \infty} \left(\frac{1}{2} - \frac{n+1}{2^{n+1}} \right)$$

$$= \frac{1}{2}$$

37. $S_N = \displaystyle\sum_{n=1}^{N} \frac{\ln\left[\frac{n^{n+1}}{(n+1)^n}\right]}{n(n+1)}$

$$= \sum_{n=1}^{N} \left(\frac{\ln n}{n} - \frac{\ln(n+1)}{n+1} \right) \quad \textit{Partial fractions}$$

$$= \left(\frac{\ln 1}{1} - \frac{\ln 2}{2} \right) + \left(\frac{\ln 2}{2} - \frac{\ln 3}{3} \right)$$

$$+ \cdots + \left(\frac{\ln N}{N} - \frac{\ln(N+1)}{N+1} \right)$$

$$= 0 - \frac{\ln(N+1)}{N+1}$$

$$S = \lim_{N \to \infty} S_N$$

$$= \lim_{N \to \infty} \frac{\ln(N+1)}{N+1}$$

$$= 0$$

39. $\displaystyle\sum_{k=0}^{\infty} (2a_k + 2^{-k}) = 2\sum_{k=0}^{\infty} a_k + \sum_{k=0}^{\infty} \frac{1}{2^k}$

$$= 2(3.57) + \frac{1}{1 - \frac{1}{2}}$$

$$= 9.14$$

41. $\displaystyle\sum_{k=0}^{\infty} \left(2^{-k} + 3^{-k} \right)^2$

$$= \sum_{k=0}^{\infty} \left[4^{-k} + 2\left(6^{-k}\right) + 9^{-k} \right]$$

$$= \frac{1}{1 - \frac{1}{4}} + 2\left(\frac{1}{1 - \frac{1}{6}} \right) + \frac{1}{1 - \frac{1}{9}}$$

$$= \frac{4}{3} + 2\left(\frac{6}{5} \right) + \frac{9}{8}$$

$$= \frac{583}{120}$$

43. $P = 2{,}000e^{-0.015} + 2{,}000\left(e^{-0.015}\right)^2$

$$+ 2{,}000\left(e^{-0.015}\right)^3 + \cdots$$

$$= 2{,}000e^{-0.015}\left(1 - e^{-0.015}\right)^{-1}$$

$$\approx 132355.83$$

Need to invest \$132,355.83.

45. $N = 500 + \dfrac{2}{3}(500) + \left(\dfrac{2}{3} \right)^2 (500) +$

$$\cdots + \left(\frac{2}{3} \right)^k (500) + \cdots$$

$$= 500 \left(\frac{1}{1 - \frac{2}{3}} \right)$$

$$= 1{,}500$$

The flywheel will make 1,500 revolutions.

47. $D = h + 2[(0.75)(h)] + 2\left[(0.75)^2(h)\right] + \cdots$

$$= h + 2h(0.75)\left[1 + 0.75 + 0.75^2 + \cdots\right]$$

$$= h + 0.75(2h)\left(\frac{1}{1 - 0.75} \right)$$

$$= h + 0.75(2h)(4)$$

$$= 7h$$

If $7h = 21$, then $h = 3$ ft.

49. Let D be the total amount of deprecation.

$$D = 10{,}000(0.2) + 10{,}000(1 - 0.2)(0.2)$$

$$+ 10{,}000(1 - 0.2)^2(0.2)$$

$$+ 10{,}000(1 - 0.2)^3(0.2) + \cdots$$

$$= 10{,}000(0.2)\left[1 + 0.8 + 0.8^2 + \cdots\right]$$

$$= 10{,}000(0.2)\left(\frac{1}{1-0.8}\right)$$

$$= 10{,}000$$

The total depreciation is $10,000.

51. Just before the second injection there will be $20e^{-1/2}$ units; before the third, $20e^{-2/2} + 20e^{-1/2}$, and so on. Just before the nth injection there will be

$$S = 20e^{-1/2} + 20e^{-1} + 20e^{-3/2} + \cdots$$

$$= 20e^{-1/2}\left(1 + e^{-1/2} + \cdots\right)$$

$$= 20e^{-1/2}\left(\frac{1}{1-e^{-1/2}}\right)$$

$$= 30.8$$

The patient will have about 30.8 units of the drug.

53. Let a_n be the number of trustees on December 31 of the nth year.

$$N = 6 + 6e^{-0.2} + 6e^{-0.2(2)} + 6e^{-0.2(3)} + \cdots$$

$$= \frac{6}{1-e^{-0.2}}$$

$$\approx 33$$

In the long run there will be 33 members on the board.

55. a. Each train travels 2.5 miles before the collision, and since they travel at 10 mi/h, it takes

$$T = \frac{2.5\text{ mi}}{10\text{ mi/hr}} = 0.25\text{ hr}$$

before they collide. Since the fly travels at 16 mi/hr, it travels a total distance of

$$s = \left(16\,\frac{\text{mi}}{\text{hr}}\right)\left(\frac{1}{4}\,\text{hr}\right) = 4\text{ mi}$$

before collision.

b. Let t_1 be the time it takes the fly to fly from the first train to the second. Then

$$\frac{\text{DISTANCE BY FLY} +}{\text{DISTANCE BY TRAIN I}} = 5$$

$$16t_1 + 10t_1 = 5$$

$$t_1 = \frac{5}{26}$$

The distance traveled by the fly is

$$d_1 = 16\left(\frac{5}{26}\right) = \frac{80}{26}$$

If t_2 is the time it takes for the fly to return to Train I, then

$$10(t_1 + t_2) + 16t_2 = 5 - 10t_1$$

$$10t_1 + 10t_2 + 16t_2 = 5 - 10t_1$$

$$20t_1 + 26t_2 = 5$$

$$20\left(\frac{5}{26}\right) + 26t_2 = 5$$

$$26t_2 = \frac{130}{26} - \frac{100}{26}$$

$$t_2 = \frac{30}{26^2}$$

The distance traveled by the fly is

$$d_2 = 16\left(\frac{30}{26^2}\right) = \frac{6}{26}\left(\frac{80}{26}\right)$$

Similarly, the time needed for the fly to fly from the present position of Train I to Train II is

$$t_3 = \frac{180}{26^3}$$

and the distance traveled by the fly on

Page 300

this leg of its journey is

$$d_3 = 16t_3 = \left(\frac{6}{26}\right)^2\left(\frac{80}{26}\right)$$

In general, the distance traveled by the fly on the kth leg is

$$d_k = \left(\frac{6}{26}\right)^{k-1}\left(\frac{80}{26}\right)$$

and the total distance traveled is

$$s = \sum_{k=1}^{\infty} d_k$$

$$= \sum_{k=1}^{\infty}\left(\frac{6}{26}\right)^{k-1}\left(\frac{80}{26}\right)$$

$$= \left(\frac{80}{26}\right)\left[\frac{1}{1-\frac{6}{26}}\right]$$

$$= \left(\frac{80}{26}\right)\left[\frac{26}{20}\right]$$

$$= 4$$

The total distance traveled is 4 miles.

57. Suppose $\Sigma(a_k - b_k)$ converges. Then, so does

$$\Sigma a_k - \Sigma(a_k - b_k) = \Sigma[a_k - (a_k - b_k)]$$
$$= \Sigma b_k$$

contrary to the hypotheses. Thus,

$$\Sigma(a_k - b_k) = \Sigma a_k - \Sigma b_k$$

diverges.

59. $A = 1$ and $B = -3$

$$S_n = \sum_{k=1}^{n}\left[\frac{3^k}{4^k - 3^k} - \frac{3^{k+1}}{4^{k+1} - 3^{k+1}}\right]$$

$$= \left(\frac{3}{4-3} - \frac{3^2}{4^2 - 3^2}\right) + \left(\frac{3^2}{4^2 - 3^2} - \frac{3^3}{4^3 - 3^3}\right)$$

$$+ \cdots + \left(\frac{3^n}{4^n - 3^n} - \frac{3^{n+1}}{4^{n+1} - 3^{n+1}}\right)$$

To find the sum, we find

$$\lim_{n\to\infty}\left[\frac{3}{4-3} - \frac{3^{n+1}}{4^{n+1} - 3^{n+1}}\right]$$

$$= 3 - \lim_{n\to\infty}\frac{\left(\frac{3}{4}\right)^{n+1}}{1 - \left(\frac{3}{4}\right)^{n+1}}$$

$$= 3$$

Note: This is interesting because technology will likely be inadequate.

8.3 The Integral Test, page 612

1. The p-series is one of the form $\displaystyle\sum_{k=1}^{\infty}\frac{1}{k^p}$

(with p a constant).

3. $p = 3$; converges

5. $p = \frac{1}{3}$; diverges

7. $S = \displaystyle\sum_{k=1}^{\infty}\frac{1}{(2+3k)^2}$

$f > 0$, continuous on $[1,\infty)$, and decreasing.

$$I = \int_1^{\infty}\frac{dx}{(2+3x)^2}\,dx$$

$$= \lim_{b\to\infty}\int_1^{b}(2+3x)^{-2}\,dx$$

$$= -\frac{1}{3}\lim_{b\to\infty}(2+3x)^{-1}\bigg|_1^b$$

$$= \frac{1}{15}$$

I converges, so *S* converges.

9. $S = \sum_{k=2}^{\infty} \dfrac{\ln k}{k}$

$f > 0$, continuous on $[2, \infty)$, and decreasing for $k \geq 3$ (but the term with $k = 2$ cannot change the divergence or convergence).

$$I = \int_2^{\infty} \frac{\ln x}{x}\,dx$$

$$= \lim_{b \to \infty} \int_2^b \ln x\,dx$$

$$= \frac{1}{2} \lim_{b \to \infty} (\ln x)^2 \Big|_2^b$$

$$= \infty$$

I diverges, so *S* diverges.

11. $S = \sum_{k=1}^{\infty} \dfrac{(\tan^{-1} k)^2}{1 + k^2}$

$f > 0$, continuous on $[1, \infty)$, and decreasing.

$$I = \int_1^{\infty} \frac{(\tan^{-1} x)^2}{1 + x^2}\,dx$$

$$= \lim_{b \to \infty} \int_1^b \frac{(\tan^{-1} x)^2}{1 + x^2}\,dx$$

$$= \frac{1}{3} \lim_{b \to \infty} (\tan^{-1} x)^3 \Big|_1^b$$

$$= \frac{\pi^3}{24} - \frac{\pi^3}{192}$$

I converges, so *S* converges.

13. $S = \sum_{k=1}^{\infty} \dfrac{k}{k + 1}$ diverges by the divergence test since $\lim_{k \to \infty} \dfrac{k}{k + 1} = 1 \neq 0$.

15. $\sum_{k=2}^{\infty} \dfrac{k}{\sqrt{k^2 - 1}}$ diverges by the divergence test since $\lim_{k \to \infty} \dfrac{k}{\sqrt{k^2 - 1}} = 1 \neq 0$.

17. $\sum_{k=0}^{\infty} \cos(k\pi)$ diverges by the divergence test since $\lim_{k \to \infty} \cos(k\pi)$ does not exist (divergence by oscillation).

19. $\sum_{k=1}^{\infty} \dfrac{\ln k}{k^2}$; apply the integral test:

$$I = \lim_{n \to \infty} \int_1^n \frac{\ln x}{x^2}\,dx \quad \boxed{u = \ln x; \; dv = \frac{dx}{x^2}}$$

$$= \lim_{n \to \infty} \left[-\frac{\ln x}{x} \Big|_1^n + \int_1^n x^{-2}\,dx \right]$$

$$= \lim_{n \to \infty} \left[-\frac{1}{x}(\ln x + 1) \right] \Big|_1^n$$

$$= \lim_{n \to \infty} \left(-\frac{\ln n}{n} - \frac{1}{n} + 1 \right)$$

$$= 1$$

The series converges.

21. $\sum_{k=1}^{\infty} \left(2 + \dfrac{3}{k} \right)^k$

$\lim_{k \to \infty} \left(2 + \dfrac{3}{k} \right)^k = \infty$ because $\lim_{k \to \infty} 2^k = \infty$, so the series diverges.

23. $\sum_{k=1}^{\infty} \dfrac{1}{k^4}$; this is a *p*-series where $p = 4 > 1$, so *S* converges.

25. $\sum_{k=1}^{\infty} k^{-3/4}$; this is a *p*-series where $p = \frac{3}{4} < 1$, so *S* diverges.

27. $\displaystyle\sum_{k=1}^{\infty} \frac{k - \sin k}{3k + 2\sin k}$ diverges by the divergence test:

$$\lim_{k\to\infty} \frac{k - \sin k}{3k + 2\sin k} = \lim_{k\to\infty} \frac{1 - \frac{\sin k}{k}}{3 + \frac{2\sin k}{k}}$$
$$= \frac{1}{3} \neq 0$$

29. $\displaystyle\sum_{k=2}^{\infty} \frac{\ln\sqrt{k}}{\sqrt{k}}$ diverges by the integral test:

$$\int_2^\infty \frac{\ln\sqrt{x}}{\sqrt{x}}\,dx = \lim_{n\to\infty} \int_2^n \frac{\ln\sqrt{x}}{\sqrt{x}}\,dx$$
$$= \lim_{n\to\infty} \left[\left. \left(2\sqrt{x}\ln\sqrt{x}\right)\right|_2^n - \int_2^n \frac{dx}{\sqrt{x}} \right]$$

$$\boxed{u = \ln\sqrt{x};\ dv = \frac{dx}{\sqrt{x}}}$$

$$= \lim_{n\to\infty} \left[\sqrt{x}(\ln x - 2) \right]\Big|_2^n$$
$$= \infty$$

31. $\displaystyle\sum_{k=1}^{\infty} \frac{k}{e^k}$

$$I = \lim_{n\to\infty} \int_1^n \frac{x}{e^x}\,dx \quad \boxed{u = x;\ dv = e^{-x}dx}$$
$$= \lim_{n\to\infty} \left[-\frac{x}{e^x}\Big|_1^n + \int_1^n e^{-x}\,dx \right]$$
$$= \lim_{n\to\infty} \left(-\frac{n}{e^n} - \frac{1}{e^n} + \frac{2}{e} \right)$$
$$= \frac{2}{e}$$

The series converges.

33. $\displaystyle\sum_{k=1}^{\infty} \frac{\tan^{-1}k}{1 + k^2}$

$$I = \lim_{n\to\infty} \int_1^n \frac{\tan^{-1}x\,dx}{1 + x^2}$$

$$= \lim_{n\to\infty} \frac{1}{2}\left(\tan^{-1}x\right)^2 \Big|_1^n$$
$$= \frac{\pi^2}{8} - \frac{\pi^2}{32}$$

The series converges.

35. $\displaystyle\sum_{k=1}^{\infty} \tan^{-1}k$ diverges by the divergence test:

$$\lim_{n\to\infty} \tan^{-1}k = \frac{\pi}{2} \neq 0$$

37. $\displaystyle\sum_{k=1}^{\infty} k\sin\frac{1}{k}$ diverges by the divergence test:

$$\lim_{k\to\infty} k\sin\frac{1}{k} = \lim_{u\to0} \frac{\sin u}{u} = 1 \neq 0$$

Let $u = \dfrac{1}{k}$ so that $u \to 0$ as $k \to \infty$.

39. $S = \displaystyle\sum_{k=1}^{\infty} \frac{k-1}{k+1}$ diverges because

$$\lim_{k\to\infty} \frac{k-1}{k+1} = 1 \neq 0$$

The necessary condition for convergence is not satisfied.

41. $S = \displaystyle\sum_{k=1}^{\infty} \frac{k^2 + 1}{k^3} = \sum_{k=1}^{\infty} \left(\frac{1}{k} + \frac{1}{k^2} \right)$

The first term is the divergent harmonic series, and the second terms is a convergent p-series, so S diverges (by Theorem 8.7).

43. $S = \displaystyle\sum_{k=1}^{\infty} \frac{1}{k^2}$

This is a p-series with $p = 2 > 1$, so S converges.

45. $S = \displaystyle\sum_{k=1}^{\infty} \frac{k^{\sqrt{3}} + 1}{k^{2.7321}} = \sum_{k=1}^{\infty} \left[\frac{1}{k^{2.7321-\sqrt{3}}} + \frac{1}{k^{2.7321}} \right]$

Both of these are p-series with $p > 1$, so the series converges.

SURVIVAL HINT: *Compare Problems 45 and 46, and notice how they relate to the p-series test (Theorem 8.12).*

47. $S = \displaystyle\sum_{k=1}^{\infty} \left[\dfrac{1}{k} + \dfrac{k+1}{k+2} \right]$

$\displaystyle\lim_{k \to \infty} \left[\dfrac{1}{k} + \dfrac{k+1}{k+2} \right] = 1 \neq 0$

The necessary condition for convergence is not satisfied, so S diverges.

49. $S = \displaystyle\sum_{k=1}^{\infty} \left[\dfrac{1}{2^k} - \dfrac{1}{k} \right] = \sum_{k=1}^{\infty} \dfrac{1}{2^k} - \sum_{k=1}^{\infty} \dfrac{1}{k}$

The first series converges (geometric series, $\frac{1}{2} < 1$) and the second series is a divergent p-series (harmonic series), so S diverges.

51. $\displaystyle\sum_{k=2}^{\infty} \dfrac{k}{(k^2-1)^p}$

$I = \displaystyle\lim_{n \to \infty} \dfrac{1}{2} \int_2^n (x^2-1)^{-p}(2x\,dx)$

$= \displaystyle\lim_{n \to \infty} \dfrac{1}{2} \dfrac{(x^2-1)^{-p+1}}{-p+1} \Big|_2^n$

$= \dfrac{1}{2(1-p)} \displaystyle\lim_{n \to \infty} \left[(n^2-1)^{1-p} - 3^{1-p} \right]$

The improper integral converges if $p > 1$, so the series also converges for $p > 1$, and diverges if $p \leq 1$.

53. $\displaystyle\sum_{k=3}^{\infty} \dfrac{1}{k^p \ln k}$

If $p \leq 1$, the series diverges since

$\displaystyle\int_3^{\infty} \dfrac{dx}{x^p \ln x} \geq \int_3^{\infty} \dfrac{dx}{x \ln x} = \infty$

If $p > 1$, the series converges since

$\displaystyle\int_3^{\infty} \dfrac{dx}{x^p \ln x} \leq \int_3^{\infty} \dfrac{dx}{x^p} = \dfrac{3^{1-p}}{p-1}$

Thus, the series converges if $p > 1$ and diverges if $p \leq 1$.

55. $S = \displaystyle\sum_{k=2}^{\infty} \dfrac{1}{k(\ln k)^p}$

$I = \displaystyle\lim_{N \to \infty} \int_2^N (\ln x)^{-p} \dfrac{dx}{x}$

$= \displaystyle\lim_{N \to \infty} \dfrac{(\ln x)^{-p+1}}{1-p} \Big|_2^N$

$= \begin{cases} \infty & \text{if } p \leq 1 \\ \dfrac{1}{(p-1)(\ln 2)^{p-1}} & \text{if } p > 1 \end{cases}$

I converges if $p > 1$ and diverges if $p \leq 1$, so S converges if $p > 1$ and diverges if $p \leq 1$.

57. It is not true, so we provide a counterexample.

Let $a_k = \dfrac{1}{k^{1.1}}$ and $b_k = \dfrac{1}{k^{0.1}}$ so Σa_k converges ($p = 1.1 > 1$) and Σb_k diverges ($p = 0.1 \leq 1$). We also see that

$$\lim_{k \to \infty} a_k = \lim_{k \to \infty} b_k = 0$$

Now,

$$\sum a_k b_k = \sum \dfrac{1}{k^{1.1}} \dfrac{1}{k^{0.1}} = \sum k^{-1.2}$$

converges since $1.2 > 1$, which contradicts the premise that it must diverge.

59. From the proof of the integral test, we have

$$\int_1^{n+1} f(x)\,dx < S_n < a_1 + \int_1^n f(x)\,dx$$

where $S_n = \displaystyle\sum_{k=1}^n a_k$ is the nth partial sum of

$\sum\limits_{k=1}^{\infty} a_k$. Since $\sum\limits_{k=1}^{\infty} a_k$ converges, we know that $\lim\limits_{n\to\infty} S_n$ exists (and is finite). Thus, by the squeeze theorem

$$\lim_{n\to\infty}\int_1^{n+1} f(x)\,dx \le \lim_{n\to\infty} S_n \le a_1 + \lim_{n\to\infty}\int_1^{n} f(x)\,dx$$

$$\int_1^{\infty} f(x)\,dx \le \sum_{k=1}^{\infty} a_k \le a_1 + \int_1^{\infty} f(x)\,dx$$

8.4 Comparison Tests, page 618

1. The geometric series converges when $|r| < 1$ and diverges when $|r| \ge 1$.

3. $\sum\limits_{k=1}^{\infty}\cos^k\left(\dfrac{\pi}{6}\right)$ is a convergent geometric series with $r = \cos\frac{\pi}{6} < 1$.

5. $\sum\limits_{k=0}^{\infty}(1.5)^k$ is a divergent geometric series with $r = 1.5 > 1$.

7. $\sum\limits_{k=1}^{\infty}\dfrac{1}{k}$ is a divergent p-series since $p = 1 \le 1$.

9. $\sum\limits_{k=1}^{\infty}\dfrac{1}{k^{3/2}}$ is a convergent p-series since $p = 1.5 > 1$.

11. $\sum\limits_{k=0}^{\infty} 1^k$ is a divergent geometric series with $r = 1 \ge 1$.

13. $\sum\limits_{k=1}^{\infty}\dfrac{1}{k^2 + k}$ is dominated by the convergent p-series $\sum\limits_{k=1}^{\infty}\dfrac{1}{k^2}$, so the series converges.

15. $\sum\limits_{k=1}^{\infty}\dfrac{1}{\sqrt{k}} = \sum\limits_{k=1}^{\infty}\dfrac{1}{k^{1/2}}$ is the divergent p-series with $p = \frac{1}{2}$.

17. $\sum\limits_{k=1}^{\infty}\dfrac{1}{\sqrt{2k+3}}$ behaves like the divergent p-series $\sum\limits_{k=1}^{\infty}\dfrac{1}{k^{1/2}}$.

19. $\sum\limits_{k=1}^{\infty}\dfrac{1}{\sqrt{k^3+2}}$ behaves like the convergent p-series $\sum\limits_{k=1}^{\infty}\dfrac{1}{k^{3/2}}$.

21. $\sum\limits_{k=1}^{\infty}\dfrac{2k^2}{k^4 - 4}$ behaves like the convergent p-series $\sum\limits_{k=1}^{\infty}\dfrac{1}{k^2}$.

23. $\sum\limits_{k=1}^{\infty}\dfrac{(k+2)(k+3)}{k^{7/2}}$ behaves like the convergent p-series $\sum\limits_{k=1}^{\infty}\dfrac{1}{k^{3/2}}$.

25. $\sum\limits_{k=1}^{\infty}\dfrac{2k+3}{k^2 + 3k + 2}$ behaves like the divergent p-series $\sum\limits_{k=1}^{\infty}\dfrac{2}{k}$.

27. $\sum\limits_{k=1}^{\infty}\dfrac{k}{(k+2)2^k}$ behaves like the convergent geometric series $\sum\limits_{k=1}^{\infty}\dfrac{1}{2^k}$.

29. $\sum\limits_{k=1}^{\infty}\dfrac{1}{k(k+2)}$ behaves like the convergent p-series $\sum\limits_{k=1}^{\infty}\dfrac{1}{k^2}$.

31. $\displaystyle\sum_{k=1}^{\infty}\frac{1}{\sqrt{k}\,2^k}$ is dominated by $\displaystyle\sum_{k=1}^{\infty}\frac{1}{2^{k+1}}$ since

$\dfrac{1}{\sqrt{k}2^k}<\dfrac{1}{2^{k+1}}$ for $k>4$. Since $\displaystyle\sum_{k=1}^{\infty}\frac{1}{2^{k+1}}$

converges (geometric series with $r=\frac{1}{2}<1$), the series also converges.

33. $\displaystyle\sum_{k=2}^{\infty}\frac{1}{\sqrt{k}\ln k}$ dominates

$$\sum_{k=2}^{\infty}\frac{1}{\sqrt{k}\left(\sqrt{k}\right)}=\sum_{k=2}^{\infty}\frac{1}{k}$$

since $\dfrac{1}{\ln k}>\dfrac{1}{\sqrt{k}}$. Since $\displaystyle\sum_{k=2}^{\infty}\frac{1}{k}$ diverges, the

series also diverges.

35. $\displaystyle\sum_{k=1}^{\infty}\frac{2k^3+k+1}{k^3+k^2+1}$ diverges since

$\displaystyle\lim_{k\to\infty}\frac{2k^3+k+1}{k^3+k^2+1}=2\neq0$; the necessary

condition for convergence is not satisfied.

37. $\displaystyle\sum_{k=1}^{\infty}\frac{|\sin(k!)|}{k^2}$ is dominated by the convergent

p-series $\displaystyle\sum_{k=1}^{\infty}\frac{1}{k^2}$.

39. $\displaystyle\sum_{k=1}^{\infty}\frac{6k^2+2k+1}{k^{1.1}(4k^2+k+4)}$ behaves like the

convergent p-series $\displaystyle\sum_{k=1}^{\infty}\frac{1}{k^{1.1}}$.

> **SURVIVAL HINT:** The crux of Problems 39 and 40 is the exponent on k. Note that in Problem 39 it is $1.1>1$ and in Problem 40 it is $0.9<1$. What does this tell you? You will see this pattern many times in the following problems.

41. $\displaystyle\sum_{k=1}^{\infty}\frac{\sqrt[6]{k}}{\sqrt[4]{k^3+2}\,\sqrt[8]{k}}=\sum_{k=1}^{\infty}\frac{k^{4/24}}{(k^3+2)^{6/24}k^{3/24}}$

$$=\sum_{k=1}^{\infty}\frac{k^{1/24}}{(k^3+2)^{6/24}}$$

This last expression behaves like the

divergent p-series $\displaystyle\sum_{k=1}^{\infty}\frac{1}{k^{17/24}}$.

43. $\displaystyle\sum_{k=1}^{\infty}\frac{1}{k^3+4}$ behaves like the convergent

p-series $\displaystyle\sum_{k=1}^{\infty}\frac{1}{k^3}$.

45. $\displaystyle\sum_{k=1}^{\infty}\frac{\ln(k+1)}{(k+1)^3}$ behaves like the convergent

series $\Sigma(\ln k)/k^3$ (q-log series with $q=3>1$), so the series converges.

47. $\displaystyle\sum_{k=2}^{\infty}\frac{1}{(k+3)(\ln k)^{1.1}}$ behaves like

$\displaystyle\sum_{k=1}^{\infty}\frac{1}{k(\ln k)^{1.1}}$ which converges by the

integral test since

$$\int_{2}^{\infty}\frac{dx}{x(\ln x)^{1.1}}=10\ln 2^{-0.1}\approx10.37$$

Thus, the series converges.

49. $\displaystyle\sum_{k=1}^{\infty}k^{(1-k)/k}=\sum_{k=1}^{\infty}\frac{1}{k^{1-1/k}}$

Compare with a divergent harmonic series ($p=1-k^{-1}\leq1$). Thus, the series diverges.

51. Compare $\displaystyle\sum_{k=1}^{\infty}\frac{k^2}{(k+3)!}$ with the convergent series $\displaystyle\sum_{k=1}^{\infty}\frac{1}{(k+1)!}$ (see Example 3). Since

$$\lim_{k\to\infty}\frac{\frac{k^2}{(k+3)!}}{\frac{1}{(k+1)!}}=\lim_{k\to\infty}\frac{k^2(k+1)!}{(k+3)(k+2)(k+1)!}=1$$

We see the series also converges by the limit comparison test.

53. $\displaystyle\sum_{k=2}^{\infty}\frac{1}{(\ln k)^{\ln k}}=\sum_{k=2}^{N}\frac{1}{(\ln k)^{\ln k}}+\sum_{k=N+1}^{\infty}\frac{1}{(\ln k)^{\ln k}}$

for any $N>2$. Since $\ln k>e^2$ for $k>e^{e^2}\approx 1{,}620$, we have

$$\sum_{k=2}^{\infty}\frac{1}{(\ln k)^{\ln k}}<\sum_{k=2}^{1{,}620}\frac{1}{(\ln k)^{\ln k}}+\sum_{k=1{,}621}^{\infty}\frac{1}{(e^2)^{\ln k}}$$

The first sum on the right is a finite number a and the second is the convergent series

$$\sum_{k=1{,}621}^{\infty}\frac{1}{(e^2)^{\ln k}}=\sum_{k=1{,}621}^{\infty}\frac{1}{k^2}$$

so the given series converges by the direct comparison test.

55. Let $\displaystyle\lim_{k\to\infty}b_k=L$ where $0<L<\infty$. Then,

$$\lim_{k\to\infty}\frac{a_kb_k}{a_k}=\lim_{k\to\infty}b_k=L$$

implies Σa_kb_k and Σa_k either both converge or both diverge by the limit comparison test.

57. If Σa_k converges, then $\displaystyle\lim_{k\to\infty}a_k=0$,

$$\lim_{k\to\infty}a_k^{-1}\neq 0$$

so Σa_k^{-1} diverges.

59. Since $\displaystyle\lim_{k\to\infty}\frac{a_k}{b_k}=0$, it follows that there is an

integer N so that

$$\left|\frac{a_k}{b_k}-0\right|<1$$

if $k>N$ (definition of limit with $\epsilon=1$). Thus, $a_k<b_k$ for $k>N$ and

$$\sum_{k=N}^{\infty}a_k<\sum_{k=N}^{\infty}b_k$$

Thus, Σa_k converges by direct comparison with the convergent series Σb_k.

8.5 The Ratio Test and the Root Test, page 625

1. Given the series Σa_k with $a_k>0$, suppose that

$$\lim_{k\to\infty}\frac{a_{k+1}}{a_k}=L$$

The ratio test states the following:
If $L<1$, then Σa_k converges.
If $L>1$ or if L is infinite, then Σa_k diverges.
If $L=1$, the test is inconclusive.

SURVIVAL HINT: *The ratio test and the root tests are two of the most useful convergence tests.*

3. $\displaystyle\sum_{k=1}^{\infty}\frac{1}{k!}$; $a_k=\dfrac{1}{k!}$, use the ratio test.

$$\lim_{k\to\infty}\frac{\frac{1}{(k+1)!}}{\frac{1}{k!}}=\lim_{k\to\infty}\frac{k!}{(k+1)!}$$
$$=\lim_{k\to\infty}\frac{1}{k+1}$$
$$=0\leq 1$$

The series converges.

5. $\displaystyle\sum_{k=1}^{\infty}\frac{k!}{2^{3k}}$; $a_k=\dfrac{k!}{2^{3k}}$, use the ratio test.

$$\lim_{k\to\infty}\frac{(k+1)!2^{3k}}{2^{3k+3}k!}=\frac{1}{8}\lim_{k\to\infty}(k+1)$$
$$=\infty>1$$

The series diverges.

7. $\displaystyle\sum_{k=1}^{\infty}\frac{k}{2^k}$; $a_k=\dfrac{k}{2^k}$, use the ratio test.

$$\lim_{k\to\infty}\frac{(k+1)2^k}{2^{k+1}k}=\frac{1}{2}<1$$

The series converges.

9. $\displaystyle\sum_{k=1}^{\infty}\frac{k^{100}}{e^k}$; $a_k=\dfrac{k^{100}}{e^k}$, use the ratio test.

$$\lim_{k\to\infty}\frac{(k+1)^{100}e^k}{e^{k+1}k^{100}}=e^{-1}<1$$

The series converges.

11. $\displaystyle\sum_{k=1}^{\infty}k\left(\frac{4}{3}\right)^k$; $a_k=k\left(\dfrac{4}{3}\right)^k$, diverges since

$$\lim_{k\to\infty}k\frac{4^k}{3^k}=\infty\neq0$$

Following the directions, we use the ratio test:

$$\lim_{k\to\infty}\frac{(k+1)4^{k+1}3^k}{3^{k+1}k4^k}=\frac{4}{3}>1$$

The series diverges.

13. $\displaystyle\sum_{k=1}^{\infty}\left(\frac{2}{k}\right)^k$; $a_k=\left(\dfrac{2}{k}\right)^k$, use the root test.

$$\lim_{k\to\infty}\sqrt[k]{\left(\frac{2}{k}\right)^k}=\lim_{k\to\infty}\frac{2}{k}=0<1$$

The series converges.

15. $\displaystyle\sum_{k=1}^{\infty}\frac{k^5}{10^k}$; $a_k=\dfrac{k^5}{10^k}$, use the ratio test.

$$\lim_{k\to\infty}\frac{10^k(k+1)^5}{k^5 10^{k+1}}=\frac{1}{10}\lim_{k\to\infty}\frac{(k+1)^5}{k^5}$$
$$=\frac{1}{10}<1$$

The series converges.

17. $\displaystyle\sum_{k=1}^{\infty}\left(\frac{k}{3k+1}\right)^k$; $a_k=\left(\dfrac{k}{3k+1}\right)^k$, use the root test.

$$\lim_{k\to\infty}\sqrt[k]{\left(\frac{k}{3k+1}\right)^k}=\lim_{k\to\infty}\frac{k}{3k+1}$$
$$=\frac{1}{3}<1$$

The series converges.

19. $\displaystyle\sum_{k=1}^{\infty}\frac{k!}{(k+2)^4}$; $a_k=\dfrac{k!}{(k+2)^4}$, use the ratio test.

$$\lim_{k\to\infty}\frac{(k+2)^4(k+1)!}{k!(k+3)^4}=\lim_{k\to\infty}\frac{(k+2)^4(k+1)}{(k+3)^4}$$
$$=\infty>1$$

the series diverges.

21. $\displaystyle\sum_{k=1}^{\infty}\frac{(k!)^2}{(2k)!}$; $a_k=\dfrac{(k!)^2}{(2k)!}$, use the ratio test.

$$\lim_{k\to\infty} \frac{(2k)![(k+1)!]^2}{(k!)^2(2k+2)!} = \lim_{k\to\infty} \frac{(k+1)^2}{(2k+2)(2k+1)}$$
$$= \frac{1}{4} < 1$$

the series converges.

23. $\displaystyle\sum_{k=1}^{\infty} k^2 2^{-k};\ a_k = k^2 2^{-k}$, use the ratio test.

$$\lim_{k\to\infty} \frac{2^k(k+1)^2}{k^2 2^{k+1}} = \frac{1}{2}\lim_{k\to\infty}\frac{(k+1)^2}{k^2}$$
$$= \frac{1}{2} < 1$$

The series converges.

25. $\displaystyle\sum_{k=1}^{\infty}\left(\frac{k-2}{k}\right)^{k^2};$

$a_k = \left(\dfrac{k-2}{k}\right)^{k^2}$, use the root test.

$$\lim_{k\to\infty}\left[\left(\frac{k-2}{k}\right)^{k^2}\right]^{1/k} = \lim_{k\to\infty}\left(\frac{k-2}{k}\right)^{k}$$
$$= \lim_{k\to\infty}\left(1-\frac{2}{k}\right)^{k}$$
$$= e^{-2} \quad \textit{l'Hôpital's rule}$$
$$< 1$$

The series converges.

27. $\displaystyle\sum_{k=1}^{\infty}\frac{1,000}{k}$ diverges by direct comparison

with the divergent p-series $\displaystyle\sum_{k=1}^{\infty}\frac{1}{k}$.

Note: the ratio test is inconclusive.

29. $\displaystyle\sum_{k=1}^{\infty}\frac{5k+2}{k2^k};$ use the ratio test.

$$\lim_{k\to\infty}\frac{(5k+7)(k)2^k}{(k+1)2^{k+1}(5k+2)}$$
$$= \frac{1}{2}\lim_{k\to\infty}\frac{(5k+7)(k)}{(k+1)(5k+2)}$$
$$= \frac{1}{2} < 1$$

The given series converges.

31. $\displaystyle\sum_{k=1}^{\infty}\frac{\sqrt{k!}}{2^k};$ use the ratio test.

$$\lim_{k\to\infty}\frac{\sqrt{(k+1)!}2^k}{2^{k+1}\sqrt{k!}} = \frac{1}{2}\lim_{k\to\infty}\sqrt{k+1}$$
$$= \infty > 1$$

The given series diverges.

33. $\displaystyle\sum_{k=1}^{\infty}\frac{2^k k!}{k^k};$ use the ratio test.

$$\lim_{k\to\infty}\frac{2^{k+1}(k+1)!k^k}{(k+1)^{k+1}2^k k!} = 2\lim_{k\to\infty}\frac{k^k}{(k+1)^k}$$
$$= 2\lim_{k\to\infty}\frac{1}{\left(1+\frac{1}{k}\right)^k}$$
$$= \frac{2}{e} < 1$$

The given series converges.

35. $\displaystyle\sum_{k=1}^{\infty}\frac{\sqrt{k+1}}{k^{k+0.5}};$ use the ratio test.

$$\lim_{k\to\infty}\frac{\sqrt{k+2}k^{k+0.5}}{(k+1)^{k+1.5}\sqrt{k+1}}$$
$$= \lim_{k\to\infty}\frac{1}{\left(1+\frac{1}{k}\right)^k}\frac{k^{0.5}}{(k+1)^{1.5}}\sqrt{\frac{k+2}{k+1}}$$
$$= \left(\frac{1}{e}\right)(0)(1)$$
$$= 0 < 1$$

The given series converges.

37. $\displaystyle\sum_{k=1}^{\infty} \frac{k!}{(k+1)!} = \sum_{k=1}^{\infty} \frac{1}{k+1}$

is the harmonic series, so it diverges.

39. $\displaystyle\sum_{k=1}^{\infty} \frac{k}{4k^3-5}$ behaves like the convergent p-

series $\displaystyle\sum_{k=1}^{\infty} \frac{1}{k^2}$ so the series converges.

41. $\displaystyle\sum_{k=1}^{\infty} \left(1+\frac{1}{k}\right)^{-k^2}$; use the root test.

$$\lim_{k\to\infty} \sqrt[k]{\left(1+\frac{1}{k}\right)^{-k^2}} = \lim_{k\to\infty} \left(1+\frac{1}{k}\right)^{-k}$$

$$= \lim_{k\to\infty} \frac{1}{\left(1+\frac{1}{k}\right)^k}$$

$$= \frac{1}{e} < 1$$

The given series converges.

43. Compare $\displaystyle\sum_{k=1}^{\infty} \left|\frac{\cos k}{2^k}\right|$ directly with the

convergent geometric series $\displaystyle\sum_{k=1}^{\infty} \frac{1}{2^k}$:

$$\sum_{k=1}^{\infty} \left|\frac{\cos k}{2^k}\right| \le \sum_{k=1}^{\infty} \frac{1}{2^k} = 1$$

The given series converges.

45. $\displaystyle\sum_{k=2}^{\infty} \left(\frac{\ln k}{k}\right)^k$; use the root test.

$$\lim_{k\to\infty} \sqrt[k]{\left(\frac{\ln k}{k}\right)^k} = \lim_{k\to\infty} \frac{\ln k}{k}$$

$$= \lim_{k\to\infty} \frac{\frac{1}{k}}{1}$$

$$= 0 < 1$$

The given series converges.

47. $\displaystyle\sum_{k=1}^{\infty} k^2 x^k$; use the ratio test.

$$\lim_{k\to\infty} \frac{(k+1)^2 x^{k+1}}{k^2 x^k} = x$$

By the ratio test, the given series converges when $x < 1$ and diverges when $x > 1$. Since the ratio test fails when the ratio equals 1, investigate

$$\sum_{k=1}^{\infty} k^2$$

separately to see that it diverges at $x = 1$. Thus, the given series converges for $0 \le x \le 1$.

49. $\displaystyle\sum_{k=1}^{\infty} \frac{(x+0.5)^k}{k\sqrt{k}}$; use the ratio test.

$$\lim_{k\to\infty} \frac{\frac{(x+0.5)^{k+1}}{(k+1)^{3/2}}}{\frac{(x+0.5)^k}{k^{3/2}}} = \lim_{k\to\infty} \frac{(x+0.5)k^{3/2}}{(k+1)^{3/2}} = x + 0.5$$

By the ratio test, the given series converges when $x + 0.5 < 1$ or $x < 0.5$ and diverges when $x > 0.5$. Since the ratio test fails when the ratio equals 1, investigate

$$\sum_{k=1}^{\infty} k^{-3/2}$$

separately to see that it converges if $x = 0.5$. Thus, the given series converges for $0 \le x \le 0.5$.

51. $\displaystyle\sum_{k=1}^{\infty} \frac{x^k}{k!}$; use the ratio test.

$$\lim_{k\to\infty} \frac{x^{k+1}k!}{(k+1)!x^k} = \lim_{k\to\infty} \frac{x}{k+1} = 0 < 1$$

for all x. Thus, the given series converges for all $x \ge 0$.

53. $\sum_{k=1}^{\infty}(ax)^k$; use the ratio test.

$$\lim_{k\to\infty}\frac{(ax)^{k+1}}{(ax)^k}=ax$$

By the ratio test, the given series converges when $ax<1$ or when $x<a^{-1}$ and diverges when $x>a^{-1}$. Since the ratio test fails when the ratio is 1, investigate

$$\sum_{k=1}^{\infty}1^k$$

separately to see that this series diverges at $x=a^{-1}$. Thus, the given series converges for $0\le x<a^{-1}$.

55. $\sum_{k=1}^{\infty}k^pe^{-k}$; use the root test.

$$\lim_{k\to\infty}\frac{k^{p/k}}{e}=e^{-1}<1$$

for all pThus, the given series converges for all p, and the integral test shows that

$$\int_1^{\infty}x^pe^{-x}\,dx$$

also converges for all p.

57. a. Since $L<1$ and

$$\lim_{k\to\infty}\frac{a_{k+1}}{a_k}=L$$

the series converges and the necessary condition for convergence is satisfied; that is,

$$\lim_{k\to\infty}a_k=0$$

b. For $\sum_{k=1}^{\infty}\frac{x^k}{k!}$ use the ratio test.

$$\lim_{k\to\infty}\frac{x^{k+1}k!}{(k+1)!x^k}=\lim_{k\to\infty}\frac{x}{k+1}=0$$

This series converges for all x so that

$$\lim_{k\to\infty}\frac{x^k}{k!}=0$$

59. a. Since

$$\lim_{n\to\infty}\sqrt[n]{a_n}=R<1$$

for all but a finite number of terms in the sequence $\{\sqrt[n]{a_n}\}$ must be "close" to R. Hence, if $R<x<1$, there is an N such that $0\le\sqrt[n]{a_n}\le x$ for all $n>N$, and

$$\sum_{n=N}^{\infty}a_n=\sum_{n=N}^{\infty}\left(\sqrt[n]{a_n}\right)^n\le\sum_{n=N}^{\infty}x^n$$

Since $|x|<1$, the geometric series on the right converges, and by comparison, so does $\sum_{n=N}^{\infty}a_n$. This, in turn, implies the convergence of $\sum_{n=1}^{\infty}a_n$.

b. If $R>1$, then all but a finite number of terms in the sequence $\{\sqrt[n]{a_n}\}$ satisfy $\sqrt[n]{a_n}>1$. Hence, $a_n>1$ for infinitely many values of n, so $\{a_n\}$ cannot tend to 0. It follows from the divergence test that $\sum_{n=1}^{\infty}a_n$ must diverge.

c. $\sum_{n=1}^{\infty}\frac{1}{n}$ diverges and $\sum_{n=1}^{\infty}\frac{1}{n^2}$ converges, yet

$$\lim_{n\to\infty}\sqrt[n]{a_n}=1$$

for both.

8.6 Alternating Series; Absolute and Conditional Convergence, page 636

1. Consider the series $A = \Sigma a_k$. If $\Sigma |a_k|$ converges, then A converges absolutely. If Σa_k diverges, then A may converge conditionally.

3. $\displaystyle\sum_{k=1}^{\infty} \frac{(-1)^{k+1}k}{k^2+1}$ does not converge absolutely

   since $\displaystyle\sum_{k=1}^{\infty} \frac{k}{k^2+1}$ behaves like the divergent

   p-series $\Sigma(1/k)$. To apply the alternating series test, first note that $\left\{ \dfrac{k}{k^2+1} \right\}$ is a decreasing sequence since

   $$\frac{d}{dx}\left(\frac{x}{x^2+1} \right) = \frac{-(x^2-1)}{(x^2+1)^2} < 0$$

   for $x > 1$. Since

   $$\lim_{k\to\infty} \frac{k}{k^2+1} = 0$$

   (l'Hôpital's rule), it follows that S converges conditionally.

5. $\displaystyle\sum_{k=1}^{\infty} \frac{(-1)^{k+1}k}{2k+1}$ diverges because

   $$\lim_{k\to\infty} \frac{k}{2k+1} = \frac{1}{2} \neq 0$$

7. $\displaystyle\sum_{k=1}^{\infty} \frac{(-1)^{k+1}}{k^{3/2}}$ converges absolutely because

   $$T = \sum_{k=1}^{\infty} \frac{1}{k^{3/2}}$$

   is a convergent p-series.

9. $\displaystyle\sum_{k=1}^{\infty} (-1)^{k+1} \frac{k^2}{e^k}$

   Apply the ratio test to the series of absolute values:

   $$\lim_{k\to\infty} \frac{(k+1)^2 e^k}{k^2 e^{k+1}} = \frac{1}{e} < 1$$

   The given series is absolutely convergent.

11. $\displaystyle\sum_{k=1}^{\infty} (-1)^k \frac{(1+k^2)}{k^3}$ does not converge

    absolutely since $\displaystyle\sum_{k=1}^{\infty} \frac{1+k^2}{k^2}$ behaves like the

    divergent harmonic series $\Sigma(1/k)$. To apply the alternating series test, note that
    $$\frac{d}{dx}\left(\frac{1+x^2}{x^3} \right) = \frac{-(x^2+3)}{x^4} < 0$$

    and

    $$\lim_{k\to\infty} \frac{1+k^2}{k^3} = 0$$

    Thus, the given series converges conditionally.

13. $\displaystyle\sum_{k=2}^{\infty} (-1)^k \frac{k!}{\ln k}$ diverges by the divergence

    test because for $k > 2$, $\dfrac{k!}{\ln k} > \dfrac{k}{\ln k}$

    $$\lim_{k\to\infty} \frac{k}{\ln k} = \infty \neq 0$$

15. $\displaystyle\sum_{k=1}^{\infty} (-1)^{k+1} \frac{2^k}{k!}$ Apply the ratio test to the

    series of absolute values:

    $$\lim_{k\to\infty} \frac{2^{k+1}k!}{(k+1)!2^k} = \lim_{k\to\infty} \frac{2}{k+1} = 0 < 1$$

    The given series is absolutely convergent.

17. $\displaystyle\sum_{k=1}^{\infty}(-1)^{k+1}\frac{2^{2k+1}}{k!}$ Apply the ratio test to the series of absolute values:

$$\lim_{k\to\infty}\frac{2^{2k+3}k!}{(k+1)!2^{2k+1}}=\lim_{k\to\infty}\frac{4}{k+1}=0<1$$

it follows that the given series is absolutely convergent.

19. $\displaystyle\sum_{k=1}^{\infty}\frac{(-1)^{k+1}k}{(k+1)(k+2)}$ does not converge absolutely since $\displaystyle\sum_{k=1}^{\infty}\frac{k}{(k+1)(k+2)}$

behaves like the divergent harmonic series $\Sigma(1/k)$. To apply the alternating series test, note that

(1) $\displaystyle\lim_{k\to\infty}\frac{k}{(k+1)(k+2)}=0$

(2) $\left\{\dfrac{k}{(k+1)(k+2)}\right\}$ is decreasing

because for $k>1$, we have $k+1>0$, $k+3>0$, and $\dfrac{1}{k+2}>0$, so we need:

$$\frac{k+1}{(k+2)(k+3)}\le\frac{k}{(k+1)(k+2)}$$
$$\frac{k+1}{k+3}\le\frac{k}{k+1}$$
$$(k+1)^2\le k(k+3)$$
$$k^2+2k+1\le k^2+3k$$
$$1\le k$$

This is a true statement, so the given series converges conditionally.

21. $\displaystyle\sum_{k=2}^{\infty}\frac{(-1)^{k+1}}{\ln(\ln k)}$ does not converge absolutely

since $\displaystyle\sum_{k=2}^{\infty}\frac{1}{\ln(\ln k)}$ diverges. To see this, use the integral test:

$$\int_{2}^{\infty}\frac{dx}{\ln(\ln x)}>\int_{2}^{\infty}\frac{dx}{x\ln x}=\infty$$

Since the sequence $\left\{\dfrac{1}{\ln(\ln k)}\right\}$ is decreasing because

$$\frac{d}{dx}\left[\frac{1}{\ln(\ln x)}\right]=\frac{-1}{x\ln x(\ln(\ln x))^2}<0$$

and

$$\lim_{k\to\infty}\frac{1}{\ln(\ln k)}=0$$

Thus the given series converges conditionally.

23. $\displaystyle\sum_{k=1}^{\infty}\frac{(-1)^{k+1}\ln k}{k}$ does not converge

absolutely since $\displaystyle\sum_{k=1}^{\infty}\frac{\ln k}{k}$ diverges by the integral test:

$$\int_{1}^{\infty}\ln x\,\frac{dx}{x}=\frac{1}{2}\lim_{N\to\infty}(\ln x)^2\Big|_{1}^{N}=\infty$$

To apply the alternating series test, note that

$$\frac{d}{dx}\left(\frac{\ln x}{x}\right)=\frac{-(\ln x-1)}{x^2}<0$$

and

$$\lim_{k \to \infty} \frac{\ln k}{k} = \lim_{k \to \infty} \frac{\frac{1}{k}}{1} = 0 \qquad \text{l'Hôpital's rule}$$

Thus, the given series converges conditionally.

25. $\displaystyle\sum_{k=1}^{\infty} (-1)^{k+1} \frac{\ln k}{k^2}$; use the integral test:

$$I = \int_{1}^{\infty} \frac{\ln x}{x^2}\, dx$$

$$= \lim_{N \to \infty} \left[-\frac{\ln x}{x} - \frac{1}{x} \right] \Big|_{1}^{N}$$

$$= 1$$

Thus, the given series is absolutely convergent.

27. $\displaystyle\sum_{k=1}^{\infty} (-1)^{k+1} \left(\frac{k}{k+1} \right)^k$

$$\lim_{k \to \infty} \left(\frac{k}{k+1} \right)^k = \lim_{k \to \infty} \frac{1}{\left(1 + \frac{1}{k} \right)^k} = \frac{1}{e} \neq 0$$

This series diverges since the necessary condition for convergence is not satisfied.

29. $\displaystyle\sum_{k=1}^{\infty} (-1)^{k+1} \left(\frac{1}{k} \right)^{1/k}$

$$\lim_{k \to \infty} \frac{1}{k^{1/k}} = \lim_{k \to \infty} \exp\left(\ln k^{-1/k} \right)$$

$$= \lim_{k \to \infty} \exp\left(-\frac{\ln k}{k} \right)$$

$$= \lim_{k \to \infty} e^{-1/k} \quad \text{l'Hôpital's rule}$$

$$= 1 \neq 0$$

This series diverges because the necessary condition for convergence is not satisfied.

31. $S = \displaystyle\sum_{k=1}^{\infty} \frac{(-1)^{k+1}}{2^{2k-2}}$

   a. $S_4 = 1 - \frac{1}{4} + \frac{1}{16} - \frac{1}{64} = \frac{51}{64}$

      $|S - S_4| < a_5 = \frac{1}{256} \approx 0.0039$

   b. $\dfrac{1}{2^{2n-2}} < 0.0005$

      $2^{2n-2} > 2{,}000$

      $2n - 2 > \log_2 2{,}000$

      $n > 6.48$

      Choose $n = 7$; $\ S_7 = \frac{3{,}277}{4{,}096} \approx 0.800$.

33. $S = \displaystyle\sum_{k=1}^{\infty} \frac{(-1)^{k}}{k^2}$

   a. $S_4 = -1 + \frac{1}{4} - \frac{1}{9} + \frac{1}{16} = -\frac{115}{144}$

      $|S - S_4| < a_5 = \frac{1}{25} = 0.04$

   b. $\dfrac{1}{n^2} < 0.0005$

      $n^2 > 2{,}000$

      $n > \sqrt{2{,}000}$

      Choose $n = 45$; we use technology to find

      $S_{45} \approx -0.823$.

35. $S = \displaystyle\sum_{k=1}^{\infty} \frac{(-1)^{k+1}}{k^3}$

   a. $S_4 = 1 - \frac{1}{8} + \frac{1}{27} - \frac{1}{64} = \frac{1{,}549}{1{,}728}$

      $|S - S_4| < a_5 = \frac{1}{125} \approx 0.008$

   b. $\dfrac{1}{n^3} < 0.0005$

      $n^3 > 2{,}000$

      $n > 12.599$

      Choose $n = 13$; we use technology to find $S_{13} \approx 0.902$.

37. $\displaystyle\sum_{k=1}^{\infty} \frac{x^k}{k}$; use generalized ratio test.

$$\lim_{k\to\infty}\left|\frac{x^{k+1}k}{(k+1)x^k}\right| = |x|$$

This series converges if $|x| < 1$ and diverges if $|x| > 1$.

For $x = 1$, $\displaystyle\sum_{k=1}^{\infty}\frac{1}{k}$ diverges.

For $x = -1$, $\displaystyle\sum_{k=1}^{\infty}\frac{(-1)^k}{k}$ converges.

The interval of convergence is $[-1, 1)$.

39. $\displaystyle\sum_{k=1}^{\infty}\frac{2^k x^k}{k!}$; use generalized ratio test.

$$\lim_{k\to\infty}\left|\frac{2^{k+1}x^{k+1}k!}{(k+1)!2^k x^k}\right| = \lim_{k\to\infty}\frac{2|x|}{(k+1)} = 0 < 1$$

Converges for all x. You can also answer by saying the interval of convergence is $(-\infty, \infty)$.

41. $\displaystyle\sum_{k=1}^{\infty}(-1)^{k+1}\left(\frac{x}{k}\right)^k$; use the generalized ratio test.

$$\lim_{k\to\infty}\left|\frac{x^{k+1}k^k}{(k+1)^{k+1}x^k}\right| = 0 < 1$$

Converges for all x.

43. $E_{\max} = \dfrac{1}{6} \approx 0.1667$

45. $E_{\max} = \dfrac{1}{\ln 8} \approx 0.4809$

47. $\displaystyle\sum_{k=2}^{\infty}\frac{(-1)^{k+1}}{k(\ln k)^p}$ converges absolutely for $p > 1$ since by the integral test.

$$I = \int_2^{\infty}\frac{dx}{x(\ln x)^p} = \lim_{N\to\infty}\left.\frac{1}{(1-p)(\ln x)^{p-1}}\right|_2^N$$

which converges when $p > 1$. Testing for

conditional convergence, we note that for all p,

$$\frac{d}{dx}\left[\frac{1}{x(\ln x)^p}\right] = \frac{-(p + \ln x)}{x^2(\ln x)^{p+1}} < 0$$

and

$$\lim_{k\to\infty}\frac{1}{k(\ln k)^p} = 0$$

(l'Hôpital's rule is needed for $p < 0$). To summarize, the given series converges absolutely for $p > 1$ and conditionally for $p \le 1$.

49. $\displaystyle\sum_{k=1}^{\infty}\frac{x^k}{k!}$; use the generalized ratio test.

$$\lim_{k\to\infty}\left|\frac{x^{k+1}k!}{(k+1)x^k}\right| = \lim_{k\to\infty}\frac{|x|}{k+1} = 0 < 1$$

for all x, so the given series converges for all x. Since it is convergent, the necessary condition must be satisfied. Thus,

$$\lim_{k\to\infty}\frac{x^k}{k!} = 0$$

for all x.

51. a. $S_{2m} = \displaystyle\sum_{k=1}^{2m}\frac{(-1)^{k+1}}{k}$

$$= 1 - \frac{1}{2} + \frac{1}{3} - \frac{1}{4} + \cdots - \frac{1}{2^{2m}}$$

$$= 1 + \left(\frac{1}{2} - 1\right) + \frac{1}{3} + \left(\frac{1}{4} - \frac{1}{2}\right)$$

$$+ \cdots + \left(\frac{1}{2m} - \frac{1}{m}\right)$$

$$= H_{2m} - H_m$$

b. $S_{2m} = H_{2m} - H_m$

$$= \Big\{[H_{2m} - \ln(2m)] - [H_m - \ln m]$$

$$+ \ln(2m) - \ln m\Big\}$$

$$S = \lim_{2m \to \infty} S_{2m}$$

$$= \lim_{m \to \infty} \left\{ [H_{2m} - \ln(2m)] - [H_m - \ln m] \right.$$

$$\left. + \ln(2m) - \ln m \right\}$$

$$= \gamma - \gamma + \ln 2$$

53. Follow the outline given in Problem 52 to find the sum. The complete solution can be found on page 89 of the 1983 issue of *School Science and Mathematics*.

55. Conditionally convergent series should not be rearranged.

57. Σk^{-1} diverges and Σk^{-2} converges.

59. $\displaystyle\sum_{k=1}^{\infty} a_k$　By the ratio test,

$$\lim_{k \to \infty} \frac{a_{k+1}}{a_k} = L < 1$$

since the series converges, by hypothesis. Then,

$$T = \sum_{k=1}^{\infty} a_k^2$$

converges because

$$\lim_{k \to \infty} \frac{a_{k+1}^2}{a_k^2} = L^2 < 1$$

8.7 Power Series, page 647

1. $L = \lim\limits_{k \to \infty} \left| \dfrac{\frac{(k+1)x^{k+1}}{k+2}}{\frac{kx^k}{k+1}} \right|$

$$= \lim_{k \to \infty} \left| \frac{(k+1)^2}{(k+2)k} \right| |x|$$

$$= |x|$$

The interval of absolute convergence is $|x| < 1$.

Endpoints:

$x = 1$: $\displaystyle\sum_{k=1}^{\infty} \frac{k}{k+1}$ diverges by the divergence test.

$x = -1$: $\displaystyle\sum_{k=1}^{\infty} \frac{(-1)^k k}{k+1}$ does not exist, so it diverges.

The convergence set is $(-1, 1)$.

3. $L = \lim\limits_{k \to \infty} \left| \dfrac{\frac{(k+1)(k+2)x^{k+1}}{k+3}}{\frac{k(k+1)x^k}{k+2}} \right|$

$$= \lim_{k \to \infty} \left| \frac{(k+1)(k+2)^2}{(k+3)(k+1)k} \right| |x|$$

$$= |x|$$

The interval of absolute convergence is $|x| < 1$.

Endpoints:

$x = 1$: $\displaystyle\sum_{k=1}^{\infty} \frac{k(k+1)}{k+2}$ diverges by the divergence test.

$x = -1$: $\displaystyle\sum_{k=1}^{\infty} \frac{(-1)^k k(k+1)}{k+2}$ diverges by the divergence test.

The convergence set is $(-1, 1)$.

5. $L = \lim\limits_{k \to \infty} \left| \dfrac{(k+1)^2 3^{k+1}(x-3)^{k+1}}{k^2 3^k (x-3)^k} \right|$

$$= \lim_{k \to \infty} \left| \frac{(k+1)^2 3^{k+1}}{k^2 3^k} \right| |x - 3|$$

$$= 3|x - 3|$$

The interval of absolute convergence is

$$3|x - 3| < 1$$
$$|x - 3| < \frac{1}{3}$$
$$-\frac{1}{3} < x - 3 < \frac{1}{3}$$
$$\frac{8}{3} < x < \frac{10}{3}$$

Endpoints:

$x = \frac{8}{3}$: $\displaystyle\sum_{k=1}^{\infty}(-1)^k k^2$ diverges by the

divergence test.

$x = \frac{10}{3}$: $\displaystyle\sum_{k=1}^{\infty} k^2$ diverges by the

divergence test.

The convergence set is $\left(\frac{8}{3}, \frac{10}{3}\right)$.

7. $\displaystyle L = \lim_{k\to\infty}\left|\frac{\frac{3^{k+1}(x+3)^{k+1}}{4^{k+1}}}{\frac{3^k(x+3)^k}{4^k}}\right|$

$\displaystyle = \lim_{k\to\infty}\left|\frac{3^{k+1}4^k}{4^{k+1}3^k}\right||x + 3|$

$\displaystyle = \frac{3}{4}|x + 3|$

The interval of absolute convergence is
$$\frac{3}{4}|x + 3| < 1$$
$$|x + 3| < \frac{4}{3}$$
$$-\frac{4}{3} < x + 3 < \frac{4}{3}$$
$$-\frac{13}{3} < x < -\frac{5}{3}$$

Endpoints:

$x = -\frac{13}{3}$: $\displaystyle\sum_{k=0}^{\infty}\frac{(-1)^k 3^k\left(\frac{4}{3}\right)^k}{4^k} = \sum_{k=0}^{\infty}(-1)^k$

diverges by the divergence test.

$x = -\frac{5}{3}$: $\displaystyle\sum_{k=0}^{\infty} 1$ diverges by the

divergence test.

The convergence set is $\left(-\frac{13}{3}, -\frac{5}{3}\right)$.

9. $\displaystyle L = \lim_{k\to\infty}\left|\frac{\frac{(k+1)!(x-1)^{k+1}}{5^{k+1}}}{\frac{k!(x-1)^k}{5^k}}\right|$

$\displaystyle = \lim_{k\to\infty}\left|\frac{(k+1)!5^k}{5^{k+1}k!}\right||x - 1|$

$\displaystyle = \lim_{k\to\infty}\left|\frac{k+1}{5}\right||x - 1|$

This limit does not exist unless $x = 1$. The convergence set is the single point $x = 1$.

11. $\displaystyle L = \lim_{k\to\infty}\left|\frac{\frac{(k+1)^2(x-1)^{k+1}}{2^{k+1}}}{\frac{k^2(x-1)^k}{2^k}}\right|$

$\displaystyle = \lim_{k\to\infty}\left|\frac{(k+1)^2 2^k}{2^{k+1}k^2}\right||x - 1|$

$\displaystyle = \frac{1}{2}|x - 1|$

The interval of absolute convergence is
$$\frac{1}{2}|x - 1| < 1$$
$$-2 < x - 1 < 2$$
$$-1 < x < 3$$

Endpoints:

$x = -1$: $\displaystyle\sum_{k=1}^{\infty}(-1)^k k^2$ diverges by the

divergence test.

$x = 3$: $\displaystyle\sum_{k=1}^{\infty} k^2$ diverges by the

divergence test.

The convergence set is $(-1, 3)$.

13. $L = \lim\limits_{k \to \infty} \left| \dfrac{\frac{(k+1)(3x-4)^{k+1}}{(k+2)^2}}{\frac{k(3x-4)^k}{(k+1)^2}} \right|$

$= \lim\limits_{k \to \infty} \left| \dfrac{(k+1)^3}{(k+2)k} \right| |3x-4|$

$= |3x-4|$

The interval of absolute convergence is

$$|3x-4| < 1$$
$$-1 < 3x - 4 < 1$$
$$3 < 3x < 5$$
$$1 < x < \frac{5}{3}$$

Endpoints:

$x = 1$: $\displaystyle\sum_{k=1}^{\infty} \dfrac{(-1)^k k}{(k+1)^2}(-1)^k$ converges

by the alternating series test:

$$\lim\limits_{k \to \infty} \dfrac{k}{(k+1)} = 0 \text{ and}$$

$$f(x) = \dfrac{x}{(x+1)^2}$$

$$f'(x) = \dfrac{-(x-1)}{(x+1)^3} < 0 \,(\text{for } x > 1)$$

$x = \frac{5}{3}$: $\displaystyle\sum_{k=1}^{\infty} \dfrac{k}{(k+1)^2}$ diverges by the

comparison test; behaves like

$\displaystyle\sum_{k=1}^{\infty} \dfrac{1}{k}$ (the harmonic series).

The convergence set is $[1, \frac{5}{3})$.

15. $L = \lim\limits_{k \to \infty} \left| \dfrac{\frac{(k+1)x^{k+1}}{7^{k+1}}}{\frac{kx^k}{7^k}} \right|$

$= \lim\limits_{k \to \infty} \left| \dfrac{(k+1)7^k}{7^{k+1}k} \right| |x|$

$= \lim\limits_{k \to \infty} \left| \dfrac{k+1}{7k} \right| |x|$

$= \dfrac{1}{7}|x|$

The interval of absolute convergence is

$$\frac{1}{7}|x| < 1$$
$$-7 < x < 7$$

Endpoints:

$x = -7$: $\displaystyle\sum_{k=1}^{\infty}(-1)^k k$ diverges by

the divergence test.

$x = 7$: $\displaystyle\sum_{k=1}^{\infty} k$ diverges by the

divergence test.

The convergence set is $(-7, 7)$.

17. $L = \lim\limits_{k \to \infty} \left| \dfrac{\frac{[(k+1)!]^2 x^{k+1}}{(k+1)^{k+1}}}{\frac{(k!)^2 x^k}{k^k}} \right|$

$= \lim\limits_{k \to \infty} \left| \dfrac{[(k+1)!]^2 k^k}{(k+1)^{k+1}(k!)^2} \right| |x|$

$= \lim\limits_{k \to \infty} \left[\dfrac{k+1}{(1+\frac{1}{k})^k} \right] |x|$

$= \infty$

The convergence set is the point $x = 0$.

19. $L = \lim\limits_{k \to \infty} \left| \dfrac{\frac{(-1)^{k+1}x^{k+1}}{(k+1)\left[\ln(k+1)^2\right]}}{\frac{(-1)^k x^k}{k(\ln k)^2}} \right|$

$= \lim\limits_{k \to \infty} \left| \dfrac{k(\ln k)^2}{(k+1)[\ln(k+1)]^2} \right| |x|$

$= |x|$ *l'Hôpital's rule*

The interval of absolute convergence is
$$-1 < x < 1$$

Endpoints:

$x = -1$: $\displaystyle\sum_{k=2}^{\infty} \frac{(-1)^k(-1)^k}{k(\ln k)^2} = \sum_{k=2}^{\infty} \frac{1}{k(\ln k)^2}$

converges by the integral test

where $f(x) = \dfrac{1}{x(\ln x)^2}$ is

positive, continuous, and
decreasing on $(2, \infty)$.

$$\int_2^\infty \frac{dx}{x(\ln x)^2} = \lim_{N\to\infty} \left.\frac{-1}{\ln x}\right|_2^N$$
$$= \frac{1}{\ln 2}$$

$x = 1$: $\displaystyle\sum_{k=2}^{\infty} \frac{(-1)^k}{k(\ln k)^2}$ converges

absolutely by the integral test.
The convergence set is $[-1, 1]$.

21. $L = \displaystyle\lim_{k\to\infty} \left| \dfrac{\frac{(2x)^{2(k+1)}}{(k+1)!}}{\frac{(2x)^{2k}}{k!}} \right|$

$= \displaystyle\lim_{k\to\infty} \left| \dfrac{2^{2k+2}k!}{(k+1)!2^{2k}} \right| |x|^2$

$= \displaystyle\lim_{k\to\infty} \left| \dfrac{2^2}{k+1} \right| |x|^2$

$= 0$

The power series converges for all x. The
convergence set is $(-\infty, \infty)$.

23. $L = \displaystyle\lim_{k\to\infty} \left| \dfrac{\frac{(k+1)!(3x)^{3(k+1)}}{2^{k+1}}}{\frac{k!(3x)^{3k}}{2^k}} \right|$

$= \displaystyle\lim_{k\to\infty} \left| \dfrac{(k+1)!3^{3k+3}2^k}{2^{k+1}k!3^{3k}} \right| |x|^3$

$= \displaystyle\lim_{k\to\infty} \left| \dfrac{k+1}{2} \right| |3x|^3$

$= \infty$

The power series converges only for $x = 0$.

25. $L = \displaystyle\lim_{k\to\infty} \left| \dfrac{\frac{2^{k+1}(2x-1)^{2(k+1)}}{(k+1)!}}{\frac{2^k(2x-1)^{2k}}{k!}} \right|$

$= \displaystyle\lim_{k\to\infty} \left| \dfrac{2^{k+1}(2x-1)^{2(k+1)}k!}{(k+1)!2^k(2x-1)^{2k}} \right|$

$= \displaystyle\lim_{k\to\infty} \dfrac{2}{k+1}|2x-1|^2$

$= 0$

The series converges for all x. The
convergence set is $(-\infty, \infty)$.

27. $L = \displaystyle\lim_{k\to\infty} \left| \dfrac{\frac{x^{k+1}}{(k+1)^{3/2}}}{\frac{x^k}{k^{3/2}}} \right|$

$= \displaystyle\lim_{k\to\infty} \left| \dfrac{k\sqrt{k}}{(k+1)\sqrt{k+1}} \right| |x|$

$= |x|$

The interval of absolute convergence is
$$-1 < x < 1$$

Endpoints:

$x = -1$: $\displaystyle\sum_{k=1}^{\infty} \frac{(-1)^k}{k\sqrt{k}}$ converges absolutely

(p-series with $p = \frac{3}{2} > 1$).

$x = 1$: $\displaystyle\sum_{k=1}^{\infty} \frac{1^k}{k\sqrt{k}}$ converges absolutely

(p-series with $p = \frac{3}{2} > 1$).

The convergence set is $[-1, 1]$.

29. $\displaystyle\lim_{k\to\infty} \left| \dfrac{(k+1)^2(x+1)^{2k+3}}{k^2(x+1)^{2k+1}} \right| = |x+1|^2 < 1$

so $R = 1$.

31. $\displaystyle\lim_{k\to\infty}\left|\dfrac{\frac{(k+1)!x^{k+1}}{(k+1)^k}}{\frac{k!x^k}{k^k}}\right|$

$\displaystyle=\lim_{k\to\infty}\left|\dfrac{(k+1)!k^kx^{k+1}}{k!(k+1)^{k+1}x^k}\right|$

$\displaystyle=\lim_{k\to\infty}\left|\dfrac{(k+1)xk^k}{(k+1)^{k+1}}\right|$

$\displaystyle=\lim_{k\to\infty}\left|\dfrac{1}{\left(1+\frac{1}{k}\right)^k}\right||x|$

$\displaystyle=\dfrac{1}{e}|x|$

$\dfrac{1}{e}|x|<1$ when $|x|<e$.

Thus, $R=e$.

33. $\displaystyle\lim_{k\to\infty}\left|\dfrac{\frac{[(k+1)!]^2x^{k+1}}{[2(k+1)]!}}{\frac{(k!)^2x^k}{(2k)!}}\right|$

$\displaystyle=\lim_{k\to\infty}\left|\dfrac{[(k+1)!]^2(2k)!x^{k+1}}{(k!)^2(2k+2)!x^k}\right|$

$\displaystyle=\lim_{k\to\infty}\left|\dfrac{(k+1)^2x}{(2k+2)(2k+1)}\right|$

$\displaystyle=\lim_{k\to\infty}\left|\dfrac{k+1}{2(2k+1)}\right||x|$

$\displaystyle=\dfrac{1}{4}|x|$

$\dfrac{1}{4}|x|<1$ when $|x|<4$.

Thus, $R=4$.

35. $\displaystyle\lim_{k\to\infty}\left|\dfrac{(k+1)(ax)^{k+1}}{k(ax)^k}\right|=|ax|$

$|ax|<1$ when $|x|<\dfrac{1}{|a|}$.

Thus, $R=\dfrac{1}{|a|}$.

37. $f(x)=1+\dfrac{x}{2}+\dfrac{x^2}{4}+\dfrac{x^3}{8}+\dfrac{x^4}{16}+\cdots$

$f'(x)=(1)\dfrac{1}{2}+(2)\dfrac{x}{4}+(3)\dfrac{x^2}{8}+(4)\dfrac{x^3}{16}+\cdots$

$\displaystyle=\sum_{k=1}^{\infty}\dfrac{kx^{k-1}}{2^k}$

39. $f(x)=2+3x+4x^2+5x^3+\cdots$

$f'(x)=3+8x+15x^2+\cdots$

$\displaystyle=\sum_{k=1}^{\infty}k(k+2)x^{k-1}$

41. $f(x)=1+x+x^2+x^3+\cdots$

$f'(x)=1+2x+3x^2+\cdots$

$\displaystyle=\sum_{k=1}^{\infty}kx^{k-1}$

43. $f(x)=1+\dfrac{x}{2}+\dfrac{x^2}{4}+\dfrac{x^3}{8}+\dfrac{x^4}{16}+\cdots$

$\displaystyle F(x)=\int_0^x f(u)\,du$

$\displaystyle=x+\dfrac{x^2}{2(2)}+\dfrac{x^3}{3(2)^2}+\dfrac{x^4}{4(2)^3}+\cdots$

$\displaystyle=\sum_{k=1}^{\infty}\dfrac{x^k}{k(2)^{k-1}}$

Alternatively, we can write

$\displaystyle f(x)=\sum_{k=0}^{\infty}\left(\dfrac{x}{2}\right)^k$

$\displaystyle F(x)=\sum_{k=0}^{\infty}\int_0^x\dfrac{u^k}{2^k}\,du=\sum_{k=0}^{\infty}\dfrac{x^{k+1}}{(k+1)2^k}$

45. $\displaystyle f(x)=\sum_{k=0}^{\infty}(k+2)x^k$

$$F(x) = \sum_{k=0}^{\infty} \int_0^x (k+2)\, u^k \, du$$

$$= \sum_{k=0}^{\infty} \frac{(k+2)x^{k+1}}{(k+1)}$$

47. $f(x) = \sum_{k=0}^{\infty} \dfrac{x^k}{k!}$

$$F(x) = \sum_{k=0}^{\infty} \int_0^x \frac{u^k}{k!} \, du = \sum_{k=0}^{\infty} \frac{x^{k+1}}{(k+1)!}$$

49. Since

$$\frac{1}{1-x} = \sum_{k=0}^{\infty} x^k$$

for $|x| < 1$, we can differentiate term by term to obtain

$$\frac{1}{(1-x)^2} = \sum_{k=1}^{\infty} k x^{k-1}$$

for $|x| < 1$. And, we can differentiate again to obtain

$$\frac{2}{(1-x)^3} = \sum_{k=2}^{\infty} k(k-1) x^{k-2}$$

for $|x| < 1$. Thus, a power series for

$$f(x) = \frac{1}{(1-x)^3}$$

is

$$\sum_{k=2}^{\infty} \frac{1}{2} k(k-1) x^{k-2}$$

$$= 1 + 3x + 6x^2 + 10x^3 + \cdots$$

This is absolutely convergent for $|x| < 1$.
Endpoints:

$x = -1$: $\sum_{k=2}^{\infty} \frac{1}{2}(-1)^{k-2} k(k-1)$ diverges

$x = 1$: $\sum_{k=2}^{\infty} \frac{1}{2} k(k-1)$ diverges

The interval of convergence is $(-1, 1)$.

51. Let $S = \sum_{k=1}^{\infty} \dfrac{(k+3)! x^k}{k!(k+4)!}$

Use the generalized ratio test.

$$L = \lim_{k \to \infty} \left| \frac{[(k+4)!]^2 k! x^{k+1}}{(k+1)!(k+3)!(k+5)! x^k} \right|$$

$$= \lim_{k \to \infty} \frac{k+4}{(k+1)(k+5)} |x|$$

$$= 0 < 1$$

The radius of convergence is $R = \infty$.

53. $f(x) = \sum_{k=0}^{\infty} \dfrac{(-1)^k x^{2k+1}}{(2k+1)!}$

$$= x - \frac{x^3}{3!} + \frac{x^5}{5!} - \frac{x^7}{7!} + \cdots$$

$$f'(x) = 1 - \frac{x^2}{2!} + \frac{x^4}{4!} - \frac{x^6}{6!} + \cdots$$

$$f''(x) = -x + \frac{x^3}{3!} - \frac{x^5}{5!} + \cdots$$

$$= -f(x)$$

55. $S = \sum_{k=1}^{\infty} \left| \dfrac{\sin(k! x)}{k^2} \right| \leq \sum_{k=1}^{\infty} \dfrac{1}{k^2}$

S converges absolutely. Let

$$T = S' = \sum_{k=1}^{\infty} \frac{k! \cos(k! x)}{k^2}$$

$\lim_{k \to \infty} \dfrac{k! \cos(k! x)}{k^2} \neq 0$ since the limit does not exist. Thus, T is always divergent. Theorem 8.23 does not apply to S since S is not a power series.

57. $S = \displaystyle\sum_{k=1}^{\infty} a_k x^{kp}$

By the ratio test,

$$\lim_{k\to\infty}\left|\frac{a_{k+1}x^{kp+p}}{a_k x^{kp}}\right| = |x|^p \lim_{k\to\infty}\left|\frac{a_{k+1}}{a_k}\right|$$

$$= |x|^p\left(\frac{1}{R}\right)$$

$|x|^p\left(\dfrac{1}{R}\right) < 1$ if $|x|^p < R$ or $|x| < R^{1/p}$

Thus, the radius of convergence $R^{1/p}$.

59. Let $S = \displaystyle\sum_{k=1}^{\infty}\frac{k}{x^k}$ and let $t = x^{-1}$ so

$S = \displaystyle\sum_{k=1}^{\infty} kt^k$ which converges by the

ratio test if $|t| < 1$ and diverges if $|t| \geq 1$.
This means S converges if $|x| > 1$.

8.8 Taylor and Maclaurin Series, page 661

1. Suppose there is an open interval I containing c throughout which the function f and all its derivatives exist. Then the power series

$$f(c) + \frac{f'(c)}{1!}(x-c) + \frac{f''(c)}{2!}(x-c)^2 \text{ is}$$

$$+ \frac{f'''(c)}{3!}(x-c)^3 + \cdots$$

called the Taylor series of f at c. The special case where $c = 0$ is called the Maclaurin series of f:

$$f(0) + \frac{f'(0)}{1!}x + \frac{f''(0)}{2!}x^2 + \frac{f'''(0)}{3!}x^3 + \cdots$$

3. $e^{2x} = 1 + 2x + \dfrac{1}{2!}(2x)^2 + \cdots + \dfrac{1}{k!}(2x)^k + \cdots$

$$= \sum_{k=0}^{\infty}\frac{(2x)^k}{k!}$$

5. $e^{x^2} = 1 + x^2 + \dfrac{1}{2!}\left(x^2\right)^2 + \cdots + \dfrac{1}{k!}\left(x^2\right)^k + \cdots$

$$= \sum_{k=0}^{\infty}\frac{x^{2k}}{k!}$$

7. $\sin x^2 = x^2 - \dfrac{1}{3!}\left(x^2\right)^3 + \dfrac{1}{5!}\left(x^2\right)^5 - \cdots$

$$= \sum_{k=0}^{\infty}\frac{(-1)^k x^{4k+2}}{(2k+1)!}$$

9. $\sin ax = ax - \dfrac{1}{3!}(ax)^3 + \dfrac{1}{5!}(ax)^5 - \cdots$

$$= \sum_{k=0}^{\infty}\frac{(-1)^k (ax)^{2k+1}}{(2k+1)!}$$

11. $\cos 2x^2 = 1 - \dfrac{1}{2!}\left(2x^2\right)^2 + \dfrac{1}{4!}\left(2x^2\right)^4 - \cdots$

$$= \sum_{k=0}^{\infty}\frac{(-1)^k (2x^2)^{2k}}{(2k)!}$$

13. $x^2 \cos x = x^2 - \dfrac{1}{2!}(x)^4 + \dfrac{1}{4!}(x)^6 - \cdots$

$$= \sum_{k=0}^{\infty}\frac{(-1)^k (x)^{2k+2}}{(2k)!}$$

15. $x^2 + 2x + 1$ is its own Maclaurin series.

17. $xe^x = x + x^2 + \dfrac{1}{2!}x^3 + \cdots$

$$= \sum_{k=0}^{\infty}\frac{x^{k+1}}{k!}$$

19. $e^x + \sin x = \left[1 + x + \dfrac{x^2}{2!} + \dfrac{x^3}{3!} + \cdots\right]$

$$+ \left[x - \frac{x^3}{3!} + \frac{x^5}{5!} + \cdots\right]$$

$$= 1 + 2x + \frac{x^2}{2!} + \frac{x^4}{4!} + \frac{2x^5}{5!}$$
$$+ \frac{x^6}{6!} + \frac{2x^9}{9!} + \cdots$$

21. $\dfrac{1}{1+4x} = \dfrac{1}{1-(-4x)}$
$$= 1 + (-4x) + (-4x)^2 + \cdots$$
$$= \sum_{k=0}^{\infty} (-4)^k x^k$$

23. $\dfrac{1}{a+x} = \dfrac{1}{a\left(1 + \frac{x}{a}\right)}$
$$= \frac{1}{a\left(1 - \left[-\frac{x}{a}\right]\right)}$$
$$= \frac{1}{a} \sum_{k=0}^{\infty} \left(-\frac{x}{a}\right)^k$$
$$= \sum_{k=0}^{\infty} \frac{(-1)^k x^k}{a^{k+1}}$$

25. $\ln(3+x) = \ln 3\left(1 + \dfrac{x}{3}\right)$
$$= \ln 3 + \ln\left(1 + \frac{x}{3}\right)$$
$$= \ln 3 + \frac{1}{3}x - \frac{1}{2(3^2)}x^2$$
$$+ \frac{1}{3(3^3)}x^3 - \frac{1}{4(3^4)}x^4 + \cdots$$
$$= \ln 3 + \sum_{k=0}^{\infty} \frac{(-1)^k x^{k+1}}{(k+1)3^{k+1}}$$

27. $e^x \approx e + e(x-1) + \dfrac{1}{2!}e(x-1)^2$
$$+ \frac{1}{3!}e(x-1)^3$$

29. $\cos x \approx \cos\dfrac{\pi}{3} - \left(x - \dfrac{\pi}{3}\right)\sin\dfrac{\pi}{3}$
$$- \frac{\left(x-\frac{\pi}{3}\right)^2}{2!}\cos\frac{\pi}{3} + \frac{\left(x-\frac{\pi}{3}\right)^3}{3!}\sin\frac{\pi}{3}$$

31. $f(x) = \tan x$; $f(0) = 0$
$$f'(x) = \sec^2 x;\ f'(0) = 1$$
$$f''(x) = 2\sec^2 x \tan x;\ f''(0) = 0$$
$$f'''(x) = 2\sec^2 x(\sec^2 x) + 4\sec x(\sec x \tan x)\tan x$$
$$= 2\sec^4 x + 4\sec^2 x \tan^2 x;\ f'''(0) = 2$$
$$\tan x \approx 0 + 1 \cdot x + \frac{0x^2}{2!} + \frac{2x^3}{3!}$$

33. $f(x) = (2-x)^{-1}$; $f(5) = -\dfrac{1}{3}$
$$f'(x) = (2-x)^{-2};\ f'(5) = \frac{1}{9}$$
$$f''(x) = 2(2-x)^{-3};\ f''(5) = -\frac{2}{27}$$
$$f'''(x) = 6(2-x)^{-4};\ f'''(5) = \frac{6}{81}$$
$$\frac{1}{2-x} \approx -\frac{1}{3} + \frac{1}{9}(x-5) - \frac{1}{27}(x-5)^2$$
$$+ \frac{1}{81}(x-5)^3$$

> **SURVIVAL HINT:** *Contrast the solutions for Problems 33 and 35. These two problems show different, but equally correct, solutions for this type of problem.*

35. $f(x) = \dfrac{3}{2x-1}$
$$= \frac{3}{2(x-2)+3}$$
$$= \frac{1}{1 + \frac{2}{3}(x-2)}$$
$$= 1 + \left[-\frac{2}{3}(x-2)\right] + \left[-\frac{2}{3}(x-2)\right]^2$$
$$+ \left[-\frac{2}{3}(x-2)\right]^3 + \cdots$$
$$\approx 1 - \frac{2}{3}(x-2) + \frac{4}{9}(x-2)^2 - \frac{8}{27}(x-2)^3$$

37. $\sqrt{1+x}$

$= (1+x)^{1/2}$

$= 1 + \dfrac{1}{2}x + \dfrac{\frac{1}{2}\left(\frac{1}{2}-1\right)x^2}{2!}$

$\qquad + \dfrac{\frac{1}{2}\left(\frac{1}{2}-1\right)\left(\frac{1}{2}-2\right)x^3}{3!} + \cdots$

$= 1 + \dfrac{1}{2}x - \dfrac{1}{8}x^2 + \dfrac{1}{16}x^3 - \dfrac{5}{128}x^4 + \cdots$

Since the exponent, $\frac{1}{2}$, is greater than 0 and not an integer, the interval of convergence is $[-1, 1]$.

39. $(1+x)^{2/3} = 1 + \dfrac{2}{3}x + \dfrac{\frac{2}{3}\left(-\frac{1}{3}\right)}{2!}x^2$

$\qquad + \dfrac{\frac{2}{3}\left(-\frac{1}{3}\right)\left(-\frac{4}{3}\right)}{3!}x^3 + \cdots$

$\qquad = 1 + \dfrac{2}{3}x - \dfrac{1}{9}x^2 + \dfrac{4}{81}x^3 + \cdots$

Since the exponent, $\frac{2}{3}$, is greater than 0 and not an integer, the interval of convergence is $[-1, 1]$.

41. $\dfrac{x}{\sqrt{1-x^2}} = x\left(1-x^2\right)^{-1/2}$

$= x\left[1 - \left(-\dfrac{1}{2}\right)x^2 + \dfrac{\left(-\frac{1}{2}\right)\left(-\frac{3}{2}\right)x^4}{2!}\right.$

$\qquad \left. - \dfrac{\left(-\frac{1}{2}\right)\left(-\frac{3}{2}\right)\left(-\frac{5}{2}\right)x^6}{3!} + \cdots\right]$

$= x + \dfrac{1}{2}x^3 + \dfrac{3}{8}x^5 + \dfrac{5}{16}x^7 + \cdots$

Since the exponent, $-\frac{1}{2}$, is between -1 and 1, the interval of convergence is

$$-1 < -x^2 < 1$$
$$-1 < \ x^2 < 1$$

$x^2 < 1$ implies $-1 < x < 1$, so the interval of convergence is $(-1, 1)$.

43. $\sinh x = \dfrac{1}{2}\left(e^x - e^{-x}\right)$

$= \dfrac{1}{2}\left[\left(1 + x + \dfrac{1}{2!}x^2 + \dfrac{1}{3!}x^3 + \cdots\right)\right.$

$\qquad \left. - \left(1 - x + \dfrac{1}{2!}x^2 - \dfrac{1}{3!}x^3 + \cdots\right)\right]$

$= x + \dfrac{1}{3!}x^3 + \dfrac{1}{5!}x^5 + \cdots$

$= \displaystyle\sum_{k=0}^{\infty} \dfrac{x^{2k+1}}{(2k+1)!}$

45. $R_n\left(\dfrac{1}{2}\right) \leq \dfrac{\exp(z_n)}{(n+1)!}\left(\dfrac{1}{3}\right)^{n+1}$ for $0 < z_n < \dfrac{1}{3} < 1$

$\dfrac{\exp(z_n)}{(n+1)3^{n+1}} < \dfrac{3}{3^{n+1}(n+1)!} < 0.0005$

Thus, $3^{n+1}(n+1)! > 6{,}000$ or choose $n = 4$ (five terms). Thus,

$e^{1/3} \approx \dfrac{\left(\frac{1}{3}\right)^0}{0!} + \dfrac{\left(\frac{1}{3}\right)^1}{1!} + \dfrac{\left(\frac{1}{3}\right)^2}{2!} + \dfrac{\left(\frac{1}{3}\right)^3}{3!} + \dfrac{\left(\frac{1}{3}\right)^4}{4!}$

$\qquad \approx 1.3956$

47. $f(x) = \dfrac{1}{x^2 - 3x + 2}$

$= \dfrac{1}{(x-2)(x-1)}$

$= \dfrac{1}{1-x} - \dfrac{\frac{1}{2}}{1 - \frac{x}{2}}$

$= \left[1 + x + x^2 + \cdots\right]$

$\qquad - \dfrac{1}{2}\left[1 + \dfrac{1}{2}x + \left(\dfrac{1}{2}x\right)^2 + \cdots\right]$

$= \displaystyle\sum_{k=0}^{\infty}\left[1 - \dfrac{1}{2^{k+1}}\right]x^k$

49. $f(x) = \sin x \cos x$

$$= \frac{1}{2}\sin 2x$$

$$= \frac{1}{2}\sum_{k=0}^{\infty}(-1)^k\frac{(2x)^{2k+1}}{(2k+1)!}$$

$$= \sum_{k=0}^{\infty}\frac{(-1)^k 2^{2k}x^{2k+1}}{(2k+1)!}$$

51. $f(x) = \ln\left[\dfrac{1+2x}{1-3x+2x^2}\right]$

$$= \ln(1+2x) - \ln(1-2x) - \ln(1-x)$$

$$= \sum_{k=0}^{\infty}(-1)^k\left[\frac{2^{k+1} - (-2)^{k+1} - (-1)^{k+1}}{k+1}\right]x^{k+1}$$

$$= \sum_{k=0}^{\infty}\left[(-1)^k 2^{k+1} + 2^{k+1} + 1\right]\frac{x^{k+1}}{k+1}$$

53. a. $f(x) = x + \sin x$

$$= x + \left(x - \frac{x^3}{3!} + \frac{x^5}{5!} - \cdots\right)$$

Thus,

$$\lim_{x\to 0}\frac{x + \sin x}{x}$$

$$= \lim_{x\to 0}\left(1 + 1 - \frac{x^2}{3!} + \frac{x^4}{5!} - \cdots\right)$$

$$= 2$$

b. $g(x) = e^x - 1$

$$= x + \frac{x^2}{2!} + \frac{x^3}{3!} + \cdots$$

Thus,

$$\lim_{x\to 0}\frac{e^x - 1}{x}$$

$$= \lim_{x\to 0}\left[1 + \frac{x}{2!} + \frac{x^2}{3!} + \cdots\right] = 1$$

55. Let $f(x) = \ln(x+1)$. Then

$$f^{(n+1)}(x) = \frac{(-1)^n n!}{(x+1)^{n+1}}$$

and for $0 < c < x < 1$

$$|R_n(x)| = \left|\frac{f^{(n+1)}(c)}{(n+1)!}x^{n+1}\right|$$

$$\leq \frac{|x|^{n+1}}{n+1}$$

since $(c+1)^{n+1} \geq 1$.

57. Maclaurin was stating the first derivative test.

59. $J_0 = \displaystyle\sum_{k=0}^{\infty}\frac{(-1)^k x^{2k}}{(k!)^2 2^{2k}}$

$J_0' = \displaystyle\sum_{k=1}^{\infty}\frac{(-1)^k 2k x^{2k-1}}{(k!)^2 2^{2k}}$

$J_0'' = \displaystyle\sum_{k=1}^{\infty}\frac{(-1)^k 2k(2k-1)x^{2k-2}}{(k!)^2 2^{2k}}$

$x^2 J_0'' + x J_0' + x^2 J_0$

$$= \sum_{k=1}^{\infty}\frac{(-1)^k 2k(2k-1)x^{2k}}{(k!)^2 2^{2k}}$$

$$+ \sum_{k=1}^{\infty}\frac{(-1)^k 2k x^{2k}}{(k!)^2 2^{2k}}$$

$$+ \sum_{k=0}^{\infty}\frac{(-1)^k x^{2k+2}}{(k!)^2 2^{2k}}$$

$$= \sum_{k=1}^{\infty}\left[\frac{(-1)^k 2k(2k-1)}{(k!)^2 2^{2k}}\right.$$

$$\left. + \frac{(-1)^k 2k}{(k!)^2 2^{2k}} + \frac{(-1)^{k-1}}{[(k-1)!]^2 2^{2(k-1)}}\right]x^{2k}$$

$$= \sum_{k=1}^{\infty}(-1)^k\left[\frac{2k(2k-1) + 2k - 2^2 k^2}{(k!)^2 2^{2k}}\right]x^{2k}$$

$$= 0$$

Chapter 8 Review

 Studying for a chapter examination is a personal process, one which nobody else can do for you. Simply take the time to review what you have done.

SURVIVAL HINT: Work all of Chapter 8 problems in the Proficiency Examination (whether they are assigned or not). Work through all of the problems before looking at the answers, and *then* correct each of the problems. The answers to all these problems are given in the answer section at the back of the text. If you worked the problem correctly, move on to the next problem, but if you did not work it correctly (or you did not know what to do), then look at the solutions below, look back in the chapter to study the procedure, or ask your instructor.

Finally, go back over the homework problems you have been assigned. If you worked a problem correctly, move on to the next problem, but if you missed it on your homework, then you should look back in the book or talk to your instructor about how to work the problem.

If you follow these steps, you should be successful with your review of this chapter.

Proficiency Examination, page 664

1. A sequence is a function whose domain is a set of nonnegative integers. That is, it is a succession of numbers that are listed according to a given prescription or rule.

2. If the terms of the sequence $\{a_n\}$ approach the number L as n increases without bound, we say that the sequence $\{a_n\}$ *converges to the limit L* and write $L = \lim\limits_{n\to\infty} a_n$.

3. A sequence converges if the limit of the nth element is finite (and unique). If not, it diverges.

4. **a.** The terms of a bounded sequence lie within a finite range.
   b. A sequence is monotonic if it is either nondecreasing or nonincreasing.
   c. A sequence is strictly monotonic if it is increasing or decreasing.

5. A monotonic sequence $\{a_n\}$ converges if it is bounded and diverges otherwise.

6. An infinite series is a sum of infinitely many terms.

7. A sequence $\{a_n\}$ converges if $\lim\limits_{n\to\infty} a_n = L$ is finite (and unique). A series $\sum\limits_{k=0}^{\infty} a_k$ converges if the sequence $\{S_n\}$ of partial sums $S_n = \sum\limits_{k=1}^{n} a_k$ converges.

8. The middle terms of a telescoping series vanish (by addition and subtraction of the same numbers). Specifically, the series $S = \Sigma a_k$ telescopes if $a_k = b_k - b_{k-1}$.

9. The harmonic series is a p-series with $p = 1$. It diverges, but the alternating harmonic series converges.

10. The ratio of consecutive terms of a geometric series is a constant, r. That is, a geometric series is one of the form Σar^k. It diverges if $|r| \geq 1$ and converges to
$$S = \frac{a}{1 - r} \text{ if } |r| < 1.$$

11. **a.** $S = \sum\limits_{k=0}^{\infty} a_k$ diverges if $\lim\limits_{k\to\infty} a_k \neq 0$.
   b. If $a_k = f(k)$ for $k = 1, 2, \cdots$, where f is a positive, continuous, and decreasing function of x for $x \geq 1$, then

Page 326

$$\sum_{k=1}^{\infty} a_k \quad \text{and} \quad \int_1^{\infty} f(x)\,dx$$

either both converge or both diverge.

c. $\sum_{k=1}^{\infty} \dfrac{1}{k^p}$ is a convergent p-series if $p > 1$.
It diverges when $p \le 1$.

d. Let $0 \le a_k \le c_k$ for all k. If $\sum_{k=1}^{\infty} c_k$

converges, then $\sum_{k=1}^{\infty} a_k$ also converges. Let

$0 \le d_k \le a_k$ for all k. If $\sum_{k=1}^{\infty} d_k$ diverges,

then $\sum_{k=1}^{\infty} a_k$ also diverges.

e. Suppose $a_k > 0$ and $b_k > 0$ for all
sufficiently large k and that

$$\lim_{k \to \infty} \frac{a_k}{b_k} = L$$

where L is finite and positive
$(0 < L < \infty)$. Then Σa_k and Σb_k
either both converge or both diverge.

f. Suppose $a_k > 0$ and $b_k > 0$ for all
sufficiently large k. Then, if

$$\lim_{k \to \infty} \frac{a_k}{b_k} = 0$$

and Σb_k converges, the series Σa_k
converges. If

$$\lim_{k \to \infty} \frac{a_k}{b_k} = \infty$$

and Σb_k diverges, the series Σa_k
diverges.

g. Given the series Σa_k with $a_k > 0$,
suppose that

$$\lim_{k \to \infty} \frac{a_{k+1}}{a_k} = L$$

The ratio test states the following:
If $L < 1$, then Σa_k converges.
If $L > 1$ or if L is infinite, then Σa_k
diverges.
If $L = 1$, the test is inconclusive.

h. Given the series Σa_k with $a_k \ge 0$,
suppose that

$$\lim_{k \to \infty} \sqrt[k]{a_k} = L$$

The root test states the following:
If $L < 1$, then Σa_k converges.
If $L > 1$ or if L is infinite, then Σa_k
diverges.
If $L = 1$, the root test is inconclusive.

i. If $a_k > 0$, then an alternating series

$$\sum_{k=1}^{\infty} (-1)^k a_k \quad \text{or} \quad \sum_{k=1}^{\infty} (-1)^{k+1} a_k$$

converges if both of the following two
conditions are satisfied:
1. $\lim_{k \to \infty} a_k = 0$
2. $\{a_k\}$ is a decreasing sequence; that is,
$a_{k+1} < a_k$ for all k.

12. Suppose an alternating series

$$\sum_{k=1}^{\infty} (-1)^k a_k \quad \text{or} \quad \sum_{k=1}^{\infty} (-1)^{k+1} a_k$$

satisfies the conditions of the alternating
series test; namely

$$\lim_{k \to \infty} a_k = 0$$

and $\{a_k\}$ is a decreasing sequence

$(a_{k+1} < a_k)$. If the series has sum S, then $|S - S_n| < a_{n+1}$, where S_n is the nth partial sum of the series.

13. A series of real numbers Σa_k must converge if the related absolute value series $\Sigma |a_k|$ converges.

14. The series Σa_k is absolutely convergent if the related series $\Sigma |a_k|$ converges. The series Σa_k is conditionally convergent if it converges but $\Sigma |a_k|$ diverges.

15. For the series Σa_k, suppose $a_k \neq 0$ for $k \geq 1$ and that

$$\lim_{k \to \infty} \left| \frac{a_{k+1}}{a_k} \right| = L$$

where L is a real number or ∞. Then:
If $L < 1$, the series Σa_k converges absolutely and hence converges.
If $L > 1$ or if L is infinite, the series Σa_k diverges.
If $L = 1$, the test is inconclusive.

16. **a.** An infinite series of the form

$$\sum_{k=0}^{\infty} a_k (x - c)^k$$

$$= a_0 + a_1(x - c) + a_2(x - c)^2 + \cdots$$

is called a power series in $(x - c)$.

b. Let $\Sigma a_k u^k$ be a power series, and consider

$$L = \lim_{k \to \infty} \left| \frac{a_{k+1}}{a_k} \right|$$

Then:
  If $L = \infty$, the power series converges only at $u = 0$.
  If $L = 0$, the power series converges for all real u.

If $0 < L < \infty$, let $R = 1/L$. Then the power series *converges absolutely* for $|u| < R$ (or $-R < u < R$) and *diverges* for $|u| > R$. This is called the *interval of absolute convergence*.

Finally, check for convergence at the endpoints $u = -R$ and $u = R$. The resulting set (with convergent endpoints) is the *convergence set*.

17. The interval of convergence of a power series consists of those values of x for which the series converges. If the interval is $-R < x < R$, then R is the radius of convergence.

18. **a.** $P_n(x) = \sum_{k=1}^{n} a_k (x - c)^k$ is a Taylor polynomial if $a_k = \dfrac{f^{(k)}(c)}{k!}$.

b. If f and all its derivatives exist in an open interval I containing c, then for each x in I

$$f(x) = f(c) + \frac{f'(c)}{1!}(x - c)$$
$$+ \frac{f''(c)}{2!}(x - c)^2 + \frac{f'''(c)}{3!}(x - c)^3 + \cdots$$
$$+ \frac{f^{(n)}(c)}{n!}(x - c)^n + R_n(x)$$

where the remainder function $R_n(x)$ is given by

$$R_n(x) = \frac{f^{(n+1)}(z_n)}{(n + 1)!}(x - c)^{n+1}$$

for some z_n that depends on x and lies between c and x.

19. The Taylor series is $T = \sum_{k=1}^{\infty} a_k (x - c)^k$ and is a Maclaurin series if $c = 0$.

20. The binomial function $(1 + x)^p$ is represented by its Maclaurin series

$$(1 + x)^p = 1 + px + \frac{p(p - 1)}{2!} x^2$$
$$+ \frac{p(p - 1)(p - 2)}{3!} x^3 + \cdots$$
$$= \sum_{k=0}^{\infty} \binom{p}{k} x^k$$

$-1 < x < 1$ if $p \le -1$;
$-1 < x \le 1$ if $-1 < p < 0$;
$-1 \le x \le 1$ if $p \ge 0$, p not an integer;
for all x if p is a nonnegative integer.

21. This is the definition of e (assuming that $n \to \infty$).

22. a. Assume $n \to \infty$. The sequence has an upper bound of 4, a lower bound of 0, and after $n = 2$ is monotone decreasing:

$$\lim_{n \to \infty} \frac{\frac{e^{n+1}}{(n+1)!}}{\frac{e^n}{n!}} = \lim_{n \to \infty} \frac{e}{n + 1} = 0$$

Thus, the sequence converges to 0 by the BMCT.

b. $S = \sum_{k=1}^{\infty} \frac{e^k}{k!}$; use ratio test.

$$\lim_{k \to \infty} \frac{e^{k+1} k!}{(k + 1)! e^k} = \lim_{k \to \infty} \frac{e}{k + 1} = 0 < 1$$

S converges.

c. Answers vary. In part **a**, we consider the convergence of a sequence, and in part **b** we consider the convergence of a series.

23. $\sum_{k=2}^{\infty} \frac{1}{k \ln k}$; use the integral test. Let

$$f(x) = \frac{1}{x \ln x}; \text{ it is continuous, positive,}$$
and decreasing for $x > 1$.

$$\lim_{b \to \infty} \int_2^b (x \ln x)^{-1} \, dx = \lim_{b \to \infty} \ln|\ln x| \Big|_2^b = \infty$$

The series diverges.

24. $\sum_{k=1}^{\infty} \frac{\pi^k k!}{k^k}$; use the ratio test.

$$\lim_{k \to \infty} \frac{\pi^{k+1} (k + 1)! k^k}{(k + 1)^{k+1} \pi^k k!} = \pi \lim_{k \to \infty} \frac{1}{\left(1 + \frac{1}{k}\right)^k} = \frac{\pi}{e} > 1$$

so the series diverges.

25. $\sum_{k=2}^{\infty} \frac{1}{(\ln k)^{1/k}}$; use divergence test.

$$\lim_{k \to \infty} (\ln k)^{-1/k} = \lim_{k \to \infty} \exp\left[\ln(\ln k)^{-1/k}\right]$$
$$= \lim_{k \to \infty} \exp\left(\frac{-\ln(\ln k)}{k}\right)$$
$$= 1 \ne 0$$

The series diverges.

26. $\sum_{k=1}^{\infty} \frac{3k^2 - k + 1}{(1 - 2k)k}$; check the necessary condition:

$$\lim_{k \to \infty} \frac{3k^2 - k + 1}{(1 - 2k)k} = -\frac{3}{2} \ne 0$$

The series diverges.

27. $\sum_{k=1}^{\infty} \frac{(-1)^{k+1}}{k^2}$ converges absolutely when compared with the convergent p-series $\Sigma(1/k^2)$. (We see $p = 2 > 1$).

28. $\sum_{k=0}^{\infty} (-1)^k (k + 1) u^k$

$$L = \lim_{k \to \infty} \left| \frac{(-1)^{k+1}(k+2)u^{k+1}}{(-1)^k(k+1)u^k} \right|$$

$$= \lim_{k \to \infty} \frac{k+2}{k+1}|u|$$

$$= |u|$$

The series converges absolutely for $-1 < u < 1$.

Endpoints:

$u = -1$: $\displaystyle\sum_{k=1}^{\infty}(-1)^{k+1}k(-1)^{k-1} = \sum_{k=1}^{\infty}k$

diverges by the divergence test.

$u = 1$: $\displaystyle\sum_{k=1}^{\infty}(-1)^{k+1}k$ diverges by

the divergence test.

The convergence set is $(-1, 1)$.

29. $\displaystyle\sin x = \sum_{k=0}^{\infty}\frac{(-1)^k x^{2k+1}}{(2k+1)!}$

$\displaystyle\sin 2x = \sum_{k=0}^{\infty}\frac{(-1)^k(2x)^{2k+1}}{(2k+1)!}$

30. $f(x) = \dfrac{1}{x-3}$ at $c = \dfrac{1}{2}$

$$= \frac{1}{x - \frac{1}{2} - \frac{5}{2}}$$

$$= \frac{-\frac{2}{5}}{1 - \frac{2}{5}\left(x - \frac{1}{2}\right)}$$

$$= -\frac{2}{5}\left[1 + \frac{2}{5}\left(x - \frac{1}{2}\right) + \cdots\right]$$

$$= -\frac{2}{5}\sum_{k=0}^{\infty}\left(\frac{2}{5}\right)^k\left(x - \frac{1}{2}\right)^k$$

$$= -\sum_{k=0}^{\infty}\left(\frac{2}{5}\right)^{k+1}\left(x - \frac{1}{2}\right)^k$$

Supplementary Problems, page 665

1. $\displaystyle\lim_{n \to \infty}\frac{2^n}{n^2+1} = \lim_{n \to \infty}\frac{\ln 2(2^n)}{2n}$ *l'Hôpital's rule*

$$= \lim_{n \to \infty}\frac{\ln 2^2(2^n)}{2}$$

$$= \infty$$

Thus, $\displaystyle\lim_{n \to \infty}\frac{(-2)^n}{n^2+1}$ does not exits.

3. $\displaystyle\lim_{n \to \infty}\left(1 - \frac{2}{n}\right)^n = \lim_{n \to \infty}\left[\left(1 + \frac{-2}{n}\right)^{-n/2}\right]^{-2}$

$$= e^{-2}$$

The sequence converges.

5. $\displaystyle\lim_{n \to \infty}\frac{n + (-1)^n}{n} = \lim_{n \to \infty}\left[1 + (-1)^n n^{-1}\right]$

$$= 1 + 0$$

$$= 1$$

The sequence converges.

7. $\displaystyle\lim_{n \to \infty}\frac{5n^4 - n^2 - 700}{3n^4 - 10n^2 + 1}$

$$= \lim_{n \to \infty}\frac{5 - n^{-2} - 700n^{-4}}{3 - 10n^{-2} + n^{-4}}$$

$$= \frac{5}{3}$$

The sequence converges.

9. $\displaystyle\lim_{n \to \infty}\left(\sqrt{n+1} - \sqrt{n}\right)$

$$= \lim_{n \to \infty}\frac{\left(\sqrt{n+1} - \sqrt{n}\right)\left(\sqrt{n+1} + \sqrt{n}\right)}{\sqrt{n+1} + \sqrt{n}}$$

$$= \lim_{n \to \infty}\frac{n+1-n}{\sqrt{n+1} + \sqrt{n}}$$

$$= 0$$

The sequence converges.

11. $\displaystyle\lim_{n \to \infty}\frac{\ln n}{n} = \lim_{n \to \infty}\frac{1}{n}$ *l'Hôpital's rule*

$$= 0$$

The sequence converges.

13. $\lim\limits_{n\to\infty}\left(1+\dfrac{4}{n}\right)^n = \lim\limits_{n\to\infty}\left[\left(1+\dfrac{1}{\frac{n}{4}}\right)^{n/4}\right]^4$

$= e^4$

The sequence converges.

15. $\lim\limits_{n\to\infty}\dfrac{n^{3/4}\sin n^2}{n+4} = \lim\limits_{n\to\infty}\dfrac{\sin n^2}{n^{1/4}+4n^{-3/4}} = 0$

since $|\sin n^2| \le 1$ while $n^{1/4}$ becomes infinite and $n^{-3/4}\to 0$. The sequence converges.

17. $\displaystyle\sum_{k=-123,456,788}^{123,456,789}\dfrac{k}{370,370,367} = \dfrac{123,456,789}{370,370,367}$

$= \dfrac{1}{3}$

The series converges.

SURVIVAL HINT: *Note that a finite series always converges.*

---

19. $\displaystyle\sum_{k=1}^{\infty}\left(\dfrac{e}{3}\right)^k = \left(\dfrac{e}{3}\right)\dfrac{1}{1-\frac{e}{3}} = \dfrac{e}{3-e}$

21. $S = \displaystyle\sum_{k=1}^{\infty}\dfrac{1}{4k^2-1}$

$S_n = \displaystyle\sum_{k=1}^{n}\dfrac{1}{4k^2-1}$

$= \displaystyle\sum_{k=1}^{\infty}\left[\dfrac{\frac{1}{2}}{2k-1} - \dfrac{\frac{1}{2}}{2k+1}\right]$

$= \dfrac{1}{2}\left[\left(\dfrac{1}{1}-\dfrac{1}{3}\right)+\left(\dfrac{1}{3}-\dfrac{1}{5}\right)+\cdots\right.$

$\left.+\left(\dfrac{1}{2n-1}-\dfrac{1}{2n+1}\right)\right]$

$= \dfrac{1}{2}\left(1-\dfrac{1}{2n+1}\right)$

$S = \lim\limits_{n\to\infty}S_n$

$= \dfrac{1}{2}\lim\limits_{n\to\infty}\left(1-\dfrac{1}{2n+1}\right) = \dfrac{1}{2}$

23. $\displaystyle\sum_{k=0}^{\infty}\dfrac{e^k+3^{k-1}}{6^{k+1}}$

$= \displaystyle\sum_{k=0}^{\infty}\left[\dfrac{1}{6}\left(\dfrac{e}{6}\right)^k + \dfrac{1}{36}\left(\dfrac{3}{6}\right)^{k-1}\right]$

$= \dfrac{\frac{1}{6}}{1-\frac{e}{6}} + \dfrac{1}{36}\left[2+1+\dfrac{1}{2}+\cdots+\left(\dfrac{1}{2}\right)^{k-1}+\cdots\right]$

$= \dfrac{1}{6-e} + \dfrac{\frac{1}{18}}{1-\frac{1}{2}}$

$= \dfrac{1}{6-e} + \dfrac{1}{9} = \dfrac{15-e}{9(6-e)}$

25. $S = \displaystyle\sum_{k=1}^{\infty}\dfrac{k}{(k+1)(k+2)(k+3)}$

$S_n = \dfrac{1}{2}\displaystyle\sum_{k=1}^{n}\left(\dfrac{-1}{k+1}+\dfrac{4}{k+2}+\dfrac{-3}{k+3}\right)$

$= \dfrac{1}{2}\left[\left(-\dfrac{1}{2}+\dfrac{4}{3}-\dfrac{3}{4}\right)+\left(-\dfrac{1}{3}+\dfrac{4}{4}-\dfrac{3}{5}\right)\right.$

$+\left(-\dfrac{1}{4}+\dfrac{4}{5}-\dfrac{3}{6}\right)+\cdots$

$\left.+\left(\dfrac{-1}{n+1}+\dfrac{4}{n+2}-\dfrac{3}{n+3}\right)\right]$

$= \dfrac{1}{2}\left[-\dfrac{1}{2}-\dfrac{1}{3}-\left(\dfrac{1}{4}+\dfrac{1}{5}+\cdots+\dfrac{1}{n+1}\right)\right.$

$+\dfrac{4}{3}+4\left(\dfrac{1}{4}+\dfrac{1}{5}+\cdots+\dfrac{1}{n+1}\right)$

$+\dfrac{4}{n+2}-3\left(\dfrac{1}{4}+\dfrac{1}{5}+\cdots+\dfrac{1}{n+1}\right)$

$\left.-3\left(\dfrac{1}{n+2}+\dfrac{1}{n+3}\right)\right]$

$= \dfrac{1}{2}\left[\left(-\dfrac{1}{2}+\dfrac{4}{3}-\dfrac{1}{3}\right)\right.$

$\left.+\left(\dfrac{-3}{n+2}+\dfrac{1}{n+2}-\dfrac{3}{n+3}\right)\right]$

$= \dfrac{1}{2}\left[\dfrac{1}{2}+\dfrac{1}{n+2}-\dfrac{3}{n+3}\right]$

$$S = \frac{1}{2} \lim_{n \to \infty} \left[\frac{1}{2} + \frac{1}{n+2} - \frac{3}{n+3} \right]$$

$$= \frac{1}{4}$$

27. $\displaystyle\sum_{k=1}^{\infty} \frac{1}{\sqrt{k^2+4}}$; use limit comparison with

$\displaystyle\sum_{k=1}^{\infty} \frac{1}{k}$ which diverges.

29. $\displaystyle\sum_{k=1}^{\infty} \frac{1}{2k-1}$; use limit comparison with

$\displaystyle\sum_{k=1}^{\infty} \frac{1}{k}$ which diverges.

31. $\displaystyle\sum_{k=1}^{\infty} \frac{k^3}{k!}$; use the ratio test.

$$\lim_{k \to \infty} \frac{(k+1)^3 k!}{(k+1)! k^3} = \lim_{k \to \infty} \frac{(k+1)^2}{k^3} = 0 < 1$$

The series converges.

33. $\displaystyle\sum_{k=1}^{\infty} \frac{k}{(2k-1)!}$; use the ratio test.

$$\lim_{k \to \infty} \frac{(k+1)(2k-1)!}{(2k+1)! k} = \lim_{k \to \infty} \frac{k+1}{k(2k+1)(2k)}$$
$$= 0 < 1$$

The series converges.

35. $\displaystyle\sum_{k=0}^{\infty} \frac{k^2}{3^k}$; use the ratio test.

$$\lim_{k \to \infty} \frac{(k+1)^2 3^k}{3^{k+1} k^2} = \frac{1}{3} < 1$$

The series converges.

37. $\displaystyle\sum_{k=2}^{\infty} \frac{1}{k(\ln k)^2}$; use integral test.

$$I = \lim_{b \to \infty} \int_2^b (\ln x)^{-2} \frac{dx}{x}$$

$$= \lim_{b \to \infty} \frac{-1}{\ln x} \Big|_2^b$$

$$= \frac{1}{\ln 2}$$

Since I converges, so does the series.

39. $\displaystyle\sum_{k=1}^{\infty} \frac{1}{\sqrt{k(k+1)}}$ behaves like $\displaystyle\sum_{k=1}^{\infty} \frac{1}{k}$ which diverges.

41. $\displaystyle\sum_{k=1}^{\infty} \frac{k!}{k^2(k+1)^2}$; use the ratio test.

$$\lim_{k \to \infty} \frac{(k+1)! k^2 (k+1)^2}{(k+1)^2 (k+2)^2 k!} = \lim_{k \to \infty} \frac{(k+1)k^2}{(k+2)^2}$$
$$= \infty > 1$$

The series diverges.

43. $\displaystyle\sum_{k=1}^{\infty} \frac{k^2 3^k}{k!}$; use the ratio test.

$$\lim_{k \to \infty} \frac{(k+1)^2 3^{k+1} k!}{(k+1)! k^2 3^k} = 3 \lim_{k \to \infty} \frac{k+1}{k^2} = 0 < 1$$

The series converges.

45. $\displaystyle\sum_{k=1}^{\infty} \frac{k}{k^2+1}$; use the integral test.

$$I = \lim_{n \to \infty} \int_1^n \frac{x \, dx}{x^2+1}$$

$$= \lim_{n \to \infty} \frac{1}{2} \ln(x^2+1) \Big|_1^n$$

$$= \lim_{n \to \infty} \frac{1}{2} \left[\ln(n^2+1) - \ln 2 \right]$$

$$= \infty$$

The series diverges.

47. $\dfrac{1}{1 \cdot 2} - \dfrac{1}{2 \cdot 3} + \dfrac{1}{3 \cdot 4} - \dfrac{1}{4 \cdot 5} + \cdots$

$= \displaystyle\sum_{k=1}^{\infty} \dfrac{(-1)^{k+1}}{k(k+1)}$

The terms of $\left\{ \dfrac{1}{k(k+1)} \right\}$ are decreasing

$\left(\text{because } \dfrac{1}{(k+1)(k+2)} < \dfrac{1}{k(k+1)} \right)$

and

$$\lim_{k \to \infty} \dfrac{k}{k(k+1)} = 0$$

so the series converges by the alternating

series test. Since $\displaystyle\sum_{k=0}^{\infty} \dfrac{1}{k(k+1)}$ converges

(compare with the p-series with $p = 2$), the
given series converges absolutely.

49. $1 - \dfrac{1}{3} + \dfrac{1}{9} - \dfrac{1}{27} + \dfrac{1}{81} - \cdots = \displaystyle\sum_{k=0}^{\infty} \dfrac{(-1)^k}{3^k}$

This series converges because it is a
geometric series with ratio $|r| = \frac{1}{3} < 1$; we
conclude the series converges absolutely.

51. $-1 + \dfrac{1}{\sqrt{2}} - \dfrac{1}{\sqrt[3]{3}} + \dfrac{1}{\sqrt[4]{4}} - \cdots = \displaystyle\sum_{k=1}^{\infty} \dfrac{(-1)^k}{\sqrt[k]{k}}$

Diverges by the divergence test since

$$\lim_{k \to \infty} \dfrac{1}{\sqrt[k]{k}} = 1 \neq 0$$

53. $\displaystyle\sum_{k=1}^{\infty} (-1)^k \left(\dfrac{3k + 85}{4k + 1} \right)^k$; compare with

$\displaystyle\sum_{k=1}^{\infty} \left(\dfrac{3k + 85}{4k + 1} \right)^k$ which behaves like

$\displaystyle\sum_{k=1}^{\infty} \left(\dfrac{3}{4} \right)^k$ which is a convergent geometric

series ($r = \frac{3}{4} < 1$). Thus, the given series
converges absolutely.

55. $\displaystyle\sum_{k=1}^{\infty} \dfrac{(-1)^{k(k+1)/2}}{2^k}$ converges absolutely since

$\displaystyle\sum_{k=1}^{\infty} \dfrac{1}{2^k}$ is a convergent geometric series.

57. $\displaystyle\lim_{k \to \infty} \left| \dfrac{(k+1)(x-1)^{k+1}}{k(x-1)^k} \right| = |x - 1|$

$|x - 1| < 1$ if $0 < x < 2$. At $x = 0$ and at
$x = 2$, the series diverges by the divergence
test. The interval of convergence is $(0, 2)$.

59. $\displaystyle\lim_{k \to \infty} \left| \dfrac{\frac{\ln(k+1)(x^2)^{k+1}}{\sqrt{k+1}}}{\frac{\ln k (x^2)^k}{\sqrt{k}}} \right|$

$= \displaystyle\lim_{x \to \infty} \left| \dfrac{\ln(k+1) x^{2k+2} \sqrt{k}}{\sqrt{k+1}(\ln k) x^{2k}} \right|$

$= x^2$

$x^2 < 1$ if $-1 < x < 1$. At $x = \pm 1$ (note
$x^2 = 1$), the series diverges because

$$\sum_{k=1}^{\infty} \dfrac{\ln k}{\sqrt{k}} \geq \sum_{k=1}^{\infty} \dfrac{1}{k^{1/2}}$$

The interval of convergence is $(-1, 1)$.

61. Use the root test:

$$\lim_{k \to \infty} \sqrt[k]{\left| \dfrac{(-1)^k x^k}{k^k} \right|} = \lim_{k \to \infty} \dfrac{|x|}{k} = 0 < 1$$

The series converges absolutely for all x.

63. $\displaystyle\lim_{k \to \infty} \left| \dfrac{(x+2)^{k+1} k \ln(k+1)}{(k+1)\ln(k+2)(x+2)^k} \right|$

$= \displaystyle\lim_{k \to \infty} \dfrac{k \ln(k+1)}{(k+1)\ln(k+2)} |x + 2|$

$= |x + 2| < 1$

if $-3 < x < -1$.

Endpoints:

$x = -1$: $\displaystyle\sum_{k=1}^{\infty} \frac{1}{k \ln(k+1)}$ diverges by the integral test:

$$\int_{1}^{\infty} [\ln(x+1)]^{-1} \frac{dx}{x} = \infty$$

$x = -3$: $\displaystyle\sum_{k=1}^{\infty} \frac{(-1)^k}{k \ln(k+1)}$ converges by the alternating series test:

$$\lim_{k\to\infty} \frac{1}{k \ln(k+1)} = 0$$

and

$$f(x) = \frac{1}{x \ln(x+1)}$$

f decreases because if you increase the denominator then you decrease fractions with constant numerators. The interval of convergence is $[-3, -1)$.

65. $\displaystyle\lim_{k\to\infty} \left| \frac{(3k+3)! x^{k+1} k!}{(k+1)!(3k)! x^k} \right|$

$= \displaystyle\lim_{k\to\infty} [3(3k+2)(3k+1)]|x|$

$= \infty$

The interval of convergence is reduced to a single point, $x = 0$.

67. $\displaystyle\lim_{k\to\infty} \left| \frac{x^{2k+1}(2k-1)!}{(2k+1)! x^{2k-1}} \right| = \lim_{k\to\infty} \frac{|x|^2}{(2k+1)(2k)}$

$= 0$

The series converges for all x.

69. $f(x) = x^3 \sin x$

$= x^3 \displaystyle\sum_{k=0}^{\infty} \frac{(-1)^k x^{2k+1}}{(2k+1)!}$

$= \displaystyle\sum_{k=0}^{\infty} \frac{(-1)^k x^{2k+4}}{(2k+1)!}$

71. $f(x) = x^2 + \tan^{-1} x$

$= x^2 + \left[x - \dfrac{x^3}{3} + \dfrac{x^5}{5} - \dfrac{x^7}{7} + \cdots \right]$

$= x + x^2 + \displaystyle\sum_{k=1}^{\infty} \frac{(-1)^k x^{2k+1}}{2k+1}$

73. $f(x) = \dfrac{11x - 1}{2 + x - 3x^2}$

$= \dfrac{11x - 1}{(2 + 3x)(1 - x)}$

$= \dfrac{-5}{3x + 2} + \dfrac{2}{1 - x}$

$= -\dfrac{5}{2} \displaystyle\sum_{k=0}^{\infty} (-1)^k \left(\frac{3}{2}x \right)^k + 2 \displaystyle\sum_{k=0}^{\infty} x^k$

$= \displaystyle\sum_{k=0}^{\infty} \left[2 + \frac{5(-1)^{k+1}}{2} \left(\frac{3}{2} \right)^k \right] x^k$

75. $|S - S_8| \leq a_9 = \dfrac{8}{9!} \approx 0.000022$

77. See Problem 31, Section 8.8

$\tan x = x + \dfrac{x^3}{3} + \dfrac{2x^5}{15} + \cdots$

$\displaystyle\int \tan x \, dx = \int \left[x + \dfrac{x^3}{3} + \dfrac{2x^5}{15} + \cdots \right] dx$

$\ln|\cos x| = -\dfrac{1}{2} x^2 - \dfrac{1}{12} x^4 - \dfrac{1}{45} x^6 + \cdots$

79. **a.** By looking for patterns, it looks like the a_k term should be

$a_k = (-1)^k (4k+1) \left[\dfrac{1 \cdot 3 \cdot 5 \cdots (2k-1)}{2 \cdot 4 \cdot 6 \cdots (2k)} \right]^3$

There is an interesting discussion of this identity, as well as biographical information on Ramanujan, in the article, "Ramanujan the phenomenon," by S. G.Gindikin, *Quantum*, pp. 4-9, April, 1998.

b. By calculator

$$1 - 5\left(\frac{1}{2}\right)^3 + 9\left(\frac{1\cdot 3}{2\cdot 4}\right)^3 - 13\left(\frac{1\cdot 3\cdot 5}{2\cdot 4\cdot 6}\right)^3 + \cdots$$
$$\approx 0.4528808594$$

and

$$\frac{2}{\pi} \approx 0.6366197724$$

It looks like we need more terms. Using a spreadsheet for $k = 20$ we find the sum is approximately 0.690061.

81. Let $A = \sum_{k=1}^{\infty} a_k, \quad A_n = \sum_{k=1}^{n} a_k,$

$$B = \sum_{k=1}^{\infty} b_k, \quad B_n = \sum_{k=1}^{n} B_k,$$

Given $\epsilon > 0$,

$|A - A_n| < \dfrac{\epsilon}{2}$ for $n > N_1$

$|B - B_n| < \dfrac{\epsilon}{2}$ for $n > N_2$

$|(A + B) - (A_n + B_n)|$
$\quad = |(A - A_n) + (B - B_n)|$
$\quad \leq |A - A_n| + |B - B_n|$
$\quad < \epsilon$

which is true if $n > \max(N_1, N_2)$.

83. a. $\sum_{k=1}^{n}(a_k - a_{k+2})$

$= (a_1 - a_3) + (a_2 - a_4) + (a_3 - a_5)$
$\quad + \cdots + (a_{n-1} - a_{n+1}) + (a_n - a_{n+2})$

$= a_1 - a_3 + a_2 - a_4 + a_3 - a_5 + \cdots$
$\quad + a_{n-2} - a_n + a_{n-1} - a_{n+1} + a_n - a_{n+2}$
$= (a_1 + a_2) + 0 + \cdots + 0 - (a_{n+1} + a_{n+2})$

b. $\sum_{k=1}^{\infty}(a_k - a_{k+2})$

$= \lim_{n\to\infty}\sum_{k=1}^{n}(a_k - a_{k+2})$

$= a_1 + a_2 - \lim_{n\to\infty} a_{n+1} - \lim_{n\to\infty} a_{n+2}$

$= a_1 + a_2 - A - A$

$= a_1 + a_2 - 2A$

c. Let $a_k = k^{1/k}$ and $L = \lim_{k\to\infty} k^{1/k}$.

Then

$$\ln L = \lim_{k\to\infty}\frac{\ln k}{k} = \lim_{k\to\infty}\frac{\frac{1}{k}}{1} = 0$$

so

$$A = \lim_{k\to\infty} a_k = \lim_{k\to\infty} k^{1/k} = e^0 = 1$$

and by part **b.**

$$\sum_{k=1}^{\infty}(a_k - a_{k+2}) = a_1 + a_2 - 2A$$
$$= 1 + 2^{1/2} - 2(1)$$
$$= \sqrt{2} - 1$$

d. $\sum_{k=2}^{\infty}\dfrac{1}{k^2 - 1} = \sum_{k=1}^{\infty}\dfrac{1}{k(k+2)}$

$$= \frac{1}{2}\lim_{n\to\infty}\sum_{k=1}^{n}\left[\frac{1}{k} - \frac{1}{k+2}\right]$$

Let $a_k = \dfrac{1}{k}$, so $\lim_{k\to\infty} a_k = 0$ and

$$\sum_{k=2}^{\infty} \frac{1}{k^2 - 1} = \frac{1}{2}\left[1 + \frac{1}{2} - 2(0)\right] = \frac{3}{4}$$

85. $f(x) = \cos x$
$f'(x) = -\sin x$
$f''(x) = -\cos x$
$f'''(x) = \sin x$
$\vdots$

$\cos x = \cos c - (x - c)\sin c$
$\qquad - \frac{1}{2!}(\cos c)(x - c)^2$
$\qquad + \frac{1}{3!}(\sin c)(x - c)^3 + \cdots$

87. $\sqrt{x^3 + 1} = \left(1 + x^3\right)^{1/2}$

$\qquad = 1 + \frac{1}{2}x^3 + \frac{\frac{1}{2}\left(-\frac{1}{2}\right)}{2!}\left(x^3\right)^2 + \cdots$

$\qquad = 1 + \frac{1}{2}x^3 - \frac{1}{8}x^6 + \frac{1}{16}x^9 + \cdots$

$\int \sqrt{x^3 + 1}\, dx$

$\quad = \int \left[1 + \frac{1}{2}x^3 - \frac{1}{8}x^6 + \frac{1}{16}x^9 + \cdots\right] dx$

$\quad = x + \frac{1}{8}x^4 - \frac{1}{56}x^7 + \frac{1}{160}x^{10} + \cdots$

89. $f(x) = a^x,\ f(0) = 1;$
$f'(x) = a^x \ln a,\ f'(0) = \ln a$
$f''(x) = a^x(\ln a)^2;\ f''(0) = (\ln a)^2$
$a^x = 1 + (\ln a)x + \frac{1}{2!}(\ln a)^2 x^2 + \frac{1}{3!}(\ln a)^3 x^3 + \cdots$

$\quad = \sum_{k-0}^{\infty} \frac{1}{k!}(\ln a)^k x^k$

91. $\quad \sin u = u - \frac{1}{3!}u^3 + \frac{1}{5!}u^5 - \cdots$

$\quad \sin 0.2 = 0.2 - \frac{1}{3!}(0.2)^3 + \frac{1}{5!}(0.2)^5 - \cdots$

$\qquad \approx 0.1986693$

with error no greater than

$$\frac{1}{7!}(0.2)^7 \approx 2.5 \times 10^{-9}$$

93. $e^{2x} \approx 1 + 2x + \frac{1}{2!}(2x)^2 + \frac{1}{3!}(2x)^3$

$\quad R_3(x) = \frac{2^4 x^4}{4!}$

$\qquad \leq \frac{2}{3}\left(10^{-3}\right)^4$

$\qquad = 6.6\left(10^{-13}\right)$

$\qquad < 10^{-12}$

95. $\cos x = 1 - \frac{1}{2!}x^2 + R_3(x)$

$\quad |R_3(x)| \leq \frac{1}{4!}x^4 \leq 0.00005$

$\quad$ so $|x| < 0.187.$

97. This is Putnam Problem A2 in the morning session of 1984. Let S_n denote the nth partial sum of the series. Then

$$S_n = \sum_{k=1}^{n} \frac{6^k}{(3^{k+1} - 2^{k+1})(3^k - 2^k)}$$

$$= \sum_{k=1}^{n} \left[\frac{3^k}{3^k - 2^k} - \frac{3^{k+1}}{3^{k+1} - 2^{k+1}}\right]$$

$$= 3 - \frac{3^{n+1}}{3^{n+1} - 2^{n+1}}$$

and the series converges to

$$\lim_{n \to \infty} S_n = 3 - 1 = 2$$

99. This is Putnam Problem A-2 in the morning session of 1982. Since $x = \log_n 2 > 0,$

$$B_n(x) = 1^x + 2^x + 3^x + \cdots + n^x$$
$$\leq n(n^x)$$

and

$$0 \le \frac{B_n(\log_n 2)}{(n\log_2 n)^2} \le \frac{n\left(n^{\log_n 2}\right)}{(n\log_2 n)^2}$$

As

$$\sum_{n=2}^{\infty} \frac{2}{n(\log_2 n)^2}$$

converges by the integral test, the given series converges by the comparison test.

Cumulative Review
Chapters 6-8, page 671

1. a.   The work done by the variable force $F(x)$ in moving an object along the x-axis from $x = a$ to $x = b$ is given by

$$W = \int_a^b F(x)\,dx$$

   b.   Suppose a flat surface (a plate) is submerged vertically in a fluid of weight density ρ (lb/ft^3) and that the submerged portion of the plate extends from $x = a$ to $x = b$ on a vertical axis. Then the total force F exerted by the fluid is given by

$$F = \int_a^b \rho\, h(x) L(x)\,dx$$

   where $h(x)$ is the depth at x and $L(x)$ is the corresponding length of a typical horizontal approximating strip.

3. Let $f(x) = P(x)/D(x)$, where $P(x)$ and $D(x)$ are polynomials with no common factors and $D(x) \ne 0$.
   Step 1 If the degree of P is greater than or equal to the degree of D, use long (or synthetic) division to express $\dfrac{P(x)}{D(x)}$ as the sum of a polynomial and a fraction $\dfrac{R(x)}{D(x)}$ in which the degree of the remainder polynomial $R(x)$ is less than the degree of the denominator polynomial $D(x)$.
   Step 2 Factor the denominator $D(x)$ into the product of linear and irreducible quadratic powers.
   Step 3 Express $\dfrac{P(x)}{D(x)}$ as a cascading sum of partial fractions of the form

$$\frac{A_i}{(x-r)^n} \quad \text{and} \quad \frac{A_k x + B_k}{(x^2 + sx + t)^m}$$

5. First, find the interval of convergence for the series and then check the convergence at each of the endpoints.

7. $\displaystyle \int \ln\sqrt{x}\,dx = \frac{1}{2}\int \ln x\,dx$ $\boxed{u = \ln x,\ du = \frac{1}{x}dx}$

$$= \frac{1}{2}\left[x\ln x - \int x\left(\frac{1}{x}\right)dx\right]$$

$$= \frac{1}{2}x\ln x - \frac{1}{2}x + C$$

9. $\displaystyle \int \frac{\tan^{-1}x}{1+x^2}\,dx = \int u\,du$

$$\boxed{u = \tan^{-1}x,\ du = (1+x^2)^{-1}dx}$$

$$= \frac{1}{2}u^2 + C$$

$$= \frac{1}{2}\left(\tan^{-1}x\right)^2 + C$$

11. $\displaystyle \int \sqrt{4 - x^2}\,dx$

$$= \int \sqrt{4 - 4\sin^2\theta}\,(2\cos\theta\,d\theta)$$

$$\boxed{x = 2\sin\theta,\ dx = 2\cos\theta\,d\theta}$$

$$= \int 4\sqrt{1 - \sin^2\theta}\,\cos\theta\,d\theta$$

$$= \int 4\cos^2\theta\,d\theta$$

$$= 4\int \frac{1 + \cos 2\theta}{2}\,d\theta$$

$$= 2\theta + \sin 2\theta + C$$

$$= 2\sin^{-1}\left(\frac{x}{2}\right) + 2\left(\frac{x}{2}\right)\left(\frac{1}{2}\sqrt{4 - x^2}\right) + C$$

$$= 2\sin^{-1}\left(\frac{x}{2}\right) + \frac{1}{2}x\sqrt{4 - x^2} + C$$

13. $\displaystyle\int \frac{(3 + 2\sin t)}{\cos t}\,dt$

$$= \int 3\sec t\,dt + \int 2\tan t\,dt$$

$$= 3\ln|\sec t + \tan t| + 2\ln|\sec t| + C$$

15. $\displaystyle\int \frac{dx}{x^2 + x + 1}$

$$= \int \frac{dx}{\left(x + \frac{1}{2}\right)^2 + \frac{3}{4}}$$

$$= \int \frac{du}{u^2 + \frac{3}{4}}$$

$$= \frac{2}{\sqrt{3}}\tan^{-1}\frac{2u}{\sqrt{3}} + C$$

$$= \frac{2}{\sqrt{3}}\tan^{-1}\frac{2}{\sqrt{3}}\left(x - \frac{1}{2}\right) + C$$

$$= \frac{2}{\sqrt{3}}\tan^{-1}\left(\frac{2x + 1}{\sqrt{3}}\right) + C$$

17. $\displaystyle\int \cos^{-1}(-x)\,dx$ $\boxed{\text{Parts or Formula 168}}$

$$= -\left[(-x)\cos^{-1}(-x) - \sqrt{1 - x^2}\right] + C$$

$$= x\cos^{-1}(-x) + \sqrt{1 - x^2} + C$$

19. $\displaystyle\int \frac{e^{-x} - e^x}{e^{2x} + e^{-2x} + 2}\,dx$

$\boxed{u = e^x + e^{-x},\ du = (e^x - e^{-x})\,dx}$

$$= \int \frac{(-1)du}{u^2}$$

$$= \frac{1}{u} + C$$

$$= \frac{1}{e^x + e^{-x}} + C$$

21. $\displaystyle\int_0^\infty xe^{-2x}\,dx = \lim_{N\to\infty}\int_0^N xe^{-2x}\,dx$

$$= \lim_{N\to\infty}\left[-e^{-2x}\left(\frac{x}{2} + \frac{1}{4}\right)\right]\Big|_0^N$$

$$= \lim_{N\to\infty}\left[-e^{-2N}\left(\frac{N}{2} + \frac{1}{4}\right) + \frac{1}{4}\right]$$

$$= \frac{1}{4}$$

23. $\displaystyle\int_0^\infty x^2 e^{-x}\,dx$ $\boxed{\text{Parts or Formula 185}}$

$$= \lim_{N\to\infty}\frac{e^{-x}}{-1}\left(x^2 - \frac{2x}{-1} + \frac{2}{1}\right)\Big|_0^N$$

$$= \lim_{N\to\infty}\left[2 - e^{-N}\left(N^2 + 2N + 2\right)\right]$$

$$= 2$$

25. a.

$$\frac{x^2 + 17x - 8}{(2x + 3)(x - 1)^2} = \frac{A}{x - 1} + \frac{B}{(x - 1)^2} + \frac{C}{2x + 3}$$

$$x^2 + 17x - 8 = A(x - 1)(2x + 3) + B(2x + 3) + C(x - 1)^2$$

Thus,

$$\begin{cases} 2A + C = 1 \\ A + 2B - 2C = 17 \\ -3A + 3B + C = -8 \end{cases}$$

Solving, we obtain $A = 3$, $B = 2$, and $C = -5$.

b. $\displaystyle\int \frac{x^2 + 17x - 8}{(2x + 3)(x - 1)^2}dx$

$\displaystyle= \int\left[\frac{3}{x - 1} + \frac{2}{(x - 1)^2} + \frac{-5}{2x + 3}\right]dx$

$\displaystyle= 3\ln|x - 1| - \frac{2}{x - 1} - \frac{5}{2}\ln|2x + 3| + C$

c. $\displaystyle\int_2^\infty \frac{x^2 + 17x - 8}{(2x + 3)(x - 1)^2}dx$

$\displaystyle= \lim_{N\to\infty}\left[3\ln|x - 1| - \frac{2}{x - 1} - \frac{5}{2}\ln|2x + 3|\right]\Big|_2^N$

$= \infty$

The integral diverges because two of the three pieces diverge.

27. $\displaystyle\sum_{k=0}^\infty \frac{k^2 3^{k+1}}{4^k}$; use the ratio test.

$\displaystyle\lim_{k\to\infty}\frac{\frac{(k+1)^2 3^{k+2}}{4^{k+1}}}{\frac{k^2 3^{k+1}}{4^k}} = \lim_{k\to\infty}\frac{(k+1)^2 3^{k+2} 4^k}{k^2 3^{k+1} 4^{k+1}}$

$\displaystyle= \lim_{k\to\infty}\frac{(k+1)^2(3)}{k^2(4)}$

$\displaystyle= \frac{3}{4} < 1$

The series converges.

29. $\displaystyle\sum_{k=1}^\infty \frac{(-1)^{k+1}k}{k + 1}$ diverges by the divergence

test: $\displaystyle\lim_{k\to\infty}\frac{(-1)^{k+1}k}{k + 1} \neq 0$

31. $\displaystyle\sum_{k=1}^\infty ke^{-k^2}$; use the integral test.

$\displaystyle\int_0^\infty xe^{-x^2}dx = \lim_{t\to\infty}\left[-\frac{1}{2}e^{-t^2} + \frac{1}{2}\right]$

$\displaystyle= \frac{1}{2}$

The series converges.

33. $\displaystyle\sum_{k=1}^\infty 5\left(\frac{-2}{3}\right)^{2k-1}$

$\displaystyle= 5\left[\frac{-2}{3} + \left(\frac{-2}{3}\right)^3 + \left(\frac{-2}{3}\right)^5 + \cdots\right]$

$\displaystyle= -\frac{10}{3}\left[1 + \left(\frac{2}{3}\right)^2 + \left(\frac{2}{3}\right)^4 + \cdots\right]$

$\displaystyle= \frac{-10}{3}\left[\frac{1}{1 - \frac{4}{9}}\right]$

$= -6$

35. $\displaystyle\lim_{k\to\infty}\left|\frac{\frac{x^{k+1}}{(k+1)^3}}{\frac{x^k}{k^3}}\right| = \lim_{k\to\infty}\left|\frac{k^3}{(k+1)^3}\right||x|$

$= |x| < 1$

Endpoints:

$x = 1$: $\displaystyle\sum_{k=1}^\infty \frac{1}{k^3}$ converges (*p*-series with

$p = 3 > 1$).

$x = -1$: $\displaystyle\sum_{k=1}^\infty \frac{(-1)^k}{k^3}$ converges absolutely.

Convergence set is $-1 \leq x \leq 1$.

37. $\displaystyle y' - \frac{y}{x} = \frac{x}{1 + x^2}$

$\displaystyle u = e^{\int(-1/x)dx} = e^{-\ln x} = \frac{1}{x}$

$\displaystyle y = \frac{1}{\frac{1}{x}}\left[\int \frac{1}{x}\left(\frac{x}{1 + x^2}\right)dx + C\right]$

$\displaystyle= x\left[\tan^{-1}x + C\right]$

$= x\tan^{-1}x + Cx$

39. $f(x) = \sin x^2$

$\displaystyle= x^2 - \frac{x^6}{3!} + \frac{x^{10}}{5!} - \cdots$

$$= \sum_{k=0}^{\infty} \frac{(-1)^k x^{4k+2}}{(2k+1)!}$$

41. $f(x) = \dfrac{3x-2}{5+2x}$

$$= \frac{3}{2} - \frac{\frac{19}{10}}{1+\frac{2}{5}x} \qquad \textit{By long division}$$

$$= \frac{3}{2} - \frac{19}{10}\sum_{k=0}^{\infty}\left(-\frac{2}{5}x\right)^k$$

43. Let (x,y) measure the distant to a point on the representative strip.

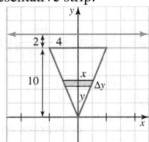

By similar triangles,

$$\frac{x}{4} = \frac{y}{10}$$

$$x = \frac{2}{5}y$$

$$\Delta F = \underbrace{62.4\,\pi x^2 \Delta y}_{\text{Force (Weight)}}\underbrace{(12-y)}_{\text{Distance}}$$

$$F = \int_0^6 64.4\pi\left(\frac{2}{3}y\right)^2 (12-y)\,dy$$

$$\approx 16{,}937 \text{ ft-lb}$$

45. a. Use disks

$$V = \int_3^8 \pi\left(\frac{1}{\sqrt{x+1}}\right)^2 dx$$

$$= \pi\int_3^8 \frac{dx}{x+1}$$

$$= \pi\ln|x+1|\big|_3^8$$

$$= \pi\ln\frac{9}{4}$$

b. Use shells

$$V = \int_3^8 2\pi x\left(\frac{1}{\sqrt{x+1}}\right) dx \quad \boxed{\begin{array}{l}u=x+1\\ du=dx\end{array}}$$

$$= 2\pi\int_4^9 \frac{u-1}{\sqrt{u}}\,du$$

$$= 2\pi\left[\frac{2}{3}u^{3/2} - 2u^{1/2}\right]\Big|_4^9$$

$$= \frac{64\pi}{3}$$

c. Use washers

$$V = \int_3^8 \pi\left[\left(\frac{1}{\sqrt{x+1}}+1\right)^2 - 1^2\right] dx$$

$$= \pi\int_3^8\left[\frac{1}{x+1} + \frac{2}{\sqrt{x+1}}\right] dx$$

$$= \pi\left[\ln|x+1| + 4\sqrt{x+1}\right]\Big|_3^8$$

$$= \pi\left(\ln\frac{9}{4} + 4\right)$$

47. $A = \displaystyle\int_0^4 \frac{2x\,dx}{\sqrt{x^2+9}} = 2\sqrt{x^2+9}\,\Big|_0^4 = 4$

49. $S_n = \displaystyle\sum_{k=0}^{n} \frac{1}{(a+k)(a+k+1)}$

$$= \sum_{k=0}^{n}\left(\frac{1}{a+k} - \frac{1}{a+k+1}\right)$$

$$= \left(\frac{1}{a+0} - \frac{1}{a+1}\right) + \left(\frac{1}{a+1} - \frac{1}{a+2}\right)$$
$$+ \cdots + \left(\frac{1}{a+n} - \frac{1}{a+n+1}\right)$$
$$= \frac{1}{a} - \frac{1}{a+n+1}$$
$$S = \lim_{n \to 0} S_n = \frac{1}{a}$$

51. $\displaystyle\sum_{k=0}^{\infty} (a_k - b_k)^2$

$$= \sum_{k=0}^{\infty} \left(a_k^2 - 2a_k b_k + b_k^2\right)$$
$$= \sum_{k=0}^{\infty} a_k^2 - 2\sum_{k=0}^{\infty} a_k b_k + \sum_{k=0}^{\infty} b_k^2$$
$$= 4 - 2(3) + 4$$
$$= 2$$

53. False; for a counterexample, let $x_k = \dfrac{1}{k^2}$.

$$S = \sum_{k=1}^{\infty} \frac{1}{k^2} \text{ converges, but}$$
$$T = \sum_{k=1}^{\infty} \frac{1}{k} \text{ diverges.}$$

55. $s(t) = \displaystyle\int_0^{\pi/3} \left|\sin t + \sin^2 t \cos^3 t\right| dt$

$$= \int_0^{\pi/3} \left[\sin t + \sin^2 t (1 - \sin^2 t) \cos t\right] dt$$
$$= \left[-\cos t + \frac{1}{3}\sin^3 t - \frac{1}{5}\sin^5 t\right]\Big|_0^{\pi/3}$$
$$= \frac{1}{2} + \frac{11\sqrt{3}}{160}$$

57. $V = \pi \displaystyle\int_0^3 \left(x\sqrt{9 - x^2}\right)^2 dx$

$$= \pi \int_0^3 \left(9x^2 - x^4\right) dx$$
$$= \pi \left(3x^3 - \frac{x^5}{5}\right)\Big|_0^3$$
$$= \frac{162\pi}{5}$$

59. For the centroid, it is assumed that $\rho = 1$.

$$m = \int_0^1 x^2 e^{-x}\, dx$$
$$= \left[-x^2 e^{-x} - 2x e^{-x} - 2e^{-x}\right]\Big|_0^1$$
$$= 2 - 5e^{-1}$$

$$M_y = \int_0^1 x\left(x^2 e^{-x}\right) dx$$
$$= 6 - 16e^{-1}$$

$$M_x = \frac{1}{2}\int_0^1 \left(x^2 e^{-x}\right)^2 dx$$
$$= \frac{3}{8} - \frac{21e^{-2}}{8}$$

$$\overline{x} = \frac{1}{m}M_y \qquad\qquad \overline{y} = \frac{1}{m}M_x$$
$$= \frac{6 - 16e^{-1}}{2 - 5e^{-1}} \qquad = \frac{\frac{3-21e^{-2}}{8}}{2 - 5e^{-1}}$$
$$\approx 0.71 \qquad\qquad \approx 0.12$$

The centroid $(\overline{x}, \overline{y}) \approx (0.71, 0.12)$

CHAPTER 9 Vectors in the Plane and in Space

Chapter Overview
We live in a three dimensional world, and, for the most part, all of your work in this book up to now has been two-dimensional. The gateway idea for higher dimensions is the idea of a vector. In this chapter, we focus on various algebraic and geometric aspects of vector representations.

9.1 Vectors in $\mathbb{R}^3$, page 684

1.   **3.**

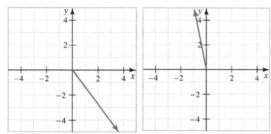

5. $\langle 4, 3 \rangle$; $\|\mathbf{PQ}\| = 5$ **7.** $\langle -5, 0 \rangle$; $\|\mathbf{PQ}\| = 5$

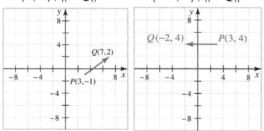

SURVIVAL HINT: *In the book we use boldface to represent vectors as in $4\mathbf{i} + 2\mathbf{j} + 3\mathbf{k}$ or $\mathbf{PQ}$ In your work, it is difficult to write boldface, so instead you should use arrow overbars instead, as in $4\overrightarrow{i} + 2\overrightarrow{j} + 3\overrightarrow{k}$ $\mathbf{k}$ or $\overrightarrow{PQ}$.*

9. $\mathbf{PQ} = (1+1)\mathbf{i} + (-2+2)\mathbf{j} = 2\mathbf{i}$
$\|\mathbf{PQ}\| = \sqrt{(1+1)^2 + (-2+2)^2} = 2$

11. $\mathbf{PQ} = (0+4)\mathbf{i} + (-1+3)\mathbf{j} = 4\mathbf{i} + 2\mathbf{j}$
$\|\mathbf{PQ}\| = \sqrt{(0+4)^2 + (-1+3)^2} = 2\sqrt{5}$
13. Let $\mathbf{v} = \mathbf{i} + \mathbf{j}$; $\|\mathbf{v}\| = \sqrt{1^2 + 1^2} = \sqrt{2}$
$\mathbf{u} = \dfrac{\mathbf{v}}{\|\mathbf{v}\|} = \dfrac{1}{\sqrt{2}}(\mathbf{i} + \mathbf{j})$
15. Let $\mathbf{v} = 3\mathbf{i} - 4\mathbf{j}$; $\|\mathbf{v}\| = \sqrt{3^2 + (-4)^2} = 5$
$\mathbf{u} = \dfrac{\mathbf{v}}{\|\mathbf{v}\|} = \dfrac{1}{5}(3\mathbf{i} - 4\mathbf{j}) = \dfrac{3}{5}\mathbf{i} - \dfrac{4}{5}\mathbf{j}$
17. $s\mathbf{u} + t\mathbf{v} = \langle -3s + t, 4s - t \rangle$ so that
$$\begin{cases} -3s + t = 6 \\ 4s - t = 0 \end{cases}$$

Solving simultaneously, $s = 6, t = 24$.
19. $s\mathbf{u} + t\mathbf{v} = \langle -3s + t, 4s - t \rangle$ so that
$$\begin{cases} -3s + t = -2 \\ 4s - t = 1 \end{cases}$$
Solving, $s = -1, t = -5$.
21. $2\mathbf{u} + 3\mathbf{v} - \mathbf{w}$
$= (6 + 12 - 1)\mathbf{i} + (-8 - 9 - 1)\mathbf{j}$
$= 17\mathbf{i} - 18\mathbf{j}$

23. $\|\mathbf{v}\| = \sqrt{4^2 + (-3)^2} = 5;$
$\|\mathbf{u}\| = \sqrt{3^2 + (-4)^2} = 5$
$\|\mathbf{v}\|\mathbf{u} + \|\mathbf{u}\|\mathbf{v}$
$= 5(3\mathbf{i} - 4\mathbf{j}) + 5(4\mathbf{i} - 3\mathbf{j})$
$= 35\mathbf{i} - 35\mathbf{j}$

25. $\begin{cases} x - y - 1 = 0 \\ 2x + 3y - 12 = 0 \end{cases}$
Solving, $x = 3, y = 2.$

27. $\begin{cases} x^2 + y^2 = 20 \\ y = x + 2 \end{cases}$
Solving simultaneously, we have $(2, 4)$ and $(-4, -2).$

29. $\mathbf{u} = (\cos 30°)\mathbf{i} + (\sin 30°)\mathbf{j}$
$= \dfrac{\sqrt{3}}{2}\mathbf{i} + \dfrac{1}{2}\mathbf{j}$

31. Let $\mathbf{v} = -4\mathbf{i} + \mathbf{j}$ so $\|\mathbf{v}\| = \sqrt{17}$; then
$\mathbf{u} = -\dfrac{1}{\sqrt{17}}(-4\mathbf{i} + \mathbf{j})$
$= \dfrac{4}{\sqrt{17}}\mathbf{i} - \dfrac{1}{\sqrt{17}}\mathbf{j}$

33. $\mathbf{u} + \mathbf{v} = 5\mathbf{i} + \mathbf{j};$ $\|\mathbf{u} + \mathbf{v}\| = \sqrt{26};$
The desired unit vector is
$\dfrac{5}{\sqrt{26}}\mathbf{i} + \dfrac{1}{\sqrt{26}}\mathbf{j}$

35. $(x_2 + 2)\mathbf{i} + (y_2 - 3)\mathbf{j} = 5\mathbf{i} + 7\mathbf{j}$, so
$x_2 + 2 = 5$ or $x_2 = 3$; and $y_2 - 3 = 7$ or
$y_2 = 10.$ The terminal point is $(3, 10).$

37. a. The midpoint of $\overline{PQ}$ is
$$M = \left(\dfrac{-3 + 9}{2}, \dfrac{-8 - 2}{2}\right) = (3, -5)$$
If the initial point of this vector is at $P,$ then its terminal point is at: $(3, -5),$ so the vector is $6\mathbf{i} + 3\mathbf{j}.$

b. $PQ = (9 + 3)\mathbf{i} + (-2 + 8)\mathbf{j}$
$= 12\mathbf{i} + 6\mathbf{j}$
Then $\frac{5}{6}\langle 12, 6 \rangle = \langle 10, 5 \rangle = 10\mathbf{i} + 5\mathbf{j}.$
This vector has initial point at $P,$ and terminal point at $(7, -3).$

39. $\|\mathbf{v}\| = \sqrt{\cos^2\theta + \sin^2\theta} = 1$

41. Not necessarily equal. Equal magnitudes say nothing about their direction.

43. a. $\|\mathbf{u} - \mathbf{u}_0\| = \sqrt{(x - x_0)^2 + (y - y_0)^2}$
$= 1$
This is the set of points on the circle with center (x_0, y_0) and radius 1.

b. $\|\mathbf{u} - \mathbf{u}_0\| \le r$ is the set of points on or interior to the circle with center (x_0, y_0) and radius $r.$

45. $a\mathbf{u} + b(\mathbf{u} - \mathbf{v}) + c(\mathbf{u} + \mathbf{v}) = \mathbf{0}$
$(a + b + c)\mathbf{u} = (b - c)\mathbf{v}$
Since $\mathbf{u}$ and $\mathbf{v}$ are not parallel, we must have
$$\begin{cases} a + b + c = 0 \\ b - c = 0 \end{cases}$$
Solving, $a = -2t$, $b = t$, and $c = t$ for any number $t.$

47. Let $\mathbf{F}_3 = a\mathbf{i} + b\mathbf{j}.$ We want
$$\mathbf{F}_1 + \mathbf{F}_2 + \mathbf{F}_3 = \mathbf{0}$$
Thus, $3\mathbf{i} + 4\mathbf{j} + 3\mathbf{i} - 7\mathbf{j} + a\mathbf{i} + b\mathbf{j} = 0\mathbf{i} + 0\mathbf{j}$
so that $a = -6$ and $b = 3$; $\mathbf{F}_3 = -6\mathbf{i} + 3\mathbf{j}$

49. The boat travels east to west, a distance of 2.1 mi in 0.5 hr, so its velocity relative to still water is $\mathbf{B} = 4.2(-\mathbf{i}).$ The velocity of the river is $\mathbf{R} = -3.1\mathbf{j}$, so the actual velocity of the boat is
$$\mathbf{v} = \mathbf{B} + \mathbf{R} = -4.2\mathbf{i} - 3.1\mathbf{j}$$
The actual speed is

$$\|\mathbf{v}\| = \sqrt{(-4.2)^2 + (-3.1)^2}$$
$$\approx 5.22 \text{ mi/h}$$

Let θ be the angle between the direction of the boat and still water. Then

$$\theta = \tan^{-1}\frac{3.1}{4.2} \approx 36.4°$$

so the direction is $90 - 36.4° = 53.6°$; that is N53.6°W.

51. Let P, Q, R, and S, be consecutive vertices in counterclockwise order of a parallelogram with diagonals $\overline{PR}$ and $\overline{QS}$. The point T is where the diagonals intersect. Then, for positive constants a and b,

$$\mathbf{PT} = a\mathbf{PR} \quad \textit{These vectors have the same direction.}$$
$$\mathbf{QT} = b\mathbf{QS} \quad \textit{These vectors have the same direction.}$$
$$\mathbf{PT} + \mathbf{TQ} = \mathbf{PQ} \quad \textit{Definition of vector addition.}$$

We see that,
$$a\mathbf{PR} - b\mathbf{QS} = \mathbf{PQ}$$

Note that
$$\mathbf{RT} = \mathbf{RP} - \mathbf{TP}$$
$$= \mathbf{RP} + \mathbf{PT}$$
$$= \mathbf{RP} + a\mathbf{PR}$$
$$= (1-a)\mathbf{RP}$$

and
$$\mathbf{TS} = \mathbf{QS} - \mathbf{QT}$$
$$= \mathbf{QS} - b\mathbf{QS}$$
$$= (1-b)\mathbf{QS}$$

Thus, we see
$$\mathbf{RS} = \mathbf{RT} + \mathbf{TS}$$
$$= (1-a)\mathbf{RP} + (1-b)\mathbf{QS}$$

Since $\mathbf{PQ} = \mathbf{SR}$, then
$$a\mathbf{PR} - b\mathbf{QS} = (1-a)\mathbf{PR} - (1-b)\mathbf{QS}$$

it follows that $a = (1-a)$ or $a = \frac{1}{2}$ and

$b = (1-b)$ or $b = \frac{1}{2}$, which means that the point T bisects the diagonals of the parallelogram.

53. a.
$$\mathbf{CN} = \mathbf{CA} + \mathbf{AN} \qquad \mathbf{BM} = \mathbf{BA} + \mathbf{AM}$$
$$= -\mathbf{AC} + \frac{1}{2}\mathbf{AB} \qquad = -\mathbf{AB} + \frac{1}{2}\mathbf{AC}$$
$$= \frac{1}{2}\mathbf{AB} - \mathbf{AC} \qquad = \frac{1}{2}\mathbf{AC} - \mathbf{AB}$$

b.
$$\mathbf{CP} = r\mathbf{CN} = r\left[\frac{1}{2}\mathbf{AB} - \mathbf{AC}\right]$$
$$\mathbf{BP} = s\mathbf{BM} = s\left[\frac{1}{2}\mathbf{AC} - \mathbf{AB}\right]$$
$$\mathbf{CB} = \mathbf{CP} + \mathbf{PB}$$
$$= \mathbf{CP} - \mathbf{BP}$$
$$= \frac{r}{2}\mathbf{AB} - r\mathbf{AC} - \frac{s}{2}\mathbf{AC} + s\mathbf{AB}$$
$$= \left(\frac{r}{2} + s\right)\mathbf{AB} + \left(-r - \frac{s}{2}\right)\mathbf{AC}$$

Since $\mathbf{CB} = \mathbf{CA} + \mathbf{AB}$ we see

$$\begin{cases} \dfrac{r}{2} + s = 1 \\ -r - \dfrac{s}{2} = -1 \end{cases}$$

Solving we find $r = \frac{2}{3}$, $s = \frac{2}{3}$. The same procedure applies to any other pair of medians.

c. Let $P(\overline{x}, \overline{y})$ be the centroid. The midpoint of $\overline{AB}$ is

$$N\left(\frac{x_1 + x_2}{2}, \frac{y_1 + y_2}{2}\right)$$

From part **b**, we have $\mathbf{CP} = \frac{2}{3}\mathbf{CN}$, so

$$\overline{x} - x_3 = \frac{2}{3}\left[\left(\frac{x_1 + x_2}{2}\right) - x_3\right]$$
$$\overline{x} = \frac{1}{3}(x_1 + x_2 + x_3)$$

and similarly,

$$\bar{y} = \frac{1}{3}(y_1 + y_2 + y_3)$$

55. a. Given $a\mathbf{u} = b\mathbf{v}$; assume $ab \neq 0$, so that
$ab = k$ or $b = \frac{k}{a}$.

$a\mathbf{u} = \frac{k}{a}\mathbf{v}$ or $\mathbf{u} = k\mathbf{v}$. This says that
$\mathbf{u}$ and $\mathbf{v}$ are parallel. This is
contrary to the assumption that they
are not parallel. Thus, we must
conclude that $a = b = 0$.

b. $a_1\mathbf{i} + b_1\mathbf{j} = a_2\mathbf{i} + b_2\mathbf{j}$
$(a_1 - a_2)\mathbf{i} = (b_1 - b_2)\mathbf{j}$

where $\mathbf{i}$ and $\mathbf{j}$ are linearly independent;
thus, $a_1 = a_2$ and $b_1 = b_2$ by part **a.**

57. a. $\mathbf{u} + \mathbf{0} = \langle u_1, u_2 \rangle + \langle 0, 0 \rangle$
$= \langle u_1 + 0, u_2 + 0 \rangle$
$= \langle u_1, u_2 \rangle$
$= \mathbf{u}$

b. $\mathbf{u} + (-\mathbf{u}) = \langle u_1, u_2 \rangle + \langle -u_1, -u_2 \rangle$
$= \langle u_1 - u_1, u_2 - u_2 \rangle$
$= \langle 0, 0 \rangle$
$= \mathbf{0}$

59. $\mathbf{AB} = \mathbf{AM} + \mathbf{MN} + \mathbf{NB}$
$\mathbf{AD} = \mathbf{AM} + \mathbf{MN} + \mathbf{ND}$
$\mathbf{CB} = \mathbf{CM} + \mathbf{MN} + \mathbf{NB}$
$\mathbf{CD} = \mathbf{CM} + \mathbf{MN} + \mathbf{ND}$
Adding, we obtain
$\mathbf{AB} + \mathbf{AD} + \mathbf{CB} + \mathbf{CD}$
$= 4\mathbf{MN} + 2(\mathbf{AM} + \mathbf{NB} + \mathbf{ND} + \mathbf{CM})$
Since M and N are the midpoints of $\overline{AC}$
and $\overline{BD}$, then $\mathbf{AM} = -\mathbf{CM}$ and
$\mathbf{NB} = -\mathbf{ND}$. Thus,
$\mathbf{AB} + \mathbf{AD} + \mathbf{CB} + \mathbf{CD} = 4\mathbf{MN} + \mathbf{0}$
$$\mathbf{MN} = \frac{1}{4}(\mathbf{AB} + \mathbf{AD} + \mathbf{CB} + \mathbf{CD})$$

9.2 Coordinates and Vectors in $\mathbb{R}^3$, page 694

1. a. $\mathbf{u} + \mathbf{v} = \langle 4 + (-2), -3 + 5, 1 + 3 \rangle$
$= \langle 2, 2, 4 \rangle$

b. $\mathbf{u} - \mathbf{v} = \langle 4 - (-2), -3 - 5, 1 - 3 \rangle$
$= \langle 6, -8, -2 \rangle$

c. $\frac{5}{2}\mathbf{u} = \frac{5}{2}\langle 4, -3, 1 \rangle$
$= \left\langle 10, -\frac{15}{2}, \frac{5}{2} \right\rangle$

d. $2\mathbf{u} + 3\mathbf{v} = 2\langle 4, -3, 1 \rangle + 3\langle -2, 5, 3 \rangle$
$= \langle 2, 9, 11 \rangle$

3. a. $\mathbf{u} + \mathbf{v} = \langle 1 + 0, -2 + (-1), 5 + 3 \rangle$
$= \langle 1, -3, 8 \rangle$

b. $\mathbf{u} - \mathbf{v} = \langle 1 - 0, -2 - (-1), 5 - 3 \rangle$
$= \langle 1, -1, 2 \rangle$

c. $\frac{5}{2}\mathbf{u} = \frac{5}{2}\langle 1, -2, 5 \rangle$
$= \left\langle \frac{5}{2}, -5, \frac{25}{2} \right\rangle$

d. $2\mathbf{u} + 3\mathbf{v} = 2\langle 1, -2, 5 \rangle + 3\langle 0, -1, 3 \rangle$
$= \langle 2, -7, 19 \rangle$

SURVIVAL HINT: *You should spend some time thinking in three dimensions. Pay attention to the drawing lessons which are designed to help you think in three dimensions. Take your time, and don't be afraid to experiment and use an eraser when your drawing "does not look right."*

5.

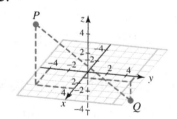

$$d = \sqrt{(1-3)^2 + (5+4)^2 + (-3-5)^2}$$
$$= \sqrt{149}$$

7.

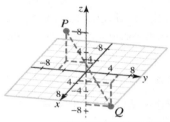

$$d = \sqrt{(3+3)^2 + (6+5)^2 + (-7-8)^2}$$
$$= \sqrt{382}$$

9. $(x-0)^2 + (y-0)^2 + (z-0)^2 = 1$
$$x^2 + y^2 + z^2 = 1$$

11. $(x-0)^2 + (y-4)^2 + (z+5)^2 = 9$
$$x^2 + (y-4)^2 + (z+5)^2 = 9$$

13. $x^2 + y^2 + z^2 - 2y + 2z - 2 = 0$
$$x^2 + (y^2 - 2y + 1^2) + (z^2 + 2z + 1^2)$$
$$= 2 + 1 + 1$$
$$x^2 + (y-1)^2 + (z+1)^2 = 4$$
$$C(0, 1, -1); \ r = 2$$

15. $x^2 + y^2 + z^2 - 6x + 2y - 2z + 10 = 0$
$$(x^2 - 6x + 3^2) + (y^2 + 2y + 1^2)$$
$$+ (z^2 - 2z + 1^2) = -10 + 9 + 1 + 1$$
$$(x-3)^2 + (y+1)^2 + (z-1)^2 = 1$$
$$C(3, -1, 1); \ r = 1$$

17.   **19.**

          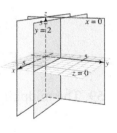

21. $\mathbf{PQ} = (-1-1)\mathbf{i} + [1-(-1)]\mathbf{j} + (4-3)\mathbf{k}$
$$= -2\mathbf{i} + 2\mathbf{j} + \mathbf{k}$$
$$\|\mathbf{PQ}\| = \sqrt{(-2)^2 + (2)^2 + (1)^2} = 3$$

23. $\mathbf{PQ} = -4\mathbf{i} - 4\mathbf{j} - 4\mathbf{k}$
$$\|\mathbf{PQ}\| = 4\sqrt{1+1+1} = 4\sqrt{3}$$

25. $\mathbf{u} + \mathbf{v} - 2\mathbf{w}$
$$= \langle 2, -1, 3 \rangle + \langle 1, 1, -5 \rangle - 2\langle 5, 0, 7 \rangle$$
$$= \langle -7, 0, -16 \rangle$$
This is $-7\mathbf{i} - 16\mathbf{k}$.

27. $4\mathbf{u} + \mathbf{w} = 4\langle 2, -1, 3 \rangle + \langle 5, 0, 7 \rangle$
$$= \langle 13, -4, 19 \rangle$$
This is $13\mathbf{i} - 4\mathbf{j} + 19\mathbf{k}$.

29. $\dfrac{\mathbf{v}}{\|\mathbf{v}\|} = \dfrac{\langle 3, -2, 1 \rangle}{\sqrt{9+4+1}}$
$$= \dfrac{1}{\sqrt{14}} \langle 3, -2, 1 \rangle$$

31. $\dfrac{\mathbf{v}}{\|\mathbf{v}\|} = \dfrac{\langle -5, 3, 4 \rangle}{\sqrt{25+9+16}}$
$$= \dfrac{1}{5\sqrt{2}} \langle -5, 3, 4 \rangle$$

33.   **35.**

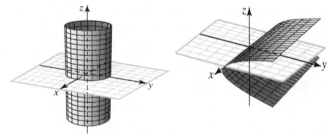

37. The center is the midpoint of $\mathbf{AB}$; namely
$M\left(-\frac{1}{2}, \frac{5}{2}, 0\right)$; the diameter is
$$\|\mathbf{AB}\| = \sqrt{(-3)^2 + 1^2 + 6^2} = \sqrt{46}$$

$r^2 = \frac{46}{4} = \frac{23}{2}$; the equation of the sphere is

$$\left(x + \frac{1}{2}\right)^2 + \left(y - \frac{5}{2}\right)^2 + z^2 = \frac{23}{2}$$
$$(2x + 1)^2 + (2y - 5)^2 + 4z^2 = 46$$

39. $\|\mathbf{i} - \mathbf{j} + \mathbf{k}\| = \sqrt{1 + 1 + 1} = \sqrt{3}$

41. $\|2(\mathbf{i} - \mathbf{j} + \mathbf{k}) - 3(2\mathbf{i} + \mathbf{j} - \mathbf{k})\|^2$
$= \|-4\mathbf{i} - 5\mathbf{j} + 5\mathbf{k}\|^2$
$= 16 + 25 + 25$
$= 66$

43. $\|\mathbf{v}\|\mathbf{w} = \sqrt{1 + 4 + 4}\,(2\mathbf{i} + 4\mathbf{j} - \mathbf{k})$
$= 6\mathbf{i} + 12\mathbf{j} - 3\mathbf{k}$

45. $\|\mathbf{v} - \mathbf{w}\|(\mathbf{v} + \mathbf{w})$
$= \|-\mathbf{i} - 6\mathbf{j} + 3\mathbf{k}\|(3\mathbf{i} + 2\mathbf{j} + \mathbf{k})$
$= \sqrt{46}(3\mathbf{i} + 2\mathbf{j} + \mathbf{k})$

47. $\mathbf{v} = -2\langle 1, -3, 5\rangle \neq s\mathbf{u}$ so $\mathbf{v}$ is not parallel to $\mathbf{u}$.

49. $\mathbf{v} = -\frac{1}{2}\langle 2, -3, 5\rangle = -\frac{1}{2}\mathbf{u}$, so $\mathbf{v}$ is parallel to $\mathbf{u}$.

51. $\|\mathbf{AB}\|^2 = 4 + 4 + 1 = 9$
$\|\mathbf{AC}\|^2 = 4 + 16 + 16 = 36$
$\|\mathbf{BC}\|^2 = 0 + 36 + 9 = 45$
$\|\mathbf{AB}\|^2 + \|\mathbf{AC}\|^2 = \|\mathbf{BC}\|^2$
$\triangle ABC$ is right, but not isosceles.

53. $\|\mathbf{AB}\|^2 = 25 + 4 + 49 = 78$
$\|\mathbf{AC}\|^2 = 64 + 16 + 169 = 249$
$\|\mathbf{BC}\|^2 = 9 + 36 + 36 = 81$
$\triangle ABC$ is neither right nor isosceles.

55. Let A, B, and C be the points $A(2, 3, 2)$, $B(-1, 4, 0)$, and $C(-4, 5, -2)$. Then
$$\mathbf{AB} = \langle -3, 1, -2\rangle$$
and

$$\mathbf{AC} = \langle -6, 2, -4\rangle$$

Since $\mathbf{AC}$ is a multiple of $\mathbf{AB}$, it follows that A, B, and C lie on the same line.

57. Let $\mathbf{F}_1$, $\mathbf{F}_2$, and $\mathbf{F}_3$ be the forces in the three legs, where for constants a, b, c, we have

$$\mathbf{F}_1 = a\mathbf{PA} = a\langle 0, -2, -6\rangle$$
$$\mathbf{F}_2 = b\mathbf{PB} = b\langle -\sqrt{3}, 1, -6\rangle$$
$$\mathbf{F}_3 = c\mathbf{PB} = c\langle \sqrt{3}, 1, -6\rangle$$

The total force exerted is $\mathbf{F} = -150\mathbf{k}$, so we have

$$\mathbf{F} = \mathbf{F}_1 + \mathbf{F}_2 + \mathbf{F}_3$$

$$-150\langle 0, 0, 1\rangle = a\langle 0, -2, -6\rangle + b\langle -\sqrt{3}, 1, -6\rangle + c\langle \sqrt{3}, 1, -6\rangle$$

Equating coefficients we find that

$$\begin{cases} -\sqrt{3}b + \sqrt{3}c = 0 \\ -2a + b + c = 0 \\ -6a - 6b - 6c = -150 \end{cases}$$

Solving, we obtain $a = b = c = \dfrac{25}{3}$, so

$$\mathbf{F}_1 = \frac{25}{3}\langle 0, -2, -6\rangle$$
$$\mathbf{F}_2 = \frac{25}{3}\langle -\sqrt{3}, 1, -6\rangle$$
$$\mathbf{F}_3 = \frac{25}{3}\langle \sqrt{3}, 1, -6\rangle$$

59. The midpoint M of $\overline{BC}$ is $M(1, 2, 2)$. Let $P(x, y, z)$ so that

$\mathbf{AP} = (x+1)\mathbf{i} + (y-3)\mathbf{j} + (z-9)\mathbf{k}$

$\mathbf{AM} = 2\mathbf{i} - \mathbf{j} - 7\mathbf{k}$

$\mathbf{AP} = \dfrac{2}{3}\mathbf{AM}$ so $3\mathbf{AP} = 2\mathbf{AM}$

or

$3(x+1) = 4$ or $x = \dfrac{1}{3}$

$3(y-3) = -2$ or $y = \dfrac{7}{3}$

$3(z-9) = -14$ or $z = \dfrac{13}{3}$

The desired point is $P\left(\dfrac{1}{3}, \dfrac{7}{3}, \dfrac{13}{3}\right)$.

9.3 The Dot Product, page 703

1. If $\mathbf{u} = \langle u_1, u_2, u_3\rangle$ and $\mathbf{v} = \langle v_1, v_2, v_3\rangle$, then $\mathbf{u}\cdot\mathbf{v} = u_1v_1 + u_2v_2 + u_3v_3$. Work can be expressed as a dot product of force and displacement vectors.

3. $\mathbf{v}\cdot\mathbf{w} = 3(2) + (-2)(-1) + 4(-6)$
   $= -16$

5. $\mathbf{v}\cdot\mathbf{w} = 2(-3) + 3(5) + (-1)(4)$
   $= 5$

7. $\mathbf{v}\cdot\mathbf{w} = 0$; orthogonal

9. $\mathbf{v}\cdot\mathbf{w} = 3(6) + (-2)(9) = 0$; orthogonal

11. $(\mathbf{v}+\mathbf{w})\cdot(\mathbf{v}-\mathbf{w})$
    $= 4(2) + (-1)(-3) + (0)(2)$
    $= 11$

13. $(\|\mathbf{v}\|\,\mathbf{w})\cdot(\|\mathbf{w}\|\,\mathbf{v})$
    $= \left[\left(\sqrt{9+4+1}\right)(\mathbf{i}+\mathbf{j}-\mathbf{k})\right]$
    $\quad \cdot \left[\sqrt{1+1+1}(3\mathbf{i}-2\mathbf{j}+\mathbf{k})\right]$
    $= \sqrt{14}\sqrt{3}(3-2-1)$
    $= 0$

15. $2\mathbf{v} - 3\mathbf{w} = 2(\mathbf{i}-2\mathbf{j}+2\mathbf{k}) - 3(2\mathbf{i}+4\mathbf{j}-\mathbf{k})$
    $= -4\mathbf{i} - 16\mathbf{j} + 7\mathbf{k}$

17. $\|2\mathbf{v}-3\mathbf{w}\|$
    $= \|2(\mathbf{i}-2\mathbf{j}+2\mathbf{k}) - 3(2\mathbf{i}+4\mathbf{j}-\mathbf{k})\|$
    $= \|-4\mathbf{i}-16\mathbf{j}+7\mathbf{k}\|$
    $= \sqrt{(-4)^2 + (-16)^2 + 7^2}$
    $= \sqrt{16+256+49}$
    $= \sqrt{321}$

19. $\|\mathbf{i}+\mathbf{j}+\mathbf{k}\| = \sqrt{1^2+1^2+1^2}$
    $= \sqrt{3}$

21. $\|2\mathbf{i}+\mathbf{j}-3\mathbf{k}\|^2 = 2^2 + 1^2 + (-3)^2$
    $= 4+1+9$
    $= 14$

23. $\cos\theta = \dfrac{\mathbf{v}\cdot\mathbf{w}}{\|\mathbf{v}\|\|\mathbf{w}\|}$
    $= \dfrac{1-1+1}{\sqrt{1^2+1^2+1^2}\sqrt{1^2+1^2+1^2}}$
    $= \dfrac{1}{3}$

    Thus, $\theta \approx 71°$.

25. $\cos\theta = \dfrac{\mathbf{v}\cdot\mathbf{w}}{\|\mathbf{v}\|\|\mathbf{w}\|}$
    $= \dfrac{0+0-2}{\sqrt{0^2+2^2+1^2}\sqrt{1^2+0^2+(-2)^2}}$
    $= \dfrac{-2}{\sqrt{5}\sqrt{5}}$
    $= -\dfrac{2}{5}$

    Thus, $\theta \approx 114°$.

27. The scalar projection: $\dfrac{\mathbf{v}\cdot\mathbf{w}}{\|\mathbf{w}\|} = \dfrac{2}{2} = 1$

    The vector projection:

$$\left(\frac{\mathbf{v}\cdot\mathbf{w}}{\mathbf{w}\cdot\mathbf{w}}\right)\mathbf{w}=\left(\frac{2}{4}\right)\mathbf{w}=\mathbf{k}$$

29. The scalar projection: $\dfrac{\mathbf{v}\cdot\mathbf{w}}{\|\mathbf{w}\|}=\dfrac{0}{3}=0$

The vector projection:

$$\left(\frac{\mathbf{v}\cdot\mathbf{w}}{\mathbf{w}\cdot\mathbf{w}}\right)\mathbf{w}=\left(\frac{0}{9}\right)\mathbf{w}=\mathbf{0}$$

31. For $\mathbf{u}=\langle u_1,u_2,u_3\rangle$ to be orthogonal to $\mathbf{v}=\langle 1,1,-1\rangle$ and $\mathbf{w}=\langle -1,1,1\rangle$, we must have

$$\begin{cases}\mathbf{u}\cdot\mathbf{v}=u_1+u_2-u_3=0\\ \mathbf{u}\cdot\mathbf{w}=-u_1+u_2+u_3=0\end{cases}$$

Solving, we obtain $\mathbf{u}=\langle a,0,a\rangle$ for an arbitrary a. Two unit vectors of this form are $\mathbf{u}_1=\left\langle \frac{\sqrt{2}}{2},0,\frac{\sqrt{2}}{2}\right\rangle$ and

$\mathbf{u}_2=\left\langle -\frac{\sqrt{2}}{2},0,-\frac{\sqrt{2}}{2}\right\rangle.$

33. $\mathbf{u}=\dfrac{-\mathbf{v}}{\|\mathbf{v}\|}$

$$=\frac{-(2\mathbf{i}+3\mathbf{j}-2\mathbf{k})}{\sqrt{2^2+3^2+(-2)^2}}$$

$$=-\frac{1}{\sqrt{17}}(2\mathbf{i}+3\mathbf{j}-2\mathbf{k})$$

35. $x(\mathbf{i}+\mathbf{j}+\mathbf{k})+y(\mathbf{i}-\mathbf{j}+2\mathbf{k})+z(\mathbf{i}+\mathbf{k})$
    $=2\mathbf{i}+\mathbf{k}$

so

$$\begin{cases}x+y+z=2\\ x-y=0\\ x+2y+z=1\end{cases}$$

Solving, $x=-1$, $y=-1$, and $z=4$.

37. The two vectors, call them $\mathbf{u}$ and $\mathbf{v}$, must have a dot product of 0.

$$\mathbf{u}\cdot\mathbf{v}=6-2a-2a=0\text{ or }a=\tfrac{3}{2}$$

39. $\|\mathbf{v}\|=\sqrt{2^2+(-3)^2+(-5)^2}=\sqrt{38}$
    $v_1=2,v_2=-3,v_3=-5$

$$\cos\alpha=\frac{2}{\sqrt{38}},\ \alpha\approx 1.24\text{ or }71°$$

$$\cos\beta=\frac{-3}{\sqrt{38}},\ \beta\approx 2.08\text{ or }119°$$

$$\cos\gamma=\frac{-5}{\sqrt{38}},\ \gamma\approx 2.52\text{ or }144°$$

41. $\|\mathbf{v}\|=\sqrt{5^2+(-4)^2+(3)^2}=5\sqrt{2}$
    $v_1=5,v_2=-4,v_3=3$

$$\cos\alpha=\frac{5}{5\sqrt{2}}=\frac{1}{\sqrt{2}},\ \alpha\approx 0.79\text{ or }45°$$

$$\cos\beta=\frac{-4}{5\sqrt{2}},\ \beta\approx 2.17\text{ or }124°$$

$$\cos\gamma=\frac{3}{5\sqrt{2}},\ \gamma\approx 1.13\text{ or }65°$$

43. **a.** $\mathbf{v}\cdot\mathbf{w}=8+18=26$

   b. $\cos\theta=\dfrac{\mathbf{v}\cdot\mathbf{w}}{\|\mathbf{v}\|\,\|\mathbf{w}\|}=\dfrac{26}{(7)(5)}=\dfrac{26}{35}$

   c. $\mathbf{v}\cdot(\mathbf{v}-s\mathbf{w})=0$

$$s=\frac{\|\mathbf{v}\|^2}{\mathbf{v}\cdot\mathbf{w}}$$

$$=\frac{49}{26}$$

   d. $(\mathbf{v}+t\mathbf{w})\cdot\mathbf{w}=\mathbf{v}\cdot\mathbf{w}+t\|\mathbf{w}\|^2=0$

$$t=-\frac{\mathbf{v}\cdot\mathbf{w}}{\mathbf{w}\cdot\mathbf{w}}=\frac{-26}{25}$$

45. $\cos\theta=\dfrac{\mathbf{v}\cdot\mathbf{w}}{\|\mathbf{v}\|\,\|\mathbf{w}\|}=\dfrac{2-1-2}{6}=-\dfrac{1}{6}$

The vector projection of $\mathbf{v}$ onto $\mathbf{w}$ is

$$\mathbf{u} = \left(\frac{\mathbf{v} \cdot \mathbf{w}}{\mathbf{w} \cdot \mathbf{w}}\right)\mathbf{w}$$
$$= -\frac{1}{6}\mathbf{w}$$
$$= -\frac{1}{6}(2\mathbf{i} + \mathbf{j} - \mathbf{k})$$

47. $\mathbf{PQ} = (3 - 1)\mathbf{i} + (1 - 0)\mathbf{j} + [2 - (-1)]\mathbf{k}$
$$= 2\mathbf{i} + \mathbf{j} + 3\mathbf{k}$$
$$W = \mathbf{F} \cdot \mathbf{PQ} = 4 + 3 + 3 = 10$$

49. Consider a Cartesian coordinate system with the end of the log at the origin and units in feet. The position vector, $\mathbf{f}$, along Fred's rope is
$$\mathbf{f} = \sqrt{8^2 - 1^2 - 2^2}\mathbf{i} - (1)\mathbf{j} + 2\mathbf{k}$$
$$= \sqrt{59}\mathbf{i} - \mathbf{j} + 2\mathbf{k}$$
The position vector, $\mathbf{s}$, along Sam's rope is
$$\mathbf{s} = \sqrt{8^2 - 1^2 - 1^2}\mathbf{i} + (1)\mathbf{j} + (1)\mathbf{k}$$
$$= \sqrt{62}\mathbf{i} + \mathbf{j} + \mathbf{k}$$
The resultant force, $\mathbf{F}$, on the log is
$$\mathbf{F} = 30\left(\frac{\mathbf{f}}{\|\mathbf{f}\|}\right) + 20\left(\frac{\mathbf{s}}{\|\mathbf{s}\|}\right)$$
$$= \frac{30}{8}\left(\sqrt{59}\mathbf{i} - \mathbf{j} + 2\mathbf{k}\right) + \frac{20}{8}\left(\sqrt{62}\mathbf{i} + \mathbf{j} + \mathbf{k}\right)$$
$$\approx 48.49\mathbf{i} - 1.25\mathbf{j} + 10\mathbf{k}$$
This resultant force has magnitude $\|\mathbf{F}\| \approx 49.53$ lb and points in the direction of the unit vector $\langle 0.979, -0.025, 0.202 \rangle$

51. a. The force vector is
$$\mathbf{F} = 50[(\cos\tfrac{\pi}{3})\mathbf{i} + (\sin\tfrac{\pi}{3})\mathbf{j}]$$
and the "drag" vector is $\mathbf{D} = 20\mathbf{i}$. The work is:
$$W = \mathbf{F} \cdot \mathbf{D}$$
$$= 50\left(\cos\frac{\pi}{3}\right)(20)$$
$$= 500 \text{ ft} \cdot \text{lb}$$

b. The force vector is
$$\mathbf{F} = 50[(\cos\frac{\pi}{4})\mathbf{i} + (\sin\frac{\pi}{4})\mathbf{j}]$$
and the "drag" vector is $\mathbf{D} = 20\mathbf{i}$. The work is:
$$W = \mathbf{F} \cdot \mathbf{D}$$
$$= 50\left(\cos\frac{\pi}{4}\right)(20)$$
$$= 500\sqrt{2} \text{ ft} \cdot \text{lb}$$

53. a. $\displaystyle\int_{-b}^{b} x^2(x^3 - 5x)\, dx$
$$= \left(\frac{x^6}{6} - \frac{5x^4}{4}\right)\Bigg|_{-b}^{b}$$
$$= 0$$

b.
$$\int_{-b}^{b} \sin kx \sin nx\, dx$$
$$= \frac{1}{2}\int_{-\pi}^{\pi}[\cos(k - n)x]\, dx$$
$$\quad - \frac{1}{2}\int_{-\pi}^{\pi}[\cos(k + n)x]\, dx$$
$$= \frac{1}{2}\left[\frac{\sin(k - n)x}{k - n} - \frac{\sin(k + n)x}{k + n}\right]\Bigg|_{-\pi}^{\pi}$$
$$= 0$$
Therefore, the given functions are orthogonal.

55. The vector $\mathbf{B}$ bisects the angle between $\mathbf{u}$ and $\mathbf{v}$. To prove this, let α be the angle between $\mathbf{B}$ and $\mathbf{u}$, and let β be the angle between $\mathbf{B}$ and $\mathbf{v}$. Then
$$\mathbf{u} \cdot \mathbf{B} = \mathbf{u} \cdot (\|\mathbf{v}\|\,\mathbf{u} + \|\mathbf{u}\|\,\mathbf{v})$$
$$= \|\mathbf{v}\|\,(\mathbf{u} \cdot \mathbf{u}) + \|\mathbf{u}\|\,(\mathbf{u} \cdot \mathbf{v})$$
$$= \|\mathbf{v}\|\|\mathbf{u}\|^2 + \|\mathbf{u}\|\,(\mathbf{u} \cdot \mathbf{v})$$

$$\cos \alpha = \frac{\mathbf{u} \cdot \mathbf{B}}{\|\mathbf{u}\| \|\mathbf{B}\|}$$

$$= \frac{\|\mathbf{v}\| \|\mathbf{u}\|^2 + \|\mathbf{u}\|(\mathbf{u} \cdot \mathbf{v})}{\|\mathbf{u}\| \|\mathbf{B}\|}$$

$$= \frac{\|\mathbf{v}\| \|\mathbf{u}\| + (\mathbf{u} \cdot \mathbf{v})}{\|\mathbf{B}\|}$$

Similarly,

$$\mathbf{v} \cdot \mathbf{B} = \|\mathbf{u}\| \|\mathbf{v}\|^2 + \|\mathbf{v}\| (\mathbf{u} \cdot \mathbf{v})$$

$$\cos \beta = \frac{\mathbf{v} \cdot \mathbf{B}}{\|\mathbf{v}\| \|\mathbf{B}\|}$$

$$= \frac{\|\mathbf{v}\| (\mathbf{v} \cdot \mathbf{u}) + \|\mathbf{u}\| \|\mathbf{v}\|^2}{\|\mathbf{v}\| \|\mathbf{B}\|}$$

$$= \frac{(\mathbf{v} \cdot \mathbf{u}) + \|\mathbf{u}\| \|\mathbf{v}\|}{\|\mathbf{B}\|}$$

Since $\cos \alpha = \cos \beta$, it follows that $\alpha = \beta$ and the vector $\mathbf{B}$ bisects the angle between $\mathbf{u}$ and $\mathbf{v}$.

57. a. $(\mathbf{v} + \mathbf{w}) \cdot (\mathbf{v} + \mathbf{w})$
$$= \mathbf{v} \cdot \mathbf{v} + \mathbf{v} \cdot \mathbf{w} + \mathbf{w} \cdot \mathbf{v} + \mathbf{w} \cdot \mathbf{w}$$
$$= \|\mathbf{v}\|^2 + 2\mathbf{v} \cdot \mathbf{w} + \|\mathbf{w}\|^2$$

b. $\|\mathbf{v} + \mathbf{w}\|^2$
$$= (\mathbf{v} + \mathbf{w}) \cdot (\mathbf{v} + \mathbf{w})$$
$$= \|\mathbf{v}\|^2 + \|\mathbf{w}\|^2 + 2\|\mathbf{v}\| \|\mathbf{w}\| \cos \theta$$
$$\leq \|\mathbf{v}\|^2 + \|\mathbf{w}\|^2 + 2\|\mathbf{v}\| \|\mathbf{w}\|$$
$$= (\|\mathbf{v}\| + \|\mathbf{w}\|)^2$$
Thus, $\|\mathbf{v} + \mathbf{w}\| \leq \|\mathbf{v}\| + \|\mathbf{w}\|$

59. If $\mathbf{v} = v_1\mathbf{i} + v_2\mathbf{j} + v_3\mathbf{k}$, then
$$\cos \alpha = \frac{v_1}{\|\mathbf{v}\|}, \cos \beta = \frac{v_2}{\|\mathbf{v}\|}, \cos \gamma = \frac{v_3}{\|\mathbf{v}\|}$$
so
$$\cos^2\alpha + \cos^2\beta + \cos^2\gamma$$
$$= \frac{v_1^2}{\|\mathbf{v}\|^2} + \frac{v_2^2}{\|\mathbf{v}\|^2} + \frac{v_3^2}{\|\mathbf{v}\|^2}$$

$$= \frac{v_1^2 + v_2^2 + v_3^2}{\|\mathbf{v}\|^2}$$
$$= \frac{\|\mathbf{v}\|^2}{\|\mathbf{v}\|^2}$$
$$= 1$$

9.4 The Cross Product, page 714

SURVIVAL HINT: *The are two important vector operations: (1) dot product (a number) and (2) cross product (a vector). Sometimes a dot product is called a scalar product and cross product is called a vector product to emphasize the nature of each product.*

1. $\mathbf{v} \times \mathbf{w} = \mathbf{i} \times \mathbf{j} = \mathbf{k}$

3. $\mathbf{v} \times \mathbf{w} = \begin{vmatrix} \mathbf{i} & \mathbf{j} & \mathbf{k} \\ 3 & 0 & 2 \\ 2 & 1 & 0 \end{vmatrix} = -2\mathbf{i} + 4\mathbf{j} + 3\mathbf{k}$

5. $\mathbf{v} \times \mathbf{w} = \begin{vmatrix} \mathbf{i} & \mathbf{j} & \mathbf{k} \\ 3 & -2 & 4 \\ 1 & 4 & -7 \end{vmatrix} = -2\mathbf{i} + 25\mathbf{j} + 14\mathbf{k}$

7. $\mathbf{v} \times \mathbf{w} = \begin{vmatrix} \mathbf{i} & \mathbf{j} & \mathbf{k} \\ 3 & -1 & 2 \\ 2 & 3 & -4 \end{vmatrix} = -2\mathbf{i} + 16\mathbf{j} + 11\mathbf{k}$

9. $\mathbf{v} \times \mathbf{w} = \begin{vmatrix} \mathbf{i} & \mathbf{j} & \mathbf{k} \\ 1 & -6 & 10 \\ -1 & 5 & -6 \end{vmatrix} = -14\mathbf{i} - 4\mathbf{j} - \mathbf{k}$

11. $\mathbf{v} \times \mathbf{w} = \begin{vmatrix} \mathbf{i} & \mathbf{j} & \mathbf{k} \\ 1 & 0 & 1 \\ 1 & 1 & 0 \end{vmatrix} = -\mathbf{i} + \mathbf{j} + \mathbf{k}$
$$\|\mathbf{v} \times \mathbf{w}\| = \sqrt{1 + 1 + 1} = \sqrt{3}$$
$$\|\mathbf{v}\| = \sqrt{1 + 1} = \sqrt{2}$$
$$\|\mathbf{w}\| = \sqrt{1 + 1} = \sqrt{2}$$

$$\sin\theta = \frac{\|\mathbf{v}\times\mathbf{w}\|}{\|\mathbf{v}\|\|\mathbf{w}\|}$$

$$= \frac{\sqrt{3}}{\sqrt{2}\sqrt{2}}$$

$$= \frac{\sqrt{3}}{2}$$

13. $\mathbf{v}\times\mathbf{w} = \begin{vmatrix} \mathbf{i} & \mathbf{j} & \mathbf{k} \\ 0 & 1 & 1 \\ 1 & 0 & 1 \end{vmatrix} = \mathbf{i}+\mathbf{j}-\mathbf{k}$

$$\|\mathbf{v}\times\mathbf{w}\| = \sqrt{1+1+1} = \sqrt{3}$$

$$\|\mathbf{v}\| = \sqrt{1+1} = \sqrt{2}$$

$$\|\mathbf{w}\| = \sqrt{1+1} = \sqrt{2}$$

$$\sin\theta = \frac{\|\mathbf{v}\times\mathbf{w}\|}{\|\mathbf{v}\|\|\mathbf{w}\|}$$

$$= \frac{\sqrt{3}}{\sqrt{2}\sqrt{2}}$$

$$= \frac{\sqrt{3}}{2}$$

15. $\mathbf{v}\times\mathbf{w} = \begin{vmatrix} \mathbf{i} & \mathbf{j} & \mathbf{k} \\ 2 & 0 & 1 \\ 1 & -1 & -1 \end{vmatrix}$

$$= \mathbf{i}+3\mathbf{j}-2\mathbf{k}$$

$$\|\mathbf{v}\times\mathbf{w}\| = \sqrt{1+9+4} = \sqrt{14}$$

A unit vector is

$$\frac{1}{\sqrt{14}}\mathbf{i}+\frac{3}{\sqrt{14}}\mathbf{j}-\frac{2}{\sqrt{14}}\mathbf{k}$$

so a unit vector is either of

$$\pm\left(\frac{1}{\sqrt{14}}\mathbf{i}+\frac{3}{\sqrt{14}}\mathbf{j}-\frac{2}{\sqrt{14}}\mathbf{k}\right)$$

17. $\mathbf{v}\times\mathbf{w} = \begin{vmatrix} \mathbf{i} & \mathbf{j} & \mathbf{k} \\ 1 & 1 & 1 \\ 3 & 12 & -4 \end{vmatrix}$

$$= -16\mathbf{i}+7\mathbf{j}+9\mathbf{k}$$

$$\|\mathbf{v}\times\mathbf{w}\| = \sqrt{256+49+81} = \sqrt{386}$$

A unit vector is

$$\frac{-16}{\sqrt{386}}\mathbf{i}+\frac{7}{\sqrt{386}}\mathbf{j}+\frac{9}{\sqrt{386}}\mathbf{k}$$

so a unit vector is either of

$$\pm\left(\frac{-16}{\sqrt{386}}\mathbf{i}+\frac{7}{\sqrt{386}}\mathbf{j}+\frac{9}{\sqrt{386}}\mathbf{k}\right)$$

19. $\mathbf{v}\times\mathbf{w} = \begin{vmatrix} \mathbf{i} & \mathbf{j} & \mathbf{k} \\ 3 & 4 & 0 \\ 1 & 1 & -1 \end{vmatrix}$

$$= -4\mathbf{i}+3\mathbf{j}-\mathbf{k}$$

$$\|\mathbf{v}\times\mathbf{w}\| = \sqrt{16+9+1} = \sqrt{26}$$

$$A = \|\mathbf{v}\times\mathbf{w}\|$$

$$= \sqrt{26}$$

21. $\mathbf{v}\times\mathbf{w} = \begin{vmatrix} \mathbf{i} & \mathbf{j} & \mathbf{k} \\ 4 & -1 & 1 \\ 2 & 3 & -1 \end{vmatrix}$

$$= -2\mathbf{i}+6\mathbf{j}+14\mathbf{k}$$

$$\|\mathbf{v}\times\mathbf{w}\| = \sqrt{4+36+196} = 2\sqrt{59}$$

$$A = \|\mathbf{v}\times\mathbf{w}\|$$

$$= 2\sqrt{59}$$

23. $\mathbf{PQ} = (1-0)\mathbf{i}+(1-1)\mathbf{j}+(0-1)\mathbf{k}$

$$= \mathbf{i}-\mathbf{k}$$

$$\mathbf{PR} = (1-0)\mathbf{i}+(0-1)\mathbf{j}+(1-1)\mathbf{k}$$

$$= \mathbf{i}-\mathbf{j}$$

$$\mathbf{PQ}\times\mathbf{PR} = \begin{vmatrix} \mathbf{i} & \mathbf{j} & \mathbf{k} \\ 1 & 0 & -1 \\ 1 & -1 & 0 \end{vmatrix}$$

$$= -\mathbf{i}-\mathbf{j}-\mathbf{k}$$

$$A = \tfrac{1}{2}\|\mathbf{PQ}\times\mathbf{PR}\|$$

$$= \tfrac{1}{2}\sqrt{1+1+1}$$

$$= \tfrac{1}{2}\sqrt{3}$$

25. $\mathbf{PQ} = (2-1)\mathbf{i} + (3-2)\mathbf{j} + (1-3)\mathbf{k}$
$= \mathbf{i} + \mathbf{j} - 2\mathbf{k}$
$\mathbf{PR} = (3-1)\mathbf{i} + (1-2)\mathbf{j} + (2-3)\mathbf{k}$
$= 2\mathbf{i} - \mathbf{j} - \mathbf{k}$

$$\mathbf{PQ} \times \mathbf{PR} = \begin{vmatrix} \mathbf{i} & \mathbf{j} & \mathbf{k} \\ 1 & 1 & -2 \\ 2 & -1 & -1 \end{vmatrix}$$
$$= -3\mathbf{i} - 3\mathbf{j} - 3\mathbf{k}$$
$$A = \frac{1}{2}\|\mathbf{PQ} \times \mathbf{PR}\|$$
$$= \frac{1}{2}\sqrt{9+9+9}$$
$$= \frac{3}{2}\sqrt{3}$$

27. **a.** Does not exist because it is a cross product of a vector (**u**) with a scalar (**v** · **w**).

b. Scalar since it is a dot product of vectors.

29. **a.** Scalar since it is a dot product of vectors.

b. Vector since it is a cross product of vectors.

31. $\begin{vmatrix} 0 & 1 & 1 \\ 2 & 1 & 2 \\ 5 & 0 & 0 \end{vmatrix} = 5;\ V = 5$

33. $\begin{vmatrix} 2 & 1 & -1 \\ 3 & 0 & 1 \\ 0 & 1 & 1 \end{vmatrix} = -8;\ V = 8$

35. The dot product of two vectors is a scalar. The dot product of two vectors is 0 if and only if the vectors are orthogonal. It can also be used to find work and the projection of a vector onto another, including finding direction cosines. The cross product of two vectors is a vector normal to the plane formed by the given vectors. Its magnitude can be used to find the area of the parallelogram formed by these vectors.

37. $\mathbf{u} = \mathbf{i}$, $\mathbf{v} = \mathbf{i} + \mathbf{j} + \mathbf{k}$, and $\mathbf{w} = \mathbf{i} + 2\mathbf{j} + s\mathbf{k}$; next, let

$$\mathbf{n} = \mathbf{u} \times \mathbf{v} = \begin{vmatrix} \mathbf{i} & \mathbf{j} & \mathbf{k} \\ 1 & 0 & 0 \\ 1 & 1 & 1 \end{vmatrix} = -\mathbf{j} + \mathbf{k}$$

n must be orthogonal to every vector in the plane determined by **u** and **v** so
$$\mathbf{n} \cdot \mathbf{w} = 0$$
$$0 - 2 + s = 0$$
$$s = 2$$

39. First,

$$\mathbf{u} \times \mathbf{v} = \begin{vmatrix} \mathbf{i} & \mathbf{j} & \mathbf{k} \\ 1 & 1 & 0 \\ 2 & -1 & 1 \end{vmatrix} = \mathbf{i} - \mathbf{j} - 3\mathbf{k}$$

$$(\mathbf{u} \times \mathbf{v}) \times \mathbf{w} = \begin{vmatrix} \mathbf{i} & \mathbf{j} & \mathbf{k} \\ 1 & -1 & -3 \\ 3 & 0 & 0 \end{vmatrix} = -9\mathbf{j} + 3\mathbf{k}$$

Next,

$$\mathbf{v} \times \mathbf{w} = \begin{vmatrix} \mathbf{i} & \mathbf{j} & \mathbf{k} \\ 2 & -1 & 1 \\ 3 & 0 & 0 \end{vmatrix} = 3\mathbf{j} + 3\mathbf{k}$$

$$\mathbf{u} \times (\mathbf{v} \times \mathbf{w}) = \begin{vmatrix} \mathbf{i} & \mathbf{j} & \mathbf{k} \\ 1 & 1 & 0 \\ 0 & 3 & 3 \end{vmatrix} = 3\mathbf{i} - 3\mathbf{j} + 3\mathbf{k}$$

Thus, $(\mathbf{u} \times \mathbf{v}) \times \mathbf{w} \neq \mathbf{u} \times (\mathbf{v} \times \mathbf{w})$. Cross product is not an associative operation.

41. Since $\mathbf{v} \times \mathbf{w}$ is orthogonal to both **v** and **w**, we can have $\mathbf{v} \times \mathbf{w} = \mathbf{w}$ only if $\mathbf{w} \cdot \mathbf{w} = 0$ because

$$0 = (\mathbf{v} \times \mathbf{w}) \cdot \mathbf{w} = \mathbf{w} \cdot \mathbf{w}$$

Thus, $\mathbf{w} = \mathbf{0}$.

43. Note $30° = \frac{\pi}{6}$; let $\mathbf{F} = \mathbf{u} = -40\mathbf{k}$ and

$$\mathbf{PQ} = 2[(\cos\frac{\pi}{6})\mathbf{j} + (\sin\frac{\pi}{6})\mathbf{k}]$$

$$= \sqrt{3}\mathbf{j} + \mathbf{k}$$

The torque is $\mathbf{T} = \mathbf{PQ} \times \mathbf{F}$

$$= \begin{vmatrix} \mathbf{i} & \mathbf{j} & \mathbf{k} \\ 0 & \sqrt{3} & 1 \\ 0 & 0 & -40 \end{vmatrix}$$

$$= -40\sqrt{3}\mathbf{i}$$

45. Let $\mathbf{v} = 2\mathbf{i} - \mathbf{j} + \mathbf{k}$,

$\mathbf{u} = \mathbf{PQ} = -2\mathbf{i} + 4\mathbf{j}$,

$\mathbf{w} = \mathbf{PR} = 4\mathbf{j} - 6\mathbf{k}$; $\mathbf{n}$ is normal to the plane of $\mathbf{u}$ and $\mathbf{w}$ if

$$\mathbf{n} = \mathbf{u} \times \mathbf{w}$$

$$= \begin{vmatrix} \mathbf{i} & \mathbf{j} & \mathbf{k} \\ -2 & 4 & 0 \\ 0 & 4 & -6 \end{vmatrix}$$

$$= -24\mathbf{i} - 12\mathbf{j} - 8\mathbf{k}$$

$$\cos\theta = \frac{\mathbf{n} \cdot \mathbf{v}}{\|\mathbf{n}\|\|\mathbf{v}\|}$$

$$= \frac{-48 + 12 - 8}{\sqrt{24^2 + 12^2 + 8^2}\sqrt{4+1+1}}$$

$$= \frac{-44}{\sqrt{784}\sqrt{6}}$$

$$= \frac{-11}{7\sqrt{6}}$$

$$\theta \approx 130°$$

The acute angle is $50°$, so the angle between the vector and the plane is $90° - 50° = 40°$.

47. First, we find

$$D = \begin{vmatrix} x_1 & y_1 & 1 \\ x_2 & y_2 & 1 \\ x_3 & y_3 & 1 \end{vmatrix}$$

$$= (x_2 y_3 - x_3 y_2) - (x_1 y_3 - x_3 y_1) + (x_1 y_2 - x_2 y_1)$$

$$= x_2 y_3 - x_3 y_2 - x_1 y_3 + x_3 y_1 + x_1 y_2 - x_2 y_1$$

Let $\mathbf{v} = \mathbf{PR} = \langle x_3 - x_1, y_3 - y_1, 0 \rangle$
and $\mathbf{u} = \mathbf{PQ} = \langle x_2 - x_1, y_2 - y_1, 0 \rangle$.

$$A = \frac{1}{2}\|\mathbf{u} \times \mathbf{v}\|$$

$$= \frac{1}{2}\left\| \begin{vmatrix} \mathbf{i} & \mathbf{j} & \mathbf{k} \\ x_2 - x_1 & y_2 - y_1 & 0 \\ x_3 - x_1 & y_3 - y_1 & 0 \end{vmatrix} \right\|$$

$$= \frac{1}{2}\|[(x_2 - x_1)(y_3 - y_1) - (x_3 - x_1)(y_2 - y_1)]\mathbf{k}\|$$

$$= \frac{1}{2}\sqrt{[(x_2 - x_1)(y_3 - y_1) - (x_3 - x_1)(y_2 - y_1)]^2}$$

$$= \frac{1}{2}|(x_2 - x_1)(y_3 - y_1) - (x_3 - x_1)(y_2 - y_1)|$$

$$= \frac{1}{2}\Big[x_2 y_3 - x_2 y_1 - x_1 y_3 + x_1 y_1 - x_3 y_2$$

$$+ x_3 y_1 + x_1 y_2 - x_1 y_1\Big]$$

$$= \frac{1}{2}|x_2 y_3 - x_3 y_2 - x_1 y_3 + x_3 y_1 + x_1 y_2 - x_2 y_1|$$

$$= \frac{1}{2}|D|$$

49. $(\mathbf{u} \times \mathbf{v}) \cdot \mathbf{w}$

$$= \begin{vmatrix} \mathbf{i} & \mathbf{j} & \mathbf{k} \\ u_1 & u_2 & u_3 \\ v_1 & v_2 & v_3 \end{vmatrix} \cdot (w_1\mathbf{i} + w_2\mathbf{j} + w_3\mathbf{k})$$

$$= \left(\begin{vmatrix} u_2 & u_3 \\ v_2 & v_3 \end{vmatrix}\mathbf{i} - \begin{vmatrix} u_1 & u_3 \\ v_1 & v_3 \end{vmatrix}\mathbf{j} \right.$$

$$\left. + \begin{vmatrix} u_1 & u_2 \\ v_1 & v_2 \end{vmatrix}\mathbf{k} \right) \cdot (w_1\mathbf{i} + w_2\mathbf{j} + w_3\mathbf{k})$$

$$= \begin{vmatrix} u_1 & u_2 & u_3 \\ v_1 & v_2 & v_3 \\ w_1 & w_2 & w_3 \end{vmatrix}$$

51.
$$\mathbf{u} + \mathbf{v} + \mathbf{w} = \mathbf{0}$$
$$(\mathbf{u} + \mathbf{v} + \mathbf{w}) \times \mathbf{v} = \mathbf{0}$$
$$\mathbf{u} \times \mathbf{v} + \mathbf{v} \times \mathbf{v} + \mathbf{w} \times \mathbf{v} = \mathbf{0}$$
$$\mathbf{u} \times \mathbf{v} + \mathbf{w} \times \mathbf{v} = \mathbf{0}$$
$$\mathbf{u} \times \mathbf{v} = -\mathbf{w} \times \mathbf{v}$$
$$\mathbf{u} \times \mathbf{v} = \mathbf{v} \times \mathbf{w}$$

Repeat the same steps by crossing both sides with $\mathbf{w}$.
$$\mathbf{u} + \mathbf{v} + \mathbf{w} = \mathbf{0}$$
$$(\mathbf{u} + \mathbf{v} + \mathbf{w}) \times \mathbf{w} = \mathbf{0}$$
$$\mathbf{u} \times \mathbf{w} + \mathbf{v} \times \mathbf{w} + \mathbf{w} \times \mathbf{w} = \mathbf{0}$$
$$\mathbf{u} \times \mathbf{w} + \mathbf{v} \times \mathbf{w} = \mathbf{0}$$
$$\mathbf{v} \times \mathbf{w} = -\mathbf{u} \times \mathbf{w}$$
$$\mathbf{v} \times \mathbf{w} = \mathbf{w} \times \mathbf{u}$$

53. Recall $\cos\theta = \dfrac{\mathbf{v} \cdot \mathbf{w}}{\|\mathbf{v}\|\|\mathbf{w}\|}$ and

$$\sin\theta = \frac{\|\mathbf{v} \times \mathbf{w}\|}{\|\mathbf{v}\|\|\mathbf{w}\|} \text{ so}$$

$$\tan\theta = \frac{\sin\theta}{\cos\theta}$$
$$= \frac{\|\mathbf{v} \times \mathbf{w}\|}{\|\mathbf{v}\|\|\mathbf{w}\|} \cdot \frac{\|\mathbf{v}\|\|\mathbf{w}\|}{\mathbf{v} \cdot \mathbf{w}}$$
$$= \frac{\|\mathbf{v} \times \mathbf{w}\|}{\mathbf{v} \cdot \mathbf{w}}$$

55. Let $\mathbf{a} = \langle a_1, a_2, a_3 \rangle$, $\mathbf{b} = \langle b_1, b_2, b_3 \rangle$, and $\mathbf{c} = \langle c_1, c_2, c_3 \rangle$.

$$\mathbf{i} \times (\mathbf{b} \times \mathbf{c}) = \mathbf{i} \times \begin{vmatrix} \mathbf{i} & \mathbf{j} & \mathbf{k} \\ b_1 & b_2 & b_3 \\ c_1 & c_2 & c_3 \end{vmatrix}$$
$$= c_1\mathbf{b} - b_1\mathbf{c}$$

$$\mathbf{j} \times (\mathbf{b} \times \mathbf{c}) = \mathbf{j} \times \begin{vmatrix} \mathbf{i} & \mathbf{j} & \mathbf{k} \\ b_1 & b_2 & b_3 \\ c_1 & c_2 & c_3 \end{vmatrix}$$
$$= c_2\mathbf{b} - b_2\mathbf{c}$$

$$\mathbf{k} \times (\mathbf{b} \times \mathbf{c}) = \mathbf{k} \times \begin{vmatrix} \mathbf{i} & \mathbf{j} & \mathbf{k} \\ b_1 & b_2 & b_3 \\ c_1 & c_2 & c_3 \end{vmatrix}$$
$$= c_3\mathbf{b} - b_3\mathbf{c}$$

$$\mathbf{a} \times (\mathbf{b} \times \mathbf{c})$$
$$= (a_1\mathbf{i} + a_2\mathbf{j} + a_3\mathbf{k}) \times (\mathbf{b} \times \mathbf{c})$$
$$= a_1[\mathbf{i} \times (\mathbf{b} \times \mathbf{c})] + a_2[\mathbf{j} \times (\mathbf{b} \times \mathbf{c})] + a_3[\mathbf{k} \times (\mathbf{b} \times \mathbf{c})]$$
$$= a_1[c_1\mathbf{b} - b_1\mathbf{c}] + a_2[c_2\mathbf{b} - b_2\mathbf{c}] + a_3[c_3\mathbf{b} - b_3\mathbf{c}]$$
$$= (a_1c_1 + a_2c_2 + a_3c_3)\mathbf{b} - (a_1b_1 + a_2b_2 + a_3b_3)\mathbf{c}$$
$$= (c_1a_1 + c_2a_2 + c_3a_3)\mathbf{b} - (b_1a_1 + b_2a_2 + b_3a_3)\mathbf{c}$$
$$= (\mathbf{c} \cdot \mathbf{a})\mathbf{b} - (\mathbf{b} \cdot \mathbf{a})\mathbf{c}$$

57. $\mathbf{u} \times (\mathbf{v} \times \mathbf{w}) + \mathbf{v} \times (\mathbf{w} \times \mathbf{u}) + \mathbf{w} \times (\mathbf{u} \times \mathbf{v})$
$$= (\mathbf{u} \cdot \mathbf{w})\mathbf{v} - (\mathbf{u} \cdot \mathbf{v})\mathbf{w} + (\mathbf{v} \cdot \mathbf{u})\mathbf{w} - (\mathbf{v} \cdot \mathbf{w})\mathbf{u}$$
$$+ (\mathbf{w} \cdot \mathbf{v})\mathbf{u} - (\mathbf{w} \cdot \mathbf{u})\mathbf{v}$$

$$= (\mathbf{v} \cdot \mathbf{u})\mathbf{w} - (\mathbf{u} \cdot \mathbf{v})\mathbf{w} - (\mathbf{v} \cdot \mathbf{w})\mathbf{u}$$
$$+ (\mathbf{w} \cdot \mathbf{v})\mathbf{u} + (\mathbf{u} \cdot \mathbf{w})\mathbf{v} - (\mathbf{w} \cdot \mathbf{u})\mathbf{v}$$
$$= \mathbf{0}$$

59. $\mathbf{u} \times (\mathbf{u} \times \mathbf{v}) = (\mathbf{v} \cdot \mathbf{u})\mathbf{u} - (\mathbf{u} \cdot \mathbf{u})\mathbf{v}$ and
$$\mathbf{u} \times [\mathbf{u} \times (\mathbf{u} \times \mathbf{v})]$$
$$= \mathbf{u} \times (\mathbf{v} \cdot \mathbf{u})\mathbf{u} - \mathbf{u} \times (\mathbf{u} \cdot \mathbf{u})\mathbf{v}$$
$$= \mathbf{0} - (\mathbf{u} \cdot \mathbf{u})(\mathbf{u} \times \mathbf{v})$$
$$= -\|\mathbf{u}\|^2(\mathbf{u} \times \mathbf{v})$$

Therefore,
$$\mathbf{u} \times [\mathbf{u} \times (\mathbf{u} \times \mathbf{v})] \cdot \mathbf{w}$$
$$= -\|\mathbf{u}\|^2[(\mathbf{u} \times \mathbf{v}) \cdot \mathbf{w}]$$
$$= -\|\mathbf{u}\|^2[\mathbf{u} \cdot (\mathbf{v} \times \mathbf{w})]$$

9.5 Parametric Representation of Curves; Lines in $\mathbb{R}^3$, page 724

1. A parameter is an arbitrary constant or a variable in a mathematical expression, which distinguishes various specific cases.

3. $x = 1 + 3t$, $y = -1 - 2t$, $z = -2 + 5t$
$$\frac{x - 1}{3} = \frac{y + 1}{-2} = \frac{z + 2}{5}$$

5. $\mathbf{v} = (2-1)\mathbf{i} + (1+1)\mathbf{j} + (3-2)\mathbf{k}$
$= \mathbf{i} + 2\mathbf{j} + \mathbf{k}$
$$\frac{x-1}{1} = \frac{y+1}{2} = \frac{z-2}{1}$$
$x = 1+t,\ y = -1+2t,\ z = 2+t$

7. $\dfrac{x-1}{1} = \dfrac{y+3}{-3} = \dfrac{z-6}{-5}$
$x = 1+t,\ y = -3-3t,\ z = 6-5t$

9. $\dfrac{x}{11} = \dfrac{y-4}{-3} = \dfrac{z+3}{5}$
$x = 11t,\ y = 4-3t,\ z = -3+5t$

11. $x = 3+t,\ y = -1,\ z = 0$

13. $x = 4+4t,\ y - -3+3t,\ z = -2+t$
$x = 0;\ t = -1$ for point $(0, -6, -3)$;
$y = 0;\ t = 1$ for point $(8, 0, -1)$;
$z = 0;\ t = 2$ for point $(12, 3, 0)$

15. $x = 6-2t,\ y = 1+t,\ z = 3t$
$x = 0;\ t = 3$ for point $(0, 4, 9)$;
$y = 0;\ t = -1$ for point $(8, 0, -3)$;
$z = 0;\ t = 0$ for point $(6, 1, 0)$

17. A vector parallel to the first line is
$\mathbf{v}_1 = 2\mathbf{i} - 3\mathbf{j} + 5\mathbf{k}$. A vector parallel to
the second line is
$\mathbf{v}_2 = 4\mathbf{i} - 6\mathbf{j} + 10\mathbf{k} = 2(2\mathbf{i} - 3\mathbf{j} + 5\mathbf{k})$;
since these vectors have the same direction
numbers the lines are either coincident or
parallel; since $P(4, 6, -2)$ is a point on
the first line, but not on the second line,
$\mathbf{v}_1$, is parallel to $\mathbf{v}_2$ and the first line is
parallel to the second.

19. A vector parallel to the first line is
$\mathbf{v}_1 = 3\mathbf{i} - 4\mathbf{j} - 7\mathbf{k}$. A vector parallel to
the second line is $\mathbf{v}_2 = 3\mathbf{i} - 4\mathbf{j} - 7\mathbf{k}$.
Since $\mathbf{v}_2 = \mathbf{v}_1$, the lines are either
coincident or parallel; since $P(3, 1, -4)$ is
a point on the first line, but not the second,
they are parallel.

21. A vector parallel to the first line is
$\mathbf{v}_1 = 2\mathbf{i} - \mathbf{j} + \mathbf{k}$. A vector parallel to the
second line is $\mathbf{v}_2 = 3\mathbf{i} - \mathbf{j} + \mathbf{k}$. Since these
lines do not have proportional direction
numbers, the lines are not parallel or
coincident. If they intersect, we must have:
for x: $3 + 2t_1 = -2 + 3t_2$ or $2t_1 - 3t_2 = -5$
for y: $1 - t_1 = 3 - t_2$ or $t_1 - t_2 = -2$
for z: $4 + t_1 = 2 + t_2$ or $t_1 - t_2 = -2$
Solving this system simultaneously we find
$t_1 = -1,\ t_2 = 1$. Thus, $x = 3 - 2 = 1$,
$y = 1 - (-1) = 2$, and $z = 4 + (-1) = 3$.
The point of intersection is $(1, 2, 3)$.

23. $t = x$, so
$y = x - 1$
$0 \le x \le 3$

25. $t = \dfrac{x}{60}$
$y = \dfrac{80x}{60} - \dfrac{16x^2}{60^2}$
$= \dfrac{4}{3}x - \dfrac{1}{225}x^2$
$0 \le x \le 180$

27. $t = x^{1/3}$
$x^{1/3} = y^{1/2}$
$y = x^{2/3},\ x \ge 0$

29. $\dfrac{x}{3} = \cos\theta$

$\dfrac{y}{3} = \sin\theta$

$x^2 + y^2 = 9$

$-3 \leq x \leq 3$

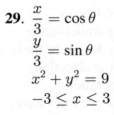

31. $\sin t = x - 1$

$\cos t = y + 2$

$(x-1)^2 + (y+2)^2 = 1$

$0 \leq x \leq 2$

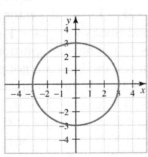

33. $\dfrac{x}{4} = \tan 2t$

$\dfrac{y}{3} = \sec 2t$

$\dfrac{x^2}{16} + 1 = \dfrac{y^2}{9}$

$\dfrac{y^2}{9} - \dfrac{x^2}{16} = 1$

$-\infty < x < \infty$

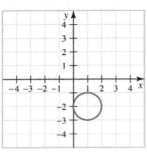

35. $t = x^{1/3}$

$y = 3\ln x^{1/3}$

$\quad = \ln x$

$x > 0$

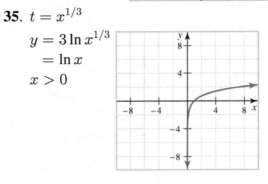

37. $\pm\dfrac{1}{\sqrt{21}}(4\mathbf{i} + 2\mathbf{j} + \mathbf{k})$

39. The equation of the circle is

$$x^2 + y^2 = 3^2$$
$$\left(\dfrac{x}{3}\right)^2 + \left(\dfrac{y}{3}\right)^2 = 1$$

Use the identity $\cos^2 t + \sin^2 t = 1$. Let

$\cos t = \dfrac{x}{3}$ or $x = 3\cos t$;

$\sin t = \dfrac{y}{3}$ or $y = 3\sin t$

The orientation is counterclockwise because t is measured in a positive direction for $0 \leq t < 2\pi$.

41. The equation of the ellipse is

$$\left(\dfrac{x}{3}\right)^2 + \left(\dfrac{y}{2}\right)^2 = 1$$

Let $\cos t = \dfrac{x}{3}$ or $x = 3\cos t$;

$\sin t = \dfrac{y}{2}$ or $y = 2\sin t$

The orientation is counterclockwise because t is measured in a positive direction for $0 \leq t < 2\pi$.

43. The equation of the hyperbola is

$$\left(\dfrac{x}{4}\right)^2 - \left(\dfrac{y}{3}\right)^2 = 1$$

Use the identity $1 + \tan^2 t = \sec^2 t$ or $\sec^2 t - \tan^2 t = 1$.

Let $\sec t = \dfrac{x}{4}$ or $x = \dfrac{4}{\cos t}$;

$\tan t = \dfrac{y}{3}$ or $y = 3\tan t$ for $0 \leq t < 2\pi$.

45. $y = \cos 2\pi t$

$\quad = 1 - 2\sin^2 \pi t$

$\quad = 1 - 2x^2$

This is a parabolic arc where $0 \le x \le 1$; $-1 \le y \le 1$.

47. Since $\cos^2 t + \sin^2 t = 1$,

$$\frac{y}{b} + \frac{x^2}{16a^2} = 1$$

$$y = \frac{b}{16a^2}\left(16a^2 - x^2\right)$$

49. $\mathbf{PQ} = 2\mathbf{i} + \mathbf{j} - 2\mathbf{k}$;
    $\mathbf{v} \cdot \mathbf{PQ} = 6 - 4 - 2 = 0$,
    so $\mathbf{v}$ and $\mathbf{PQ}$ are orthogonal.

51. The lines are parallel to each other since they are both parallel to $\mathbf{v} = \langle 2, 4, 1 \rangle$.

    $L_1 : \ x = a + 2t \quad L_2 : \ x = 2 - 4s$
    $ y = 1 + 4t \qquad y = b - 8s$
    $ z = -2 + t \qquad z = -1 - 2s$

    Thus, the lines will coincide if they intersect. Solving

    $$\begin{cases} a + 2t = 2 - 4s \\ 1 + 4t = b - 8s \\ -2 + t = -1 - 2s \end{cases}$$

    From the third equation, $t = 1 - 2s$, so the first two equations are

    $$\begin{cases} a + 2(1 - 2s) = 2 - 4s \\ 1 + 4(1 - 2s) = b - 8s \end{cases}$$

    We solve this system to find that the lines intersect when $a = 0$, $b = 5$.

53. To show that lines L_1 and L_2 intersect, solve

    $$\begin{cases} -1 + 2t = -2 - s \\ 3 - t = 5 + 2s \\ 2 + 2t = -2s \end{cases}$$

    to obtain $t = 0$ and $s = -1$. Thus, the point of intersection is $(-1, 3, 2)$.

To find the angle θ between the lines, note $\mathbf{v}_1 = \langle 2, -1, 2 \rangle$ and $\mathbf{v}_2 = \langle -1, 2, -2 \rangle$ are vectors parallel to L_1 and L_2, respectively, so

$$\mathbf{v}_1 \cdot \mathbf{v}_2 = \langle 2, -1, 2 \rangle \cdot \langle -1, 2, -2 \rangle = -8$$

and

$$\begin{aligned} \cos \theta &= \frac{\mathbf{v}_1 \cdot \mathbf{v}_2}{\|\mathbf{v}_1\| \, \|\mathbf{v}_2\|} \\ &= \frac{-8}{\sqrt{4 + 1 + 4}\sqrt{1 + 4 + 4}} \\ &= -\frac{8}{9} \\ \theta &\approx 152.7° \end{aligned}$$

Thus, the corresponding acute angle is

$$180° - 152.7° \approx 27°$$

55. The given line is parallel to the vector $\mathbf{v} = \langle -1, -2, 1 \rangle$ and contains the point $Q(2, 1, 5)$. Then $\mathbf{QP} = \langle -3, 2, -4 \rangle$ is the vector from Q to the given point $P(-1, 3, 1)$, and the cross product $\mathbf{QP} \times \mathbf{v} = \langle -6, 7, 8 \rangle$ is orthogonal to both the given line and the line we seek. This means the cross product

$$\mathbf{v} \times (\mathbf{QP} \times \mathbf{v}) = \langle -23, 2, -19 \rangle$$

is parallel to the required line, which thus has parametric equations

$$x = -1 - 23t, y = 3 + 2t, z = 1 - 19t$$

57. Vectors parallel to the given lines are $\mathbf{v}_1 = \langle a_1, b_1, c_1 \rangle$, and $\mathbf{v}_2 = \langle a_2, b_2, c_2 \rangle$. If $\mathbf{v}_1 \cdot \mathbf{v}_2 = a_1 a_2 + b_1 b_2 + c_1 c_2 = 0$, then the

vectors are orthogonal. In this case, the lines are perpendicular and intersect at $P(x_0, y_0, z_0)$.

59. Consider a fixed circle of radius a with center at the origin O. Let E be the point with coordinates $(a, 0)$. A ray makes an angle θ with the positive x-axis and contains the center A of a moving circle of radius R which makes contact with the fixed circle at D. A ray from A to a point $P(x, y)$—drawn to the right and below A for convenience—on this moving circle also makes an angle ϕ with respect to $\overline{OD}$. This point P was originally at E, before the second circle started moving. Arcs $\overline{DP}$ and $\overline{DE}$ are the same length. We have

$$a\theta = \phi R \text{ or } \phi = \frac{a\theta}{R}$$

Consider a right triangle with $\overline{AP}$ as hypotenuse and third vertex above P and to the right of A (for convenience). Label the acute angle at A, α, and label the legs x_1 and y_1, respectively. Then,

$$\alpha = \phi - \theta = \left(\frac{a - R}{R}\right)\theta$$

$$x_1 = R\cos\alpha = R\cos\frac{(a - R)\theta}{R}$$

$$y_1 = R\sin\alpha = R\sin\frac{(a - R)\theta}{R}$$

The distance x_2 from O to the projection of A on the x-axis is $x_2 = (a - R)\cos\theta$. The vertical distance y_1 from A to the x-axis is $y_2 = (a - R)\sin\theta$. Putting all this together, we have

$$x = x_2 + x_1$$
$$= (a - R)\cos\theta + R\cos\frac{(a - R)\theta}{R}$$
$$y = y_2 - y_1$$
$$= (a - R)\sin\theta - R\sin\frac{(a - R)\theta}{R}$$

9.6 Planes in $\mathbb{R}^3$, page 733

1. The plane containing $P_0(x_0, y_0, z_0)$ with normal vector $A\mathbf{i} + B\mathbf{j} + C\mathbf{k}$ has the equation

$$A(x - x_0) + B(y - y_0) + C(z - z_0) = 0$$

Conversely, if $Ax + By + Cz + D = 0$ is the equation of a plane, a vector normal to the plane is given by $\mathbf{N} = A\mathbf{i} + B\mathbf{j} + C\mathbf{k}$. Note that the coefficients of the variables are the direction numbers of the normal.

3. $-2(x + 1) + 4(y - 3) - 8z = 0$
$$-2x - 2 + 4y - 12 - 8z = 0$$
$$-2x + 4y - 8z - 14 = 0$$
$$x - 2y + 4z + 7 = 0$$

5. $5(x - 2) - 3(y + 2) + 4(z + 3) = 0$
$$5x - 3y + 4z - 4 = 0$$

7.   **9.**

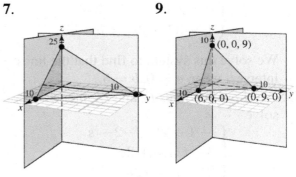

11. $-(x - 0) + 0(y + 7) + (z - 1) = 0$
$$x - z + 1 = 0$$

13. $-(x - 1) - 2(y - 1) + 3(z + 1) = 0$
$$x + 2y - 3z - 6 = 0$$

15. $x = 0$

17. $\pm \dfrac{1}{\sqrt{38}}(5\mathbf{i} - 3\mathbf{j} + 2\mathbf{k})$

19. $d = \dfrac{|-10|}{\sqrt{2^2 + 3^2 + 5^2}}$

$\quad = \dfrac{10}{\sqrt{38}}$

21. $d = \dfrac{\left|3a - 4a + 3a + \frac{1}{a}\right|}{\sqrt{9 + 4 + 1}}$

$\quad = \dfrac{\left|2a + \frac{1}{a}\right|}{\sqrt{14}}$

$\quad = \dfrac{2a^2 + 1}{|a|\sqrt{14}}$

23. A normal to the plane is
$$\mathbf{N} = \mathbf{AB} \times \mathbf{AC}$$

$\quad = \begin{vmatrix} \mathbf{i} & \mathbf{j} & \mathbf{k} \\ 5 & 2 & 6 \\ 4 & -2 & -1 \end{vmatrix}$

$\quad = 10\mathbf{i} + 29\mathbf{j} - 18\mathbf{k}$

The equation of the plane containing the given points is $10x + 29y - 18z - 1 = 0$, so

$$d = \frac{|-10 + 58 - 18 - 1|}{\sqrt{10^2 + 29^2 + 18^2}} = \frac{29}{\sqrt{1{,}265}}$$

25. The equation of the plane normal to the given vector is $2x - y + 2z - 4 = 0$, so

$$d = \frac{|-2 - 2 + 2 - 4|}{\sqrt{4 + 1 + 4}} = \frac{6}{3} = 2$$

27. $d = \dfrac{|27 + 12 + 8|}{\sqrt{3^2 + 4^2}} = \dfrac{47}{5}$

29. $d = \dfrac{|48 - 15 - 2|}{\sqrt{12^2 + 5^2}} = \dfrac{31}{13}$

31. The given line is parallel to the vector
$\mathbf{v} = 3\mathbf{i} + \mathbf{j} + 2\mathbf{k}$ and contains the point
$Q(2, -1, 1)$. Thus, $\mathbf{PQ} = \mathbf{i} - \mathbf{j} + 2\mathbf{k}$ and

$\mathbf{PQ} \times \mathbf{v} = \begin{vmatrix} \mathbf{i} & \mathbf{j} & \mathbf{k} \\ 1 & -1 & 2 \\ 3 & 1 & 2 \end{vmatrix}$

$\quad = -4\mathbf{i} + 4\mathbf{j} + 4\mathbf{k}$

$d = \dfrac{\|\mathbf{PQ} \times \mathbf{v}\|}{\|\mathbf{v}\|}$

$\quad = \dfrac{\sqrt{(-4)^2 + 4^2 + 4^2}}{\sqrt{9 + 1 + 4}}$

$\quad = \dfrac{4\sqrt{3}}{\sqrt{14}}$

33. The first line is parallel to
$\mathbf{v}_1 = -\mathbf{i} + 2\mathbf{j} + 3\mathbf{k}$ and contains
$P_1(2, 5, 0)$, while the second is parallel to
$\mathbf{v}_2 = 2\mathbf{i} - \mathbf{j} + 2\mathbf{k}$ and contains
$P_2(0, -1, 1)$. Then $\mathbf{P}_1\mathbf{P}_2 = -2\mathbf{i} - 6\mathbf{j} + \mathbf{k}$;
$$\mathbf{N} = \mathbf{v}_1 \times \mathbf{v}_2$$

$\quad = \begin{vmatrix} \mathbf{i} & \mathbf{j} & \mathbf{k} \\ -1 & 2 & 3 \\ 2 & -1 & 2 \end{vmatrix}$

$\quad = 7\mathbf{i} + 8\mathbf{j} - 3\mathbf{k}$

$d = \dfrac{|\mathbf{P}_1\mathbf{P}_2 \cdot \mathbf{N}|}{\|\mathbf{N}\|}$

$\quad = \dfrac{|-14 - 48 - 3|}{\sqrt{49 + 64 + 9}}$

$\quad = \dfrac{65}{\sqrt{122}}$

35. The first line is parallel to $\mathbf{v}_1 = \mathbf{i} - 2\mathbf{j}$ and contains $P_1(-1, 0, 3)$, while the second is parallel to $\mathbf{v}_2 = \mathbf{i} - \mathbf{j} + \mathbf{k}$ and contains $P_2(0, -1, 2)$. Then $\mathbf{P}_1\mathbf{P}_2 = \mathbf{i} - \mathbf{j} - \mathbf{k}$ and

$$\mathbf{N} = \mathbf{v}_1 \times \mathbf{v}_2$$

$$= \begin{vmatrix} \mathbf{i} & \mathbf{j} & \mathbf{k} \\ 1 & -2 & 0 \\ 1 & -1 & 1 \end{vmatrix}$$

$$= -2\mathbf{i} - \mathbf{j} + \mathbf{k}$$

$$d = \frac{|\mathbf{P}_1\mathbf{P}_2 \cdot \mathbf{N}|}{\|\mathbf{N}\|}$$

$$= \frac{|-2 + 1 - 1|}{\sqrt{4 + 1 + 1}}$$

$$= \frac{2}{\sqrt{6}}$$

37. a. A vector parallel to the given line is $\mathbf{v} = 3\mathbf{i} - 2\mathbf{j} + \mathbf{k}$ and the normal to the given plane is $\mathbf{N} = \mathbf{i} + 2\mathbf{j} + \mathbf{k}$.

$$\mathbf{N} \cdot \mathbf{v} = 3 - 4 + 1 = 0$$

so the vectors are orthogonal which means that the line is parallel to the given plane.

b. The point $Q(1, 0, -1)$ is on the line, and the distance from the line to the plane is the same as the distance from Q to the plane, namely

$$d = \frac{|1 + 0 - 1 - 1|}{\sqrt{6}} = \frac{1}{\sqrt{6}}$$

39. The distance from $P(x, y, z)$ to the plane $2x - 5y + 3z = 7$ is

$$d = \frac{|2x - 5y + 3z - 7|}{\sqrt{4 + 25 + 9}}$$

Also,

$$\|\mathbf{P}_0\mathbf{P}\|^2 = (x + 1)^2 + (y - 2)^2 + (z - 4)^2$$

Thus, $\|\mathbf{P}_0\mathbf{P}\| = d$ when

$$(x + 1)^2 + (y - 2)^2 + (z - 4)^2$$

$$= \frac{1}{38}(2x - 5y + 3z - 7)^2$$

41. a. $\mathbf{N} = a\mathbf{i} + b\mathbf{j} + c\mathbf{k}$ is normal to the plane, and $\mathbf{v} = A\mathbf{i} + B\mathbf{j} + C\mathbf{k}$ is parallel to the line. The cosine of the acute angle, ϕ between the vectors is

$$\cos \phi = \frac{|\mathbf{v} \cdot \mathbf{N}|}{\|\mathbf{v}\| \, \|\mathbf{N}\|}$$

$$= \frac{aA + bB + cC}{\sqrt{a^2 + b^2 + c^2}\sqrt{A^2 + B^2 + C^2}}$$

b. $\cos \phi = \dfrac{2 + 3 - 1}{\sqrt{3}\sqrt{14}}$

$$\phi \approx 52°$$

Therefore, $\theta \approx 38°$.

43. $\mathbf{N}_1 = 2\mathbf{i} + 3\mathbf{j}, \mathbf{N}_2 = 3\mathbf{i} - \mathbf{j} + \mathbf{k}$

$$\mathbf{N}_1 \times \mathbf{N}_2 = \begin{vmatrix} \mathbf{i} & \mathbf{j} & \mathbf{k} \\ 2 & 3 & 0 \\ 3 & -1 & 1 \end{vmatrix}$$

$$= 3\mathbf{i} - 2\mathbf{j} - 11\mathbf{k}$$

45. $\mathbf{N}_1 = \mathbf{i} + \mathbf{j} + \mathbf{k}, \mathbf{N}_2 = 2\mathbf{i} + 3\mathbf{j} - \mathbf{k}$

$$\mathbf{N}_1 \times \mathbf{N}_2 = \begin{vmatrix} \mathbf{i} & \mathbf{j} & \mathbf{k} \\ 1 & 1 & 1 \\ 2 & 3 & -1 \end{vmatrix}$$
$$= -4\mathbf{i} + 3\mathbf{j} + \mathbf{k}$$

$\|\mathbf{N}_1 \times \mathbf{N}_2\|^2 = 16 + 9 + 1 = 26;$

$\cos\alpha = \dfrac{-4}{\sqrt{26}}$ so $\alpha \approx 2.47$ or $142°;$

$\cos\beta = \dfrac{3}{\sqrt{26}}$ so $\beta \approx 0.94$ or $54°;$

$\cos\gamma = \dfrac{1}{\sqrt{26}}$ so $\gamma \approx 1.37$ or $79°.$

47. A vector normal to the given plane is
$\mathbf{N} = 2\mathbf{i} - 3\mathbf{j} + \mathbf{k}$, and this vector is parallel to the desired line. Thus,

$$\frac{x-1}{2} = \frac{y+5}{-3} = \frac{z-3}{1}$$

49. A vector normal to the first plane is
$\mathbf{N}_1 = \mathbf{i} + \mathbf{j} + \mathbf{k}$ and a vector normal to the second plane is $\mathbf{N}_2 = \mathbf{i} - \mathbf{j} + \mathbf{k}$. Then

$$\mathbf{N}_1 \times \mathbf{N}_2 = \begin{vmatrix} \mathbf{i} & \mathbf{j} & \mathbf{k} \\ 1 & 1 & 1 \\ 1 & -1 & 1 \end{vmatrix}$$
$$= 2\mathbf{i} - 2\mathbf{k}$$

The unit vectors are $\mathbf{N} = \pm \dfrac{1}{\sqrt{2}}(\mathbf{i} - \mathbf{k})$.

51. Two parallel planes have the same normal vectors, so the desired equation is

$$2x - y + 3z + D = 0$$

The plane passes through $P(1, 2, -1)$, so
$$2 - 2 - 3 + D = 0$$
$$D = 3$$
Thus, $2x - y + 3z + 3 = 0$.

53. Parametric equations for the given line are
$x = 1 + 2t, y = -1 - t, z = 3t$. This line intersects the plane when

$$3(1 + 2t) + 2(-1 - t) - 3t = 5$$
$$6t + 3 - 2t - 2 - 3t = 5$$
$$t = 4$$

Thus, the given point is $P(9, -5, 12)$.

55. The vectors $\mathbf{N}_1 = \mathbf{i} + 2\mathbf{j} - 3\mathbf{k}$,
$\mathbf{N}_2 = \mathbf{i} - 2\mathbf{j} + \mathbf{k}$ are normal to the given planes, so the cross-product

$$\mathbf{N}_1 \times \mathbf{N}_2 = \begin{vmatrix} \mathbf{i} & \mathbf{j} & \mathbf{k} \\ 1 & 2 & -3 \\ 1 & -2 & 1 \end{vmatrix}$$
$$= -4\mathbf{i} - 4\mathbf{j} - 4\mathbf{k}$$

is parallel to the required line. Since the line passes through $P(2, 3, 1)$, its equation is

$$\frac{x-2}{1} = \frac{y-3}{1} = \frac{z-1}{1}$$

57. Since $\mathbf{N}_1 = A_1\mathbf{i} + B_1\mathbf{j} + C_1\mathbf{k}$ is normal to the first plane and $\mathbf{N}_2 = A_2\mathbf{i} + B_2\mathbf{j} + C_2\mathbf{k}$ is normal to the second, the angle between the planes is $\pi/2$ if and only if $\mathbf{N}_1 \cdot \mathbf{N}_2 = 0$; that is, $A_1 A_2 + B_1 B_2 + C_1 C_2 = 0$.

59. $\mathbf{N}_1 = \mathbf{v}_1 \times \mathbf{w}_1$ is normal to p_1 and $\mathbf{N}_2 = \mathbf{v}_2 \times \mathbf{w}_2$ is normal to p_2. Then, $\mathbf{v} = \mathbf{N}_1 \times \mathbf{N}_2$ is parallel to both planes, or aligned with their line of intersection.

9.7 Quadratic Surfaces, page 740

1. A quadric surface is the graph of an equation of the form
$Ax^2 + By^2 + Cz^2 + Dxy + Exz + Fyz$
$+ Gx + Hy + Iz + j = 0$
where $A, B, C, D, E, F, G, H, I,$ and J are constants.

3. Circular cone, B

5. Sphere, A

7. Paraboloid, z-axis, K

9. Paraboloid, G

11. Hyperboloid of two sheets, I

13. Ellipsoid, C

15.   **17.**

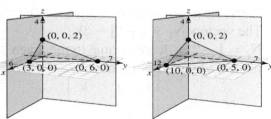

19.   **21.**

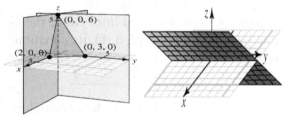

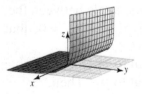

23.   **25.**

27.   **29.**

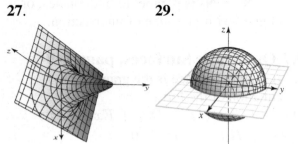

31. Ellipsoid

xy-plane: ellipse

yz-plane: ellipse

xz-plane: ellipse

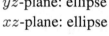
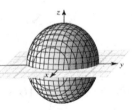

33. Hyperboloid of two sheets

parallel to the xy-plane:

$|z| > 1$: ellipse

yz-plane: hyperbola

xz-plane: hyperbola

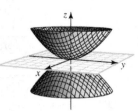

35. Elliptic cone

parallel to the xy-plane; $z > 0$: ellipse

parallel to the yz-plane: hyperbola

parallel to the xz-plane: hyperbola

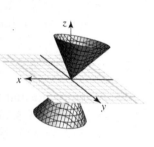

37. Hyperbolic paraboloid

xy-plane: hyperbola

yz-plane: parabola

xz-plane: parabola

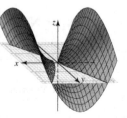

39. Elliptic cone

parallel to the xy-plane; $z \neq 0$: ellipse

parallel to the yz-plane: $x \neq 0$: hyperbola parallel to the xz-plane:

$y \neq 0$: hyperbola

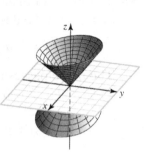

41. Elliptic paraboloid
xy-plane; parabola
yz-plane; parabola
xz-plane; ellipse

43. Elliptic cone
xy-plane; hyperbola
yz-plane; ellipse
xz-plane; hyperbola

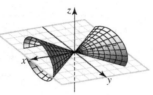

45. Ellipsoid centered at $(1, -2, 3)$ with semi-axes 2, 1, and 3.

47.
$$7x^2 + y^2 + 3z^2 - 9x + 4y + 7z = 1$$
$$7\left[x^2 - \frac{9}{7}x\right] + [y^2 + 4y] + 3\left[z^2 + \frac{7}{3}z\right] = 1$$
$$7\left(x - \frac{9}{14}\right)^2 + (y + 2)^2 + 3\left(z + \frac{7}{6}\right)^2 = \frac{503}{42}$$

This is an ellipsoid centered at $\left(\frac{9}{14}, -2, -\frac{7}{6}\right)$ with semi-axes approximately 1.31, 3.46, and 2.00.

49.
$$x^2 - x - y + z^2 + z + 5 = 0$$
$$\left(x - \frac{1}{2}\right)^2 - y + \left(z + \frac{1}{2}\right)^2 = -5 + \frac{1}{4} + \frac{1}{4}$$
$$y - \frac{9}{2} = \left(x - \frac{1}{2}\right)^2 + \left(z + \frac{1}{2}\right)^2$$

This is an elliptic paraboloid with vertex $\left(\frac{1}{2}, \frac{9}{2}, -\frac{1}{2}\right)$ and the y-axis as the axis of symmetry.

51. The ellipsoid $\dfrac{x^2}{9} + \dfrac{y^2}{4} + \dfrac{z^2}{5} = 1$ and the elliptic cone $z^2 = \dfrac{x^2}{3} + \dfrac{y^2}{2}$ intersect where

$$\frac{x^2}{9} + \frac{y^2}{4} + \frac{1}{5}\left(\frac{x^2}{3} + \frac{y^2}{2}\right) = 1$$
$$\frac{8}{45}x^2 + \frac{7}{20}y^2 = 1$$

This is an ellipse in the xy-plane.

53. Each cross-section perpendicular to the z-axis (for $|z| < c$) is an ellipse with equation

$$\frac{x^2}{a^2} + \frac{y^2}{b^2} = 1 - \frac{z^2}{c^2}$$
$$\frac{x^2}{\alpha^2} + \frac{y^2}{\beta^2} = 1$$

where $\alpha^2 = a^2\left(1 - \dfrac{z^2}{c^2}\right)$ and

$\beta^2 = b^2\left(1 - \dfrac{z^2}{c^2}\right)$. The area of this cross section is

$$\pi\alpha\beta = \pi\left[a\sqrt{1 - \frac{z^2}{c^2}}\right]\left[b\sqrt{1 - \frac{z^2}{c^2}}\right]$$
$$= \pi a b\left(1 - \frac{z^2}{c^2}\right)$$

Integrating with respect to z, we find that the volume of the ellipsoid is

$$V = \int_{-c}^{c} \pi a b\left(1 - \frac{z^2}{c^2}\right) dz$$
$$= \frac{\pi a b}{c^2}\left[c^2 z - \frac{z^3}{3}\right]_{-c}^{c}$$
$$= \frac{4}{3}\pi a b c$$

55. The intersection of the surfaces is found by solving the equations simultaneously:

$$x^2 + y^2 = 1 - y^2$$
$$x^2 + 2y^2 = 1$$

This is the equation of an ellipse.

57. a. The equatorial radius is $a = 6{,}378.2$ and the polar radius is $b = 6{,}356.5$.

b. The volume is

$$V = \frac{4}{3}\pi a^2 b$$
$$= \frac{4}{3}\pi (6{,}378.2)^2(6{,}356.5)$$
$$\approx 1.08319 \times 10^{12} \text{ km}^3$$

Chapter 9 Review

Studying for a chapter examination is a personal process, one which nobody else can do for you. Simply take the time to review what you have done.

59. If the point $P(x, y, z)$ is on the required surface, then

$$\sqrt{(x-1)^2 + y^2 + z^2} + \sqrt{(x+1)^2 + y^2 + z^2} = 3$$
$$\sqrt{(x-1)^2 + y^2 + z^2} = 3 - \sqrt{(x+1)^2 + y^2 + z^2}$$
$$(x-1)^2 + y^2 + z^2 = 9 - 6\sqrt{(x+1)^2 + y^2 + z^2}$$
$$+ (x+1)^2 + y^2 + z^2$$
$$6\sqrt{(x+1)^2 + y^2 + z^2} = 9 + 4x$$
$$6\left[(x+1)^2 + y^2 + z^2\right] = 81 + 72x + 16x^2$$
$$20x^2 + 36y^2 + 36z^2 = 45$$

This is an ellipsoid.

Proficiency Examination, page 742

1. a. A vector is a directed line segment.

   b. A scalar is a real number.

   c. If $\mathbf{v} = \langle a_1, a_2, a_3 \rangle$, then
   $s\mathbf{v} = \langle sa_1, sa_2, sa_3 \rangle$.
   Geometrically, $s\mathbf{v}$ is a directed line segment $s\|\mathbf{v}\|$ units long and points along $\mathbf{v}$ if $s > 0$ or in the opposite direction if $s < 0$.

   d. If two vectors are arranged so that their initial points coincide, then the sum of the vectors is the diagonal of the parallelogram formed by the two vectors.

2. a. $\mathbf{u} + \mathbf{v} = \mathbf{v} + \mathbf{u}$

   b. $(\mathbf{u} + \mathbf{v}) + \mathbf{w} = \mathbf{u} + (\mathbf{v} + \mathbf{w})$

   c. $\mathbf{u} + \mathbf{0} = \mathbf{u}$

   d. $\mathbf{u} + (-\mathbf{u}) = \mathbf{0}$

Page 366

e. $\|\mathbf{v}\| = (\mathbf{v} \cdot \mathbf{v})^{1/2}$

f. $\mathbf{v} \cdot \mathbf{w} = \mathbf{w} \cdot \mathbf{v}$

g. $c(\mathbf{v} \cdot \mathbf{w}) = (c\mathbf{v}) \cdot \mathbf{w} = \mathbf{v} \cdot (c\mathbf{w})$

h. $\mathbf{u} \cdot (\mathbf{v} + \mathbf{w}) = \mathbf{u} \cdot \mathbf{v} + \mathbf{u} \cdot \mathbf{w}$

i. $\mathbf{v} \times \mathbf{0} = \mathbf{0} \times \mathbf{v} = \mathbf{0}$

j. $\mathbf{v} \times \mathbf{w} = -(\mathbf{w} \times \mathbf{v})$

k. $\mathbf{u} \times (\mathbf{v} + \mathbf{w}) = (\mathbf{u} \times \mathbf{v}) + (\mathbf{u} \times \mathbf{w})$

or

$(\mathbf{u} + \mathbf{v}) \times \mathbf{w} = (\mathbf{u} \times \mathbf{w}) + (\mathbf{v} \times \mathbf{w})$

3. a. If $\mathbf{v} = a_1\mathbf{i} + a_2\mathbf{j} + a_3\mathbf{k}$ and $\mathbf{w} = b_1\mathbf{i} + b_2\mathbf{j} + b_3\mathbf{k}$, the cross product, written $\mathbf{v} \times \mathbf{w}$, is the vector

$$\mathbf{v} \times \mathbf{w} = (a_2b_3 - a_3b_2)\mathbf{i} + (a_3b_1 - a_1b_3)\mathbf{j} + (a_1b_2 - a_2b_1)\mathbf{k}$$

These terms can be obtained by using a determinant

$$\mathbf{v} \times \mathbf{w} = \begin{vmatrix} \mathbf{i} & \mathbf{j} & \mathbf{k} \\ a_1 & a_2 & a_3 \\ b_1 & b_2 & b_3 \end{vmatrix}$$

b. If $\mathbf{v}$ and $\mathbf{w}$ are nonzero vectors in $\mathbb{R}^3$, that are not multiples of one another, then $\mathbf{v} \times \mathbf{w}$ is orthogonal to both $\mathbf{v}$ and $\mathbf{w}$.

c. If $\mathbf{v}$ and $\mathbf{w}$ are nonzero vectors in $\mathbb{R}^3$ with θ the angle between $\mathbf{v}$ and $\mathbf{w}$ $(0 \le \theta \le \pi)$, then

$$\|\mathbf{v} \times \mathbf{w}\| = \|\mathbf{v}\|\|\mathbf{w}\|\sin\theta$$

d. If $\mathbf{u} = s\mathbf{v}$, then $\mathbf{u} \times \mathbf{v} = \mathbf{0}$.

4. a. $\|\mathbf{u}\| = \|a\mathbf{i} + b\mathbf{j} + c\mathbf{k}\|$
$= \sqrt{a^2 + b^2 + c^2}$

b. A unit vector is a vector with length 1.

c. $\mathbf{u} = \dfrac{\mathbf{v}}{\|\mathbf{v}\|}$

d. $\|\mathbf{u} + \mathbf{v}\| \le \|\mathbf{u}\| + \|\mathbf{v}\|$

e. $\mathbf{i}, \mathbf{j}, \mathbf{k}$

5. a. $(x-a)^2 + (y-b)^2 + (z-c)^2 = r^2$

b. Draw a planar curve. Define a direction (generatrix). Move a line (directrix) along the curve parallel to the generatrix. The resulting surface is a cylinder.

6. a. $\|\mathbf{v} \times \mathbf{w}\|^2 = \|\mathbf{v}\|^2\|\mathbf{w}\|^2 - (\mathbf{v} \cdot \mathbf{w})^2$

b. An object moving with displacement $\mathbf{R}$ with a constant force $\mathbf{F}$ does work $W = \mathbf{F} \cdot \mathbf{R}$.

7. a. The dot product of vectors $\mathbf{v} = a_1\mathbf{i} + a_2\mathbf{j} + a_3\mathbf{k}$ and $\mathbf{w} = b_1\mathbf{i} + b_2\mathbf{j} + b_3\mathbf{k}$ is the scalar denoted by $\mathbf{v} \cdot \mathbf{w}$ and given by

$$\mathbf{v} \cdot \mathbf{w} = a_1b_1 + a_2b_2 + a_3b_3$$

Alternately, $\mathbf{u} \cdot \mathbf{v} = \|\mathbf{u}\|\,\|\mathbf{v}\|\cos\theta$ where θ is the angle between the vectors $\mathbf{u}$ and $\mathbf{v}$.

b. If $\mathbf{v} = a_1\mathbf{i} + a_2\mathbf{j} + a_3\mathbf{k}$ is a nonzero vector, then the direction cosines of $\mathbf{v}$ are $\cos\alpha = \dfrac{a_1}{\|\mathbf{v}\|}$; $\cos\beta = \dfrac{a_2}{\|\mathbf{v}\|}$; $\cos\gamma = \dfrac{a_3}{\|\mathbf{v}\|}$.

8. a. If θ is the angle between the nonzero vectors $\mathbf{v}$ and $\mathbf{w}$, then

$$\cos\theta = \frac{\mathbf{v} \cdot \mathbf{w}}{\|\mathbf{v}\|\|\mathbf{w}\|}$$

b. $\mathbf{u}$ and $\mathbf{v}$ are orthogonal vectors if the lines determined by those vectors are perpendicular.

c. Algebraically, $\mathbf{u} \cdot \mathbf{v} = 0$.

9. a. The vector projection of $\mathbf{AB}$ onto $\mathbf{AC}$ is the vector from $\mathbf{A}$ to a point D on $\mathbf{AC}$ so $\overline{BD}$ is perpendicular to the line formed by $\mathbf{AC}$.

b. The formula for the vector projection of
v in the direction of **w** is the vector

$$\text{proj}_\mathbf{w}\mathbf{v} = \left(\frac{\mathbf{v} \cdot \mathbf{w}}{\mathbf{w} \cdot \mathbf{w}}\right)\mathbf{w}$$

c. The scalar projection of **AB** onto **AC** is
the component of **v** in the direction of **w**.

d. It is the scalar

$$\text{comp}_\mathbf{w}\mathbf{v} = \frac{\mathbf{v} \cdot \mathbf{w}}{\|\mathbf{w}\|}$$

10. Place the little finger of your right hand
along the positive x-axis and wrap your
fingers around the positive y-axis. Then
your thumb points in the direction of the
positive z-axis.

11. a. If $\mathbf{u} = a_1\mathbf{i} + a_2\mathbf{j} + a_3\mathbf{k}$,
$\mathbf{v} = b_1\mathbf{i} + b_2\mathbf{j} + b_3\mathbf{k}$, and
$\mathbf{w} = c_1\mathbf{i} + c_2\mathbf{j} + c_3\mathbf{k}$, then the triple scalar
product can be found by evaluating the
determinant

$$(\mathbf{u} \times \mathbf{v}) \cdot \mathbf{w} = \begin{vmatrix} a_1 & a_2 & a_3 \\ b_1 & b_2 & b_3 \\ c_1 & c_2 & c_3 \end{vmatrix}$$

b. The volume of a parallelepiped formed
by **u**, **v**, and **w** is the absolute value of
the triple scalar product, $|(\mathbf{u} \times \mathbf{v}) \cdot \mathbf{w}|$.

12. Suppose the line passes through the point
(x_0, y_0, z_0) and is parallel to
$\mathbf{v} = A\mathbf{i} + B\mathbf{j} + C\mathbf{k}$.

a. $x = x_0 + tA$, $y = y_0 + tB$, $z = z_0 + tC$

b. $\dfrac{x - x_0}{A} = \dfrac{y - y_0}{B} = \dfrac{z - z_0}{C}$

13. a. The plane $Ax + By + Cz + D = 0$ has
normal vector $\mathbf{N} = A\mathbf{i} + B\mathbf{j} + C\mathbf{k}$

b. The plane containing the point
$P_0(x_0, y_0, z_0)$ with normal vector

$\mathbf{N} = A\mathbf{i} + B\mathbf{j} + C\mathbf{k}$ has *point-normal*
form
$A(x - x_0) + B(y - y_0) + C(z - z_0) = 0$

c. The standard form of a plane with
normal vector $\mathbf{N} = A\mathbf{i} + B\mathbf{j} + C\mathbf{k}$ is
$Ax + By + Cz + D = 0$.

14. a. The distance $|P_1 P_2|$ between
$P_1(x_1, y_1, z_1)$ and $P_2(x_2, y_2, z_2)$ is
$|P_1 P_2|$
$$= \sqrt{(x_2 - x_1)^2 + (y_2 - y_1)^2 + (z_2 - z_1)^2}$$

b. The distance from the point (x_0, y_0, z_0)
to the plane $Ax + By + Cz + D = 0$ is
given by

$$d = \frac{|Ax_0 + By_0 + Cz_0 + D|}{\sqrt{A^2 + B^2 + C^2}}$$

c. The (shortest) distance from the point P
to the line L is given by the formula

$$d = \frac{\|\mathbf{v} \times \mathbf{QP}\|}{\|\mathbf{v}\|}$$

where **v** is a vector aligned with L and Q
is any point on L.

15. A quadric surface is the set of all (x, y, z)
that satisfy the equation
$Ax^2 + By^2 + Cz^2 + Dxy + Exz + Fyz$
$+ Gx + Hy + Iz + J = 0$
where $A, B, C, D, E, F, G, H, I$, and J are
constants.

16. a. $2\mathbf{v} + 3\mathbf{w} = [2(2) + 3(3)]\mathbf{i} + [2(-3) + 3(-2)]\mathbf{j}$
$+ [2(1) + 3(0)]\mathbf{k}$
$= 13\mathbf{i} - 12\mathbf{j} + 2\mathbf{k}$

b. $\|\mathbf{v}\|^2 - \|\mathbf{w}\|^2 = [2^2 + (-3)^2 + 1^2]$
$- [3^2 + (-2)^2]$
$= 14 - 13$
$= 1$

$$= \frac{1}{9}\|\mathbf{AB}\|^2 + \|\mathbf{AB}\|^2 - 2\left[\frac{1}{3}\|\mathbf{AB}\|^2\left(\frac{1}{2}\right)\right]$$
$$= \frac{7}{9}\|\mathbf{AB}\|^2$$

91. The result (from Problems 89 and 90) is still true for nonequilateral triangles, but the proof is more complicated.

93. Let a, b, c satisfy $\boldsymbol{\gamma} = a\mathbf{u} + b\boldsymbol{\alpha} + c\boldsymbol{\beta}$; then
$$\boldsymbol{\gamma}\cdot\mathbf{u} = a(\mathbf{u}\cdot\mathbf{u}) + b(\boldsymbol{\alpha}\cdot\mathbf{u}) + c(\boldsymbol{\beta}\cdot\mathbf{u})$$
$$= a(\mathbf{u}\cdot\mathbf{u}) + 0 + 0$$
$$a = \frac{\boldsymbol{\gamma}\cdot\mathbf{u}}{\mathbf{u}\cdot\mathbf{u}} = \frac{\boldsymbol{\gamma}\cdot\mathbf{u}}{\|\mathbf{u}\|^2}$$

Similarly,
$$\boldsymbol{\gamma}\cdot\boldsymbol{\alpha} = b(\boldsymbol{\alpha}\cdot\boldsymbol{\alpha}) \quad \boldsymbol{\gamma}\cdot\boldsymbol{\beta} = c(\boldsymbol{\beta}\cdot\boldsymbol{\beta})$$
$$b = \frac{\boldsymbol{\gamma}\cdot\boldsymbol{\alpha}}{\|\boldsymbol{\alpha}\|^2} \qquad c = \frac{\boldsymbol{\gamma}\cdot\boldsymbol{\beta}}{\|\boldsymbol{\beta}\|^2}$$

Thus,
$$\boldsymbol{\gamma} = \frac{(\boldsymbol{\gamma}\cdot\mathbf{u})\mathbf{u}}{\|\mathbf{u}\|^2} + \frac{(\boldsymbol{\gamma}\cdot\boldsymbol{\alpha})\boldsymbol{\alpha}}{\|\boldsymbol{\alpha}\|^2} + \frac{(\boldsymbol{\gamma}\cdot\boldsymbol{\beta})\boldsymbol{\beta}}{\|\boldsymbol{\beta}\|^2}$$

95. Let s and t be scalars such that
$$\mathbf{AP} = s\mathbf{AM} \text{ and } \mathbf{CP} = t\mathbf{CR}_k. \text{ Then,}$$
$$\mathbf{CR}_k = \mathbf{AR}_k - \mathbf{AC}$$
$$= \frac{1}{2k+1}\mathbf{AB} - \mathbf{AC}$$
and since $\mathbf{CM} = \frac{1}{2}\mathbf{CD} = \frac{1}{2}\mathbf{AB}$,
$$\mathbf{AM} = \mathbf{AC} + \mathbf{CM} = \mathbf{AC} + \frac{1}{2}\mathbf{AB}$$
Substituting into the vector equation,
$$\mathbf{AP} = \mathbf{AC} + \mathbf{CP}$$
we obtain
$$s\left(\mathbf{AC} + \frac{1}{2}\mathbf{AB}\right) = \mathbf{AC} + t\left(\frac{1}{2k+1}\mathbf{AB} - \mathbf{AC}\right)$$
$$(s+t-1)\mathbf{AC} = \left(\frac{t}{2k+1} - \frac{1}{2}s\right)\mathbf{AB}$$

Since $\mathbf{AC}$ and $\mathbf{AB}$ are not parallel, we have
$$\begin{cases} s+t-1 = 0 \\ \dfrac{t}{2k+1} - \dfrac{1}{2}s = 0 \end{cases}$$

Solving $(s,t) = \left(\dfrac{2}{2k+3}, \dfrac{2k+1}{2k+3}\right)$.

Finally using proportionalities in the similar triangles $\triangle ACR_k$ and $\triangle PR_kR_{k+1}$, we find
$$\frac{\|\mathbf{AR}_{k+1}\|}{\|\mathbf{AR}_k\|} = \frac{\|\mathbf{CP}\|}{\|\mathbf{CR}_k\|} = \frac{2k+1}{2k+3}$$
$$\|\mathbf{AR}_{k+1}\| = \left(\frac{2k+1}{2k+3}\right)\|\mathbf{AR}_k\|$$
$$= \left(\frac{2k+1}{2k+3}\right)\left(\frac{1}{2k+1}\right)\|\mathbf{AB}\|$$
$$\mathbf{AR}_{k+1} = \left(\frac{1}{2k+3}\right)\mathbf{AB}$$

97. This is Putnam Problem 4 of the morning session of 1939. Suppose the required line L meets the given lines at points $A, B, C,$ and D, respectively. Then $A(1,0,a)$, $B(b,1,0)$, $C(0,c,1)$, $D(6d,6d,-d)$ for some $a, b, c,$ and d. Treat $A, B, C,$ and D as vectors. The condition that they be collinear is that the vectors
$$\mathbf{B} - \mathbf{A} = \langle b-1, 1, -a\rangle,$$
$$\mathbf{C} - \mathbf{A} = \langle -1, c, 1-a\rangle,$$
$$\mathbf{D} - \mathbf{A} = \langle 6d-1, 6d, -d-a\rangle$$
The proportionality of the first two tells us that
$$c = \frac{1}{1-b} = \frac{a-1}{a}$$
while the first and third give
$$6d = \frac{1-6d}{1-b} = \frac{a+d}{a}$$
Rewrite the middle member here using the first string of equations
$$6d = (1-6d)\left(\frac{a-1}{a}\right) = \frac{a+d}{a}$$
Clearing fractions $6ad = a+d$,
$$a + 6d - 1 - 6ad = a + d$$

Adding these equations, we find $4d = a + 1$, so

$$6a(a+1) = 24ad = 4(a+d) = 5a + 1$$

This quadratic equation has roots $a = \frac{1}{3}$, $-\frac{1}{2}$ and the corresponding values of the other unknowns are

$$b = \tfrac{3}{2}, \tfrac{2}{3}; \qquad c = -2, 3; \qquad d = \tfrac{1}{3}, \tfrac{1}{8}$$

The direction vectors of the lines (proportional to $\mathbf{B} - \mathbf{A}$, $\mathbf{C} - \mathbf{A}$, and $\mathbf{D} - \mathbf{A}$) in the two cases are $(3, 6, -2)$ and $(-2, 6, 3)$, respectively. The two lines are given parametrically by

$$L_1 \colon s \to \left(1, 0, \frac{1}{3}\right) + s(3, 6, -2)$$

$$L_2 \colon t \to \left(1, 0, -\frac{1}{2}\right) + t(-2, 6, 3)$$

These lines cross the given lines (in order) for $s = 0, \frac{1}{6}, -\frac{1}{3}, \frac{1}{3}$ and $t = 0, \frac{1}{6}, \frac{1}{2}, \frac{1}{8}$. In nonparametric form, L_1 is given by

$$y = 2(x-1) = 1 - 3z$$

and L_2 is given by

$$y = 3(1-x) = 2z + 1$$

For a general treatment of this problem, see D. M. Y. Sommerville, *Analytic Geometry of Three Dimensions*, Cambridge, 1934, p. 184.

99. This is Putnam Problem 2 of the morning session of 1983. Let $\mathbf{OA}$ be the long hand and $\mathbf{OB}$ be the short hand. We can think of $\mathbf{OA}$ as fixed and $\mathbf{OB}$ as rotating at constant speed. Let $\mathbf{v}$ be the vector $\mathbf{OB}$ under this assumption. The rate of change of the distance between A and B is the component of $\mathbf{v}$ in the direction of $\mathbf{AB}$. Since $\mathbf{v}$ is orthogonal to $\mathbf{OB}$ and the magnitude of $\mathbf{v}$ is constant, this component is maximal when $\angle OAB$ has closed to

form a triangle; that is, when the distance $\mathbf{AB}$ is $\sqrt{4^2 - 3^2} = \sqrt{7}$. Alternately, let x be the distance $\mathbf{AB}$ and $\theta = \angle AOB$. By the law of cosines,

$$x^2 = 3^2 + 4^2 - 2(3)(4) \cos \theta$$
$$= 25 - 24 \cos \theta$$

Now,

$$2x \frac{dx}{dt} = 24 \sin \theta$$
$$\frac{dx}{d\theta} = \frac{12 \sin \theta}{\sqrt{25 - 24 \cos \theta}}$$

Since dx/dt is an odd function of θ, $|dx/dt|$ is a minimum when $dx/d\theta$ is a maximum or a minimum. Since dx/dt is a periodic differentiable function of θ, $d^2x/ds^2 = 0$ at the extremes for dx/dt. For such θ,

$$12 \cos \theta = x \frac{d^2x}{d\theta^2} + \left(\frac{dx}{d\theta}\right)^2$$
$$= \left(\frac{dx}{d\theta}\right)^2$$
$$= \frac{144 \sin^2 \theta}{x^2}$$
$$x^2 = \frac{12 \sin^2 \theta}{\cos \theta}$$
$$= \frac{12 - 12 \cos^2 \theta}{\cos \theta}$$
$$= 25 - 24 \cos \theta$$

It follows that

$$12 \cos^2 \theta - 25 \cos \theta + 12 = 0$$

The only allowable solution for $\cos \theta$ is $\cos \theta = \frac{3}{4}$ and hence

$$x = \sqrt{25 - 24 \cos \theta}$$
$$= \sqrt{25 - 18}$$
$$= \sqrt{7}$$

CHAPTER 10 Vector-Valued Functions

Chapter Overview
You have now completed a course in single-variable calculus, and at this point should be comfortable with the three new ideas, limit, derivative, and integral. In the last chapter, you learned about vectors, and in this chapter we join together the ideas of calculus and vectors into what is known as **vector calculus**. In this chapter, we will see that vector-valued functions behave much like the functions studied earlier in this text.

10.1 Introduction to Vector Functions, page 758

SURVIVAL HINT: *The first two problems in this chapter make sure you understand the new concept necessary for working in this chapter.*

1. A vector-valued function **F** with domain D assigns a unique vector $\mathbf{F}(t)$ to each scalar t in the set D. The set of all vectors **v** of the form $\mathbf{v} = \mathbf{F}(t)$ for t in D is the range of **F**. In this text we consider vector functions whose range are vectors in $\mathbb{R}^2$ or $\mathbb{R}^3$; that is,

$$\mathbf{F}(t) = f_1(t)\mathbf{i} + f_2(t)\mathbf{j} \text{ in } \mathbb{R}^2 \text{ (plane)}$$
$$\mathbf{F}(t) = f_1(t)\mathbf{i} + f_2(t)\mathbf{j} + f_3(t)\mathbf{k} \text{ in } \mathbb{R}^3 \text{ (space)}$$

where $f_1, f_2,$ and f_3 are real-valued (scalar-valued) functions of the real number t defined on the domain set D. In this context f_1, f_2, and f_3 are called the components of **F**.

3. $t \neq 0$

5. $t \neq \dfrac{(2n+1)\pi}{2}$
   n an integer

7. $t \neq \dfrac{n\pi}{2}$
   n an integer

9. $t > 0$

11. parabolic cylinder; in $\mathbb{R}^2$, the graph is a parabola in the xy-plane

13. cylinder; in $\mathbb{R}^2$, the graph is a a circle in the xy-plane

15. plane; in $\mathbb{R}^2$, the graph is a line in the xz-plane parallel to and four units below the x-axis

17. a circular helix; in $\mathbb{R}^2$, the graph is a circle in the xy-plane

19. the curve is in the intersection of the parabolic cylinder $y = (1-x)^2$ with the plane $x + z = 1$; in $\mathbb{R}^2$, the graph is a parabola in the xy-plane

SURVIVAL HINT: *In addition to the basic operations of addition, subtraction, and scalar multiplication, we work with the operations of dot and cross product of vector functions.*

21. $2\mathbf{F}(t) - 3\mathbf{G}(t)$
$$= [2(2t) - 3(1-t)]\mathbf{i} + [2(-5) - 3(0)]\mathbf{j}$$
$$+ \left[2(t^2) - 3\left(\frac{1}{t}\right)\right]\mathbf{k}$$

$$= (7t - 3)\mathbf{i} - 10\mathbf{j} + \left(2t^2 - \frac{3}{t}\right)\mathbf{k}$$

23. $\mathbf{F}(t) \cdot \mathbf{G}(t)$

$$= (2t\mathbf{i} - 5\mathbf{j} + t^2\mathbf{k}) \cdot \left[(1 - t)\mathbf{i} + \left(\frac{1}{t}\right)\mathbf{k}\right]$$

$$= (2t - 2t^2) + t$$

$$= 3t - 2t^2$$

25. $\mathbf{G}(t) \cdot \mathbf{H}(t)$

$$= \left[(1 - t)\mathbf{i} + \left(\frac{1}{t}\right)\mathbf{k}\right] \cdot (\sin t\,\mathbf{i} + e^t\mathbf{j})$$

$$= (1 - t)\sin t$$

27. $\mathbf{F}(t) \times \mathbf{H}(t) = \begin{vmatrix} \mathbf{i} & \mathbf{j} & \mathbf{k} \\ 2t & -5 & t^2 \\ \sin t & e^t & o \end{vmatrix}$

$$= -t^2 e^t\,\mathbf{i} + (t^2 \sin t\,)\mathbf{j} + (2te^t + 5\sin t)\mathbf{k}$$

29. $2e^t\mathbf{F}(t) + t\mathbf{G}(t) + 10\mathbf{H}(t)$

$$= [2e^t(2t) + t(1 - t) + 10\sin t]\mathbf{i}$$
$$+ [2e^t(-5) + 10e^t]\mathbf{j}$$
$$+ [2e^t t^2 + 1]\mathbf{k}$$

$$= (4te^t - t^2 + t + 10\sin t)\mathbf{i}$$
$$+ (2e^t t^2 + 1)\mathbf{k}$$

31. $\mathbf{G}(t) \cdot [\mathbf{H}(t) \times \mathbf{F}(t)] = \begin{vmatrix} 1 - t & 0 & \frac{1}{t} \\ \sin t & e^t & 0 \\ 2t & -5 & t^2 \end{vmatrix}$

$$= t^2 e^t - t^3 e^t - 2e^t - \frac{5}{t}\sin t$$

33. $x = 2\sin t,\ y = 2\sin t,\ z = \sqrt{8}\cos t$

$$x^2 + y^2 + z^2 = 4\sin^2 t + 4\sin^2 t + 8\cos^2 t$$
$$= 8(\sin^2 t + \cos^2 t)$$
$$= 8$$

This is a sphere with center at the origin and a radius of $2\sqrt{2}$.

35. $\mathbf{F}(t) = t\mathbf{i} + t^2\mathbf{j} + 2\mathbf{k}$

37. $\mathbf{F}(t) = 2t\,\mathbf{i} + (1 - t)\mathbf{j} + (\sin t)\mathbf{k}$

39. $\mathbf{F}(t) = t^2\mathbf{i} + t\mathbf{j} + \sqrt{9 - t^2 - t^4}\mathbf{k}$

41. $\lim\limits_{t \to 1} [2t\mathbf{i} - 3\mathbf{j} + e^t\mathbf{k}] = 2\mathbf{i} - 3\mathbf{j} + e\mathbf{k}$

43. $\lim\limits_{t \to 0} \left[\dfrac{(\sin t)\mathbf{i} - t\mathbf{k}}{t^2 + t - 1}\right] = \mathbf{0}$

45. $\lim\limits_{t \to 0} \left[\dfrac{te^t}{1 - e^t}\mathbf{i} + \dfrac{e^{t-1}}{\cos t}\mathbf{j}\right]$

$$= \left[\lim\limits_{t \to 0} \frac{te^t + e^t}{-e^t}\right]\mathbf{i} + \left[\lim\limits_{t \to 0} \frac{e^{t-1}}{\cos t}\right]\mathbf{j}$$

$$= \left[\lim\limits_{t \to 0}(-t - 1)\right]\mathbf{i} + e^{-1}\mathbf{j}$$

$$= -\mathbf{i} + e^{-1}\mathbf{j}$$

47. $\lim\limits_{t \to 0^+} \left[\dfrac{\sin 3t}{\sin 2t}\mathbf{i} + \dfrac{\ln(\sin t)}{\ln(\tan t)}\mathbf{j} + (t\ln t)\mathbf{k}\right]$

$$= \lim\limits_{t \to 0^+} \left[\left(\frac{3\cos 3t}{2\cos 2t}\right)\mathbf{i} + \left(\frac{\cos t \tan t}{\sin t \sec^2 t}\right)\mathbf{j} + \left(\frac{\frac{1}{t}}{-\frac{1}{t^2}}\right)\mathbf{k}\right]$$

$$= \frac{3}{2}\mathbf{i} + \mathbf{j}$$

49. continuous for all t

51. continuous for $t \neq 0,\ t \neq -1$

53. continuous for all $t \neq 0$

55. We show $\mathbf{R}(t) = \left\langle t, \dfrac{1 - t}{t}, \dfrac{1 - t^2}{t} \right\rangle$ lies on a plane by finding constants A, B, C, and D so that $Ax + By + Cz = D$. Thus,

$$At + B\left(\frac{1 - t}{t}\right) + C\left(\frac{1 - t^2}{t}\right) = D$$

$$(-B - D) + (A - C)t + (B + C)\left(\frac{1}{t}\right) = 0$$

Solving the system of equations
$$\begin{cases} -B - D = 0 \\ A - C = 0 \\ B + C = 0 \end{cases}$$

we obtain $A = C = -B = D$. Choosing $A = 1$, $B = -1, C = 1$, and $D = 1$, we conclude that $\mathbf{R}(t)$ lies on the plane
$$x - y + z = 1$$

57. a. $\lim\limits_{t \to 0} e^t \mathbf{F}(t) = \lim\limits_{t \to 0} e^t(t\mathbf{i} + t^2\mathbf{j} + t^3\mathbf{k}) = \mathbf{0}$

b. $\lim\limits_{t \to 1} \mathbf{F}(t) \cdot \mathbf{G}(t)$

$$= \lim\limits_{t \to 1} \left[(t\mathbf{i} + t^2\mathbf{j} + t^3\mathbf{k}) \cdot \left(\frac{1}{t}\mathbf{i} - e^t\mathbf{j} \right) \right]$$

$$= \lim\limits_{t \to 1} \left(1 - t^2 e^t \right)$$

$$= 1 - e$$

$$\left[\lim\limits_{t \to 1} \mathbf{F}(t) \right] \cdot \left[\lim\limits_{t \to 1} \mathbf{G}(t) \right]$$

$$= \left[\lim\limits_{t \to 1} (t\mathbf{i} + t^2\mathbf{j} + t^3\mathbf{k}) \right] \cdot \left[\lim\limits_{t \to 1} \left(\frac{1}{t}\mathbf{i} - e^t\mathbf{j} \right) \right]$$

$$= (\mathbf{i} + \mathbf{j} + \mathbf{k}) \cdot (\mathbf{i} - e\mathbf{j})$$

$$= 1 - e$$

c. $\lim\limits_{t \to 1} [\mathbf{F}(t) \times \mathbf{G}(t)]$

$$= \lim\limits_{t \to 1} \begin{vmatrix} \mathbf{i} & \mathbf{j} & \mathbf{k} \\ t & t^2 & t^3 \\ \frac{1}{t} & -e^t & 0 \end{vmatrix}$$

$$= \lim\limits_{t \to 1} \left[t^3 e^t \mathbf{i} + t^2\mathbf{j} + (-te^t - t)\mathbf{k} \right]$$

$$= e\mathbf{i} + \mathbf{j} + (-e - 1)\mathbf{k}$$

$$\left[\lim\limits_{t \to 1} \mathbf{F}(t) \right] \times \left[\lim\limits_{t \to 1} \mathbf{G}(t) \right]$$

$$= \begin{vmatrix} \mathbf{i} & \mathbf{j} & \mathbf{k} \\ 1 & 1 & 1 \\ 1 & -e & 0 \end{vmatrix}$$

$$= e\mathbf{i} + \mathbf{j} + (-e - 1)\mathbf{k}$$

59. $\lim\limits_{t \to t_0} [\mathbf{F}(t) + \mathbf{G}(t)]$

$$= \lim\limits_{t \to t_0} \Big[f_1(t)\mathbf{i} + f_2(t)\mathbf{j} + f_3(t)\mathbf{k}$$

$$+ g_1(t)\mathbf{i} + g_2(t)\mathbf{j} + g_3(t)\mathbf{k} \Big]$$

$$= \lim\limits_{t \to t_0} \left[f_1(t)\mathbf{i} + f_2(t)\mathbf{j} + f_3(t)\mathbf{k} \right]$$

$$+ \lim\limits_{t \to t_0} \left[g_1(t)\mathbf{i} + g_2(t)\mathbf{j} + g_3(t)\mathbf{k} \right]$$

$$= \lim\limits_{t \to t_0} \mathbf{F}(t) + \lim\limits_{t \to t_0} \mathbf{G}(t)$$

10.2 Differentiation and Integration of Vector Functions, page 769

1. A smooth curve has no corners; the first derivative vector $\mathbf{R}'(t)$ is continuous.

3. $\mathbf{F}'(t) = \mathbf{i} + 2t\mathbf{j} + (1 + 3t^2)\mathbf{k}$

5. $\mathbf{F}'(s) = \dfrac{1}{s}(s\mathbf{i} + 5\mathbf{j} - e^s\mathbf{k}) + (\ln s)(\mathbf{i} - e^s\mathbf{k})$

$$= (1 + \ln s)\mathbf{i} + 5s^{-1}\mathbf{j} - e^s(\ln s + s^{-1})\mathbf{k}$$

7. $\mathbf{F}'(t) = 2t\mathbf{i} - t^{-2}\mathbf{j} + 2e^{2t}\mathbf{k}$

$\mathbf{F}''(t) = 2\mathbf{i} + 2t^{-3}\mathbf{j} + 4e^{2t}\mathbf{k}$

9. $\mathbf{F}'(s) = (\cos s)\mathbf{i} - (\sin s)\mathbf{j} + 2s\mathbf{k}$

$\mathbf{F}''(s) = (-\sin s)\mathbf{i} - (\cos s)\mathbf{j} + 2\mathbf{k}$

11. $f(x) = 2x^2 + (x + 1)\left(-3x^2 \right)$

$$= -3x^3 - x^2$$

$$f'(x) = -9x^2 - 2x$$

13. $g(x) = \sqrt{(\sin x)^2 + (-2x)^2 + (\cos x)^2}$

$$= \sqrt{1 + 4x^2}$$

$$g'(x) = \frac{4x}{\sqrt{1 + 4x^2}}$$

15. $\mathbf{V}(t) = \mathbf{R}'(t) = \mathbf{i} + 2t\mathbf{j} + 2\mathbf{k};$

$\mathbf{V}(1) = \mathbf{i} + 2\mathbf{j} + 2\mathbf{k}$

$\mathbf{A}(t) = \mathbf{V}'(t) = \mathbf{R}''(t) = 2\mathbf{j};$

$\mathbf{A}(1) = 2\mathbf{j}$

speed $= \|\mathbf{V}(1)\| = \sqrt{1^2 + 2^2 + 2^2} = 3$

Direction of motion is that of the unit vector

$$\frac{\mathbf{V}}{\|\mathbf{V}\|} = \frac{1}{3}\mathbf{i} + \frac{2}{3}\mathbf{j} + \frac{2}{3}\mathbf{k}$$

17. $\mathbf{V}(t) = \mathbf{R}'(t) = -\sin t\,\mathbf{i} + \cos t\,\mathbf{j} + 3\mathbf{k};$

$\mathbf{V}(\frac{\pi}{4}) = -\frac{\sqrt{2}}{2}\mathbf{i} + \frac{\sqrt{2}}{2}\mathbf{j} + 3\mathbf{k}$

$\mathbf{A}(t) = \mathbf{V}'(t) = \mathbf{R}''(t) = -\cos t\,\mathbf{i} - \sin t\mathbf{j} + 3\mathbf{k}$

$\mathbf{A}(\frac{\pi}{4}) = -\frac{\sqrt{2}}{2}\mathbf{i} - \frac{\sqrt{2}}{2}\mathbf{j}$

speed $= \|\mathbf{V}(\frac{\pi}{4})\| = \sqrt{\frac{2}{4} + \frac{2}{4} + 9} = \sqrt{10}$

Direction of motion is that of the unit vector

$$\frac{\mathbf{V}}{\|\mathbf{V}\|} = -\frac{1}{2\sqrt{5}}\mathbf{i} + \frac{1}{2\sqrt{5}}\mathbf{j} + \frac{3}{\sqrt{10}}\mathbf{k}$$

19. $\mathbf{V}(t) = \mathbf{R}'(t) = e^t\mathbf{i} - e^{-t}\mathbf{j} + 2e^{2t}\mathbf{k}$;

$\mathbf{V}(\ln 2) = 2\mathbf{i} - \frac{1}{2}\mathbf{j} + 8\mathbf{k}$

$\mathbf{A}(t) = \mathbf{V}'(t) = \mathbf{R}''(t) = e^t\mathbf{i} + e^{-t}\mathbf{j} + 4e^{2t}\mathbf{k}$

$\mathbf{A}(\ln 2) = 2\mathbf{i} + \frac{1}{2}\mathbf{j} + 16\mathbf{k}$

speed $= \|\mathbf{V}(\ln 2)\|$

$$= \sqrt{2^2 + \left(\frac{1}{2}\right)^2 + 8^2}$$

$$= \frac{\sqrt{273}}{2}$$

Direction of motion is that of the unit vector

$$\frac{\mathbf{V}}{\|\mathbf{V}\|} = \frac{1}{\sqrt{273}}(4\mathbf{i} - \mathbf{j} + 16\mathbf{k})$$

21. $\mathbf{F}'(t) = 2t\mathbf{i} + 2\mathbf{j} + (3t^2 + 2t)\mathbf{k}$

$\mathbf{F}'(0) = 2\mathbf{j}$

$\mathbf{F}'(1) = 2\mathbf{i} + 2\mathbf{j} + 5\mathbf{k}$

$\mathbf{F}'(-1) = -2\mathbf{i} + 2\mathbf{j} + \mathbf{k}$

23. $\mathbf{F}'(t) = 2t\,\mathbf{i} - \sin t\,\mathbf{j} + (-t^2\sin t + 2t\cos t)\mathbf{k}$

$\mathbf{F}'(0) = \mathbf{0}$

$\mathbf{F}'(\frac{\pi}{2}) = \pi\mathbf{i} - \mathbf{j} - \frac{\pi^2}{4}\mathbf{k}$

25. The point corresponding to $t_0 = -1$ is $P_0 = \langle -1, 1, -1\rangle$. The required line is parallel to $\mathbf{F}'(t) = \left\langle \frac{-3}{t^4}, \frac{-2}{t^3}, \frac{-1}{t^2}\right\rangle$ evaluated at $t_0 = -1$; that is, parallel to

$\mathbf{F}'(-1) = \langle -3, 2, -1\rangle$. Thus, the line has parametric equations

$x = -1 - 3t,\ y = 1 + 2t,\ z = -1 - t$

27. $\displaystyle\int \langle t, -e^{3t}, 3\rangle\,dt = \frac{t^2}{2}\mathbf{i} - \frac{e^{3t}}{3}\mathbf{j} + 3t\mathbf{k} + \mathbf{C}$

29. $\displaystyle\int \langle \ln t, -t, 3\rangle\,dt$

$= (t\ln t - t)\mathbf{i} - \frac{1}{2}t^2\mathbf{j} + 3t\mathbf{k} + \mathbf{C}$

31. $\displaystyle\int \langle t\ln t, -\sin(1 - t), t\rangle\,dt$

$= \frac{t^2}{2}\left(\ln t - \frac{1}{2}\right)\mathbf{i} - \cos(1 - t)\mathbf{j} + \frac{t^2}{2}\mathbf{k} + \mathbf{C}$

33. $\mathbf{R}(t) = \displaystyle\int \mathbf{V}(t)\,dt$

$= \displaystyle\int \left\langle t^2, -e^{2t}, \sqrt{t}\right\rangle\,dt$

$= \left\langle \frac{1}{3}t^3, -\frac{1}{2}e^{2t}, \frac{2}{3}t^{3/2}\right\rangle + \mathbf{C}$

$\mathbf{R}(0) = \left\langle 0, -\frac{1}{2}, 0\right\rangle + \langle c_1, c_2, c_3\rangle$

$= \langle 1, 4, -1\rangle$

so $\mathbf{C} = \left\langle 1, \frac{9}{2}, -1\right\rangle$

$\mathbf{R}(t) = \left(\frac{1}{3}t^3 + 1\right)\mathbf{i} + \left(-\frac{1}{2}e^{2t} + \frac{9}{2}\right)\mathbf{j}$

$\qquad + \left(\frac{2}{3}t^{3/2} - 1\right)\mathbf{k}$

35. $\mathbf{R}(t) = \displaystyle\int \mathbf{V}(t)\,dt$

$= \displaystyle\int \left\langle 2t^{1/2}, \cos t, 0\right\rangle\,dt$

$= \left\langle \frac{4}{3}t^{3/2}, \sin t, 0\right\rangle + \mathbf{C}$

$\mathbf{R}(0) = \langle 0, 0, 0\rangle + \langle c_1, c_2, c_3\rangle$

$= \langle 1, 1, 0\rangle$

so $\mathbf{C} = \left\langle 1, 1, 0\right\rangle$

$$\mathbf{R}(t) = \left(\frac{4}{3}t^{3/2} + 1\right)\mathbf{i} + (\sin t + 1)\mathbf{j}$$

37. $\mathbf{V}(t) = \displaystyle\int \langle \cos t, 0, -t \sin t \rangle \, dt$

$$= \langle \sin t, 0, t \cos t - \sin t \rangle + \mathbf{C}_1$$
$$\mathbf{V}(0) = \langle 0, 0, 0 \rangle + \mathbf{C}_1 = \langle 2, 0, 3 \rangle;$$
$$\mathbf{C}_1 = \langle 2, 0, 3 \rangle$$
$$\mathbf{V}(t) = \langle 2 + \sin t, 0, 3 + t \cos t - \sin t \rangle$$
$$\mathbf{R}(t) = \int (\sin t + 2, 0, 3 + t \cos t - \sin t) \, dt$$
$$= \langle 2t - \cos t, 0, 3t + 2\cos t + t \sin t \rangle + \mathbf{C}_2$$
$$\mathbf{R}(0) = \langle -1, 0, 2 \rangle + \mathbf{C}_2 = \langle 1, -2, 1 \rangle$$
$$\mathbf{C}_2 = \langle 2, -2, -1 \rangle$$
$$\mathbf{R}(t) = \langle 2 + 2t - \cos t, -2,$$
$$-1 + 3t + 2\cos t + t \sin t \rangle$$

39. $\mathbf{V}(t) = \dfrac{d\mathbf{R}}{dt} = e^t\mathbf{i} + t^2\mathbf{j}$

$$\mathbf{R}(t) = \left\langle e^t, \frac{t^3}{3} \right\rangle + \mathbf{C}$$
$$\mathbf{R}(0) = \langle 1, 0 \rangle + \mathbf{C} = \langle 1, -1 \rangle$$
Thus, $\mathbf{C} = \langle 0, -1 \rangle$ so
$$\mathbf{R}(t) = e^t\mathbf{i} + \left(\frac{1}{3}t^3 - 1\right)\mathbf{j}$$

41. Let $f(t) = \mathbf{v} \cdot t^4\mathbf{w}$

$$= 2t^4 - 2t^4 - 15t^4$$
$$= -15t^4$$
$$\frac{d}{dt}f(t) = -60t^3$$
$$\frac{d^2}{dt^2}f(t) = -180t^2$$

43. $\mathbf{v} \times \mathbf{w} = \begin{vmatrix} \mathbf{i} & \mathbf{j} & \mathbf{k} \\ 2 & -1 & 5 \\ 1 & 2 & -3 \end{vmatrix}$

$$= -7\mathbf{i} + 11\mathbf{j} + 5\mathbf{k}$$
Let $f(t) = t\mathbf{v} \times t^2\mathbf{w} = t^3\mathbf{v} \times \mathbf{w}$, so
$$\frac{d}{dt}f(t) = 3t^2(-7\mathbf{i} + 11\mathbf{j} + 5\mathbf{k})$$

45. $(\mathbf{F} \cdot \mathbf{G})'(t) = \left[(3 + t^2)\sin(2 - t) - t^{-1}e^{2t}\right]'$

$$= -(3 + t^2)\cos(2 - t) + 2t \sin(2 - t)$$
$$- 2t^{-1}e^{2t} + t^{-2}e^{2t}$$

Also,
$$\mathbf{F}'(t) = 2t\mathbf{i} + 3\sin 3t\,\mathbf{j} - t^{-2}\mathbf{k}$$
$$\mathbf{G}'(t) = -\cos(2 - t)\mathbf{i} - 2e^{2t}\mathbf{k}$$
and
$(\mathbf{F}' \cdot \mathbf{G})(t) + (\mathbf{F} \cdot \mathbf{G}')(t)$
$$= [2t\mathbf{i} + 3\sin 3t\,\mathbf{j} - t^{-2}\mathbf{k}] \cdot [\sin(2 - t)\mathbf{i} - e^{2t}\mathbf{k}]$$
$$+ [(3 + t^2)\mathbf{i} - \cos 3t\,\mathbf{j} + t^{-1}\mathbf{k}] \cdot [-\cos(2 - t)\mathbf{i} - 2e^{2t}\mathbf{k}]$$
$$= 2t\sin(2 - t) + t^{-2}e^{2t} - (3 + t^2)\cos(2 - t) - 2t^{-1}e^{2t}$$

Therefore,
$$(\mathbf{F} \cdot \mathbf{G})'(t) = (\mathbf{F}' \cdot \mathbf{G})(t) + (\mathbf{F} \cdot \mathbf{G}')(t)$$

47. $\displaystyle\int_0^{2a} [\cos t\,\mathbf{i} + \sin t\,\mathbf{j} + \sin t \cos t\,\mathbf{k}] \, dt$

$$= \left[\sin t\,\mathbf{i} - \cos t\,\mathbf{j} + \left(\frac{1}{2}\sin^2 t\right)\mathbf{k}\right]\Bigg|_0^{2a}$$
$$= (\sin 2a)\mathbf{i} + (1 - \cos 2a)\mathbf{j} + \left(\frac{1}{2}\sin^2 2a\right)\mathbf{k}$$

Solve
$$(\sin 2a)\mathbf{i} + (1 - \cos 2a)\mathbf{j} + \left(\frac{1}{2}\sin^2 2a\right)\mathbf{k}$$
$$= \mathbf{i} + \mathbf{j} + \frac{1}{2}\mathbf{k}$$
This give rise to the system
$$\begin{cases} \sin 2a = 1 \\ 1 - \cos 2a = 1 \\ \dfrac{1}{2}\sin^2 2a = \dfrac{1}{2} \end{cases}$$
From the first component
$$\sin 2a = 1$$
$$a = \frac{\pi}{4}$$
Checking, we see this value also satisfies the second and third equations.

Page 383

49. $\mathbf{F}'(t) = k(-\sin(kt)\,\mathbf{i} + \cos(kt)\,\mathbf{j})$
$\mathbf{F}''(t) = -k^2[\cos(kt)\,\mathbf{i} + \sin(kt)\,\mathbf{j}]$
$\qquad = -k^2\mathbf{F}(t)$
Thus, $\mathbf{F}''(t)$ is parallel to $\mathbf{F}(t)$.

51. $\mathbf{F}'(t) = \left\langle e^t - 1, \frac{3}{2}t^{1/2}, -\sin t \right\rangle = \mathbf{0}$ when
$t = 0$, so $\mathbf{F}(t)$ is not smooth on $[-1, 2]$.

53. $\mathbf{F} = \langle t, 0, 0 \rangle$ and $\mathbf{G} = \langle -t, 0, 0 \rangle$ are both
smooth on $[1, 2]$, but $\mathbf{F} + \mathbf{G} = \langle 0, 0, 0 \rangle$ is
not.

55. Let $\mathbf{F}(t) = u(t)\mathbf{i} + v(t)\mathbf{j} + w(t)\mathbf{k}$; then
$[h(t)\mathbf{F}(t)]'$
$= [h(t)u(t)]'\mathbf{i} + [h(t)v(t)]'\mathbf{j} + [h(t)w(t)]'\mathbf{k}$
$= [h(t)u'(t) + h'(t)u(t)]\mathbf{i}$
$\qquad + [h(t)v'(t) + h'(t)v(t)]\mathbf{j}$
$\qquad + [h(t)w'(t) + h'(t)w(t)]\mathbf{k}$
$= [h(t)u'(t)\mathbf{i} + h(t)v'(t)\mathbf{j} + h(t)w'(t)\mathbf{k}]$
$\qquad + [h'(t)u(t)\mathbf{i} + h'(t)v(t)\mathbf{j} + h'(t)w(t)\mathbf{k}]$
$= h(t)\mathbf{F}'(t) + h'(t)\mathbf{F}(t)$

57. Let $\mathbf{F}(t) = u_1(t)\mathbf{i} + v_1(t)\mathbf{j} + w_1(t)\mathbf{k}$ and
$\mathbf{G}(t) = u_2(t)\mathbf{i} + v_2(t)\mathbf{j} + w_2(t)\mathbf{k}$
$[\mathbf{F} \times \mathbf{G}]'(t)$

$= \dfrac{d}{dt}\begin{vmatrix} \mathbf{i} & \mathbf{j} & \mathbf{k} \\ u_1(t) & v_1(t) & w_1(t) \\ u_2(t) & v_2(t) & w_2(t) \end{vmatrix}$

$= [v_1(t)w_2(t) - v_2(t)w_1(t)]'\mathbf{i}$
$\quad - [u_1(t)w_2(t) - u_2(t)w_1(t)]'\mathbf{j}$
$\quad + [u_1(t)v_2(t) - u_2(t)v_1(t)]'\mathbf{k}$
$= [v_1'(t)w_2(t) + v_1(t)w_2'(t)$
$\qquad - v_2(t)w_1'(t) - v_2'(t)w_1(t)]\mathbf{i}$
$\quad - [u_1'(t)w_2(t) + u_1(t)w_2'(t)$
$\qquad - u_2(t)w_1'(t) - u_2'(t)w_1(t)]\mathbf{j}$
$\quad + [u_1'(t)v_2(t) + u_1(t)v_2'(t)$
$\qquad - u_2(t)v_1'(t) - u_2'(t)v_1(t)]\mathbf{k}$
Also,

$(\mathbf{F}' \times \mathbf{G})(t) + (\mathbf{F} \times \mathbf{G}')(t)$

$= \begin{vmatrix} \mathbf{i} & \mathbf{j} & \mathbf{k} \\ u_1'(t) & v_1'(t) & w_1'(t) \\ u_2(t) & v_2(t) & w_2(t) \end{vmatrix}$

$\quad + \begin{vmatrix} \mathbf{i} & \mathbf{j} & \mathbf{k} \\ u_1(t) & v_1(t) & w_1(t) \\ u_2'(t) & v_2'(t) & w_2'(t) \end{vmatrix}$

$= [v_1'(t)w_2(t) - v_2(t)w_1'(t)]\mathbf{i}$
$\quad - [u_1'(t)w_2(t) - u_2(t)w_1'(t)]\mathbf{j}$
$\quad + [u_1'(t)v_2(t) - u_2(t)v_1'(t)]\mathbf{k}$
$\quad + [v_1(t)w_2'(t) - v_2'(t)w_1(t)]\mathbf{i}$
$\quad - [u_1(t)w_2'(t) - u_2'(t)w_1(t)]\mathbf{j}$
$\quad + [u_1(t)v_2'(t) - u_2'(t)v_1(t)]\mathbf{k}$
Thus,
$[\mathbf{F} \times \mathbf{G}]'(t) = [\mathbf{F}' \times \mathbf{G}](t) + [\mathbf{F} \times \mathbf{G}'](t)$

59. $\dfrac{d}{dt}(\|\mathbf{R}(t)\|) = \dfrac{d}{dt}\left[(\mathbf{R} \cdot \mathbf{R})^{1/2}\right]$
$\qquad = \dfrac{1}{2}(\mathbf{R} \cdot \mathbf{R})^{-1/2}(2\mathbf{R} \cdot \mathbf{R}')$
$\qquad = \dfrac{1}{\|\mathbf{R}\|}\,\mathbf{R} \cdot \mathbf{R}'$

10.3 Modeling Ballistics and Planetary Motion, page 779

> **SURVIVAL HINT:** *After practicing differentiation and integration of vector functions, we are able to do some real world mathematics! In this section we investigate some topics pointing to the need for the study of calculus.*

1. $T_f = \dfrac{2}{g}v_0 \sin\alpha \qquad\qquad R_f = \dfrac{v_0^2}{g}\sin 2\alpha$

$\quad = \dfrac{2}{32}(128)\sin 35° \qquad = \dfrac{128^2}{32}\sin 2(35°)$

$\quad \approx 4.6 \text{ sec} \qquad\qquad\quad \approx 481 \text{ ft}$

3. $T_f = \dfrac{2}{g}v_0 \sin\alpha$　　　$R_f = \dfrac{v_0^2}{g}\sin 2\alpha$

$\quad = \dfrac{2}{9.8}(850)\sin 48.5°$　$= \dfrac{850^2}{9.8}\sin 2(48.5°)$

$\quad \approx 129.9 \text{ sec}$　　　$\approx 73{,}175 \text{ m}$

5. $T_f = \dfrac{2}{g}v_0 \sin\alpha$　　　$R_f = \dfrac{v_0^2}{g}\sin 2\alpha$

$\quad = \dfrac{2}{9.8}(23.3)\sin 23.74°$　$= \dfrac{23.3^2}{9.8}\sin 2(23.74°)$

$\quad \approx 1.9 \text{ sec}$　　　$\approx 41 \text{ m}$

7. $T_f = \dfrac{2}{g}v_0 \sin\alpha$　　　$R_f = \dfrac{v_0^2}{g}\sin 2\alpha$

$\quad = \dfrac{2}{32}(100)\sin 14.11°$　$= \dfrac{100^2}{32}\sin 2(14.11°)$

$\quad \approx 1.5 \text{ sec}$　　　$\approx 148 \text{ ft}$

9. $T_f = \dfrac{2}{g}v_0 \sin\alpha$　　　$R_f = \dfrac{v_0^2}{g}\sin 2\alpha$

$\quad = \dfrac{2}{32}(450)\sin 49.31°$　$= \dfrac{450^2}{32}\sin 2(49.31°)$

$\quad \approx 21.3 \text{ sec}$　　　$\approx 6{,}257 \text{ ft}$

11. $r = \|\mathbf{R}(t)\| = \sqrt{(2t)^2 + t^2} = t\sqrt{5}$

$\dfrac{dr}{dt} = \sqrt{5}; \ \dfrac{d^2 t}{dt^2} = 0;$

$\theta = \tan^{-1}\left(\dfrac{t}{2t}\right) = \tan^{-1}\dfrac{1}{2}$

$\dfrac{d\theta}{dt} = \dfrac{d^2\theta}{dt^2} = 0$

$\mathbf{V}(t) = \sqrt{5}\,\mathbf{u}_r; \ \mathbf{A}(t) = \mathbf{0}$

13. $r = \|\mathbf{R}(t)\| = \sqrt{(\cos t)^2 + (\sin t)^2} = 1$

$\dfrac{dr}{dt} = \dfrac{d^2 r}{dt^2} = 0;$

$\theta = \tan^{-1}\left(\dfrac{\sin t}{\cos t}\right) = t$

$\dfrac{d\theta}{dt} = 1; \ \dfrac{d^2\theta}{dt^2} = 0$

$\mathbf{V}(t) = (0)\mathbf{u}_r + (1)(1)\mathbf{u}_\theta = \mathbf{u}_\theta$

$\mathbf{A}(t) = \left[0 - (1)(1)^2\right]\mathbf{u}_r$

$\qquad + \left[1(0) + 2(0)(1)\right]\mathbf{u}_\theta$

$\qquad = -\mathbf{u}_r$

15. $r = \sin 2t$

$\dfrac{dr}{dt} = 2\cos 2t; \ \dfrac{d^2 r}{dt^2} = -4\sin 2t$

$\dfrac{d\theta}{dt} = 2; \ \dfrac{d^2\theta}{dt^2} = 0$

$\mathbf{V}(t) = \dfrac{dr}{dt}\mathbf{u}_r + r\dfrac{d\theta}{dt}\mathbf{u}_\theta$

$\qquad = (2\cos 2t)\mathbf{u}_r + (2\sin 2t)\mathbf{u}_\theta$

$\mathbf{A}(t) = \left[\dfrac{d^2 r}{dt^2} - r\left(\dfrac{d\theta}{dt}\right)^2\right]\mathbf{u}_r$

$\qquad + \left[r\dfrac{d^2\theta}{dt^2} + 2\dfrac{dr}{dt}\dfrac{d\theta}{dt}\right]\mathbf{u}_\theta$

$\qquad = [-4\sin 2t - (\sin 2t)(4)]\mathbf{u}_r$

$\qquad + [(\sin 2t)(0) + 2(2)(2\cos 2t)]\mathbf{u}_\theta$

$\qquad = (-8\sin 2t)\mathbf{u}_r + (8\cos 2t)\mathbf{u}_\theta$

17. $r = 5 + 5[\cos(2t+1)]$

$\dfrac{dr}{dt} = -10\sin(2t+1)$

$\dfrac{d^2 r}{dt^2} = -20\cos(2t+1)$

$\dfrac{d\theta}{dt} = 2; \ \dfrac{d^2\theta}{dt^2} = 0$

$\mathbf{V}(t) = \dfrac{dr}{dt}\mathbf{u}_r + r\dfrac{d\theta}{dt}\mathbf{u}_\theta$

$\qquad = -10\sin(2t+1)\mathbf{u}_r$

$\qquad + 10[1 + \cos(2t+1)]\mathbf{u}_\theta$

$\mathbf{A}(t) = \left[\dfrac{d^2 r}{dt^2} - r\left(\dfrac{d\theta}{dt}\right)^2\right]\mathbf{u}_r$

$\qquad + \left[r\dfrac{d^2\theta}{dt^2} + 2\dfrac{dr}{dt}\dfrac{d\theta}{dt}\right]\mathbf{u}_\theta$

$\qquad = \Big\{-20\cos(2t+1)$

$\qquad - 20[1 + \cos(2t+1)]\Big\}\mathbf{u}_r$

$\qquad + 4[-10\sin(2t+1)]\mathbf{u}_\theta$

$\qquad = [-40\cos(2t+1) - 20]\mathbf{u}_r$

$\qquad - 40\sin(2t+1)\mathbf{u}_\theta$

19. $\mathbf{V} = \dfrac{d\mathbf{R}}{dt}$

$\underline{= (-a\omega \sin \omega t)\mathbf{i} + (a\omega \cos \omega t)\mathbf{j}}$

21. $\|\mathbf{V}\| = \sqrt{(-a\omega \sin \omega t)^2 + (a\omega \cos \omega t)^2}$

$= |a\omega|$

23. The acceleration vector (from Problem 22) is $-a\omega^2 \mathbf{R}$. Since $\mathbf{R}$ points away from the center, $\mathbf{A} = -a\omega^2 \mathbf{R}$ points toward the center.

25. $\alpha = 45°$, so we can use the formula for maximum range: $R_m = v_0^2/g$.

$$2,000 = \frac{v_0^2}{9.8}$$

$$v_0 = 140 \text{ m/sec}$$

27. $\sin 2\alpha = \dfrac{R_f g}{v_0^2}$

$= \dfrac{600(32)}{(167.1)^2}$

≈ 0.6876

$\alpha \approx 21.7°$

29. The maximum height is reached when $y'(t) = 0$. We can use this equation to find the time at which this occurs, then use that time in the equation for $y(t)$ to find the height.

$$y(t) = -16t^2 + (V_0 \sin \alpha)t + s_0$$

$$= -16t^2 + (144)\left(\frac{1}{2}\right)t + 4$$

$$= -16t^2 + 72t + 4$$

$$y'(t) = -32t + 72$$

$y'(t) = 0$ when $t = 2.25$ sec.

$$y(2.25) = -16(2.25)^2 + 72(2.25) + 4$$

$$= 85 \text{ ft}$$

The ball will land when $y = 0$. We can use $y(t) = 0$ to find the time of flight, and then use that time to find $x(t)$.

$-16t^2 + 72t + 4 = 0$

$$t = \frac{-72 \pm \sqrt{72^2 - 4(16)(4)}}{2(-16)}$$

$t \approx 4.5549$ sec   (reject negative solution)

$x(t) = (v_0 \cos \alpha)t = \frac{\sqrt{3}}{2}(144)t = 72\sqrt{3}t$

$x(4.5549) \approx 568$ ft

To find the distance to the fence we will find t for $y(t) = 5$, then use that time in the equation $x(t)$. $-16t^2 + 72t + 4 = 5$ when $t \approx 4.49$ sec (again reject negative solution).

$x(4.49) = 72\sqrt{3}(4.49) \approx 560$ ft.

31. Let α be the nozzle angle. Then,

$x = (50 \cos \alpha)t$, $y = -16t^2 + (50 \sin \alpha)t$

For the range, $y = 0$ or $16t^2 = 50t \sin \alpha$.

$$T_f = \frac{50 \sin \alpha}{16} \text{ and } t_h = \frac{25 \sin \alpha}{16}$$

where t_h is the time at which the water reaches its highest point. We know $y_{\max} = 5$, so

$$-16\left(\frac{25 \sin \alpha}{16}\right)^2 + 50 \sin \alpha \left(\frac{25 \sin \alpha}{16}\right) = 5$$

$$80 = 625 \sin^2 \alpha$$

$$\alpha \approx 21°$$

The time of travel is

$$T_f = \frac{25 \sin 21°}{8} \approx 1.12 \text{ seconds}$$

and $x \approx 50(\cos 21°)(1.12) \approx 52.3$ ft

33. Since $\alpha = 90°$, $\sin \alpha = 1$ and $\cos \alpha = 0$. Thus, $x(t) = 0$, $y(t) = -\frac{1}{2}gt^2 + v_0 t + s_0$.

35. $T_f = \dfrac{2v_0 \sin \alpha}{g}$

$= \dfrac{2(125)\sqrt{2}}{32(2)} \approx 5.5$ sec

37. Evel's "range" was at least 1,700 ft. Using the formula $R = v_0^2 \sin 2\alpha/g$, we find

$$\frac{v_0^2 \sin 2(45°)}{32} = 1,700$$

$$v_0 \approx 233.238 \text{ ft/s}$$

(reject negative value)

This is about 159 mi/h.

39. $\mathbf{V} = (v_0 \cos \alpha)\mathbf{i} + (-gt + v_0 \sin \alpha)\mathbf{j}$

$\mathbf{R} = [(v_0 \cos \alpha)t]\mathbf{i}$

$$+ \left[-\frac{gt^2}{2} + (v_0 \sin \alpha)t + s_0 \right]\mathbf{j}$$

$v_0 = 8$, $g = 32$, $s_0 = 120$, $\alpha = -\frac{\pi}{6}$

$$-\frac{1}{2}gt^2 + (v_0 \sin \alpha)t + s_0 = y(t)$$

$$-16t^2 - 4t + 120 = 0$$

$$4t^2 + t - 30 = 0$$

$t \approx 2.6165$ sec (disregard negative value).

$$x = 4\sqrt{3}(2.6165) \approx 18.13 \text{ ft}$$

41. $\mathbf{V} = \dfrac{dr}{dt}\mathbf{u}_r + r\dfrac{d\theta}{dt}\mathbf{u}_\theta$

$\quad = (-3 \sin t)\mathbf{u}_r + (4 + 3\cos t)(3t^2)\mathbf{u}_\theta$

$\mathbf{A} = \left[\dfrac{d^2r}{dt^2} - r\left(\dfrac{d\theta}{dt}\right)^2 \right]\mathbf{u}_r + \left[r\dfrac{d^2\theta}{dt^2} + 2\dfrac{dr}{dt}\dfrac{d\theta}{dt} \right]\mathbf{u}_\theta$

$\quad = \left[-3\cos t - (4 + 3\cos t)(3t^2)^2 \right]\mathbf{u}_r$

$\qquad + \left[(4 + 3\cos t)(6t) + 2(-3\sin t)3t^2 \right]\mathbf{u}_\theta$

$\quad = \left[-3\cos t - (4 + 3\cos t)9t^4 \right]\mathbf{u}_r$

$\qquad + \left[(4 + 3\cos t)6t - 18t^2\sin t \right]\mathbf{u}_\theta$

43. $R = \dfrac{550^2}{32} \sin 2(22°) \approx 6,566.69$

This means

$$6,566.69 \approx \frac{550^2}{32}\sin 2\alpha$$

$$\alpha \approx 21.8°$$

45. We can find when the first shell is 50 ft above ground by solving the following equation.

$$50 = -16t^2 + (750 \sin 25°)t$$

$$t \approx 19.65 \text{ seconds}$$

At that time, its horizontal distance from the second gun is

$$s_1 = 20,000 - 750(\cos 25°)(19.65)$$

$$\approx 6,643.29 \text{ ft}$$

Since the second gun fires 2 seconds after the first, its shell will be in the air only 17.65 seconds before the collision. Thus,

$$-16(17.65)^2 + (v_0 \sin \alpha)(17.65) = 50$$

$$(v_0 \cos \alpha)(17.65) = 6,643.29$$

Solving, we obtain $v_0 \sin \alpha = 285.23$ and $v_0 \cos \alpha = 376.39$ so

$$\alpha = \tan^{-1}\left(\frac{285.23}{376.39} \right) \approx 37.2°$$

$$v_0 = \frac{285.23}{\sin 37.2°} \approx 471.77 \text{ ft/s}$$

47. **a.** The area of a triangle is $\frac{1}{2}bh$, where b is the base and h is the altitude. If the triangles are sufficiently thin, the sum of the lengths of the bases approximates C, the circumference, and $h \approx r$, the radius. Hence $A = \frac{1}{2}Cr$. Note that if one knows the formula for the circumference of a circle $\frac{1}{2}Cr = \frac{1}{2}(2\pi r)r = \pi r^2$.

b. A sphere can be thought of being composed of cones with the vertices at the center. The volume, V, of a cone is $\frac{1}{3}Bh$, where B is the area of the base and h is the altitude. The sum of the surface areas of the cones equals the surface area of the sphere, and the altitude of the sphere is r. Hence,

$V = \frac{1}{3}rS$, where S is the surface area of the sphere. Note that if one knows the formula for the surface area of the sphere,

$$V = \tfrac{1}{3}rS = \tfrac{1}{3}(4\pi r^2)h = \tfrac{4}{3}\pi r^3$$

c. From part **a** we know the area of a unit circle is $\frac{1}{2}C = \pi$. If the unit circle is stretched by a factor of a in the x-direction and by a factor of b in the y-direction, with $a \geq b$, so that the semimajor axis is a and the semi-minor axis is b, then the area is ab times the area of the unit circle, or πab.

49. From Kepler's first law, the orbit is an ellipse with rectangular equation

$$\frac{x^2}{a^2} + \frac{y^2}{b^2} = 1$$

with $c^2 = a^2 - b^2$. To prove Kepler's third law, we use the polar-form equation of an ellipse, namely

$$r = \frac{\frac{h^2}{k}}{1 + e\cos\theta} \quad \text{where } e = \frac{c}{a}$$

is the eccentricity of the orbit. The semimajor axis, or *mean distance*, is $\frac{1}{2}$ the sum of the least and greatest values of r. These values correspond to $\theta = 0$ and $\theta = \pi$. Substitute these values of θ into the polar equation to find (after some simplification) $b^2 = \dfrac{h^2 a}{k}$. Define T to be the period; that is, the time required for one complete revolution. Since the area of the ellipse is πab from Kepler's second law, $T = \dfrac{2\pi ab}{h}$ so

$$T^2 = \frac{4\pi^2 a^2 b^2}{h^2} = \left(\frac{4\pi^2}{k}\right)a^3$$

Thus, the square of the period of revolution of the planet is proportional to the cube of its mean distances.

51. They are the same since the planets have the same semi-minor axis.

53. The velocity vector is
$$\mathbf{V}(t) = \langle 76.6, -9.8t + 64.28 \rangle$$
so the velocity at impact is
$$\mathbf{V}(18.81) = \langle 76.6, -120.06 \rangle$$
with speed
$$s = \|\mathbf{V}(18.81)\| = 142.41 \text{ m/s}$$

55. $R_f = v_0 \cos\alpha \sec\beta \left[\dfrac{v_0 \sin\alpha + v_0 \cos\alpha \tan\beta}{\frac{1}{2}g} \right]$

$$\frac{dR_f}{d\alpha} = \frac{2}{g} v_0^2 \sec\beta \left[\cos 2\alpha + 2\tan\beta \cos\alpha(-\sin\alpha)\right]$$

This derivative is 0 when

$$\cos 2\alpha = \sin 2\alpha \tan\beta$$
$$\alpha = \frac{1}{2}\tan^{-1}(\cot\beta)$$
$$= \frac{1}{2}\left(\frac{\pi}{2} - \beta\right)$$
$$= \frac{\pi}{4} - \frac{\beta}{2}$$

57. Differentiating the equation in Problem 57 implicitly with respect to α, we obtain
$$g[2\sec\alpha(\sec\alpha\tan\alpha)]R_f^2 = 2g\sec^2\alpha\, R_f\, R_f'$$
$$- 2v_0^2 \sec^2\alpha\, R_f - 2v_0^2 \tan\alpha\, R_f' = 0$$
The maximum range R_m occurs when $R_f' = 0$ and $\alpha = \alpha_m$, so
$$g[2\sec^2\alpha_m \tan\alpha_m]R_m^2 = 2v_0^2 \sec^2\alpha_m\, R_m$$
$$R_m \tan\alpha_m = \frac{v_0^2}{g}$$

$\|\mathbf{R}'(t)\| = 2$

$\mathbf{R}' \cdot \mathbf{R}'' = -4\cos t \sin t + 4\sin t \cos t = 0$

$\mathbf{R}' \times \mathbf{R}'' = \begin{vmatrix} \mathbf{i} & \mathbf{j} & \mathbf{k} \\ -2\cos t & -2\sin t & 0 \\ 2\sin t & -2\cos t & 0 \end{vmatrix} = 4\mathbf{k}$

$\|\mathbf{R}' \times \mathbf{R}''\| = 4$

$A_T = \dfrac{\mathbf{R}' \cdot \mathbf{R}''}{\|\mathbf{R}'\|} = 0$

$A_N = \dfrac{\|\mathbf{R}' \times \mathbf{R}''\|}{\|\mathbf{R}'\|} = 2$

31. $\mathbf{R}(t) = (e^t \cos t)\mathbf{i} + (e^t \sin t)\mathbf{j} + e^t \mathbf{k}$

$\mathbf{R}'(t) = e^t[(-\sin t + \cos t)\,\mathbf{i}$
$\qquad\qquad + (\cos t + \sin t)\,\mathbf{j} + \mathbf{k}]$

$\mathbf{R}''(t) = e^t[-2\sin t\,\mathbf{i} + 2\cos t\,\mathbf{j} + \mathbf{k}]$

$\|\mathbf{R}'(t)\| = \sqrt{3e^{2t}}$

$\mathbf{R}' \cdot \mathbf{R}'' = 3e^{2t}$

$\mathbf{R}' \times \mathbf{R}''$
$\quad = e^{2t}\langle \sin t - \cos t,\ -\sin t - \cos t,\ 2\rangle$

$\|\mathbf{R}' \times \mathbf{R}''\| = e^t \sqrt{3}$

$A_T = \dfrac{\mathbf{R}' \cdot \mathbf{R}''}{\|\mathbf{R}'\|} = \dfrac{3e^{2t}}{e^t\sqrt{3}} = \sqrt{3}\,e^t$

$A_N = \dfrac{\|\mathbf{R}' \times \mathbf{R}''\|}{\|\mathbf{R}'\|} = \dfrac{e^{2t}\sqrt{6}}{e^t\sqrt{3}} = \sqrt{2}\,e^t$

33. $\mathbf{R}'(t) = \langle 8\cos 2t, 6\sin 2t\rangle$

$\text{speed} = \dfrac{ds}{dt} = \|\mathbf{R}'(t)\|$

$\qquad = \sqrt{64\cos^2 2t + 36\sin^2 2t}$

$\qquad = \sqrt{28\cos^2 2t + 36}$

$\dfrac{d^2 s}{dt^2} = \dfrac{-28\sin 4t}{\sqrt{28\cos^2 2t + 36}}$

This second derivative is 0 when $t = 0, \frac{\pi}{4}, \frac{\pi}{2}, \cdots$. The maximum speed is 8 at $t = \frac{n\pi}{2}$, n any integer, and the minimum speed is 6 at $t = \frac{(2n+1)\pi}{4}$.

35. $\mathbf{R}(t) = (1 + \cos 2t)\mathbf{i} + (\sin 2t)\mathbf{j}$

$\mathbf{V}(t) = \mathbf{R}'(t) = (-2\sin 2t)\mathbf{i} + (2\cos 2t)\mathbf{j}$

$\mathbf{A}(t) = \mathbf{V}'(t) = (-4\cos 2t)\mathbf{i} - (4\sin 2t)\mathbf{j}$

$\dfrac{ds}{dt} = \|\mathbf{V}(t)\| = 2$

$A_T = \dfrac{d^2 s}{dt^2} = 0$

$A_N = \sqrt{\|\mathbf{A}\|^2 - A_T^2} = 4$

37. Parameterize the curve by $x = t, y = 4t^2$

$\mathbf{R}(t) = t\,\mathbf{i} + 4t^2\,\mathbf{j}$

$\mathbf{V}(t) = \mathbf{i} + 8t\,\mathbf{j}$

$\mathbf{A}(t) = 8\,\mathbf{j};\ \|\mathbf{A}\| = 8$

$\dfrac{ds}{dt} = \|\mathbf{V}(t)\| = \sqrt{1 + 64t^2}$

If $\dfrac{ds}{dt} = 20$, $1 + 64t^2 = 400$

$\qquad\qquad\qquad t = \dfrac{\sqrt{399}}{8}$

$A_T = \dfrac{d^2 s}{dt^2} = \dfrac{64t}{\sqrt{1 + 64t^2}}$

$A_N = \sqrt{\|\mathbf{A}\|^2 - A_T^2}$

$\quad = \sqrt{64 - \dfrac{(64t)^2}{1 + 64t^2}}$

$\quad = \dfrac{8}{\sqrt{1 + 64t^2}}$

At $t = \dfrac{\sqrt{399}}{8}$, $A_T = \dfrac{8\sqrt{399}}{20} = \dfrac{2\sqrt{399}}{5}$

$A_N = \dfrac{8}{\sqrt{1 + 64\left(\frac{399}{64}\right)}} = \dfrac{2}{5}$

39. Since the pail moves in a vertical plane with angular velocity ω rev/s, we have

$\mathbf{R}(t) = \langle 3\cos 2\pi\omega t, 3\sin 2\pi\omega t\rangle$

$\mathbf{V}(t) = \langle -6\pi\omega \sin 2\pi\omega t, 6\pi\omega \cos 2\pi\omega t\rangle$

$\mathbf{A}(t) = \langle -12\pi^2\omega^2 \cos 2\pi\omega t,$
$\qquad\qquad\qquad -12\pi^2\omega^2 \sin 2\pi\omega t\rangle$

$\dfrac{ds}{dt} = \|\mathbf{V}(t)\| = 6\pi\omega;$

$\|\mathbf{A}(t)\| = 12\pi^2\omega^2;\ A_T = \|\mathbf{V}\|' = 0;$

$A_N = \sqrt{\|\mathbf{A}\|^2 - A_T^2} = 12\pi^2\omega^2$

The force due to the motion of the pail with mass 2/32 slugs is

$$F_N = mA_N = \frac{22}{32}\,(12\pi^2\omega^2)$$
$$\approx 7.4\,\omega^2\ \text{lbs}$$

The force on the bottom of the pail is greatest at the lowest point of the swing where it equals $7.4\omega^2 + 2$. The force is $7.4\omega^2 - 2$ lbs at the top of the swing, and to keep the water from spilling, we must have

$$7.4\omega^2 - 2 = 0$$
$$\omega \approx 0.52\ \text{rev/s (or 31.2 rev/min)}$$

41. $A_N = \kappa\left(\dfrac{ds}{dt}\right)^2 = \left(\dfrac{ds}{dt}\right)^2\dfrac{1}{\rho}$

Since $\dfrac{ds}{dt} = 45\left(\dfrac{5,280}{3,600}\right) = 66$ ft/s , we want

$$2.4 = (66)^2\left(\frac{1}{\rho}\right)$$
$$\rho = 1,815\ \text{ft}$$

43. Since the car moves in a circle, its tangential acceleration is 0. The force required to keep the car moving in a circle has magnitude

$$m\kappa\left(\frac{ds}{dt}\right)^2 = \frac{m}{\rho}\left(\frac{ds}{dt}\right)^2$$

where $m = W/g$ is the car's mass and

$\rho = 1/\kappa$. We want this force to be balanced by the frictional force keeping the car on the road, so

$$\frac{m}{\rho}\left(\frac{ds}{dt}\right)^2 = \mu W = \mu(mg)$$

Since m's cancel, the weight does not matter.

45. From Problem 43,

$$\frac{m}{\rho}\left(\frac{ds}{dt}\right)^2 = \mu W = \mu(mg)$$

Since m's cancel, the weight does not matter.

47. For the optimal safe speed to be 55 mi/h (80.667 ft/s) we want θ to satisfy

$$(80.667)^2 = \frac{150(32)(\sin\theta + 0.47\cos\theta)}{\cos\theta - 0.47\sin\theta}$$
$$\theta \approx 28.4°$$

49. The speed at the volunteer's location 15 ft from the center of the wheel is

$$\frac{ds}{dt} = 2\pi(15)\omega/60 = \frac{\pi\omega}{2}$$

feet per second, so the normal component of the volunteer's weight is

$$F_N = mA_N = \frac{W}{g}\kappa\left(\frac{ds}{dt}\right)^2 = \frac{W}{g}\left(\frac{1}{\rho}\right)\left(\frac{ds}{dt}\right)^2$$
$$= \frac{W}{g}\left(\frac{1}{15}\right)\left(\frac{\pi\omega}{2}\right)^2 = \frac{\pi^2\omega^2}{(32)(60)}W = \frac{\pi^2\omega^2}{1920}W$$

51. The curve $x = \dfrac{y^2}{120}$ has parametric form:

$$x = \frac{t^2}{120},\ y = t.$$

$x' = \dfrac{t}{60}$, $y' = 1$; $x'' = \dfrac{1}{60}$, $y'' = 0$

The curvature is

$$\kappa = \frac{|x'y'' - y'x''|}{[(x')^2 + (y')^2]^{3/2}}$$

$$= \frac{\left|\frac{t}{60}(0) - (1)\frac{1}{60}\right|}{\left[\left(\frac{t}{60}\right)^2 + 1\right]^{3/2}}$$

$$= \frac{60^2}{[t^2 + 60^2]^{3/2}}$$

At $t = 0$, $\kappa(0) = \frac{1}{60}$ and

$$A_N = \kappa \left(\frac{ds}{dt}\right)^2 = \frac{1}{60}\left(\frac{ds}{dt}\right)^2$$

Thus, the maximum safe speed $\dfrac{ds}{dt}$ satisfies

$$A_N = \frac{1}{60}\left(\frac{ds}{dt}\right)^2 = 30$$

$$\frac{ds}{dt} = 42.43 \text{ units/s}$$

53. $\mathbf{R}(t) = x\mathbf{i} + f(x)\mathbf{j}$

$\mathbf{R}'(t) = \mathbf{i} + f'(x)\mathbf{j}$

$\mathbf{R}''(t) = f''(x)\mathbf{j}$

$\|\mathbf{R}'(t)\| = \sqrt{1 + [f'(x)]^2}$

$\mathbf{R}' \cdot \mathbf{R}'' = f'(x)f''(x)$

$$\mathbf{R}' \times \mathbf{R}'' = \begin{vmatrix} \mathbf{i} & \mathbf{j} & \mathbf{k} \\ 1 & f'(x) & 0 \\ 0 & f''(x) & 0 \end{vmatrix}$$

$$= f''(x)\mathbf{k}$$

$\|\mathbf{R}' \times \mathbf{R}''\| = |f''(x)|$

$$A_T = \frac{\mathbf{R}' \cdot \mathbf{R}''}{\|\mathbf{R}'\|} = \frac{f'(x)f''(x)}{\sqrt{1 + [f'(x)]^2}}$$

$$A_N = \frac{\|\mathbf{R}' \times \mathbf{R}''\|}{\|\mathbf{R}'\|} = \frac{|f''(x)|}{\sqrt{1 + [f'(x)]^2}}$$

55. $\mathbf{A} = -g\mathbf{j}$,

$\mathbf{V} = (v_0 \cos\alpha)\mathbf{i} + (-gt + v_0\sin\alpha)\mathbf{j}$

$\|\mathbf{V}\| = \sqrt{v_0^2 \cos^2\alpha + (-gt + v_0 \sin\alpha)^2}$

The maximum height occurs when the $\mathbf{j}$ component of velocity is 0:

$$-gt_m + v_0 \sin\alpha = 0$$

$$t_m = \frac{v_0}{g}\sin\alpha$$

We find $\mathbf{V} \cdot \mathbf{A} = g^2 t - gv_0\sin\alpha$ and

$\mathbf{V} \times \mathbf{A} = -(gv_0\cos\alpha)\mathbf{k}$

At time $t = t_m$, we have

$$\|\mathbf{V}(t_m)\| = |v_0 \cos\alpha|$$

$\mathbf{V} \cdot \mathbf{A} = 0$; $\|\mathbf{V} \times \mathbf{A}\| = |gv_0\cos\alpha|$, so that

$A_T = \dfrac{\mathbf{V} \cdot \mathbf{A}}{\|\mathbf{V}\|} = 0$ and

$$A_N = \frac{\|\mathbf{V} \times \mathbf{A}\|}{\|\mathbf{V}\|} = \frac{|gv_0\cos\alpha|}{|v_0\cos\alpha|} = g$$

57. a. Let $R_s = R + R_e$ be the distance of the satellite from the center of the earth. From Example 4,

$$v = \sqrt{\frac{GM}{R_s}}; \quad T = \frac{2\pi R_s}{v} = \frac{2\pi R_s^{3/2}}{\sqrt{GM}}$$

b. $T = (24)(3,600) = \dfrac{2\pi R_s^{3/2}}{\sqrt{398,600}}$

$R_s^{3/2} = \dfrac{(24)(3,600)\sqrt{398,600}}{2\pi}$

$= 8681655$

$R_s \approx 42241.08$

Thus, $R \approx 42,241 - 6,440 \approx 36,000$ km.

c. The speed is

$$v = \sqrt{\dfrac{398,600}{35,801}} \approx 3.3367 \text{ km/s}$$

This is about 12,000 km/h.

59. Answers vary.

Chapter 10 Review

Studying for a chapter examination is a personal process, one which nobody else can do for you. Simply take the time to review what you have done.

SURVIVAL HINT: Work all of Chapter 10 problems in the Proficiency Examination (whether they are assigned or not). Work through all of the problems before looking at the answers, and *then* correct each of the problems. The answers to all these problems are given in the answer section at the back of the text. If you worked the problem correctly, move on to the next problem, but if you did not work it correctly (or you did not know what to do), then look at the solutions below, look back in the chapter to study the procedure, or ask your instructor.

Finally, go back over the homework problems you have been assigned. If you worked a problem correctly, move on to the next problem, but if you missed it on your homework, then you should look back in the book or talk to your instructor about how to work the problem.

If you follow these steps, you should be successful with your review of this chapter.

Proficiency Examination, page 805

1. A vector-valued function (or, simply, a vector function) **F** with domain D assigns a unique vector $\mathbf{F}(t)$ to each scalar t in the set D. The set of all vectors **v** of the form $\mathbf{v} = \mathbf{F}(t)$ for t in D is the range of **F**.

2. $\mathbf{F}(t) = f_1(t)\mathbf{i} + f_2(t)\mathbf{j}$ in $\mathbb{R}^2$ (plane) or $\mathbf{F}(t) = f_1(t)\mathbf{i} + f_2(t)\mathbf{j} + f_3(t)\mathbf{k}$ in $\mathbb{R}^3$ (space) where f_1, f_2, and f_3 are real-valued (scalar-valued) functions of the real number t defined on the domain set

D. In this context f_1, f_2, and f_3 are called the components of **F**.

3. The graph of

$$\mathbf{F}(t) = f_1(t)\mathbf{i} + f_2(t)\mathbf{j} + f_3(t)\mathbf{k}$$

in $\mathbb{R}^3$ (space) is the graph of the parametric equations $x = f_1(t)$, $y = f_2(t)$, $z = f_3(t)$.

4. The limit of a vector function consists of the vector sum of the limits of its individual components.

5. $\mathbf{F}'(t) = f_1'(t)\mathbf{i} + f_2'(t)\mathbf{j} + f_3'(t)\mathbf{k}$

6. $\displaystyle\int \mathbf{F}(t)\,dt$

$$= \int f_1(t)\,dt\,\mathbf{i} + \int f_2(t)\,dt\,\mathbf{j} + \int f_3(t)\,dt\,\mathbf{k}$$

7. The graph of the vector function defined by $\mathbf{F}(t)$ is said to be *smooth* on any interval of t where $\mathbf{F}'$ is continuous and $\mathbf{F}'(t) \neq 0$.

8. a. Linearity rule
$$(a\mathbf{F} + b\mathbf{G})'(t) = a\mathbf{F}'(t) + b\mathbf{G}'(t)$$
for constants a, b

b. Scalar multiple
$$(h\mathbf{F})'(t) = h'(t)\mathbf{F}(t) + h(t)\mathbf{F}'(t)$$

c. Dot product rule
$$(\mathbf{F}\cdot\mathbf{G})'(t) = (\mathbf{F}'\cdot\mathbf{G})(t) + (\mathbf{F}\cdot\mathbf{G}')(t)$$

d. Cross product rule
$$(\mathbf{F}\times\mathbf{G})'(t)$$
$$= (\mathbf{F}'\times\mathbf{G})(t) + (\mathbf{F}\times\mathbf{G}')(t)$$

e. Chain rule
$$[\mathbf{F}(h(t))]' = h'(t)\mathbf{F}'(h(t))$$

9. If the nonzero vector function $\mathbf{F}(t)$ is differentiable and has constant length, then $\mathbf{F}(t)$ is orthogonal to the derivative vector $\mathbf{F}'(t)$.

10. $\mathbf{R}(t) = x(t)\mathbf{i} + y(t)\mathbf{j} + z(t)\mathbf{k}$ is the position;

$$\mathbf{V}(t) = \frac{d\mathbf{R}(t)}{dt} = \mathbf{R}'(t) \text{ is the velocity;}$$

$$\mathbf{A}(t) = \frac{d\mathbf{V}(t)}{dt} = \frac{d^2\mathbf{R}(t)}{dt^2} \text{ is the acceleration}$$

11. The speed is $\|\mathbf{V}(t)\|$.

12. Consider a projectile that travels in a vacuum in a coordinate plane, with the x-axis along level ground. If the projectile is fired from a height of s_0 with initial speed v_0 and angle of elevation α, then at time t

($t \geq 0$) it will be at the point $(x(t), y(t))$, where

$$x(t) = (v_0\cos\alpha)t$$
$$y(t) = -\frac{1}{2}gt^2 + (v_0\sin\alpha)t + s_0$$

13. A projectile fired from ground level has time of flight T_f and range R given by the equations

$$T_f = \frac{2}{g}v_0\sin\alpha \text{ and } R = \frac{v_0^2}{g}\sin 2\alpha$$

The maximal range is $R_m = \frac{v_0^2}{g}$, and it occurs when $\alpha = \frac{\pi}{4}$.

14. First law
The planets move about the sun in elliptical orbits, with the sun at one focus.
Second law
The radius vector joining a planet to the sun sweeps over equal areas in equal intervals of time.
Third law
The square of the time of one complete revolution of a planet about its orbit is proportional to the cube of the orbit's semimajor axis.

15. $\mathbf{u}_r = (\cos\theta)\mathbf{i} + (\sin\theta)\mathbf{j}$ and
$\mathbf{u}_\theta = (-\sin\theta)\mathbf{i} + (\cos\theta)\mathbf{j}$

16. If $\mathbf{R}(t)$ is a vector function that defines a smooth graph ($\mathbf{R}'(t) \neq 0$), then at each point a unit tangent is

$$\mathbf{T}(t) = \frac{\mathbf{R}'(t)}{\|\mathbf{R}'(t)\|}$$

and the principal unit normal vector is

$$\mathbf{N}(t) = \frac{\mathbf{T}'(t)}{\|\mathbf{T}'(t)\|}$$

17. Suppose the position of a moving object is $\mathbf{R}(t)$, where $\mathbf{R}'(t)$ is continuous on the interval $[a, b]$. Then the object has speed

$$\|\mathbf{V}(t)\| = \|\mathbf{R}'(t)\| = \frac{ds}{dt}$$

for $a \le t \le b$.

18. If C is a smooth curve defined by $\mathbf{R}(t) = x(t)\mathbf{i} + y(t)\mathbf{j} + z(t))\mathbf{k}$ on an interval $[a, b]$, then the arc length of C is given by

$$s = \int_a^b \|\mathbf{R}'(t)\| \, dt$$

$$= \int_a^b \sqrt{[x'(t)]^2 + [y'(t)]^2 + [z'(t)]^2} \, dt$$

19. The curvature of a graph is an indication of how quickly the graph changes direction.

$$\kappa = \left\| \frac{d\mathbf{T}}{ds} \right\|$$

where $\mathbf{T}$ is a unit vector.

20. If $\mathbf{V} = \mathbf{R}'$, $\mathbf{A} = \mathbf{R}''$, then $\kappa = \dfrac{\|\mathbf{V} \times \mathbf{A}\|}{\|\mathbf{V}\|^3}$.

21. $\mathbf{T} = \dfrac{d\mathbf{R}}{ds}$ and $\mathbf{N} = \dfrac{1}{\kappa}\left(\dfrac{d\mathbf{T}}{ds}\right)$.

22. The radius of curvature is the reciprocal of curvature $\rho = 1/\kappa$.

23. The acceleration $\mathbf{A}$ of a moving object can be written as $\mathbf{A} = A_T\mathbf{T} + A_N\mathbf{N}$, where

$A_T = \dfrac{d^2 s}{dt^2}$ is the tangential component; and

$A_N = \kappa\left(\dfrac{ds}{dt}\right)^2$ is the normal component.

24. This is a helix of radius of 3, climbing in a counter-clockwise direction.

$$\mathbf{R} = (3\cos t)\mathbf{i} + (3\sin t)\mathbf{j} + t\mathbf{k},$$

$$\mathbf{R}' = (-3\sin t)\mathbf{i} + (3\cos t)\mathbf{j} + \mathbf{k},$$
$$\|\mathbf{R}'\| = \sqrt{9\sin^2 t + 9\cos^2 t + 1} = \sqrt{10}$$

$$s = \int_0^{2\pi} \|\mathbf{R}'\| \, dt$$

$$= \int_0^{2\pi} \sqrt{10} \, dt$$

$$= 2\pi\sqrt{10}$$

25. $\mathbf{F}' = \dfrac{1}{(1+t)^2}\mathbf{i} + \dfrac{t\cos t - \sin t}{t^2}\mathbf{j} + (-\sin t)\mathbf{k}$

$\mathbf{F}'' = -\dfrac{2}{(1+t)^3}\mathbf{i}$

$\qquad + \dfrac{-t^2\sin t - 2t\cos t + 2\sin t}{t^3}\mathbf{j} - (\cos t)\mathbf{k}$

26. $\begin{vmatrix} \mathbf{i} & \mathbf{j} & \mathbf{k} \\ 3t & 0 & 3 \\ 0 & \ln t & -t^2 \end{vmatrix}$

$= (-3\ln t)\mathbf{i} + (3t^3)\mathbf{j} + (3t\ln t)\mathbf{k}$

$\displaystyle\int_1^2 [(-3\ln t)\mathbf{i} + (3t^3)\mathbf{j} + (3t\ln t\,\mathbf{k}] \, dt$ ⟶ Formulas 196 and 199

$= \left\{-3(t\ln t - t)\mathbf{i} + \dfrac{3t^4}{4}\mathbf{j} + 3\left[\dfrac{t^2}{2}\left(\ln t - \dfrac{1}{2}\right)\right]\mathbf{k}\right\}\Big|_1^2$

$= (6 - 6\ln 2)\mathbf{i} + 12\mathbf{j} + (6\ln 2 - 3)\mathbf{k} - \left(3\mathbf{i} + \dfrac{3}{4}\mathbf{j} - \dfrac{3}{4}\mathbf{k}\right)$

$= (3 - 6\ln 2)\mathbf{i} + \dfrac{45}{4}\mathbf{j} + \left(6\ln 2 - \dfrac{9}{4}\right)\mathbf{k}$

27. $\mathbf{F}'' = e^t\mathbf{i} - t^2\mathbf{j} + 3\mathbf{k}$,

$\mathbf{F}' = e^t\mathbf{i} - \dfrac{t^3}{3}\mathbf{j} + 3t\mathbf{k} + \mathbf{C}_1$;

but $\mathbf{F}'(0) = 3\mathbf{k}$ so

$\mathbf{F}' = (e^t - 1)\mathbf{i} - \dfrac{t^3}{3}\mathbf{j} + (3t + 3)\mathbf{k}$

$\mathbf{F} = (e^t - t)\mathbf{i} - \dfrac{t^4}{12}\mathbf{j} + (\dfrac{3t^2}{2} + 3t)\mathbf{k} + \mathbf{C}_2$;

but $\mathbf{F}(0) = \mathbf{i} - 2\mathbf{j}$ so

$\mathbf{F} = (e^t - t)\mathbf{i} - \left(\dfrac{t^4}{12} + 2\right)\mathbf{j} + \left(\dfrac{3t^2}{2} + 3t\right)\mathbf{k}$

28. $\mathbf{R} = t\mathbf{i} + 2t\mathbf{j} + te^t\mathbf{k}$

$\mathbf{V} = \mathbf{R}' = \mathbf{i} + 2\mathbf{j} + e^t(t+1)\mathbf{k}$

The speed is:

$$\frac{ds}{dt} = \|\mathbf{V}\| = \sqrt{1 + 4 + e^{2t}(t+1)^2}$$

$$= \sqrt{5 + e^{2t}(t+1)^2}$$

$\mathbf{A} = \mathbf{V}' = e^t(t+2)\mathbf{k}$

29. $\mathbf{R} = t^2\mathbf{i} + 3t\mathbf{j} - 3t\mathbf{k}$

$$\mathbf{T} = \frac{\mathbf{R}'}{\|\mathbf{R}'\|} = \frac{2t\mathbf{i} + 3\mathbf{j} - 3\mathbf{k}}{\sqrt{4t^2 + 18}}$$

$$\mathbf{T}' = \frac{36\mathbf{i} - 12t\mathbf{j} + 12t\mathbf{k}}{(4t^2 + 18)^{3/2}}$$

$$\mathbf{N} = \frac{\mathbf{T}'}{\|\mathbf{T}'\|} = \frac{3\mathbf{i} - t\mathbf{j} + t\mathbf{k}}{(2t^2 + 9)^{1/2}}$$

$\mathbf{A} = \mathbf{R}'' = 2\mathbf{i}$

$$\mathbf{R}' \times \mathbf{R}'' = \begin{vmatrix} \mathbf{i} & \mathbf{j} & \mathbf{k} \\ 2t & 3 & -3 \\ 2 & 0 & 0 \end{vmatrix} = -6\mathbf{j} - 6\mathbf{k}$$

$$\kappa = \frac{\|\mathbf{R}' \times \mathbf{R}''\|}{\|\mathbf{R}'\|^3}$$

$$= \frac{\sqrt{36 + 36}}{(4t^2 + 18)^{3/2}}$$

$$= \frac{\sqrt{72}}{(4t^2 + 18)^{3/2}}$$

$$= \frac{3}{(2t^2 + 9)^{3/2}}$$

$$A_T = \frac{d^2s}{dt^2} = \frac{4t}{(4t^2 + 18)^{1/2}} = \frac{2\sqrt{2}t}{\sqrt{2t^2 + 9}}$$

$$A_N = \kappa \left(\frac{ds}{dt} \right)^2$$

$$= \frac{3}{(2t^2 + 9)^{3/2}}(4t^2 + 18)$$

$$= \frac{6}{\sqrt{2t^2 + 9}}$$

30. a. $\mathbf{R} = [(v_0\cos\alpha)t]\mathbf{i}$

$+ [(v_0\sin\alpha)t - \frac{1}{2}gt^2 + s_0]\mathbf{j};$

In our case:

$\mathbf{R}(t) = (25\sqrt{3}t)\mathbf{i} + (25t - 16t^2)\mathbf{j}$

$\mathbf{R}'(t) = (25\sqrt{3})\mathbf{i} + (25 - 32t)\mathbf{j}$

The maximum height is reached when the $\mathbf{j}$ component of $\mathbf{R}'$ is 0; that is, when

$25 - 32t = 0; t = \frac{25}{32}$

$y_{\max} = 25\left(\frac{25}{32}\right) - 16\left(\frac{25}{32}\right)^2 \approx 9.77$ ft

b. $T_f = \frac{2}{g}v_0\sin\alpha = \frac{1}{16}(50)\left(\frac{1}{2}\right) = \frac{25}{16}$ sec

$\text{Range} = \frac{v_0^2}{g}\sin 2\alpha = \frac{50^2}{32}\sin 60°$

$$= \frac{625\sqrt{3}}{16} \approx 67.7 \text{ ft}$$

Supplementary Problems, page 806

1. $\lim\limits_{t\to 0}\left[t^2\mathbf{i} - 3te^t\mathbf{j} + \frac{\sin 2t}{t}\mathbf{k}\right] = 2\mathbf{k}$

3. $\lim\limits_{t\to 0}\langle t, 0, 5\rangle \cdot \langle\sin t, 3t, -(1-t)\rangle$

$= (5\mathbf{k}) \cdot (-\mathbf{k}) = -5$

5. Let $N = \lim\limits_{t\to 0^+}\left(1 + \frac{1}{t}\right)^t$ so

$\ln N = \lim\limits_{t\to 0^+}\left[t\ln\left(1 + \frac{1}{t}\right)\right]$

so $e^0 = N$, which implies $N = 1$.

Also, $\lim\limits_{t\to 0^+}\dfrac{\sin t}{t} = 1$ and $\lim\limits_{t\to 0^+}t = 0$.

$$\lim\limits_{t\to 0^+}\left[\left(1 + \frac{1}{t}\right)^t\mathbf{i} - \left(\frac{\sin t}{t}\right)\mathbf{j} - t\mathbf{k}\right]$$

$= \mathbf{i} - \mathbf{j}$

7. $\mathbf{F}(t) = te^t\mathbf{i} + t^2\mathbf{j}$

$\mathbf{F}'(t) = (1 + t)e^t\mathbf{i} + 2t\mathbf{j}$

$\mathbf{F}''(t) = (2 + t)e^t\mathbf{i} + 2\mathbf{j}$

9. $\mathbf{F}(t) = \langle t^2, t^{3/2}, t^{-3}\rangle$

$\mathbf{F}'(t) = \langle 2t, \frac{3}{2}t^{1/2}, -3t^{-4}\rangle$

$\mathbf{F}''(t) = \langle 2, \frac{3}{4}t^{-1/2}, 12t^{-5} \rangle$

11. $\mathbf{F}(t) = \langle 2t^{-1}, -2t, te^{-t} \rangle$
$\mathbf{F}'(t) = \langle -2t^{-2}, -2, (1-t)e^{-t} \rangle$
$\mathbf{F}''(t) = \langle 4t^{-3}, 0, (t-2)e^{-t} \rangle$

13. $\mathbf{F}(t) = (t^2 + e^t)\mathbf{i} + (te^{-t})\mathbf{j} + (e^{t+1})\mathbf{k}$
$\mathbf{F}'(t) = (2t + e^t)\mathbf{i} + (1-t)e^{-t}\mathbf{j} + e^{t+1}\mathbf{k}$
$\mathbf{F}''(t) = (2 + e^t)\mathbf{i} + (t-2)e^{-t}\mathbf{j} + e^{t+1}\mathbf{k}$

15.   17.

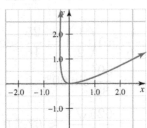
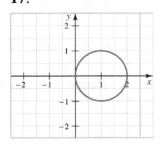

19. The graph is a circular helix of radius 2, traversed clockwise. It begins $(t = 0)$ at $(0, 2, 0)$ and rises 10π units with each revolution.

21. $x = t$, $z = -5t$ is the line $z = -5x$ in the plane $y = 3$.

23. The graph is the intersection of the cylinder $y = x^2 + 1$ with the plane $y = z + 1$; in $\mathbb{R}^2$, the graph is a parabola in the xy-plane.

25. The graph is the intersection of the plane $z = -2x$ and the paraboloid $y = x^2$.

27. **a.** $x^2 + y^2 = 1 \neq z$; no

    b. $x^2 + y^2 = e^{-2t}(\cos^2 t + \sin^2 t)$
$$= e^{-2t} = z$$
    yes

    c. $x^2 + y^2 = \left(1 + \frac{1}{t}\right)^2 + \left(1 - \frac{1}{t}\right)^2$
$$= 2 + \frac{2}{t^2} = z$$
    yes

29. $\dfrac{d}{dt}[\mathbf{F}(t) \cdot \mathbf{F}(t)] = 2\mathbf{F} \cdot \mathbf{F}'$

31. $\dfrac{d}{dt}\left[\dfrac{\mathbf{F}(t)}{\|\mathbf{F}(t)\|}\right] = \dfrac{d}{dt}\dfrac{\mathbf{F}}{\sqrt{\mathbf{F} \cdot \mathbf{F}}}$

$= \dfrac{\sqrt{\mathbf{F} \cdot \mathbf{F}}\,\mathbf{F}' - (\mathbf{F} \cdot \mathbf{F})^{-1/2}(\mathbf{F} \cdot \mathbf{F}')\mathbf{F}}{\mathbf{F} \cdot \mathbf{F}}$

$= \dfrac{\|\mathbf{F}\|^2\mathbf{F}' - (\mathbf{F} \cdot \mathbf{F}')\mathbf{F}}{\|\mathbf{F}\|^3}$

33. $\dfrac{d}{dt}[\mathbf{F}(t) \times \mathbf{F}(t)] = \mathbf{F} \times \mathbf{F}' + \mathbf{F}' \times \mathbf{F}$

$\qquad\qquad = \mathbf{F} \times \mathbf{F}' - \mathbf{F} \times \mathbf{F}'$

$\qquad\qquad = \mathbf{0}$

> **SURVIVAL HINT:**
> *It is easier to note that* $\mathbf{F} \times \mathbf{F} = \mathbf{0}$ *for any vector,*
> so, of course, $\dfrac{d}{dt}(\mathbf{F} \times \mathbf{F}) = \mathbf{0}$.

35. $\displaystyle\int_{-1}^{1} (e^{-t}\mathbf{i} + t^3\mathbf{j} + 3\mathbf{k})\, dt$

$= \left(-e^{-t}\mathbf{i} + \dfrac{t^4}{4}\mathbf{j} + 3t\mathbf{k}\right)\Big|_{-1}^{1}$

$= (e - e^{-1})\mathbf{i} + 6\mathbf{k}$

37. $\displaystyle\int [te^t\mathbf{i} - (\sin 2t)\mathbf{j} + t^2\mathbf{k}]\, dt$

$= (te^t - e^t)\mathbf{i} + \dfrac{\cos 2t}{2}\mathbf{j} + \dfrac{t^3}{3}\mathbf{k} + \mathbf{C}$

39. $\displaystyle\int t[e^t\mathbf{i} + (\ln t)\mathbf{j} + 3\mathbf{k}]\, dt$

$= (te^t - e^t)\mathbf{i} + \dfrac{1}{4}t^2(2\ln t - 1)\mathbf{j} + \dfrac{3t^2}{2}\mathbf{k} + \mathbf{C}$

41. $\mathbf{R}(t) = t\mathbf{i} + (3 - t)\mathbf{j} + 2\mathbf{k}$
$\mathbf{V}(t) = \mathbf{R}'(t) = \mathbf{i} - \mathbf{j}$
$\dfrac{ds}{dt} = \|\mathbf{V}(t)\| = \sqrt{2}$
$\mathbf{A}(t) = \mathbf{R}''(t) = \mathbf{0}$

43. $\mathbf{R}(t) = \langle t\sin t, te^{-t}, -(1 - t) \rangle$

$$\mathbf{V}(t) = \mathbf{R}'(t) = (t\cos t + \sin t)\mathbf{i}$$
$$+ (1-t)e^{-t}\mathbf{j} + \mathbf{k}$$

$$\frac{ds}{dt} = \|\mathbf{V}(t)\|$$
$$= \sqrt{t^2\cos^2 t + t\sin 2t + \sin^2 t + (1-t)^2 e^{-2t} + 1}$$

$$\mathbf{A}(t) = \mathbf{R}''(t)$$
$$= (2\cos t - t\sin t)\mathbf{i} + e^{-1}(t-2)\mathbf{j}$$

45. $\mathbf{R}(t) = t\mathbf{i} - t^2\mathbf{j}$

$\mathbf{V}(t) = \mathbf{i} - 2t\mathbf{j}$

$$\mathbf{T}(t) = \frac{\mathbf{i} - 2t\mathbf{j}}{\sqrt{4t^2 + 1}}$$

$$\frac{d\mathbf{T}}{dt} = \frac{-4t\mathbf{i} - 2\mathbf{j}}{(4t^2 + 1)^{3/2}}$$

$$\mathbf{N}(t) = \frac{-2t\mathbf{i} - \mathbf{j}}{\sqrt{4t^2 + 1}}$$

47. $\mathbf{R}(t) = \langle 4\cos t, -3t, 4\sin t \rangle$

$\mathbf{V}(t) = (-4\sin t)\mathbf{i} - 3\mathbf{j} + (4\cos t)\mathbf{k}$

$\|\mathbf{V}(t)\| = \sqrt{16 + 9} = 5$

$$\mathbf{T}(t) = \frac{1}{5}(-4\sin t\,\mathbf{i} - 3\mathbf{j} + 4\cos t\,\mathbf{k})$$

$$\frac{d\mathbf{T}}{dt} = \frac{1}{5}(-4\cos t\,\mathbf{i} - 4\cos t\,\mathbf{k})$$

$\mathbf{N}(t) = -\cos t\,\mathbf{i} - \sin t\,\mathbf{j}$

49. $\mathbf{R}(t) = t^2\mathbf{i} + 2t\mathbf{j} + e^t\mathbf{k}$

$\mathbf{V}(t) = \mathbf{R}'(t) = 2t\mathbf{i} + 2\mathbf{j} + e^t\mathbf{k}$

$\mathbf{A}(t) = \mathbf{R}''(t) = 2\mathbf{i} + e^t\mathbf{k};\ \|\mathbf{A}(t)\| = \sqrt{4 + e^{2t}}$

$$\frac{ds}{dt} = \|\mathbf{V}(t)\| = \sqrt{4t^2 + 4 + e^{2t}}$$

$$A_T = \frac{d^2 s}{dt^2} = \frac{4t + e^{2t}}{\sqrt{4 + 4t^2 + e^{2t}}}$$

$$A_N = \sqrt{\|\mathbf{A}\|^2 - A_T^2}$$

$$= \sqrt{4 + e^{2t} - \frac{16t^2 + 8te^{2t} + e^{4t}}{4 + 4t^2 + e^{2t}}}$$

$$= 2\sqrt{\frac{te^{2t} - 2te^{2t} + 2e^{2t} + 4}{4 + 4t^2 + e^{2t}}}$$

$$\kappa = \frac{\|\mathbf{V} \times \mathbf{A}\|}{\|\mathbf{V}\|^3} = \frac{A_N}{\|\mathbf{V}\|^2}$$

$$= \frac{2\sqrt{te^{2t} - 2te^{2t} + 2e^{2t} + 4}}{(4 + 4t^2 + e^{2t})^{3/2}}$$

51. $\mathbf{R}(t) = (4\sin t)\mathbf{i} + (4\cos t)\mathbf{j} + 4t\mathbf{k}$

$\mathbf{V}(t) = \mathbf{R}'(t) = (4\cos t)\mathbf{i} - (4\sin t)\mathbf{j} + 4\mathbf{k}$

$\mathbf{A}(t) = \mathbf{R}''(t) = 4(-\sin t\,\mathbf{i} - \cos t\,\mathbf{j})$

$\|\mathbf{A}(t)\| = 4$

$$\frac{ds}{dt} = \|\mathbf{V}(t)\| = \sqrt{16 + 16} = 4\sqrt{2}$$

$$A_T = \frac{d^2 s}{dt^2} = 0$$

$$A_N = \sqrt{\|\mathbf{A}\|^2 - A_T^2} = 4$$

$$\kappa = \frac{\|\mathbf{V} \times \mathbf{A}\|}{\|\mathbf{V}\|^3} = \frac{A_N}{\|\mathbf{V}\|^2} = \frac{4}{\left(4\sqrt{2}\right)^2} = \frac{1}{8}$$

53. $r = 4\cos\theta;\ r' = -4\sin\theta;\ r'' = -4\cos\theta$

At $\theta = \pi/3$, $r = 2$, $r' = -2\sqrt{3}$, $r'' = -2$

$$\kappa = \frac{\left|2^2 + 2(-2\sqrt{2})^2 - 2(-2)\right|}{[2^2 + (-2\sqrt{3})^2]^{3/2}} = \frac{1}{2}$$

55. $r = e^{-\theta};\ r' = -e^{-\theta};\ r'' = e^{-\theta}$

At $\theta = 1$, $r = e^{-1}$, $r' = -e^{-1}$, $r'' = e^{-1}$

$$\kappa = \frac{|e^{-2} + 2(e^{-2})^2 - e^{-2}|}{[e^{-2} + e^{-2}]^{3/2}} = \frac{e}{\sqrt{2}}$$

57. $r = 4\cos 3\theta;\ r' = -12\sin 3\theta;$

$r'' = -36\cos 3\theta$

At $\theta = \pi/6$, $r = 0$, $r' = -12$, $r'' = 0$

$$\kappa = \frac{|0^2 + 2(-12)^2 - 0(0)|}{[0^2 + (-12)^2]^{3/2}} = \frac{1}{6}$$

59. $\mathbf{F}(x) = \displaystyle\int_1^x [t\mathbf{i} - t^{-1}\mathbf{j} + e^{-t}\mathbf{k}]\,dt$

$\mathbf{F}'(x) = x\mathbf{i} - x^{-1}\mathbf{j} - e^{-x}\mathbf{k}$

61. $\mathbf{F}(x) = \displaystyle\int_1^x [(\sin t)\mathbf{i} - (\cos 2t)\mathbf{j} + e^{-t}\mathbf{k}]\,dt$

$\mathbf{F}'(x) = (\sin x)\mathbf{i} - (\cos 2x)\mathbf{j} + e^{-x}\mathbf{k}$

63. $\mathbf{F}(t) = \left(\dfrac{t^2}{2} + C_1\right)\mathbf{i} + \left(\dfrac{t^2}{2} + C_2\right)\mathbf{j} - \left(\dfrac{t^2}{2} + C_3\right)\mathbf{k}$

Since $\mathbf{F}(0) = \mathbf{i} + 2\mathbf{j} - 3\mathbf{k}$,

$\mathbf{F}(t) = \left(\dfrac{t^2}{2} + 1\right)\mathbf{i} + \left(\dfrac{t^2}{2} + 2\right)\mathbf{j} - \left(\dfrac{t^2}{2} + 3\right)\mathbf{k}$

65. $\mathbf{F}(t) = \left[\left(-\dfrac{1}{2}\right)\cos 2t + C_1\right]\mathbf{i}$

$\quad + \left[\left(\dfrac{e^t}{2}\right)(\sin t + \cos t) + C_2\right]\mathbf{j}$

$\quad - \left[3\ln|t + 1| + C_3\right]\mathbf{k}$

Since $\mathbf{F}(0) = \mathbf{i} - 3\mathbf{k}$,

$\mathbf{F}(t) = \left[\left(\dfrac{1}{2}\right)(3 - \cos 2t)\right]\mathbf{i}$

$\quad + \left[\left(\dfrac{e^t}{2}\right)(\sin t + \cos t) - \dfrac{1}{2}\right]\mathbf{j}$

$\quad - \left[3\ln|t + 1| + 3\right]\mathbf{k}$

67. $\mathbf{F}(t)$ is continuous for $t \neq 1$.

69. $\mathbf{F}'''(t) = (\cos t)\mathbf{i} + (\sin t)\mathbf{j} + \dfrac{t}{\pi}\mathbf{k}$

$\mathbf{F}''(t) = (\sin t + C_1)\mathbf{i} + (-\cos t + C_2)\mathbf{j}$

$\quad + \left(\dfrac{t^2}{2\pi} + C_3\right)\mathbf{k}$

Since $\mathbf{F}''(0) = 2\mathbf{j} + \mathbf{k}$,

$\mathbf{F}''(t) = \sin t\,\mathbf{i} + (3 - \cos t)\mathbf{j} + \left(\dfrac{t^2}{2\pi} + 1\right)\mathbf{k}$

$\mathbf{F}'(t) = (-\cos t + C_4)\mathbf{i} + (3t - \sin t + C_5)\mathbf{j}$

$\quad + \left(\dfrac{t^3}{6\pi} + t + C_6\right)\mathbf{k}$

Since $\mathbf{F}'(0) = \mathbf{i}$,

$\mathbf{F}'(t) = (2 - \cos t)\mathbf{i} + (3t - \sin t)\mathbf{j}$

$\quad + \left(\dfrac{t^3}{6\pi} + t\right)\mathbf{k}$

$\mathbf{F}(t) = (2t - \sin t + C_7)\mathbf{i} + \left(\dfrac{3t^2}{2} + \cos t + C_8\right)\mathbf{j}$

$\quad + \left(\dfrac{t^4}{24\pi} + \dfrac{t^2}{2} + C_9\right)\mathbf{k}$

Since $\mathbf{F}(0) = \mathbf{i}$,

$\mathbf{F}(t) = (2t - \sin t + 1)\mathbf{i} + \left(\dfrac{3t^2}{2} + \cos t - 1\right)\mathbf{j}$

$\quad + \left(\dfrac{t^4}{24\pi} + \dfrac{t^2}{2}\right)\mathbf{k}$

71. a. We have $\mathbf{R}'(t) = \mathbf{i} + \mathbf{j} + 3t^2\mathbf{k}$, so
$\mathbf{R}(1) = \mathbf{i} + \mathbf{j} + \mathbf{k}$ and
$\mathbf{R}'(1) = \mathbf{i} + \mathbf{j} + 3\mathbf{k}$.
Thus, parametric equations for the
tangent line are $x = 1 + s$,
$y = 1 + s$, $z = 1 + 3s$.

b. The tangent line in part **a** intersects
the graph of $\mathbf{R}(t)$ when $1 + s = t$,
$1 + s = t$, $1 + 3s = t^3$. Solving, we
obtain $s = 0, -3$ (or $t = 1, -2$). The
point of tangency $P(1, 1, 1)$
corresponds to $s = 0$. When
$s = -3$, we obtain $x = -2$,
$y = -2$, $z = -8$. The required point
is $(-2, -2, -8)$.

73. $\mathbf{A}(t) = \langle 24t^2, 4\rangle$

$\mathbf{V}(t) = \displaystyle\int \mathbf{A}(t)\,dt = \langle 8t^3, 4t\rangle + \mathbf{C}_1$

$\mathbf{V}(0) = \langle 0, 0\rangle + \mathbf{C}_1 = \langle 0, 0\rangle;\ \mathbf{C}_1 = \langle 0, 0\rangle$

$\mathbf{R}(t) = \displaystyle\int \mathbf{V}(t)\,dt = \langle 2t^4, 2t^2\rangle + \mathbf{C}_2$

$\mathbf{R}(0) = \langle 0, 0\rangle + \mathbf{C}_2 = \langle 1, 2\rangle;\ \mathbf{C}_2 = \langle 1, 2\rangle$

$\mathbf{R}(t) = \langle 2t^4 + 1, 2t^2 + 2\rangle$

75. $\mathbf{R}(s) = \left(a\cos\dfrac{s}{a}\right)\mathbf{i} + \left(a\sin\dfrac{s}{a}\right)\mathbf{j} + 2s\,\mathbf{k}$

$\mathbf{R}'(s) = \left(-\sin\dfrac{s}{a}\right)\mathbf{i} + \left(\cos\dfrac{s}{a}\right)\mathbf{j} + 2\mathbf{k}$

$\|\mathbf{R}'(s)\| = \sqrt{5}$

$\mathbf{T}(s) = \dfrac{1}{\sqrt{5}}\left[\left(-\sin\dfrac{s}{a}\right)\mathbf{i} + \left(\cos\dfrac{s}{a}\right)\mathbf{j} + 2\mathbf{k}\right]$

$\dfrac{d\mathbf{T}}{ds} = \dfrac{1}{a\sqrt{5}}\left[\left(-\cos\dfrac{s}{a}\right)\mathbf{i} - \left(\sin\dfrac{s}{a}\right)\mathbf{j}\right]$

$\mathbf{N}(s) = -\left(\cos\dfrac{s}{a}\right)\mathbf{i} - \left(\sin\dfrac{s}{a}\right)\mathbf{j}$

77. $W = 3{,}000$ lb; $mg = 3{,}000$, so

$m = 3{,}000/32$ slugs

$v = \dfrac{60(5{,}280)}{3{,}600} = 88$ ft/s; $\rho = 160$,

$\kappa = \dfrac{1}{160}$; $A_N = \dfrac{88^2}{160}$;

$\|\mathbf{F}\| = mA_N = \dfrac{88^2}{160}\dfrac{3{,}000}{32} = 4{,}537.5$ lb

79. a. $\mathbf{R}(t) = (e^{-t}\cos t)\mathbf{i} + (e^{-t}\sin t)\mathbf{j} + e^{-t}\mathbf{k}$

$\mathbf{V}(t) = e^{-t}(-\cos t - \sin t)\mathbf{i}$
$\qquad\qquad + e^{-t}(-\sin t + \cos t)\mathbf{j} - e^{-t}\mathbf{k}$

Speed is:

$\|\mathbf{V}(t)\|$
$= e^{-t}\sqrt{2\cos t \sin t + 1 + 1 - 2\sin t \cos t + 1}$
$= e^{-t}\sqrt{3}$

$\mathbf{A}(t) = e^{-t}(2\sin t)\mathbf{i}$
$\qquad\qquad + e^{-t}(-2\cos t)\mathbf{j} + e^{-t}\mathbf{k}$

b. $\mathbf{V} \times \mathbf{A}$

$= e^{-2t}\begin{vmatrix} \mathbf{i} & \mathbf{j} & \mathbf{k} \\ -\cos t - \sin t & -\sin t + \cos t & -1 \\ 2\sin t & -2\cos t & 1 \end{vmatrix}$

$= e^{-2t}[-(\sin t + \cos t)\mathbf{i} - (\sin t - \cos t)\mathbf{j} + 2\mathbf{k}]$

$\|\mathbf{V} \times \mathbf{A}\| = e^{-2t}\sqrt{6}$

$\kappa = \dfrac{\sqrt{6}e^{-2t}}{3\sqrt{3}e^{-3t}} = \dfrac{\sqrt{2}}{3}e^{t}$

81. Using the *cab-bac* formula,

$\mathbf{F} \times (\mathbf{G} \times \mathbf{H}) = (\mathbf{H} \cdot \mathbf{F})\mathbf{G} - (\mathbf{G} \cdot \mathbf{F})\mathbf{H}$

Thus,

$[\mathbf{F} \times (\mathbf{G} \times \mathbf{H})]' = [(\mathbf{H} \cdot \mathbf{F})\mathbf{G}]' - [(\mathbf{G} \cdot \mathbf{F})\mathbf{H}]'$

83. $r = 1 + \cos at$; $\theta = e^{-at}$;

$\dfrac{dr}{dt} = -a\sin at$; $\dfrac{d\theta}{dt} = -ae^{-at}$;

$\dfrac{d^2 r}{dt^2} = -a^2\cos at$; $\dfrac{d^2\theta}{dt^2} = a^2 e^{-at}$;

$\mathbf{V}(t) = \dfrac{dr}{dt}\mathbf{u}_r + r\dfrac{d\theta}{dt}\mathbf{u}_\theta$
$\qquad = -a\sin at\,\mathbf{u}_r - ae^{-at}(1 + \cos at)\mathbf{u}_\theta$

$\mathbf{A}(t) = \left[\dfrac{d^2 r}{dt^2} - r\left(\dfrac{d\theta}{dt}\right)^2\right]\mathbf{u}_r$
$\qquad + \left[r\dfrac{d^2\theta}{dt^2} + 2\left(\dfrac{dr}{dt}\right)\left(\dfrac{d\theta}{dt}\right)\right]\mathbf{u}_\theta$
$= [-a^2\cos at - a^2 e^{-2at}(1 + \cos at)]\mathbf{u}_r$
$\quad + [a^2 e^{-at}(1 + \cos at) + 2a^2 e^{-at}\sin at]\mathbf{u}_\theta$

85.

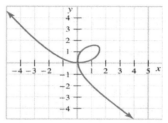

$\mathbf{R}(t) = \left(\dfrac{3t}{1 + t^3}\right)\mathbf{i} + \left(\dfrac{3t^2}{1 + t^3}\right)\mathbf{j}$

$\dfrac{d\mathbf{R}}{dt} = \left[\dfrac{3 - 6t^3}{(1 + t^3)^2}\right]\mathbf{i} + \left[\dfrac{6t - 3t^4}{(t^3 + 1)^2}\right]\mathbf{j}$

To find the tangent line when $t = 2$, first note that the point of tangency is $P_0(\frac{2}{3}, \frac{4}{3})$, and is parallel to the vector $\mathbf{R}'(2) = -\frac{5}{9}\mathbf{i} - \frac{4}{9}\mathbf{j}$, so its parametric form is $x = \frac{2}{3} - \frac{5}{9}t$, $y = \frac{4}{3} - \frac{4}{9}t$.

87. $\mathbf{R} = 10^{-8}\cos t\,\mathbf{i} + 10^{-8}\sin t\,\mathbf{j} + \dfrac{3(10^{-8})t}{2\pi}\mathbf{k}$

$\mathbf{V} = 10^{-8}(-\sin t\,\mathbf{i} + \cos t\,\mathbf{j} + \dfrac{3}{2\pi}\mathbf{k})$

$\|\mathbf{V}\| = 10^{-8}\sqrt{1 + \dfrac{9}{4\pi^2}}$

Range of t: $0 < t < 6\pi(10^8)$ since there are 3×10^8 turns at 2π radians/turn

$s = (10^{-8})\int_0^{6\pi \times 10^8} \sqrt{1 + \frac{9}{4\pi^2}}\, dt \approx 20.888\ \mu\text{m}$

89. $r = 10t$; $r' = 10$; $r'' = 0$;

$\theta = 2\pi t$; $\theta' = 2\pi$; $\theta'' = 0$

$\mathbf{V}(t) = \dfrac{dr}{dt}\mathbf{u}_r + r\dfrac{d\theta}{dt}\mathbf{u}_\theta = 10\mathbf{u}_r + 20\pi t\mathbf{u}_\theta$

a. At $t = 0$, $\mathbf{V}(0) = 10\mathbf{u}_r$

b. At $t = 0.25$, $\mathbf{V}(0.25) = 10\mathbf{u}_r + 5\pi\mathbf{u}_\theta$

91. Answers vary. Try to find some "new" facts about Galileo.

93. Answers vary. Do not just translate, but put your explanation in some context.

95. a. $3\mathbf{F}(t_0) + 5\mathbf{G}(t_0)$ exists and

$$\lim_{t \to t_0}[3\mathbf{F}(t) + 5\mathbf{G}(t)]$$

$$= 3\lim_{t \to t_0}\mathbf{F}(t) + 5\lim_{t \to t_0}\mathbf{G}(t)$$

$$= 3\mathbf{F}(t_0) + 5\mathbf{G}(t_0)$$

The function is continuous at t_0.

b. $\mathbf{F}(t_0) \cdot \mathbf{G}(t_0)$ exists and

$$\lim_{t \to t_0} [\mathbf{F}(t_0) \cdot \mathbf{G}(t_0)]$$

$$= \left[\lim_{t \to t_0} \mathbf{F}(t)\right] \cdot \left[\lim_{t \to t_0} \mathbf{G}(t_0)\right]$$

$$= \mathbf{F}(t_0) \cdot \mathbf{G}(t_0)$$

The function is continuous at t_0.

c. Let $h(t) = t$; for $t_0 = 0$; $(h(t_0))^{-1}\mathbf{F}(t_0)$ is not continuous.

d. $\mathbf{F}(t_0) \times \mathbf{G}(t_0)$ exists and

$$\lim_{t \to t_0} [\mathbf{F}(t) \times \mathbf{G}(t)]$$

$$= \left[\lim_{t \to t_0} \mathbf{F}(t)\right] \times \left[\lim_{t \to t_0} \mathbf{G}(t)\right]$$

$$= \mathbf{F}(t_0) \times \mathbf{G}(t_0)$$

The function is continuous at t_0.

97. This is Putnam Problem 6ii of the morning session of 1939. Choose rectangular coordinates with the y-axis vertical, the origin at the position of the gun, and the airplane over a point of the positive x-axis. Then the coordinates of the airplane are (u, h) where $u \geq 0$. If the gun is fired at time $t = 0$ with muzzle velocity V and angle of elevation α, then (neglecting air resistance), the shell's position at time t is given by

$$x = Vt\cos\alpha, \quad y = Vt\sin\alpha - \tfrac{1}{2}gt^2$$

Since it is given that the shell strikes the airplane,

$$u = Vt\cos\alpha \quad \text{and} \quad h = Vt\sin\alpha - \tfrac{1}{2}gt^2$$

for some t and α. Hence,

$$u^2 + \left(h + \frac{1}{2}gt^2\right)^2 = V^2t^2$$

so that

$$\frac{1}{4}g^2t^4 + (gh - V^2)t^2 + h^2 + u^2 = 0$$

In order for this equation to have a real root, it is necessary that

$$(gh - V^2)^2 \geq g^2(h^2 + u^2)$$

$$g^2u^2 \leq V^2(V^2 - 2gh)$$

$$V^2 \geq 2gh \quad \text{and} \quad u \leq \frac{V}{g}\sqrt{V^2 - 2gh}$$

This shows that the gun is within distance

$$\frac{V}{g}\sqrt{V^2 - 2gh}$$

from the point directly below the airplane when it was hit.

99. This is Putnam Problem 6 of the afternoon session of 1946. Suppose the circle has radius r. If Cartesian coordinates are chosen with origin at the center of the circle, then the coordinates of the particle are $(r\cos\theta, r\sin\theta)$ where θ is a function of time t. Differentiating this twice, we see that the acceleration vector is

$$r\frac{d\omega}{dt}\langle -\sin\theta, \cos\theta\rangle + r\omega^2\langle -\cos\theta, -\sin\theta\rangle$$

where $\omega = d\theta/dt$ is the angular velocity. Since $\langle -\sin\theta, \cos\theta\rangle$ and $\langle -\cos\theta, -\sin\theta\rangle$ are orthogonal unit vectors in the direction of the tangent and the inward normal, we see that the two terms of the acceleration vector are, respectively, the tangential and normal components of the acceleration. Since $\omega \neq 0$ at any time during the motion, the normal component of acceleration, and hence the acceleration vector itself, is never 0. Since $\omega = 0$ at the start and at the finish, by Rolle's theorem there is an intermediate time at which $d\omega/dt = 0$. At that time, the acceleration vector points inward along the radius because $r\omega^2 > 0$. *Remark*: We have interpreted "coming to rest" to mean "having velocity 0." If "coming to rest" means "remaining stationary through some time interval," then the first statement is false, for it is certainly possible that ω and $d\omega/dt$ vanish simultaneously, but not on an interval. The second statement is true, however, because the usual proof of Rolle's theorem shows that $d\omega/dt = 0$ at some point where $\omega \neq 0$, unless $\omega = 0$ identically, which is ruled out.

Cumulative Review
Chapters 1-10, page 816

1. **Step 1.** Simplify;
   Step 2. Use basic formulas;
   Step 3. Substitute;
   Step 4. Classify.
   Step 5: Try again. See Integration Strategy in Section 7.5, p. 541.
3. A vector is a directed line segment. A vector function is a function whose range is a set of vectors for each point in its domain. Vector calculus involves differentiation and integration of vector functions.

5. $\lim\limits_{x \to \infty} \dfrac{(2x^2 - 3x + 1)(1 - 5x)}{5x^3 + 4x - 9}$
$$= \lim\limits_{x \to \infty} \frac{\left(2 - \frac{3}{x} + \frac{1}{x^2}\right)\left(\frac{1}{x} - 5\right)}{5 + \frac{4}{x^2} - \frac{9}{x^3}}$$
$$= -\frac{10}{5} = -2$$

7. Let $L = \lim\limits_{x \to \infty}\left(1 - \dfrac{2}{x}\right)^{3x}$
$$\ln L = \ln \lim\limits_{x \to \infty}\left(1 - \frac{2}{x}\right)^{3x}$$
$$= \lim\limits_{x \to \infty} 3x \ln\left(1 - \frac{2}{x}\right)$$
$$= \lim\limits_{x \to \infty} \frac{3\ln\left(1 - \frac{2}{x}\right)}{\frac{1}{x}}$$
$$= 3\lim\limits_{x \to \infty} \frac{\left(\frac{1}{1-\frac{2}{x}}\right)\left(\frac{2}{x^2}\right)}{\left(\frac{-1}{x^2}\right)} \quad \text{l'Hôpital's rule}$$
$$= 3(-2)$$
$$= -6$$
Thus, $L = e^{-6}$.

9. $\lim\limits_{x \to 0} \dfrac{\cos x - 1}{e^x - x - 1} = \lim\limits_{x \to 0} \dfrac{-\sin x}{e^x - 1} \quad$ l'Hôpital's rule
$$= \lim\limits_{x \to 0} \frac{-\cos x}{e^x} \quad \text{l'Hôpital's rule}$$
$$= -1$$

11. $y = \sin^3 x + 2\tan x$
$y' = 3\sin^2 x \cos x + 2\sec^2 x$

13. $y = \sin^{-1}x + 2\tan^{-1}\dfrac{1}{x}$
$$= \frac{1}{\sqrt{1 - x^2}} + \frac{2}{1 + \left(\frac{1}{x}\right)^2}\left(-\frac{1}{x^2}\right)$$
$$= \frac{1}{\sqrt{1 - x^2}} - \frac{2}{x^2 + 1}$$

15.
$$xy^3 + x^2 e^{-y} = 4$$
$$x(3y^2 y') + y^3 + x^2(-e^{-y} y') + 2xe^{-y} = 0$$
$$y' = \frac{-(y^3 + 2xe^{-y})}{3xy^2 - x^2 e^{-y}}$$

17. $\displaystyle\int \sin^2 x \cos^3 x \, dx$

$$= \int \sin^2 x (1 - \sin^2 x) \cos x \, dx$$

$$= \int \sin^2 x (\cos x \, dx) - \int \sin^4 x (\cos \, dx)$$

$$= \frac{1}{3} \sin^3 x - \frac{1}{5} \sin^5 x + C$$

19. $\displaystyle\int \frac{dx}{1 + \cos x} = \int \frac{1 - \cos x}{1 - \cos^2 x} dx$

$$= \int \frac{dx}{\sin^2 x} - \int \frac{\cos x \, dx}{\sin^2 x}$$

$$= \int \csc^2 x \, dx - \int \cot x \csc x \, dx$$

$$= -\cot x + \csc x + C$$

21. $\displaystyle\int \frac{dx}{\sqrt{2x - x^2}} = \int \frac{dx}{\sqrt{-(x^2 - 2x + 1) + 1}}$

$$= \int \frac{dx}{\sqrt{1 - (x - 1)^2}}$$

$$= \int \frac{du}{\sqrt{1 - u^2}}$$

$$= \sin^{-1} u + C$$

$$= \sin^{-1}(x - 1) + C$$

23. a. Yes; since $f'(x)$ can be negative and $f(0) = 0$, we know that f can decrease for $x \geq 0$.

b. Yes; since $f'(x)$ can be positive and $f(0) = 0$, we know that f can be negative at $x = -2$.

c. No, it need not have a critical point. For example, if $f(x) = x$, then the conditions of this problem are satisfied.

25. The curves intersect at $x = \frac{\pi}{4}$.

$$A = \int_0^{\pi/4} (\cos x - \sin x) \, dx$$

$$+ \int_{\pi/4}^1 (\sin x - \cos x) \, dx$$

$$= (\sin x + \cos x)\big|_0^{\pi/4}$$

$$- (\cos x + \sin x)\big|_{\pi/4}^1$$

$$= 2\sqrt{2} - 1 - \cos 1 - \sin 1$$

27. $\displaystyle V = \pi \int_1^2 \left[2(3x - 2)^{-1/2}\right]^2 dx$

$$\boxed{\text{Let } u = 3x - 2;\ du = 3\, dx}$$

$$= \pi \int_1^4 4u^{-1} \frac{du}{3}$$

$$= \frac{4\pi}{3} \ln|u|\Big|_1^4$$

$$= \frac{4\pi}{3} \ln 4$$

29. a. Let $A(5, 1, 2)$, $B(3, 1, -2)$, $C(3, 2, 5)$
$$\mathbf{AB} = -2\mathbf{i} - 4\mathbf{k};\ \mathbf{AC} = -2\mathbf{i} + \mathbf{j} + 3\mathbf{k}$$

$$\mathbf{N} = \begin{vmatrix} \mathbf{i} & \mathbf{j} & \mathbf{k} \\ -2 & 0 & -4 \\ -2 & 1 & 3 \end{vmatrix} = 4\mathbf{i} + 14\mathbf{j} - 2\mathbf{k}$$

The desired plane is
$$4(x - 5) + 14(y - 1) - 2(z - 2) = 0$$
$$2x + 7y - z - 15 = 0$$

b. $P(-2, -1, 4)$; a vector normal to the plane is parallel to the line, so
$\mathbf{N} = 2\mathbf{i} + 5\mathbf{j} - 2\mathbf{k}$. The desired plane is
$$2(x + 2) + 5(y + 1) - 2(z - 4) = 0$$
$$2x + 5y - 2z + 17 = 0$$

31. $\displaystyle\sum_{k=1}^{\infty} \frac{1}{k4^k}$ converges by the ratio test

$$\lim_{k\to\infty} \frac{k4^k}{(k+1)4^{k+1}} = \frac{1}{4} < 1$$

33. $\displaystyle\sum_{k=1}^{\infty} \frac{k!}{2^k k}$ diverges by the ratio test

$$\lim_{k\to\infty} \frac{(k+1)!2^k k}{(k+1)2^{k+1} k!} = \lim_{k\to\infty} \frac{(k+1)k}{2(k+1)} = \infty$$

35. The related absolute value series is $\displaystyle\sum \frac{1}{k^3}$

which converges (*p*-series, $p = 3 > 1$). Thus, the given series converges by the absolute convergence test.

37. a. $\displaystyle\int_1^\infty x^2 e^{-x}\,dx$ $\boxed{\text{Formula 185}}$

$$= \lim_{t\to\infty} \int_1^t x^2 e^{-x}\,dx$$

$$= \lim_{t\to\infty} \frac{e^{-x}}{-1}\left(x^2 + 2x + 2\right)\Bigg|_1^t$$

$$= \frac{1}{e}(1 + 2 + 2) = \frac{5}{e}; \text{ converges}$$

b. $\displaystyle\int_0^2 \frac{dx}{\sqrt{4-x^2}} = \lim_{t\to 2^-} \int_0^t \frac{dx}{\sqrt{4-x^2}}$

$$= \lim_{t\to 2^-1} \sin^{-1}\frac{x}{2}\Bigg|_0^t$$

$$= \frac{\pi}{2}; \text{ converges}$$

39. $\mathbf{F}(t) = 2t\,\mathbf{i} + e^{-3t}\mathbf{j} + t^4\mathbf{k}$
$\mathbf{F}'(t) = 2\mathbf{i} - 3e^{-3t}\mathbf{j} + 4t^3\mathbf{k}$
$\mathbf{F}''(t) = 9e^{-3t}\mathbf{j} + 12t^2\mathbf{k}$

41. $\mathbf{R}(t) = 2(\sin 2t)\mathbf{i} + (2 + 2\cos 2t)\mathbf{j} + 6t\mathbf{k}$
$\mathbf{R}'(t) = 4(\cos 2t)\mathbf{i} - (4\sin 2t)\mathbf{j} + 6\mathbf{k}$
$\|\mathbf{R}'\| = \sqrt{16 + 36} = 2\sqrt{13}$

$$\mathbf{T} = \frac{1}{\sqrt{13}}[2(\cos 2t)\mathbf{i} - 2(\sin 2t)\mathbf{j} + 3\mathbf{k}]$$

$$\frac{d\mathbf{T}}{dt} = \frac{1}{\sqrt{13}}[-4(\sin 2t)\mathbf{i} - 4(\cos 2t)\mathbf{j}]$$

$$\mathbf{N} = -\sin 2t\,\mathbf{i} - \cos 2t\,\mathbf{j}$$

43. a. $e^x = \displaystyle\sum_{k=0}^\infty \frac{x^k}{k!}$, so

$$e^{-x^2} = \sum_{k=0}^\infty \frac{(-x^2)^k}{k!} = \sum_{k=0}^\infty \frac{(-1)^k x^{2k}}{k!}$$

and

$$x^2 e^{-x^2} = x^2\sum_{k=0}^\infty \frac{(-1)^k x^{2k}}{k!} = \sum_{k=0}^\infty \frac{(-1)^k x^{2k+2}}{k!}$$

b.

$$\int_0^1 x^2 e^{-x^2}\,dx = \sum_{k=0}^\infty \frac{(-1)^k}{k!}\int_0^1 x^{2k+2}\,dx$$

$$= \left[\frac{x^3}{3} - \frac{x^5}{5(1!)} + \frac{x^7}{7(2!)} - \frac{x^9}{9(3!)} + \cdots\right]\Bigg|_0^1$$

$$\approx \frac{1}{3} - \frac{1}{5} + \frac{1}{14} - \frac{1}{54} + \frac{1}{264} - \frac{1}{1,560} + \frac{1}{10,800}$$

$$\approx 0.189$$

45. $\dfrac{dy}{dx} + 2y = x^2$; the integrating factor is
$I = e^{\int 2\,dx} = e^{2x}$

$$y = \frac{1}{e^{2x}}\left[\int x^2 e^{2x}\,dx + C\right]$$

$$= \frac{1}{e^{2x}}\left[\frac{e^{2x}}{4}\left(2x^2 - 2x + 1\right) + C\right]$$

If $x = 0$, then $y = 2$, so

$$y = \frac{x^2}{2} - \frac{x}{2} + \frac{1}{4} + \frac{7}{4}e^{-2x}$$

47. $x = t^2 - 2t - 1$, $y = t^4 - 4t^2 + 2$

$$\frac{dx}{dt} = 2t - 2; \quad \frac{dy}{dt} = 4t^3 - 8t$$

$$\mathbf{R}'(t) = \langle 2t - 2, 4t^3 - 8t\rangle$$

At $t = 1$, the point is $P(-2, -1)$, and the tangent line is parallel to the vector

$\mathbf{R}'(1) = \langle 0, -4 \rangle$. Thus, the tangent line is vertical with equation $x = -2$.

49. The element of rope dx is lifted through a distance $50 - x$, so $dF_r = 0.25 \, dx$ and

$$W_r = \int_0^{50} 0.25(50 - x) \, dx$$

$$= -\frac{0.25}{2}(50 - x)^2 \Big|_0^{50}$$

$$= 312.5 \text{ ft-lb}$$

For the bucket, $W_b = 50(25) = 1{,}250$ ft-lb.
Thus, $W = W_b + W_r = 1{,}562.5$ ft-lb.

51. $f(x) = (2x + x^{-1})^3$
$f'(x) = 3(2x + x^{-1})^2(2 - x^{-2})$
$f'(1) = 3(3^2)(1) = 27$, so $m = 27$
The equation of the tangent line is

$$y - 27 = 27(x - 1)$$

$$27x - y = 0$$

The equation of the normal line is

$$y - 27 = -\frac{1}{27}(x - 1)$$

$$x + 27y - 730 = 0$$

53. Let $u = x^2 + x$, then

$$\frac{d}{dx} f(x^2 + x) = \left[\frac{d}{du} f(u)\right] \frac{du}{dx}$$

$$= f'(u)(2x + 1)$$

$$= (2x + 1)\left[\left(x^2 + x\right)^2 + x^2 + x\right]$$

$$= (2x + 1)\left(x^4 + 2x^3 + 2x^2 + x\right)$$

55. $V = \frac{4}{3}\pi r^3$; $\dfrac{dV}{dt} = 4\pi r^2 \dfrac{dr}{dt}$; $S = 4\pi r^2$ so

at the specified instant when $4\pi r^2 = 4\pi$, we have $r = 1$ and since $dr/dt = 2$

$$\frac{dV}{dt} = 4\pi(1)^2(2) = 8\pi$$

Thus, the volume is changing at the rate of $8\pi \text{ cm}^3/\text{s}$.

57. a.

$$\frac{dv}{dt} = a(t) = 32 - 0.08v$$

$$\int \frac{dv}{32 - 0.08v} = \int dt$$

$$-\frac{1}{0.8} \ln|32 - 0.08v| = t + C_1$$

$$32 - 0.08v = C_2 e^{-0.08t}$$

$$v = \frac{1}{0.08}\left[32 - C_2 e^{-0.08t}\right]$$

When $t = 0$, $v = 5$, so $C_2 = 31.6$, and

$$v(t) = \frac{1}{0.08}\left[32 - 31.6e^{-0.08t}\right]$$

$$= 400 - 395e^{-0.08t}$$

b. $\displaystyle\lim_{t \to \infty} (400 - 395e^{-0.08}) = 400$ ft/s

59. Answers vary.

CHAPTER 11 Partial Differentiation

Chapter Overview

In the past two chapters you have been introduced to mathematically thinking in three dimensions. Now, in this chapter, we are ready to begin the study of three dimensional calculus. The notion of limits, continuity, derivatives, and chain rules are all familiar topics from single variable calculus, and in this chapter these concepts are extended to $\mathbb{R}^3$. The last three sections give some important applications of these ideas.

11.1 Functions of Several Variables, page 826

1. A (real) function of two variables associates a real number with each point in the two-dimensional xy-plane; examples vary.

3. $f(x, y) = x^2 y + xy^2$

   a. $\quad f(0, 0) = 0 + 0 = 0$

   b. $\quad f(-1, 0) = 0 + 0 = 0$

   c. $\quad f(0, -1) = 0 + 0 = 0$

   d. $\quad f(1, 1) = (1)^2(1) + (1)(1)^2 = 2$

   e. $\quad f(2, 4) = (2)^2(4) + 2(4)^2 = 48$

   f. $\quad f(t, t) = t^2 t + t t^2 = 2t^3$

   g. $\quad f(t, t^2) = t^2(t^2) + t(t^2)^2 = t^4 + t^5$

   h. $\quad f(1 - t, t) = (1 - t)^2 t + (1 - t)t^2$

$$= t - t^2$$

	Domain	Range
5.	$x - y \geq 0$	$f \geq 0$
7.	$uv \geq 0$	$f \geq 0$
9.	$y - x > 0$	$\mathbb{R}$
11.	$\mathbb{R}^2$	$f \geq 0$
13.	$x^2 - y^2 > 0$	$f > 0$

15.   **17.**

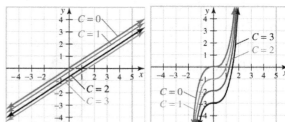

19.

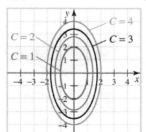

21. This is a cylinder, $y^2 + z^2 = 1$, which has the x-axis as its axis.

23. This is a plane, $x + y - z = 1$; its trace in the xy-plane is the line $x + y = 1$; its trace in the xz-plane is $x - z = 1$; and its trace in the yz-plane is $y - z = 1$.

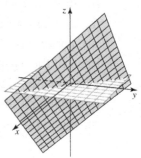

Page 413

25. This is a sphere,
$(x + 1)^2 + (y - 2)^2 + (z - 3)^2 = 4$
its cross sections in the planes
$x = -1$, $y = 2$, and $z = 3$ are circles.

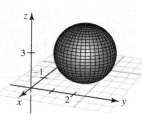

27. This is an ellipsoid;
traces are ellipses
in all three
coordinate planes.

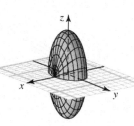

29. This is a hyperboloid
of one sheet; the trace
in the xy-plane is an
ellipse; in the xz- and
yz-planes the traces are
hyperbolas.

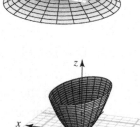

31. This is an elliptic
paraboloid; its traces in
the xz- and yz-planes
are parabolas; the
trace in the xy-plane is
a point (the origin).
In a plane parallel to the
xy-plane it is an ellipse.

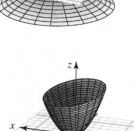

33. This is an elliptic cone;
the traces in the xz- and
yz-planes are pairs of
lines; the trace in the
xy-plane is the origin if
$z = 0$ and an ellipse
if $z \neq 0$.

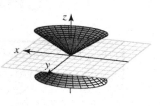

SURVIVAL HINT: *Problems 35-40 are designed
to help you to use level curves to "visualize"
three dimensional graphs.*

35. D   **37.** B   **39.** A

41. Plane parallel to
the xy-plane.

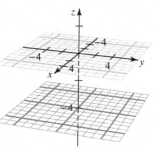

43. Cylinder which
bends upward from
the xy-plane.

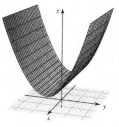

45. Plane which intersects the xy-plane
along the line
$2x - 3y = 0$.

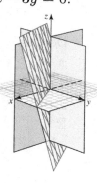

47. Elliptic paraboloid with vertex at the origin and which rises upward from the xy-plane.

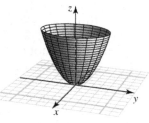

49. A surface that lies close to the xy-plane as it moves away from the origin, but "peels" upward close to the x-axis.

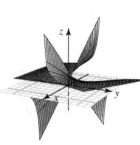

51. A paraboloid with vertex at $(0, 0, 2)$ and it rises upward from that point.

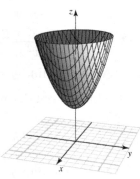

53. Equipotential curves are ellipses.

$E = 1:$ $x^2 + 2y^2 = 46$

$E = 2:$ $x^2 + 2y^2 = \dfrac{37}{4}$

$E = 3:$ $x^2 + 2y^2 = \dfrac{22}{9}$

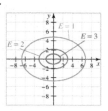

55. $\dfrac{1}{d_o} + \dfrac{1}{d_i} = \dfrac{1}{L}$

$$L = \frac{d_o d_i}{d_1 + d_o} \qquad d_o > 0, d_i > 0$$

The level curves $L + K$ satisfy

$$\frac{d_o d_i}{d_i + d_o} = K$$

$$d_o = \frac{K d_i}{d_i - R}$$

A typical level curve for (d_o, d_i) is a hyperbola with vertical asymptote $d_i = K$ and horizontal asymptote $d_o = K$.

57. $10xy = 1,000$ or $y = \dfrac{100}{x}$ (a hyperbola)

59. x machines sold at $60 - \dfrac{x}{5} + \dfrac{y}{20}$ thousand dollars apiece and y machines at $50 - \dfrac{x}{10} + \dfrac{y}{20}$ thousand dollars apiece yield revenue

$$R = \left(60 - \frac{x}{5} + \frac{y}{20}\right)x + \left(50 - \frac{x}{10} + \frac{y}{20}\right)y$$

11.2 Limits and Continuity, page 836

1. Pick a positive value ϵ. You are given the surface $z = f(x, y)$. Consider a fixed point (a, b) such that $f(a, b) = L$. Draw a circle around (a, b) with radius $\delta > 0$. If the vertical distance from $z = f(x, y)$ to $z = L$ is less than ϵ for every point in the circle for delta sufficiently small with the possible exception of the center (a, b), then

$$\lim_{(x,y) \to (a,b)} f(x, y) = L$$

If no such circle exists, then the limit does not exist. Formally,

$$\lim_{(x,y) \to (a,b)} f(x, y) = L$$

means that for any given $\epsilon > 0$ there exists a $\delta > 0$ so that

$$|f(x, y) - L| < \epsilon$$

whenever (x, y) is a point inside the punctured disk with radius δ.

3. $\lim\limits_{(x,y)\to(-1,0)} (xy^2 + x^3y + 5)$

$= (-1)0^2 + (-1)^30 + 5 = 5$

5. $\lim\limits_{(x,y)\to(1,3)} \dfrac{x+y}{x-y} = \dfrac{1+3}{1-3} = -2$

7. $\lim\limits_{(x,y)\to(1,0)} e^{xy} = e^0 = 1$

9. $\lim\limits_{(x,y)\to(0,1)} \left[e^{x^2+x}\ln(ey^2) \right] = e^0\ln e = 1$

11. $\lim\limits_{(x,y)\to(0,0)} \dfrac{x^2 - 2xy + y^2}{x-y} = \lim\limits_{(x,y)\to(0,0)} (x-y) = 0$

13. $\lim\limits_{(x,y)\to(0,0)} \dfrac{e^x\tan^{-1}y}{y} = \lim\limits_{y\to0} \dfrac{\tan^{-1}y}{y}$

$= \lim\limits_{y\to0} \dfrac{1}{1+y^2} = 1$

15. $\lim\limits_{(x,y)\to(0,0)} \dfrac{\sin(x+y)}{x+y} = \lim\limits_{t\to0} \dfrac{\sin t}{t} = 1$

17. $\lim\limits_{(x,y)\to(5,5)} \dfrac{x^4 - y^4}{x^2 - y^2}$

$= \lim\limits_{(x,y)\to(5,5)} \dfrac{(x^2+y^2)(x^2-y^2)}{x^2 - y^2}$

$= \lim\limits_{(x,y)\to(5,5)} (x^2 + y^2) = 50$

19. $\lim\limits_{(x,y)\to(2,1)} \dfrac{x^2 - 4y^2}{x - 2y} = \lim\limits_{(x,y)\to(2,1)} (x + 2y)$

$= 4$

21. $\lim\limits_{(x,y)\to(2,1)} (xy^2 + x^3y) = 2(1) + 8(1) = 10$

23. Along the line $x = 0$:

$\lim\limits_{(x,y)\to(0,0)} \dfrac{x+y}{x-y} = \lim\limits_{y\to0} \dfrac{y}{(-y)} = \lim\limits_{y\to0}(-1) = -1$

Along the line $y = 0$:

$\lim\limits_{(x,y)\to(0,0)} \dfrac{x+y}{x-y} = \lim\limits_{x\to0} \dfrac{x}{x} = \lim\limits_{x\to0} 1 = 1$

Limit does not exist.

25. $\lim\limits_{(x,y)\to(0,0)} e^{xy} = e^0 = 1$

27. $\lim\limits_{(x,y)\to(0,0)} (\sin x - \cos y) = 0 - 1 = -1$

29. $\lim\limits_{(x,y)\to(0,0)} \left[\dfrac{\sin(x^2 + y^2)}{x^2 + y^2} \right]$

$= \lim\limits_{t\to0} \dfrac{\sin t}{t} = 1$

31. $f(x, y) = \dfrac{x^2y^2}{x^4 + y^4}$

Along the line $y = mx$, we have

$\lim\limits_{x\to0} \dfrac{m^2x^4}{(1+m^4)x^4} = \dfrac{m^2}{1+m^4}$ which is not unique (the result varies with m). Thus, the limit does not exist.

33. $f(x, y) = \dfrac{x - y^2}{x^2 + y^2}$

Along the line $y = 0$, we have $\lim\limits_{x\to0} \dfrac{x}{x^2}$, which is not defined, so the limit does not exist.

35. $f(x, y) = \dfrac{xy^3}{x^2 + y^6}$

Along the line $y = kx$,

$\lim\limits_{x\to0} \dfrac{k^3x^4}{x^2 + k^6x^6} = \lim\limits_{x\to0} \dfrac{k^3x^2}{1 + k^6x^4} = 0$

while along the path $x = y^3$,

$\lim\limits_{y\to0} \dfrac{y^6}{y^6 + y^2} = \dfrac{1}{2}$

Since these path limits are not the same, $f(x, y)$ has no limiting value as

$(x, y) \to (0, 0)$. Thus, f is not continuous at $(0, 0)$.

37. $f(x, y) = \dfrac{x^2 + 2y^2}{x^2 + y^2}$

a. $\lim\limits_{(x,y)\to(3,1)} \dfrac{x^2 + 2y^2}{x^2 + y^2} = \dfrac{9+2}{9+1} = \dfrac{11}{10}$

b. Along the line $x = 0$,
$$\lim_{y\to 0} \frac{2y^2}{y^2} = 2$$
Along the line $y = 0$,
$$\lim_{x\to 0} \frac{x^2}{x^2} = 1$$
Since these path limits are not the same, the limit does not exist.

39. Along the line $y = 0$
$$\lim_{(x,y)\to(0,0)} \frac{3x^3 - 3y^3}{x^2 - y^2} = \lim_{x\to 0} \frac{3x^2}{x^2} = 0$$
Since f is to be continuous at $(0, 0)$, we must have $A = 0$.

41. In polar coordinates,
$$\lim_{(x,y)\to(0,0)} \frac{2x^2 - x^2y^2 + 2y^2}{x^2 + y^2}$$
$$= \lim_{r\to 0} \frac{2r^2 - r^4\cos^2\theta\sin^2\theta}{r^2}$$
$$= 2$$
Thus, $C = 2$.

43. $\lim\limits_{(x,y)\to(0,0)} \dfrac{xy^3}{x^2 + y^2} = 0 = f(0, 0)$

Thus, the function is continuous at $(0, 0)$.

45. Approach the origin along the line $x = t$, $y = t, z = t$:
$$\lim_{(x,y,z)\to(0,0,0)} \frac{xyz}{x^2 + y^2 + z^2} = \lim_{t\to 0} \frac{t^3}{3t^2} = 0$$

47. $\lim\limits_{(x,y)\to(0,0)} (1 + x^2 + y^2)^{1/(x^2+y^2)}$
$$= \lim_{r\to 0} (1 + r^2)^{1/r^2}$$
$$= L$$
Thus, $\ln L = \lim\limits_{r\to 0} \dfrac{\ln(1 + r^2)}{r^2}$
$$= 1 \qquad \text{l'Hôpital's rule}$$
Then $L = e$.

49. $\lim\limits_{(x,y)\to(0,0)} x\ln\sqrt{x^2 + y^2} = \lim\limits_{r\to 0} (r\cos\theta)\ln r$
$$= \cos\theta \lim_{r\to 0} \frac{\ln r}{r^{-1}}$$
$$= 0 \qquad \text{l'Hôpital's rule}$$

51. False; let $f(x, y) = \dfrac{x^2 y^2}{x^4 + y^4}$.

Then $f(x, y)$ is continuous for all $(x, y) \neq (0, 0)$ but $\lim\limits_{(x,y)\to(0,0)} f(x, y)$ does not exist.

53. $|2x^2 + 3y^2 - 0| \leq 3|x^2 + y^2| < 3\sqrt{x^2 + y^2}$ for $x^2 + y^2 < 1$, for a point (x, y) near $(0, 0)$. Thus, if $\sqrt{x^2 + y^2} < \delta$, we have
$$|(2x^2 + 3y^2) - 0| < 3\sqrt{x^2 + y^2} < 3\delta$$
and
$$|(2x^2 + 3y^2) - 0| < \epsilon$$
if $\delta = \dfrac{\epsilon}{3}$.

55. $\left|\dfrac{x^2 - y^2}{x + y} - 0\right| = |x - y|$ if $x \neq -y$
$$\leq |x| + |y|$$
$$\leq \sqrt{2}\sqrt{x^2 + y^2}$$
Thus, if $\sqrt{x^2 + y^2} < \delta$, we have
$$|x - y| \leq \sqrt{2}\sqrt{x^2 + y^2} \leq \sqrt{2}\,\delta$$
and

$$\left| \frac{x^2 - y^2}{x + y} - 0 \right| < \epsilon$$

if $\delta = \dfrac{\epsilon}{\sqrt{2}}$.

57. Let $\epsilon = \dfrac{f(a, b)}{2}$; since f is continuous at (a, b), there exists a $\delta > 0$ so that
$$|f(x, y) - f(a, b)| < \epsilon$$
whenever
$$0 < \sqrt{(x - a)^2 + (y - b)^2} < \delta$$
Thus, if (x, y) is any point inside the punctured disk

$$0 < \sqrt{(x - a)^2 + (y - b)^2} < \delta$$

we must have

$$-\epsilon < f(x, y) - f(a, b) < \epsilon$$

which implies

$$f(a, b) - \epsilon < f(x, y)$$

and since $\epsilon = \dfrac{f(a, b)}{2}$, we have

$$f(x, y) > f(a, b) - \frac{f(a, b)}{2} = \frac{1}{2}f(a, b) > 0$$

59. Let $\epsilon > 0$ be given; there exist $\delta_1 > 0$ and $\delta_2 > 0$ so that $|f(x, y) - L| < \frac{\epsilon}{2}$ if

$$0 < \sqrt{(x - x_0)^2 + (y - y_0)^2} < \delta_1$$

$|g(x, y) - M| < \frac{\epsilon}{2}$ if

$$0 < \sqrt{(x - x_0)^2 + (y - y_0)^2} < \delta_2$$

Let $\delta = \min(\delta_1, \delta_2)$; then if

$$0 < \sqrt{(x - x_0)^2 + (y - y_0)^2} < \delta$$

we have

$$|f(x, y) + g(x, y) - (L + M)|$$
$$\leq |f(x, y) - L| + |g(x, y) - M|$$
$$< \frac{\epsilon}{2} + \frac{\epsilon}{2}$$
$$= \epsilon$$

Thus,

$$\lim_{(x,y) \to (x_0, y_0)} [f(x, y) + g(x, y)] = L + m$$

11.3 Partial Derivatives, page 846

1. Given a function of two (or more) variables, hold all but one independent variable(s) constant and differentiate with respect to the remaining variable.

> **SURVIVAL HINT:** *To "take a derivative" of a function of several variables means to treat one of the variables as the variable, and the other variables as constants.*

3. $f(x, y) = x^3 + x^2y + xy^2 + y^3$
$f_x = 3x^2 + 2xy + y^2$
$f_y = x^2 + 2xy + 3y^2$
$f_{xx} = 6x + 2y$
$f_{yx} = 2x + 2y$

5. $f(x, y) = \dfrac{x}{y}$

$f_x = \dfrac{1}{y}$

$f_y = \dfrac{-x}{y^2}$

$f_{xx} = 0$

$$f_{yx} = \frac{-1}{y^2}$$

7. $f(x, y) = \ln(2x + 3y)$

$$f_x = \frac{2}{2x + 3y}$$

$$f_y = \frac{3}{2x + 3y}$$

$$f_{xx} = \frac{-4}{(2x + 3y)^2}$$

$$f_{xy} = \frac{-6}{(2x + 3y)^2}$$

9. a. $f(x, y) = (\sin x^2)(\cos y)$

$f_x = (2x \cos x^2)(\cos y)$

$f_y = -(\sin x^2)(\sin y)$

b. $f(x, y) = \sin(x^2 \cos y)$

$f_x = \cos(x^2 \cos y)(2x \cos y)$

$f_y = \cos(x^2 \cos y)(-x^2 \sin y)$

11. $f(x, y) = \sqrt{3x^2 + y^4}$

$$f_x = \frac{3x}{(3x^2 + y^4)^{1/2}}$$

$$f_y = \frac{2y^3}{(3x^2 + y^4)^{1/2}}$$

13. $f(x, y) = x^2 e^{x+y} \cos y$

$f_x = (\cos y)(x^2 e^{x+y} + 2x e^{x+y})$

$\quad = x e^{x+y}(x + 2)(\cos y)$

$f_y = -x^2 e^{x+y} \sin y + x^2 e^{x+y} \cos y$

$\quad = x^2 e^{x+y}(\cos y - \sin y)$

15. $f(x, y) = \sin^{-1}(xy)$

$$f_x = \frac{y}{\sqrt{1 - x^2 y^2}}$$

$$f_y = \frac{x}{\sqrt{1 - x^2 y^2}}$$

17. $f(x, y, z) = xy^2 + yz^3 + xyz$

$f_x = y^2 + yz$

$f_y = 2xy + z^3 + xz$

$f_z = 3yz^2 + xy$

19. $f(x, y, z) = \dfrac{x + y^2}{z}$

$$f_x = \frac{1}{z}$$

$$f_y = \frac{2y}{z}$$

$$f_z = -\frac{x + y^2}{z^2}$$

21. $f(x, y, z) = \ln(x + y^2 + z^3)$

$$f_x = \frac{1}{x + y^2 + z^3}$$

$$f_y = \frac{2y}{x + y^2 + z^3}$$

$$f_z = \frac{3z^2}{x + y^2 + z^3}$$

23. $\dfrac{2x}{9} + z z_x = 0 \qquad$ and $\qquad -\dfrac{y}{2} + z z_y = 0$

$$z_x = -\frac{2x}{9z} \qquad\qquad z_y = \frac{y}{2z}$$

25. $6xy + y^3 z_x - 2z z_x = 0$

$(y^3 - 2z)z_x = -6xy$

$$z_x = \frac{-6xy}{y^3 - 2z}$$

$$\quad = \frac{6xy}{2z - y^3}$$

$3x^2 + y^3 z_y + 3y^2 z - 2z z_y = 0$

$(y^3 - 2z)z_y = -3x^2 - 3y^2 z$

$$z_y = -\frac{3x^2 + 3y^2 z}{y^3 - 2z}$$

$$\quad = \frac{3x^2 + 3y^2 z}{2z - y^3}$$

27. $\dfrac{1}{2\sqrt{x}} + (\cos xz)(z + x z_x) = 0$

$$z_x = -\frac{\frac{1}{2\sqrt{x}} + z \cos xz}{x \cos xz}$$

$$= -\frac{1}{2x^{3/2}\cos xz} - \frac{z}{x}$$

Now, with respect to y:

$$2y + (\cos xz)(xz_y) = 0$$

$$z_y = -\frac{2y}{x\cos xz}$$

29. a. $f_x = y^3 + 3x^2 y;\ f_x(1, -1) = -4$
b. $f_y = 3xy^2 + x^3;\ f_y(1, -1) = 4$

31. a. $f_x = x^2\cos(x+y) + 2x\sin(x+y)$

$$f_x\left(\frac{\pi}{2}, \frac{\pi}{2}\right) = -\frac{\pi^2}{4}$$

b. $f_y = x^2\cos(x+y)$

$$f_y\left(\frac{\pi}{2}, \frac{\pi}{2}\right) = -\frac{\pi^2}{4}$$

33. $f(x, y) = \int_x^y \left(t^2 + 2t + 1\right) dt$

$$= \int_0^y \left(t^2 + 2t + 1\right) dt$$

$$- \int_0^x \left(t^2 + 2t + 1\right) dt$$

$$f_x = -(x^2 + 2x + 1)$$
$$f_y = y^2 + 2y + 1$$

35. $f_x = 6xy;\ f_{xx} = 6y;\ f_y = 3x^2 - 3y^2;$
$f_{yy} = -6y;$ thus

$$f_{xx} + f_{yy} = 0$$

so f is harmonic.

37. $f_x = e^x\sin y;\ f_{xx} = e^x\sin y$
$f_y = e^x\cos y;\ f_{yy} = -e^x\sin y$
Thus

$$f_{xx} + f_{yy} = 0$$

so f is harmonic.

39. $f_x = -y^2\sin xy^2$
$\quad f_{xy} = -y^2(2xy)\cos xy^2 - 2y\sin xy^2$
$\quad\quad = -2xy^3\cos xy^2 - 2y\sin xy^2$
$\quad f_y = -2xy\sin xy^2$

$f_{yx} = -2y\left[(\sin xy^2) + x(\cos xy^2)(y^2)\right]$
$\quad\quad = -2xy^3\cos xy^2 - 2y\sin xy^2$
They are the same.

41. $f_x = 2x - 2y\cos z$
$f_{xz} = 2y\sin z$
$f_y = 2y - 2x\cos z$
$f_{yz} = 2x\sin z$
$f_{yzz} = 2\cos z$
$f_{xzy} = 2\sin z$

$$f_{xzy} - f_{yzz} = 2(\sin z - x\cos z)$$

43. a. $\dfrac{\partial D_1}{\partial p_1} < 0$ and $\dfrac{\partial D_2}{\partial p_2} < 0$

since the demand for a commodity decreases as its price increases.

b. $\dfrac{\partial D_1}{\partial p_2} < 0$ and $\dfrac{\partial D_2}{\partial p_2} < 0$

since the demand for Q_1 decreases as the price for the complementary commodity Q_2 increases; likewise the demand for Q_2 increases as the price for Q_1 decreases.

c. paint and paint brushes

45. a. $C_m = \sigma(T - t)\left(-0.67m^{-1.67}\right)$

$$= -0.67\sigma(T - t)m^{-1.67}$$

b. $C_T = \sigma m^{-0.67}$
c. $C_t = -\sigma m^{-0.67}$

47. $Q = 120K^{2/3}L^{2/5}$

a. $\dfrac{\partial Q}{\partial K} = 80K^{-1/3}L^{2/5}$

$$\dfrac{\partial Q}{\partial L} = 48K^{2/3}L^{-3/5}$$

b. $\dfrac{\partial^2 Q}{\partial K^2} = -\dfrac{80}{3}K^{-4/3}L^{2/5} < 0$

$$\dfrac{\partial^2 Q}{\partial L^2} = -\dfrac{144}{5}K^{2/3}L^{-8/5} < 0$$

Interpretation: Marginal productivity of capital and labor are both decreasing functions. This is called the *law of eventually diminishing marginal productivity.*

49. a. Moving parallel to **j** means y is changing, and x is constant.

$$\frac{\partial T}{\partial y} = 4xy + 1 \qquad \frac{\partial T}{\partial y}(2,1) = 9$$

b. In this case x is changing and y is constant.

$$\frac{\partial T}{\partial x} = 3x^2 + 2y^2$$

$$\frac{\partial T}{\partial x}(2,1) = 3(2)^2 + 2(1) = 14$$

51. a. $\quad \dfrac{\partial u}{\partial x} = -e^{-x}\cos y \qquad \dfrac{\partial u}{\partial y} = -e^{-x}\sin y$

$\quad \dfrac{\partial v}{\partial x} = -e^{-x}\sin y \qquad \dfrac{\partial v}{\partial y} = e^{-x}\cos y$

Thus, $\dfrac{\partial u}{\partial x} \neq \dfrac{\partial v}{\partial y}$ and $\dfrac{\partial u}{\partial y} \neq -\dfrac{\partial v}{\partial x}$

so the Cauchy-Riemann equations are not satisfied.

b. $\quad \dfrac{\partial u}{\partial x} = 2x \qquad \dfrac{\partial u}{\partial y} = 2y$

$\quad \dfrac{\partial v}{\partial x} = 2y \qquad \dfrac{\partial v}{\partial y} = 2x$

Thus, $\dfrac{\partial u}{\partial x} = \dfrac{\partial v}{\partial y}$ but $\dfrac{\partial u}{\partial y} \neq -\dfrac{\partial v}{\partial x}$

so the Cauchy-Riemann equations are not satisfied.

c. $\quad \dfrac{\partial u}{\partial x} = \dfrac{2x}{x^2 + y^2} \qquad \dfrac{\partial u}{\partial y} = \dfrac{2y}{x^2 + y^2}$

$\quad \dfrac{\partial v}{\partial x} = \dfrac{-2y}{x^2 + y^2} \qquad \dfrac{\partial v}{\partial y} = \dfrac{2x}{x^2 + y^2}$

Thus, $\dfrac{\partial u}{\partial x} = \dfrac{\partial v}{\partial y}$ and $\dfrac{\partial u}{\partial y} = -\dfrac{\partial v}{\partial x}$

so the Cauchy-Riemann equations are satisfied.

53. If $P(L, K) = L^\alpha K^\beta$, then

$$\frac{\partial P}{\partial L} = \alpha L^{\alpha-1} K^\beta \text{ and } \frac{\partial P}{\partial K} = \beta L^\alpha K^{\beta-1},$$

so

$$L\frac{\partial P}{\partial L} + K\frac{\partial P}{\partial K} = L\left(\alpha L^{\alpha-1} K^\beta\right) + K\left(\beta L^\alpha K^{\beta-1}\right)$$
$$= \alpha L^\alpha K^\beta + \beta L^\alpha K^\beta$$
$$= (\alpha + \beta)P$$

55. $P(x, t) = P_0 + P_1 e^{-kx}\sin\left(At - kx\right)$

$$\frac{\partial P}{\partial x} = -P_1 k e^{-kx}[\sin(At - kx) + \cos(At - kx)]$$

$$\frac{\partial^2 P}{\partial x^2} = 2k^2 P_1 e^{-kx}\cos(At - kx)$$

$$\frac{\partial P}{\partial t} = A P_1 e^{-kx}\cos(At - kx)$$

so $\dfrac{\partial P}{\partial t} = c^2\dfrac{\partial^2 P}{\partial x^2}$ where $c^2 = \dfrac{A}{2k^2}$

57. Using the definition of the derivative, we find

$$f_x(0, 0) = \lim_{h \to 0}\frac{f(0 + h, 0) - f(0, 0)}{h}$$

$$= \lim_{h \to 0}\frac{h^2\sin\left(\frac{1}{h^2}\right) - 0}{h}$$

$$= 0$$

since $\left|\sin\frac{1}{h}\right| \leq 1$ for all $h \neq 0$.

$$f_y(0, 0) = \lim_{h \to 0}\frac{f(0, 0 + h) - f(0, 0)}{h}$$

$$= \lim_{h \to 0}\frac{h\sin\left(\frac{1}{h^2}\right)}{h}$$

$$= \lim_{h \to 0}\sin\left(\frac{1}{h^2}\right)$$

This limit does not exist.

59. a. $\dfrac{\partial A}{\partial a} = \dfrac{1}{2}b\sin\gamma;\ \dfrac{\partial A}{\partial b} = \dfrac{1}{2}a\sin\gamma;$

$\dfrac{\partial A}{\partial \gamma} = \dfrac{1}{2}ab\cos\gamma$

b. $a = \dfrac{2A}{b\sin\gamma};\ \dfrac{\partial a}{\partial \gamma} = -\dfrac{2A\cos\gamma}{b\sin^2\gamma}$

11.4 Tangent Planes, Approximations, and Differentiability, page 857

> **SURVIVAL HINT:** *Theorem 11,4 presents the essential idea of the chain rule for a differential function of more than one variable. Read this theorem, then read it again. Look at the "What this says" box and make sure that you understand what it says before moving on. If you do not, ask you instructor or a fellow student.*

1. A function $f(x, y)$ is differentiable at (x_0, y_0) if the increment of f can be expressed as

$\Delta f = f_x(x_0, y_0)\Delta x + f_y(x_0, y_0)\Delta y + \epsilon_1\Delta x + \epsilon_2\Delta y$

where $\epsilon_1 \to 0$ and $\epsilon_2 \to 0$ as both $\Delta x \to 0$ and $\Delta y \to 0$.

3. $z = \left(x^2 + y^2\right)^{1/2}$

$z_x = \dfrac{1}{2}\left(x^2 + y^2\right)^{-1/2}(2x) = \dfrac{x}{\sqrt{x^2 + y^2}}$

$z_y = \dfrac{y}{\sqrt{x^2 + y^2}}$

At P_0, $z_x = \dfrac{3}{\sqrt{10}}$, $z_y = \dfrac{1}{\sqrt{10}}$.

The equation of the tangent plane is:

$z - \sqrt{10} = \dfrac{3}{\sqrt{10}}(x - 3) + \dfrac{1}{\sqrt{10}}(y - 1)$

$3x + y - \sqrt{10}z = 0$

5. $f(x, y) = x^2 + y^2 + \sin xy$

$f_x = 2x + y\cos xy$

$f_y = 2y + x\cos xy$

At P_0,

$f_x(0, 2) = 2(0) + 2\cos 0 = 2$

$f_y(0, 2) = 2(2) + 0\cos 0 = 4$

The equation of the tangent plane is:

$z - 4 = 2(x - 0) + 4(y - 2)$

$2x + 4y - z - 4 = 0$

7. $z = \tan^{-1}\dfrac{y}{x}$

$z_x = \dfrac{-\dfrac{y}{x^2}}{1 + \left(\dfrac{y}{x}\right)^2} = \dfrac{-y}{x^2 + y^2}$

$z_y = \dfrac{\dfrac{1}{x}}{1 + \left(\dfrac{y}{x}\right)^2} = \dfrac{x}{x^2 + y^2}$

At P_0, $z_x = -\dfrac{1}{4}$, $z_y = \dfrac{1}{4}$

The equation of the tangent plane is:

$z - \dfrac{\pi}{4} = -\dfrac{1}{4}(x - 2) + \dfrac{1}{4}(y - 2)$

$x - y + 4z - \pi = 0$

9. $df = 10xy^3 dx + 15x^2y^2 dy$

11. $df = y(\cos xy)\,dx + x(\cos xy)\,dy$

13. $df = -\dfrac{y}{x^2}\,dx + \dfrac{1}{x}\,dy$

15. $df = ye^x dx + e^x dy$

17. $df = 9x^2 dx - 8y^3 dy + 5\,dz$

19. $df = 2z^2\cos(2x - 3y)\,dx$
$\qquad - 3z^2\cos(2x - 3y)\,dy$
$\qquad + 2z\sin(2x - 3y)\,dz$

21. $f(x, y) = xy^3 + 3xy^2$

$f_x = y^3 + 3y^2$

$f_y = 3xy^2 + 6xy$

Since $f, f_x,$ and f_y are all continuous, we know f is differentiable.

23. $f(x, y) = e^{2x+y^2}$

$f_x = 2e^{2x+y^2};\ f_y = 2ye^{2x+y^2}$
Since f, f_x, and f_y are all continuous, we
know f is differentiable.

25. $f(x, y) = \cos(2x - 3y)$
$f_x = -2\sin(2x - 3y)$
$f_y = 3\sin(2x - 3y)$
Since f, f_x, and f_y are all continuous, we
know f is differentiable.

27. $f_x = 1 + 2y + \lambda y$
$f_y = 2x + \lambda x$
$f_\lambda = xy - 10$

29. $f_x = 2x - 3\lambda$
$f_y = 2y - 2\lambda$
$f_\lambda = 6 - 3x - 2y$

31. $x_0 = 1,\ \Delta x = 0.01$
$y_0 = 2,\ \Delta y = 0.03$
$f(1, 2) = 3(1)^4 + 2(2)^4$
$\qquad = 35$
$f_x = 12x^3;\ f_x(1, 2) = 12$
$f_y = 8y^3;\ f_y(1, 2) = 64$
$f(1.01, 2.03) \approx f(1, 2) + f_x(1, 2)\Delta x + f_y(1, 2)\Delta y$
$\qquad = 35 + 12(0.01) + 64(0.03)$
$\qquad = 37.04$
By calculator,
$f(1.01, 2.03) \approx 37.08544565$

33. $x_0 = \frac{\pi}{2},\ \Delta x = 0.01$
$y_0 = \frac{\pi}{2},\ \Delta y = -0.01$
$f\left(\frac{\pi}{2}, \frac{\pi}{2}\right) = \sin\left(\frac{\pi}{2} + \frac{\pi}{2}\right)$
$\qquad\qquad = 0$
$f_x = \cos(x + y);\ f_x\left(\frac{\pi}{2}, \frac{\pi}{2}\right) = -1$
$f_y = \cos(x + y);\ f_y\left(\frac{\pi}{2}, \frac{\pi}{2}\right) = -1$
$f\left(\frac{\pi}{2} + 0.01, \frac{\pi}{2} - 0.01\right)$
$\approx f\left(\frac{\pi}{2}, \frac{\pi}{2}\right) + f_x\left(\frac{\pi}{2}, \frac{\pi}{2}\right)\Delta x + f_y\left(\frac{\pi}{2}, \frac{\pi}{2}\right)\Delta y$
$= 0 + (-1)(0.01) + (-1)(-0.01) = 0$
By calculator, $f\left(\frac{\pi}{2} + 0.01, \frac{\pi}{2} - 0.01\right) \approx 0$

35. $x_0 = 1,\ \Delta x = 0.01$
$y_0 = 1,\ \Delta y = -0.02$
$f(1, 1) = e^{(1)(1)}$
$\qquad = e$
$f_x = ye^{xy};\ f_x(1, 1) = e$
$f_y = xe^{xy};\ f_y(1, 1) = e$
$f(1.01, 0.98) \approx f(1, 1) + f_x(1, 1)\Delta x + f_y(1, 1)\Delta y$
$\qquad\qquad = e + e(0.01) + e(-0.02)$
$\qquad\qquad = 2.691$

By calculator, $f(1.01, 0.98) \approx 2.6906963$

37. $z_x = -2x$ is equal to 0 when $x = 0$
$z_y = -2y + 4$ is equal to 0 when $y = 2$
The tangent plane is horizontal when
$z_x = z_y = 0$: Since

$$z(0, 2) = 5 - 0^2 - 2^2 + 4(2) = 9$$

the equation of the horizontal plane is
$z = 9$.

39.   Since x and y are very small, we can
take $\Delta x = x$ and $\Delta y = y$. Then

$$f(x, y) \approx f(0, 0) + xf_x(0, 0) + yf_y(0, 0)$$

41. $f(x, y) = \dfrac{1}{(x + 1)^2 + (y + 1)^2};\ f(0, 0) = \dfrac{1}{2}$
$f_x(x, y) = \dfrac{-2(x + 1)}{\left[(x + 1)^2 + (y + 1)^2\right]^2};\ f_x(0, 0) = -\dfrac{1}{2}$
$f_y(x, y) = \dfrac{-2(y + 1)}{\left[(x + 1)^2 + (y + 1)^2\right]^2};\ f_y(0, 0) = -\dfrac{1}{2}$
Thus,

$$f(x, y) \approx f(0, 0) + xf_x(0, 0) + yf_y(0, 0)$$
$$\approx \frac{1}{2} - \frac{1}{2}x - \frac{1}{2}y$$
$$= \frac{1}{2}(1 - x - y)$$

43. Let x, y, and z be the length, width, and

height of the box, respectively. The total cost is

$$C = 2xy + 1.50(xy + 2xz + 2yz)$$
$$= 3.5xy + 3xz + 3yz$$

$$C_x = 3.5y + 3z; \; C_y = 3.5x + 3z;$$
$$C_z = 3x + 3y$$

$$\Delta C \approx (3.5y + 3z)\Delta x + (3.5x + 3z)\Delta y$$
$$+ (3x + 3y)\Delta z$$

Since $x = 2$, $y = 4$, $z = 3$ and $|\Delta x| \leq 0.02$, $|\Delta y| \leq 0.02$, and $|\Delta z| \leq 0.02$, we have

$$|\Delta C| \leq |3.5(4) + 3(3)|(0.02)$$
$$+ |3.5(2) + 3(3)|(0.02)$$
$$+ |3(2) + 3(4)|(0.02)$$
$$= 1.14$$

Thus, the maximum possible error will cost $1.14.

45. a. $\quad x = \dfrac{4{,}000 - p}{500}; \; y = \dfrac{3{,}000 - q}{450}$

$$R = x(p) \cdot p + y(q) \cdot q$$
$$= \left(\dfrac{4{,}000 - p}{500}\right)p + \left(\dfrac{3{,}000 - q}{450}\right)q$$

b. $\quad R_p = 8 - \dfrac{2p}{500}; \; R_q = \dfrac{20}{3} - \dfrac{2q}{450}$

$$\Delta R \approx R_p \Delta p + R_q \Delta q$$
$$= \left(8 - \dfrac{2p}{500}\right)\Delta p + \left(\dfrac{20}{3} - \dfrac{2q}{450}\right)\Delta q$$

When $p = 500$, $q = 750$, $\Delta p = 20$, and $\Delta q = 18$, we have

$$\Delta R \approx \left[8 - \dfrac{2(500)}{500}\right](20) + \left[\dfrac{20}{3} - \dfrac{2(750)}{450}\right](18)$$
$$= 180$$

That is, revenue is increased by approximately $180.

47. $\quad dR = \dfrac{c\,dx}{r^4} - \dfrac{4cx}{r^5}\,dr$

$$\dfrac{dR}{R} = \dfrac{cr^4 dx}{cxr^4} - \dfrac{4cr^4\,dx}{cxr^5}$$
$$= \dfrac{dx}{x} - \dfrac{4\,dr}{r}$$

This implies
$$\dfrac{\Delta R}{R} \approx \dfrac{\Delta x}{x} - 4\dfrac{\Delta r}{r}$$
$$\approx 0.03 - 4(-0.02)$$
$$= 0.11$$

Thus, R increases by approximately 11%.

49. $\quad F(x, y) = \dfrac{1.786xy}{1.798x + y}$

$$F_x = \dfrac{(1.798x + y)(1.786y) - (1.786xy)(1.798)}{(1.798x + y)^2}$$

$$F_y = \dfrac{(1.798x + y)(1.786x) - (1.786xy)(1)}{(1.798x + y)^2}$$

$$x = 5, y = 4, \Delta x = 0.1, \Delta y = 0.04$$
$$F_x(5, 4) \approx 0.1693; \; F_y(5, 4) \approx 0.4758$$

$$\Delta F \approx F_x dx + F_y dy$$
$$\approx (0.1693)(0.1) + (0.4758)(0.04)$$
$$\approx 0.0360 \text{ cal}$$

51. $f(x, y) = 10xy^{1/2}; \; x = 30, y = 36, dx = 1$
$$10xy^{1/2} = C$$
$$\left(\dfrac{1}{2}\right)xy^{-1/2}dy + y^{1/2}dx = 0$$

$$dy = \dfrac{-\sqrt{y}\,dx(2)\sqrt{y}}{x}$$

$$= -\frac{2y\,dx}{x}$$

$$= -\frac{2(36)(1)}{30}$$

$$= -2.4$$

The manufacturer should decrease the level of unskilled labor by about 2.4 hours. Alternatively, we can work with approximations

$$\Delta f \approx df = \left(10y^{1/2}\right)dx + \left(5xy^{-1/2}\right)dy$$

For f to stay constant when $x = 30$, $y = 36$, $dx = 1$, we want $\Delta f = 0$. Thus, dy must satisfy

$$10(36)^{1/2}(1) + 5(30)(36)^{-1/2}dy = 0$$

$$dy \approx \frac{-10(36)}{5(3)}$$

$$= -2.4$$

The manufacturer should decrease the level of unskilled labor about 2.4 hours.

53. $T = 2\pi\sqrt{\dfrac{L}{g}} = 2\pi L^{1/2}g^{-1/2}$

$$T_L = 2\pi\left(\frac{1}{2}L^{-1/2}\right)g^{-1/2}$$

$$= \frac{\pi}{\sqrt{Lg}}$$

$$T_g = 2\pi\left(-\frac{1}{2}g^{-3/2}\right)L^{1/2}$$

$$= -\frac{\pi}{g}\sqrt{\frac{L}{g}}$$

For $L = 4$, $g = 32$, $dL = 0.03$, $dg = 0.2$

$$\Delta T \approx dT = T_L dL + T_g dg$$

$$= \left[\frac{\pi}{\sqrt{(4)(32)}}\right](0.03) + \left[\frac{-\pi\sqrt{4}}{(32)^{3/2}}\right](0.2)$$

$$\approx 0.0014$$

The correct period is approximately 0.0014 seconds more than the computed period.

55. $S = \dfrac{x}{x-y}$; $S_x = \dfrac{-y}{(x-y)^2}$; $S_y = \dfrac{x}{(x-y)^2}$

$$\Delta S = \left[\frac{-y}{(x-y)^2}\right]\Delta x + \left[\frac{x}{(x-y)^2}\right]\Delta y$$

The maximum will occur when $\Delta x = 0$ and $\Delta y = 0.01$. Thus, for $x = 1.2$, $y = 0.5$, we have

$$\Delta S \leq \left[\frac{1.2}{(1.2 - 0.5)^2}\right]\Delta y$$

$$= \left(\frac{120}{49}\right)(0.01)$$

$$= \frac{12}{490}$$

The maximum possible error in the measurement of S is about $\frac{12}{490}$ lb.

57. Approach along the line $y = mx$;

$$\lim_{x\to 0}\frac{x(mx)}{x^2 + (mx)^2} = \lim_{x\to 0}\frac{m}{m^2 + 1} \neq 0$$

Thus, f is not continuous at $(0,0)$, so it cannot be differentiable at $(0,0)$ either.

59. $A = hb = ab\sin\frac{\pi}{6} = \frac{1}{2}ab$;

$$\frac{\Delta a}{a} = 0.04, \quad \frac{\Delta b}{b} = -0.03$$

$$\Delta A = \frac{1}{2}b\,\Delta a + \frac{1}{2}a\,\Delta b$$

$$\frac{\Delta A}{A} = \frac{b\,\Delta a}{ba} + \frac{a\,\Delta b}{ab}$$

$$= \frac{da}{a} + \frac{db}{b}$$

$$= 0.04 - 0.03$$

$$= 0.01$$

Thus, A increases by about 1%.

11.5 Chain Rules, page 866

1. Let $f(x, y)$ be a differentiable function of x and y, and let $x = x(t)$ and $y = y(t)$ be differentiable functions of t. Then $z = f(x, y)$ is a differentiable function of t, and

$$\frac{dz}{dt} = \frac{\partial z}{\partial x}\frac{dx}{dt} + \frac{\partial z}{\partial y}\frac{dy}{dt}$$

Recall the chain rule for a single variable:

$$\frac{dy}{dx} = \frac{dy}{du}\frac{du}{dx}$$

The corresponding rule for two variables is *essentially* the same, the formula involves *both* variables. For two parameters, suppose $z = f(x, y)$ is differentiable at (x, y) and that the partial derivatives of $x = x(u, v)$ and $y = y(u, v)$ exist at $(u.v)$. Then the composite function

$$z = f[x(u, v), y(u, v)]$$

is differentiable at (u, v) with

$$\frac{\partial z}{\partial u} = \frac{\partial z}{\partial x}\frac{\partial x}{\partial u} + \frac{\partial z}{\partial y}\frac{\partial y}{\partial u}$$

and

$$\frac{\partial z}{\partial v} = \frac{\partial z}{\partial x}\frac{\partial x}{\partial v} + \frac{\partial z}{\partial y}\frac{\partial y}{\partial v}$$

The chain rule allows us to compute $\dfrac{dz}{dt}$ without determining $z(t)$ explicitly.

3. $z = f(x, y)$, $x = x(u, v, w)$, $y = y(u, v, w)$,

$$\frac{\partial z}{\partial u} = \frac{\partial f}{\partial x}\frac{\partial x}{\partial u} + \frac{\partial f}{\partial y}\frac{\partial y}{\partial u}$$

$$\frac{\partial z}{\partial v} = \frac{\partial f}{\partial x}\frac{\partial x}{\partial v} + \frac{\partial f}{\partial y}\frac{\partial y}{\partial v}$$

$$\frac{\partial z}{\partial w} = \frac{\partial f}{\partial x}\frac{\partial x}{\partial w} + \frac{\partial f}{\partial y}\frac{\partial y}{\partial w}$$

5. a. $f(t) = \left(4 + e^{6t}\right)e^{2t}$

$$= 4e^{2t} + e^{8t}$$

$$f'(t) = 8e^{2t} + 8e^{8t}$$

$$= 8e^{2t}\left(1 + e^{6t}\right)$$

b. $f'(t) = \dfrac{\partial z}{\partial x}\dfrac{dx}{dt} + \dfrac{\partial z}{\partial y}\dfrac{dy}{dt}$

$$= \left(4 + y^2\right)2e^{2t} + (2xy)3e^{3t}$$

In terms of t:

$$= \left(4 + e^{6t}\right)2e^{2t} + \left(2e^{2t}e^{3t}\right)3e^{3t}$$

$$= 8e^{2t}\left(1 + e^{6t}\right)$$

7. a. $f(t) = (\cos 3t)\left(\tan^2 3t\right)$

$$= \frac{\sin^2 3t}{\cos 3t}$$

$$f'(t) = \frac{(\cos 3t)(6\sin 3t)(\cos 3t) - \sin^2 3t(-3\sin 3t)}{\cos^2 3t}$$

$$= \frac{(6\sin 3t)(1 - \sin^2 3t) + 3\sin^3 3t}{\cos^2 3t}$$

$$= \frac{6\sin 3t - 3\sin^3 3t}{\cos^2 3t}$$

b. $f'(t) = \dfrac{\partial}{\partial x}\left(xy^2\right)\dfrac{d}{dt}(\cos 3t) + \dfrac{\partial}{\partial y}\left(xy^2\right)\dfrac{d}{dt}(\tan 3t)$

$\qquad = y^2(-3\sin 3t) + 2xy\left(3\sec^2 t\right)$

$\qquad = -3\sin 3t \tan^2 3t + 6\cos 3t \tan 3t \sec^2 3t$

$\qquad = \dfrac{-3\sin^3 3t + 6\sin 3t}{\cos^2 3t}$

9. a. $F(u,v) = u^2\sin^2 v + u^2 - 4uv + 4v^2$

$\qquad \dfrac{\partial F}{\partial u} = 2u\sin^2 v + 2u - 4v$

$\qquad \dfrac{\partial F}{\partial v} = 2u^2\sin v \cos v - 4u + 8v$

b. $\dfrac{\partial F}{\partial u} = \dfrac{\partial F}{\partial x}\dfrac{\partial x}{\partial u} + \dfrac{\partial F}{\partial y}\dfrac{\partial y}{\partial u}$

$\qquad = 2x\sin v + 2y(1)$

$\qquad = 2u\sin^2 v + 2(u - 2v)$

$\qquad \dfrac{\partial F}{\partial v} = \dfrac{\partial F}{\partial x}\dfrac{\partial x}{\partial v} + \dfrac{\partial F}{\partial y}\dfrac{\partial y}{\partial v}$

$\qquad = 2xu\cos v + (2y)(-2)$

$\qquad = 2u^2\sin v \cos v - 4(u - 2v)$

$\qquad = 2u^2\sin v \cos v - 4u + 8v$

11. a. $F(u,v) = \ln\left[e^{uv^2 + uv}\right] = uv^2 + uv$

$\qquad \dfrac{\partial F}{\partial u} = v^2 + v$

$\qquad \dfrac{\partial F}{\partial v} = 2uv + u$

b. $\dfrac{\partial F}{\partial u} = \dfrac{\partial F}{\partial x}\dfrac{\partial x}{\partial u} + \dfrac{\partial F}{\partial y}\dfrac{\partial y}{\partial u}$

$\qquad = \dfrac{1}{x}\left[e^{uv^2}v^2\right] + \dfrac{1}{y}\left[e^{uv}v\right]$

$\qquad = \dfrac{e^{uv^2}}{e^{uv^2}}v^2 + \dfrac{e^{uv}}{e^{uv}}v$

$\qquad = v^2 + v$

$\dfrac{\partial F}{\partial v} = \dfrac{\partial F}{\partial x}\dfrac{\partial x}{\partial v} + \dfrac{\partial F}{\partial y}\dfrac{\partial y}{\partial v}$

$\qquad = \dfrac{1}{x}\left[e^{uv^2}(2uv)\right] + \dfrac{1}{y}\left[e^{uv}u\right]$

$\qquad = 2uv + u$

13. $\dfrac{\partial w}{\partial s} = \dfrac{\partial w}{\partial x}\dfrac{\partial x}{\partial s} + \dfrac{\partial w}{\partial y}\dfrac{\partial y}{\partial s} + \dfrac{\partial w}{\partial z}\dfrac{\partial z}{\partial s}$

$\quad\;\; \dfrac{\partial w}{\partial t} = \dfrac{\partial w}{\partial x}\dfrac{\partial x}{\partial t} + \dfrac{\partial w}{\partial y}\dfrac{\partial y}{\partial t} + \dfrac{\partial w}{\partial z}\dfrac{\partial z}{\partial t}$

15. $\dfrac{\partial w}{\partial s} = \dfrac{\partial w}{\partial x}\dfrac{\partial x}{\partial s} + \dfrac{\partial w}{\partial y}\dfrac{\partial y}{\partial s} + \dfrac{\partial w}{\partial z}\dfrac{\partial z}{\partial s}$

$\quad\;\; \dfrac{\partial w}{\partial t} = \dfrac{\partial w}{\partial x}\dfrac{\partial x}{\partial t} + \dfrac{\partial w}{\partial y}\dfrac{\partial y}{\partial t} + \dfrac{\partial w}{\partial z}\dfrac{\partial z}{\partial t}$

$\quad\;\; \dfrac{\partial w}{\partial u} = \dfrac{\partial w}{\partial x}\dfrac{\partial x}{\partial u} + \dfrac{\partial w}{\partial y}\dfrac{\partial y}{\partial u} + \dfrac{\partial w}{\partial z}\dfrac{\partial z}{\partial u}$

17. $\dfrac{dw}{dt} = \dfrac{\partial w}{\partial x}\dfrac{dx}{dt} + \dfrac{\partial w}{\partial y}\dfrac{dy}{dt} + \dfrac{\partial w}{\partial z}\dfrac{dz}{dt}$

$\quad\;\; = [yz\cos(xyz)](-3)$

$\qquad + [xz\cos(xyz)]\left(-e^{1-t}\right) + [xy\cos(xyz)](4)$

$\quad\;\; = \cos(xyz)[-3yz - e^{1-t}xz + 4xy]$

19. $\dfrac{dw}{dt} = \dfrac{\partial w}{\partial x}\dfrac{dx}{dt} + \dfrac{\partial w}{\partial y}\dfrac{dy}{dt} + \dfrac{\partial w}{\partial z}\dfrac{dz}{dt}$

$\quad\;\; = e^{x^3 + yz}\left[3x^2\left(\dfrac{-2}{t^2}\right) + \dfrac{2z}{2t - 3} + y(2t)\right]$

$\quad\;\; = \left(e^{x^3 + yz}\right)\left[\dfrac{-6x^2}{t^2} + 2ty + \dfrac{2z}{2t - 3}\right]$

21. $\dfrac{\partial w}{\partial r} = \dfrac{\partial w}{\partial x}\dfrac{\partial x}{\partial r} + \dfrac{\partial w}{\partial y}\dfrac{\partial y}{\partial r} + \dfrac{\partial w}{\partial z}\dfrac{\partial z}{\partial r}$

$\quad\;\; = \dfrac{2s}{2 - z} + \dfrac{t\cos rt}{2 - z} + 0$

$\quad\;\; = \dfrac{2s + t\cos rt}{2 - z}$

$\dfrac{\partial w}{\partial t} = \dfrac{\partial w}{\partial x}\dfrac{\partial x}{\partial t} + \dfrac{\partial w}{\partial y}\dfrac{\partial y}{\partial t} + \dfrac{\partial w}{\partial z}\dfrac{\partial z}{\partial t}$

$$= 0 + \frac{r\cos rt}{2-z} + \frac{(x+y)(2st)}{(2-z)^2}$$
$$= \frac{(2-z)r\cos rt + 2st(x+y)}{(2-z)^2}$$

23. $F(x,y) = (x^2-y)^{3/2} + x^2 y - 2$
$F_x = \frac{3}{2}(x^2-y)^{1/2}(2x) + 2xy$
$F_y = \frac{3}{2}(x^2-y)^{1/2}(-1) + x^2$
$\frac{dy}{dx} = \frac{-F_x}{F_y}$
$$= \frac{-\left[3x(x^2-y)^{1/2} + 2xy\right]}{-\frac{3}{2}(x^2-y)^{1/2} + x^2}$$

25. $F(x,y) = x\cos y + y\tan^{-1}x - x$
$F_x = \cos y + \frac{y}{1+x^2} - 1$
$F_y = -x\sin y + \tan^{-1}x$
$\frac{dy}{dx} = \frac{-F_x}{F_y} = \frac{-\left[\cos y + \frac{y}{1+x^2} - 1\right]}{-x\sin y + \tan^{-1}x}$
$$= \frac{(1-\cos y)(1+x^2) - y}{(1+x^2)(-x\sin y + \tan^{-1}x)}$$

27. $F(x,y) = \tan^{-1}\left(\frac{x}{y}\right) - \tan^{-1}\left(\frac{y}{x}\right)$
$F_x = \frac{1}{1+\left(\frac{x}{y}\right)^2}\left(\frac{1}{y}\right) - \frac{1}{1+\left(\frac{y}{x}\right)^2}\left(-\frac{y}{x^2}\right)$
$= \frac{y}{x^2+y^2} + \frac{y}{x^2+y^2}$
$= \frac{2y}{x^2+y^2}$
$F_y = \frac{1}{1+\left(\frac{x}{y}\right)^2}\left(\frac{-x}{y^2}\right) - \frac{1}{1+\left(\frac{y}{x}\right)^2}\left(\frac{1}{x}\right)$
$= \frac{-x}{x^2+y^2} - \frac{x}{x^2+y^2}$
$= \frac{-2x}{x^2+y^2}$

$$\frac{dy}{dx} = \frac{-F_x}{F_y} = \frac{-\left(\frac{2y}{x^2+y^2}\right)}{\left(\frac{-2x}{x^2+y^2}\right)} = \frac{y}{x}$$

29. a. $z = \frac{2}{xy}$; $z_x = -\frac{2}{x^2 y}$; $z_y = \frac{-2}{xy^2}$
$\frac{\partial^2 z}{\partial x \partial y} = \frac{\partial}{\partial x}\left(\frac{-2}{xy^2}\right)$
$= \frac{-2}{y^2}\left(\frac{-1}{x^2}\right)$
$= \frac{2}{x^2 y^2}$
$= \frac{z^2}{2}$

b. $z_{xx} = \frac{4}{x^3 y}$

c. $z_{yy} = \frac{4}{xy^3}$

31. a. $-x^{-2} - z^{-2}z_x = 0$
$z_x = -\frac{z^2}{x^2}$
$-y^{-2} - z^{-2}z_y = 0$
$z_y = \frac{-z^2}{y^2}$
$z_{yx} = \frac{\partial}{\partial x}\left(\frac{-z^2}{y^2}\right)$
$= \frac{-1}{y^2}(2zz_x)$
$= \frac{-2z}{y^2}\left(\frac{-z^2}{x^2}\right)$
$= \frac{2z^3}{x^2 y^2}$

b. $\dfrac{\partial^2 z}{\partial x^2} = -\dfrac{x^2\left(2z\frac{\partial z}{\partial x}\right) - z^2(2x)}{x^4}$

$\qquad = \dfrac{2z^2(x+z)}{x^4}$

c. $\dfrac{\partial^2 z}{\partial y^2} = -\dfrac{y^2\left(2z\frac{\partial z}{\partial y}\right) - z^2(2y)}{y^4}$

$\qquad = \dfrac{2z^2(y+z)}{y^4}$

33. a. $z^2 = \tan y - \sin x$

$\qquad 2zz_x = -\cos x \qquad 2zz_y = \sec^2 y$

$\qquad z_x = -\dfrac{\cos x}{2z} \qquad z_y = \dfrac{\sec^2 y}{2z}$

$\qquad z_{yx} = \dfrac{\partial}{\partial x}\left(\dfrac{\sec^2 y}{2z}\right)$

$\qquad\quad = \dfrac{\sec^2 y}{2}\left(\dfrac{-1}{z^2}z_x\right)$

$\qquad\quad = \dfrac{-\sec^2 y}{2z^2}\left(\dfrac{-\cos x}{2z}\right)$

$\qquad\quad = \dfrac{\sec^2 y \cos x}{4z^3}$

b. $z_{xx} = -\dfrac{1}{2}\dfrac{-z\sin x - z_x\cos x}{z^2}$

$\qquad\quad = \dfrac{2z^2\sin x - \cos^2 x}{4z^3}$

c. $z_{yy} = \dfrac{1}{2z^2}\left(2z\sec^2 y \tan y - \sec^2 y\, z_y\right)$

$\qquad\quad = \dfrac{4z^2\sec^2 y\tan y - \sec^4 y}{4z^3}$

35. $\dfrac{\partial z}{\partial u} = \dfrac{\partial z}{\partial x}\dfrac{\partial x}{\partial u} + \dfrac{\partial z}{\partial y}\dfrac{\partial y}{\partial u}$

$\qquad = \dfrac{\partial z}{\partial x}\cdot a + \dfrac{\partial z}{\partial y}\cdot 0$

$\qquad = a\dfrac{\partial z}{\partial x}$

$\dfrac{\partial z}{\partial v} = \dfrac{\partial z}{\partial x}\dfrac{\partial x}{\partial v} + \dfrac{\partial z}{\partial y}\dfrac{\partial y}{\partial v}$

$\qquad = \dfrac{\partial z}{\partial x}\cdot 0 + \dfrac{\partial z}{\partial y}\cdot b$

$\qquad = b\dfrac{\partial z}{\partial y}$

$\dfrac{\partial^2 z}{\partial u^2} = a\dfrac{\partial}{\partial u}z_x$

$\qquad = a\left[(z_x)_x x_u + (z_x)_y y_u\right]$

$\qquad = a^2 z_{xx}$

$\dfrac{\partial^2 z}{\partial v^2} = b\dfrac{\partial}{\partial v}z_y$

$\qquad = b\left[(z_y)_x x_v + (z_y)_y y_v\right] = b^2 z_{yy}$

37. $V = \ell w h$

$\dfrac{dV}{dt} = (\ell w)\dfrac{dh}{dt} + (\ell h)\dfrac{dw}{dt} + (wh)\dfrac{d\ell}{dt}$

When $t = 5$, $h = 20 - 3(5) = 5$;

$\ell = 10 + 2(5) = 20$; $w = 8 + 2(5) = 18$

$V'(5) = 20(18)(-3) + (20)(5)(2) + (18)(5)(2)$

$\qquad = -700 < 0$ (V is decreasing)

For the surface area, $S = 2(\ell h + \ell w + hw)$;

$\dfrac{dS}{dt} = 2\left[(\ell + w)\dfrac{dh}{dt} + (\ell + h)\dfrac{dw}{dt} + (h + w)\dfrac{d\ell}{dt}\right]$

$S'(5) = 2[(20 + 18)(-3) + (20 + 5)(2) + (5 + 18)(2)]$

$\qquad = -36 < 0$ (S is also decreasing)

39. a. $\dfrac{\partial C}{\partial a} = \dfrac{(b-a)(-te^{-at}) - (e^{-at} - e^{-bt})(-1)}{(b-a)^2}$

$\qquad = \dfrac{[1 - t(b-a)]e^{-at} - e^{-bt}}{(b-a)^2}$

$\dfrac{\partial C}{\partial b} = \dfrac{(b-a)(te^{-bt}) - (e^{-at} - e^{-bt})(1)}{(b-a)^2}$

$\qquad = \dfrac{[(b-a)t + 1]e^{-bt} - e^{-at}}{(b-a)^2}$

$$\frac{\partial C}{\partial t} = \frac{1}{b-a}\left[-ae^{-at} + be^{-bt}\right]$$

b. Since $a = \dfrac{\ln b}{t}$, b constant, we have

$$e^{-at} = \frac{1}{b}, \; \frac{db}{dt} = 0, \; \frac{da}{dt} = \frac{-\ln b}{t^2}, \text{ and}$$

$$(b-a)t = \left(b - \frac{\ln b}{t}\right)t = bt - \ln b$$

Thus,

$$\frac{dC}{dt} = \frac{\partial C}{\partial a}\frac{da}{dt} + \frac{\partial C}{\partial b}\frac{db}{dt} + \frac{\partial C}{\partial t}\frac{dt}{dt}$$

$$= \left[\frac{(1 - bt + \ln b)(\frac{1}{b}) - e^{-bt}}{(b - \frac{\ln b}{t})^2}\right]\left(\frac{-\ln b}{t^2}\right)$$

$$+ \left(\frac{\partial C}{\partial b}\right)(0) + \frac{1}{b-a}\left[-ae^{-at} + be^{-bt}\right](1)$$

$$= \left[\frac{(1 - bt + \ln b)(\frac{1}{b}) - e^{-bt}}{(bt - \ln b)^2}\right](-\ln b)$$

$$+ \frac{t}{bt - \ln b}\left[\frac{-\ln b}{bt} + be^{-bt}\right]$$

41. $Q = 240 - 21\sqrt{x} + 4(0.2y + 12)^{3/2}$

$$Q_x = -21\left(\frac{1}{2}\right)\frac{1}{\sqrt{x}} = \frac{-10.5}{\sqrt{x}}$$

$$Q_y = 4\left(\frac{3}{2}\right)(0.2y + 12)^{1/2}(0.2)$$

$$= 1.2(0.2y + 12)^{1/2}$$

$$\frac{dx}{dt} = 6; \; \frac{dy}{dt} = \frac{10}{\sqrt{t}}$$

$$\frac{dQ}{dt} = Q_x\left(\frac{dx}{dt}\right) + Q_y\left(\frac{dy}{dt}\right)$$

$$= \left(\frac{-10.5}{\sqrt{x}}\right)(6) + 1.2(0.2y + 12)^{1/2}\left(\frac{10}{\sqrt{t}}\right)$$

When $t = 4$,

$x = 120 + 6(4)$ and $y = 380 + 10\sqrt{4(4)}$

$= 144$ $= 420$

$$\left.\frac{dQ}{dt}\right|_{t=4} = \frac{-10.5(6)}{\sqrt{144}} + \frac{1.2(10)}{\sqrt{4}}[0.2(420) + 12]^{1/2}$$

$$\approx 53.54$$

Thus, the monthly demand for bicycles will be increasing at the rate of about 54 bicycles per month (4 months from now).

43. $\quad \dfrac{1}{R} = \dfrac{1}{R_1} + \dfrac{1}{R_2} + \dfrac{1}{R_3}$

$$-\frac{1}{R^2}\frac{\partial R}{\partial R_1} = -\frac{1}{R_1^2}$$

$$\frac{\partial R}{\partial R_1} = \left(\frac{R}{R_1}\right)^2$$

Similarly,

$$\frac{\partial R}{\partial R_2} = \left(\frac{R}{R_2}\right)^2; \; \frac{\partial R}{\partial R_3} = \left(\frac{R}{R_3}\right)^2$$

When $R_1 = 100$, $R_2 = 200$, $R_3 = 300$,

$$\frac{1}{R} = \frac{1}{100} + \frac{1}{200} + \frac{1}{300}$$

$$R \approx 54.545$$

With $\dfrac{dR_1}{dt} = -1.5, \; \dfrac{dR_2}{dt} = 2; \; \dfrac{dR_3}{dt} = -1.5$

$$\frac{dR}{dt} = \left(\frac{R}{R_1}\right)^2\frac{dR_1}{dt} + \left(\frac{R}{R_2}\right)^2\frac{dR_2}{dt} + \left(\frac{R}{r_3}\right)^2\frac{dR_3}{dt}$$

$$\approx \left(\frac{54.545}{100}\right)^2(-1.5) + \left(\frac{54.545}{200}\right)^2(2)$$

$$+ \left(\frac{54.545}{300}\right)^2(-1.5)$$

$$\approx -0.3471$$

The joint resistance is decreasing at the approximate rate of 0.3471 ohms/second.

45. Let $x = u - v$, $y = v - u$, so $z = f(w)$.

$$\frac{\partial z}{\partial u} = \frac{dz}{dx}\frac{\partial x}{\partial u} + \frac{\partial z}{\partial y}\frac{\partial y}{\partial u} = z_x(1) + z_y(-1)$$

$$\frac{\partial z}{\partial v} = \frac{dz}{dx}\frac{\partial x}{\partial v} + \frac{\partial z}{\partial y}\frac{\partial y}{\partial v} = z_x(-1) + z_y(1)$$

Thus,

$$z_u + z_v = (z_x - z_y) + (z_y - z_x)$$
$$= 0$$

47. Let $u = \dfrac{r-s}{s} = \dfrac{r}{s} - 1$, so $w = f(u)$.

$$\frac{\partial w}{\partial r} = \frac{df}{du}\frac{\partial u}{\partial r} = [f'(u)](s^{-1})$$

$$\frac{\partial w}{\partial s} = \frac{df}{du}\frac{\partial u}{\partial s} = f'(u)(-rs^{-2})$$

Thus,

$$r\frac{\partial w}{\partial r} + s\frac{\partial w}{\partial s} = r\left[\frac{1}{s}f'(u)\right] + s\left[\frac{-r}{s^2}f'(u)\right]$$
$$= 0$$

49. $\dfrac{\partial w}{\partial x} = \dfrac{df}{dt}\dfrac{\partial t}{\partial x}$

$$= [f'(t)]\left(\frac{1}{2}\right)(x^2+y^2+z^2)^{-1/2}(2x)$$

$$\frac{\partial w}{\partial y} = \frac{df}{dt}\frac{\partial t}{\partial y}$$

$$= [f'(t)]\left(\frac{1}{2}\right)(x^2+y^2+z^2)^{-1/2}(2y)$$

$$\frac{\partial w}{\partial z} = \frac{df}{dt}\frac{\partial t}{\partial z}$$

$$= [f'(t)]\left(\frac{1}{2}\right)(x^2+y^2+z^2)^{-1/2}(2z)$$

$$\left(\frac{\partial w}{\partial x}\right)^2 + \left(\frac{\partial w}{\partial y}\right)^2 + \left(\frac{\partial w}{\partial z}\right)^2$$

$$= [f'(t)]^2(x^2+y^2+z^2)^{-1}(x^2+y^2+z^2)$$

$$= [f'(t)]^2$$
$$= \left(\frac{dw}{dt}\right)^2$$

51. $\dfrac{dz}{d\theta} = \dfrac{\partial f}{\partial x}\dfrac{dx}{d\theta} + \dfrac{\partial f}{\partial y}\dfrac{dy}{d\theta}$

$$= (-\sin\theta)f_x + (\cos\theta)f_y$$
$$= -yf_x + xf_y$$

$$\frac{d^2z}{d\theta^2} = \frac{d}{d\theta}\left(\frac{dz}{d\theta}\right)$$

$$= \frac{d}{d\theta}[-yf_x + xf_y]$$

$$= \frac{\partial}{\partial x}[-yf_x + xf_y]\frac{dx}{d\theta}$$

$$+ \frac{\partial}{\partial y}[-yf_x + xf_y]\frac{dy}{d\theta}$$

$$= (-yf_{xx} + xf_{yx} + f_y)(-\sin\theta)$$
$$+ (-yf_{xy} - f_x + xf_{yy})(\cos\theta)$$

$$= y^2 f_{xx} - yxf_{yz} - yf_y - xyf_{xy} - xf_x + x^2 f_{yy}$$

$$= y^2 f_{xx} + x^2 f_{yy} - 2xyf_{xy} - xf_x - yf_y$$

53. $\dfrac{\partial z}{\partial r} = \dfrac{\partial z}{\partial x}\dfrac{\partial x}{\partial r} + \dfrac{\partial z}{\partial y}\dfrac{\partial y}{\partial r}$

$$= f_x(e^r\cos\theta) + f_y(e^r\sin\theta)$$
$$= xf_x + yf_y$$

$$\frac{\partial^2 z}{\partial r^2} = \frac{\partial}{\partial r}\left(\frac{\partial z}{\partial r}\right)$$

$$= \frac{\partial}{\partial r}(xf_x + yf_y)$$

$$= \frac{\partial}{\partial x}(xf_x + yf_y)\frac{\partial x}{\partial r} + \frac{\partial}{\partial y}(xf_x + yf_y)\frac{\partial y}{\partial r}$$

$$= (xf_{xx} + yf_{yx} + f_x)(e^r\cos\theta)$$
$$+ (xf_{xy} + f_y + yf_{yy})(e^r\sin\theta)$$

$$= x^2 f_{xx} + y^2 f_{yy} + 2xyf_{xy} + xf_x + yf_y$$

$$\frac{\partial z}{\partial\theta} = \frac{\partial z}{\partial x}\frac{\partial x}{\partial\theta} + \frac{\partial z}{\partial y}\frac{\partial y}{\partial\theta}$$

$$= f_x(-e^r\sin\theta) + f_y(e^r\cos\theta)$$

$$= -yf_x + xf_y$$

$$\frac{\partial^2 z}{\partial \theta^2} = \frac{\partial}{\partial \theta}\left(\frac{\partial z}{\partial \theta}\right)$$

$$= \frac{\partial}{\partial \theta}(-yf_x + xf_y)$$

$$= (-yf_{xx} + xf_{yx} + f_y)(-e^r \sin \theta)$$
$$+ (-yf_{xy} - f_x + xf_{yy})(e^r \cos \theta)$$

$$= y^2 f_{xx} + x^2 f_{yy} - 2xy f_{xy} - xf_x - yf_y$$

We have

$$\frac{\partial^2 z}{\partial r^2} + \frac{\partial^2 z}{\partial \theta^2} = (x^2 + y^2)f_{xx} + (x^2 + y^2)f_{yy}$$
$$+ (2xy - 2xy)f_{xy}$$
$$+ (x - x)f_x + (y - y)f_y$$

$$= (x^2 + y^2)f_{xx} + (x^2 + y^2)f_{yy}$$

$$= e^{2r}(f_{xx} + f_{yy})$$

$$f_{xx} + f_{yy} = e^{-2r}\left(\frac{\partial^2 z}{\partial r^2} + \frac{\partial^2 z}{\partial \theta^2}\right)$$

$$\frac{\partial^2 z}{\partial x^2} + \frac{\partial^2 z}{\partial y^2} = e^{-2r}\left(\frac{\partial^2 z}{\partial r^2} + \frac{\partial^2 z}{\partial \theta^2}\right)$$

55. By the chain rule

$$\frac{\partial u}{\partial r} = \frac{\partial u}{\partial x}\cos \theta + \frac{\partial u}{\partial y}\sin \theta$$

$$\frac{\partial v}{\partial \theta} = -\frac{\partial v}{\partial x}(r \sin \theta) + \frac{\partial v}{\partial y}(r \cos \theta)$$

Substituting,

$$\frac{\partial u}{\partial x} = \frac{\partial v}{\partial y} \quad \text{and} \quad \frac{\partial u}{\partial y} = -\frac{\partial v}{\partial x}$$

into the first equation, we obtain

$$\frac{\partial u}{\partial r} = \frac{\partial v}{\partial y}\cos \theta - \frac{\partial v}{\partial x}\sin \theta$$

$$= \frac{1}{r}\left[\frac{\partial v}{\partial y}(r \cos \theta) - \frac{\partial v}{\partial x}(r \sin \theta)\right]$$

$$\qquad\qquad\qquad \textit{Multiply by } \frac{r}{r} = 1$$

$$= \frac{1}{r}\frac{\partial v}{\partial \theta} \qquad \textit{Substitute from the second equation.}$$

The equation $\dfrac{\partial v}{\partial r} = -\dfrac{1}{r}\dfrac{\partial u}{\partial \theta}$ is obtained

by computing $\dfrac{\partial v}{\partial r}$ and $\dfrac{\partial u}{\partial \theta}$ and

substituting $\dfrac{\partial u}{\partial x} = \dfrac{\partial v}{\partial y}$ and $\dfrac{\partial u}{\partial y} = -\dfrac{\partial v}{\partial x}$.

57. For the equation $F(x, y, z) = 4$, where
$F(x, y, z) = x^2 + 2xyz + y^3 + e^z$, we have

$$F_x = 2x + 2yz$$
$$F_y = 2xz + 3y^2$$
$$F_z = 2xy + e^z$$

so

$$\frac{\partial z}{\partial x} = \frac{-F_x}{F_z} = \frac{-(2x + 2yz)}{2xy + e^z}$$

$$\frac{\partial z}{\partial y} = \frac{-F_y}{F_z} = \frac{-(2xz + 3y^2)}{2xy + e^z}$$

59. a. If $f(x, y) = x^2 y + 2y^3$, then

$$f(tx, ty) = (tx)^2(ty) + 2(ty)^3$$
$$= t^3(x^2 y + 2y^3)$$
$$= t^3 f(x, y)$$

Thus, the degree is $n = 3$.

b. Let $u = tx$ and $v = ty$, so
$f(u, v) = t^n f(x, y)$. Then,

$$\frac{\partial f}{\partial u}\frac{du}{dt} + \frac{\partial f}{\partial v}\frac{dv}{dt} = nt^{n-1}f(x, y)$$

$$x\frac{\partial f}{\partial u} + y\frac{\partial f}{\partial v} = nt^{n-1}f(x, y)$$

$$x\frac{\partial f}{\partial x} + y\frac{\partial f}{\partial y} = nf(x, y) \qquad \textit{Let } t = 1.$$

11.6 Directional Derivatives and the Gradient, page 881

1. $\nabla f = f_x \mathbf{i} + f_y \mathbf{j} = (2x - 2y)\mathbf{i} - 2x\mathbf{j}$

3. $\nabla F = (-\frac{y}{x^2} + \frac{1}{y})\mathbf{i} + + \left(\frac{1}{x} - \frac{x}{y^2}\right)\mathbf{j}$

5. $\nabla f = e^{3-y}(\mathbf{i} - x\mathbf{j})$

7. $\nabla f = \cos(x + 2y)(\mathbf{i} + 2\mathbf{j})$

9. $\nabla f = e^{y+3z}(\mathbf{i} + x\mathbf{j} + 3x\mathbf{k})$

11. $\mathbf{u} = \frac{1}{\sqrt{2}}\mathbf{i} + \frac{1}{\sqrt{s}}\mathbf{j};$
$$\nabla f = (2x + y)\mathbf{i} + x\mathbf{j}$$
$$\nabla f(1, -2) = \mathbf{j}$$
$$D_{\mathbf{u}}f = \frac{1}{\sqrt{2}}$$

13. $\mathbf{u} = \frac{\sqrt{2}}{2}\mathbf{i} + \frac{\sqrt{2}}{2}\mathbf{j};$
$$\nabla f = \frac{2x}{x^2 + 3y}\mathbf{i} + \frac{3}{x^2 + 3y}\mathbf{j}$$
$$\nabla f(1, 1) = \frac{1}{2}\mathbf{i} + \frac{3}{4}\mathbf{j}$$
$$D_{\mathbf{u}}f = \frac{\sqrt{2}}{4} + \frac{3\sqrt{2}}{8} = \frac{5\sqrt{2}}{8}$$

15. $\mathbf{u} = -\frac{1}{\sqrt{10}}\mathbf{i} - \frac{3}{\sqrt{10}}\mathbf{j};$
$$\nabla f = \sec\left(xy - y^3\right)\tan\left(xy - y^3\right)[y\mathbf{i} + (x - 3y^2)\mathbf{j}]$$
$$\nabla f(2, 0) = \mathbf{0}$$
$$D_{\mathbf{u}}f = 0$$

17. $\mathbf{N} = \nabla f = 2x\mathbf{i} + 2y\mathbf{j} + 2z\mathbf{k}$
$$= 2(x\mathbf{i} + y\mathbf{j} + z\mathbf{k})$$
$$\mathbf{N}(1, -1, 1) = 2(\mathbf{i} - \mathbf{j} + \mathbf{k})$$
$$\mathbf{N_u} = \pm\frac{\mathbf{N}}{2\sqrt{3}} = \pm\frac{\sqrt{3}}{3}(\mathbf{i} - \mathbf{j} + \mathbf{k})$$
The tangent plane is:
$$(x - 1) - (y + 1) + (z - 1) = 0$$
$$x - y + z - 3 = 0$$

19. $\mathbf{N} = \nabla f$
$$= \cos(x + y)\mathbf{i} + \cos(x + y)\mathbf{j} + (\sin z)\mathbf{k}$$

$$\mathbf{N}\left(\frac{\pi}{2}, \frac{\pi}{2}, \frac{\pi}{2}\right) = -\mathbf{i} - \mathbf{j} + \mathbf{k}$$
$$\mathbf{N_u} = \pm\frac{\mathbf{N}}{\sqrt{3}} = \pm\frac{\sqrt{3}}{3}(-\mathbf{i} - \mathbf{j} + \mathbf{k})$$
The tangent plane is:
$$-\left(x - \frac{\pi}{2}\right) - \left(y - \frac{\pi}{2}\right) + \left(z - \frac{\pi}{2}\right) = 0$$
$$x + y - z - \frac{\pi}{2} = 0$$

21. Write the function as $\ln x - \ln(y - z) = 0$.
$$\mathbf{N} = \nabla f = \frac{1}{x}\mathbf{i} - \frac{1}{y - z}\mathbf{j} + \frac{1}{y - z}\mathbf{k}$$
$$\mathbf{N}(2, 5, 3) = \frac{1}{2}\mathbf{i} - \frac{1}{2}\mathbf{j} + \frac{1}{2}\mathbf{k} = \frac{1}{2}(\mathbf{i} - \mathbf{j} + \mathbf{k})$$
$$\mathbf{N_u} = \pm\frac{\sqrt{3}}{3}(\mathbf{i} - \mathbf{j} + \mathbf{k})$$

The tangent plane is:
$$(x - 2) - (y - 5) + (z - 3) = 0$$
$$x - y + z = 0$$

23. $\mathbf{N} = \nabla f = e^{x+2y}(z\mathbf{i} + 2z\mathbf{j} + \mathbf{k})$
$$\mathbf{N}(2, -1, 3) = 3\mathbf{i} + 6\mathbf{j} + \mathbf{k}$$
$$\mathbf{N_u} = \pm\frac{\mathbf{N}}{\sqrt{46}}$$
$$= \pm\frac{1}{\sqrt{46}}(3\mathbf{i} + 6\mathbf{j} + \mathbf{k})$$
The tangent plane is:
$$3x + 6y + z - 3 = 0$$

25. $\nabla f = 3\mathbf{i} + 2\mathbf{j}$
$\nabla f(1, -1) = 3\mathbf{i} + 2\mathbf{j}; \|\nabla f\| = \sqrt{13}$

27. $\nabla f = 3x^2\mathbf{i} + 3y^2\mathbf{j}$
$\nabla f(3, -3) = 27(\mathbf{i} + \mathbf{j}); \|\nabla f\| = 27\sqrt{2}$

29. $\nabla f = \dfrac{1}{(x^2 + y^2)}(x\mathbf{i} + y\mathbf{j}) = \dfrac{1}{x^2 + y^2}(x\mathbf{i} + y\mathbf{j})$

$\nabla f(1,2) = \dfrac{1}{5}(\mathbf{i} + 2\mathbf{j}); \|\nabla f\| = \dfrac{1}{\sqrt{5}}$

31. $\nabla f = [2(x+y) + 2(x+z)]\mathbf{i}$
$+ [2(x+y) + 2(y+z)]\mathbf{j}$
$+ [2(x+z) + 2(y+z)]\mathbf{k}$

$\nabla f(2,-1,2) = 2(5\mathbf{i} + 2\mathbf{j} + 5\mathbf{k});$

$\|\nabla f\| = \sqrt{216} = 6\sqrt{6}$

33. Let $f(x,y) = ax + by - c$, so $f(x,y) = 0$.

$\nabla f = a\mathbf{i} + b\mathbf{j}; \mathbf{u} = \pm\dfrac{a\mathbf{i} + b\mathbf{j}}{\sqrt{a^2 + b^2}}$

35. Let $f(x,y) = \dfrac{x^2}{a^2} + \dfrac{y^2}{b^2} - 1$, so $f(x,y) = 0$.

$\nabla f = \dfrac{2x}{a^2}\mathbf{i} + \dfrac{2y}{b^2}\mathbf{j}$

$= 2a^{-2}b^{-2}(b^2 x\mathbf{i} + a^2 y\mathbf{j})$

$\mathbf{u} = \pm\dfrac{b^2 x_0\mathbf{i} + a^2 y_0\mathbf{j}}{\sqrt{b^4 x_0^2 + a^4 y_0^2}}$

37. $\mathbf{u} = \cos\frac{\pi}{6}\mathbf{i} + \sin\frac{\pi}{6}\mathbf{j} = \frac{1}{2}(\sqrt{3}\,\mathbf{i} + \mathbf{j})$

$\nabla f = \nabla(x^2 + y^2)$

$= 2(x\mathbf{i} + y\mathbf{j})$

$\nabla f(1,1) = 2(\mathbf{i} + \mathbf{j});$

$D_\mathbf{u}f = \sqrt{3} + 1$

39. $\nabla f = e^{x^2 y^2}(2xy^2\mathbf{i} + 2x^2 y\mathbf{j})$

$\nabla f(1,-1) = 2e(\mathbf{i} - \mathbf{j});$

$\mathbf{V} = \langle 2-1, 3-(-1)\rangle = \langle 1, 4\rangle;$

$\mathbf{u} = \left\langle \dfrac{1}{\sqrt{17}}, \dfrac{4}{\sqrt{17}}\right\rangle$

$D_\mathbf{u}f = -2e\left(\dfrac{3}{\sqrt{17}}\right) = -\dfrac{6e}{\sqrt{17}}$

41. $\nabla f = \nabla xyz = yz\mathbf{i} + xz\mathbf{j} + xy\mathbf{k}$

$\nabla f(1,-1,2) = -2\mathbf{i} + 2\mathbf{j} - \mathbf{k}$

$\mathbf{v} \times \mathbf{w} = -\mathbf{i} + 7\mathbf{j} + 5\mathbf{k}$

$\mathbf{u} = \dfrac{1}{\sqrt{75}}(-\mathbf{i} + 7\mathbf{j} + 5\mathbf{k})$

$D_\mathbf{u}f = \dfrac{2 + 14 - 5}{5\sqrt{3}} = \dfrac{11\sqrt{3}}{15}$

43. $\nabla T = (y+z)\mathbf{i} + (x+z)\mathbf{j} + (x+y)\mathbf{k}$

$\nabla T_0 = \nabla T(1,1,1) = 2(\mathbf{i} + \mathbf{j} + \mathbf{k})$

Maximum rate of temperature change is

$\|\nabla \mathbf{T}_0\| = 2\sqrt{3}$ in the direction of

$\mathbf{u} = \dfrac{\sqrt{3}}{3}(\mathbf{i} + \mathbf{j} + \mathbf{k})$

45. $\nabla z = -6x\mathbf{i} - 5y\mathbf{j}$

$\nabla z_0\left(\frac{1}{4}, -\frac{1}{2}\right) = -\frac{3}{2}\mathbf{i} + \frac{5}{2}\mathbf{j}$

For the most rapid decrease, she should head in the direction

$-\nabla z_0 = \dfrac{3}{2}\mathbf{i} - \dfrac{5}{2}\mathbf{j}$

Practically speaking, she should head in the direction the water is running since streams run perpendicular to the contours.

47. To find the directional derivative of f in the direction of $\mathbf{u}$ we need f_x and f_y, which we can find with a system of equations.

$\begin{cases} (f_x\mathbf{i} + f_y\mathbf{j}) \cdot \left(\dfrac{3\mathbf{i} - 4\mathbf{j}}{5}\right) = 8 \\ (f_x\mathbf{i} + f_y\mathbf{j}) \cdot \left(\dfrac{12\mathbf{i} + 5\mathbf{j}}{13}\right) = 1 \end{cases}$

$\begin{cases} 3f_x - 4f_y = 40 \\ 12f_x + 5f_y = 13 \end{cases}$

Solving simultaneously: $f_x = 4$, $f_y = -7$.

Now for $\mathbf{v} = 3\mathbf{i} - 5\mathbf{j}$:

$(f_x\mathbf{i} + f_y\mathbf{j}) \cdot \left(\dfrac{3\mathbf{i} - 5\mathbf{j}}{\sqrt{34}}\right) = (4\mathbf{i} - 7\mathbf{j}) \cdot \left(\dfrac{3\mathbf{i} - 5\mathbf{j}}{\sqrt{34}}\right)$

$= \dfrac{12 + 35}{\sqrt{34}}$

≈ 8.06

49. Let $\nabla f_0 = f_x \mathbf{i} + f_y \mathbf{j}$ be the gradient of f at $P_0(1, 2)$. The unit vector from P_0 toward $Q(3, -4)$ is $\mathbf{u} = \frac{1}{\sqrt{10}}(\mathbf{i} - 3\mathbf{j})$. Since the maximal directional derivative points in the direction of ∇f_0 and has magnitude 50, we want to have

$$f_x \mathbf{i} + f_y \mathbf{j} = 50\mathbf{u} = \frac{50}{\sqrt{10}}(\mathbf{i} - 3\mathbf{j})$$

Thus,

$$\nabla f(1, 2) = \frac{50(\mathbf{i} - 3\mathbf{j})}{\sqrt{10}}$$
$$= 5\sqrt{10}(\mathbf{i} - 3\mathbf{j})$$

51. a. Write $\dfrac{x^2}{a^2} + \dfrac{y^2}{b^2} + \dfrac{z^2}{c^2} = 1$ as $F(x, y, z) = 1$, where

$$F(x, y, z) = \frac{x^2}{a^2} + \frac{y^2}{b^2} + \frac{z^2}{c^2}$$

Then,

$$F_x = \frac{2x}{a^2}, \ F_y = \frac{2y}{b^2}, \text{ and } F_z = \frac{2z}{c^2}$$

so the tangent plane at $P_0(x_0, y_0, z_0)$ has the equation

$$\frac{2x_0}{a^2}(x - x_0) + \frac{2y_0}{b^2}(y - y_0) + \frac{2z_0}{c^2}(z - z_0) = 0$$
$$\frac{x_0 x}{a^2} - \frac{x_0^2}{a^2} + \frac{y_0 y}{b^2} - \frac{y_0^2}{b^2} + \frac{z_0 z}{c^2} - \frac{z_0^2}{c^2} = 0$$
$$\frac{x_0 x}{a^2} + \frac{y_0 y}{b^2} + \frac{z_0 z}{c^2} = 1$$

b. As in part **a**, let

$$F(x, y, z) = \frac{x^2}{a^2} + \frac{y^2}{b^2} - \frac{z^2}{c^2}$$

and obtain $F_x = \dfrac{2x}{a^2}$, $F_y = \dfrac{2y}{b^2}$, and $F_z = -\dfrac{2z}{c^2}$. Thus, the tangent plane at $P_0(x_0, y_0, z_0)$ has the equation

$$\frac{2x_0}{a^2}(x - x_0) + \frac{2y_0}{b^2}(y - y_0) - \frac{2z_0}{c^2}(z - z_0) = 0$$
$$\frac{x_0 x}{a^2} - \frac{x_0^2}{a^2} + \frac{y_0 y}{b^2} - \frac{y_0^2}{b^2} - \frac{z_0 z}{c^2} + \frac{z_0^2}{c^2} = 0$$
$$\frac{x_0 x}{a^2} + \frac{y_0 y}{b^2} - \frac{z_0 z}{c^2} = 1$$

c. Write

$$\frac{z}{c} = \frac{x^2}{a^2} + \frac{y^2}{b^2}$$

as $F(x, y, z) = 0$ where

$$F(x, y, z) = \frac{x^2}{a^2} + \frac{y^2}{b^2} - \frac{z}{c}$$

Then, $F_x = \dfrac{2x}{a^2}$, $F_y = \dfrac{2y}{b^2}$, and $F_z = \dfrac{-1}{c}$ so the tangent plane at $P_0(x_0, y_0, z_0)$ has the equation

$$\frac{2x_0}{a^2}(x - x_0) + \frac{2y_0}{b^2}(y - y_0) - \frac{1}{c}(z - z_0) = 0$$
$$\frac{2x_0 x}{a^2} - \frac{2x_0^2}{a^2} + \frac{2y_0 y}{b^2} - \frac{2y_0^2}{b^2} - \frac{z}{c} + \frac{z_0}{c} = 0$$
$$\frac{2x_0 x}{a^2} + \frac{2y_0 y}{b^2} - \frac{z}{c} = \frac{z_0}{c}$$

53. a. Note that $r_1 + r_2 = C$ (a constant) is a level curve of the function $f = r_1 + r_2$. By the normal property of the gradient, we know that $\nabla(r_1 + r_2)$ is a normal to $r_1 + r_2 = C$. So, $\mathbf{T} \cdot \nabla(r_1 + r_2) = 0$.

b. From part **a**,

$$\mathbf{T} \cdot \nabla(r_1 + r_2) = 0$$

Let $\mathbf{R}_1 = \mathbf{PF}_1$ and $\mathbf{R}_2 = \mathbf{PF}_2$ be the vectors from P to the two foci, so that

$$r_1 = \|\mathbf{R}_1\| \text{ and } r_2 = \|\mathbf{R}_2\|$$

By direct computation, it can be shown that

$$\nabla r_1 = \frac{\mathbf{R}_1}{r_1} \text{ and } \nabla r_2 = \frac{\mathbf{R}_2}{r_2}$$

and by substituting into the vector equation $\mathbf{T} \cdot \nabla(r_1 + r_2) = 0$, we have

$$\mathbf{T} \cdot \nabla r_1 = -\mathbf{T} \cdot \nabla r_2$$

$$\mathbf{T} \cdot \left(\frac{\mathbf{R}_1}{r_1}\right) = -\mathbf{T} \cdot \left(\frac{\mathbf{R}_2}{r_2}\right)$$

or

$$\frac{\|\mathbf{T}\|\|\mathbf{R}_1\|}{r_1}\cos(\pi - \theta_1) = -\frac{\|\mathbf{T}\|\|\mathbf{R}_2\|}{r_2}\cos\theta_2$$

so that

$$\cos(\pi - \theta_1) = -\cos\theta_2$$

$$\cos\theta_1 = \cos\theta_2$$

$$\theta_1 = \theta_2$$

55. $\nabla f = f_x\mathbf{i} + f_y\mathbf{j}$

$D_{\mathbf{u}}f = f_x\cos\theta + f_y\sin\theta$ for $f = xy^2e^{x-2y}$

$\nabla f = e^{x-2y}[(y^2 + xy^2)\mathbf{i} + (2xy - 2xy^2)\mathbf{j}]$

$\nabla f(-1, 3) = 12e^{-7}\mathbf{j}$

$D_{\mathbf{u}}f = 12e^{-7}\sin\frac{\pi}{6} = 6e^{-7}$

57. True; since $\nabla f = f_x\mathbf{i} + f_y\mathbf{j} = \mathbf{0}$, it follows that $f_x(x, y) = f_y(x, y) = 0$ throughout the disk, so $f(x, y)$ must be a constant function.

59. a. $\nabla r = \dfrac{(x\mathbf{i} + y\mathbf{j} + z\mathbf{k})}{\sqrt{x^2 + y^2 + z^2}} = \dfrac{\mathbf{R}}{\|\mathbf{R}\|}$

b. $\nabla r^n = \nabla(x^2 + y^2 + z^2)^{n/2}$

$= \dfrac{n}{2}(r^2)^{n/2-1}(2x\mathbf{i} + 2y\mathbf{j} + 2z\mathbf{k})$

$= nr^{n-2}(x\mathbf{i} + y\mathbf{j} + z\mathbf{k})$

$= nr^{n-2}\mathbf{R}$

11.7 Extrema of Functions of Two Variables, page 893

1. Solve $f_x = 0$, $f_y = 0$ to find critical points, and compute $D(x_0, y_0) = f_{xx}f_{yy} - f_{xy}^2$ at each critical point (x_0, y_0). Then:
  (i) (x_0, y_0) is a relative maximum if $D > 0$ and $f_{xx}(x_0, y_0) < 0$;
  (ii) (x_0, y_0) is a relative minimum if $D > 0$ and $f_{xx}(x_0, y_0) > 0$;
  (iii) (x_0, y_0) is a saddle point if $D < 0$.
  (iv) further analysis is necessary if $D = 0$.

3. $f(x, y) = \dfrac{9x}{x^2 + y^2 + 1}$

$f_x = \dfrac{9(x^2 + y^2 + 1 - 2x^2)}{(x^2 + y^2 + 1)^2}$

$= \dfrac{9(y^2 + 1 - x^2)}{(x^2 + y^2 + 1)^2}$

$f_y = -\dfrac{9x(2y)}{(x^2 + y^2 + 1)^2}$

$= \dfrac{-18xy}{(x^2 + y^2 + 1)^2}$

$f_x = f_y = 0$ only when $x = \pm 1, y = 0$

$f_{xx} = \dfrac{18x^3 - 54xy^2 - 54x}{(x^2 + y^2 + 1)^3}$

$f_{xy} = \dfrac{54x^2y - 18y^3 - 18y}{(x^2 + y^2 + 1)^3}$

$f_{yy} = \dfrac{54xy^2 - 18x^3 - 18x}{(x^2 + y^2 + 1)^3}$

x	y	f_{xx}	f_{xy}	f_{yy}	D	Classify
1	0	$-\frac{9}{2}$	0	$-\frac{9}{2}$	+	rel maximum
−1	0	$\frac{9}{2}$	0	$\frac{9}{2}$	+	rel minimum

SURVIVAL HINT: *Good organization is important when working a complicated problem. Look at Theorem 11.13, and look at the summary in the margin. This provided a hint about how to set up your solution for Problems 3-22 illustrates this organization.*

5. $f(x, y) = (x - 2)^2 + (y - 3)^4$
$f_x = 2(x - 2)$
$f_y = 4(y - 3)^3$
$f_x = f_y = 0$, when $x = 2$, $y = 3$;
$f_{yy} = 12(y - 3)^2$, $f_{xx} = 2$, $f_{xy} = 0$

x	y	f_{xx}	f_{xy}	f_{yy}	D	Classify
2	3	2	0	0	0	inclusive

Note that f is a sum of squares, which implies that a minimum occurs at $(2, 3)$.

7. $f(x, y) = (1 + x^2 + y^2)e^{1-x^2-y^2}$
$f_x = e^{1-x^2-y^2}\left[(-2x)(1 + x^2 + y^2) + 2x\right]$
$\quad = -2x(x^2 + y^2)e^{1-x^2-y^2}$
$f_y = e^{1-x^2-y^2}\left[(-2y)(1 + x^2 + y^2) + 2y\right]$
$\quad = -2y(x^2 + y^2)e^{1-x^2-y^2}$
$f_x = f_y = 0$ only when $x = y = 0$
$f_{xx} = 2(2x^4 + 2x^2y^2 - y^2 - 3x^2)e^{1-x^2-y^2}$
$f_{yy} = 2(2y^4 + 2x^2y^2 - x^2 - 3y^2)e^{1-x^2-y^2}$
$f_{xy} = 4xy(x^2 + y^2 - 1)e^{1-x^2-y^2}$

x	y	f_{xx}	f_{xy}	f_{yy}	D	Classify
0	0	0	0	0	0	inclusive

Examination of the graph of $f(x, y)$ suggests that a maximum occurs at $(0, 0)$.

9. $f(x, y) = x^2 + xy + y^2$
$f_x = 2x + y$
$f_y = x + 2y$
$f_x = f_y = 0$ only when $x = 0$, $y = 0$.
$f_{xx} = 2$; $f_{xy} = 1$; $f_{yy} = 2$

x	y	f_{xx}	f_{xy}	f_{yy}	D	Classify
0	0	2	1	2	+	rel minimum

11. $f(x, y) = -x^3 + 9x - 4y^2$
$f_x = -3x^2 + 9$
$f_y = -8y$
$f_x = f_y = 0$ when $x = \pm\sqrt{3}$, $y = 0$.
$f_{xx} = -6x$; $f_{yy} = -8$; $f_{xy} = 0$

x	y	f_{xx}	f_{xy}	f_{yy}	D	Classify
$\sqrt{3}$	0	$-6\sqrt{3}$	0	-8	+	rel maximum
$-\sqrt{3}$	0	$6\sqrt{3}$	0	-8	$-$	saddle point

13. $f(x, y) = (x^2 + 2y^2)e^{1-x^2-y^2}$
$f_x = -2x(x^2 + 2y^2 - 1)e^{1-x^2-y^2}$
$f_y = -2y(x^2 + 2y^2 - 2)e^{1-x^2-y^2}$
$f_x = f_y = 0$ when $x = 0$, $y = 0$ or when $x = \pm 1$, $y = 0$; $x = 0$, $y = \pm 1$. Note that $x^2 + 2y^2 = 1$, $x^2 + 2y^2 = 2$ cannot both be simultaneously true.
$f_{xx} = 2(2x^4 + 4x^2y^2 - 5x^2 - 2y^2 + 1)e^{1-x^2-y^2}$
$f_{yy} = 2(2x^2y^2 - x^2 + 4y^4 - 10y^2 + 2)e^{1-x^2-y^2}$
$f_{xy} = 4xy(x^2 + 2y^2 - 3)e^{1-x^2-y^2}$

x	y	f_{xx}	f_{xy}	f_{yy}	D	Classify
0	0	$2e$	0	$4e$	+	rel minimum
1	0	-4	0	2	$-$	saddle point
-1	0	-4	0	2	$-$	saddle point
0	1	-2	0	-8	+	rel maximum
0	-1	-2	0	-8	+	rel maximum

15. $f(x, y) = x^{-1} + y^{-1} + 2xy$
$f_x = -\dfrac{1}{x^2} + 2y$
$f_y = -\dfrac{1}{y^2} + 2x$
$f_x = f_y = 0$ when $y = \dfrac{1}{2x^2}$ and $x = \dfrac{1}{2y^2}$.
Solving, we obtain $x = y$ and
$x = \dfrac{1}{2x^2}$ or $x^3 = \dfrac{1}{2}$. Similarly, $y^3 = \dfrac{1}{2}$.
Thus, $\left(\dfrac{1}{\sqrt[3]{2}}, \dfrac{1}{\sqrt[3]{2}}\right)$ is the critical point, since $x \neq 0$.

$$f_{xx} = \frac{2}{x^3};\ f_{yy} = \frac{2}{y^3};\ f_{xy} = 2$$

x	y	f_{xx}	f_{xy}	f_{yy}	D	Classify
$2^{-1/3}$	$2^{-1/3}$	4	2	4	+	rel minimum

17. $f(x,y) = x^3 + y^3 + 3x^2 - 18y^2 + 81y + 5$
$f_x = 3x^2 + 6x = 3x(x+2)$
$f_y = 3y^2 - 36y + 81 = 3(y-3)(y-9)$
$f_x = f_y = 0$ when $x = 0, x = -2$, and
$y = 3, y = 9$; critical points are $(0,3)$,
$(0,9)$, $(-2,3)$, and $(-2,9)$.
$f_{xx} = 6x + 6;\ f_{xy} = 0;\ f_{yy} = 6y - 36$

x	y	f_{xx}	f_{xy}	f_{yy}	D	Classify
0	3	6	0	-18	-	saddle point
0	9	6	0	18	+	rel minimum
-2	9	-6	0	18	-	saddle point
-2	3	-6	0	-18	+	rel maximum

19. $f(x,y) = x^2 + y^2 - 6xy + 9x + 5y + 2$
$f_x = 2x - 6y + 9$
$f_y = 2y - 6x + 5$
$f_x = f_y = 0$ when $x = \frac{3}{2}, y = 2$.
$f_{xx} = 2;\ f_{xy} = -6;\ f_{yy} = 2$

x	y	f_{xx}	f_{xy}	f_{yy}	D	Classify
$\frac{3}{2}$	2	2	-6	2	-	saddle point

21. $f(x,y) = x^2 + y^3 + \dfrac{768}{x+y},\ x \neq -y$

$f_x = 2x - \dfrac{768}{(x+y)^2};\ f_y = 3y^2 - \dfrac{768}{(x+y)^2}$

$f_x = f_y = 0$ when

$$2x(x+y)^2 = 3y^2(x+y)^2 = 768$$

Since $x \neq -y$, we have $2x = 3y^2$ and

$$2x\left(x \pm \sqrt{\frac{2x}{3}}\right)^2 = 768$$

$x = 6, 8.985$ *Use technology*
For these x-values, we find $y = 2$ and
$y \approx -2.447$, respectively.

$$f_{xx} = 2 + \frac{1{,}536}{(x+y)^3};\ f_{xy} = \frac{1{,}536}{(x+y)^3};$$

$$f_{yy} = 6y + \frac{1{,}536}{(x+y)^3}$$

x	y	f_{xx}	f_{xy}	f_{yy}	D	Classify
6	2	5	3	15	+	rel minimum
8.985	-2.447	7.5	5.5	-9.2	-	saddle point

23. $f(x,y) = x^2 + xy + y^2$
$f_x = 2x + y;\ f_y = x + 2y$
$f_x = f_y = 0$ only at $(0,0)$.
On the boundary $x^2 + y^2 = 1$, let $x = \cos t$,
$y = \sin t$ for $0 \leq t \leq 2\pi$. Then,
$$F(t) = f(\cos t, \sin t) = 1 + \cos t \sin t$$
$$F'(t) = -\sin^2 t + \cos^2 t = \cos 2t$$
$F'(t) = 0$ when

$$t = \frac{\pi}{4} \qquad \left(\frac{\sqrt{2}}{2}, \frac{\sqrt{2}}{2}\right)$$
$$t = \frac{3\pi}{4} \qquad \left(-\frac{\sqrt{2}}{2}, \frac{\sqrt{2}}{2}\right)$$
$$t = \frac{5}{4} \qquad \left(-\frac{\sqrt{2}}{2}, -\frac{\sqrt{2}}{2}\right)$$
$$t = \frac{7\pi}{4} \qquad \left(\frac{\sqrt{2}}{2}, -\frac{\sqrt{2}}{2}\right)$$

$f(0,0) = 0$; this is a minimum since a minimum must occur somewhere and this is the only possibility.

$$f\left(\frac{\sqrt{2}}{2}, \frac{\sqrt{2}}{2}\right) = f\left(-\frac{\sqrt{2}}{2}, -\frac{\sqrt{2}}{2}\right) = \frac{3}{2};\ \text{maximum}$$

$$f\left(-\frac{\sqrt{2}}{2}, \frac{\sqrt{2}}{2}\right) = f\left(-\frac{\sqrt{2}}{2}, -\frac{\sqrt{2}}{2}\right) = \frac{1}{2}$$

25. $f(x,y) = xy - 2x - 5y$
$f_x = y - 2$
$f_y = x - 5$
$f_x = f_y = 0$ only at $(5,2)$. The boundary
consists of three lines:
On $y = 0, x = t$ for $0 \leq t \leq 7$, we have

$F_1(t) = f(t,0) = 0 - 2t - 0$
$F_1'(t) = -2 \neq 0$
On $x = 7$, $y = t$ for $0 \leq t \leq 7$
　$F_2(t) = f(7,t) = 7t - 14 - 5t$
　$F_2'(t) = 2 \neq 0$
On $x = t$, $y = t$ for $0 \leq t \leq 7$
　$F_3(t) = f(t,t) = t^2 - 7t$
　$F_3'(t) = 2t - 7$ with boundary
　critical point $\left(\dfrac{7}{2}, \dfrac{7}{2}\right)$ where $t = \dfrac{7}{2}$.

Thus, we have an interior critical point $(5,2)$, a boundary critical point $\left(\frac{7}{2}, \frac{7}{2}\right)$, and three boundary endpoints $(0,0)$, $(7,0)$, and $(7,7)$.

x	y	$f(x,y)$	Classify
5	2	-10	
$\frac{7}{2}$	$\frac{7}{2}$	-12.25	
0	0	0	maximum
7	0	-14	minimum
7	7	0	maximum

The largest value of f on S is 0 and the smallest is -14.

27. $f(x,y) = 2\sin x + 5\cos y$
$f_x = 2\cos x$; $f_y = -5\sin y$
$f_x = f_y = 0$ when $x = \frac{\pi}{2}$, $y = 0, \pi$
Critical points are $\left(\frac{\pi}{2}, 0\right)$, $\left(\frac{\pi}{2}, \pi\right)$.
On $x = t$, $y = 0$ for $0 \leq t \leq 2$;
$F_1(t) = f(t,0) = 2\sin t + 5$
$F_1'(t) = 2\cos t = 0$
　　when $t = \dfrac{\pi}{2}$; point $\left(\dfrac{\pi}{2}, 0\right)$
On $x = 2$, $y = t$ for $0 \leq t \leq 5$;
$F_2(t) = f(2,t) = 2\sin 2 + 5\cos t$
$F_1'(t) = -5\sin t = 0$
when $t = 0, \pi$; points $(2,0)$, $(2,\pi)$

On $y = 5$, $x = t$ for $0 \leq t \leq 2$;

$F_3(t) = f(t,5) = 2\sin t + 5\cos 5$
$F_3'(t) = 2\cos t = 0$
　　when $t = \dfrac{\pi}{2}$; point $\left(\dfrac{\pi}{2}, 5\right)$

On $x = 0$, $y = t$ for $0 \leq t \leq 5$;

　$F_4(t) = f(0,t) = 5\cos t$
　$F_4'(t) = -5\sin t = 0$
　　　when $t = 0, \pi$
　　points $(0,0)$, $(0,\pi)$

There is one interior critical point, $\left(\frac{\pi}{2}, \pi\right)$; four boundary critical points $\left(\frac{\pi}{2}, 0\right)$, $(2,\pi)$, $\left(\frac{\pi}{2}, 5\right)$, $(0,\pi)$; and four boundary endpoints $(0,0)$, $(2,0)$, $(2,5)$, $(0,5)$.

x	y	$f(x,y)$	Classify
$\frac{\pi}{2}$	0	7	maximum
2	π	-3.18	
$\frac{\pi}{2}$	5	3.42	
0	π	-5	minimum
0	0	5	
2	0	6.82	
2	5	3.24	
0	5	1.42	
$\frac{\pi}{2}$	π	-3	

The largest value of f on S is 7 and the smallest is -5.

29. $m = \dfrac{5(25) - (1)(5)}{5(15) - 1^2} = \dfrac{60}{37} \approx 1.62$

$b = \dfrac{15(5) - (1)(25)}{5(15) - 1^2} = \dfrac{25}{37} \approx 0.68$

$y = 1.62x + 0.68$

31. $m = \dfrac{5(155.68) - (28.75)(27.14)}{5(184.47) - (28.75)^2} \approx -0.02$

$b = \dfrac{184.47(27.14) - (28.75)(155.68)}{95.79} \approx 5.54$

$y = -0.02x + 5.54$

33. Minimize $S = x^2 + y^2 + z^2$ subject to $y^2 = 4 + xz$. We have,

$$S(x,z) = x^2 + (4 + xz) + z^2$$

$S_x = 2x + z$; $S_z = x + 2z$ and
$S_x = S_z = 0$ when $z = -2x = -\frac{1}{2}x$;
$x = 0$, $z = 0$, $y = \pm 2$. The closest
points are $(0, 2, 0)$ and $(0, -2, 0)$.

35. Let $x, y,$ and z be the dimensions of the
box. The volume is $V = xyz$ or
$z = \dfrac{V}{xy}$, and the surface area is
$S = xy + 2xz + 2yz$. We wish to
minimize

$$S = xy + 2xz + 2yz$$
$$= xy + \frac{2xV}{xy} + \frac{2yV}{xy}$$
$$= xy + \frac{2V}{y} + \frac{2V}{x}$$

$S_x = y - \dfrac{2V}{x^2}$ and $S_y = x - \dfrac{2V}{y^2}$. Then
$S_x = S_y = 0$ when $x = y = \sqrt[3]{2V}$,
$z = \sqrt[3]{0.25V}$.

$$S_{xx} = \frac{4V}{x^3},\ S_{xy} = 1,\ S_{yy} = \frac{4V}{y^3}$$

$D > 0$, so the dimensions for the
minimum construction are $x = \sqrt[3]{2V}$,
$y = \sqrt[3]{2V}$, $z = \sqrt[3]{0.25V}$.

37. If $x, y,$ and z are the numbers we wish to
maximize $P = xyz$ subject to

$$x + y + z = 54$$

Since $z = 54 - x - y$, we have

$$P = xy(54 - x - y)$$
$$= 54xy - x^2y - xy^2$$
and
$$P_x = 54y - 2xy - y^2$$
$$P_y = 54x - x^2 - 2xy$$
Since $x > 0$, $y > 0$, it follows that
$P_x = P_y = 0$ when $x = y = 18$. Since

$$D = P_{xx}P_{yy} - P_{xy}^2$$
$$= (-2y)(-2x) - (54 - 2x - 2y)^2$$

We have $P_{xx}(18, 18) < 0$ and
$D(18, 18) > 0$, so a relative maximum
occurs at the critical point $(18, 18)$. Thus,
the product is maximized when $x = 18$,
$y = 18$, and $z = 54 - 18 - 18 = 18$.

39. First, find the critical points for $T(x, y)$:

$$T_x = 4x - y = 0$$
$$T_y = -x + 2y - 2 = 0$$

The only critical point is $\left(\frac{2}{7}, \frac{8}{7}\right)$. There are
three boundary lines:
I. $x = -1$
$$T_1 = 2 + y + y^2 - 2y + 1$$
$$= y^2 - y + 3$$
$$T_1' = 2y - 1$$
$$T_1' = 0 \text{ when } y = \frac{1}{2} \text{ or at } \left(-1, \frac{1}{2}\right)$$

II. $y = 2$
$$T_2 = 2x^2 - 2x + 4 - 4 + 1$$
$$= 2x^2 - 2x + 1$$
$$T_2' = 4x - 2$$
$$T_2' = 0 \text{ when } x = \frac{1}{2} \text{ or at } \left(\frac{1}{2}, 2\right)$$

III. Line through $(-1, -2)$ and $(3, 2)$

$$\frac{y-2}{x-3} = \frac{4}{4} \text{ or } y = x - 1$$

$$T_3 = 2x^2 - x(x-1) + (x-1)^2$$
$$\qquad - 2(x-1) + 1$$
$$\quad = 2x^2 - 3x + 4$$
$$T_3' = 4x - 3$$

$$T_3' = 0 \text{ when } x = \frac{3}{4} \text{ or at } \left(\frac{3}{4}, -\frac{1}{4}\right)$$

Evaluation	(x,y)	$T(x,y)$
interior critical point:	$\left(\frac{2}{7}, \frac{8}{7}\right)$	$-\frac{1}{7}$
boundary critical points:	$\left(-1, \frac{1}{2}\right)$	$\frac{11}{4}$
	$\left(\frac{1}{2}, 2\right)$	$\frac{1}{2}$
	$\left(\frac{3}{4}, -\frac{1}{4}\right)$	$\frac{23}{8}$
boundary end points:	$(-1, 2)$	5
	$(-1, -2)$	9
	$(3, 2)$	13

We see the minimum is $-\frac{1}{7}$ and the maximum is 13. The temperature is the greatest $(13°C)$ at $(3, 2)$ and is least $\left(-\frac{1}{7}°C\right)$ at $\left(\frac{2}{7}, \frac{8}{7}\right)$.

41. $R_x = -2x + 2y + 8 = 0$
$R_y = -4y + 2x + 5 = 0$
Solving simultaneously, $x = \frac{21}{2}$, $y = \frac{13}{2}$
$R_{xx} = -2$, $R_{xy} = 2$, $R_{yy} = -4$, $D > 0$,
so the revenue is maximized at $\left(\frac{21}{2}, \frac{13}{2}\right)$.
In other words, the revenue is maximized when 10,500 units of A and 6,500 units of B are produced.

43. The profit for each bottle of California water is $x - 2$, and for each bottle of New York water is $y - 2$. The profit is

$$P(x, y) = (x - 2)(40 - 50x + 40y)$$
$$\qquad + (y - 2)(20 + 60x - 70y)$$
$$\qquad = -50x^2 + 100xy - 70y^2 + 20x + 80y - 120$$

$$P_x = -100x + 100y + 20$$
$$P_y = 100x - 140y + 80$$

Solving simultaneously, $x = 2.7$, $y = 2.5$.
$P_{xx} = -100$; $P_{xy} = 100$; $P_{yy} = -140$
Then $D > 0$, $P_{xx} < 0$, so the profit is maximized at $(2.7, 2.5)$. The owner should charge \$2.70 for California water and \$2.50 for New York water.

45. The domestic profit is:

$$D = x(60 - 0.2x + 0.05y) - 10x$$
$$\quad = x(50 - 0.2x + 0.05y)$$

The foreign profit is:

$$F = y(50 - 0.1y + 0.05x) - 10y$$
$$\quad = y(40 - 0.1y + 0.05x)$$

$$P(x, y) = x(50 - 0.2x + 0.05y)$$
$$\qquad + y(40 - 0.1y + 0.05x)$$
$$\qquad = -0.2x^2 + 0.1xy - 0.1y^2 + 50x + 40y$$
$$P_x = -0.4x + 0.1y + 50 = 0$$
$$P_y = 0.1x - 0.2y + 40 = 0$$
Solving simultaneously, $x = 200$, $y = 300$.
$P_{xx} = -0.4$; $P_{xy} = 0.1$; $P_{yy} = -0.2$
$D > 0$ and $D_{xx} < 0$, so $(200, 300)$ is a maximum. That is, 200 machines should be supplied to the domestic market and 300 to the foreign market.

47. $y = 2.12x + 0.065$
$k = m \approx 2.12$

49. a.

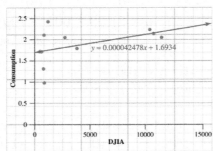

b. $y = 0.000041971x + 1.6978$ where x is the DJIA and y is the wine consumption.

c. Answers vary. For example, in 2014 compare the DJIA with the wine consumption and compare it the equation given in part **b.** To find the wine consumption you can check with the *Wine Institute* at **www.wineinstitute.org**.

51. a. We want $W = kx^m$ so

$$\ln W = \ln k + m \ln x$$

We do linear regression with $\ln x$ and $\ln W$ instead of with x and W. Using technology, we find

$$\ln W = 3.4323 + 0.5718X$$

so $W = e^{3.4323}(x^{0.5718})$. This means $m = 0.5718$, $k = e^{3.4323} \approx 30.951$.

b. The poor agreement of m with the expected value is largely due to the exceptional performance of the 60 kg lifter. You will notice this if you graph the least squares approximation and the data points you will see that all the data points cluster except for the outlier 60 kg. It is the outlier that causes $m \approx 0.57$

to differ from the expected value of approximately 0.67.

53. $f(x, y) = 4x^2 e^y - 2x^4 - e^{4y}$
$f_x = 8xe^y - 8x^3$
$f_y = 4x^2 e^y - 4e^{4y}$
Then, $f_x = 0$ and $f_y = 0$ when
$$8xe^y - 8x^3 = 0 \quad \text{and} \quad 4x^2 e^y - 4e^{4y} = 0$$
$$8x(e^y - x^2) = 0 \qquad 4e^y(x^2 - e^{3y}) = 0$$
$$e^y = x^2 \qquad\qquad e^{3y} = x^2$$
Note: in the first equation, if $x = 0$, then $f_y \neq 0$, and in the second equation $e^y = 0$ is impossible. Thus, $e^y = e^{3y}$, so $y = 0$, $x = \pm 1$. The critical points are $(1, 0)$ and $(-1, 0)$. The discriminant is

$$D = f_{xx}f_{yy} - f_{xy}^2$$
$$= \left(8e^y - 24x^2\right)\left(4x^2 e^y - 16e^{4y}\right) - \left(8xe^y\right)^2$$

and $D > 0$ with $f_{xx} < 0$ at both $(1, 0)$ and $(-1, 0)$, so they both correspond to relative maxima.

55. Tom travels $\sqrt{x^2 + 1.2^2}$ miles, Dick travels $\sqrt{y^2 + 2.5^2}$ miles, and Mary travels $4.3 - x - y$ miles. Thus, the total time of travel is

$$T = \frac{\sqrt{x^2 + 1.2^2}}{2} + \frac{\sqrt{y^2 + 2.5^2}}{4} + \frac{4.3 - x - y}{6}$$

Then,

$$T_x = \frac{1}{2}\frac{x}{\sqrt{x^2 + 1.2^2}} - \frac{1}{6} = 0$$

$$T_y = \frac{1}{4}\frac{y}{\sqrt{y^2 + 2.5^2}} - \frac{1}{6} = 0$$

which has the unique solution $x \approx 0.4243$ and $y \approx 2.2361$.

Boundary cases

If $x = 0$, then the time of travel is

$$T_1 = \frac{1.2}{2} + \frac{\sqrt{y^2 + 2.5^2}}{4} + \frac{4.3 - y}{6}$$

$$T_1' = \frac{1}{4} \frac{y}{\sqrt{y^2 + 2.5^2}} - \frac{1}{6}$$

$T_1' = 0$ when $y \approx 2.2361$ and we find $T_1 \approx 1.7825$.

If $y = 0$, then the time of travel is

$$T_2 = \frac{\sqrt{x^2 + 1.2^2}}{2} + \frac{2.5}{4} + \frac{4.3 - x}{6}$$

$$T_2' = \frac{1}{2} \frac{x}{\sqrt{x^2 + 1.2^2}} - \frac{1}{6}$$

$T_2' = 0$ when $x \approx 0.4243$ and we find $T_2 \approx 1.9074$.

If Tom trudges directly to B and Dick rows from B to A, then the time of travel is

$$T_3 = \frac{1.2}{2} + \frac{2.5}{4} + \frac{4.3}{6} \approx 1.9417$$

Finally, if Tom trudges in a direct line toward F, and Dick rows in a direct line toward F, then

$$T_4: \quad \frac{x}{1.2} = \frac{4.3}{3.7} \text{ or } x \approx 1.3946$$

$$\text{and } \frac{y}{2.5} = \frac{4.3}{3.7} \text{ or } y \approx 2.9054$$

We now evaluate the times:

Time	x	y	Time (in hours)
T	0.4243	2.2361	1.7482
T_1	0	2.2361	1.7825
T_2	0.4243	0	1.9074
T_3	0	0	1.9417
T_4	1.3946	2.9054	1.8781

The minimum time of travel occurs when Dick waits 0.4243 miles from the line $\overline{AS}$ and Mary waits

$$4.3 - 0.4243 - 2.2361 = 1.6396$$

miles from the finish line.

57. a. $z = f(x, y)$ has a minimum of 0 at the origin because as (x, y) moves toward the origin, the values of f drop toward 0.

   b. $z = f(x, y)$ has a saddle point at the origin. In the first and third quadrants, f approaches 0 from above, while in the second and fourth quadrants f approaches 0 from below as $(x, y) \rightarrow (0, 0)$. In the plane $y = x$, the cross section is parabola-like with a low point at $(0, 0)$. In the plane $y = -x$, the cross section is parabola-like with a high point at $(0, 0)$.

59. $F(m, b) = \sum_{k=1}^{n} [y_k - (mx_k + b)]^2$

$$F_m(m, b) = 2 \sum_{k=1}^{n} [y_k - (mx_k + b)](-x_k)$$

$$F_b(m, b) = 2 \sum_{k=1}^{n} [y_k - (mx_k + b)](-1)$$

Then $F_m = F_b = 0$ when

$$\left(\sum_{k=1}^{n} x_k^2\right)m + \left(\sum_{k=1}^{n} x_k\right)b = \sum_{k=1}^{n} x_k y_k$$

$$\left(\sum_{k=1}^{n} x_k\right)m + \left(\sum_{k=1}^{n} 1\right)b = \sum_{k=1}^{n} y_k$$

or equivalently

$$Am + Bb = D$$
$$Bm + nb = C$$

where

$$A = \sum_{k=1}^{n} x_k^2, \; B = \sum_{k=1}^{n} x_k, \; C = \sum_{k=1}^{n} y_k$$

and $D = \sum_{k=1}^{n} x_k y_k$. Solving this system, we have

$$m = \frac{Dn - BC}{An - B^2} \qquad b = \frac{AC - BD}{An - B^2}$$

Thus,

$$m = \frac{n\sum_{k=1}^{n} x_k y_k - \left(\sum_{k=1}^{n} x_k\right)\left(\sum_{k=1}^{n} y_k\right)}{n\sum_{k=1}^{n} x_k^2 - \left(\sum_{k=1}^{n} x_k\right)^2}$$

and

$$b = \frac{\left(\sum_{k=1}^{n} x_k^2\right)\left(\sum_{k=1}^{n} y_k\right) - \left(\sum_{k=1}^{n} x_k\right)\left(\sum_{k=1}^{n} x_k y_k\right)}{n\sum_{k=1}^{n} x_k^2 - \left(\sum_{k=1}^{n} k_k\right)^2}$$

11.8 Lagrange Multipliers, page 904

SURVIVAL HINT: *Before working Problems 1-18, look at the procedure box on "Method of Lagrange Multiplies" on page 898.*

1. $f(x, y) = xy; g(x, y) = 2x + 2y$
$f_x = y, \; f_y = x, \; g_x = 2, \; g_y = 2$
Solve the system
$$\begin{cases} y = 2\lambda \\ x = 2\lambda \\ 2x + 2y = 5 \end{cases}$$
to find $x = y = \frac{5}{4}$. The constrained maximum is $f\left(\frac{5}{4}, \frac{5}{4}\right) = \frac{25}{16}$.

3. $f(x, y) = x^2 + y^2; g(x, y) = x + y$
$f_x = 2x, \; f_y = 2y, \; g_x = 1, \; g_y = 1$
Solve the system
$$\begin{cases} 2x = \lambda \\ 2y = \lambda \\ x + y = 24 \end{cases}$$
to find $x = y = 12$. The constrained minimum is $f(12, 12) = 288$.

5. $f(x, y) = x^2 + y^2; g(x, y) = x + y$
$f_x = 2x, \; f_y = 2y, \; g_x = 1, \; g_y = 1$
Solve the system
$$\begin{cases} 2x = \lambda \\ 2y = \lambda \\ x + y = 9 \end{cases}$$
to find $x = y = \frac{9}{2}$. The constrained minimum is $f\left(\frac{9}{2}, \frac{9}{2}\right) = \frac{81}{2}$.

7. $f(x, y) = x^2 + y^2; g(x, y) = xy$
$f_x = 2x, \; f_y = 2y, \; g_x = y, \; g_y = x$
Solve the system
$$\begin{cases} 2x = \lambda y \\ 2y = \lambda x \\ xy = 1 \end{cases}$$
to find $x = y = 1$ and $x = y = -1$. The constrained minimum is $f(\pm 1, \pm 1) = 2$.

9. $f(x, y) = x^2 + y^2 - xy - 4$;
$g(x, y) = x + y$

$f_x = 2x - y$, $f_y = 2y - x$, $g_x = 1$, $g_y = 1$

Solve the system
$$\begin{cases} 2x - y = \lambda \\ 2y - x = \lambda \\ x + y = 6 \end{cases}$$
to find $x = y = 3$. The constrained minimum is $f(3,3) = 5$.

11.
$f(x,y) = x^2 - y^2$; $g(x,y) = x^2 + y^2$
$f_x = 2x$, $f_y = -2y$, $g_x = 2x$, $g_y = 2y$

Solve the system
$$\begin{cases} 2x = 2\lambda x \\ -2y = 2\lambda y \\ x^2 + y^2 = 4 \end{cases}$$
to find $x = 0$, $y = \pm 2$ or $x = \pm 2$, $y = 0$. The constrained minimum is $f(0, \pm 2) = -4$.

13. $f(x,y) = \cos x + \cos y$;
$g(x,y) = y - x$
$f_x = -\sin x$, $f_y = -\sin y$,
$g_x = -1$, $g_y = 1$

Solve the system
$$\begin{cases} -\sin x = -\lambda \\ -\sin y = \lambda \\ y = x + \dfrac{\pi}{4} \end{cases}$$
to find
$$x = \frac{(8n - 1)\pi}{8}, \quad y = \frac{(8n + 1)\pi}{8}$$
If $n = 1$, $f\left(\frac{7\pi}{8}, \frac{9\pi}{8}\right) \approx -1.8478$;
if $n = 2$, $f\left(\frac{15\pi}{8}, \frac{17\pi}{8}\right) \approx 1.8478$;
the constrained maximum is approximately 1.8478 for
$f\left(-\frac{\pi}{8} + n\pi, \frac{\pi}{8} + n\pi\right)$ with n even.

15. $f(x,y) = \ln(xy^2)$; $g(x,y) = 2x^2 + 3y^2$
$f_x = \frac{1}{x}$, $f_y = \frac{2}{y}$,

$g_x = 4x$, $g_y = 6y$

Solve the system
$$\begin{cases} \dfrac{1}{x} = (4x)\lambda \\ \dfrac{2}{y} = (6y)\lambda \\ 2x^2 + 3y^2 = 8 \end{cases}$$
to find $x = \sqrt{\frac{4}{3}}$ (negative value not in the domain), $y = \frac{4}{3}$.

The constrained maximum is
$f\left(\sqrt{\frac{4}{3}}, \frac{4}{3}\right) = \ln\left[\sqrt{\frac{4}{3}}\left(\frac{4}{3}\right)^2\right] = \frac{5}{2}\ln\frac{4}{3}$.

17. $f(x,y,z) = x^2 + y^2 + z^2$;
$g(x,y,z) = x - 2y + 3z$
$f_x = 2x$, $f_y = 2y$, $f_z = 2z$
$g_x = 1$, $g_y = -2$, $g_z = 3$

Solve the system
$$\begin{cases} 2x = \lambda \\ 2y = -2\lambda \\ 2z = 3\lambda \\ x - 2y + 3z = 4 \end{cases}$$
to find $\lambda = \frac{4}{7}$ and then $x = \frac{2}{7}$, $y = -\frac{4}{7}$, $z = \frac{6}{7}$. The constrained minimum is
$f\left(\frac{2}{7}, -\frac{4}{7}, \frac{6}{7}\right) = \frac{8}{7}$.

19. $f(x,y,z) = 2x^2 + 4y^2 + z^2$;
$g(x,y,z) = 4x - 8y + 2z$
$f_x = 4x$, $f_y = 8y$, $f_z = 2z$
$g_x = 4$, $g_y = -8$, $g_z = 2$

Solve the system
$$\begin{cases} 4x = 4\lambda \\ 8y = -8\lambda \\ 2z = 2\lambda \\ 4x - 8y + 2z = 10 \end{cases}$$
to find $\lambda = x = -y = z$ and then $x = \frac{5}{7}$, $y = -\frac{5}{7}$, $z = \frac{5}{7}$. The constrained minimum is

$f\left(\frac{5}{7}, -\frac{5}{7}, \frac{5}{7}\right) = \frac{25}{7}$. By using opposite
values for any two variables in the
constraint equation, the third variable
can be made arbitrarily large, thus f can
be made arbitrarily large and does not
have a maximum.

21. $f(x, y, z) = x - y + z$;
$g(x, y, z) = x^2 + y^2 + z^2$
$f_x = 1,\ f_y = -1,\ f_z = 1$
$g_x = 2x,\ g_y = 2y,\ g_z = 2z$
Solve the system
$$\begin{cases} 1 = 2\lambda x \\ -1 = 2\lambda y \\ 1 = 2\lambda z \\ x^2 + y^2 + z^2 = 100 \end{cases}$$
to find $x = z = \pm\dfrac{10}{\sqrt{3}},\ y = \mp\dfrac{10}{\sqrt{3}}$

The constrained maximum is
$$f\left(\frac{10}{\sqrt{3}}, -\frac{10}{\sqrt{3}}, \frac{10}{\sqrt{3}}\right) = \frac{30}{\sqrt{3}} = 10\sqrt{3}.$$
The constrained minimum is
$$f\left(-\frac{10}{\sqrt{3}}, \frac{10}{\sqrt{3}}, -\frac{10}{\sqrt{3}}\right) = -\frac{30}{\sqrt{3}}$$
$$= -10\sqrt{3}$$

23. Minimize the square of the distance.
$f(x, y, z) = x^2 + y^2 + z^2$, subject to
$g(x, y, z) = Ax + By + Cz = D$.
$f_x = 2x,\ f_y = 2y,\ f_z = 2z$;
$g_x = A,\ g_y = B,\ g_z = C$.
Solve the system
$$\begin{cases} 2x = A\lambda \\ 2y = B\lambda \\ 2z = C\lambda \\ Ax + By + Cz = D \end{cases}$$

to find $\lambda = \dfrac{2D}{A^2 + B^2 + C^2}$ and then
$x = \dfrac{AD}{A^2 + B^2 + C^2},\ y = \dfrac{BD}{A^2 + B^2 + C^2}$,
$z = \dfrac{CD}{A^2 + B^2 + C^2}$; let
$H = A^2 + B^2 + C^2$. The minimum
distance is
$$S = \left[f\left(\frac{AD}{H}, \frac{BD}{H}, \frac{CD}{H}\right)\right]^{1/2} = \frac{|D|}{\sqrt{A^2 + B^2 + C^2}}$$

25. Minimize the square of the distance.
$f(x, y, z) = x^2 + y^2 + z^2$, subject to
$g(x, y, z) = 2x + y + z = 1$.
$f_x = 2x,\ f_y = 2y,\ f_z = 2z$;
$g_x = 2,\ g_y = 1,\ g_z = 1$.
Solve the system
$$\begin{cases} 2x = 2\lambda \\ 2y = \lambda \\ 2x + y + z = 1 \end{cases}$$
to find $x = \frac{1}{3},\ y = \frac{1}{6},\ z = \frac{1}{6}$. The
nearest point is $\left(\frac{1}{3}, \frac{1}{6}, \frac{1}{6}\right)$ and the
minimum distance is
$$S = \left[f\left(\frac{1}{3}, \frac{1}{6}, \frac{1}{6}\right)\right]^{1/2} = \frac{1}{\sqrt{6}}$$

27. $f(x, y, z) = xy^2z$
$g(x, y, z) = x + y + z = 12$
$f_x = y^2z,\ f_y = 2xyz,\ f_z = xy^2$
$g_x = 1,\ g_y = 1,\ g_z = 1$
Solve the system
$$\begin{cases} y^2z = \lambda \\ xy^2 = \lambda \\ 2xyz = \lambda \\ x + y + z = 12 \end{cases}$$
to find $x = z = 3,\ y = 6$. The largest
product is $f(3, 6, 3) = 324$.

29. Minimize
$T(x, y, z) = 100 - xy - xz - yz$,
subject to $g(x, y, z) = 10$, where
$g(x, y, z) = x + y + z$.
$T_x = -y - z$, $T_y = -x - z$, $T_z = -x - y$
$g_x = 1$, $g_y = 1$, $g_z = 1$.
Solve the system
$$\begin{cases} -y - z = \lambda \\ -x - z = \lambda \\ -x - y = \lambda \\ x + y + z = 10 \end{cases}$$
to obtain $x = y = z = -\frac{\lambda}{2}$, and then
find $x = y = z = \frac{10}{3}$. The lowest
temperature is
$$T\left(\frac{10}{3}, \frac{10}{3}, \frac{10}{3}\right) = \frac{200}{3}$$

31. Let x and y denote the sides of the
field. We wish to maximize $A = xy$
subject to $F(x, y) = 2x + 2y = 320$.
$A_x = y$, $A_y = x$, $F_x = 2$, $F_y = 2$.
Solve the system
$$\begin{cases} y = 2\lambda \\ x = 2\lambda \\ 2x + 2y = 320 \end{cases}$$
to find $x = y = 80$ and the maximum
value of A is $(80)(80) = 6{,}400$ yd^2.

33. Let x and y be the radius and height of
the cylinder. Minimize the cost
$f(x, y) = 2(2\pi x^2) + 2\pi xy$; subject to
$g(x, y) = 4\pi$ where
$g(x, y) = \pi x^2 y = 4\pi$
$f_x = 8\pi x + 2\pi y$
$f_y = 2\pi x$, $g_x = 2\pi xy$, $g_y = \pi x^2$
Solve the system

$$\begin{cases} 8\pi x + 2\pi y = 2\lambda \pi xy \\ 2\pi x = \lambda \pi x^2 \\ x^2 y \pi = 4\pi \end{cases}$$
to obtain $y = 4x$, and then find the radius
$x = 1$ in. and the height $y = 4$ in.

35. Maximize $f(x, y) = 50x^{1/2}y^{3/2}$,
subject to $g(x, y) = 8$ where
$g(x, y) = x + y$.
$f_x = 25x^{-1/2}y^{3/2}$, $f_y = 75x^{1/2}y^{1/2}$,
$g_x = 1$, $g_y = 1$.
Solve the system
$$\begin{cases} 25x^{-1/2}y^{3/2} = \lambda \\ 75x^{1/2}y^{1/2} = \lambda \\ x + y = 8 \end{cases}$$
to find $x = 2$, $y = 6$. \$2,000 to
development and \$6,000 to promotion
gives the maximum sales of
$$f(2, 6) = 50\sqrt{2}\left(6\sqrt{6}\right) = 600\sqrt{3};$$
this is about 1,039 units.

37. Let s be the length of
the living space and
y the depth.

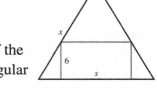

The height of the
equilateral triangular
face is
$h = \dfrac{\sqrt{3}}{2}x$ so by similar triangles,

$$\frac{\frac{\sqrt{3}}{2}x - 6}{\frac{\sqrt{3}}{2}x} = \frac{\frac{s}{2}}{\frac{x}{2}}$$

$$s = x - 4\sqrt{3}$$

The volume of the livable space is

$$V = 6(x - 4\sqrt{3})y = 6xy - 24\sqrt{3}\,y$$

and the surface area of the building is

$$S = 2\left[\frac{1}{2}\left(\frac{\sqrt{3}}{2}x^2\right)\right] + 3xy = \frac{\sqrt{3}}{2}x^2 + 3xy$$

The problem is to maximize
$V = 6\left(x - 4\sqrt{3}\right)y$ subject to

$S = \dfrac{\sqrt{3}}{2}x^2 + 3xy = 500$. We have,

$V_x = 6y$, $V_y = 6x - 24\sqrt{3}$,
$S_x = \sqrt{3}x + 3y$, $S_y = 3x$, so we must
solve the system

$$\begin{cases} 6y = \lambda\left(\sqrt{3}x + 3y\right) \\ 6x - 24\sqrt{3} = \lambda(3x) \\ \dfrac{\sqrt{3}}{2}x^2 + 3xy = 500 \end{cases}$$

We find that

$$\lambda = \frac{6y}{\sqrt{3}x + 3y} = \frac{6x - 24\sqrt{3}}{3x}$$

so that $y = \dfrac{x^2}{12} - \dfrac{\sqrt{3}}{3}x$, and then

$$\frac{\sqrt{3}}{2}x^2 + 3x\left(\frac{x^2}{12} - \frac{\sqrt{3}}{3}x\right) = 500$$

$x \approx 13.866$ and $y \approx 8.017$.

39. Minimize

$$F = \frac{8m}{k^2}E$$
$$= \frac{1}{x^2} + \frac{1}{y^2} + \frac{1}{z^2}$$

subject to $V = xyz = C$. We have

$$F_x = \frac{-2}{x^3}; \; F_y = \frac{-2}{y^3}; \; F_z = \frac{-2}{z^3}$$

$V_x = yz$; $V_y = xz$; $V_z = xy$.
We must solve the system of equations
(for nonnegative x, y, and z):

$$\begin{cases} \dfrac{-2}{x^3} = \lambda(yz) \\ \dfrac{-2}{y^3} = \lambda(xz) \\ \dfrac{-2}{z^3} = \lambda(xy) \\ xyz = C \end{cases}$$

Solving, we obtain $x = y = z = \sqrt[3]{C}$.

41. Proceeding as in the solution to Problem
40, we must solve the system of equations:

$$\begin{cases} 2x = 20\lambda \\ 4y = 12\lambda \\ 20x + 12y = 100 \end{cases}$$

Solving, we obtain $\lambda = \frac{100}{236}$ and then
$x \approx 4.24$, $y \approx 1.27$. In this case, the farmer
should apply 4.24 acre-ft of water and 1.27
lb of fertilizer to maximize the yield.

43. Maximize $F = A^2 = s(s - a)(s - b)(s - c)$.
Since $s = \frac{1}{2}(a + b + c)$, we substitute this into
the formula to find

$$F = \frac{1}{2}(a + b + c)\frac{1}{2}(b + c - a)\frac{1}{2}(a + c - b)\frac{1}{2}(a + b - c)$$
$$= \frac{1}{16}\left[2a^2b^2 + 2a^2c^2 + 2b^2c^2 - a^4 - b^4 - c^4\right]$$

subject to $P = a + b + c = P_0$. We must
solve the system of equations:

$$\begin{cases} F_a = \dfrac{1}{16}\left[-4a^3 + 4ab^2 + 4ac^2\right] = \lambda(1) \\ F_b = \dfrac{1}{16}\left[-4b^3 + 4ba^2 + 4bc^2\right] = \lambda(1) \\ F_c = \dfrac{1}{16}\left[-4c^3 + 4ca^2 + 4cb^2\right] = \lambda(1) \\ a + b + c = P_0 \end{cases}$$

Solving, we obtain $a = b = c = \frac{1}{3}P_0$, so the
triangle with maximum area is equilateral.

45. $f(x, y, z) = x^2 + y^2 + z^2$
$g(x, y, z) = x + y = 4$
$h(x, y, z) = y + z = 6$
$f_x = 2x, \ f_y = 2y, \ f_z = 2z$
$g_x = 1, \ g_y = 1, \ g_z = 0$
$h_x = 0, \ h_y = 1, \ h_z = 1$
Solve the system

$$\begin{cases} 2x = \lambda \\ 2y = \lambda + \mu \\ 2z = \mu \\ x + y = 4 \\ y + z = 6 \end{cases}$$

to find $x = \frac{2}{3}, y = \frac{10}{3}, z = \frac{8}{3}$.
The minimum is $f\left(\frac{2}{3}, \frac{10}{3}, \frac{8}{3}\right) = \frac{56}{3}$.

47. $f(x, y, z) = xy + xz$
$g(x, y, z) = 2x + 3z = 5$
$h(x, y, z) = xy = 4$
$f_x = y + z, \ f_y = x, \ f_z = x$
$g_x = 2, \ g_y = 0, \ g_z = 3$
$h_x = y, \ h_y = x, \ h_z = 0$
Solve the system

$$\begin{cases} y + z = 2\lambda + \mu y \\ x = \mu x \\ x = 3\lambda \\ 2x + 3z = 5 \\ xy = 4 \end{cases}$$

to find $\mu = 1, \ \lambda = \frac{5}{12}$, and then $x = \frac{5}{4}$,
$y = \frac{16}{5}, z = \frac{5}{6}$.
The maximum is $f\left(\frac{5}{4}, \frac{16}{5}, \frac{5}{6}\right) = \frac{121}{24}$.

49. PROFIT = REVENUE − COST
$$P(x, y) = \left(\frac{320y}{y+2} + \frac{160x}{x+4}\right)150$$
$$- \left[\left(\frac{320y}{y+2} + \frac{160x}{x+4}\right)50 + 1{,}000(x + y)\right]$$

a. $g(x, y) = x + y - 8$
$$P_x = \frac{100(160)(4)}{(x+4)^2} - 1{,}000$$
$$P_y = \frac{100(320)(2)}{(y+2)^2} - 1{,}000$$
$g_x = 1, g_y = 1$
Since $P_x = P_y = \lambda$,

$$\frac{100(160)(4)}{(x+4)^2} - 1{,}000 = \frac{100(320)(2)}{(y+2)^2} - 1{,}000$$
$$(x+4)^2 = (y+2)^2$$
$$x + 4 = \pm(y + 2)$$

Reject the negative solution as leading to negative spending. Substituting $y = x + 2$ in the constraint equation $x + y = 8$ leads to $x = 3$ thousand dollars for development and $y = 5$ thousand dollars for promotion.

b. $\lambda = P_y = \dfrac{64{,}000}{49} - 1{,}000 \approx 306.122$
(for each $1,000). Since the change in this promotion/development is $100, the corresponding increase in profit is $30.61. Remember that the Lagrange multiplier is the change in maximum profit for a unit (one thousand dollar) change in the constraint. The actual increase in profit is $29.68.

c. To maximize the profit when unlimited funds are available maximize $P(x, y)$ without constraints. To do this, find the critical points by setting $P_x = 0$ and $P_y = 0$, that is

P_x: $\dfrac{64}{(x+4)^2} - 1 = 0$

$(x+4)^2 = 64$

$x = 4$ *Reject negative*

P_y: $\dfrac{64}{(y+2)^2} - 1 = 0$

$(y+2)^2 = 64$

$y = 6$ *Reject negative*

Thus, \$4,000 should be spent on development and \$6,000 should be spent on promotion to maximize profit.

d. If there were a restriction on the amount spent on development and promotion, then constraints would be $g(x, y) = x + y = k$ for some positive constant k. The corresponding Lagrange equations would be

$\dfrac{64}{(x+4)^2} - 1 = \lambda \qquad \dfrac{64}{(y+2)^2} - 1 = \lambda$

and $x + y = k$. To obtain the answer in part **c**, eliminate λ. Beginning with the Lagrange equations from part **c**, set $\lambda = 0$ to obtain

$\dfrac{64}{(x+4)^2} - 1 = 0$

$(x+4)^2 = 64$

$x = 4$ *Reject negative*

Similarly, $y = 6$, just as we found in part **c**.

51. $P = 2pq + (2p+2q)(1-p-q)$

$= -2p^2 - 2pq - 2q^2 + 2p + 2q$

$P_p = -4p - 2q + 2$

$P_q = -2p - 4q + 2$

To maximize P, we want $P_p = P_q = 0$, so

$\begin{cases} -4p - 2q + 2 = 0 \\ -2p - 4q + 2 = 0 \end{cases}$

Solving this systems of equations, we obtain $p = q = \frac{1}{3}$, and thus $r = 1 - p - q = \frac{1}{3}$. To form the discriminant, we find

$P_{pp} = -4, \; P_{qq} = -4, \; P_{pq} = -2$

so $D = (-4)(-4) - (-2)^2 > 0$ and since $P_{pp} = -4 < 0$, it follows that the critical point $\left(\frac{1}{3}, \frac{1}{3}\right)$ corresponds to a maximum. Finally,

$P_{max} = P\left(\frac{1}{3}, \frac{1}{3}, \frac{1}{3}\right)$

$= 2\left(\frac{1}{3}\right)\left(\frac{1}{3}\right) + 2\left(\frac{1}{3}\right)\left(\frac{1}{3}\right) + 2\left(\frac{1}{3}\right)\left(\frac{1}{3}\right)$

$= \frac{2}{3}$

53. Let x, y, z be the length, width, and height of one-eighth of the rectangular box, respectively.

$f(x, y, z) = xyz;$

$g(x, y, z) = \dfrac{x^2}{a^2} + \dfrac{y^2}{b^2} + \dfrac{z^2}{c^2} = 1$

$f_x = yz, \; f_y = xz, \; f_z = xy$

$g_x = \dfrac{2x}{a^2}, \; g_y = \dfrac{2y}{b^2}, \; g_z = \dfrac{2z}{c^2}$

Solve the system of equations:

$$\begin{cases} yz = \dfrac{2x}{a^2}\lambda \\[2mm] xz = \dfrac{2y}{b^2}\lambda \\[2mm] xy = \dfrac{2z}{c^2}\lambda \end{cases}$$

Solving simultaneously,

$$x = \frac{a}{\sqrt{3}}, y = \frac{b}{\sqrt{3}}, z = \frac{c}{\sqrt{3}}$$

The maximum volume is

$$8f\left(\frac{a}{\sqrt{3}}, \frac{b}{\sqrt{3}}, \frac{c}{\sqrt{3}}\right) = \frac{8abc}{3\sqrt{3}} \text{ cubic units}$$

55. The goal is to minimize cost $C = px + qy$ subject to the fixed production function $Q(x, y) = Q_0$. Since $C_x = p$ and $C_y = q$, the three Lagrange equations are

$$\begin{cases} p = \lambda Q_x \\ q = \lambda Q_y \\ Q(x, y) = Q_0 \end{cases}$$

Solving this system leads to

$$\frac{Q_x}{p} = \frac{Q_y}{q}$$

Since $Q = f(x, y)$, we have

$$\frac{f_x}{p} = \frac{f_y}{q}$$

$$yz = \lambda(1); \; xz = \lambda(1); \; xy = \lambda(1)$$

Solving this system we find $x = y = z = C/3$, and the maximum value of P is

57. Maximize $Q(x, y) = cx^\alpha y^\beta$ subject to $C(x, y) = px + qy = k$. We have $Q_x = x^{\alpha-1}y^\beta$; $Q_y = c\beta x^\alpha y^{\beta-1}$; $C_x = p$; $C_y = q$. We must solve the system of equations

$$\begin{cases} c\alpha x^{\alpha-1}y^\beta = \lambda(p) \\ c\beta x^\alpha y^{\beta-1} = \lambda(q) \\ px + qy = k \end{cases}$$

a. Solving simultaneously,

$$x = \frac{k\alpha}{p(\alpha + \beta)} = \frac{k\alpha}{p} \quad \text{and} \quad y = \frac{k\beta}{q}$$

b. If we drop the condition $\alpha + \beta = 1$, the maximum occurs at

$$x = \frac{k\alpha}{p(\alpha + \beta)}, y = \frac{k\beta}{q(\alpha + \beta)}$$

If k is increased by 1 unit, the maximum output increases by

$$\frac{dQ}{dk} = \lambda = \frac{c\alpha\left[\frac{k\alpha}{p(\alpha+\beta)}\right]^{\alpha-1}\left[\frac{k\beta}{q(\alpha+\beta)}\right]^\beta}{p}$$

$$= c\left(\frac{k}{\alpha + \beta}\right)^{\alpha+\beta-1}\left(\frac{\alpha}{p}\right)^\alpha\left(\frac{\beta}{q}\right)^\beta$$

59. a. Maximize $P = xyz$ subject to $x + y + z = C$. The Lagrange equations are

$$P\left(\frac{C}{3}, \frac{C}{3}, \frac{C}{3}\right) = \frac{1}{27}C^3$$

In other words,

$$\frac{C^3}{27} \geq xyz$$

for all x, y, z such that $x + y + z = C$, which means so that

$$\frac{(x+y+z)^3}{27} \geq xyz$$

$$\frac{x+y+z}{3} \geq \sqrt[3]{xyz}$$

$$A(x, y, z) \geq G(x, y, z)$$

Thus, $G(x, y, z) \leq A(x, y, z)$.

b. The geometric mean of n positive numbers $x_1, x_2, \cdots, x_n$ is $G = (x_1 x_1 \cdots x_n)^{1/n}$ and the arithmetic mean is $A = \frac{1}{n}(x_1 + x_2 + \cdots + x_n)$. Then,

$$G(x_1, x_2, \cdots, x_n) \leq A(x_1, x_2, \cdots, x_n)$$

for all x_i.

Chapter 11 Review

 Studying for a chapter examination is a personal process, one which nobody else can do for you. Simply take the time to review what you have done.

SURVIVAL HINT: Work all of Chapter 11 problems in the Proficiency Examination (whether they are assigned or not). Work through all of the problems before looking at the answers, and *then* correct each of the problems. The answers to all these problems are given in the answer section at the back of the text. If you worked the problem correctly, move on to the next problem, but if you did not work it correctly (or you did not know what to do), then look at the solutions below, look back in the chapter to study the procedure, or ask your instructor.

Finally, go back over the homework problems you have been assigned. If you worked a problem correctly, move on to the next problem, but if you missed it on your homework, then you should look back in the book or talk to your instructor about how to work the problem.

If you follow these steps, you should be successful with your review of this chapter.

Proficiency Examination, page 908

1. a. A function of two variables is a rule that assigns to each ordered pair (x, y) in a set D a unique number $f(x, y)$.

b. The set D is called the domain of the function, and the corresponding values of $f(x, y)$ constitute the range of f.

2. The notation

$$\lim_{(x,y)\to(x_0,y_0)} f(x, y) = L$$

means that the functional values $f(x, y)$ can be made arbitrarily close to L by choosing a point (x, y) sufficiently close (but not equal) to the point (x_0, y_0). In other words, given some $\epsilon > 0$, we wish to find a $\delta > 0$ so that for any point (x, y) in the punctured disk of radius δ centered at (x_0, y_0), the functional value $f(x, y)$ lies between $L - \epsilon$ and $L + \epsilon$.

3. Suppose $\lim_{(x,y)\to(x_0,y_0)} = L$ and

$\lim\limits_{(x,y)\to(x_0,y_0)} g(x,y) = M$, then, for a constant a:

a. $\lim\limits_{(x,y)\to(x_0,y_0)} [af](x,y) = aL$

b. $\lim\limits_{(x,y)\to(x_0,y_0)} [f+g](x,y)$

$= \left[\lim\limits_{(x,y)\to(x_0,y_0)} f(x,y)\right] + \left[\lim\limits_{(x,y)\to(x_0,y_0)} g(x,y)\right]$

$= L + M$

c. $\lim\limits_{(x,y)\to(x_0,y_0)} [fg](x,y)$

$= \left[\lim\limits_{(x,y)\to(x_0,y_0)} f(x,y)\right]\left[\lim\limits_{(x,y)\to(x_0,y_0)} g(x,y)\right]$

$= LM$

d. $\lim\limits_{(x,y)\to(x_0,y_0)} \left[\dfrac{f}{g}\right](x,y)$

$= \dfrac{\lim\limits_{(x,y)\to(x_0,y_0)} f(x,y)}{\lim\limits_{(x,y)\to(x_0,y_0)} g(x,y)}$

$= \dfrac{L}{M} \qquad (M \neq 0)$

4. The function $f(x,y)$ is continuous at the point (x_0, y_0) if and only if

1. $f(x_0, y_0)$ is defined;
2. $\lim\limits_{(x,y)\to(x_0,y_0)} f(x,y)$ exists;
3. $\lim\limits_{(x,y)\to(x_0,y_0)} f(x,y) = f(x_0, y_0)$

Also, f is continuous on a set S in its domain if it is continuous at each point in S.

5. a. If $z = f(x,y)$, then the (first) partial derivatives of f with respect to x and y are the functions f_x and f_y, respectively, defined by

$f_x(x,y) = \lim\limits_{\Delta x\to 0} \dfrac{f(x+\Delta x, y) - f(x,y)}{\Delta x}$

$f_y(x,y) = \lim\limits_{\Delta y\to 0} \dfrac{f(x, y+\Delta y) - f(x,y)}{\Delta y}$

provided the limits exist.

b. $z = f(x,y); \dfrac{\partial^2 f}{\partial x^2}, \dfrac{\partial^2 f}{\partial y^2}, \dfrac{\partial^2 f}{\partial x\partial y}, \dfrac{\partial^2 f}{\partial y\partial x}$

or $f_{xx}, f_{yy}, f_{yx}, f_{xy}$

c. Let $z = f(x,y);$

$\Delta z = \dfrac{\partial f}{\partial x}\Delta x + \dfrac{\partial f}{\partial y}\Delta y$

where $\Delta x = dx$, $\Delta y = dy$, and $\Delta z = f_x\Delta x + f_y\Delta y$.

6. The line tangent at $P_0(x_0, y_0, z_0)$ to the trace of $z = f(x,y)$ in the plane $y = y_0$ has slope $f_x(x_0, y_0)$. Likewise, the line tangent at P_0 to the trace of $z = f(x,y)$ in the plane $x = x_0$ has slope $f_y(x_0, y_0)$.

7. Suppose $f(x,y)$ is defined at each point in a circular disk that is centered at (x_0, y_0) and contains the point $(x_0 + \Delta x, y_0 + \Delta y)$. Then f is said to be differentiable at (x_0, y_0) if, the increment of f can be expressed as

$$\Delta f = f_x(x_0,y_0)\Delta x + f_y(x_0,y_0)\Delta y + \epsilon_1\Delta x + \epsilon_2\Delta y$$

where $\epsilon_1 \to 0$ and $\epsilon_2 \to 0$ as both $\Delta x \to 0$ and $\Delta y \to 0$ (and $\epsilon_1 = \epsilon_2 = 0$ when $\Delta x = \Delta y = 0$). Also, $f(x,y)$ is said to be differentiable on the region R of the plane if f is differentiable at each point in R.

8. a. If $f(x,y)$ and its partial derivatives f_x and f_y are defined in an open region R containing the point $P(x_0, y_0)$ and f_x and f_y are continuous at P, then

$\Delta f = f(x_0 + \Delta x, y_0 + \Delta y) - f(x_0, y_0)$
$\approx f_x(x_0, y_0)\Delta x + f_y(x_0, y_0)\Delta y$

so that

$$f(x_0 + \Delta x, y_0 + \Delta y)$$
$$\approx f(x_0, y_0) + f_x(x_0, y_0)\Delta x + f_y(x_0, y_0)\Delta y$$

b. If $z = f(x, y)$ and Δx and Δy are increments of x and y, respectively, and if we let $dx = \Delta x$ and $dy = \Delta y$ be differentials for x and y, respectively, then the total differential of $f(x, y)$ is

$$df = \frac{\partial f}{\partial y}dx + \frac{\partial f}{\partial y}dy$$
$$= f_x(x, y)\, dx + f_y(x, y)\, dy$$

9. a. Let $f(x, y)$ be a differentiable function of x and y, and let $x = x(t)$ and $y = y(t)$ be differentiable functions of t. Then $z = f(x, y)$ is a differentiable function of t, and

$$\frac{dz}{dt} = \frac{\partial z}{\partial x}\frac{dx}{dt} + \frac{\partial z}{\partial y}\frac{dy}{dt}$$

b. Suppose $z = f(x, y)$ is differentiable at (x, y) and that the partial derivatives of $x = x(u, v)$ and $y = y(u, v)$ exist at (u, v). Then the composite function $z = f[x(u, v), y(u, v)]$ is differentiable at (u, v) with

$$\frac{\partial z}{\partial u} = \frac{\partial z}{\partial x}\frac{\partial x}{\partial u} + \frac{\partial z}{\partial y}\frac{\partial y}{\partial u}$$

and

$$\frac{\partial z}{\partial v} = \frac{\partial z}{\partial x}\frac{\partial x}{\partial v} + \frac{\partial z}{\partial y}\frac{\partial y}{\partial v}$$

10. Let f be a function of two variables, and let $\mathbf{u} = u_1\mathbf{i} + u_2\mathbf{j}$ be a unit vector. The directional derivative of f at $P_0(x_0, y_0)$ in the direction of $\mathbf{u}$ is given by

$$D_{\mathbf{u}}f(x_0, y_0) = \lim_{h \to 0}\frac{f(x_0 + hu_1, y_0 + hu_2) - f(x_0, y_0)}{h}$$

provided the limit exists.

11. a. Let f be a differentiable function at (x, y) and let $f(x, y)$ have partial derivatives $f_x(x, y)$ and $f_y(x, y)$. Then the gradient of f, denoted by ∇f, is the vector given by

$$\nabla f(x, y) = f_x(x, y)\mathbf{i} + f_y(x, y)\mathbf{j}$$

b. If f is a differentiable function of x and y, then the directional derivative at the point $P_0(x_0, y_0)$ in the direction of the unit vector $\mathbf{u}$ is

$$D_{\mathbf{u}}f(x, y) = \nabla f \cdot \mathbf{u}$$

c. Suppose the function f is differentiable at the point P_0 and that the gradient at P_0 satisfies $\nabla f_0 \neq \mathbf{0}$. Then ∇f_0 is orthogonal to the level surface $f(x, y, z) = K$ at P_0.

12. Let f and g be differentiable functions. Then
a. $\nabla c = \mathbf{0}$ for any constant c
b. $\nabla(af + bg) = a\nabla f + b\nabla g$
c. $\nabla(fg) = f\nabla g + g\nabla f$
d. $\nabla\left(\dfrac{f}{g}\right) = \dfrac{g\nabla f - f\nabla g}{g^2}$ $g \neq 0$
e. $\nabla(f^n) = nf^{n-1}\nabla f$

13. Suppose f is differentiable and let ∇f_0 denote the gradient at P_0. Then if $\nabla f_0 \neq \mathbf{0}$:
(1) The largest value of the directional derivative of $D_{\mathbf{u}}f$ is $\|\nabla f_0\|$ and occurs when the unit vector $\mathbf{u}$ points in the direction of ∇f_0.
(2) The smallest value of $D_{\mathbf{u}}f$ is $-\|\nabla f_0\|$ and occurs when $\mathbf{u}$ points in the direction of $-\nabla f_0$.

14. Suppose the surface S has a nonzero normal vector $\mathbf{N}$ at the point P_0. Then the line through P_0 parallel to $\mathbf{N}$ is called the normal line to S at P_0, and the plane through P_0 with normal vector $\mathbf{N}$ is the tangent plane to S at P_0.

15. a. The function $f(x,y)$ is said to have an absolute maximum at (x_0, y_0) if $f(x_0, y_0) \geq f(x,y)$ for all (x,y) in the domain D of f. Similarly, f has an absolute minimum at (x_0, y_0) if $f(x_0, y_0) \leq f(x,y)$ for all (x,y) in D. Collectively, absolute maxima and minima are called absolute extrema.

b. Let f be a function defined at (x_0, y_0). Then $f(x_0, y_0)$ is a relative maximum if $f(x,y) \leq f(x_0, y_0)$ for all (x,y) in an open disk containing (x_0, y_0). $f(x_0, y_0)$ is a relative minimum if $f(x,y) \geq f(x_0, y_0)$ for all (x,y) in an open disk containing (x_0, y_0). Collectively, relative maxima and minima are called relative extrema.

c. A critical point of a function f defined on an open set S is a point (x_0, y_0) in S where either one of the following is true:
(1) $f_x(x_0, y_0) = f_y(x_0, y_0) = 0$.
(2) $f_x(x_0, y_0)$ or $f_y(x_0, y_0)$ does not exist (one or both).

16. Let $f(x,y)$ have a critical point at $P_0(x_0, y_0)$ and assume that f has continuous partial derivatives in a disk centered at (x_0, y_0). Let

$$D = f_{xx}(x_0, y_0)f_{yy}(x_0, y_0) - [f_{xy}(x_0, y_0)]^2$$

Then, a relative maximum occurs at P_0 if

$$D > 0 \text{ and } f_{xx}(x_0, y_0) < 0$$

A relative minimum occurs at P_0 if

$$D > 0 \text{ and } f_{xx}(x_0, y_0) > 0$$

A saddle point occurs at P_0 if $D < 0$. If $D = 0$, then the test is inconclusive.

17. A function of two variables $f(x,y)$ assumes an absolute extremum on any closed, bounded set S in the plane where it is continuous. Moreover, all absolute extrema must occur either on the boundary of S or at critical points in the interior of S.

18. Given a set of data points (x_k, y_k), a line $y = mx + b$, called a regression line, is obtained by minimizing the sum of squares of distances $y_k - (mx_k + b)$.

19. Assume that f and g have continuous first partial derivatives and that f has an extremum at $P_0(x_0, y_0)$ on the smooth constraint curve $g(x,y) = c$. If $\nabla g(x_0, y_0) \neq \mathbf{0}$, there is a number λ such that

$$\nabla f(x_0, y_0) = \lambda \nabla g(x_0, y_0)$$

20. Suppose f and g satisfy the hypotheses of Lagrange's theorem, and suppose that $f(x,y)$ has an extremum (minimum and/or a maximum) subject to the constraint $g(x,y) = c$. Then to find the extreme values, proceed as follows:
1. Simultaneously solve the following three equations:
$$\begin{cases} f_x(x,y) = \lambda g_x(x,y) \\ f_y(x,y) = \lambda g_y(x,y) \\ g(x,y) = c \end{cases}$$

2. Evaluate f at all points found in Step 1. The largest of these values is the maximum value of f and the smallest of these values is the minimum value of f.

21. $f(x, y) = \sin^{-1} xy$

Recall $\dfrac{d}{dx} \sin^{-1} u = \dfrac{1}{\sqrt{1-u^2}} \dfrac{du}{dx}$;

$f_x = \dfrac{y}{\sqrt{1-x^2 y^2}}$; $f_y = \dfrac{x}{\sqrt{1-x^2 y^2}}$

$f_{xy} = \dfrac{\sqrt{1-x^2 y^2}(1) - \frac{y(-2x^2 y)}{2\sqrt{1-x^2 y^2}}}{1-x^2 y^2}$

$= \dfrac{1-x^2 y^2 + x^2 y^2}{(1-x^2 y^2)^{3/2}}$

$= \dfrac{1}{(1-x^2 y^2)^{3/2}}$

$f_{yx} = \dfrac{\sqrt{1-x^2 y^2}(1) - \frac{x(-2x^2 y)}{2\sqrt{1-x^2 y^2}}}{1-x^2 y^2}$

$= \dfrac{1-x^2 y^2 + x^2 y^2}{(1-x^2 y^2)^{3/2}}$

$= \dfrac{1}{(1-x^2 y^2)^{3/2}}$

22. $\dfrac{dw}{dt} = \dfrac{\partial w}{\partial x}\dfrac{dx}{dt} + \dfrac{\partial w}{\partial y}\dfrac{dy}{dt} + \dfrac{\partial w}{\partial z}\dfrac{dz}{dt}$

$= 2xy(t\cos t + \sin t)$
$\quad + (x^2 + 2yz)(-t\sin t + \cos t) + y^2(2)$

If $t = \pi$, then $x = 0$, $y = -\pi$, $z = 2\pi^2$

and $\dfrac{dw}{dt} = 0 - 2(-\pi)(2\pi) + 2\pi^2$

$= 6\pi^2$

23. $f(x, y, z) = xy + yz + xz$ at $(1, 2, -1)$

a. $\nabla f = (y+z)\mathbf{i} + (x+z)\mathbf{j} + (y+x)\mathbf{k}$
$\nabla f = \mathbf{i} + 3\mathbf{k}$ at $P_0(1, 2, -1)$

b. $\mathbf{u} = \dfrac{P_0 Q}{\|P_0 Q\|} = \dfrac{-2\mathbf{i} - \mathbf{j}}{\sqrt{5}}$

$D_{\mathbf{u}}(f) = \nabla f \cdot \mathbf{u} = \dfrac{-2}{\sqrt{5}} = \dfrac{-2\sqrt{5}}{5}$

c. The directional derivative has its greatest value in the direction of the gradient, $\mathbf{u} = \dfrac{\mathbf{i} + 3\mathbf{k}}{\sqrt{10}}$. The magnitude is $\|\nabla f\| = \sqrt{10}$.

24. $\displaystyle\lim_{(x,y)\to(0,0)} f(x, y)$ along the line $y = x$ is

$\displaystyle\lim_{x\to 0} \dfrac{x^3}{x^3 + x^3} = \dfrac{1}{2}$. The limit does not equal $f(0, 0)$ so the function is not continuous at $(0, 0)$.

25. $f(x, y) = \ln\dfrac{y}{x} = \ln y - \ln x$.

$f_x = -\dfrac{1}{x}$, $f_y = \dfrac{1}{y}$,

$f_{xx} = \dfrac{1}{x^2}$, $f_{yy} = -\dfrac{1}{y^2}$, $f_{xy} = 0$

26. $f(x, y, z) = x^2 y + y^2 z + z^2 x$
$f_x = 2xy + z^2$, $f_y = x^2 + 2yz$,
$f_z = y^2 + 2zx$

$\dfrac{\partial f}{\partial x} + \dfrac{\partial f}{\partial y} + \dfrac{\partial f}{\partial z}$

$= 2xy + z^2 + x^2 + 2yz + y^2 + 2zx$

$= (x + y + z)^2$

27. $f(x, y) = x^4 + 2x^2 y^2 + y^4$
$\nabla f = (4x^3 + 4xy^2)\mathbf{i} + (4x^2 y + 4y^3)\mathbf{j}$
$\nabla f_0 = \nabla f(2, -2) = 64(\mathbf{i} - \mathbf{j})$
A unit vector in the direction of $\frac{2\pi}{3}$ is

$\mathbf{u} = -\frac{1}{2}\mathbf{i} + \frac{\sqrt{3}}{2}\mathbf{j}$
$D_{\mathbf{u}}(f) = \nabla f_0 \cdot \mathbf{u}$

$= -32 - 32\sqrt{3}$

28. $f(x, y) = 12xy - 2x^2 - y^4$

$f_x = 12y - 4x$; $f_y = 12x - 4y^3$
$f_x = f_y = 0$ when $x = 3y$ and $36y = 4y^3$;
critical points are $(0,0)$, $(9,3)$ and
$(-9,-3)$.
$f_{xx} = -4$; $f_{xy} = 12$; $f_{yy} = -12y^2$

x	y	f_{xx}	f_{xy}	f_{yy}	D	Classify
0	0	-4	12	0	$-$	saddle point
9	3	-4	12	-108	$+$	rel maximum
-9	-3	-4	12	-108	$+$	rel maximum

29. $f(x,y) = x^2 + 2y^2 + 2x + 3$, subject to
$g(x,y) = x^2 + y^2 = 4$.
Solve the system
$$\begin{cases} 2x + 2 = 2x\lambda \\ 4y = 2y\lambda \\ x^2 + y^2 = 4 \end{cases}$$
to find $\lambda = 2$ or $y = 0$.
If $\lambda = 2$, then $x = 1$, $y = \pm\sqrt{3}$.
If $y = 0$, then $x = \pm 2$.
Then:
$$f\left(1, \sqrt{3}\right) = 12$$
$$f\left(1, -\sqrt{3}\right) = 12$$
$$f(2,0) = 11$$
$$f(-2,0) = 3$$
$f(x,y)$ has a maximum of 12 at $\left(1, \pm\sqrt{3}\right)$
and a minimum of 3 at $(-2,0)$.

30. $f(x,y) = x^2 - 4y^2 + 3x + 6y$
$f_x = 2x + 3$; $f_y = -8y + 6$
$f_x = f_y = 0$ if $x = -\frac{3}{2}$, $y = \frac{3}{4}$
For the boundary:
On $y = 0$, $-2 \le x \le 2$, $f_1 = x^2 + 3x$
$\quad f_1' = 2x + 3$
$\quad f_1' = 0$ when $x = -\frac{3}{2}$; point $\left(-\frac{3}{2}, 0\right)$
On $x = 2$, $0 \le y \le 1$,

$f_2 = -4y^2 + 6y + 10$
$f_2' = -8y + 6$
$f_2' = 0$ when $y = \frac{3}{4}$; point $\left(2, \frac{3}{4}\right)$
On $y = 1$, $-2 \le x \le 2$,
$\quad f_3 = x^2 + 3x + 2$
$\quad f_3' = 2x + 3$
$\quad f_3' = 0$ when $x = -\frac{3}{2}$; point $\left(-\frac{3}{2}, 1\right)$
On $x = -2$, $0 \le y \le 1$,
$\quad f_4 = -4y^2 + 6y - 2$
$\quad f_4' = -8y + 6$
$\quad f_4' = 0$ when $y = \frac{3}{4}$; point $\left(-2, \frac{3}{4}\right)$
Check these points, along with the
corner points.

x	y	$f(x,y)$	Classify
$-\frac{3}{2}$	$\frac{3}{4}$	0	
$-\frac{3}{2}$	0	$-\frac{9}{4}$	minimum
2	$\frac{3}{4}$	$\frac{49}{4}$	maximum
$-\frac{3}{2}$	1	$-\frac{1}{4}$	
-2	$\frac{3}{4}$	$\frac{1}{4}$	
-2	0	-2	
-2	0	10	
2	1	12	

The largest value of f is $\frac{49}{4}$ at $\left(2, \frac{3}{4}\right)$
and the smallest is $-\frac{9}{4}$ at $\left(-\frac{3}{2}, 0\right)$.

Supplementary Problems, page 909

1. $f(x,y) = \sqrt{16 - x^2 - y^2}$
$\quad 16 - x^2 - y^2 \ge 0$
$\qquad x^2 + y^2 \le 4$
The domain consists of the circle with
center at the origin, radius 4, and its
interior.

3. $f(x,y) = \sin^{-1}x + \cos^{-1}y$
The domain is $-1 \le x \le 1$, $-1 \le y \le 1$

5. $f(x,y) = \dfrac{x^2 - y^2}{x + y} = x - y$, $x \ne y$
$\quad f_x = 1$; $f_y = -1$

33. $\displaystyle\iint_R x\ln(xy)\,dA$

$\displaystyle= \int_1^2\int_1^e x\ln(xy)\,dy\,dx$

$\displaystyle= \int_1^2 [xy(\ln xy - 1)]\big|_1^e\,dx$

$\displaystyle= \int_1^2 x[(e-1)\ln x + 1]\,dx$

$\displaystyle= \frac{1}{4}x^2[2(e-1)\ln x - (e-3)]\Big|_1^2$

$\displaystyle= 2(e-1)\ln 2 - \frac{3}{4}(e-3)$

35. $\displaystyle\iint_R \sqrt{xy}\,dA$

$\displaystyle= \int_0^1\int_0^4 x^{1/2}y^{1/2}\,dy\,dx$

$\displaystyle= \frac{2}{3}\int_0^1 x^{1/2}(8)\,dx$

$\displaystyle= \frac{32}{9}$

37. $\displaystyle\iint_R \frac{xy}{\sqrt{x^2+y^2+1}}\,dA$

$\displaystyle= \int_0^1\int_0^1 \frac{xy}{\sqrt{x^2+y^2+1}}\,dy\,dx$

$\boxed{u = x^2+y^2+1;\ du = 2y\,dy}$

$\displaystyle= \int_0^1 x\sqrt{x^2+y^2+1}\Big|_0^1\,dx$

$\displaystyle= \int_0^1 x\left[\sqrt{x^2+2} - \sqrt{x^2+1}\right]dx$

$\displaystyle= \frac{1}{3}\left[(x^2+2)^{3/2} - (x^2+1)^{3/2}\right]\Big|_0^1$

$\displaystyle= \sqrt{3} - \frac{4}{3}\sqrt{2} + \frac{1}{3}$

$\displaystyle\iint_R \frac{xy}{\sqrt{x^2+y^2+1}}\,dA$

$\displaystyle= \int_0^1\int_0^1 \frac{xy}{\sqrt{x^2+y^2+1}}\,dy\,dx$

$\boxed{u = x^2+y^2+1;\ du = 2y\,dy}$

$\displaystyle= \int_0^1 x\sqrt{x^2+y^2+1}\Big|_0^1\,dx$

$\displaystyle= \int_0^1 x\left[\sqrt{x^2+2} - \sqrt{x^2+1}\right]dx$

$\displaystyle= \frac{1}{3}\left[(x^2+2)^{3/2} - (x^2+1)^{3/2}\right]\Big|_0^1$

$\displaystyle= \sqrt{3} - \frac{4}{3}\sqrt{2} + \frac{1}{3}$

39. $\displaystyle\iint_R (x+y)^5\,dA$

$\displaystyle= \int_0^1\int_0^1 (x+y)^5\,dy\,dx$

$\displaystyle= \frac{1}{6}\int_0^1\left[(x+1)^6 - x^6\right]dx$

$\displaystyle= 3$

41. $\displaystyle\iint_R (x\cos y + y\sin x)\,dA$

$\displaystyle= \int_0^{\pi/2}\int_0^{\pi/2} (x\cos y + y\sin x)\,dy\,dx$

$\displaystyle= \int_0^{\pi/2}\left(x + \frac{\pi^2}{8}\sin x\right)dx$

$\displaystyle= \frac{\pi^2}{4}$

43. $\displaystyle T = \iint_R 20{,}000\,dA$

45. $\displaystyle M = \iint_R \rho(x,y)\,dA$

47. $\displaystyle\int_0^2 \int_0^1 x(1-x^2)^{1/2} e^{3y}\, dx\, dy$

$\displaystyle = \frac{1}{3}\int_0^2 e^{3y}\, dy$

$\displaystyle = \frac{1}{9}(e^6 - 1)$

49. $\displaystyle\int_1^3 \int_1^2 \frac{xy}{x^2+y^2}\, dy\, dx$

$\displaystyle = \frac{1}{2}\int_1^3 \left[x\ln(x^2+4) - x\ln(x^2+1)\right] dx$

$\displaystyle = \frac{1}{4}\left[(x^2+4)\ln(x^2+4) - (x^2+1)\ln(x^2+1) - 3\right]\Big|_1^3$

$\displaystyle = \frac{1}{4}[13\ln 13 - 15\ln 5 - 8\ln 2]$

51. $z = f(x,y) = 4 - x^2 - y^2$ is a paraboloid opening downward with vertex at $z = 4$ and intercepts at $(\pm 2, 0, 0))$ and $(0, \pm 2, 0)$. The integral represents the volume above the unit square in the first quadrant. The minimum value for z is $f(1,1) = 2$. Since $z \geq 2$ over the given region R, the value of the integral will be greater than the volume of the "box" with unit base and height 2.

53. Approximations may vary depending on the evaluation of z_k over each cell of the grid.

$$A \approx (0.25)^2 \sum_{k=1}^{16} z_k$$

$$\approx 0.0625(23)$$

$$\approx 1.44$$

55. From a mathematical dictionary: *Differential projective geometry* is the theory of the differential properties of configurations which are invariant under projective transformations. Projective geometry is different from Euclidean geometry, where properties of length, angle, and area are measured and compared. In projective geometry, length, shape, angles, area, similarity, and congruence are not preserved under projection, and every pair of lines in the plane intersect.

57. $\displaystyle\iint_R \frac{\partial}{\partial y}\left[\frac{\partial f(x,y)}{\partial x}\right] dA$

$\displaystyle = \int_{x_1}^{x_2}\int_{y_1}^{y_2} \frac{\partial}{\partial y}\left[\frac{\partial f(x,y)}{\partial x}\right] dy\, dx$

$\displaystyle = \int_{x_1}^{x_2} \frac{\partial}{\partial x}f(x,y)\Big|_{y=y_1}^{y=y_2} dx$

$\displaystyle = \int_{x_1}^{x_2}\left[\frac{\partial f(x,y_2)}{\partial x} - \frac{\partial f(x,y_1)}{\partial x}\right] dx$

$\displaystyle = f(x,y_2) - f(x,y_1)\big|_{x=x_1}^{x=x_2}$

$\displaystyle = f(x_2,y_2) - f(x_1,y_2) - f(x_2,y_1) + f(x_1,y_1)$

59. $\displaystyle\int_0^1\int_0^1 \frac{y-x}{(x+y)^3}\, dx\, dy = \frac{1}{2}$

$\displaystyle\int_0^1\int_0^1 \frac{y-x}{(x+y)^3}\, dy\, dx = -\frac{1}{2}$

Fubini's theorem does not apply because the integrand is not continuous at $(0,0)$.

12.2 Double Integration over Nonrectangular Regions, page 938

1. If $f(x,y) \geq 0$ on R, then the volume of the region under $z = f(x,y)$ and above R is given by

$$V = \iint_R f(x,y)\, dA$$

3. Vertically simple:

$$\int_0^4 \int_0^{4-x} xy \, dy \, dx$$

$$= \int_0^4 \frac{x}{2}(4-x)^2 \, dx$$

$$= \frac{32}{3}$$

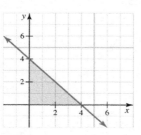

5. Horizontally simple:

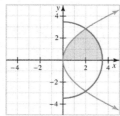

$$\int_0^{2\sqrt{3}} \int_{y^2/4}^{\sqrt{12-y^2}} dx \, dy$$

$$= \int_0^{2\sqrt{3}} \left(\sqrt{12-y^2} - \frac{y^2}{4} \right) dy \quad \boxed{\text{Formula 117}}$$

$$= \left(\frac{y}{2}\sqrt{12-y^2} + 6\sin^{-1}\frac{y}{2\sqrt{3}} - \frac{y^3}{12} \right)_0^{2\sqrt{3}}$$

$$= 3\pi - 2\sqrt{3}$$

7. Vertically simple:

$$\int_0^1 \int_0^x (x^2 + 2y^2) \, dy \, dx$$

$$= \int_0^1 \left(x^3 + \frac{2x^3}{3} \right) dx$$

$$= \frac{5}{12}$$

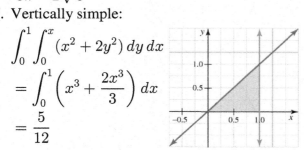

9. Vertically simple:

$$\int_0^2 \int_0^{\sin x} y \cos x \, dy \, dx$$

$$= \frac{1}{2} \int_0^2 \sin^2 x \cos x \, dx$$

$$= \frac{1}{6} \sin^3 2$$

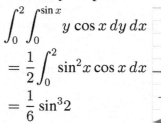
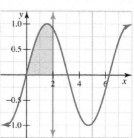

11. Horizontally simple:

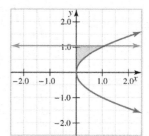

$$\int_0^{\pi/3} \int_0^{y^2} \frac{1}{y}\sin\frac{x}{y} \, dx \, dy$$

$$= \int_0^{\pi/3} \left(-\cos\frac{x}{y} \right)\Big|_0^{y^2} dy$$

$$= \int_0^{\pi/3} (1 - \cos y) \, dy$$

$$= \frac{\pi}{3} - \frac{\sqrt{3}}{2}$$

13. Horizontally simple;

$$\int_0^1 \int_0^{y^3} e^{x/y} dx \, dy = \int_0^1 y e^{x/y}\Big|_0^{y^3} dy$$

$$= \int_0^1 y \left(e^{y^2} - 1 \right) dy$$

$$= \left[\frac{1}{2} e^{y^2} - \frac{y^2}{2} \right]_0^1$$

$$= \frac{1}{2}e - 1$$

15. Vertically simple (change order);

$$\int_0^1 \int_{\tan^{-1} y}^{\pi/4} \sec x \, dx \, dy = \int_0^{\pi/4} \int_0^{\tan x} \sec x \, dy \, dx$$

$$= \int_0^{\pi/4} \sec x \tan x \, dx$$

$$= \sec x \Big|_0^{\pi/4}$$

$$= \sqrt{2} - 1$$

17. Vertically simple:

$$\iint_D (x + y) \, dA = \int_0^1 \int_x^1 (x + y) \, dy \, dx$$

$$= \int_0^1 \left(x + \frac{1}{2} - \frac{3x^2}{2} \right) dx$$

$$= \frac{1}{2}$$

19. Horizontally simple:

$$\iint_D y \, dA = \int_0^1 \int_{y^2}^{2-y} y \, dx \, dy$$

$$= \int_0^1 \left(2y - y^2 - y^3 \right) dy$$

$$= \frac{5}{12}$$

21. Horizontally simple:

$$\iint_R \frac{dA}{y^2 + 1} \, dA = \int_0^2 \int_{-y}^{2y} \frac{1}{y^2 + 1} \, dx \, dy$$

$$= \int_0^2 \frac{3y}{y^2 + 1} \, dy$$

$$= \frac{3}{2} \ln 5$$

23. Vertically simple (change order);

$$\iint_D \frac{\sin x}{x} \, dx \, dy = \int_0^\pi \int_0^x \frac{\sin x}{x} \, dy \, dx$$

$$= \int_0^\pi \frac{\sin x}{x} (x) \, dx = 2$$

25.

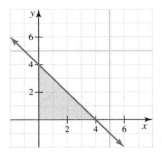

a. $\displaystyle\int_0^4 \int_0^{4-x} xy \, dy \, dx$

$$= \frac{1}{2} \int_0^4 (x^3 - 8x^2 + 16x) \, dx$$

$$= \frac{32}{3}$$

b. $\displaystyle\int_0^4 \int_0^{4-y} xy \, dx \, dy$

$$= \frac{1}{2} \int_0^4 \left(y^3 - 8y^2 + 16y \right) dy$$

$$= \frac{32}{3}$$

27.

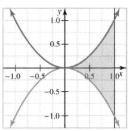

a. $\displaystyle\int_0^1 \int_{-x^2}^{x^2} dy \, dx = 2 \int_0^1 x^2 \, dx$

$$= \frac{2}{3}$$

b. $\displaystyle\int_0^1 \int_{\sqrt{y}}^1 dx \, dy + \int_{-1}^0 \int_{\sqrt{-y}}^1 dx \, dy$

$$= 2 \int_0^1 \int_{\sqrt{y}}^1 dx\, dy$$

$$= 2 \int_0^1 \left(1 - \sqrt{y}\right) dy$$

$$= \frac{2}{3}$$

29.

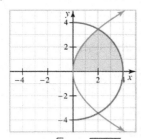

a. $\displaystyle \int_0^{2\sqrt{3}} \int_{y^2/6}^{\sqrt{16-y^2}} dx\, dy$

$$= \int_0^{2\sqrt{3}} \left(\sqrt{16-y^2} - \frac{y^2}{6}\right) dy$$

$$= \left[8\sin^{-1}\frac{y}{4} + \frac{y}{2}\sqrt{16-y^2} - \frac{y^3}{18}\right]\Bigg|_0^{2\sqrt{3}}$$

$$= \frac{8\pi}{3} + \frac{2\sqrt{3}}{3}$$

b. $\displaystyle \int_0^2 \int_0^{\sqrt{6y}} dy\, dx + \int_2^4 \int_0^{\sqrt{16-x^2}} dy\, dx$

$$= \int_0^2 \sqrt{6x}\, dx + \int_2^4 \sqrt{16-x^2}\, dx$$

$$= \frac{8\sqrt{3}}{3} + \left(-2\sqrt{3} + \frac{8\pi}{3}\right)$$

$$= \frac{1}{3}\left(8\pi + 2\sqrt{3}\right)$$

31. $\displaystyle \int_0^1 \int_0^{2y} f(x,y)\, dx\, dy$

$$\int_0^2 \int_{x/2}^1 f(x,y)\, dy\, dx$$

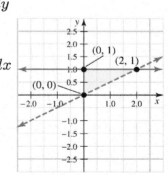

33. $\displaystyle \int_0^3 \int_{y/3}^{\sqrt{4-y}} f(x,y)\, dx\, dy$

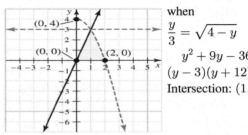

The curves intersect when
$$\frac{y}{3} = \sqrt{4-y}$$
$$y^2 + 9y - 36 = 0$$
$$(y-3)(y+12) = 0$$
Intersection: $(1,3)$

$$\int_0^1 \int_0^{3x} f(x,y)\, dy\, dx + \int_1^2 \int_0^{4-x^2} f(x,y)\, dy\, dx$$

35. $\displaystyle \int_0^7 \int_{x^2-6x}^x f(x,y)\, dy\, dx$

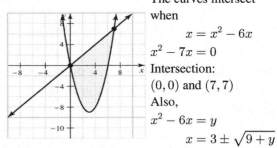

The curves intersect when
$$x = x^2 - 6x$$
$$x^2 - 7x = 0$$
Intersection:
$(0,0)$ and $(7,7)$
Also,
$$x^2 - 6x = y$$
$$x = 3 \pm \sqrt{9+y}$$

$$\int_{-9}^{0}\int_{3-\sqrt{9+y}}^{3+\sqrt{9+y}} f(x,y)\,dx\,dy$$

$$+ \int_{0}^{7}\int_{y}^{3+\sqrt{9+y}} f(x,y)\,dx\,dy$$

37. $V = \displaystyle\int_{0}^{7/3}\int_{0}^{(7-3x)/2} (7-3x-2y)\,dy\,dx$

39. Using symmetry of $x^2 + y^2 = 3$, and $x^2 + y^2 + z^2 = 7$, we have

$$V = 8\int_{0}^{\sqrt{3}}\int_{0}^{\sqrt{3-x^2}} \sqrt{7-x^2-y^2}\,dy\,dx$$

41. The projection of the ellipsoid on the xy-plane is the ellipse

$$\frac{x^2}{a^2} + \frac{y^2}{b^2} = 1$$

Using symmetry of the ellipsoid, we have

$$V = 8\int_{0}^{a}\int_{0}^{\frac{b}{a}\sqrt{a^2-x^2}} c\sqrt{1 - \frac{x^2}{a^2} - \frac{y^2}{b^2}}\,dy\,dx$$

43. The volume removed from the sphere
$$x^2 + y^2 + z^2 = 2$$
by the square cylindrical hole centered at the origin is

$$V_r = 8\int_{0}^{1}\int_{0}^{1} \sqrt{2-x^2-y^2}\,dy\,dx$$

We have used the symmetry of the sphere and square to simplify the integral. Thus, the volume that remains is

$$V = \frac{4}{3}\pi\left(\sqrt{2}\right)^3 - 8\int_{0}^{1}\int_{0}^{1} \sqrt{2-x^2-y^2}\,dy\,dx$$

45.

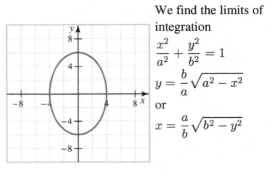

We find the limits of integration

$$\frac{x^2}{a^2} + \frac{y^2}{b^2} = 1$$

$$y = \frac{b}{a}\sqrt{a^2 - x^2}$$

or

$$x = \frac{a}{b}\sqrt{b^2 - y^2}$$

Using vertical strips and symmetry:

$$A = 4\int_{0}^{a}\int_{0}^{\frac{b}{a}\sqrt{a^2-x^2}} dy\,dx$$

Using horizontal strips:

$$A = 4\int_{0}^{b}\int_{0}^{\frac{a}{b}\sqrt{b^2-y^2}} dx\,dy$$

$$= \frac{4a}{b}\int_{0}^{b} \sqrt{b^2 - y^2}\,dy$$

$$= \frac{4a}{b}\left[\frac{b^2}{2}\sin^{-1}\frac{y}{b} + \frac{y}{2}\sqrt{b^2 - y^2}\right]\Bigg|_{0}^{b}$$

$$= \frac{4a}{b}\left[\frac{\pi b^2}{4}\right]$$

$$= \pi ab$$

47.

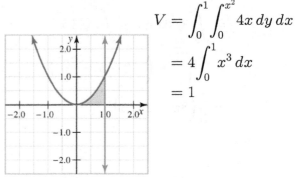

$$V = \int_{0}^{1}\int_{0}^{x^2} 4x\,dy\,dx$$

$$= 4\int_{0}^{1} x^3\,dx$$

$$= 1$$

49.

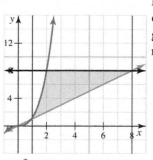

Reversing the order of integration gives a single region.

$$\int_1^8 \int_{y^{1/3}}^{y} f(x,y)\,dx\,dx$$

51.

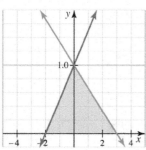

The lines intersect at $(-2,0)$, $(3,0)$, and $(0,1)$.

$$\int_0^1 \int_{2y-2}^{3-3y} \left(x^2 - xy - 1\right)dx\,dy$$
$$= \int_0^1 \left[\frac{x^3}{3} - \frac{x^2 y}{2} - x\right]\Bigg|_{2(y-1)}^{3(1-y)} dy$$
$$= \int_0^1 \left[9(1-y)^3 - \frac{9}{2}(1-y)^2 y - 3(1-y)\right.$$
$$\left. - \frac{8}{3}(y-1)^3 + 2(y-1)^2 y + 2(y-1)\right] dy$$
$$= \int_0^1 \left[-\frac{35}{3}(y-1)^3 - \frac{5}{2}(y-1)^2 y + 5(y-1)\right] dy$$
$$= \frac{5}{24}$$

53. $\int_0^3 \int_0^{\frac{1}{3}\sqrt{36-4x^2}} (3x+y)\,dy\,dx$
$$= \int_0^3 \left[2x\sqrt{9-x^2} - \frac{2}{9}(x^2 - 9)\right] dx$$
$$= 22$$

55. The integral is the volume of a cylinder of radius 3 and height 3 minus the volume of a cone of radius 3 and height 3:
$$V = \pi r^2 h - \frac{1}{3}\pi r^2 h$$
$$= \pi(3)^2(3) - \frac{\pi}{3}(3^2)(3)$$
$$= 18\pi$$

57. The integral is the volume of a cylinder of radius 4 and height 4 minus the volume of a cone of radius 4 and height 4:
$$V = \pi r^2 h - \frac{1}{3}\pi r^2 h$$
$$= \pi(4)^2(4) - \frac{\pi}{3}(4^2)(4)$$
$$= \frac{128\pi}{3}$$

$V = \pi r^2 h - \frac{1}{3}\pi r^2 h = \frac{2\pi}{3}(4^2)(4) = \frac{128\pi}{3}$

59. Let $f(x,y) = e^{y\sin x}$. Then, since $-1 \le \sin x \le 1$, we have
$$m\,dA \le \iint_D f(x,y)\,dA \le M\,dA$$
$$e^{-1}\left(\frac{3}{2}\right) \le \iint_D f(x,y)\,dA \le e^1\left(\frac{3}{2}\right)$$

Since $\frac{3}{2}e^{-1} \approx 0.55$ and $\frac{3}{2}e \approx 4.08$, we see the given integral is between 0.55 and 4.08.

12.3 Double Integrals in Polar Coordinates, page 948

1.

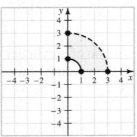

$$\int_0^{\pi/2} \int_1^3 re^{-r^2}\, dr\, d\theta$$

$$= -\frac{1}{2}\int_0^{\pi/2} \left(e^{-9} - e^{-1}\right) d\theta$$

$$= -\frac{\pi}{4}\left(e^{-9} - e^{-1}\right)$$

$$= \frac{\pi}{4}\left(\frac{1}{e} - \frac{1}{e^9}\right)$$

3.

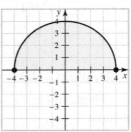

$$\int_0^{\pi} \int_0^4 r^2\sin^2\theta\, dr\, d\theta$$

$$= \frac{4^3}{3}\int_0^{\pi}\sin^2\theta\, d\theta$$

$$= \frac{32\pi}{3}$$

5.

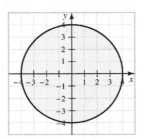

$$\int_0^{2\pi} \int_0^4 2r^2\cos\theta\, dr\, d\theta$$

$$= \frac{128}{3}\int_0^{2\pi}\cos\theta\, d\theta$$

$$= 0$$

7.

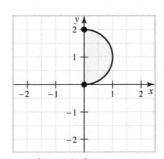

$$\int_0^{\pi/2} \int_0^{2\sin\theta} dr\, d\theta = \int_0^{\pi/2} 2\sin\theta\, d\theta$$

$$= 2$$

9. $\displaystyle\int_0^{2\pi} \int_0^4 r\, dr\, d\theta = 8\int_0^{2\pi} d\theta$

$$= 16\pi$$

11. $\displaystyle 2\int_0^{\pi} \int_0^{2(1-\cos\theta)} r\, dr\, d\theta$

$$= 4\int_0^{\pi}\left[1 - 2\cos\theta + \frac{1+\cos 2\theta}{2}\right] d\theta$$

$$= 6\pi$$

Page 474

13. $6 \displaystyle\int_0^{\pi/6} \int_0^{4\cos 3\theta} r\, dr\, d\theta = 48 \int_0^{\pi/6} \cos^2 3\theta\, d\theta$

$$= 4\pi$$

15. $8 \displaystyle\int_0^{\pi/4} \int_0^{\cos 2\theta} r\, dr\, d\theta = 4 \int_0^{\pi/4} \cos^2 2\theta\, d\theta$

$$= \frac{\pi}{2}$$

17. $2 \displaystyle\int_{\pi/6}^{\pi/2} \int_1^{2\sin\theta} r\, dr\, d\theta$

$$= \int_{\pi/6}^{\pi/2} \left(4\sin^2\theta - 1\right) d\theta$$

$$= \frac{1}{6}\left(2\pi + 3\sqrt{3}\right)$$

19. Use symmetry about the polar axis and the equation of the quarter circle.

$$2\left[\frac{1}{4}\pi(1)^2\right] + 2\int_{\pi/2}^{\pi} \int_0^{1+\cos\theta} r\, dr\, d\theta$$

$$= \frac{\pi}{2} + 2\int_{\pi/2}^{\pi} \int_0^{1+\cos\theta} r\, dr\, d\theta$$

$$= \frac{\pi}{2} + 2\int_{\pi/2}^{\pi} \left(\frac{1}{2}\cos^2\theta + \cos\theta + \frac{1}{2}\right) d\theta$$

$$= \frac{\pi}{2} + \frac{3\pi - 8}{4}$$

$$= \frac{5\pi}{4} - 2$$

21. $2 \displaystyle\int_0^{\pi/3} \int_{1+\cos\theta}^{3\cos\theta} r\, dr\, d\theta$

$$= \int_0^{\pi/3} \left[9\cos^2\theta - (1 + \cos\theta)^2\right] d\theta$$

$$= \int_0^{\pi/3} (3 + 4\cos 2\theta - 2\cos\theta)\, d\theta$$

$$= \pi$$

23. The bottom half of the loop is scanned out as θ varies from $2\pi/3$ to π. Thus, the area is

$$A = 2\int_{2\pi/3}^{\pi} \int_0^{1+2\cos\theta} r\, dr\, d\theta$$

$$= \pi - \frac{3\sqrt{3}}{2}$$

25. $\displaystyle\int_0^{2\pi} \int_0^a r^2\sin^2\theta\, dr\, d\theta = \frac{a^4}{4}\int_0^{2\pi} \frac{1 - \cos 2\theta}{2}\, d\theta$

$$= \frac{a^4}{4}\left[\frac{\theta}{2} - \frac{\sin(2\theta)}{4}\right]\Big|_0^{2\pi}$$

$$= \frac{a^4\pi}{4}$$

27. $4 \displaystyle\int_0^{\pi/2} \int_0^a \frac{r}{a^2 + r^2}\, dr\, d\theta = \int_0^{\pi/2} 2\ln 2\, d\theta$

$$= \pi\ln 2$$

29. $\displaystyle\int_0^{2\pi} \int_0^a \frac{1}{a+r}r\, dr\, d\theta = \int_0^{2\pi} \int_0^a \left(1 - \frac{a}{r+a}\right) dr\, d\theta$

$$= \int_0^{2\pi} (a - a\ln 2a + a\ln a)\, d\theta$$

$$= 2\pi a\left(1 + \ln\frac{1}{2}\right)$$

$$= 2\pi a(1 - \ln 2)$$

31. $\displaystyle\iint_D y\, dA = \int_0^{2\pi} \int_0^2 r^2\sin\theta\, dr\, d\theta$

$$= \int_0^{2\pi} \frac{8}{3}\sin\theta\, d\theta$$

$$= 0$$

33. $\displaystyle\iint_D e^{x^2+y^2}\, dA = 4\int_0^{\pi/2} \int_0^3 re^{r^2}\, dr\, d\theta$

$$= 4\int_0^{\pi/2} \frac{1}{2}\left(e^9 - 1\right) d\theta$$

$$= \pi\left(e^9 - 1\right)$$

35. $\displaystyle\iint_D \ln(x^2 + y^2 + 2)\, dA = \int_0^{\pi/2} \int_0^2 r \ln(r^2 + 2)\, dr\, d\theta$

$$= \int_0^{\pi/2} (3\ln 3 + 2\ln 2 - 2)\, d\theta$$

$$= \frac{3}{2}\pi \ln 3 + \pi \ln 2 - \pi$$

37. $\displaystyle\int_0^3 \int_0^{\sqrt{9-x^2}} x\, dy\, dx = \int_0^{\pi/2} \int_0^3 (r\cos\theta)(r\, dr\, d\theta)$

$$= \int_0^{\pi/2} \cos\theta \left[\frac{r^3}{3}\right]\Big|_0^3 d\theta$$

$$= 9\int_0^{\pi/2} \cos\theta\, d\theta$$

$$= 9$$

39. $\displaystyle\int_0^2 \int_0^{\sqrt{4-y^2}} e^{x^2+y^2}\, dx\, dy$

$$= \int_0^{\pi/2} \int_0^2 e^{r^2} r\, dr\, d\theta$$

$$= \int_0^{\pi/2} \left[\frac{1}{2} e^{r^2}\right]_0^2 d\theta$$

$$= \int_0^{\pi/2} \left[\frac{1}{2} e^4 - \frac{1}{2} e^0\right] d\theta$$

$$= \frac{\pi}{4}\left(e^4 - 1\right)$$

41. $\displaystyle\int_0^4 \int_0^{\sqrt{4y-y^2}} \frac{1}{\sqrt{x^2 + y^2}}\, dx\, dy$

$$= \int_0^{\pi/2} \int_0^{4\sin\theta} \frac{1}{r} r\, dr\, d\theta$$

$$= \int_0^{\pi/2} 4\sin\theta\, d\theta$$

$$= 4$$

43. $\displaystyle\int_0^2 \int_0^{\sqrt{2x-x^2}} \frac{x-y}{x^2+y^2}\, dy\, dx$

$$= \int_0^{\pi/2} \int_0^{2\cos\theta} \frac{r\cos\theta - r\sin\theta}{r^2}\, r\, dr\, d\theta$$

$$= \int_0^{\pi/2} [\cos\theta - \sin\theta](2\cos\theta)\, d\theta$$

$$= \left[\frac{1}{2}\sin 2\theta + \frac{1}{2}\cos 2\theta + \theta\right]\Big|_0^{\pi/2}$$

$$= \frac{\pi - 2}{2}$$

45. $\displaystyle V = \int_0^{2\pi} \int_0^2 \left(4 - r^2\right) r\, dr\, d\theta$

$$= \int_0^{2\pi} 4\, d\theta$$

$$= 8\pi$$

47.

The curves intersect at $\theta = \dfrac{\pi}{4}$ and at the pole.

$$\int_0^{\pi/4} \int_0^{4\sin\theta} r^3 \sin\theta \cos\theta\, dr\, d\theta$$

$$+ \int_{\pi/4}^{\pi/2} \int_0^{4\cos\theta} r^3 \sin\theta \cos\theta\, dr\, d\theta$$

$$= 4^3 \int_0^{\pi/4} \sin^5\theta \cos\theta\, d\theta$$

$$+ 4^3 \int_{\pi/4}^{\pi/2} \cos^5\theta \sin\theta\, d\theta$$

$$= \frac{4}{3} + \frac{4}{3}$$

$$= \frac{8}{3}$$

49. The curves intersect at $\theta = \pi/4$. Making use of the symmetry we will integrate the $\sin\theta$ curve from 0 to $\pi/4$ to get half of the desired area.

$$A = 2\int_0^{\pi/4}\int_0^{a\sin\theta} r\, dr\, d\theta$$

$$= 2\int_0^{\pi/4} \frac{a^2\sin^2\theta}{2}\, d\theta$$

$$= a^2\int_0^{\pi/4} \frac{1-\cos 2\theta}{2}\, d\theta$$

$$= \frac{a^2}{8}(\pi - 2)$$

51. $\displaystyle V = 2\int_0^{2\pi}\int_3^5 \sqrt{25-r^2}\, r\, dr\, d\theta$

$$= 2\int_0^{2\pi} \left[-\frac{1}{3}\left(25-r^2\right)^{3/2}\right]_3^5 d\theta$$

$$= 2\int_0^{2\pi} \frac{64}{3}\, d\theta$$

$$= \frac{256\pi}{3}$$

53. The circle $x^2 + y^2 = 2x$ in polar form is $r = 2\cos\theta$. Half the solid is above the xy-plane and half is below.

$$V = \iint_D z\, dA$$

$$= 4\int_0^{\pi/2}\int_0^{2\cos\theta} \sqrt{4-r^2}\, r\, dr\, d\theta$$

$$= 4\int_0^{\pi/2} \left[-\frac{8}{3}\left(\sin^3\theta - 1\right)\right] d\theta$$

$$= -\frac{32}{3}\left[-\cos\theta + \frac{\cos^3\theta}{3} - \theta\right]_0^{\pi/2}$$

$$= \frac{16}{9}(3\pi - 4)$$

55. $\displaystyle V = \iint_D z\, dA$

$$= 2\int_0^{\pi/2}\int_0^{\cos\theta} (1-r^2)\, r\, dr\, d\theta$$

$$= 2\int_0^{\pi/2} \left[\frac{\cos^2\theta}{2} - \frac{1}{4}\cos^4\theta\right] d\theta$$

$$= \frac{5\pi}{32}$$

57. The curves intersect at $\theta = \frac{\pi}{4}$. For $0 \le \theta \le \frac{\pi}{4}$, we use the curve $r = 2\sin\theta$, and for $\frac{\pi}{4} \le \theta \le \frac{\pi}{2}$ use the curve $r = 2\cos\theta$. Thus,

$$V = \iint_D x\, dA$$

$$= \iint_D r\cos\theta\, r\, dr\, d\theta$$

$$= \int_0^{\pi/4}\int_0^{2\sin\theta} r^2\cos\theta\, dr\, d\theta$$

$$+ \int_{\pi/4}^{\pi/2}\int_0^{2\cos\theta} r^2\cos\theta\, dr\, d\theta$$

$$= \int_0^{\pi/4} \frac{1}{3}(2\sin\theta)^3\cos\theta\, d\theta$$

$$+ \int_{\pi/4}^{\pi/2} \frac{1}{3}(2\cos\theta)^3\cos\theta\, d\theta$$

$$= \frac{8}{3}\frac{\sin^4\theta}{4}\Big|_0^{\pi/4} + \frac{8}{3}\left[\frac{3\theta}{8} + \frac{\sin 2\theta}{4} + \frac{\sin 4\theta}{32}\right]\Big|_{\pi/4}^{\pi/2}$$

$$= \frac{1}{6} + \frac{(3\pi - 8)}{12} = \frac{1}{4}(\pi - 2)$$

59. The first solution can't be correct because $\cos\theta$ is negative for $\pi/2 < \theta \le \pi$. The second solution is correct.

12.4 Surface Area, page 958

1. Obtain a rectangular approximating element of surface area, project onto a coordinate plane (the xy-plane) and then sum up (integrate). In particular, with $z = f(x,y)$, find f_x, f_y, to obtain

$$dS = \sqrt{f_x^2 + f_y^2 + 1}\,dA$$

Then integrate to find S.

3. The appropriate comparison is arc length to surface area, as discussed at the beginning of this section.

a. Length on the x-axis is the distance, d, between two points $P_1(a,0)$ and $P_2(b,0)$ so that

$$d = |b - a|$$
$$= \int_a^b dx$$

The arc length, L, between two points $P_1(a, f(a))$ and $P_2(b, f(b))$ is

$$L = \int_a^b \sqrt{1 + [f'(x)]^2}\,dx$$

b. The area in the xy-plane is

$$\iint_R dA$$

The surface area is

$$\iint_R dS$$
$$= \iint_R \sqrt{[f_x(x,y)]^2 + [f_y(x,y)]^2 + 1}\,dA$$

5. $z = f(x,y) = 2 - \frac{1}{2}x - \frac{1}{4}y$
$f_x = -\frac{1}{2};\ f_y = -\frac{1}{4}$
$$\sqrt{f_x^2 + f_y^2 + 1} = \sqrt{\frac{1}{4} + \frac{1}{16} + 1}$$
$$= \frac{\sqrt{21}}{4}$$
$$S = \int_0^4 \int_0^{-2x+8} \frac{\sqrt{21}}{4}\,dy\,dx$$
$$= \frac{\sqrt{21}}{4}\int_0^4 (-2x + 8)\,dx$$
$$= \frac{16\sqrt{21}}{4}$$
$$= 4\sqrt{21}$$

7. $z = f(x,y) = 2 - 2x - y$
$f_x = -2;\ f_y = -1$
$$\sqrt{f_x^2 + f_y^2 + 1} = \sqrt{4 + 1 + 1}$$
$$= \sqrt{6}$$
$$S = \int_0^1 \int_0^{2-2x} \sqrt{6}\,dy\,dx$$
$$= 2\sqrt{6}\int_0^1 (1 - x)\,dx$$
$$= \sqrt{6}$$

9. $z = f(x,y) = x^2 + y^2$
$f_x = 2x;\ f_y = 2y$

$$\sqrt{f_x^2 + f_y^2 + 1} = \sqrt{4x^2 + 4y^2 + 1}$$
$$= \sqrt{4r^2 + 1}$$

The projected region is $x^2 + y^2 \le 1$, so $r \le 1$.

$$S = \int_0^{2\pi} \int_0^1 \sqrt{4r^2 + 1}\, r\, dr\, d\theta$$
$$= \frac{1}{8} \int_0^{2\pi} \int_0^1 \sqrt{4r^2 + 1}\, (8r\, dr)\, d\theta$$
$$= \frac{1}{12} \int_0^{2\pi} \left(5\sqrt{5} - 1\right) d\theta$$
$$= \frac{\pi}{6} \left(5\sqrt{5} - 1\right)$$

11. $z = f(x,y) = \frac{1}{2}(12 - 3x - 6y)$
$f_x = -\frac{3}{2};\ f_y = -3$
$$\sqrt{f_x^2 + f_y^2 + 1} = \sqrt{\frac{9}{4} + 9 + 1}$$
$$= \frac{7}{2}$$
$$S = \int_0^1 \int_0^x \frac{7}{2}\, dy\, dx$$
$$= \frac{7}{2} \int_0^1 x\, dx$$
$$= \frac{7}{4}$$

13. $z = f(x,y) = 9 - x^2$
$f_x = -2x;\ f_y = 0$
$$\sqrt{f_x^2 + f_y^2 + 1} = \sqrt{4x^2 + 1}$$
$$S = \int_0^2 \int_0^2 \sqrt{4x^2 + 1}\, dy\, dx$$
$$= \int_0^2 \sqrt{4x^2 + 1}\, (2\, dx) \quad \boxed{\text{Formula 85}}$$
$$= \left[x\sqrt{4x^2 + 1} + \frac{1}{2}\ln\left(2x + \sqrt{4x^2 + 1}\right) \right]_0^2$$

$$= 2\sqrt{17} + \frac{1}{2}\ln\left(4 + \sqrt{17}\right)$$

15. $z = f(x,y) = x^2$
$f_x = 2x;\ f_y = 0$
$$\sqrt{f_x^2 + f_y^2 + 1} = \sqrt{4x^2 + 1}$$
$$S = \int_0^4 \int_0^4 \sqrt{4x^2 + 1}\, dy\, dx$$
$$= 2 \int_0^4 \sqrt{4x^2 + 1}\, (2\, dx) \quad \boxed{\text{Formula 85}}$$
$$= 2\left[x\sqrt{4x^2 + 1} + \frac{1}{2}\ln\left(2x + \sqrt{4x^2 + 1}\right) \right]_0^4$$
$$= 8\sqrt{65} + \ln\left(8 + \sqrt{65}\right)$$

17. $z = f(x,y) = x^2 + y^2$
$f_x = 2x;\ f_y = 2y$
$$\sqrt{f_x^2 + f_y^2 + 1} = \sqrt{4x^2 + 4y^2 + 1}$$
$$= \sqrt{4r^2 + 1}$$
If $z = 1$, $x^2 + y^2 = 1$, so $r = 1$.

$$S = 4 \int_0^{\pi/2} \int_0^1 \sqrt{4r^2 + 1}\, r\, dr\, d\theta$$
$$= \frac{1}{2} \int_0^{\pi/2} \int_0^1 \sqrt{4r^2 + 1}\, (8r\, dr)\, d\theta$$
$$= \frac{1}{2} \int_0^{\pi/2} \left(5\sqrt{5} - 1\right) \left(\frac{2}{3}\right) d\theta$$
$$= \frac{\pi}{6} \left(5\sqrt{5} - 1\right)$$

19. $z = f(x,y) = \sqrt{4 - x^2}$
$$f_x = \frac{-x}{\sqrt{4 - x^2}};\ f_y = 0$$
$$\sqrt{f_x^2 + f_y^2 + 1} = \sqrt{\frac{x^2 + 4 - x^2}{4 - x^2}}$$
$$= \frac{2}{\sqrt{4 - x^2}}$$

The projected region is the square $0 \le x \le 2$, $0 \le y \le 2$. Thus,

$$S = \int_0^2 \int_0^2 \frac{2}{\sqrt{4-x^2}} \, dy \, dx$$

$$= \int_0^2 \frac{4}{\sqrt{4-x^2}} \, dx$$

$$= 4\sin^{-1}\frac{x}{2}\Big|_0^2$$

$$= 2\pi$$

21. The intersection of the two surfaces (which is also the projection onto the xy-plane) is found by eliminating the z variable:

$$x^2 + y^2 = 4$$

$$z = f(x,y) = \sqrt{8 - x^2 - y^2}$$

$$f_x = \frac{-x}{\sqrt{8-x^2-y^2}}; \; f_y = \frac{-y}{\sqrt{8-x^2-y^2}}$$

$$\sqrt{f_x^2 + f_y^2 + 1} = \sqrt{\frac{x^2 + y^2 + 8 - x^2 - y^2}{8 - x^2 - y^2}}$$

$$= \frac{\sqrt{8}}{\sqrt{8-r^2}}$$

Since half the surface lies above the xy-plane, we have

$$S = 2\int_0^{2\pi} \int_0^2 \frac{2\sqrt{2}\,r}{\sqrt{8-r^2}} \, dr \, d\theta \quad \boxed{u = \sqrt{8-r^2}}$$

$$= 2\int_0^{2\pi} (8 - 4\sqrt{2}) \, d\theta$$

$$= 16\pi\left(2 - \sqrt{2}\right)$$

23. $z = f(x,y) = x^2 + y$

$f_x = 2x; \; f_y = 1$

$$\sqrt{f_x^2 + f_y^2 + 1} = \sqrt{4x^2 + 2}$$

$$S = \int_0^2 \int_0^5 \sqrt{4x^2 + 2} \, dy \, dx$$

$$= 5\sqrt{2}\int_0^2 \sqrt{1 + 2x^2} \, dx$$

$$= 15\sqrt{2} + \frac{5}{2}\ln\left(2\sqrt{2} + 3\right)$$

25. $z = f(x,y) = 4 - x^2 - y^2$

$f_x = -2x; \; f_y = -2y$

$$\sqrt{f_x^2 + f_y^2 + 1} = \sqrt{4x^2 + 4y^2 + 1}$$

$$= \sqrt{4r^2 + 1}$$

The projected region is $x^2 + y^2 \le 4$ or $r \le 2$.

$$S = \int_0^{2\pi} \int_0^2 \sqrt{4r^2 + 1}\, r \, dr \, d\theta$$

$$= \frac{1}{8}\int_0^{2\pi} \int_0^2 \sqrt{4r^2 + 1}\,(8r \, dr)\, d\theta$$

$$= \frac{1}{12}\int_0^{2\pi} \left(17\sqrt{17} - 1\right) d\theta$$

$$= \frac{\pi}{6}\left(17\sqrt{17} - 1\right)$$

27. $f_x = 0; \; f_y = \frac{1}{5}$;

$$\sqrt{f_x^2 + f_y^2 + 1} = \sqrt{0 + \frac{1}{25} + 1}$$

$$= \frac{1}{5}\sqrt{26}$$

$$S = \int_0^{300} \int_0^{400} \frac{1}{5}\sqrt{26} \, dy \, dx$$

$$S = \int_0^{300} \frac{400\sqrt{26}}{5} \, dx$$

$$= \frac{400(300)\sqrt{26}}{5}$$

$$= 24{,}000\sqrt{26}$$

29. $z = f(x,y) = \sqrt{a^2 - x^2}$

$$f_x = \frac{-x}{\sqrt{a^2 - x^2}}; \; f_y = 0$$

$$\sqrt{f_x^2 + f_y^2 + 1} = \sqrt{\frac{x^2}{a^2 - x^2} + 0 + 1}$$

$$= \sqrt{\frac{x^2 + a^2 - x^2}{a^2 - x^2}}$$

$$= \frac{a}{\sqrt{a^2 - x^2}}$$

$$S = 4 \int_0^a \int_0^h \frac{a \, dy \, dx}{\sqrt{a^2 - x^2}}$$

$$= 4 \int_0^a \frac{ha}{\sqrt{a^2 - x^2}} \, dx$$

$$= 4ah\sin^{-1}\frac{x}{a}\Big|_0^a$$

$$= 4ah\frac{\pi}{2}$$

$$= 2\pi ah$$

31. Solve $x^2 + y^2 + z^2 = 9z$ for z (we disregard the negative root since it corresponds to the lower part of the sphere):

$$z = \frac{9 + \sqrt{81 - 4(x^2 + y^2)}}{2}$$

$$z_x = \frac{-2x}{\sqrt{81 - 4(x^2 + y^2)}} = \frac{-2r\cos\theta}{\sqrt{81 - 4r^2}}$$

$$z_y = \frac{-2y}{\sqrt{81 - 4(x^2 + y^2)}} = \frac{-2r\sin\theta}{\sqrt{81 - 4r^2}}$$

$$\sqrt{f_x^2 + f_y^2 + 1} = \sqrt{\frac{4r^2\cos^2\theta}{81 - 4r^2} + \frac{4r^2\sin^2\theta}{81 - 4r^2} + 1}$$

$$= \frac{9}{\sqrt{81 - 4r^2}}$$

The projected region in the xy-plane is the intersection of $z^2 + r^2 = 9z$ with $r^2 = 4z$. Solving, we obtain $z = 0, 5$. (Reject $z = 0$, since $r = 0$ is a point.) The projected region is $r^2 = 4(5) = 20$, $r = \sqrt{20}$. Thus,

$$S = \int_0^{2\pi} \int_0^{\sqrt{20}} \frac{9}{\sqrt{81 - 4r^2}} r \, dr \, d\theta$$

$$= -\frac{9}{8} \int_0^{2\pi} \int_0^{\sqrt{20}} \frac{(-8r)\, dr}{\sqrt{81 - 4r^2}} \, d\theta$$

$$= -\frac{9}{4} \int_0^{2\pi} (1 - 9) \, d\theta$$

$$= 36\pi$$

33. $z = f(x, y) = \sqrt{x^2 + y^2}$

$$f_x = \frac{x}{\sqrt{x^2 + y^2}}; \quad f_y = \frac{y}{\sqrt{x^2 + y^2}}$$

$$\sqrt{f_x^2 + f_y^2 + 1} = \sqrt{\frac{x^2 + y^2}{x^2 + y^2} + 1}$$

$$= \sqrt{2}$$

$$S = \int_0^{2\pi} \int_0^h \sqrt{2} \, r \, dr \, d\theta$$

$$= \frac{\sqrt{2}}{2} \int_0^{2\pi} h^2 \, d\theta$$

$$= \sqrt{2}\pi h^2$$

35. $z = f(x, y) = \sqrt{4 - x^2}$

$$f_x = \frac{-x}{\sqrt{4 - x^2}}; \quad f_y = 0$$

$$\sqrt{f_x^2 + f_y^2 + 1} = \sqrt{\frac{x^2}{4 - x^2} + 0 + 1}$$

$$= \frac{2}{\sqrt{4 - x^2}}$$

$$S = \int_0^1 \int_0^x \frac{2}{\sqrt{4 - x^2}} \, dy \, dx$$

$$= \int_0^1 \frac{2x}{\sqrt{4 - x^2}} \, dx$$

$$= 2\left(2 - \sqrt{3}\right)$$

37. $x = f(y, z) = \sqrt{9 - z^2}$

$f_y = 0; \; f_z = \dfrac{-z}{\sqrt{9 - z^2}}$

$\sqrt{f_x^2 + f_y^2 + 1} = \sqrt{0 + \left(\dfrac{-z}{\sqrt{9 - z^2}}\right)^2 + 1}$

$= \dfrac{3}{\sqrt{9 - z^2}}$

> **SURVIVAL HINT:** It is tempting to use polar coordinates on this problem, but that approach leads to much harder integrations.

The projected region on the yz-plane is the disk, D: $y^2 + z^2 \le 9$. Since half the surface lies in front of the yz-plane, we have

$S = 2 \displaystyle\int_0^3 \int_0^{\sqrt{9 - z^2}} \dfrac{3}{\sqrt{9 - z^2}} dy \, dz$

$= 8 \displaystyle\int_0^3 3 \, dz$

$= 72$

39. $z = f(x, y) = \frac{1}{C} \sqrt{D - Ax - By}$

$f_x = -\dfrac{A}{C}; \; f_y = -\dfrac{B}{C}$

$\sqrt{f_x^2 + f_y^2 + 1} = \sqrt{\dfrac{A^2}{C^2} + \dfrac{B^2}{C^2} + 1}$

$= \sqrt{\dfrac{A^2 + B^2 + C^2}{C}}$

$S = \displaystyle\int_0^{D/A} \int_0^{-\frac{A}{B}x + \frac{D}{B}} \dfrac{\sqrt{A^2 + B^2 + C^2}}{C} dy \, dx$

$= \dfrac{\sqrt{A^2 + B^2 + C^2}}{C} \displaystyle\int_0^{D/A} \left(-\dfrac{A}{B}x + \dfrac{D}{B}\right) dx$

$= \dfrac{\sqrt{A^2 + B^2 + C^2}}{C} \left(\dfrac{D^2}{2AB}\right)$

$= \dfrac{D^2}{2ABC} \sqrt{A^2 + B^2 + C^2}$

41. $z = f(x, y) = \frac{1}{3}(12 - x - 2y)$

$f_x = -\dfrac{1}{3}; \; f_y = -\dfrac{2}{3}$

$\sqrt{f_x^2 + f_y^2 + 1} = \sqrt{\dfrac{1}{9} + \dfrac{4}{9} + 1}$

$= \dfrac{\sqrt{14}}{3}$

$S = \displaystyle\int_0^a \int_0^x \dfrac{\sqrt{14}}{3} dy \, dx$

$= \displaystyle\int_0^a \dfrac{x}{3} \sqrt{14} \, dx$

$= \dfrac{a^2}{6} \sqrt{14}$

43. $z = f(x, y) = a - x - y$

$f_x = -1; \; f_y = -1$

$\sqrt{f_x^2 + f_y^2 + 1} = \sqrt{3}$

$S = \displaystyle\int_0^{2\pi} \int_{a/2}^a \sqrt{3} \, r \, dr \, d\theta$

$= \displaystyle\int_0^{2\pi} \dfrac{3\sqrt{3}a^2}{8} d\theta$

$= \dfrac{3\sqrt{3}\pi a^2}{4}$

45. Solve $z = e^{-x} \sin y$ for x:

$x = \ln(\sin y) - \ln z$

$x_y = \dfrac{\cos y}{\sin y} = \cot y; \; x_z = \dfrac{-1}{z}$

$$\sqrt{x_y^2 + x_z^2 + 1} = \sqrt{\cot^2 y + \frac{1}{z^2} + 1}$$
$$= \sqrt{\csc^2 y + z^{-2}}$$
$$S = \int_0^1 \int_0^y \sqrt{\csc^2 y + z^{-2}}\, dz\, dy$$

47. $x = f(x, y) = \cos(x^2 + y^2)$
$f_x = -2x \sin(x^2 + y^2);$
$f_y = -2y \sin(x^2 + y^2)$
$$\sqrt{f_x^2 + f_y^2 + 1}$$
$$= \sqrt{(4x^2 + 4y^2)\sin(x^2 + y^2) + 1}$$
$$S = \int_0^{2\pi} \int_0^{\sqrt{\pi/2}} \sqrt{4r^2 \sin^2 r^2 + 1}\, r\, dr\, d\theta$$

49. $z = f(x, y) = x^2 + 5xy + y^2$
The curves intersect when $x = 1$ and $x = 5$.
$f_x = 2x + 5y;\ f_y = 2y + 5x$
$$\sqrt{f_x^2 + f_y^2 + 1}$$
$$= \sqrt{(2x + 5y)^2 + (2y + 5x)^2 + 1}$$
$$S = \int_1^5 \int_{5/x}^{6-x} \sqrt{(2x + 5y)^2 + (5x + 2y)^2 + 1}\, dy\, dx$$

51. $\mathbf{R}_u \times \mathbf{R}_v = \begin{vmatrix} \mathbf{i} & \mathbf{j} & \mathbf{k} \\ 2\sin v & 2\cos v & 2u \\ 2u\cos v & -2u\sin v & 0 \end{vmatrix}$
$$= 4u^2 \sin v\, \mathbf{i} + 4u^2 \cos v\, \mathbf{j}$$
$$+ \left(-4u\sin^2 v - 4u\cos^2 v\right)\mathbf{k}$$
$$= 4u^2 \sin v\, \mathbf{i} + 4u^2 \cos v\, \mathbf{j} - 4u\, \mathbf{k}$$

$$\|\mathbf{R}_u \times \mathbf{R}_v\| = \sqrt{16u^4 \sin^2 v + 16u^2 \cos^2 v + 16u^2}$$
$$= \sqrt{16u^4 + 16u^2}$$
$$= 4|u|\sqrt{u^2 + 1}$$

53. $\mathbf{R}_u \times \mathbf{R}_v = \begin{vmatrix} \mathbf{i} & \mathbf{j} & \mathbf{k} \\ 1 & 0 & 3u^2 \\ 0 & 2v & 0 \end{vmatrix}$
$$= -6u^2 v\, \mathbf{i} + 2v\mathbf{k}$$
$$\|\mathbf{R}_u \times \mathbf{R}_v\| = \sqrt{36u^4 v^2 + 4v^2}$$
$$= 2|v|\sqrt{9u^4 + 1}$$

55. $\mathbf{R}_u \times \mathbf{R}_v = \begin{vmatrix} \mathbf{i} & \mathbf{j} & \mathbf{k} \\ v & 1 & 1 \\ u & -1 & 1 \end{vmatrix}$
$$= 2\mathbf{i} + (u - v)\mathbf{j} - (u + v)\mathbf{k}$$
$$\|\mathbf{R}_u \times \mathbf{R}_v\| = \sqrt{4 + (u - v)^2 + (-u - v)^2}$$
$$= \sqrt{4 + 2u^2 + 2v^2}$$
$$S = \int_0^{2\pi} \int_0^1 \sqrt{4 + 2r^2}\, r\, dr\, d\theta$$
$$= \frac{1}{4}\int_0^{2\pi} \int_0^1 \sqrt{4 + 2r^2}\,(4r\, dr)\, d\theta$$
$$= \frac{1}{6}\int_0^{2\pi} \left(6\sqrt{6} - 8\right) d\theta$$
$$= \frac{2\pi}{3}\left(3\sqrt{6} - 4\right)$$

57. a. $\mathbf{R}_u = (\sin v)\mathbf{i} + (\cos v)\mathbf{j},$
$\mathbf{R}_v = (u\cos v)\mathbf{i} + (-u\sin v)\mathbf{j} + \mathbf{k}$
$$\mathbf{R}_u \times \mathbf{R}_v = \begin{vmatrix} \mathbf{i} & \mathbf{j} & \mathbf{k} \\ \sin v & \cos v & 0 \\ u\cos v & -u\sin v & 1 \end{vmatrix}$$
$$= (\cos v)\mathbf{i} - (\sin v)\mathbf{j} - u\mathbf{k}$$

b. $\|\mathbf{R}_u \times \mathbf{R}_v\| = \sqrt{1 + u^2}$
$$S = \int_0^b \int_0^a \sqrt{1 + u^2}\, du\, dv$$
$$= \int_0^b \left[\frac{\ln\left(u + \sqrt{1 + u^2}\right)}{2} + \frac{u\sqrt{1 + u^2}}{2}\right]_0^a dv$$

$$= \frac{b}{2}\left[\ln\left(a + \sqrt{1 + a^2}\right) + a\sqrt{1 + a^2}\right]$$

59. Since $F(x, y, z) = 0$, we have

$$\frac{\partial F}{\partial x} + \frac{\partial F}{\partial z}\frac{\partial z}{\partial x} = 0 \text{ and } \frac{\partial F}{\partial y} + \frac{\partial F}{\partial z}\frac{\partial z}{\partial y} = 0$$

so that

$$\frac{\partial z}{\partial x} = \frac{-F_x}{F_z} \text{ and } \frac{\partial z}{\partial y} = \frac{-F_y}{F_z}$$

and

$$z_x^2 + z_y^2 + 1 = \left(\frac{-F_x}{F_z}\right)^2 + \left(\frac{-F_y}{F_z}\right)^2 + 1$$

$$= \frac{F_x^2 + F_y^2 + F_z^2}{F_z^2}$$

Thus,

$$|A = \iint_D \frac{\sqrt{F_x^2 + F_y^2 + F_z^2}}{|F_z|}\, dA_{xy}$$

To find the surface area of a sphere, let
$f(x, y, z) = x^2 + y^2 + z^2 - R^2$
$f_x = 2x; \ f_y = 2y; \ f_z = 2z$

$$\frac{\sqrt{f_x^2 + f_y^2 + f_z^2}}{|f_z|} = \frac{2R}{2|z|}$$

$$= \frac{R}{\sqrt{R^2 - x^2 - y^2}}$$

$$S = \int_0^R \int_0^{\sqrt{R^2 - x^2}} \frac{R}{\sqrt{R^2 - x^2 - y^2}}\, dy\, dx$$

$$= 8R \int_0^{\pi/2} \int_0^R \frac{r}{\sqrt{R^2 - r^2}}\, dr\, d\theta$$

$$= 8R \int_0^{\pi/2} R\, d\theta$$

$$= 4\pi R^2$$

12.5 Triple Integrals, page 970

1. Fubini's theorem says that a triple integral of a continuous function in x, y, and z over a suitable region in space can be evaluated as an iterated integral in any of the six possible orders of integration: $xyz, xzy, yxz, yzx, zxy, zyx$.

3. $\displaystyle\int_1^4 \int_{-2}^3 \int_2^5 dx\, dy\, dz = (3)(5)(3) = 45$

5. $\displaystyle\int_{-3}^3 \int_0^1 \int_{-1}^2 dy\, dx\, dz = (3)(1)(6) = 18$

7. $\displaystyle\int_0^4 \int_1^4 \int_{-2}^3 dz\, dx\, dy = (5)(3)(4) = 60$

9. $\displaystyle\int_1^2 \int_0^1 \int_{-1}^2 8x^2 yz^3\, dx\, dy\, dz$

$$= \int_1^2 \int_0^1 24yz^3\, dy\, dz$$

$$= \int_1^2 12z^3\, dz$$

$$= 45$$

11. $\displaystyle\int_0^2 \int_0^x \int_0^{x+y} xyz\, dz\, dy\, dx$

$$= \frac{1}{2}\int_0^2 \int_0^x xy(x + y)^2\, dy\, dx$$

$$= \frac{1}{2}\int_0^2 \left(\frac{x^5}{2} + \frac{2x^5}{3} + \frac{x^5}{4}\right) dx$$

$$= \frac{1}{24}\int_0^2 17x^5\, dx$$

$$= \frac{68}{9}$$

13. $\displaystyle\int_{-1}^2 \int_0^{\pi} \int_1^4 yz\cos(xy)\, dz\, dx\, dy$

$$= \frac{15}{2} \int_{-1}^{2} \int_{0}^{\pi} y \cos xy \; dx \, dy$$

$$= \frac{15}{2} \int_{-1}^{2} \sin \pi y \, dy$$

$$= -\frac{15}{\pi}$$

15. $\int_{0}^{1} \int_{0}^{y} \int_{0}^{\ln y} e^{z+2x} dz \, dx \, dy$

$$= \int_{0}^{1} \int_{0}^{y} e^{2x}(y-1) \, dx \, dy$$

$$= \frac{1}{2} \int_{0}^{1} \left(e^{2y}-1\right)(y-1) \, dy$$

$$= \frac{1}{8}\left(5 - e^2\right)$$

17. $\int_{1}^{4} \int_{-1}^{2z} \int_{0}^{\sqrt{3x}} \frac{x-y}{x^2+y^2} dy \, dx \, dz$

$$= \int_{1}^{4} \int_{-1}^{2z} \frac{1}{2}\left[2\tan^{-1}\frac{y}{z} - \ln(x^2+y^2)\right]\Big|_{0}^{\sqrt{3x}} dx \, dz$$

$$= \int_{1}^{4} \int_{-1}^{2z} \left[\frac{\pi}{3} - \ln 2\right] dx \, dz$$

$$= \left(\frac{\pi}{3} - \ln 2\right)\int_{1}^{4} (2z+1) \, dz$$

$$= 6\pi - 18\ln 2$$

19. $\int_{1}^{3} \int_{-1}^{1} \int_{2}^{4} \left(x^2 y + y^2 z\right) dz \, dy \, dx$

$$= \int_{1}^{3} \int_{-1}^{1} \left(2x^2 y + 6y^2\right) dy \, dx$$

$$= \int_{1}^{3} 4 \, dx$$

$$= 8$$

21. $\int_{0}^{1} \int_{0}^{1-x} \int_{0}^{1-x-y} xyz \, dz \, dy \, dx$

$$= \frac{1}{2} \int_{0}^{1} \int_{0}^{1-x} xy(1-x-y)^2 \, dy \, dx$$

$$= \frac{1}{24} \int_{0}^{3} x(x-1)^4 \, dx$$

$$= \frac{1}{720}$$

23. $\int_{0}^{1} \int_{0}^{\sqrt{1-y^2}} \int_{-\sqrt{1-y^2-z^2}}^{\sqrt{1-y^2-z^2}} xyz \, dx \, dz \, dy$

$$= \int_{0}^{1} \int_{0}^{\sqrt{1-y^2}} \frac{yzx^2}{2}\Big|_{-\sqrt{1-y^2-z^2}}^{\sqrt{1-y^2-z^2}} dz \, dy$$

$$= 0$$

25. $\int_{0}^{1} \int_{0}^{x} \int_{0}^{x+y} e^z \, dz \, dy \, dx$

$$= \int_{0}^{1} \int_{0}^{x} \left(e^{x+y} - 1\right) dy \, dx$$

$$= \int_{0}^{1} \left(e^{2x} - x - e^x\right) dx$$

$$= \frac{e^2}{2} - e$$

27. $V = \int_{0}^{1} \int_{0}^{-x+1} \int_{0}^{1-x-y} dz \, dy \, dx$

$$= \int_{0}^{1} \int_{0}^{-x+1} (1-x-y) \, dy \, dx$$

$$= \int_{0}^{1} \left[(1-x)^2 - \frac{(1-x)^2}{2}\right] dx$$

$$= \int_{0}^{1} \frac{(1-x)^2}{2} \, dx$$

$$= \frac{1}{6}$$

> **SURVIVAL HINT:** This result is easily verified with the formula for the volume of a pyramid.

29. Let $A = 2 + \sqrt{1 - (x-1)^2}$ and

$B = 3 + \sqrt{1 - (x-1)^2 - (y-2)^2}$

$V = 8 \int_1^2 \int_2^A \int_3^B dz\,dy\,dx$

$= 8 \int_1^2 \int_2^A \sqrt{1 - (x-1)^2 - (y-2)^2}\,dy\,dx$

$= 8 \int_1^2 \frac{\pi}{4}(2x - x^2)\,dx$

$= \frac{4\pi}{3}$

> **SURVIVAL HINT:** This result is easily verified with the formula for a sphere.

31. The intersection of the parabolic cylinder and the elliptic paraboloid gives the region of integration in the xy-plane:

$$4 - y^2 = x^2 + 3y^2$$
$$x^2 + 4y^2 = 4$$

$V = 4 \int_0^1 \int_0^{2\sqrt{1-y^2}} \int_{x^2+3y^2}^{4-y^2} dz\,dx\,dy$

$= 4 \int_0^1 \int_0^{2\sqrt{1-y^2}} (4 - x^2 - 4y^2)\,dx\,dy$

$= 4 \int_0^1 \frac{16}{3}(1 - y^2)^{3/2}\,dy$

$= 4\pi$

33. The intersection of the two surfaces gives the region of integration in the xy-plane:

$$6 - x^2 - y^2 = 2x^2 + y^2$$
$$3x^2 + 2y^2 = 6$$

$V = 4 \int_0^{\sqrt{2}} \int_0^{\sqrt{(6-3x^2)/2}} \int_{2x^2+y^2}^{6-x^2-y^2} dz\,dy\,dx$

$= 4 \int_0^{\sqrt{2}} \int_0^{\sqrt{(6-3x^2)/2}} [6 - x^2 - y^2 - (2x^2 + y^2)]\,dy\,dx$

$= \int_0^{\sqrt{2}} \left[4\sqrt{6}(2 - x^2)^{3/2}\right] dx$

$= 3\sqrt{6}\pi$

35. $V = \int_{-1}^1 \int_{-\sqrt{1-x^2}}^{\sqrt{1-x^2}} \int_{-\sqrt{1-z^2}}^{\sqrt{1-z^2}} dy\,dz\,dx$

$= 8 \int_0^1 \int_0^{\sqrt{1-x^2}} \int_0^{\sqrt{1-z^2}} dy\,dz\,dx$

$= 8 \int_0^1 \int_0^{\sqrt{1-x^2}} \sqrt{1 - z^2}\,dz\,dx$

$= 4 \int_0^1 \left(\sin^{-1}\sqrt{1 - x^2} + x\sqrt{1 - x^2}\right) dx$

$= \frac{16}{3}$

37. $\int_0^1 \int_0^{1-y} \int_0^{1-x-y} f(x, y, z)\,dz\,dx\,dy$

39. $\int_0^1 \int_0^{1-z} \int_0^{1-y-z} f(x, y, z)\,dx\,dy\,dz$

> **SURVIVAL HINT:** *Compare (contrast) Problems 37-39 to see how rearranging the order changes the limits of integration.*

41. $\int_0^2 \int_0^{\sqrt{4-z^2}} \int_0^{\sqrt{4-y^2-z^2}} f(x, y, z)\,dx\,dy\,dz$

43. $\int_0^1 \int_{\sqrt[3]{x}}^1 \int_0^{1-y} f(x, y, z)\,dz\,dy\,dx$

45. Let $A = 3\sqrt{1 - \dfrac{x^2}{4}}$ and $B = 4\sqrt{1 - \dfrac{x^2}{4} - \dfrac{y^2}{9}}$

$$V = 8 \int_0^2 \int_0^A \int_0^B dz\, dy\, dx$$

$$= 32 \int_0^2 \int_0^A \sqrt{1 - \frac{x^2}{4} - \frac{y^2}{9}}\, dy\, dx$$

$$= 96 \int_0^2 \left[\frac{4 - x^2}{8} \frac{\pi}{2} + 0 \right] dx$$

$$= 24\pi \int_0^2 \left(1 - \frac{x^2}{4} \right) dx$$

$$= 32\pi$$

SURVIVAL HINT: *This result can be verified by using the formula for the volume of an ellipsoid.*

$$V = \frac{4}{3}\pi abc = \frac{4}{3}\pi(2)(3)(4) = 32\pi$$

---

47. The projected region of integration is found by intersecting the two surfaces

$$16 - x^2 - 2y^2 = 3x^2 + 2y^2$$
$$x^2 + y^2 = 4$$

Let D be the part of the disk $x^2 + y^2 \leq 4$ in the first quadrant.

$$V = 4 \int\!\!\int_D \int_{3x^2 + 2y^2}^{16 - x^2 - 2y^2} dz\, dy\, dx$$

$$= 4 \int\!\!\int_D \left[-4(x^2 + y^2 - 4 \right] dy\, dx$$

$$= 4 \int_0^{\pi/2} \int_0^2 4r(4 - r^2)\, dr\, d\theta$$

$$= 4 \int_0^{\pi/2} 16\, d\theta$$

$$= 32\pi$$

49. $V = 4 \int_0^3 \int_0^{\frac{1}{3}\sqrt{9-x^2}} \int_0^{\frac{x^2}{9}+y^2} dz\, dy\, dx$

$$= 4 \int_0^3 \int_0^{\frac{1}{3}\sqrt{9-x^2}} \left(\frac{x^2}{9} + y^2 \right) dy\, dx$$

$$= 4 \int_0^3 \left[\frac{x^2}{27}\sqrt{9 - x^2} + \frac{1}{81}(9 - x^2)^{3/2} \right] dx$$

$$= 4 \left(\frac{3\pi}{16} + \frac{3\pi}{16} \right)$$

$$= \frac{3\pi}{2}$$

51. $V = 8 \int_0^{\pi/2} \int_0^R \int_0^{\sqrt{R^2 - r^2}} dz\, r\, dr\, d\theta$

$$= 8 \int_0^{\pi/2} \int_0^R r\sqrt{R^2 - r^2}\, dr\, d\theta$$

$$= \frac{8}{3} \int_0^{\pi/2} R^3\, d\theta$$

$$= \frac{4}{3}\pi R^3$$

53. $z = c\left(1 - \frac{x^2}{a^2} - \frac{y^2}{b^2} \right)^{1/2}$

$$y = b\left(1 - \frac{x^2}{a^2} \right)^{1/2}$$

$$V = 8 \int_0^a \int_0^{b\sqrt{1 - \frac{x^2}{a^2}}} \int_0^{c\sqrt{1 - \frac{x^2}{a^2} - \frac{y^2}{b^2}}} dz\, dy\, dx$$

$$= 8c \int_0^a \int_0^{b\sqrt{1 - \frac{x^2}{a^2}}} \left(1 - \frac{x^2}{a^2} - \frac{y^2}{b^2} \right)^{1/2} dy\, dx$$

$$= 4bc \int_0^a \left[\left(1 - \frac{x^2}{a^2} \right) \frac{\pi}{2} \right] dx$$

$$= \frac{4\pi abc}{3}$$

55. $\displaystyle\int_a^b \int_c^d \int_r^s f(x)g(y)h(z)\,dz\,dy\,dx$

$\displaystyle = \int_a^b f(x) \int_c^d g(y) \int_r^s h(z)\,dz\,dy\,dx$

$\displaystyle = \left[\int_a^b f(x)\,dx\right]\left[\int_c^d g(y)\,dy\right]\left[\int_r^s h(z)\,dz\right]$

$\displaystyle = \iiint_B f(x)g(y)h(z)\,dV$

57. Using Problem 56,

$\displaystyle \iiint_D \sin(\pi - z)^3\,dz\,dy\,dx$

$\displaystyle = \int_0^\pi \int_0^y \int_0^x \sin(\pi - z)^3\,dz\,dx\,dy$

$\displaystyle = \frac{1}{2}\int_0^\pi (\pi - t)^2 \sin(\pi - t)^3\,dt$

$\displaystyle = \left. \frac{\cos(\pi - t)^3}{6} \right|_0^\pi$

$\displaystyle = \frac{1 - \cos \pi^3}{6}$

59. $\displaystyle\int_1^2 \int_{-1}^1 \int_0^2 \int_0^1 xyz^2 w^2\,dx\,dy\,dz\,dw$

$\displaystyle = \int_1^2 \int_{-1}^1 \int_0^2 \frac{w^2 y z^2}{2}\,dy\,dz\,dw$

$\displaystyle = \int_1^2 \int_{-1}^1 w^2 z^2\,dz\,dw$

$\displaystyle = \int_1^2 \frac{2w^2}{3}\,dw$

$\displaystyle = \frac{14}{9}$

12.6 Mass, Moments, and Probability Density Functions, page 982

1. Find the total mass. Compute the first moment with respect to the x-axis and the first moment with respect to the y-axis. Divide the first moments by the mass. The resulting coordinates are those of the center of mass.

3. A probability density function measures the probability that a continuous random variable X lies between two numbers on a number line ($\mathbb{R}$). A joint probability density function is the generalization to $\mathbb{R}^2$; that is, the probability that an ordered pair of continuous random variables (X, Y) lies within a particular region on a plane. Specifically, a joint probability density function for two random variables X and Y is a function $f(x, y)$ of two variables such that the probability that the point (X, Y) is in a region D satisfies

$$P[(x, y) \text{ is in } D] = \iint_D f(x, y)\,dA$$

5. $\displaystyle m = \int_0^3 \int_0^4 5\,dy\,dx = 60$

$\displaystyle M_x = \int_0^3 \int_0^4 5y\,dy\,dx \qquad M_y = \int_0^3 \int_0^4 5x\,dy\,dx$

$\displaystyle \quad = \frac{5}{2}\int_0^3 16\,dx \qquad\qquad = 20\int_0^3 x\,dx$

$\displaystyle \quad = 120 \qquad\qquad\qquad\quad = 90$

$\displaystyle (\overline{x}, \overline{y}) = \left(\frac{90}{60}, \frac{120}{60}\right) = \left(\frac{3}{2}, 2\right)$

7. $m = 2 \displaystyle\int_0^2 \int_{x^2}^{2x} dy\, dx = \frac{8}{3}$

$M_x = 2 \displaystyle\int_0^2 \int_{x^2}^{2x} y\, dy\, dx$ $M_y = 2 \displaystyle\int_0^2 \int_{x^2}^{2x} x\, dy\, dx$

$\qquad = \displaystyle\int_0^2 (4x^2 - x^4)\, dx$ $\qquad = 2 \displaystyle\int_0^2 x(2x - x^2)\, dx$

$\qquad = \dfrac{64}{15}$ $\qquad = \dfrac{8}{3}$

$(\overline{x}, \overline{y}) = \left(\dfrac{\frac{8}{3}}{\frac{8}{3}}, \dfrac{\frac{64}{15}}{\frac{8}{3}} \right) = \left(1, \dfrac{8}{5}\right)$

9. The parabola and line intersect when

$$2 - 3x^2 = \frac{1}{2}(1 - 3x)$$
$$6x^2 - 3x - 3 = 0$$
$$3(x - 1)(2x + 1) = 0$$
$$x = 1, -\frac{1}{2}$$

$m = \displaystyle\int_{-1/2}^{1} \int_{(1-3x)/2}^{2-3x^2} dy\, dx$

$\quad = \displaystyle\int_{-1/2}^{1} \left(-3x^2 + \frac{3}{2}x + \frac{3}{2}\right) dx$

$\quad = \dfrac{27}{16}$

$M_x = \displaystyle\int_{-\frac{1}{2}}^{1} \int_{\frac{1-3x}{2}}^{2-3x^2} y\, dy\, dx$

$\quad = \dfrac{1}{2} \displaystyle\int_{-\frac{1}{2}}^{1} \left[(2-3x^2)^2 - \frac{1}{4}(1-3x)^2\right] dx$

$\quad = \dfrac{27}{20}$

$M_y = \displaystyle\int_{-\frac{1}{2}}^{1} \int_{\frac{1-3x}{2}}^{2-3x^2} x\, dy\, dx$

$\quad = \displaystyle\int_{-\frac{1}{2}}^{1} x\left[2 - 3x^2 - \frac{1}{2}(1-3x)\right] dx$

$\quad = \dfrac{27}{64}$

$(\overline{x}, \overline{y}) = \left(\dfrac{\frac{27}{64}}{\frac{27}{16}}, \dfrac{\frac{27}{20}}{\frac{27}{16}} \right) = \left(\dfrac{1}{4}, \dfrac{4}{5}\right)$

11. $m = 4 \displaystyle\int_0^4 \int_0^{-x+4} \int_0^{4-x-y} dz\, dy\, dx$

$\quad = 4 \displaystyle\int_0^4 \int_0^{-x+4} (4 - x - y)\, dy\, dx$

$\quad = 4 \displaystyle\int_0^4 \frac{(x-4)^2}{2}\, dx$

$\quad = \dfrac{128}{3}$

$M_{yz} = \displaystyle\int_0^4 \int_0^{-x+4} \int_0^{4-x-y} 4x\, dz\, dy\, dx$

$\quad = \displaystyle\int_0^4 \int_0^{-x+4} [-4x(x + y - 4)]\, dy\, dx$

$\quad = \displaystyle\int_0^4 2x(x - 4)^2\, dx$

$\quad = \dfrac{(x-4)^3(3x+4)}{6} \bigg|_0^4$

$\quad = \dfrac{128}{3}$

$\overline{x} = \dfrac{\frac{128}{3}}{\frac{128}{3}} = 1$

By symmetry, we see $(\overline{x}, \overline{y}, \overline{z}) = (1, 1, 1)$.

13. $m = 2 \displaystyle\int_0^3 \int_0^{\sqrt{9-x^2}} (x^2 + y^2)\, dy\, dx$

$\quad = 2 \displaystyle\int_0^{\pi/2} \int_0^3 r^2\, r\, dr\, d\theta$

$\quad = 2 \displaystyle\int_o^{\pi/2} \frac{81}{4}\, dr$

$\quad = \dfrac{81\pi}{4}$

By symmetry, $\overline{x} = 0$.

$$M_x = 2 \int_0^3 \int_0^{\sqrt{9-x^2}} y(x^2 + y^2)\, dy\, dx$$

$$= 2 \int_0^3 \left[\frac{1}{2}x^2(9 - x^2) + \frac{1}{4}(81 - 18x^2 + x^4) \right] dx$$

$$= \frac{486}{5}$$

$$\overline{y} = \frac{\frac{486}{5}}{\frac{81\pi}{4}} = \frac{24}{5\pi}; \; (\overline{x}, \overline{y}) = \left(0, \frac{24}{5\pi} \right)$$

15. $m = \displaystyle\int_0^5 \int_{6y/5}^{(60-6y)/5} 7x\, dx\, dy$

$$= \int_0^5 \frac{-504}{5}(y - 5)\, dy$$

$$= 1{,}260$$

$$M_x = \int_0^6 \int_0^{5x/6} 7xy\, dy\, dx + \int_6^{12} \int_0^{-5x/6+10} 7xy\, dy\, dx$$

$$= \int_0^6 \frac{7x}{2}\left(\frac{5x}{6} \right)^2 dx + \int_6^{12} \frac{7x}{2}\left(-\frac{5x}{6} + 10 \right)^2 dx$$

$$= \int_0^6 \frac{175}{72}x^3\, dx + \int_6^{12} \left(\frac{175}{72}x^3 - \frac{175}{3}x^2 + 350x \right) dx$$

$$= \frac{1{,}575}{2} + \frac{2{,}625}{2}$$

$$= 2{,}100$$

$$M_y = 7 \int_0^5 \int_{6y/5}^{-(6/5)y+12} x^2\, dx\, dy$$

$$= \frac{7(6^3)}{3} \int_0^5 \left(-\frac{2y^3}{5^3} + \frac{6y^2}{5^2} - \frac{12y}{5} + 8 \right) dy$$

$$= 8{,}820$$

$$(\overline{x}, \overline{y}) = \left(\frac{8{,}820}{1{,}260}, \frac{2{,}100}{1{,}260} \right) = \left(7, \frac{5}{3} \right)$$

17. $m = \displaystyle\int_1^2 \int_0^{\ln x} x^{-1}\, dy\, dx = \int_1^2 \frac{\ln x}{x}\, dx = \frac{(\ln 2)^2}{2}$

$$M_x = \int_1^2 \int_0^{\ln x} \frac{y}{x}\, dy\, dx$$

$$= \frac{1}{2} \int_1^2 \frac{(\ln x)^2}{x}\, dx = \frac{(\ln 2)^3}{6}$$

$$M_y = \int_1^2 \int_0^{\ln x} dy\, dx$$

$$= \int_1^2 \ln x\, dx$$

$$= 2\ln 2 - 1$$

$$(\overline{x}, \overline{y}) = \left(\frac{2\ln 2 - 1}{\frac{(\ln 2)^2}{2}}, \frac{\frac{(\ln 2)^3}{6}}{\frac{(\ln 2)^2}{2}} \right)$$

$$= \left(\frac{4\ln 2 - 2}{(\ln 2)^2}, \frac{\ln 2}{3} \right)$$

19. a. By symmetry $\overline{x} = 0$. $\rho = kr$.

$$m = k \int_0^\pi \int_0^a (r)\, r\, dr\, d\theta$$

$$= \frac{ka^3}{3} \int_0^\pi d\theta$$

$$= \frac{k\pi a^3}{3}$$

$$M_x = k \int_0^\pi \int_0^a (r)(r\sin\theta)\, r\, dr\, d\theta$$

$$= \frac{ka^4}{4} \int_0^\pi \sin\theta\, d\theta$$

$$= \frac{ka^4}{4}(1 + 1)$$

$$= \frac{ka^4}{2}$$

$$\overline{y} = \frac{\frac{ka^4}{2}}{\frac{k\pi a^3}{3}} = \frac{3a}{2\pi}$$

The centroid is at $\left(0, \dfrac{3a}{2\pi} \right)$.

b. $\rho = k\theta$ (use $\rho = \theta$ as the k will cancel in the quotient).

$$m = \int_0^\pi \int_0^a (\theta)\, r\, dr\, d\theta$$

$$= \frac{a^2}{2} \int_0^\pi \theta \, d\theta$$

$$= \frac{\pi^2 a^2}{4}$$

$$M_x = \int_0^\pi \int_0^a (\theta)(r \sin \theta) \, r \, dr \, d\theta$$

$$= \frac{a^3}{3} \int_0^\pi \theta \sin \theta \, d\theta$$

$$= \frac{\pi a^3}{3}$$

$$M_y = \int_0^\pi \int_0^a (\theta)(r \cos \theta) \, r \, dr \, d\theta$$

$$= \frac{a^3}{3} \int_0^\pi \theta \cos \theta \, d\theta$$

$$= -\frac{2a^3}{3}$$

$$(\overline{x}, \overline{y}) = \left(\frac{-\frac{2a^3}{3}}{\frac{\pi^2 a^2}{4}}, \frac{\frac{\pi a^3}{3}}{\frac{\pi^2 a^2}{4}} \right)$$

$$= \left(-\frac{8a}{3\pi^2}, \frac{4a}{3\pi} \right)$$

21. $m = \int_1^{e^2} \int_0^{\ln x} dy \, dx$

$$= \int_1^{e^2} \ln x \, dx$$

$$= e^2 + 1$$

$$M_x = \int_1^{e^2} \int_0^{\ln x} y \, dy \, dx$$

$$= \frac{1}{2} \int_1^{e^2} (\ln x)^2 \, dx$$

$$= \frac{1}{2} \left[x \ln^2 x - 2 \int_0^{e^2} \ln x \, dx \right]$$

$$= e^2 - 1$$

$$M_y = \int_1^{e^2} \int_0^{\ln x} x \, dy \, dx$$

$$= \int_1^{e^2} x \ln x \, dx$$

$$= e^4 - \frac{e^4}{4} + \frac{1}{4}$$

$$= \frac{1}{4} \left(3e^4 + 1 \right)$$

$$(\overline{x}, \overline{y}) = \left(\frac{3e^4 + 1}{4(e^2 + 1)}, \frac{e^2 - 1}{e^2 + 1} \right)$$

23. $I_z = 4 \int_0^1 \int_0^1 (x^2 + y^2) x^2 y^2 \, dy \, dx$

$$= 4 \int_0^1 \int_0^1 \left(x^4 y^2 + x^2 y^4 \right) dy \, dx$$

$$= \frac{4}{15} \int_0^1 (5x^4 + 3x^2) \, dx$$

$$= \frac{8}{15}$$

25. $m = \int_0^{\pi/2} \int_0^{\sqrt{2 \sin 2\theta}} r \, dr \, d\theta$

$$= \int_0^{\pi/2} \sin 2\theta \, d\theta$$

$$= 1$$

$$M_y = \int_0^{\pi/2} \int_0^{\sqrt{2 \sin 2\theta}} (r \cos \theta) \, r \, dr \, d\theta$$

$$= \int_0^{\pi/2} \frac{2\sqrt{2}}{3} \cos \theta (\sin 2\theta)^{3/2} \, d\theta$$

$$\approx 0.56 \qquad \textit{Use technology to approximate.}$$

$\overline{x} \approx 0.56$, and by symmetry, $y \approx 0.56$.
$(\overline{x}, \overline{y}) = (0.56, 0.56)$.

27. $m = \int_0^a \int_0^{bx/a} kx \, dy \, dx$

$$= \frac{kb}{a} \int_0^a x^2 \, dx = \frac{a^2 bk}{3}$$

$$M_x = \int_0^a \int_0^{bx/a} kxy \, dy \, dx$$

$$= \frac{kb^2}{2a^2} \int_0^a x^3 \, dx$$

$$= \frac{a^2 b^2 k}{8}$$

$$M_y = \int_0^a \int_0^{bx/a} kx^2 \, dy \, dx$$

$$= \frac{kb}{a} \int_0^a x^3 \, dx$$

$$= \frac{a^3 bk}{4}$$

$$(\overline{x}, \overline{y}) = \left(\frac{\frac{a^3 bk}{4}}{\frac{a^2 bk}{3}}, \frac{\frac{a^2 b^2 k}{8}}{\frac{a^2 bk}{3}} \right) = \left(\frac{3a}{4}, \frac{3b}{8} \right)$$

29. The distance from the point (x, y) to the line $x = \frac{a}{2}$ through the center of the lamina is $s = |x - \frac{a}{2}|$. Thus, the moment of inertia is

$$I_L = \int_{-\pi/2}^{\pi/2} \int_0^{a\cos\theta} \left(r\cos\theta - \frac{a}{2} \right)^2 r \, dr \, d\theta$$

$$= a^4 \int_{-\pi/2}^{\pi/2} \left[\frac{\cos^6\theta}{4} - \frac{\cos^4\theta}{3} + \frac{\cos^2\theta}{8} \right] d\theta$$

$$= \frac{a^4 \pi}{64}$$

31. If $\rho = 1$ then $m = A = \pi ab$. Using symmetry:

$$I_x = 4 \int_0^a \int_0^{\frac{b}{a}\sqrt{a^2-x^2}} y^2 \, dy \, dx$$

$$= \frac{4}{3} \int_0^a \left(\frac{b}{a}\sqrt{a^2 - x^2} \right)^3 dx$$

$$= \frac{4b^3}{3a^3} \int_0^a (a^2 - x^2)^{3/2} \, dx$$

$$= \frac{b^3}{3a^3} \left(\frac{3a^4\pi}{4} \right)$$

$$= \frac{ab^3\pi}{4}$$

$$= (\pi ab)\frac{b^2}{4}$$

$$= \frac{1}{4}mb^2$$

33. $m = \dfrac{1}{8}\left(\dfrac{4}{3}\pi a^2 \right) = \dfrac{\pi a^3}{6}$

$$M_{yz} = \int_0^{\pi/2} \int_0^a \int_0^{\sqrt{a^2-r^2}} (r\cos\theta)\, r \, dz \, dr \, d\theta$$

$$= \int_0^{\pi/2} \int_0^a r^2 \cos\theta \sqrt{a^2 - r^2} \, dr \, d\theta$$

$$= \int_0^{\pi/2} \frac{\pi}{16} a^4 \cos\theta \, d\theta$$

$$= \frac{\pi a^4}{16}$$

$$\overline{x} = \frac{\frac{\pi a^4}{16}}{\frac{\pi a^3}{6}} = \frac{3a}{8}$$

By symmetry, $\overline{x} = \overline{y} = \overline{z}$, so the centroid is $\left(\dfrac{3a}{8}, \dfrac{3a}{8}, \dfrac{3a}{8} \right)$.

35. $P(X + Y \leq 1) = \displaystyle\int_0^1 \int_0^{1-x} xe^{-x}e^{-y} \, dy \, dx$

$$= -\int_0^1 (xe^{-1} - xe^{-x}) \, dx$$

$$= 1 - \frac{5}{2}e^{-1}$$

37. $P(X + Y \leq 8) = \dfrac{1}{8}\displaystyle\int_0^8 \int_0^{8-x} e^{-x/2}e^{-y/4} \, dy \, dx$

$$= -\frac{1}{2}\int_0^8 e^{-x/2}(e^{-2+x/4} - 1) \, dx$$

$$= e^{-4}(e^4 - 2e^2 + 1)$$

$$\approx 0.7476450724$$

The probability is roughly 75%.

39. $A = \int_0^3 \int_{2x/3}^{7-5x/3} y \, dy \, dx = \dfrac{63}{2}$

$A_1 = \int_0^3 \int_{2x/3}^{7-5x/3} xy \, dy \, dx = \dfrac{231}{8}$

$A_2 = \int_0^3 \int_{2x/3}^{7-5x/3} y^2 \, dy \, dx = \dfrac{469}{4}$

$\overline{x} = \dfrac{A_1}{A} \qquad \overline{y} = \dfrac{A_2}{A}$

$\quad = \dfrac{11}{12} \qquad \quad = \dfrac{67}{18}$

The center of pressure on the sail is $\left(\frac{11}{12}, \frac{67}{18}\right)$.

41. $A = \int_0^1 \int_{x^2}^1 dy \, dx$

$\quad = \int_0^1 \left(1 - x^2\right) dx$

$\quad = \dfrac{2}{3}$

The average value is:

$\dfrac{1}{A} \int_0^1 \int_{x^2}^1 e^x y^{-1/2} \, dy \, dx = \dfrac{3}{2} \int_0^1 (2e^x - 2xe^x) dx$

$\qquad\qquad\qquad\qquad\qquad = \dfrac{3}{2}(2e - 4)$

$\qquad\qquad\qquad\qquad\qquad = 3(e - 2)$

43. Since the sphere is symmetric in each of the eight octants, and xyz is positive in four and negative in four, the average value is 0.

45. $m = \dfrac{ab}{2}$

$I_z = \int_0^a \int_0^{b-bx/a} \left(x^2 + y^2\right) dy \, dx$

$\quad = \int_0^a \left[-\dfrac{bx^3}{a} - \dfrac{b^3 x^3}{3a^3} + \dfrac{b^3 x^2}{a^2} + bx^2 - \dfrac{b^3 x^3}{a} + \dfrac{b^3}{3} \right] dx$

$\quad = \dfrac{ab(a^2 + b^2)}{12}$

$d^2 = \dfrac{2ab(a^2 + b^2)}{12ab}; \quad d = \sqrt{\dfrac{a^2 + b^2}{6}}$

47. $m = \int_1^2 \int_1^{x^2} x^2 y \, dy \, dx$

$\quad = \dfrac{1}{2} \int_1^2 x^2 \left(x^4 - 1\right) dx$

$\quad = \dfrac{332}{42}$

$I_x = \int_1^2 \int_1^{x^2} x^2 y \, y^2 \, dy \, dx$

$\quad = \dfrac{1}{4} \int_1^2 x^2 \left(x^8 - 1\right) dx$

$\quad = \dfrac{1,516}{33}$

$d^2 = \dfrac{1,516(42)}{33(332)}; \quad d = \sqrt{\dfrac{5,306}{913}} \approx 2.4107$

49. a. Fix x at $x = x_0$. Then,

$\dfrac{\partial C}{\partial t}(x_0, t) = \dfrac{C_0}{\sqrt{k\pi t}} \dfrac{x_0^2}{4kt^2} e^{-x_0^2/(4kt)} - \dfrac{C_0}{2\sqrt{k\pi t^{3/2}}} e^{-x_0^2/(4kt)}$

When $\dfrac{\partial C}{\partial t}(x_0, t) = 0$, $t = t_m$ and

$\dfrac{x_0^2}{4kt_m^2} = \dfrac{1}{2t_m}$

$t_m = \dfrac{x_0^2}{2k}$ and $x_0^2 = 2kt_m$

$C_m(x_0) = \dfrac{C_0}{\sqrt{k\pi \left(\frac{x_0^2}{2k}\right)}} \exp\left(\dfrac{-x_0^2}{2x_0}\right)$

$\qquad\quad = \sqrt{\dfrac{2}{\pi}} C_0 \dfrac{e^{-1/2}}{x_0}$

b. We want x_m to satisfy

$$\sqrt{\frac{2}{\pi}} C_0 \frac{e^{-1/2}}{x_m} \geq 0.25 C_0$$

$$4\sqrt{\frac{2}{\pi}} e^{-1/2} \geq x_m$$

$$x_m \leq 1.9358$$

The danger zone is approximately 1.9 miles.

c. $\quad A = \displaystyle\int_0^{x_m} \int_0^{t_m} dt\, dx$

$\quad \approx \displaystyle\int_0^{1.9358} \int_0^{x^2/2k} dt\, dx$

$\quad = \displaystyle\int_0^{1.9358} \frac{x^2}{2k} dx$

$\quad \approx \dfrac{1.21}{k}$

The average value is:

$$\frac{1}{A} \int_0^{1.9358} \int_0^{x^2/(2k)} \frac{C_0}{\sqrt{k\pi t}} \exp\left(\frac{-x^2}{4kt}\right) dt\, dx$$

d. Answers vary. For example, we may say that the danger period at point x on the riverbank is the period of time $t_1 < t < t_2$ when the concentration exceeds $0.25 C_0$.

51. $dF = \delta\, dV; \ d = h - z$
$dW = \delta(h - z)\, dV$, so
$W = \delta \iiint (h - z)\, dV$

53. $W = \delta(15)\pi(36)(10) = 5{,}400\pi\delta$

$$W = 4\delta \int_0^6 \int_0^{\sqrt{36-x^2}} \int_{-5}^5 (15 - z)\, dz\, dy\, dx$$

$$= 4\delta \int_0^6 \int_0^{\sqrt{36-x^2}} 150\, dy\, dx$$

$$= 5{,}400\pi\delta$$

55. Sum up above ΔF.

57. Numerical integration gives

$$F = 4.39741 Gm \text{ vs } 5.0 Gm$$

59. Assume that the curve C is smooth and is described by $y = f(x)$ for $a \leq x \leq b$. Assume also that the axis of revolution is the y-axis. The distance traveled by the centroid $(\bar{x}, \bar{y})$ is $d = 2\pi h$ where $h = \bar{x}$. The area of the surface of revolution is

$$S = \int_a^b 2\pi x \sqrt{1 + [f'(x)]^2}\, dx$$

But,

$$h = \bar{x} = \frac{1}{L} \int_a^b x\, ds$$

$$= \frac{1}{L} \int_a^b x\sqrt{1 + [f'(x)]^2}\, dx$$

Thus, $S = 2\pi L h$.

12.7 Cylindrical and Spherical Coordinates, page 995

1. Cylindrical coordinates are best used in problems with axial symmetry (that is, with cylinders), spherical coordinates in problems with radial symmetry (that is, with spheres), and rectangular coordinates in all other cases.

3. a. $\left(4, \frac{\pi}{2}, \sqrt{3}\right)$ **b.** $\left(\sqrt{19}, \frac{\pi}{2}, \cos^{-1}\frac{\sqrt{3}}{\sqrt{19}}\right)$

5. a. $\left(\sqrt{5}, \tan^{-1}2, 3\right)$

b. $\left(\sqrt{14}, \tan^{-1}2, \cos^{-1}\frac{3}{\sqrt{14}}\right)$

7. a. $\left(-\frac{3}{2}, \frac{3\sqrt{3}}{2}, -3\right)$

b. $\left(\sqrt{18}, \frac{2\pi}{3}, \cos^{-1}\frac{-3}{\sqrt{18}}\right) = \left(3\sqrt{2}, \frac{2\pi}{3}, \frac{3\pi}{4}\right)$

9. a. $\left(\sqrt{2}, \sqrt{2}, \pi\right)$

  b. $\left(\sqrt{\pi^2+4}, \frac{\pi}{4}, \cos^{-1}\frac{\pi}{\sqrt{\pi^2+4}}\right)$

11. a. $\left(\frac{3}{2}, \frac{\sqrt{3}}{2}, -1\right)$ **b.** $\left(\sqrt{3}, \frac{\pi}{6}, -1\right)$

13. a. $(\sin 3\cos 2,\ \sin 3\sin 2,\ \cos 3)$

  b. $(\sin 3, 2, \cos 3)$

15. $z = r^2\cos 2\theta$ **17.** $r = \dfrac{6z}{\sqrt{4-13\cos^2\theta}}$

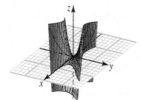 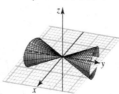

19. $\phi = \dfrac{\pi}{4}$ **21.** $\rho = \dfrac{4\cot\phi\csc\phi}{3-2\cos^2\theta}$

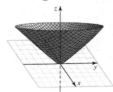 

23. $z = 2xy$ **25.** $z = x^2 - y^2$

27. $xz = 1$

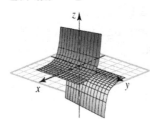

29. $\displaystyle\int_0^\pi\int_0^2\int_0^{\sqrt{4-r^2}} r\sin\theta\,dz\,dr\,d\theta$

$= \displaystyle\int_0^\pi\int_0^2 (\sin\theta)\sqrt{4-r^2}\,r\,dr\,d\theta$

$= \dfrac{8}{3}\displaystyle\int_0^\pi \sin\theta\,d\theta$

$= \dfrac{16}{3}$

31. $\displaystyle\int_0^{\pi/2}\int_0^{2\pi}\int_0^2 \cos\phi\sin\phi\,d\rho\,d\theta\,d\phi$

$= 2\displaystyle\int_0^{\pi/2}\int_0^{2\pi} \cos\phi\sin\phi\,d\theta\,d\phi$

$= 4\pi\displaystyle\int_0^{\pi/2} \cos\phi\sin\phi\,d\phi$

$= 2\pi$

33. $\displaystyle\int_0^{2\pi}\int_0^4\int_0^1 zr\,dz\,dr\,d\theta$

$= \dfrac{1}{2}\displaystyle\int_0^{2\pi}\int_0^4 r\,dr\,d\theta$

$= 4\displaystyle\int_0^{2\pi} d\theta$

$= 8\pi$

35. $\displaystyle\int_0^{\pi/2}\int_0^{\cos\theta}\int_0^{1-r^2} r\sin\theta\,dz\,dr\,d\theta$

$= \displaystyle\int_0^{\pi/2}\int_0^{\cos\theta} r(1-r^2)\sin\theta\,dr\,d\theta$

$= \displaystyle\int_0^{\pi/2}\frac{\cos^2\theta\sin\theta}{2}\,d\theta - \int_0^{\pi/2}\frac{\cos^4\theta\sin\theta}{4}\,d\theta$

$= \left[-\dfrac{\cos^3\theta}{6} + \dfrac{\cos^5\theta}{20}\right]_0^{\pi/2}$

$= \dfrac{7}{60}$

37. Set up coordinates so the axis of the cylinder is the z-axis and the origin is the centroid.

$$I_x = \int_0^{2\pi} \int_0^R \int_{-h/2}^{h/2} r^2 r\, dz\, dr\, d\theta$$

$$= h \int_0^{2\pi} \int_0^R r^3\, dr\, d\theta$$

$$= h \int_0^{2\pi} \frac{R^4}{4}\, d\theta$$

$$= \frac{1}{2} h \pi R^4$$

39. $\displaystyle\iiint_D \left(x^2 + y^2\right) dx\, dy\, dz$

There are four octants where $z > 0$.

$$4 \int_0^{1/\pi} \int_0^{\pi/2} \int_0^a r^4 r\, dr\, d\theta\, dz$$

$$= \frac{4a^6}{6} \int_0^{1/\pi} \int_0^{\pi/2} d\theta\, dz$$

$$= \frac{4a^6}{6} \int_o^{1/\pi} \frac{\pi}{2}\, dz$$

$$= \frac{a^6}{3}$$

41. $\displaystyle m = \int_0^{2\pi} \int_0^9 \int_r^9 r\, dz\, dr\, d\theta$

$$= \int_0^{2\pi} \int_0^9 \left(9r - r^2\right) dr\, d\theta$$

$$= \frac{243}{2} \int_0^{2\pi} d\theta$$

$$= 243\pi$$

By symmetry, $\overline{x} = \overline{y} = 0$.

$$M_{xy} = \int_0^{2\pi} \int_0^9 \int_r^9 rz\, dz\, dr\, d\theta$$

$$= \frac{1}{2} \int_0^{2\pi} \int_0^9 r\left(9^2 - r^2\right) dr\, d\theta$$

$$= \frac{6{,}561}{8} \int_0^{2\pi} d\theta = \frac{6{,}561\pi}{4}$$

$$\overline{z} = \frac{\frac{6{,}561\pi}{4}}{243\pi} = \frac{27}{4}$$

The centroid is $\left(0, 0, \frac{27}{4}\right)$.

43. Use cylindrical coordinates.

a. $\displaystyle m = \int_0^{2\pi} \int_0^3 \int_r^{\sqrt{9-r^2}} \left(r^2 \sin\theta \cos\theta + z\right) r\, dz\, dr\, d\theta$

b.

$$\overline{x} = \frac{1}{m} \int_0^{2\pi} \int_0^3 \int_0^{\sqrt{9-r^2}} r\cos\theta\left(r^2 \sin\theta \cos\theta + z\right) r\, dz\, dr\, d\theta$$

c. $\displaystyle I_z = \int_0^{2\pi} \int_0^3 \int_0^{\sqrt{9-r^2}} r^2 \left(r^2 \sin\theta \cos\theta + z\right) r\, dz\, dr\, d\theta$

45. $\displaystyle m = \int_0^{2\pi} \int_0^\pi \int_0^{2\sin\phi} \rho\left(\rho^2 \sin\phi\right) d\rho\, d\phi\, d\theta$

$$= \int_0^{2\pi} \int_0^\pi \sin\phi \left[\frac{(2\sin\phi)^4}{4}\right] d\phi\, d\theta$$

$$= 4 \int_0^{2\pi} \left[\frac{-\cos^5\phi}{5} + \frac{2\cos^3\phi}{3} - \cos\phi\right]_0^\pi d\theta$$

$$= \int_0^{2\pi} \frac{64}{15}\, d\theta$$

$$= \frac{128\pi}{15}$$

> **SURVIVAL HINT:** *You could expected this answer because $f(x, y, z) = x + y + z$ satisfies $f(-x, -y, -z) = -f(x, y, z)$.*

47. The sphere has volume $V = \frac{4}{3}\pi a^3$.
Conjectures vary; the average value of θ is π and of ϕ is half of that.
The average value of θ over the sphere is:

$$\frac{1}{V}\int_0^{2\pi}\int_0^{\pi}\int_0^{a}\theta\rho^2\sin\phi\,d\rho\,d\phi\,d\theta$$

$$=\frac{1}{3V}\int_0^{2\pi}\int_0^{\pi}a^3\theta\sin\phi\,d\phi\,d\theta$$

$$=\frac{2a^3}{3V}\int_0^{2\pi}\theta\,d\theta$$

$$=\frac{4\pi^2 a^3}{3V}$$

$$=\pi$$

The average value of ϕ is

$$\frac{1}{V}\int_0^{2\pi}\int_0^{\pi}\int_0^{a}\phi\,\rho^2\sin\phi\,d\rho\,d\phi\,d\theta$$

$$=\frac{1}{3V}\int_0^{2\pi}\int_0^{\pi}a^3\phi\sin\phi\,d\phi\,d\theta$$

$$=\frac{a^3}{3V}\int_0^{2\pi}\pi\,d\theta$$

$$=\frac{2\pi^2 a^3}{3V}$$

$$=\frac{\pi}{2}$$

49. $\displaystyle\iiint_D\left(x^2+y^2+z^2\right)dx\,dy\,dz$

$$=8\int_0^{\pi/2}\int_0^{\pi/2}\int_0^{\sqrt{2}}\rho^4\sin\phi\,d\rho\,d\theta\,d\phi$$

$$=\frac{32\sqrt{2}}{5}\int_0^{\pi/2}\int_0^{\pi/2}\sin\phi\,d\theta\,d\phi$$

$$=\frac{16\sqrt{2}\,\pi}{5}\int_0^{\pi/2}\sin\phi\,d\phi$$

$$=\frac{16\pi\sqrt{2}}{5}$$

51. $\displaystyle\iiint_D\frac{dx\,dy\,dz}{\sqrt{x^2+y^2+z^2}}$

$$=8\int_0^{\pi/2}\int_0^{\pi/2}\int_0^{\sqrt{3}}\rho\sin\phi\,d\rho\,d\theta\,d\phi$$

$$=12\int_0^{\pi/2}\int_0^{\pi/2}\sin\phi\,d\theta\,d\phi$$

$$=6\pi\int_0^{\pi/2}\sin\phi\,d\phi$$

$$=6\pi$$

53. The projected region is $4=x^2+y^2$

$$V=\int_0^{2\pi}\int_0^{1}\int_0^{4-r^2}r\,dz\,dr\,d\theta$$

$$=\int_0^{2\pi}\int_0^{1}r\left(4-r^2\right)dr\,d\theta$$

$$=\frac{7}{4}\int_0^{2\pi}d\theta$$

$$=\frac{7\pi}{2}$$

55. $\displaystyle V=2\int_0^{\pi/2}\int_0^{2\sin\theta}\int_0^{4-r^2}r\,dz\,dr\,d\theta$

$$=2\int_0^{\pi/2}\int_0^{2\sin\theta}r\left(4-r^2\right)dr\,d\theta$$

$$=2\int_0^{\pi/2}\left(8\sin^2\theta-4\sin^4\theta\right)d\theta$$

$$=2[4\theta-4\sin\theta\cos\theta]\big|_0^{\pi/2}$$

$$\quad-8\left[\frac{3\theta}{8}-\frac{\sin 2\theta}{4}+\frac{\sin 4\theta}{32}\right]\Big|_0^{\pi/2}$$

$$=\frac{5\pi}{2}$$

57. a. Force $=\dfrac{GmM}{R^2}=\dfrac{Gm\delta(4\pi a^3)}{3R^2}$

b. With $R=4$ and $a=3$, we obtain

$$\text{Force}=\frac{Gm\delta(9\pi)}{4}$$

c. With a rectangular mass m, we got a poor approximation using the center of mass when the separating distance was small. The approximation improves as the distance increases. With the sphere we always get perfect agreement. Apparently the symmetry of the sphere plays the key role. Either a computer with symbolic integration capability or determined work by hand will show that for the sphere the center of mass method gives the exact result.

59. $I = I_x + I_y + I_z$

$$= \iiint_S (y^2 + z^2)\,dV$$

$$+ \iiint_S (x^2 + z^2)\,dV$$

$$+ \iiint_S (x^2 + y^2)\,dV$$

$$= \iiint_S (2x^2 + 2y^2 + 2z^2)\,dV$$

$$= 2\int_0^\pi \int_0^{2\pi} \int_0^1 \rho^2 \rho^2 \sin\phi\,d\rho\,d\theta\,d\phi$$

$$= 2\int_0^{2\pi} \int_0^\pi \int_0^1 \rho^4 \sin\phi\,d\rho\,d\phi\,d\theta$$

$$= \frac{2}{5}\int_0^{2\pi} \int_0^\pi \sin\phi\,d\phi\,d\theta$$

$$= \frac{4}{5}\int_0^{2\pi} d\theta = \frac{8\pi}{5}$$

12.8 Jacobians: Change of Variables, page 1005

1. Find the Jacobian, map the region, then integrate.

3. $x = u - v,\ y = u + v$

$$\frac{\partial(x,y)}{\partial(u,v)} = \begin{vmatrix} \frac{\partial x}{\partial u} & \frac{\partial x}{\partial v} \\ \frac{\partial y}{\partial u} & \frac{\partial y}{\partial v} \end{vmatrix} = \begin{vmatrix} 1 & -1 \\ 1 & 1 \end{vmatrix} = 2$$

5. $x = u^2,\ y = u + v$

$$\frac{\partial(x,y)}{\partial(u,v)} = \begin{vmatrix} \frac{\partial x}{\partial u} & \frac{\partial x}{\partial v} \\ \frac{\partial y}{\partial u} & \frac{\partial y}{\partial v} \end{vmatrix} = \begin{vmatrix} 2u & 0 \\ 1 & 1 \end{vmatrix} = 2u$$

7. $x = e^{u+v},\ y = e^{u-v}$

$$\frac{\partial(x,y)}{\partial(u,v)} = \begin{vmatrix} \frac{\partial x}{\partial u} & \frac{\partial x}{\partial v} \\ \frac{\partial y}{\partial u} & \frac{\partial y}{\partial v} \end{vmatrix}$$

$$= \begin{vmatrix} e^{u+v} & e^{u+v} \\ e^{u-v} & -e^{u-v} \end{vmatrix}$$

$$= -2e^{u+v}e^{u-v}$$

$$= -2e^{2u}$$

9. $x = e^u \sin v,\ y = e^u \cos v$

$$\frac{\partial(x,y)}{\partial(u,v)} = \begin{vmatrix} \frac{\partial x}{\partial u} & \frac{\partial x}{\partial v} \\ \frac{\partial y}{\partial u} & \frac{\partial y}{\partial v} \end{vmatrix}$$

$$= \begin{vmatrix} e^u \sin v & e^u \cos v \\ e^u \cos v & -e^u \sin v \end{vmatrix}$$

$$= -e^{2u}\sin^2 v - e^{2u}\cos^2 v$$

$$= -e^{2u}$$

11. $x = u + v - w,\ y = 2u - v + 3w,$
$z = -u + 2v - w$

$$\frac{\partial(x,y,z)}{\partial(u,v,w)} = \begin{vmatrix} 1 & 1 & -1 \\ 2 & -1 & 3 \\ -1 & 2 & -1 \end{vmatrix} = -9$$

13. $x = u\cos v,\ y = u\sin v,\ z = we^{uv}$

$$\frac{\partial(x,y,z)}{\partial(u,v,w)} = \begin{vmatrix} \cos v & -u\sin v & 0 \\ \sin v & u\cos v & 0 \\ vwe^{uv} & uwe^{uv} & e^{uv} \end{vmatrix}$$

$$= ue^{uv}$$

15. $u = 2x - 3y,\ v = x + 4y$

$$\frac{\partial(u,v)}{\partial(x,y)} = \begin{vmatrix} 2 & -3 \\ 1 & 4 \end{vmatrix} = 11, \text{ so}$$

$$\frac{\partial(x,y)}{\partial(u,v)} = \frac{1}{11}$$

17. $u = ye^{-x}$ and $v = e^x$

$y = ue^x \qquad x = \ln v$

$\quad = uv$

$$\frac{\partial(x,y)}{\partial(u,v)} = \begin{vmatrix} 0 & \dfrac{1}{v} \\ v & u \end{vmatrix} = -1$$

19. $u = \sqrt{x^2 + y^2}, \; v = \cot^{-1}\left(\dfrac{y}{x}\right)$

$$\frac{\partial(u,v)}{\partial(x,y)} = \begin{vmatrix} \dfrac{x}{\sqrt{x^2+y^2}} & \dfrac{y}{\sqrt{x^2+y^2}} \\ \dfrac{y}{x^2+y^2} & \dfrac{-x}{x^2+y^2} \end{vmatrix}$$

$$= \frac{-x^2 - y^2}{(x^2+y^2)\sqrt{x^2+y^2}}$$

$$= \frac{-1}{\sqrt{x^2+y^2}}$$

$$= -\frac{1}{u}$$

$$\frac{\partial(x,y)}{\partial(u,v)} = -u$$

21. $u = x^2 - y^2, v = x^2 + y^2$

$$\frac{\partial(u,v)}{\partial(x,y)} = \begin{vmatrix} 2x & -2y \\ 2x & 2y \end{vmatrix}$$

$$= 4xy + 4xy$$

$$= 8xy$$

$$\frac{\partial(x,y)}{\partial(u,v)} = \frac{1}{8xy}$$

23. $A(0,5) \quad \rightarrow \quad (5,-5)$

$\; B(6,5) \quad \rightarrow \quad (11,1)$

$\; C(6,0) \quad \rightarrow \quad (6,6)$

$\; O(0,0) \quad \rightarrow \quad (0,0)$

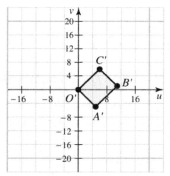

This transformation represents a rotation combined with a dilation.

25. $A(5,0) \quad \rightarrow \quad (5,0)$ Note that the new figure
$\; B(7,4) \quad \rightarrow \quad (7,-4)$ represents the same
$\; C(2,4) \quad \rightarrow \quad (2,-4)$ parallelogram reflected
$\; O(0,0) \quad \rightarrow \quad (0,0)$ with respect to the x-axis.

27. $x = u - uv, \; y = uv$

$$\frac{\partial(x,y)}{\partial(u,v)} = \begin{vmatrix} 1-v & -u \\ v & u \end{vmatrix}$$

$$= u - uv + uv$$

$$= u$$

$$dx\,dy = u\,du\,dv$$

29. Let $u = xy, \; v = \dfrac{y}{x}$. Then the transformed region has boundary lines $u = 1$, $u = 4$, and $v = 1$, $v = 4$. Thus, $x = \sqrt{\dfrac{u}{v}}, \; y = \sqrt{uv}$ and

$$\frac{\partial(x,y)}{\partial(u,v)} = \begin{vmatrix} \dfrac{1}{2\sqrt{uv}} & \dfrac{-1}{2v}\sqrt{\dfrac{u}{v}} \\ \dfrac{1}{2}\sqrt{\dfrac{v}{u}} & \dfrac{1}{2}\sqrt{\dfrac{u}{v}} \end{vmatrix}$$

$$= \frac{1}{2v}$$

The area is

$$A = \int_1^4 \int_1^4 \frac{1}{2v}\,dv\,du = 3\ln 2$$

For Problems 31-34, we have

$$u = x - y, \ v = x + y$$

so $x = \dfrac{u+v}{2}$, $y = \dfrac{v-u}{2}$. *The boundary lines* $x = 0$, $y = 0$, $x + y = 1$ *become* $-v = u$, $v = u$, $v = 1$, *respectively. The Jacobian of the transformation is*

$$\frac{\partial(x,y)}{\partial(u,v)} = \begin{vmatrix} \frac{\partial x}{\partial u} & \frac{\partial x}{\partial v} \\ \frac{\partial y}{\partial u} & \frac{\partial y}{\partial v} \end{vmatrix}$$

$$= \begin{vmatrix} \dfrac{1}{2} & \dfrac{1}{2} \\ -\dfrac{1}{2} & \dfrac{1}{2} \end{vmatrix}$$

$$= \frac{1}{2}$$

$dy \, dx = \frac{1}{2} \, du \, dv.$

31. $\displaystyle\iint_D \left(\frac{x-y}{x+y}\right)^5 dy \, dx$

$$= \int_0^1 \int_{-v}^v \frac{u^5}{v^5}\frac{1}{2} \, du \, dv$$

$$= \frac{1}{12}\int_0^1 (v - v)\, dv$$

$$= 0$$

33. $\displaystyle\iint_D (x-y)^5(x+y)^3 \, dy \, dx$

$$= \frac{1}{2}\int_0^1 \int_{-v}^v u^5 v^3 \, du \, dv$$

$$= \frac{1}{12}\int_0^1 (v^9 - v^9)\, dv$$

$$= 0$$

For Problems 35-40,

$$u = \frac{2x+y}{5}, \quad v = \frac{x-2y}{5}$$

or $y = u - 2v$ *and* $x = 2u + v$.

$$\begin{array}{lcl} A(0,0) & \rightarrow & (0,0) \\ B(1,-2) & \rightarrow & (0,1) \\ C(3,-1) & \rightarrow & (1,1) \\ D(2,1) & \rightarrow & (1,0) \end{array}$$ *and* $\dfrac{\partial(x,y)}{\partial(u,v)} = \begin{vmatrix} \frac{\partial x}{\partial u} & \frac{\partial x}{\partial v} \\ \frac{\partial y}{\partial u} & \frac{\partial y}{\partial v} \end{vmatrix}$

$$= \begin{vmatrix} 2 & 1 \\ 1 & -2 \end{vmatrix}$$

$$= -5$$

R *is the unit square in the* uv-*plane.*

35. $\displaystyle\iint_D \left(\frac{2x+y}{x-2y+5}\right)^2 dy \, dx$

$$= \int_0^1 \int_0^1 \left(\frac{u}{v+1}\right)^2 |5| \, du \, dv$$

$$= \frac{5}{3}\int_0^1 (v+1)^{-2} \, dv$$

$$= \frac{5}{6}$$

37. $\displaystyle\iint_D (2x+y)^2(x-2y) \, dy \, dx$

$$= \int_0^1 \int_0^1 (5u)^2(5v) \, |5| \, du \, dv$$

$$= \frac{5^4}{3}\int_0^1 v \, dv$$

$$= \frac{625}{6}$$

39. $\displaystyle\iint_D (2x+y)\tan^{-1}(x-2y) \, dy \, dx$

$$= \int_0^1 \int_0^1 (5u)\tan^{-1}(5v)|5| \, dv \, du$$

$$= \int_0^1 \left[25uv \tan^{-1}(5v) - \frac{5u\ln(25v^2+1)}{2} \right]_0^1 du$$

$$= \int_0^1 \left[25u \tan^{-1}5 - \frac{5u\ln 26}{2} \right] du$$

$$= \frac{25}{2}\tan^{-1}5 - \frac{5}{4}\ln 26$$

41. This problem is sufficiently simple that no transformation is necessary.

$$\iint_R e^{x+y}\, dA = \int_0^1 \int_0^{1-x} e^{x+y}\, dy\, dx$$

$$= \int_0^1 (e - e^x)\, dx$$

$$= 1$$

43. By looking at the function we see a suitable transformation can be obtained when $a = b = s = 1$ and $r = -1$.
$u = x + y$, $v = -x + y$, so
$y = \frac{1}{2}(u + v)$, $x = \frac{1}{2}(u - v)$.

$$\frac{\partial(x,y)}{\partial(u,v)} = \begin{vmatrix} \frac{\partial x}{\partial u} & \frac{\partial x}{\partial v} \\ \frac{\partial y}{\partial u} & \frac{\partial y}{\partial v} \end{vmatrix}$$

$$= \begin{vmatrix} \dfrac{1}{2} & -\dfrac{1}{2} \\ \dfrac{1}{2} & \dfrac{1}{2} \end{vmatrix}$$

$$= \frac{1}{2}$$

$$\begin{array}{ccc} A(0,0) & \to & (0,0) \\ B(1,1) & \to & (2,0) \\ C(0,2) & \to & (2,2) \\ D(-1,1) & \to & (0,2) \end{array}$$

$$dy\, dx = \frac{1}{2}\, du\, dv$$

$$\iint_R \left(\frac{x+y}{2}\right)^2 e^{(y-x)/2}\, dy\, dx$$

$$= \int_0^2 \int_0^2 \left(\frac{u}{2}\right)^2 e^{v/2} \left(\frac{1}{2}\right) dv\, du$$

$$= \frac{1}{4}(e - 1) \int_0^2 u^2\, du$$

$$= \frac{2}{3}(e - 1)$$

45. $x = ar\cos\theta$, $y = br\sin\theta$

Since $\dfrac{x^2}{a^2} + \dfrac{y^2}{b^2} = r^2$, the transformed region is $r \le 1, 0 \le \theta \le \frac{\pi}{2}$.

$$\frac{\partial(x,y)}{\partial(r,\theta)} = \begin{vmatrix} \frac{\partial x}{\partial r} & \frac{\partial x}{\partial \theta} \\ \frac{\partial y}{\partial r} & \frac{\partial y}{\partial \theta} \end{vmatrix}$$

$$= \begin{vmatrix} a\cos\theta & -ar\sin\theta \\ b\sin\theta & br\cos\theta \end{vmatrix}$$

$$= abr$$

$$dy\, dx = abr\, dr\, d\theta$$

$$\iint_{D^*} \exp\left(-\frac{x^2}{a^2} - \frac{y^2}{b^2}\right) dy\, dx$$

$$= ab \int_0^{\pi/2} \int_0^1 r\, e^{-r^2}\, dr\, d\theta$$

$$= \frac{ab}{2} \int_0^{\pi/2} \left(1 - e^{-1}\right) d\theta$$

$$= \frac{ab\pi}{4}(1 - e^{-1})$$

47. Let $\begin{cases} u = \dfrac{x}{\sqrt{5}} \\ v = \dfrac{y}{2} \end{cases}$

Then,

$$\frac{\partial(x,y)}{\partial(u,v)} = \begin{vmatrix} \sqrt{5} & 0 \\ 0 & 2 \end{vmatrix} = 2\sqrt{5}$$

$$\iint_{D^*} e^{-(4x^2+5y^2)}\, dx\, dy$$

$$= \int_{-1}^1 \int_{-\sqrt{1-v^2}}^{\sqrt{1-v^2}} e^{-4\left(\sqrt{5}u\right)^2 - 5(2v)^2} \left(2\sqrt{5}\right) du\, dv$$

$$= 8\sqrt{5} \int_0^1 \int_0^{\sqrt{1-v^2}} e^{-20(u^2+v^2)}\, du\, dv$$

$$= 8\sqrt{5}\int_0^{\pi/2}\int_0^1 e^{-20r^2} r\, dr\, d\theta$$

$$= 8\sqrt{5}\int_0^{\pi/2} \frac{(1-e^{-20})}{40}\, d\theta$$

$$= \frac{\sqrt{5}\pi}{10}(1-e^{-20})$$

49. Let $u = xy^3$, $v = \dfrac{y}{x}$, $\dfrac{u}{v^3} = \dfrac{xy^3}{\left(\frac{y}{x}\right)^3} = x^4$

so $x = \sqrt[4]{\dfrac{u}{v^3}}$; $y = vx$ so $y = \sqrt[4]{uv}$

$$\frac{\partial(u,v)}{\partial(x,y)} = \begin{vmatrix} y^3 & 3xy^2 \\ \dfrac{-y}{x^2} & \dfrac{1}{x} \end{vmatrix}$$

$$= \frac{y^3}{x} + \frac{3xy^3}{x^2}$$

$$= \frac{4y^3}{x}$$

$$\frac{\partial(x,y)}{\partial(u,v)} = \frac{x}{4y^3} = \frac{1}{4}\frac{\sqrt[4]{\frac{u}{v^3}}}{(\sqrt[4]{uv})^3} = \frac{1}{4\sqrt{uv^3}}$$

The area of D^* is

$$A = \int_1^2\int_3^6 \frac{1}{4\sqrt{uv^3}}\, du\, dv$$

$$= \frac{(3\sqrt{2}-4)\sqrt{3}}{2}$$

$$\approx 0.2101$$

$$\bar{x} = \frac{1}{A}\int\!\!\int_{D^*} x\, dA$$

$$= \frac{1}{A}\int_1^2\int_3^6 \sqrt[4]{\frac{u}{v^3}}\left(\frac{1}{4\sqrt{uv^3}}\right) du\, dv$$

$$= \frac{1}{A}\int_1^2\int_3^6 \frac{1}{4}u^{-1/4}v^{-9/4}\, du\, dv$$

$$\approx \frac{0.2401}{A}$$

$$\approx 1.14$$

$$\bar{y} = \frac{1}{A}\int\!\!\int_{D^*} y\, dA$$

$$= \frac{1}{A}\int_1^2\int_3^6 \sqrt[4]{uv}\left(\frac{1}{4\sqrt{uv^3}}\right) du\, dv$$

$$= \frac{1}{A}\int_1^2\int_3^6 \frac{1}{4}u^{-1/4}v^{-5/4}\, du\, dv$$

$$\approx \frac{0.3297}{A}$$

$$\approx 1.57$$

Thus, the centroid is located at $(1.14, 1.57)$ in the xy-plane.

51. $x = au$, $y = bv$, $z = cw$.

$$\frac{\partial(x,y,z)}{\partial(u,v,w)} = \begin{vmatrix} a & 0 & 0 \\ 0 & b & 0 \\ 0 & 0 & c \end{vmatrix} = abc$$

$$dy\, dx\, dz = abc\, du\, dv\, dw$$

$$\frac{x^2}{a^2} + \frac{y^2}{b^2} + \frac{z^2}{c^2} = u^2 + v^2 + w^2$$

Thus, the volume of the ellipsoid is equal to abc times the volume of sphere $x^2 + y^2 + z^2 = 1$:

$$abc\left[\frac{4}{3}\pi(1)\right] = \frac{4}{3}\pi abc$$

53. Let $u = x - y$, $v = x + y$ so that $x = \dfrac{u+v}{2}$, $y = \dfrac{-u+v}{2}$ and

$$\frac{\partial(x,y)}{\partial(u,v)} = \begin{vmatrix} \dfrac{1}{2} & \dfrac{1}{2} \\ -\dfrac{1}{2} & \dfrac{1}{2} \end{vmatrix} = \frac{1}{2}$$

The given region R is bounded by the lines $x - 3y = 1$, $x + y = 1$, and $x = 4$ which transform into $2u - v = 1$, $v = 1$, and $u + v = 8$.

$$\iint\limits_R \ln\left(\frac{x-y}{x+y}\right) dy\, dx$$

$$= \frac{1}{2}\int_1^5 \int_{(v+1)/2}^{8-v} \ln\left(\frac{u}{v}\right) du\, dv$$

$$= \frac{1}{4}\int_1^5 \left[-(v+1)\ln\frac{v+1}{2v}\right.$$

$$\left. + 2(8-v)\ln\frac{8-v}{v} + 3(v-5)\right] dv$$

$$= \frac{1}{4}\left(49\ln 7 - \frac{75}{2}\ln 5 - 27\ln 3 + 6\right)$$

55. We find $A = \frac{1}{5}$, $B = \frac{6}{5}$, so the transformed region is $\frac{1}{5}u^2 + \frac{6}{5}v^2 = 1$. Since $u = x + 2y$, $v = 2x - y$, we have

$$x = \frac{u + 2v}{5}, \quad y = \frac{2u - v}{5}$$

$$\frac{\partial(x,y)}{\partial(u,v)} = \begin{vmatrix} \dfrac{1}{5} & \dfrac{2}{5} \\ \dfrac{2}{5} & -\dfrac{1}{5} \end{vmatrix} = -\frac{1}{5}$$

Thus, the area is

$$A = \iint\limits_R dy\, dx$$

$$= 4\int_0^{\sqrt{5/6}} \int_0^{\sqrt{5-6v^2}} \frac{1}{5}\, du\, dv$$

$$= \frac{4}{5}\int_0^{\sqrt{5/6}} \sqrt{5 - 6v^2}\, dv = \frac{\pi}{\sqrt{6}}$$

57. Let $u = x + y$, $v = 2y$, $x = u - v/2$, $y = v/2$. The boundary $x + y = 0$ becomes $u = 0$; $x + y = 2$ becomes $u = 2$; $y = 0$ becomes $v = 0$; and $y = 1$ becomes $v = 2$.

$$\frac{\partial(x,y)}{\partial(u,v)} = \begin{vmatrix} 1 & -\dfrac{1}{2} \\ 0 & \dfrac{1}{2} \end{vmatrix} = \frac{1}{2}$$

$$dy\, dx = \tfrac{1}{2}\, du\, dv$$

$$\iint\limits_R f(x+y)\, dy\, dx$$

$$= \frac{1}{2}\int_0^2 \int_0^2 f(u)\, dv\, du$$

$$= \int_0^2 f(u)\, du$$

$$= \int_0^2 f(t)\, dt$$

59. For the circumference, solve the equation

$$\frac{x^2}{a^2} + \frac{y^2}{b^2} = 1$$

for y to obtain $y = b\sqrt{1 - \dfrac{x^2}{a^2}}$. Then

$$\frac{dy}{dx} = \frac{-bx}{a\sqrt{a^2 - x^2}}$$

and the circumference, C, of the ellipse is given by the improper integral

$$C = 4\int_0^a \sqrt{1 + \left(\frac{-bx}{a\sqrt{a^2 - x}}\right)}\, dx$$

$$= 4\int_0^a \frac{1}{a\sqrt{a^2 - x^2}}\sqrt{a^4 + (b^2 - a^2)x^2}\, dx$$

Chapter 12 Review

 Studying for a chapter examination is a personal process, one which nobody else can do for you. Simply take the time to review what you have done.

SURVIVAL HINT: Work all of Chapter 12 problems in the Proficiency Examination (whether they are assigned or not). Work through all of the problems before looking at the answers, and *then* correct each of the problems. The answers to all these problems are given in the answer section at the back of the text. If you worked the problem correctly, move on to the next problem, but if you did not work it correctly (or you did not know what to do), then look at the solutions below, look back in the chapter to study the procedure, or ask your instructor.

Finally, go back over the homework problems you have been assigned. If you worked a problem correctly, move on to the next problem, but if you missed it on your homework, then you should look back in the book or talk to your instructor about how to work the problem.

If you follow these steps, you should be successful with your review of this chapter.

Proficiency Examination, page 1008

1. If f is defined on a closed, bounded region R in the xy-plane, then the double integral of f over R is defined by

$$\iint_R f(x,y)\, dA = \lim_{\|P\| \to 0} \sum_{k=1}^{N} f(x_k^*, y_k^*) \Delta A_k$$

provided this limit exists. If the limit exists, we say that f is integrable over R.

2. If $f(x,y)$ is continuous over the rectangle $R: a \le x \le b, c \le y \le c$, then the double integral

$$\iint_R f(x,y)\, dA$$

may be evaluated by either iterated integral; that is,

$$\iint_R f(x,y)\, dA = \int_c^d \int_a^b f(x,y)\, dx\, dy$$

$$= \int_a^b \int_c^d f(x,y)\, dy\, dx$$

3. A type I region contains points (x,y) such that for each fixed x between constants a and b, y varies from $g_1(x)$ to $g_2(x)$, where g_1

and g_2 are continuous functions. This is vertically simple.

$$\iint_D f(x,y)\, dA = \int_a^b \int_{g_1(x)}^{g_2(x)} f(x,y)\, dy\, dx$$

whenever both integrals exist.

4. A type II region contains points (x,y) such that for each fixed y between constants c and d, x varies from $h_1(y)$ to $h_2(y)$, where h_1 and h_2 are continuous functions. This is horizontally simple.

$$\iint_D f(x,y)\, dA = \int_c^d \int_{h_1(y)}^{h_2(y)} f(x,y)\, dx\, dy$$

whenever both integrals exist.

5. The area of the region D in the xy-plane is given by

$$A = \iint_D dA$$

6. If f is continuous and $f(x,y) \ge 0$ on the region D, the volume of the solid under the surface $z = f(x,y)$ above the region D is given by

$$V = \int\!\!\int_D dA$$

7. **a.** Linearity rule: for constants a and b,

$$\int\!\!\int_D [af(x,y) + bg(x,y)]\, dA$$

$$= a\int\!\!\int_D f(x,y)\, dA + b\int\!\!\int_D g(x,y)\, dA$$

b. Dominance rule: If $f(x,y) \geq g(x,y)$ throughout a region D, then

$$\int\!\!\int_D f(x,y)\, dA \geq \int\!\!\int_D g(x,y)\, dA$$

c. Subdivision rule: If the region of integration D can be subdivided into two subregions D_1 and D_2, then

$$\int\!\!\int_D f(x,y)\, dA$$

$$= \int\!\!\int_{D_1} f(x,y)\, dA + \int\!\!\int_{D_2} f(x,y)\, dA$$

8. If f is continuous in the polar region D such that for each fixed θ between α and β, r varies between $h_1(\theta)$ and $h_2(\theta)$, then

$$\int\!\!\int_D f(r,\theta)\, dA = \int_\alpha^\beta \int_{h_1(\theta)}^{h_2(\theta)} f(r,\theta)\, r\, dr\, d\theta$$

9. Assume that the function $f(x,y)$ has continuous partial derivatives f_x and f_y in a region R of the xy-plane. Then the portion of the surface $z = f(x,y)$ that lies over R has surface area

$$S = \int\!\!\int_R \sqrt{[f_x(x,y)]^2 + [f_y(x,y)]^2 + 1}\, dA$$

10. Let D be a region in the xy-plane on which x, y, z and their partial derivatives with respect to u and v are continuous. Also, let S be a surface defined by a vector function

$$\mathbf{R}(u.v) = x(u,v)\mathbf{i} + y(u,v)\mathbf{j} + z(u,v)\mathbf{k}$$

Then the surface area is defined by

$$S = \int\!\!\int_D \|\mathbf{R}_u(u,v) \times \mathbf{R}_v(u,v)\|\, du\, dv$$

11. If $f(x,y,z)$ is continuous over a rectangular solid R: $a \leq x \leq b, c \leq y \leq d, r \leq z \leq s$, then the triple integral may be evaluated by the iterated integral

$$\int\!\!\int\!\!\int_R f(x,y,z)\, dV = \int_r^s \int_c^d \int_a^b f(x,y,z)\, dx\, dy\, dz$$

The iterated integration can be performed in any order (with appropriate adjustments) to the limits of integration: $dx\, dy\, dz$, $dx\, dz\, dy$, $dz\, dx\, dy$, $dy\, dx\, dz$, $dy\, dz\, dx$, $dz\, dy\, dx$.

12. If V is the volume of the solid region S, then

$$\int\!\!\int\!\!\int_S dV$$

13. If ρ is a continuous density function on the lamina corresponding to a plane region R, then the mass m of the lamina is given by

$$m = \int\!\!\int_R \rho(x,y)\, dA$$

Page 505

14. If ρ is a continuous density function on a lamina corresponding to a plane region R, then the moments of mass with respect to the x-axis is

$$M_x = \int\int_R \rho(x,y)\, y\, dA$$

15. If m is the mass of the lamina, the center of mass is $(\overline{x}, \overline{y}))$, where

$$\overline{x} = \frac{M_y}{m} \quad \text{and} \quad \overline{y} = \frac{M_x}{m}$$

If the density ρ is constant, the point $(\overline{x}, \overline{y})$ is called the centroid of the region.

16. The moments of inertia of a lamina of variable density ρ about the x- and y-axes, respectively, are

$$I_x = \int\int_R \rho(x,y)\, y^2\, dA$$

and

$$I_y = \int\int_R \rho(x,y)\, x^2\, dA$$

17. A joint probability density function for the random variables X and Y is a continuous, nonnegative function $f(x, y)$ such that

$$P[(X, Y) \text{ in } R] = \int\int_R f(x,y)\, dy\, dx$$

and

$$\int_{-\infty}^{\infty}\int_{-\infty}^{\infty} f(x,y)\, dy\, dx = 1$$

where $P[(X, Y)$ in $R]$ denotes the probability that (X, Y) is in the region R in the xy-plane.

18. Rectangular to cylindrical:

$$y = \sqrt{x^2 + y^2};\ \tan\theta = \frac{y}{x};\ z = z$$

Rectangular to spherical:

$$\rho = \sqrt{x^2 + y^2 + z^2};\ \tan\theta = \frac{y}{x}$$

$$\phi = \cos^{-1}\left(\frac{z}{\sqrt{x^2 + y^2 + z^2}}\right)$$

Cylindrical to rectangular:

$$z = r\cos\theta;\ y = r\sin\theta;\ z = z$$

Cylindrical to spherical:

$$\rho = \sqrt{r^2 + z^2};\ \theta = \theta;\ \phi = \cos^{-1}\left(\frac{z}{\sqrt{r^2 + z^2}}\right)$$

Spherical to rectangular:

$$x = \rho\sin\phi\cos\theta;\ y = \rho\sin\phi\sin\theta;\ z = \rho\cos\phi$$

Spherical to cylindrical:

$$r = \rho\sin\phi;\ \theta = \theta;\ z = \rho\cos\phi$$

19. Let f be a continuous function on the bounded, solid region S. Then the triple integral of f over S is given by:

a. $$\int\int\int_S f(r, \theta, z)\, r\, dz\, dr\, d\theta$$

in cylindrical coordinates

b. $$\int\int\int_S f(\rho, \theta, \phi)\, \rho^2 \sin\phi\, d\rho\, d\theta\, d\phi$$

in spherical coordinates

20. $\dfrac{\partial(x,y)}{\partial(u,v)} = \begin{vmatrix} \frac{\partial x}{\partial u} & \frac{\partial x}{\partial v} \\ \frac{\partial y}{\partial u} & \frac{\partial y}{\partial v} \end{vmatrix}$

$$= \frac{\partial x}{\partial u}\frac{\partial y}{\partial v} - \frac{\partial y}{\partial u}\frac{\partial x}{\partial v}$$

21. Let f be a continuous function on a region D, and let T be a one-to-one transformation that maps the region D^* in the uv-plane onto a region D in the xy-plane under the change of variables $x = x(u,v)$, $y = y(u,v)$ where g and h are continuously differentiable on D^*. Then

$$\iint\limits_{D} f(x,y)\, dy\, dx$$

$$= \iint\limits_{D^*} f[x(u,v),\, y(u,v]\, |J(u,v|\, du\, dv$$

22. $\displaystyle\int_0^{\pi/3}\int_0^{\sin y} e^{-x}\cos y\, dx\, dy$

$$= \int_0^{\pi/3}\left[\left(e^{-\sin y}\right)(-\cos y) + \cos y\right] dy$$

$$= e^{-\sqrt{3}/2} + \frac{\sqrt{3}}{2} - 1$$

23. $\displaystyle\int_{-1}^1\int_0^z\int_y^{y-z} (x+y-z)\, dx\, dy\, dz$

$$= \int_{-1}^1\int_0^z\left[\frac{(y-z)^2}{2} + (y-z)y\right.$$

$$\left. - (y-z)z - \frac{y^2}{2} - y^2 + yz\right] dy\, dz$$

$$= \int_{-1}^1\int_0^z\left(-2yz + \frac{3z^2}{2}\right) dy\, dz$$

$$= \int_{-1}^1 z^3\, dz$$

$$= 0$$

24. $\displaystyle 2\int_0^3\int_0^{9-x^2} dy\, dx = 2\int_0^3 \left(9 - x^2\right) dx$

$$= 36$$

25. $A = \displaystyle\int_0^{\pi/2}\int_0^1 \cos r^2\, r\, dr\, d\theta$

$$= \frac{1}{2}\int_0^{\pi/2} \sin 1\, d\theta$$

$$= \frac{\pi}{4}\sin 1$$

26. $z = c\left(1 - \dfrac{x}{a} - \dfrac{y}{b}\right)$; $y = b\left(1 - \dfrac{x}{a}\right)$ and the projected region on the xy-plane is the triangle bounded by $x = 0$, $y = 0$, and

$$y = b\left(1 - \frac{x}{a}\right)$$

$$V = \int_0^a\int_0^{b(1-x/a)}\int_0^{c(1-x/a-y/b)} dz\, dy\, dx$$

$$= -\frac{c}{ab}\int_0^a\int_0^{b(1-x/a)} (bx + ay - ab)\, dy\, dx$$

$$= \frac{bc}{2a^2}\int_0^a (x-a)^2\, dx$$

$$= \frac{abc}{6}$$

27. The appliance fails during the first year if both components fail in that time; that is, if (X,Y) lies in the square $0 \leq x \leq 1$, $0 \leq y \leq 1$. The probability of this occurring is

$$P[0 \leq x \leq 1,\, 0 \leq y \leq 1]$$

$$= \int_0^1\int_0^1 \frac{1}{4}e^{-x/2}e^{-y/2}\, dy\, dx$$

$$= \frac{1 - e^{-1/2}}{2}\int_0^1 e^{-x/2}\, dx$$

$$= \left(1 - e^{-1/2}\right)^2$$

Thus, the probability of product failure is about 15%.

28. The density is $\rho = r$ and the projected region is $x^2 + y^2 = 4$. In cylindrical coordinates,

$$m = \int_0^{2\pi} \int_0^2 \int_{r^2}^4 (r) r \, dz \, dr \, d\theta$$

$$= \int_0^{2\pi} \int_0^2 r^2(4 - r^2) \, dr \, d\theta$$

$$= \frac{64}{15} \int_0^{2\pi} d\theta = \frac{128\pi}{15}$$

29. First find the intersection of the plane and the paraboloid:

$$\begin{cases} x^2 + 2y^2 = 4x \\ (x-2)^2 + 2y^2 = 4 \end{cases}$$

$$\frac{(x-2)^2}{4} + \frac{y^2}{2} = 1$$

This is a translated ellipse centered at $(2, 0)$ with x intercepts of 0 and 4.

$$V = \int_0^4 \int_0^{\sqrt{4x-x^2}/\sqrt{2}} \int_{x^2+2y^2}^{4x} dz \, dy \, dx$$

$$= \int_0^4 \int_0^{\sqrt{4x-x^2}/\sqrt{2}} (4x - x^2 - 2y^2) \, dy \, dx$$

$$= \int_0^4 \left\{ \frac{(4x-x^2)^{3/2}}{\sqrt{2}} - \frac{2}{3}\left[\frac{(4x-x^2)^{3/2}}{2\sqrt{2}} \right] \right\} dx$$

$$= \frac{2}{3\sqrt{2}} \int_0^4 (4x - x^2)^{3/2} \, dx$$

$$= \frac{\sqrt{2}}{3} \int_0^4 \left[2^2 - (x-2)^2 \right]^{3/2} dx$$

$$= 2\pi\sqrt{2}$$

30. Let $u = x + y$ and $v = x - 2y$, so that $x = \frac{1}{3}(2u + v)$, $y = \frac{1}{3}(u - v)$.

$$\frac{\partial(x,y)}{\partial(u,v)} = \begin{vmatrix} \frac{2}{3} & \frac{1}{3} \\ \frac{1}{3} & -\frac{1}{3} \end{vmatrix} = -\frac{1}{3}$$

The region R is bounded by the lines $y = 0$,

$y = 2 - x$, $y = x$ which transform into the lines $u = v$, $u = 2$, and $u = -2v$.

$$\iint_R (x + y) e^{x-2y} \, dy \, dx$$

$$= \int_0^2 \int_{-u/2}^u u e^v \left| -\frac{1}{3} \right| dv \, du$$

$$= \frac{1}{3} \int_0^2 \left(u e^u - u e^{-u/2} \right) du$$

$$= \frac{1}{3}(e^2 + 8e^{-1} - 3)$$

Supplementary Problems, page 1010

1.

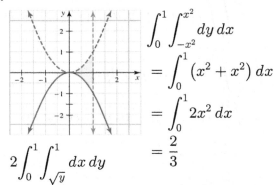

$$\int_0^1 \int_{-x^2}^{x^2} dy \, dx$$

$$= \int_0^1 (x^2 + x^2) \, dx$$

$$= \int_0^1 2x^2 \, dx$$

$$= \frac{2}{3}$$

$$2 \int_0^1 \int_{\sqrt{y}}^1 dx \, dy$$

3.

$$\int_1^4 \int_0^{4-y} (x + y) \, dx \, dy$$

$$= -\frac{1}{2} \int_1^4 (y^2 - 16) \, dy$$

$$= \frac{27}{2}$$

$$\int_0^3 \int_1^{4-x} (x + y) \, dy \, dx$$

5.

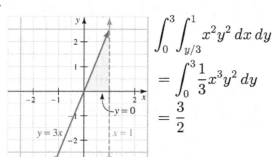

$$\int_0^3 \int_{y/3}^1 x^2 y^2 \, dx \, dy$$

$$= \int_0^3 \frac{1}{3} x^3 y^2 \, dy$$

$$= \frac{3}{2}$$

$$\int_0^1 \int_0^{3x} x^2 y^2 \, dy \, dx$$

7.

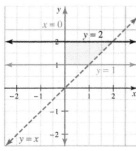

Evaluate the original integral:

$$\int_1^2 \int_0^y \frac{1}{x^2 + y^2} \, dx \, dy$$

$$= \int_1^2 \frac{1}{y} \tan^{-1} \frac{x}{y} \bigg|_0^y \, dy$$

$$= \int_1^2 \frac{\pi}{4y} \, dy$$

$$= \frac{\pi}{4} \ln 2$$

$$\int_0^1 \int_1^2 \frac{1}{x^2 + y^2} \, dy \, dx$$
$$+ \int_1^2 \int_x^2 \frac{1}{x^2 + y^2} \, dy \, dx$$

9. Both integrals are the same.

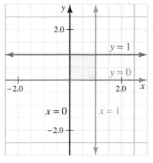

Evaluate the original integral:

$$\int_0^1 \int_0^1 x\sqrt{x^2 + y} \, dx \, dy$$

$$= \frac{1}{3} \int_0^1 \left[(1+y)^{3/2} - y^{3/2} \right] dy$$

$$= \frac{8\sqrt{2} - 4}{15}$$

11.

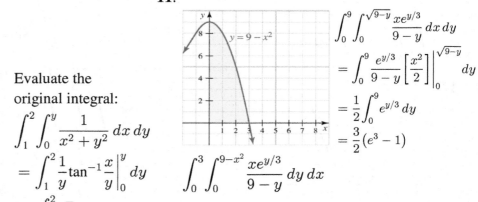

$$\int_0^9 \int_0^{\sqrt{9-y}} \frac{xe^{y/3}}{9-y} \, dx \, dy$$

$$= \int_0^9 \frac{e^{y/3}}{9-y} \left[\frac{x^2}{2} \right] \bigg|_0^{\sqrt{9-y}} \, dy$$

$$= \frac{1}{2} \int_0^9 e^{y/3} \, dy$$

$$= \frac{3}{2} (e^3 - 1)$$

$$\int_0^3 \int_0^{9-x^2} \frac{xe^{y/3}}{9-y} \, dy \, dx$$

13.

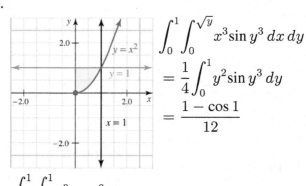

$$\int_0^1 \int_0^{\sqrt{y}} x^3 \sin y^3 \, dx \, dy$$

$$= \frac{1}{4} \int_0^1 y^2 \sin y^3 \, dy$$

$$= \frac{1 - \cos 1}{12}$$

$$\int_0^1 \int_{x^2}^1 x^3 \sin y^3 \, dy \, dx$$

15.

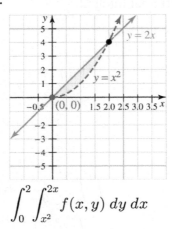

$$\int_0^2 \int_{x^2}^{2x} f(x, y)\, dy\, dx$$

17.

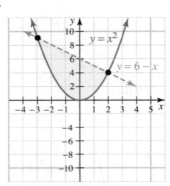

$$\int_0^4 \int_{-\sqrt{y}}^{\sqrt{y}} f(x, y)\, dx\, dy$$

$$+ \int_4^9 \int_{-\sqrt{y}}^{6-y} f(x, y)\, dx\, dy$$

19. $\displaystyle \int_0^1 \int_{\sqrt{x}}^1 e^{y^3}\, dy\, dx = \int_0^1 \int_0^{y^2} e^{y^3}\, dx\, dy$

$$= \int_0^1 y^2 e^{y^3}\, dy$$

$$= \frac{1}{3}(e - 1)$$

21. $\displaystyle \int_0^1 \int_1^4 \int_x^y z\, dz\, dy\, dx = \frac{1}{2} \int_0^1 \int_1^4 (y^2 - x^2)\, dy\, dx$

$$= \frac{1}{2} \int_0^1 (21 - 3x^2)\, dx$$

$$= 10$$

23. $\displaystyle \int_0^{\pi/4} \int_0^{2\pi} \int_0^{\theta} r^2 \sin\phi\, dr\, d\theta\, d\phi$

$$= \frac{1}{3} \int_0^{\pi/4} \int_0^{2\pi} \theta^3 \sin\phi\, d\theta\, d\phi$$

$$= \frac{4\pi^4}{3} \int_0^{\pi/4} \sin\phi\, d\phi$$

$$= \frac{2\pi^4}{3} (2 - \sqrt{2})$$

25. $\displaystyle \int_1^2 \int_x^{x^2} \int_0^{\ln x} xe^z\, dz\, dy\, dx$

$$= \int_1^2 \int_x^{x^2} x(x - 1)\, dy\, dx$$

$$= \int_1^2 x^2(x - 1)^2\, dx$$

$$= \frac{31}{30}$$

27. $\displaystyle \iint_D x^3 \sqrt{4 - y^2}\, dA = 0$

since the region D is symmetric with respect to the origin and $f(-x, -y) = -f(x, y)$ where $f(x, y) = x^3 \sqrt{4 - y^2}$.

29. $\displaystyle \iint_D (x^2 + y^2 + 1)\, dA$

$$= \int_0^{2\pi} \int_0^2 (r^2 + 1)\, r\, dr\, d\theta$$

$$= 2\pi \int_0^2 (r^3 + r)\, dr$$

$$= 12\pi$$

Page 510

31. $\displaystyle\iint_D (x^2 + y^2)^n \, dA = \int_0^{2\pi} \int_0^2 r^{2n} r \, dr \, d\theta$

$\displaystyle = \int_0^{2\pi} \frac{2^{2n+2}}{2n+2} \, d\theta$

$\displaystyle = \frac{2^{2n+2}}{n+1} \pi$

33. $\displaystyle\iint_D 2x \, dA$

$\displaystyle = \int_0^1 \int_0^x 2x \, dy \, dx + \int_1^2 \int_0^{1/x^2} 2x \, dy \, dx$

$\displaystyle = \int_0^1 2x^2 \, dx + \int_1^2 2x^{-1} \, dx$

$\displaystyle = \frac{2}{3} + 2\ln 2$

35. $\displaystyle\iint_D x \, dy \, dx = \int_{-1}^3 \int_{x^2}^{2x+3} x \, dy \, dx$

$\displaystyle = \int_{-1}^3 x(2x+3-x^2) \, dx$

$\displaystyle = \frac{32}{3}$

37. $\displaystyle\int_{-\sqrt{3}}^{\sqrt{3}} \int_{-\sqrt{3-y^2}}^{\sqrt{3-y^2}} \int_{(x^2+y^2)^2}^9 y^2 dz \, dx \, dy$

$\displaystyle = \int_0^{2\pi} \int_0^{\sqrt{3}} \int_{r^4}^9 r^2 \sin^2\theta \, r \, dz \, dr \, d\theta$

$\displaystyle = \int_0^{2\pi} \int_0^{\sqrt{3}} r^2 \sin^2\theta (9 - r^4) \, r \, dr \, d\theta$

$\displaystyle = \int_0^{2\pi} \int_0^{\sqrt{3}} \sin^2\theta (9r^3 - r^7) \, dr \, d\theta$

$\displaystyle = \frac{81}{16} \int_0^{2\pi} (1 - \cos 2\theta) \, d\theta = \frac{81\pi}{8}$

39. $\displaystyle\iint_D xy^2 \, dA = \int_{-1}^1 \int_{(y-1)/2}^1 xy^2 \, dx \, dy$

$\displaystyle = \frac{1}{2} \int_{-1}^1 \left[\frac{3y^2}{4} - \frac{y^4}{4} + \frac{y^2}{2} \right] dy$

$\displaystyle = \frac{1}{5}$

41. $\displaystyle\iint_R \frac{\partial^2 f}{\partial x \partial y} \, dx \, dy$

$\displaystyle = \int_c^d \int_a^b \frac{\partial}{\partial x} \frac{\partial f}{\partial y} \, dx \, dy$

$\displaystyle = \int_c^d \left[\frac{\partial f(b,y)}{\partial y} - \frac{\partial f(a,y)}{\partial y} \right] dy$

$\displaystyle = f(b,d) - f(b,c) - f(a,d) + f(a,c)$

$\displaystyle = 5 - 1 + 3 + 4$

$\displaystyle = 11$

43. $\displaystyle\iiint_H z^2 \, dV$

$\displaystyle = \int_0^{\pi/2} \int_0^{2\pi} \int_0^1 \rho^4 \cos^2\phi \sin\phi \, d\rho \, d\theta \, d\phi$

$\displaystyle = \frac{1}{5} \int_0^{\pi/2} \int_0^{2\pi} \cos^2\phi \sin\phi \, d\theta \, d\phi$

$\displaystyle = \frac{2\pi}{5} \int_0^{\pi/2} \cos^2\phi \sin\phi \, d\phi$

$\displaystyle = \frac{2\pi}{15}$

45. The intersection of the surfaces is $2z + z^2 = 8$, so $z = 2$ (reject $z = -2$) and the projected region on the xy-plane is $x^2 + y^2 \le 4$.

$$\iiint_D \frac{dV}{x^2 + y^2 + z^2}$$

$$= \int_0^{2\pi} \int_0^2 \int_{r^2/2}^{\sqrt{8-r^2}} \frac{r}{r^2 + z^2} \, dz \, dr \, d\theta$$

$$= \int_0^{2\pi} \int_0^2 \tan^{-1}\left(\frac{\sqrt{8-r^2}}{r}\right) - \tan^{-1}\left(\frac{r}{2}\right) dr \, d\theta$$

$$= \int_0^{2\pi} \left(\ln 2 + 2\sqrt{2} - 2\right) d\theta$$

$$= 2\pi \left(\ln 2 + 2\sqrt{2} - 2\right)$$

47. The curves intersect when $2 = 4\sin 2\theta$ or
$\theta = \dfrac{\pi}{12}, \dfrac{5\pi}{12}$; let ρ be the density.

$$m = \int_{\pi/12}^{5\pi/12} \int_{\sqrt{2a}}^{2a\sqrt{\sin 2\theta}} \rho r \, dr \, d\theta$$

49.
$$\int_0^{\pi/2} \int_0^{2a\cos\theta} r \sin 2\theta \, dr \, d\theta$$

$$= \int_0^{2a} \int_0^{\sqrt{2ax-x^2}} \frac{2xy}{x^2 + y^2} \, dy \, dx$$

$$= \int_0^{2a} x(\ln 2ax - 2\ln x) \, dx$$

$$= a^2$$

51.
$$\int_0^1 \int_0^y f(x,y) \, dx \, dy + \int_1^4 \int_0^{(4-y)/3} f(x,y) \, dx$$

$$= \int_0^1 \int_x^{4-3x} f(x,y) \, dy \, dx$$

53. The circle and the cardioid intersect where

$$\cos\theta = 1 - \cos\theta$$
$$\cos\theta = \frac{1}{2}$$
$$\theta = \frac{\pi}{3}, -\frac{\pi}{3}$$

$$2\int_0^{\pi/3} \int_{1-\cos\theta}^{\cos\theta} r \, dr \, d\theta = \int_0^{\pi/3} (2\cos\theta - 1) \, d\theta$$
$$= \sqrt{3} - \frac{\pi}{3}$$

55. The limaçon passes through the pole (origin) if $\sin\theta = -\frac{1}{2}$. This happens at $\theta = \frac{7\pi}{6}$ and $\theta = \frac{11\pi}{6}$, but we do only half the limaçon and multiply by 2, so we discard $\frac{11\pi}{6}$.

$$2\int_{\pi/2}^{7\pi/6} \int_0^{1+\sin\theta} r \, dr \, d\theta - 2\int_{7\pi/6}^{3\pi/2} \int_0^{1+2\sin\theta} r \, dr \, d\theta$$

$$= \int_{\pi/2}^{7\pi/6} (1 + 2\sin\theta)^2 \, d\theta - \int_{7\pi/6}^{3\pi/2} (1 + 2\sin\theta)^2 d\theta$$

$$= \pi + 3\sqrt{3}$$

57.
$$\int_0^1 \int_0^1 (x^2 + y^2) \, dy \, dx = \int_0^1 \left(x^2 + \frac{1}{3}\right) dx$$
$$= \frac{2}{3}$$

59. The plane and paraboloid intersect where

$$4 - x^2 - y^2 = 4 - 2x$$
$$(x-1)^2 + y^2 = 1$$

The polar form of the projected region is $r = 2\cos\theta$.

$$V = \int_{-\pi/2}^{\pi/2} \int_0^{2\cos\theta} \int_{4-2x}^{4-x^2-y^2} r \, dz \, dr \, d\theta$$

$$= \int_{-\pi/2}^{\pi/2} \int_0^{2\cos\theta} \int_{4-2r\cos\theta}^{4-r^2} r \, dz \, dr \, d\theta$$

$$= \int_{-\pi/2}^{\pi/2} \int_0^{2\cos\theta} \left(4 - r^2 - 4 + 2r\cos\theta\right) r \, dr \, d\theta$$

$$= \frac{4}{3} \int_{-\pi/2}^{\pi/2} \cos^4\theta \, d\theta = \frac{\pi}{2}$$

61. The projected region is $x^2 + y^2 = 9$. We have $z_x = 2x$, $z_y = 2y$, and
$$z_x^2 + z_y^2 + 1 = 4x^2 + 4y^2 + 1$$

$$S = \int_0^{2\pi} \int_0^3 \sqrt{4r^2 + 1}\, r\, dr\, d\theta$$

$$= \int_0^{2\pi} \frac{1}{12}\left(37\sqrt{37} - 1\right) d\theta$$

$$= \frac{\pi}{6}\left(37\sqrt{37} - 1\right)$$

63. $\begin{cases} y^2 + z^2 = 2x \\ x + y = 1 \end{cases} \Rightarrow z^2 + (y+1)^2 = 3$

If $z = 0$, $y = -1 \pm \sqrt{3}$.

$$2\int_{-1-\sqrt{3}}^{-1+\sqrt{3}} \int_0^{\sqrt{3-(y+1)^2}} \int_{(y^2+z^2)/2}^{1-y} dx\, dz\, dy$$

$$= 2\int_{-1-\sqrt{3}}^{-1+\sqrt{3}} \int_0^{\sqrt{3-(y+1)^2}} \left(1 - y - \frac{y^2}{2} - \frac{z^2}{2}\right) dz\, dy$$

$$= \frac{2}{3}\int_{-1-\sqrt{3}}^{-1+\sqrt{3}} \left[(2 - y^2 - 2y)^{3/2}\right] dy$$

$$= \frac{9\pi}{4}$$

65. $4\int_0^{\pi/2} \int_0^{\cos\theta} 2r\sqrt{1-r^2}\, dr\, d\theta$

$$= \frac{8}{3}\int_0^{\pi/2} \left(1 - \sin^3\theta\right) d\theta$$

$$= \frac{8}{3}\int_0^{\pi/2} \left[1 - (1 - \cos^2\theta)\sin\theta\right] d\theta$$

$$= \frac{4(3\pi - 4)}{9}$$

67. The sphere intersects the cone where

$$z^2 + z^2 = 2az$$

$$z^2 - az = 0$$

$$z(z - a) = 0$$

$$z = a$$

Thus, the projected region in the xy-plane is the circle $x^2 + y^2 = a^2$. In spherical coordinates, the equation of the cone is

$\phi = \dfrac{\pi}{4}$ and the equation of the sphere is

$$\rho^2 = 2az$$

$$= 2a\rho\cos\phi$$

$$\rho^2 - 2a\rho\cos\phi = 0$$

$$\rho = 2a\cos\phi$$

$$V = \int_0^{\pi/4} \int_0^{2\pi} \int_0^{2a\cos\phi} \rho^2\sin\phi\, d\rho\, d\theta\, d\phi$$

$$= \frac{8a^3}{3}\int_0^{\pi/4} \int_0^{2\pi} \sin\phi\cos^3\phi\, d\theta\, d\phi$$

$$= \frac{16\pi a^3}{3}\int_0^{\pi/4} \sin\phi\cos^3\phi\, d\phi$$

$$= \pi a^3$$

69. $\int_0^{\phi_0} \int_0^{2\pi} \int_0^a \rho^2\sin\phi\,(\rho^2\sin\phi)\, d\rho\, d\theta\, d\phi$

$$= \int_0^{\phi_0} \int_0^{2\pi} \frac{a^5\sin^3\phi}{5}\, d\theta\, d\phi$$

$$= \frac{2\pi a^5}{5}\int_0^{\phi_0} \sin\phi\,(1 - \cos^2\phi)\, d\phi$$

$$= \frac{2\pi a^5}{15}\left(\cos^3\phi_0 - 3\cos\phi_0 + 2\right)$$

71. $m = \int_0^1 \int_0^{1-x} (x^2 + y^2)\, dy\, dx$

$$= \int_0^1 \left[x^2(1-x) + \frac{1}{3}(1-x)^3\right] dx$$

$$= \frac{1}{6}$$

$$M_y = \int_0^1 \int_0^{1-x} x(x^2 + y^2)\, dy\, dx$$

$$= \int_0^1 \left(-\frac{4}{3}x^4 + 2x^3 - x^2 + \frac{1}{3}x\right) dx$$

$$= \frac{1}{15}$$

$$\bar{x} = \frac{\frac{1}{15}}{\frac{1}{6}} = \frac{6}{15} = \frac{2}{5}$$

Page 513

73. By symmetry, $\bar{x} = \bar{y} = 0$. The volume of the cone is $V = \dfrac{\pi R^2 H}{3}$, and its equation in spherical coordinates is $\phi_0 = \tan^{-1}\dfrac{R}{H}$. The "top" of the cone is $z = H = \rho \cos\phi$, so $\rho = H \sec\phi$.

$$M_{xy} = \int_0^{2\pi} \int_0^{\tan^{-1}(R/H)} \int_0^{H\sec\phi} (\rho\cos\phi)\,\rho^2 \sin\phi \, d\rho \, d\phi \, d\theta$$

$$= \int_0^{2\pi} \int_0^{\tan^{-1}(R/H)} \frac{1}{4}(H\sec\phi)^4 \cos\phi \sin\phi \, d\phi$$

$$= \frac{1}{4} H^3 \int_0^{2\pi} \frac{R^2}{2H} \, d\theta$$

$$= \frac{\pi H^2 R^2}{4}$$

Thus, $\bar{z} = \dfrac{M_{xy}}{m} = \dfrac{\pi H^2 R^2}{4} \cdot \dfrac{3}{\pi R^2 H} = \dfrac{3}{4}H$

75. $\dfrac{\partial(u, v, w)}{\partial(x, y, z)} = \begin{vmatrix} \frac{\partial u}{\partial x} & \frac{\partial u}{\partial y} & \frac{\partial u}{\partial z} \\ \frac{\partial v}{\partial x} & \frac{\partial v}{\partial y} & \frac{\partial v}{\partial z} \\ \frac{\partial w}{\partial x} & \frac{\partial w}{\partial y} & \frac{\partial w}{\partial z} \end{vmatrix}$

$$= \begin{vmatrix} 2x & 2y & 2z \\ 0 & 4y & 2z \\ 0 & 0 & 4z \end{vmatrix}$$

$$= 32xyz$$

77. $m = \displaystyle\int_0^{\pi/6} \int_0^{\cos 3\theta} (r\theta)\, r \, dr \, d\theta$

$$= \int_0^{\pi/6} \frac{1}{3}\theta \cos^3(3\theta) \, d\theta$$

$$= \frac{3\pi - 7}{243}$$

79. a. $u = x + y,\ v = x - y;$
$u \geq 0$ becomes $y \geq -x,$
$v \geq 0$ becomes $y \leq x;$ thus, $-x \leq y \leq x$
$u + v \geq 1$ becomes $2x \geq 1$ or $x \geq \frac{1}{2}$
$u + v \leq 2$ becomes $2x \leq 2$ or $x \leq 1;$
thus $\frac{1}{2} \leq x \leq 1.$

b. $\dfrac{\partial(u, v)}{\partial(x, y)} = \begin{vmatrix} \frac{\partial u}{\partial x} & \frac{\partial u}{\partial y} \\ \frac{\partial v}{\partial x} & \frac{\partial v}{\partial y} \end{vmatrix}$

$$= \begin{vmatrix} 1 & 1 \\ 1 & -1 \end{vmatrix}$$

$$= -2$$

$$\iint_D (u + v)\, du\, dv = 2\int_{1/2}^x \int_{-x}^x 2x \, dy \, dx$$

$$= 4 \int_{1/2}^1 x(x + x)\, dx$$

$$= \frac{7}{3}$$

81. $f(x, y, z) = Ax + By + Cz - 1$
$\nabla f = A\mathbf{i} + B\mathbf{j} + C\mathbf{k}$, and the unit normal vector is $\mathbf{N} = \dfrac{A\mathbf{i} + B\mathbf{j} + C\mathbf{k}}{\sqrt{A^2 + B^2 + C^2}}$

$$\|\mathbf{N} \cdot \mathbf{k}\| = \frac{C}{\sqrt{A^2 + B^2 + C^2}}$$

$$m = \frac{\sqrt{A^2 + B^2 + C^2}}{C} \iint_S dS$$

$$= \frac{\sqrt{A^2 + B^2 + C^2}}{2ABC}$$

$$M_{yz} = \frac{\sqrt{A^2 + B^2 + C^2}}{C} \int_0^{1/A} \int_0^{(-Ax+1)/B} x \, dx \, dy$$

$$= \frac{\sqrt{A^2 + B^2 + C^2}}{BC} \int_0^{1/A} (-Ax^2 + x)\, dx$$

$$= \frac{\sqrt{A^2 + B^2 + C^2}}{6A^2 BC}$$

$$\bar{x} = \frac{2ABC}{6A^2 BC} = \frac{1}{3A}$$

Similarly, $\bar{y} = \dfrac{1}{3B}$, $\bar{z} = \dfrac{1}{3C}$

83. a. $y = 1 - x^2,\ \rho = xy$

$$m = \int_0^1 \int_0^{1-x^2} xy\, dy\, dx$$

$$= \frac{1}{2} \int_0^1 x\left(1 - x^2\right)^2 dx$$

$$= \frac{1}{12}$$

$$M_x = \int_0^1 \int_0^{1-x^2} y(xy)\, dy\, dx$$

$$= \frac{1}{3} \int_0^1 x\left(1 - x^2\right)^3 dx$$

$$= \frac{1}{24}$$

$$M_y = \int_0^1 \int_0^{1-x^2} x(xy)\, dy\, dx$$

$$= \frac{1}{2} \int_0^1 x^2 \left(1 - 2x^2 + x^4\right) dx$$

$$= \frac{4}{105}$$

$$(\overline{x}, \overline{y}) = \left(\frac{4(12)}{105}, \frac{12}{24}\right) = \left(\frac{16}{35}, \frac{1}{2}\right)$$

b. $I_z = \displaystyle\int_0^1 \int_0^{1-x^2} xy\left(x^2 + y^2\right) dy\, dx$

$$= \frac{1}{4} \int_0^1 \left(x^9 - 2x^7 + 2x^5 - 2x^3 + x\right) dx$$

$$= \frac{11}{240}$$

85. $V = 2 \displaystyle\int_0^{\pi/2} \int_0^{2a\cos\theta} \int_0^r r\, dz\, dr\, d\theta$

$$= 2 \int_0^{\pi/2} \int_0^{2a\cos\theta} r^2\, dr\, d\theta$$

$$= \frac{16a^3}{3} \int_0^{\pi/2} \cos^3\theta\, d\theta$$

$$= \frac{32a^3}{9}$$

87. $\dfrac{x^2}{a^2} + \dfrac{y^2}{b^2} = R^2\sin^2\phi$ and $\dfrac{z^2}{c^2} = R^2\cos^2\phi$ so

$$\frac{x^2}{a^2} + \frac{y^2}{b^2} + \frac{z^2}{c^2} = R^2$$

89. $x^{2/3} + y^{2/3} + z^{2/3} = a^{2/3}$

Let $x = ar^3$, $y = as^3$, $z = at^3$, so that
$r^2 + s^2 + t^2 = 1$.

$$\frac{\partial(x, y, z)}{\partial(r, s, t)} = \begin{vmatrix} \frac{\partial x}{\partial r} & \frac{\partial x}{\partial s} & \frac{\partial x}{\partial t} \\ \frac{\partial y}{\partial r} & \frac{\partial y}{\partial s} & \frac{\partial y}{\partial t} \\ \frac{\partial z}{\partial r} & \frac{\partial z}{\partial s} & \frac{\partial z}{\partial t} \end{vmatrix}$$

$$= \begin{vmatrix} 3ar^2 & 0 & 0 \\ 0 & 3as^2 & 0 \\ 0 & 0 & 3at^2 \end{vmatrix}$$

$$= 27a^3 r^2 s^2 t^2$$

Now the spherical coordinates are
$r = \rho \sin\phi \cos\theta$, $s = \rho \sin\phi \sin\theta$,
$t = \rho \cos\phi$

$$V = \iint\limits_{\text{sphere}} \int 27a^3 r^2 s^2 t^2\, dr\, ds\, dt$$

$$= 8(27a^3) \int_0^{\pi/2} \int_0^{\pi/2} \int_0^1 (\rho \sin\phi \cos\theta)^2$$
$$\cdot (\rho \sin\phi \sin\theta)^2 (\rho \cos\phi)^2 \rho^2 \sin\phi\, d\rho\, d\theta\, d\phi$$

$$= 216a^3 \int_0^{\pi/2} \int_0^{\pi/2} \int_0^1 \rho^8 \sin^5\phi \cos^2\phi \cos^2\theta \sin^2\theta\, d\rho\, d\theta\, d\phi$$

$$= 24a^3 \int_0^{\pi/2} \int_0^{\pi/2} \cos^2\theta \sin^2\theta \sin^5\phi \cos^2\phi\, d\theta\, d\phi$$

$$= \frac{3a^3\pi}{2} \int_0^{2\pi} \sin^5\phi \cos^2\phi\, d\phi$$

$$= \frac{4a^3\pi}{35}$$

91. Projecting onto the yz-plane, we see that the region of integration is the circle
$y^2 + z^2 = 1$.

$$m = 8 \int_0^1 \int_0^{\sqrt{1-z^2}} \int_0^{\sqrt{1-z^2}} dx\, dy\, dz$$

$$= 8 \int_0^1 \int_0^{\sqrt{1-z^2}} \sqrt{1-z^2}\, dy\, dz$$

$$= 8 \int_0^1 \left(1-z^2\right) dz$$

$$= \frac{16}{3}$$

$$M_{xy} = \int_{-1}^{1} \int_{-\sqrt{1-x^2}}^{\sqrt{1-z^2}} \int_{-\sqrt{1-z^2}}^{\sqrt{1-z^2}} z\, dx\, dy\, dz = 0$$

$$= \int_{-1}^{1} \int_{-\sqrt{1-x^2}}^{\sqrt{1-z^2}} 2z\sqrt{1-z^2}\, dy\, dz$$

$$= \int_{-1}^{1} 4z(1-z^2)\, dz$$

$$= 0$$

Similarly, $M_{yz} = M_{xz} = 0$, so the centroid is the origin $(\overline{x}, \overline{y}, \overline{z}) = (0,0,0)$.

93. In spherical coordinates, the mass is

$$m = \int_0^{\pi/2} \int_0^{\pi/2} \int_0^1 \frac{\rho^2 \sin\phi}{\rho^2+1}\, d\rho\, d\phi\, d\theta$$

$$= \int_0^{\pi/2} \int_0^{\pi/2} \left(1 - \frac{\pi}{4}\right) \sin\phi\, d\phi\, d\theta$$

$$= \int_0^{\pi/2} \left(1 - \frac{\pi}{4}\right) d\theta$$

$$= \frac{\pi}{2}\left(1 - \frac{\pi}{4}\right)$$

$$M_{xy} = \int_0^{\pi/2} \int_0^{\pi/2} \int_0^1 \frac{1}{\rho^2+1}(\rho \cos\phi)\, \rho^2 \sin\phi\, d\rho\, d\phi\, d\theta$$

$$= \int_0^{\pi/2} \int_0^{\pi/2} \left(\frac{1}{2} - \frac{\ln 2}{2}\right) \sin\phi \cos\phi\, d\phi\, d\theta$$

$$= \frac{1}{4} \int_0^{\pi/2} (1 - \ln 2)\, d\theta$$

$$= \frac{\pi}{8}(1 - \ln 2)$$

$$\overline{z} = \frac{1 - \ln 2}{4 - \pi}$$

By symmetry, $\overline{x} = \overline{y} = \overline{z}$.

95. Let $\overline{z}_s$ be the z-coordinate of the centroid of the upper hemisphere of radius 1. $\overline{z}_s = \frac{3}{8}$, and by symmetry, $\overline{z}_c = -1$.

$$m = m_s + m_c = \frac{4\pi}{6} + 8 = \frac{2\pi + 24}{3}$$

$$M_{xy} = \left(\frac{2\pi}{3}\right)\left(\frac{3}{8}\right) + 8(-1) = \frac{\pi - 32}{4}$$

$$\overline{z} = \frac{3(\pi - 32)}{4(2\pi + 24)}$$

By symmetry, $\overline{x} = \overline{y} = 0$, so the center of mass is $\left(0, 0, \dfrac{3(\pi - 32)}{8(\pi + 12)}\right)$.

97. This is Putnam Problem 5 in the morning session of 1942. Choose the axes so that the generated surface starts in the xz-plane and is revolved about the z-axis. The generated circle is half of a torus (*i.e.* a solid bounded by a torus). It is clear from symmetry that the centroid lies at the point $(0, \overline{y}, 0)$ on the y-axis, that the requirement of the problem is that $\overline{y} = b - a$. To find the centroid, we introduce polar coordinates in the xy-plane. Corresponding to the element of area $r\, dr\, d\theta$ in the plane there is the element of volume

$$2\sqrt{a^2 - (r-b)^2}\, r\, dr\, d\theta$$

which contributes

$$2r\sin\theta \sqrt{a^2 - (r-b)^2}\, r\, dr\, d\theta$$

to the moment M of the solid in the y-

direction. We have $\overline{y} = M_y/V$ where V is the volume of the half torus.

$$V = \int_0^\pi \int_{b-a}^{b+a} 2\sqrt{a^2 - (r-b)^2}\, r\, dr\, d\theta$$

$$= 2\pi \int_{b-a}^{b+a} 2\sqrt{a^2 - (r-b)^2}\, r\, dr$$

$$= 2\pi \int_{-\pi/2}^{\pi/2} \cos\phi (b + a\sin\phi)\cos\phi\, d\phi$$

$$= \pi^2 a^2 b$$

$$M_y = \int_0^\pi \int_{b-a}^{b+a} 2\sqrt{a^2 - (r-b)^2}\, r^2 \sin\theta\, dr\, d\theta$$

$$= 4\pi \int_{b-a}^{b+a} \sqrt{a^2 - (r-b)^2}\, r^2\, dr$$

$$= 4a^2\pi \int_{-\pi/2}^{\pi/2} \cos\phi (b + a\sin\phi)^2 \cos\phi\, d\phi$$

$$= 2\pi^2 a^2 b^2 + \frac{\pi a^4}{2}$$

In both integrals, we used the substitution $r = b + a\sin\phi$. Hence

$$\overline{y} = \frac{M_y}{V} = \frac{a^2 + 4b^2}{2\pi b}$$

We require $\overline{y} = b - a$, so

$$2\pi b^2 - 2\pi ab = a^2 + 4b^2$$

If $c = b/a$, then

$$(2\pi - 4)c^2 - 2\pi c - 1 = 0$$

$$c = \frac{\pi + \sqrt{\pi^2 + 2\pi - 4}}{2\pi - 4}$$

We chose the positive sign since c must be positive. Remark: the volume of the half torus could have been obtained from Pappus' theorem. The critical ratio is 2.90.

99. This is Putnam Problem 5 in the morning session of 1958. Suppose there are two continuous solutions and let g be their difference. Then g is continuous and

$$g(x, y) = \int_0^x \int_0^y g(u, v)\, du\, dv$$

Since g is continuous, it is bounded on the given square. Let M be a bound. Then

$$|g(x, y)| \leq \int_0^x \int_0^y |g(u, v)|\, du\, dv$$

$$\leq \int_0^x \int_0^y M\, du\, dv$$

$$= Mxy$$

for $0 \leq x \leq 1, 0 \leq y \leq 1$. We now prove

$$|g(x, y)| \leq M\frac{x^n y^n}{n!n!}$$

for any integer n. This has been proved for $n = 1$. Assume that it is true for $n = k$; then

$$|g(x, y)| \leq \int_0^x \int_0^y |g(u, v)|\, du\, dv$$

$$\leq \int_0^1 \int_0^y M\, du\, dv$$

$$= M\frac{x^{k+1} y^{k+1}}{(k+1)!(k+1)!}$$

Thus, the proposition is true by mathematical induction. But for any fixed x and y

$$\lim_{n\to\infty} M\frac{x^n y^n}{n!n!} = 0$$

Hence, $|g(x, y)| \leq 0$; that is, $g(x, y) = 0$. Thus, there cannot be two different continuous solutions. Remark: This problem generalizes immediately to n-dimensions. See Problem 4885, *American Mathematical Monthly*, Vol. 67 (1960), p. 87. Solution is given in Vol. 68 (1961), p. 73.

CHAPTER 13 Vector Analysis

Chapter Overview
This chapter combines what you have learned about differentiation, integration, and vectors in order to apply calculus to fluid and electrical flow as an introduction to fluid dynamics and electromagnetic theory. There are three extremely important results introduced in this chapter, Green's theorem, Stokes' theorem and divergence theorem. This chapter will be a springboard for you into the world of higher mathematics.

13.1 Properties of a Vector Field: Divergence and Curl, page 1026

1. Answers vary; the operator

$$\nabla = \frac{\partial}{\partial x}\mathbf{i} + \frac{\partial}{\partial y}\mathbf{j} + \frac{\partial}{\partial z}\mathbf{k}$$

is called the del operator and is applicable to a scalar function only:

$$\nabla f = f_x\mathbf{i} + f_y\mathbf{j} + f_z\mathbf{k}$$

The del operator is used to compute divergence by

$$\text{div } \mathbf{F} = \nabla \cdot \mathbf{F}$$

This is applicable to vector functions only and generates a scalar. It is also used to compute curl by

$$\text{curl } \mathbf{F} = \nabla \times \mathbf{F}$$

This is applicable to vector functions only and generates a vector

$$\nabla f = \frac{\partial f}{\partial x}\mathbf{i} + \frac{\partial f}{\partial y}\mathbf{j} + \frac{\partial f}{\partial z}\mathbf{k}$$

Finally, we note that $\nabla^2 f = \nabla \cdot \nabla f$.

3. $\text{div } \mathbf{F} = \nabla \cdot \mathbf{F}$
$$= \left(\frac{\partial}{\partial x}\mathbf{i} + \frac{\partial}{\partial y}\mathbf{j} + \frac{\partial}{\partial z}\mathbf{k}\right) \cdot (x^2\mathbf{i} + xy\mathbf{j} + z^3\mathbf{k})$$
$$= 2x + x + 3z^2$$
$$= 3x + 3z^2$$

$$\text{curl } \mathbf{F} = \nabla \times \mathbf{F} = \begin{vmatrix} \mathbf{i} & \mathbf{j} & \mathbf{k} \\ \frac{\partial}{\partial x} & \frac{\partial}{\partial y} & \frac{\partial}{\partial z} \\ x^2 & xy & z^3 \end{vmatrix}$$
$$= 0\mathbf{i} - 0\mathbf{j} + (y - 0)\mathbf{k}$$
$$= y\mathbf{k}$$

5. $\text{div } \mathbf{F} = \nabla \cdot \mathbf{F}$
$$= \left(\frac{\partial}{\partial x}\mathbf{i} + \frac{\partial}{\partial y}\mathbf{j} + \frac{\partial}{\partial z}\mathbf{k}\right) \cdot (2y\mathbf{j})$$
$$= 2$$

$$\text{curl } \mathbf{F} = \nabla \times \mathbf{F} = \begin{vmatrix} \mathbf{i} & \mathbf{j} & \mathbf{k} \\ \frac{\partial}{\partial x} & \frac{\partial}{\partial y} & \frac{\partial}{\partial z} \\ 0 & 2y & 0 \end{vmatrix}$$
$$= 0\mathbf{i} - 0\mathbf{j} + 0\mathbf{k}$$
$$= \mathbf{0}$$

7. $\text{div } \mathbf{F} = \nabla \cdot \mathbf{F}$
$$= \left(\frac{\partial}{\partial x}\mathbf{i} + \frac{\partial}{\partial y}\mathbf{j} + \frac{\partial}{\partial z}\mathbf{k}\right) \cdot (\mathbf{i} + \mathbf{j} + \mathbf{k})$$
$$= 0$$

$$\text{curl } \mathbf{F} = \nabla \times \mathbf{F} = \begin{vmatrix} \mathbf{i} & \mathbf{j} & \mathbf{k} \\ \frac{\partial}{\partial x} & \frac{\partial}{\partial y} & \frac{\partial}{\partial z} \\ 1 & 1 & 1 \end{vmatrix}$$

$$= \mathbf{0}$$

Note: $\mathbf{F}$ is a constant vector, so the result is the same regardless of the given point.

9. $\text{div } \mathbf{F} = \nabla \cdot \mathbf{F}$

$$= \left(\frac{\partial}{\partial x} \mathbf{i} + \frac{\partial}{\partial y} \mathbf{j} + \frac{\partial}{\partial z} \mathbf{k} \right) \cdot (xyz\mathbf{i} + y\mathbf{j} + x\mathbf{k})$$

$$= yz + 1$$

$$\text{curl } \mathbf{F} = \nabla \times \mathbf{F} = \begin{vmatrix} \mathbf{i} & \mathbf{j} & \mathbf{k} \\ \frac{\partial}{\partial x} & \frac{\partial}{\partial y} & \frac{\partial}{\partial z} \\ xyz & y & x \end{vmatrix}$$

$$= 0\mathbf{i} - (1 - xy)\mathbf{j} - xz\mathbf{k}$$

At $(1, 2, 3)$,

div $\mathbf{F} = 6 + 1 = 7$,

curl $\mathbf{F} = \mathbf{j} - 3\mathbf{k}$

11. $\text{div } \mathbf{F} = \nabla \cdot \mathbf{F}$

$$= \left(\frac{\partial}{\partial x} \mathbf{i} + \frac{\partial}{\partial y} \mathbf{j} + \frac{\partial}{\partial z} \mathbf{k} \right) \cdot (e^{-xy}\mathbf{i} + e^{xz}\mathbf{j} + e^{yz}\mathbf{k})$$

$$= -ye^{-xy} + ye^{yz}$$

curl $\mathbf{F} = \nabla \times \mathbf{F}$

$$= \begin{vmatrix} \mathbf{i} & \mathbf{j} & \mathbf{k} \\ \frac{\partial}{\partial x} & \frac{\partial}{\partial y} & \frac{\partial}{\partial z} \\ e^{-xy} & e^{xz} & e^{yz} \end{vmatrix}$$

$$= (ze^{yz} - xe^{xz})\mathbf{i} - 0\mathbf{j} + (ze^{xz} + xe^{-xy})\mathbf{k}$$

At $(3, 2, 0)$,

div $\mathbf{F} = 2 - 2e^{-6}$,

curl $\mathbf{F} = -3\mathbf{i} + 3e^{-6}\mathbf{k}$

13. $\text{div } \mathbf{F} = \nabla \cdot \mathbf{F}$

$$= = \left(\frac{\partial}{\partial x} \mathbf{i} + \frac{\partial}{\partial y} \mathbf{j} \right) \cdot [(\sin x)\mathbf{i} + (\cos y)\mathbf{j}]$$

$$= \cos x - \sin y$$

$$\text{curl } \mathbf{F} = \nabla \times \mathbf{F} = \begin{vmatrix} \mathbf{i} & \mathbf{j} & \mathbf{k} \\ \frac{\partial}{\partial x} & \frac{\partial}{\partial y} & \frac{\partial}{\partial z} \\ \sin x & \cos y & 0 \end{vmatrix}$$

$$= \mathbf{0}$$

15. $\text{div } \mathbf{F} = \nabla \cdot \mathbf{F}$

$$= \left(\frac{\partial}{\partial x} \mathbf{i} + \frac{\partial}{\partial y} \mathbf{j} \right) \cdot (x\mathbf{i} - y\mathbf{j})$$

$$= 1 - 1 = 0$$

$$\text{curl } \mathbf{F} = \nabla \times \mathbf{F} = \begin{vmatrix} \mathbf{i} & \mathbf{j} & \mathbf{k} \\ \frac{\partial}{\partial x} & \frac{\partial}{\partial y} & \frac{\partial}{\partial z} \\ x & -y & 0 \end{vmatrix}$$

$$= -1 + 1$$

$$= \mathbf{0}$$

17. $\text{div } \mathbf{F} = \nabla \cdot \mathbf{F}$

$$= \left(\frac{\partial}{\partial x} \mathbf{i} + \frac{\partial}{\partial y} \mathbf{j} \right)$$

$$\cdot \left(\frac{x}{\sqrt{x^2 + y^2}} \mathbf{i} + \frac{y}{\sqrt{x^2 + y^2}} \mathbf{j} \right)$$

$$= \frac{\sqrt{x^2 + y^2} - \frac{x^2}{\sqrt{x^2+y^2}}}{x^2 + y^2}$$

$$+ \frac{\sqrt{x^2 + y^2} - \frac{y^2}{\sqrt{x^2+y^2}}}{x^2 + y^2}$$

$$= \frac{1}{\sqrt{x^2 + y^2}}$$

curl $\mathbf{F} = \nabla \times \mathbf{F}$

$$= = \begin{vmatrix} \mathbf{i} & \mathbf{j} & \mathbf{k} \\ \frac{\partial}{\partial x} & \frac{\partial}{\partial y} & \frac{\partial}{\partial z} \\ \frac{x}{\sqrt{x^2+y^2}} & \frac{y}{\sqrt{x^2+y^2}} & 0 \end{vmatrix}$$

$$= \mathbf{0}$$

19. $\operatorname{div} \mathbf{F} = \nabla \cdot \mathbf{F}$

$$= \left(\frac{\partial}{\partial x}\mathbf{i} + \frac{\partial}{\partial y}\mathbf{j} + \frac{\partial}{\partial z}\mathbf{k} \right) \cdot \left(x^2\mathbf{i} - z^2\mathbf{k} \right)$$

$$= 2x - 2z$$

$$\operatorname{curl} \mathbf{F} = \nabla \times \mathbf{F} = \begin{vmatrix} \mathbf{i} & \mathbf{j} & \mathbf{k} \\ \frac{\partial}{\partial x} & \frac{\partial}{\partial y} & \frac{\partial}{\partial z} \\ x^2 & 0 & -z^2 \end{vmatrix}$$

$$= \mathbf{0}$$

21. $\operatorname{div} \mathbf{F} = \nabla \cdot \mathbf{F}$

$$= \left(\frac{\partial}{\partial x}\mathbf{i} + \frac{\partial}{\partial y}\mathbf{j} + \frac{\partial}{\partial z}\mathbf{k} \right) \cdot \left(y\mathbf{i} + z\mathbf{j} + x\mathbf{k} \right)$$

$$= 0$$

$$\operatorname{curl} \mathbf{F} = \nabla \times \mathbf{F}$$

$$= \begin{vmatrix} \mathbf{i} & \mathbf{j} & \mathbf{k} \\ \frac{\partial}{\partial x} & \frac{\partial}{\partial y} & \frac{\partial}{\partial z} \\ y & z & x \end{vmatrix}$$

$$= -\mathbf{i} - \mathbf{j} - \mathbf{k}$$

23. $\operatorname{div} \mathbf{F} = \nabla \cdot \mathbf{F}$

$$= \left(\frac{\partial}{\partial x}\mathbf{i} + \frac{\partial}{\partial y}\mathbf{j} + \frac{\partial}{\partial z}\mathbf{k} \right)$$

$$\cdot \left(\ln z\,\mathbf{i} + e^{xy}\mathbf{j} + \tan^{-1}\frac{x}{z}\mathbf{k} \right)$$

$$= xe^{xy} - \frac{x}{x^2 + z^2}$$

$$\operatorname{curl} \mathbf{F} = \nabla \times \mathbf{F}$$

$$= \begin{vmatrix} \mathbf{i} & \mathbf{j} & \mathbf{k} \\ \frac{\partial}{\partial x} & \frac{\partial}{\partial y} & \frac{\partial}{\partial z} \\ \ln z & e^{xy} & \tan^{-1}\frac{x}{z} \end{vmatrix}$$

$$= \left(\frac{1}{z} - \frac{z}{x^2 + z^2} \right)\mathbf{j} + (ye^{xy})\mathbf{k}$$

25. $\operatorname{div} \mathbf{F} = \nabla \cdot \mathbf{F}$

$$= \left(\frac{\partial}{\partial x}\mathbf{i} + \frac{\partial}{\partial y}\mathbf{j} + \frac{\partial}{\partial z}\mathbf{k} \right)$$

$$\cdot \left(\frac{x\mathbf{i} + y\mathbf{j} + z\mathbf{k}}{\sqrt{x^2 + y^2 + z^2}} \right)$$

$$= \frac{\sqrt{x^2 + y^2 + z^2} - \frac{x^2}{\sqrt{x^2+y^2+z^2}}}{x^2 + y^2 + z^2}$$

$$+ \frac{\sqrt{x^2 + y^2 + z^2} - \frac{y^2}{\sqrt{x^2+y^2+z^2}}}{x^2 + y^2 + z^2}$$

$$+ \frac{\sqrt{x^2 + y^2 + z^2} - \frac{z^2}{\sqrt{x^2+y^2+z^2}}}{x^2 + y^2 + z^2}$$

$$= \frac{2(x^2 + y^2 + z^2)}{(x^2 + y^2 + z^2)^{3/2}}$$

$$= \frac{2}{\sqrt{x^2 + y^2 + z^2}}$$

$$\operatorname{curl} \mathbf{F} = \nabla \times \mathbf{F}$$

$$= \begin{vmatrix} \mathbf{i} & \mathbf{j} & \mathbf{k} \\ \frac{\partial}{\partial x} & \frac{\partial}{\partial y} & \frac{\partial}{\partial z} \\ \frac{x}{\sqrt{x^2+y^2+z^2}} & \frac{y}{\sqrt{x^2+y^2+z^2}} & \frac{z}{\sqrt{x^2+y^2+z^2}} \end{vmatrix}$$

$$= \frac{-2yz + 2yz}{2(x^2 + y^2 + z^2)^{3/2}}\mathbf{i} + \frac{-2xz + 2xz}{2(x^2 + y^2 + z^2)^{3/2}}\mathbf{j}$$

$$+ \frac{-2xy + 2xy}{2(x^2 + y^2 + z^2)^{3/2}}\mathbf{k}$$

$$= \mathbf{0}$$

27. $u_x = -e^{-x}(\cos y - \sin y)$

$u_{xx} = e^{-x}(\cos y - \sin y)$

$u_y = e^{-x}(-\sin y - \cos y)$

$u_{yy} = e^{-x}(-\cos y + \sin y)$

$u_{xx} + u_{yy} = 0;\ u$ is harmonic

29. $w_x = \dfrac{-x}{(x^2 + y^2 + z^2)^{3/2}}$

$$w_{xx} = -\frac{(x^2 + y^2 + z^2)^{3/2} - 3x^2(x^2 + y^2 + z^2)^{1/2}}{(x^2 + y^2 + z^2)^3}$$

$$= \frac{2x^2 - y^2 - z^2}{(x^2 + y^2 + z^2)^{5/2}}$$

Similarly,

$$w_{yy} = \frac{2y^2 - x^2 - z^2}{(x^2 + y^2 + z^2)^{5/2}}$$

$$w_{zz} = \frac{2z^2 - x^2 - y^2}{(x^2 + y^2 + z^2)^{5/2}}$$

$w_{xx} + w_{yy} + w_{zz} = 0$; w is harmonic

31. $\mathbf{F} = \nabla f = y^3 z^2 \mathbf{i} + 3xy^2 z^2 \mathbf{j} + 2xy^3 z \mathbf{k}$
　　div $\mathbf{F} = 6xyz^2 + 2xy^3$

33. $\mathbf{F} \times \mathbf{G} = \begin{vmatrix} \mathbf{i} & \mathbf{j} & \mathbf{k} \\ 2 & 2x & 3y \\ x & -y & z \end{vmatrix}$

$$= (2xz + 3y^2)\mathbf{i} + (3xy - 2z)\mathbf{j}$$
$$+ (-2y - 2x^2)\mathbf{k}$$

curl($\mathbf{F} \times \mathbf{G}$)

$$= \begin{vmatrix} \mathbf{i} & \mathbf{j} & \mathbf{k} \\ \frac{\partial}{\partial x} & \frac{\partial}{\partial y} & \frac{\partial}{\partial z} \\ 2xz + 3y^2 & 3xy - 2z & -2y - 2x^2 \end{vmatrix}$$

$$= (-2 + 2)\mathbf{i} - (-2x - 4x)\mathbf{j} + (3y - 6y)\mathbf{k}$$
$$= 6x\mathbf{j} - 3y\mathbf{k}$$

35. From Problem 33,
　　$\mathbf{F} \times \mathbf{G} = (2xz + 3y^2)\mathbf{i} + (3xy - 2z)\mathbf{j}$
　　　　　　$+ (-2y - 2x^2)\mathbf{k}$

div($\mathbf{F} \times \mathbf{G}$)

$$= \frac{\partial(2xz + 3y^2)}{\partial x} + \frac{\partial(3xy - 2z)}{\partial y} + \frac{\partial(-2y - 2x^2)}{\partial z}$$

$$= 2z + 3x$$

37. Let $\mathbf{A} = a\mathbf{i} + b\mathbf{j} + c\mathbf{k}$

$$\mathbf{A} \times \mathbf{R} = \begin{vmatrix} \mathbf{i} & \mathbf{j} & \mathbf{k} \\ a & b & c \\ x & y & z \end{vmatrix}$$

$$= (bz - cy)\mathbf{i} - (az - cx)\mathbf{j}$$
$$+ (ay - bx)\mathbf{k}$$

div($\mathbf{A} \times \mathbf{R}$)

$$= \frac{\partial}{\partial x}(bz - cy) - \frac{\partial}{\partial y}(az - cx)$$
$$+ \frac{\partial}{\partial z}(ay - bx)$$

$$= 0$$

39. Let $\boldsymbol{\omega} = a\mathbf{i} + b\mathbf{j} + c\mathbf{k}$; then
　a.　$\mathbf{V} = \boldsymbol{\omega} \times \mathbf{R}$

$$= \begin{vmatrix} \mathbf{i} & \mathbf{j} & \mathbf{k} \\ a & b & c \\ x & y & z \end{vmatrix}$$

$$= (bz - cy)\mathbf{i} + (cx - az)\mathbf{j}$$
$$+ (ay - bx)\mathbf{k}$$

　b.　div $\mathbf{V} = 0 + 0 + 0 = 0$
　　curl $\mathbf{V}$

$$= \begin{vmatrix} \mathbf{i} & \mathbf{j} & \mathbf{k} \\ \frac{\partial}{\partial x} & \frac{\partial}{\partial y} & \frac{\partial}{\partial z} \\ bz - cy & cx - ax & ay - bx \end{vmatrix}$$

$$= 2a\mathbf{i} + 2b\mathbf{j} + 2c\mathbf{k}$$
$$= 2\boldsymbol{\omega}$$

41. Let $\mathbf{F} = f_1\mathbf{i} + f_2\mathbf{j} + f_3\mathbf{k}$ and
　　$\mathbf{G} = g_1\mathbf{i} + g_2\mathbf{j} + g_3\mathbf{k}$

$$\mathbf{F} \times \mathbf{G} = \begin{vmatrix} \mathbf{i} & \mathbf{j} & \mathbf{k} \\ f_1 & f_2 & f_3 \\ g_1 & g_2 & g_3 \end{vmatrix}$$

$$= (f_2 g_3 - f_3 g_2)\mathbf{i} + (f_3 g_1 - f_1 g_3)\mathbf{j}$$
$$+ (f_1 g_2 - f_2 g_1)\mathbf{k}$$

$$\text{div}(\mathbf{F} \times \mathbf{G}) = (f_2 g_3 - f_3 g_2)_x + (f_3 g_1 - f_1 g_3)_y$$
$$+ (f_1 g_2 - f_2 g_1)_z$$

I. (div $\mathbf{F}$)(div $\mathbf{G}$)
　$= [(f_1)_x + (f_2)_y + (f_3)_z][(g_1)_x + (g_2)_y + (g_3)_z]$
　$\neq \text{div}(\mathbf{F} \times \mathbf{G})$

II. $(\text{curl } \mathbf{F}) \cdot \mathbf{G} - \mathbf{F} \cdot (\text{curl } \mathbf{G})$

$$= \langle (f_3)_y - (f_2)_z, (f_1)_z - (f_3)_x, (f_2)_x - (f_1)_y \rangle$$
$$\cdot \langle g_1, g_2, g_3 \rangle$$
$$- \langle f_1, f_2, f_3 \rangle \cdot \langle (g_3)_y - (g_2)_z, (g_1)_z$$
$$- (g_3)_x, (g_2)_x - (g_1)_y \rangle$$
$$= g_1[(f_3)_y - (f_2)_z] + g_2[(f_1)_z - (f_3)_x]$$
$$+ g_3[(f_2)_x - (f_1)_y] - f_1[(g_3)_y - (g_2)_z]$$
$$- f_2[(g_1)_z - (g_3)_x] - f_3[(g_2)_x - (g_1)_y]$$
$$= (f_2 g_3 - f_3 g_2)_x + (f_3 g_1 - f_1 g_3)_y$$
$$+ (f_1 g_2 - f_2 g_1)_z$$
$$= \text{div}(\mathbf{F} \times \mathbf{G})$$

III. $\mathbf{F}(\text{div } \mathbf{G}) + (\text{div } \mathbf{F})\mathbf{G}$ is a vector, so it can't possibly equal $\text{div}(\mathbf{F} \times \mathbf{G})$.

IV. Note that since

$$(\text{curl } \mathbf{F}) \cdot \mathbf{G} - \mathbf{F} \cdot (\text{curl } \mathbf{G}) = \text{div}(\mathbf{F} \times \mathbf{G})$$

we cannot also have

$$(\text{curl } \mathbf{F}) \cdot \mathbf{G} + \mathbf{F} \cdot (\text{curl } \mathbf{G}) = \text{div}(\mathbf{F} \times \mathbf{G})$$

unless

$$2\mathbf{F} \cdot (\text{curl } \mathbf{G}) = 0$$

which is not generally true, as the following counterexample illustrates: $\mathbf{F} = z\mathbf{k}$, $\mathbf{G} = x\mathbf{j}$, so

$$\mathbf{F} \cdot (\text{curl } \mathbf{G}) = z\mathbf{k} \cdot \mathbf{k} = z$$

43. $\text{curl } \mathbf{F} = \begin{vmatrix} \mathbf{i} & \mathbf{j} & \mathbf{k} \\ \frac{\partial}{\partial x} & \frac{\partial}{\partial y} & \frac{\partial}{\partial z} \\ u(x,y) & v(x,y) & 0 \end{vmatrix}$

$$= 0\mathbf{i} + 0\mathbf{j} + \left(\frac{\partial}{\partial x} v(x,y) - \frac{\partial}{\partial y} u(x,y) \right)\mathbf{k}$$
$$= \mathbf{0}$$

if and only if the coefficient of $\mathbf{k}$ is zero—that is if and only if $\dfrac{\partial v}{\partial x} = \dfrac{\partial u}{\partial y}$.

45. $\text{div } \mathbf{F} + \text{div } \mathbf{G}$
$$= [(f_1)_x + (f_2)_y + (f_3)_z] + [(g_1)_x + (g_2)_y + (g_3)_z]$$
$$= [(f_1)_x + (g_1)_x] + [(f_2)_y + (g_2)_y] + [(f_3)_z + (g_3)_z]$$
$$= \text{div}(\mathbf{F} + \mathbf{G})$$

47. $\text{curl}(c\mathbf{F}) = \begin{vmatrix} \mathbf{i} & \mathbf{j} & \mathbf{k} \\ \frac{\partial}{\partial x} & \frac{\partial}{\partial y} & \frac{\partial}{\partial z} \\ cf_1 & cf_2 & cf_3 \end{vmatrix}$

$$= c \begin{vmatrix} \mathbf{i} & \mathbf{j} & \mathbf{k} \\ \frac{\partial}{\partial x} & \frac{\partial}{\partial y} & \frac{\partial}{\partial z} \\ f_1 & f_2 & f_3 \end{vmatrix}$$
$$= c \text{ curl } \mathbf{F}$$

49. Let $\mathbf{F} = M\mathbf{i} + N\mathbf{j} + P\mathbf{k}$

$$\text{div}(f\mathbf{F}) = (fM)_x + (fN)_y + (fP)_x$$
$$= f(M_x + N_y + P_z) + Mf_x + Nf_y + Pf_z$$
$$= f \text{ div } \mathbf{F} + [f_x\mathbf{i} + f_y\mathbf{j} + f_z\mathbf{k}] \cdot [M\mathbf{i} + N\mathbf{j} + P\mathbf{k}]$$
$$= f \text{ div } \mathbf{F} + \nabla f \cdot \mathbf{F}$$

51. Apply the formula in Problem 49, with $\mathbf{F} = \nabla g$. Then

$$\text{div}(f \nabla g) = f \text{ div} \nabla g + \nabla f \cdot \nabla g$$

53. Let $\mathbf{F} = M\mathbf{i} + N\mathbf{j} + P\mathbf{k}$

$\text{div}(\text{curl } \mathbf{F}) = \text{div} \begin{vmatrix} \mathbf{i} & \mathbf{j} & \mathbf{k} \\ \frac{\partial}{\partial x} & \frac{\partial}{\partial y} & \frac{\partial}{\partial z} \\ M & N & P \end{vmatrix}$

$$= \text{div}[(P_y - N_z)\mathbf{i} + (M_z - P_x)\mathbf{j} + (N_x - M_y)\mathbf{k}]$$
$$= P_{yx} - N_{zx} + M_{zy} - P_{xy} + N_{xz} - M_{yz}$$
$$= (P_{yx} - P_{xy}) + (M_{zy} - M_{yz}) + (N_{xz} - N_{zx})$$
$$= 0$$

55. $\text{div}\left(\dfrac{1}{r^3}\mathbf{R} \right) = \text{div}\left(\dfrac{1}{r^3}(x\mathbf{i} + y\mathbf{j} + z\mathbf{k}) \right)$

$$= \frac{-(2x^2 - y^2 - z^2) - (2y^2 - x^2 - z^2) - (2z^2 - x^2 - y^2)}{(x^2 + y^2 + z^2)^{5/2}}$$
$$= 0$$

57. $\text{div}(r\mathbf{R}) = \dfrac{\partial}{\partial x}(rx) + \dfrac{\partial}{\partial y}(ry) + \dfrac{\partial}{\partial z}(rz)$

$$= \left(r + \frac{x^2}{r}\right) + \left(r + \frac{y^2}{r}\right) + \left(r + \frac{z^2}{r}\right)$$

$$= \frac{3r^2 + (x^2 + y^2 + z^2)}{r}$$

$$= \frac{3r^2 + r^2}{r}$$

$$= 4r$$

59. div $\nabla(fg) = \nabla \cdot (f\nabla g + g\nabla f)$

$$= f(\nabla \cdot \nabla g) + \nabla f \cdot \nabla g$$
$$\qquad + \nabla g \cdot \nabla f + g(\nabla \cdot \nabla f)$$
$$= f\,\text{div}(\nabla g) + 2\nabla f \cdot \nabla g + g\,\text{div}(\nabla f)$$

13.2 Line Integrals, page 1036

SURVIVAL HINT: *When finding a line integral of a function $f(x, y, z)$ defined on the smooth curve C, use the symbol*

$$\int_C f(x, y, z)\, ds$$

If the curve C is a closed curve, then we often use the symbol

$$\oint_C f(x, y, z)\, ds$$

1. Let $f(x, y, z)$ where $x = x(t)$, $y = y(t)$, $z = z(t)$ over $a \le t \le b$ is given. Then

$$\int_C f\, ds$$

$$= \int_a^b f[x(t), y(t), z(t)]\sqrt{[x'(t)]^2 + [y'(t)]^2 + [z'(t)]^2}\, dt$$

and

$$\int_C f\, dx = \int_a^b f[x(t), y(t), z(t)]\, x'(t)\, dt$$

3. $x = 1 - t$, $y = 4(1-t)^2$, $0 \le t \le 1$

$$\int_C (-y\, dx + x\, dy)$$

$$= \int_0^1 \left[-[4(1 - t)^2](-dt) + (1 - t)(8t - 8)\, dt\right]$$

$$= \int_0^1 \left[-4t^2 + 8t - 4\right] dt$$

$$= -\frac{4}{3}$$

5. $x = t$, $y = \dfrac{2t - 1}{4}$, $4 \le t \le 8$

$$\int_C (x\, dy - y)\, dx$$

$$= \int_4^8 \left[t\left(\frac{1}{2}dt\right) - \left(\frac{2t - 1}{4}\right)dt\right]$$

$$= \frac{1}{4}\int_4^8 dt$$

$$= 1$$

7. C needs to be considered in two pieces: Let $x = t$, then $y = -2t$ on $-1 \le t \le 0$ and $y = 2t$ on $0 \le t \le 1$.

On $[-1, 0]$: $\quad [(x + y)^2 dx - (x - y)^2\, dy]$
$$= (-t)^2 dt - (3t)^2(-2\, dt)$$
$$= 19t^2\, dt$$

On $[0, 1]$: $\quad [(x + y)^2 dx - (x - y)^2 dy]$
$$= (3t)^2 dt - (-t)^2(2\, dt)$$
$$= 7t^2 dt$$

$$\int_C [(x + y)^2 dx - (x - y)^2 dy]$$

$$= \int_{-1}^0 19t^2\, dt + \int_0^1 7t^2\, dt$$

$$= \frac{26}{3}$$

9. $ds = \sqrt{[x'(t)]^2 + [y'(t)]^2}\, dt$

$= \sqrt{(3t^{1/2})^2 + (3)^2}\, dt$

$= \sqrt{9t + 9}\, dt$

$\int_C \dfrac{1}{3+y}\, ds = \int_0^1 \dfrac{1}{3+3t}\sqrt{9t+9}\, dt$

$= \int_0^1 \dfrac{\sqrt{t+1}}{t+1}\, dt$

$= 2\sqrt{2} - 2$

11. $ds = \sqrt{[x'(t)]^2 + [y'(t)]^2}\, dt$

$= \sqrt{\cos^2 t + \sin^2 t}\, dt$

$= dt$

$\int_C (3x - 2y)\, ds = \int_0^\pi (3\sin t - 2\cos t)\, dt$

$= 6$

13. $ds = \sqrt{[x'(t)]^2 + [y'(t)]^2}\, dt$

$= \sqrt{1 + 9}\, dt$

$= \sqrt{10}\, dt$

$\int_C \dfrac{x^2}{y^2}\, ds = \int_0^1 \dfrac{t^2}{9t^2}\sqrt{10}\, dt$

$= \dfrac{\sqrt{10}}{9}\int_0^1 dt$

$= \dfrac{\sqrt{10}}{9}$

15. a. Let $x = \cos\theta$ and $y = \sin\theta$ on
$0 \le \theta \le \frac{\pi}{2}$
$(x^2 + y^2)dx + 2xy\, dy$
$= (1)(-\sin\theta\, d\theta) + 2\cos\theta\sin\theta(\cos\theta)\, d\theta$

$\int_C [(x^2 + y^2)dx + 2xy\, dy]$

$= \int_0^{\pi/2} \left(-\sin\theta + 2\sin\theta\cos^2\theta\right) d\theta$

$= -\dfrac{1}{3}$

b. Let $x = 1 - t$ and $y = t$ on $0 \le t \le 1$
$(x^2 + y^2)dx + 2xy\, dy$
$= \left[(1-t)^2 + t^2\right](-dt) + 2(1-t)(t)\, dt$
$= \left(-4t^2 + 4t - 1\right) dt$

$\int_C [(x^2 + y^2)dx + 2xy\, dy]$

$= \int_0^1 \left(-4t^2 + 4t - 1\right) dt$

$= -\dfrac{1}{3}$

17. We must use two smooth pieces:
On $[0,2]$: $x = t, y = 0$

$[(x^2 + y^2)dx + 2xy\, dy] = t^2\, dt$

On $[0,4]$: $x = 2, y = t$

$[(x^2 + y^2)dx + 2xy\, dy] = 4t\, dt$

$\int_C [(x^2 + y^2)dx + 2xy\, dy]$

$= \int_0^2 t^2\, dt + \int_0^4 4t\, dt$

$= \dfrac{8}{3} + 32$

$= \dfrac{104}{3}$

19. C must be divided into two smooth pieces.
On $[-1,0]$: $x = t, y = 1$

$$(-xy^2\,dx + x^2\,dy) = -t\,dt$$

On $[0, \frac{\pi}{2}]$: $x = \sin\theta$ and $y = \cos\theta$

$$(-xy^2\,dx + x^2\,dy)$$
$$= \cos^2\theta - \sin\theta\,(\cos\theta\,d\theta) + \sin^2\theta(-\sin\theta\,d\theta)$$
$$= (-\sin\theta\cos^3\theta - \sin^3\theta)\,d\theta$$

$$\int_C (-xy^2\,dx + x^2\,dy)$$
$$= \int_{-1}^0 (-t)dt$$
$$+ \int_0^{\pi/2} (-\sin\theta\cos^3\theta - \sin^3\theta)\,d\theta$$
$$= \frac{1}{2} + \left(-\frac{11}{12}\right)$$
$$= -\frac{5}{12}$$

21. Let $x = 2\cos\theta$ and $y = 2\sin\theta$ on $[0, 2\pi]$.

$$(x^2 - y^2)\,dx + x\,dy$$
$$= (4\cos^2\theta - 4\sin^2\theta)(-2\sin\theta\,d\theta)$$
$$\quad + 2\cos\theta(2\cos\theta\,d\theta)$$
$$= [8(2\cos^2\theta - 1)(-\sin\theta)$$
$$\quad + 2(1 + \cos 2\theta)]\,d\theta$$

$$\oint_C (x^2 - y^2)\,dx + x\,dy$$
$$= \int_0^{2\pi} [8(2\cos^2\theta - 1)(-\sin\theta)\,d\theta$$
$$+ \int_0^{2\pi} 2(1 + \cos 2\theta)]\,dx$$
$$= 4\pi$$

23. Write the equation of the line in parametric form where $x = 2t$ and $y = t$ on $0 \le t \le 1$.

$$\mathbf{F}\cdot d\mathbf{R} = [(5(2t) + t)\mathbf{i} + 2t\mathbf{j}]\cdot(2\mathbf{i} + \mathbf{j})\,dt$$
$$= 24t\,dt$$

$$\int_0^1 24t\,dt = 12$$

25. We recognize this as the same as Problem 23.

$$\mathbf{F}\cdot d\mathbf{R} = [(5(2t) + t)\mathbf{i} + 2t\mathbf{j}]\cdot(2\mathbf{i} + \mathbf{j})\,dt$$
$$= 24t\,dt$$

$$\int_0^1 24t\,dt = 12$$

27. a. $\displaystyle\int_C (y\,dx - x\,dy + dz)$

$$= \int_0^{\pi/2} [3\cos t(3\cos t) - 3\sin t(-3\sin t) + 1]\,dt$$
$$= 5\pi$$

b. $\displaystyle\int_C (y\,dx - x\,dy + dz)$

$$= \int_0^{\pi/2} [a\cos t(a\cos t) - a\sin t(-a\sin t) + 1]\,dt$$
$$= \frac{\pi}{2}(a^2 + 1)$$

29. a. $\displaystyle\int_C (-y\,dx + x\,dy + xz\,dz)$

$$= \int_0^{2\pi} [-\sin t(-\sin t) + \cos^2 t + t\cos t]\,dt$$
$$= 2\pi$$

b. $C: x = \cos t,\ y = \sin t,\ z = 0,\ 0 \le t \le 2\pi$

$$\int_C (-y\,dx + x\,dy + xz\,dz)$$
$$= \int_0^{2\pi} [-\sin t(-\sin t) + \cos^2 t + 0]\,dt$$
$$= 2\pi$$

31. $C: \quad x^2 + 4y^2 - 8y + 3 = 0$

$\qquad x^2 + 4(y^2 - 2y + 1) = -3 + 4$

$\qquad\qquad x^2 + 4(y - 1)^2 = 1$

Let $x = \cos t$, $y = \frac{1}{2}\sin t + 1$, $z = 0$;

$0 \le t < 2\pi$

$\qquad \mathbf{F} = x\mathbf{i} + xy\mathbf{j} + x^2 yz\mathbf{k}$

$\qquad = (\cos t)\mathbf{i} + (\cos t)\left(\frac{1}{2}\sin t + 1\right)\mathbf{j}$

$\qquad\qquad + (\cos t)^2\left(\frac{1}{2}\sin t + 1\right)(0)\mathbf{k}$

$\mathbf{R} = (\cos t)\mathbf{i} + \left(\frac{1}{2}\sin t + 1\right)\mathbf{j} + 0\mathbf{k}$

$d\mathbf{R} = \left[(-\sin t)\mathbf{i} + \left(\frac{1}{2}\cos t\right)\mathbf{j}\right]dt$

$\oint_C \mathbf{F} \cdot d\mathbf{R}$

$\qquad = \int_0^{2\pi}\left[-\cos t \sin t + \frac{1}{4}\cos^2 t \sin t + \frac{1}{2}\cos^2 t\right]dt$

$\qquad = \frac{\pi}{2}$

33. $ds = \sqrt{(1)^2 + (2t)^2 + (2t^2)^2}\, dt$

$\qquad = \sqrt{1 + 4t^2 + 4t^4}\, dt$

$\qquad = (2t^2 + 1)\, dt$

$\int_C 2xy^2 z\, ds = \int_0^1 2t(t^2)^2\left(\frac{2}{3}t^3\right)(2t^2 + 1)\, dt$

$\qquad = \int_0^1 \frac{4}{3}t^8(2t^2 + 1)\, dt$

$\qquad = \frac{116}{297}$

35. $C_1: \quad x = y = t, z = 0, 0 \le t \le 1$

$\qquad \mathbf{R} = t\mathbf{i} + t\mathbf{j}$

$\qquad \mathbf{F} = t^2\mathbf{i} + t^2\mathbf{j} - t\mathbf{k}, \ d\mathbf{R} = (\mathbf{i} + \mathbf{j})\, dt$

$C_2: x = 1, y = 1 - t, z = 0, 0 \le t \le 1$

$\mathbf{R} = \mathbf{i} + (1 - t)\mathbf{j}$

$\mathbf{F} = (1 - t)^2 + \mathbf{j} - \mathbf{k}, \ d\mathbf{R} = -\mathbf{j}\, dt$

$C_3: x = 1 - t, y = 0, z = 0, 0 \le t \le 1$

$\mathbf{R} = (1 - t)\mathbf{i}$

$\mathbf{F} = (1 - t)^2\mathbf{j} - (t - 1)\mathbf{k}, \ d\mathbf{R} = -\mathbf{i}\, dt$

$\int_C \mathbf{F} \cdot d\mathbf{R} = \int_0^1 2t^2\, dt - \int_0^1 dt + \int_0^1 0\, dt = -\frac{1}{3}$

37. $C_1: (0, 0)$ to $(0, 1): x = 0, y = t, 0 \le t \le 1$

$\qquad \mathbf{F} = 2\mathbf{j}, \mathbf{R} = t\mathbf{j}, \ d\mathbf{R} = \mathbf{j}dt$

$C_2: (0, 1)$ to $(2, 1), \ x = t, y = 1, 0 \le t \le 2$

$\qquad \mathbf{F} = -t\mathbf{i} + 2\mathbf{j}, \ \mathbf{R} = t\mathbf{i} + \mathbf{j}, d\mathbf{R} = \mathbf{i}\, dt$

$C_3: (2, 1)$ to $(1, 0); x = 2 - t, y = 1 - t$,

$0 \le t \le 1$

$\mathbf{F} = -(2 - t)\mathbf{i} + 2\mathbf{j}, \mathbf{R} = (2 - t)\mathbf{i} + (1 - t)\mathbf{j}$,

$d\mathbf{R} = (-\mathbf{i} - \mathbf{j})\, dt$

$C_4: (1, 0)$ to $(0, 0), x = 1 - t, y = 0$,

$0 \le t \le 1$

$\mathbf{F} = -(1 - t)\mathbf{i} + 2\mathbf{j}$,

$\mathbf{R} = (1 - t)\mathbf{i}, d\mathbf{R} = -\mathbf{i}\, dt$

$\oint_C \mathbf{F} \cdot \mathbf{T}\, ds = \oint_C \mathbf{F} \cdot d\mathbf{R}$

$\qquad = \int_0^1 2\, dt + \int_0^2 (-t)\, dt$

$\qquad\qquad + \int_0^1 (2 - t - 2)\, dt + \int_0^1 (1 - t)\, dt$

$\qquad = 2 - 2 - \frac{1}{2} + \frac{1}{2}$

$\qquad = 0$

39. $\dfrac{dx}{dt} = 2\cos t(-\sin t);\ \dfrac{dy}{dt} = 2\sin t\cos t$

$ds = \sqrt{4\cos^2 t\sin^2 t + 4\sin^2 t\cos^2 t}\ dt$

$\quad = 2\sqrt{2}|\sin t\cos t|\ dt$

$\displaystyle\int_C (x+y)\ ds$

$\displaystyle = 2\sqrt{2}\int_{-\pi/4}^{0} (\cos^2 t + \sin^2 t)\ dx|\sin t\cos t|\ dt$

$\displaystyle = -2\sqrt{2}\int_{-\pi/4}^{0} \sin t\cos t\ dt$

$\displaystyle = \frac{\sqrt{2}}{2}$

41. For $0 \le t \le 2\pi$, $x = \cos t$, $y = \sin t$;

$dx = -\sin t\ dt;\ dy = \cos t\ dt$

$\dfrac{x\ dy - y\ dx}{x^2 + y^2} = \dfrac{\cos^2 t + \sin^2 t}{\cos^2 t + \sin^2 t} = 1\ dt$

$\displaystyle\oint_C \frac{x\ dy - y\ dx}{x^2 + y^2} = \int_0^{2\pi} 1\ dt = 2\pi$

43. $x = t$, $y = at$; $\mathbf{R}(t) = t\mathbf{i} + at\mathbf{j}$

$dW = \mathbf{F} \cdot d\mathbf{R}$

$\quad = (a\mathbf{i} + \mathbf{j}) \cdot (\mathbf{i} + a\mathbf{j})\ dt$

$\quad = 2a\ dt$

$\displaystyle W = 2a\int_a^0 dt = -2a^2$

45. $C_1\colon \mathbf{R} = \cos t\,\mathbf{i} + \sin t\,\mathbf{j};\ 0 \le t \le \pi$

$\quad C_2\colon \mathbf{R} = t\mathbf{i};\ -1 \le t \le 1$

$\displaystyle\int_0^{\pi} [(\cos^2 t + \sin^2 t)(-\sin t) + (\cos t + \sin t)(\cos t)]\ dt$

$\displaystyle + \int_{-1}^{1} t^2\ dt = -2 + \frac{\pi}{2} + \frac{2}{3}$

$\displaystyle = \frac{\pi}{2} - \frac{4}{3}$

47. $W = \displaystyle\int_C \mathbf{F} \cdot d\mathbf{R}$

$\displaystyle = \int_0^1 \left[\left(t^2\right)^2 - \left(t^3\right)^2 + 2\left(t^2\right)\left(t^3\right)(2t) \right.$

$\displaystyle \qquad\qquad \left. - t^2\left(3t^2\right)dt \right]$

$\displaystyle = \int_0^1 \left(t^4 - t^6 + 4t^6 - 3t^4 \right) dt$

$\displaystyle = \frac{1}{35}$

49. $\mathbf{R} = 2t\mathbf{i} + t\mathbf{j} + 2t\mathbf{k};\ 0 \le t \le 1$

$d\mathbf{R} = (2\mathbf{i} + \mathbf{j} + 2\mathbf{k})\ dtt;$

$\mathbf{F} = 2t\mathbf{i} + t\mathbf{j} + (4t^2 - t)\mathbf{k}$

$W = \displaystyle\int_C \mathbf{F} \cdot d\mathbf{R}$

$\displaystyle = \int_0^1 \left[(2t)(2) + t + 2\left(4t^2 - t\right) \right] dt$

$\displaystyle = \int_0^1 \left(8t^2 + 3t \right) dt$

$\displaystyle = \frac{25}{6}$

51. $ds = \sqrt{(\sqrt{2}\cos t)^2 + (-\sin t)^2 + (-\sin t)^2}\ dt$

$\quad = \sqrt{2}\ dt$

$m = \displaystyle\int_C \rho\ ds$

$\displaystyle = \int_C \left(\sqrt{2}\sin t\right)(\cos t)(\cos t)\left(\sqrt{2}\ dt\right)$

$\displaystyle = \frac{4}{3}$

$\overline{x} = \dfrac{1}{m}\displaystyle\int_C x(\rho\ ds)$

$\displaystyle = \frac{1}{m}\int_C x^2 yz\ ds$

$$= \frac{1}{m}\int_0^\pi \left(\sqrt{2}\sin t\right)^2 (\cos t)(\cos t)\sqrt{2}\,dt$$

$$= \frac{1}{m}\left(\frac{\sqrt{2}\pi}{4}\right)$$

$$= \frac{3\sqrt{2}\pi}{16}$$

$$\bar{y} = \frac{1}{m}\int_C y(\rho\,ds)$$

$$= \frac{1}{m}\int_C xy^2 z\,ds$$

$$= \frac{1}{m}\int_0^\pi \left(\sqrt{2}\sin t\right)(\cos^2 t)(\cos t)\sqrt{2}\,dt$$

$$= 0$$

$$\bar{z} = \frac{1}{m}\int_C z(\rho\,ds)$$

$$= \frac{1}{m}\int_C xyz^2\,ds$$

$$= \frac{1}{m}\int_0^\pi \left(\sqrt{2}\sin t\right)(\cos t)(\cos^2 t)\sqrt{2}\,dt$$

$$= 0$$

The center of mass is $\left(\dfrac{3\sqrt{2}\,\pi}{16}, 0, 0\right)$.

53. $ds = \sqrt{(2)^2 + (2t)^2 + (t^2)^2}\,dt$

$$= \sqrt{(t^2+2)^2}\,dt$$

$$= \left(t^2+2\right)dt$$

$$m = \int_C ds = \int_0^2 \left(t^2+2\right)dt = \frac{20}{3}$$

$$\bar{x} = \frac{1}{m}\int_C x\,ds = \frac{1}{m}\int_0^2 (2t)(t^2+2)\,dt = \frac{12}{5}$$

$$\bar{y} = \frac{1}{m}\int_C y\,ds = \frac{1}{m}\int_0^2 t^2(t^2+2)\,dt = \frac{44}{25}$$

$$\bar{z} = \frac{1}{m}\int_C z\,ds = \frac{1}{m}\int_0^2 \frac{1}{3}t^3(t^2+2)\,dt = \frac{14}{15}$$

The centroid is $\left(\dfrac{12}{5}, \dfrac{44}{25}, \dfrac{14}{15}\right)$.

55. Let t measure the rotation in radians. The laborer climbs 50 ft in $5(2\pi) = 10\pi$ radians, so z increases at the rate of $50/(10\pi) = 5/\pi$ ft/radians. The helical path is

$$\mathbf{R}(t) = \langle 10\cos t,\, 10\sin t,\, \frac{5t}{\pi}\rangle$$

and the force exerted by the laborer and the sand is

$$\mathbf{F} = \langle 0, 0, 180 + 40\rangle$$

The work done is

$$W = \int_C \mathbf{F}\cdot d\mathbf{R}$$

$$= \int_0^{10\pi} (180+40)\left(\frac{5}{\pi}\right)dt$$

$$= 11{,}000 \text{ ft-lb}$$

57. $\mathbf{R} = 5{,}000(\cos t\,\mathbf{i} + \sin t\,\mathbf{j});$
$d\mathbf{R} = 5{,}000(-\sin t\,\mathbf{i} + \cos t\,\mathbf{j})\,dt$
$\mathbf{F} = 5{,}000(5{,}000\cos t\,\mathbf{i} + 5{,}000\sin t\,\mathbf{j})$

$$W = \int_C \mathbf{F}\cdot d\mathbf{R}$$

$$= \int_0^{2\pi} (5{,}000)^3(-\cos t\sin t + \sin t\cos t)\,dt$$

$$= 0$$

59. Counterexample; $f(x) = \sin x$ and C: $x = t, y = 0$ for $0 \le t \le 2\pi$.

$$\int_C f(x)\,ds = \int_0^{2\pi} \sin t\sqrt{1+0}\,dt = 0$$

but $f(x) \ne 0$ on C.

13.3 The Fundamental Theorem and Path Independence, page 1047

1. With f continuously differentiable and $\mathbf{F} = \nabla f$

$$\int_A^B \mathbf{F} \, d\mathbf{R} = f(B) - f(A)$$

3. $\mathbf{F} = (e^{2x}\sin y)\mathbf{i} + (e^{2x}\cos y)\mathbf{j}$

$$\frac{\partial}{\partial y}(e^{2x}\sin y) = e^{2x}\cos y$$

$$\frac{\partial}{\partial x}(e^{2x}\cos y) = 2e^{2x}\cos y$$

These are not the same, so by the cross-partials test, the field is not conservative.

5. $\mathbf{F} = 2xt^3\mathbf{i} + 3x^2y^2\mathbf{j}$; since

$$\frac{\partial}{\partial y}(2xy^3) = \frac{\partial}{\partial x}(3y^2x^2) = 6xy^2$$

the field is conservative. Also,

$\dfrac{\partial f}{\partial x} = 2xy^3$ and $f(x,y) = x^2y^3 + c(y)$, so

$$\frac{\partial f}{\partial y} = 3x^2y^2 + c'(y) = 3x^2y^2$$

$c'(y) = 0$, so $c(y) = K$. If we pick $K = 0$, then $f(x,y) = x^2y^3$.

7. $\mathbf{F} = (-y + e^x\sin y)\mathbf{i} + (x+2)e^x\cos y\,\mathbf{j}$

$$\frac{\partial}{\partial y}(-y + e^x\sin y) = -1 + e^x\cos y$$

$$\frac{\partial}{\partial x}(x+2)e^x\cos y = e^x\cos y(x+3)$$

These are not the same, so by the cross-partials test, the field is not conservative.

For Problems 9-12, we have

$$\int_C \mathbf{F} \cdot d\mathbf{R} = \int_C [(3x+2y)\,dx + (2x+3y)\,dy]$$

$$\frac{\partial}{\partial y}(3x+2y) = 2 \qquad \frac{\partial}{\partial x}(2x+3y) = 2$$

These are the same, so the line integrals are all path independent.

9. Consider the path C_1: $x = 1 - t$, $y = 0$, $0 \le t \le 2$.

$$\int_C \mathbf{F} \cdot d\mathbf{R} = \int_{C_1} \mathbf{F} \cdot d\mathbf{R}$$
$$= \int_2^3 [3(1-t)(-1)]\,dt$$
$$= 0$$

11. The integral is 0 since C is a closed curve.

For Problems 13-16, we have

$$\int_C \mathbf{F} \cdot d\mathbf{R} = \int_C [2x^2y\,dx + x^3\,dy]$$

$$\frac{\partial}{\partial y}(2x^2y) = 2x^2 \qquad \frac{\partial}{\partial x}(x^3) = 3x^2$$

These are not the same, so by the cross-partials test, the field is not conservative. In other words, the line integral is path dependent.

13. Consider the path C: $x = \cos t$, $y = \sin t$, $-\frac{\pi}{2} \le t \le \frac{\pi}{2}$.

$$\int_C \mathbf{F} \cdot d\mathbf{R} = \int_C [2x^2y\,dx + x^3\,dy]$$
$$= \int_{-\pi/2}^{\pi/2} [(2\cos^2 t\sin t)(-\sin t) + \cos^3 t(\cos t)]\,dt$$
$$= \int_{-\pi/2}^{\pi/2} \cos^2 t(-2\sin^2 t + \cos^2 t)\,dt$$
$$= \frac{\pi}{8}$$

15. C_1: $x = -t$, $y = -1$; $-1 \le t \le 1$
   C_2: $x = -1$, $y = t$; $-1 \le t \le 1$
   C_3: $x = t$, $y = 1$; $-1 \le t \le 1$
   C_4: $x = 1$, $y = -t$; $-1 \le t \le 1$

$$\oint_C \left[2x^2 y\, dx + x^3\, dy \right]$$

$$= \int_{-1}^{1} \left[2(-t)^2(-1)(-dt) + (-t)^3(0) \right]$$

$$+ \int_{-1}^{1} \left[2(-1)^2(t)(0) + (-1)^3 dt \right]$$

$$+ \int_{-1}^{1} \left[2t^2(1)\, dt + t^3(0) \right]$$

$$+ \int_{-1}^{1} \left[2(1)^2(-t)(0) + (1)^3(-dt) \right]$$

$$= \int_{-1}^{1} \left[2t^2 - 1 + 2t^2 - 1 \right] dt$$

$$= -\frac{4}{3}$$

For Problems 17-20, we have

$$\int_C \mathbf{F} \cdot d\mathbf{R} = \int_C \left[2xy\, dx + x^2\, dy \right]$$

$$\frac{\partial}{\partial y}(2xy) = 2x \qquad \frac{\partial}{\partial x}(x^2) = 2x$$

These are the same, so the line integrals are all path independent.

17. The integral is 0 because the path is closed.

19. Because of path independence, it is the same as Problem 18, namely 32.

21. $u(x,y) = 2x - y$; $v(x,y) = y^2 - x$
   $\dfrac{\partial u}{\partial y} = -1$; $\dfrac{\partial v}{\partial x} = -1$; **F** is conservative
   $f(x,y) = \displaystyle\int u(x,y)\, dx$

$$= \int (2x - y)\, dx$$

$$= x^2 - yx + c(y)$$

$$\frac{\partial f}{\partial y} = -x + c'(y)$$

$$= -x + y^2$$

We see $c'(y) = y^2$, so $c(y) = \dfrac{y^3}{3}$ and the scalar potential is
$f(x,y) = x^2 - xy + \frac{1}{3}y^3$

$$\int_C \mathbf{F} \cdot d\mathbf{R} = f(1,1) - f(0,0)$$

$$= \frac{1}{3} - 0$$

$$= \frac{1}{3}$$

23. $\dfrac{\partial u}{\partial y} = 2x$; $\dfrac{\partial v}{\partial x} = 2x$; **F** is conservative.
   $f(x,y) = x^2 y + c(y)$
   $\dfrac{\partial f}{\partial y} = x^2 + c'(y)$
   $= x^2$
   $c'(y) = 0$, so $c(y) = 0$
   Scalar potential: $f(x,y) = x^2 y$

$$\int_C \mathbf{F} \cdot d\mathbf{R} = f(1,1) - f(0,0)$$

$$= 1 - 0$$

$$= 1$$

25. $\dfrac{\partial}{\partial y}\left[\dfrac{y+1}{(y+1)^2} \right] = \dfrac{-1}{(y+1)^2}$

$\dfrac{\partial}{\partial x}\left[\dfrac{-x}{(y+1)^2} \right] = \dfrac{-1}{(y+1)^2}$

F is conservative.

$$\frac{\partial f}{\partial x} = \frac{1}{y+1}$$

$$f(x,y) = \frac{x}{y+1} + c(y)$$

$$\frac{\partial f}{\partial y} = \frac{-x}{(y+1)^2} + c'(y)$$

$$= \frac{-x}{(y+1)^2}$$

$c'(y) = 0$, so $c(y) = 0$.

Scalar potential $f(x,y) = \dfrac{x}{y+1}$

$$\int_C \mathbf{F} \cdot d\mathbf{R} = f(1,1) - f(0,0)$$

$$= \frac{1}{2} - 0$$

$$= \frac{1}{2}$$

27. $\text{curl } \mathbf{F} = \begin{vmatrix} \mathbf{i} & \mathbf{j} & \mathbf{k} \\ \frac{\partial}{\partial x} & \frac{\partial}{\partial y} & \frac{\partial}{\partial z} \\ yze^{xy} & xze^{xy} & e^{xy} \end{vmatrix}$

$$= (xe^{xy} - xe^{xy})\mathbf{i} - (ye^{xy} - ye^{xy})\mathbf{j}$$
$$+ (ze^{xy} + xyze^{xy} - ze^{xy} - xyze^{xy})\mathbf{k}$$

$$= \mathbf{0}$$

F is conservative from Theorem 13.4.

$$\frac{\partial f}{\partial x} = e^{xy}yz$$

$$f = ze^{xy} + c(y,z)$$

$$\frac{\partial f}{\partial y} = e^{xy}xz + \frac{\partial c}{\partial y}$$

$$= e^{xy}xz$$

Thus, $\dfrac{\partial c}{\partial y} = 0$, so $c = c_1(z)$.

$$f = ze^{xy} + c_1(z)$$

$$\frac{\partial f}{\partial z} = e^{xy} + c_1'(z)$$

$$= e^{xy}$$

We see, $c_1'(z) = 0$, so $c_1 = 0$.

Scalar potential: $f(x,y,z) = ze^{xy}$.

29. Let $r = x^2 + y^2 + z^2$; then

$$\text{curl } \mathbf{F} = \begin{vmatrix} \mathbf{i} & \mathbf{j} & \mathbf{k} \\ \frac{\partial}{\partial x} & \frac{\partial}{\partial y} & \frac{\partial}{\partial z} \\ rx & ry & rz \end{vmatrix}$$

$$= (2yz - 2yz)\mathbf{i} - (2xz - 2xz)\mathbf{j}$$
$$+ (2xy - 2xy)\mathbf{k}$$

$$= \mathbf{0}$$

F is conservative from Theorem 13.4.

$$\frac{\partial f}{\partial x} = x(x^2 + y^2 + z^2)$$

$$f = \frac{1}{4}(x^2 + y^2 + z^2)^2 + c(y,z)$$

$$\frac{\partial f}{\partial y} = y(x^2 + y^2 + z^2) + \frac{\partial c}{\partial y}$$

$$= y(x^2 + y^2 + z^2)$$

Thus, $\dfrac{\partial c}{\partial y} = 0$, so $c = c_1(z)$.

$$f = \frac{1}{4}(x^2 + y^2 + z^2)^2 + c_1(z)$$

$$\frac{\partial f}{\partial z} = z(x^2 + y^2 + z^2) + c_1'(z)$$

$$= z(x^2 + y^2 + z^2)$$

We see, $c_1'(z) = 0$, so $c_1 = 0$.

Scalar potential:

$$f(x,y,z) = \tfrac{1}{4}(x^2 + y^2 + z^2)^2.$$

31. $\text{curl } \mathbf{F} = \begin{vmatrix} \mathbf{i} & \mathbf{j} & \mathbf{k} \\ \frac{\partial}{\partial x} & \frac{\partial}{\partial y} & \frac{\partial}{\partial z} \\ xy^2 + yz & x^2y + xz + 3y^2z & xy + y^3 \end{vmatrix}$

$$= (x + 3y^2 - x - 3y^2)\mathbf{i} - (y - y)\mathbf{j}$$
$$+ (2xy + z - 2xy - z)\mathbf{k}$$

$$= \mathbf{0}$$

F is conservative from Theorem 13.4.

$$\frac{\partial f}{\partial x} = xy^2 + yz$$

$$f = \frac{1}{2}x^2y^2 + xyz + c(y, z)$$

$$\frac{\partial f}{\partial y} = x^2y + xz + \frac{\partial c}{\partial y}$$

$$= x^2y + xz + 3y^2z$$

Thus, $\frac{\partial c}{\partial y} = 3y^2z$, so $c = y^3z + c_1(z)$.

$$f = \frac{1}{2}x^2y^2 + xyz + y^3z + c_1(z)$$

$$\frac{\partial f}{\partial z} = xy + y^3 + c_1'(z)$$

$$= xy + y^3$$

We see, $c_1'(z) = 0$, so $c_1 = 0$.

Scalar potential:

$$f(x, y, z) = \frac{1}{2}x^2y^2 + xyz + y^3z$$

33. $\mathbf{F} = \langle \sin z, -z\sin y, x\cos z + \cos y \rangle$

$$\text{curl } \mathbf{F} = \begin{vmatrix} \mathbf{i} & \mathbf{j} & \mathbf{k} \\ \frac{\partial}{\partial x} & \frac{\partial}{\partial y} & \frac{\partial}{\partial z} \\ \sin z & -z\sin y & x\cos z + \cos y \end{vmatrix}$$

$$= \mathbf{0}$$

$\mathbf{F}$ is conservative.

$$\frac{\partial f}{\partial x} = \sin z$$

$$f = x\sin z + c(y, z)$$

$$\frac{\partial f}{\partial y} = -z\sin y$$

$$= \frac{\partial}{\partial y}(x\sin z + c)$$

$$= \frac{\partial c}{\partial y}$$

Thus, $\frac{\partial c}{\partial y} = -z\sin y$, so

$$c = z\cos y + c_1(z).$$

$$f = x\sin z + z\cos y + c_1(z)$$

$$\frac{\partial f}{\partial z} = x\cos z + \cos y$$

$$= \frac{\partial}{\partial z}[x\sin z + z\cos y + c_1(z)]$$

$$= x\cos z + \cos y + c_1'(z)$$

We see, $c_1'(z) = 0$, so $c_1 = 0$.

$$f(x, y, z) = x\sin z + z\cos y$$

$$\int_C \mathbf{F} \cdot d\mathbf{R} = f(0, -1, 1) - f(1, 0, -1)$$

$$= [0 + (1)\cos(-1)] - [(1)\sin(-1) + (-1)(1)]$$

$$= \cos(-1) - \sin(-1) + 1$$

$$= \cos 1 + \sin 1 + 1$$

35. $\mathbf{F} = \left\langle \frac{y}{1+x^2} + \tan^{-1}z, \tan^{-1}x, \frac{x}{1+z^2} \right\rangle$

curl $\mathbf{F} = \mathbf{0}$, so $\mathbf{F}$ is conservative.

$$\frac{\partial f}{\partial x} = \frac{y}{1+x^2} + \tan^{-1}z$$

$$f = y\tan^{-1}x + x\tan^{-1}z + c(y, z)$$

$$\frac{\partial f}{\partial y} = \tan^{-1}x$$

$$= \frac{\partial}{\partial y}[y\tan^{-1}x + x\tan^{-1}z + c]$$

$$= \tan^{-1}x + \frac{\partial c}{\partial y}$$

Thus, $\frac{\partial c}{\partial y} = 0$, so $c = c_1(z)$.

$$f = y\tan^{-1}x + x\tan^{-1}z + c_1(z)$$

$$\frac{\partial f}{\partial z} = \frac{x}{1+z^2}$$

$$= \frac{\partial}{\partial z}[y\tan^{-1}x + x\tan^{-1}z + c_1(z)]$$

$$= \frac{x}{1+z^2} + c_1'(z)$$

We see, $c_1'(z) = 0$, so $c_1 = 0$.

$$f(x, y, z) = y\tan^{-1}x + x\tan^{-1}z$$

$$\int_C \mathbf{F} \cdot d\mathbf{R} = f(0, -1, 1) - f(1, 0, -1)$$

$$= [-\tan^{-1}0 + 0] - [0 + (1)\tan^{-1}(-1)]$$

$$= -\tan^{-1}(-1)$$

$$= \frac{\pi}{4}$$

37. $\dfrac{\partial u}{\partial y} = 2y; \dfrac{\partial v}{\partial x} = 2y;$ **F** is conservative.

$$\frac{\partial f}{\partial x} = 3x^2 + 2x + y^2$$

$$f(x, y) = x^3 + x^2 + xy^2 + c(y)$$

$$\frac{\partial f}{\partial y} = 2xy + c'(y)$$

$$= 2xy + y^3$$

We see $c'(y) = y^3$, so $c(y) = \frac{1}{4}y^4$.

$$f(x, y) = x^3 + x^2 + xy^2 + \frac{y^4}{4}$$

$$\int_C \left[(3x^2 + 2x + y^2)\, dx + (2xy + y^3)\, dy \right]$$

$$= \int_C \mathbf{F} \cdot d\mathbf{R}$$

$$= f(1, 1) - f(0, 0)$$

$$= \frac{13}{4} - 0$$

$$= \frac{13}{4}$$

39. $\dfrac{\partial u}{\partial y} = 1; \dfrac{\partial v}{\partial x} = 1;$ **F** is conservative.

$$\frac{\partial f}{\partial x} = y - x^2$$

$$f(x, y) = xy - \frac{x^3}{3} + c(y)$$

$$\frac{\partial f}{\partial y} = x + c'(y)$$

$$= x + y^2$$

We see $c'(y) = y^2$, so $c(y) = \frac{1}{3}y^3$.

$$f(x, y) = xy - \frac{x^3}{3} + \frac{y^3}{3}$$

$$\int_C \left[(y - x^2)\, dx + (x + y^2)\, dy \right]$$

$$= \int_C \mathbf{F} \cdot d\mathbf{R}$$

$$= f(0, 3) - f(-1, -1)$$

$$= 9 - 1$$

$$= 8$$

41. $\dfrac{\partial u}{\partial y} = \cos y; \dfrac{\partial v}{\partial x} = \cos y;$ **F** is conservative.

$$\frac{\partial f}{\partial x} = \sin y$$

$$f(x, y) = x \sin y + c(y)$$

$$\frac{\partial f}{\partial y} = x \cos y + c'(y)$$

$$= 3 + x \cos y$$

We see $c'(y) = 3$, so $c(y) = 3y$, so $f(x, y) = x \sin y + 3y$. When $t = 0$, $(x, y) = (0, 0)$ and when $t = 1$, $(x, y) = (-2, \frac{\pi}{2})$

$$\int_C \left[(\sin y)\, dx + (3 + x \cos y)\, dy \right]$$

$$= \int_C \mathbf{F} \cdot d\mathbf{R}$$

$$= f\left(-2, \frac{\pi}{2}\right) - f(0, 0)$$

$$= \frac{1}{2}(3\pi - 4)$$

43. $\dfrac{\partial u}{\partial y} = 1; \dfrac{\partial v}{\partial x} = 1$

F $= y\mathbf{i} + x\mathbf{j};$ **F** is conservative.

$$\frac{\partial f}{\partial x} = y; f(x, y) = xy + c(y)$$

$$c'(y) = 0, \text{ so } c(y) = 0$$

$$f(x, y) = xy$$

$$\int_C (y\mathbf{i} + x\mathbf{j}) \cdot d\mathbf{R} = f(2,4) - f(0,0)$$
$$= 8$$

45. $\dfrac{\partial u}{\partial y} = 2;\ \dfrac{\partial v}{\partial x} = 2$

$\mathbf{F} = 2y\mathbf{i} + 2x\mathbf{j};\ \mathbf{F}$ is conservative.

$\dfrac{\partial f}{\partial x} = 2y;\ f(x,y) = 2xy + c(y)$

$\dfrac{\partial f}{\partial y} = 2x + c'(y)$

$c'(y) = 0$, so $c(y) = 0$

$f(x,y) = 2xy$

$$\int_C (2y\,dx + 2x\,dy) = f(4,4) - f(0,0)$$
$$= 32$$

47. We want

$$\mathbf{G} = \langle g(x)(x^4 + y^4), -g(x)(xy^3)\rangle$$

to be conservative, so

$$\frac{\partial}{\partial y}[g(x)(x^4 + y^4)] = \frac{\partial}{\partial x}[-g(x)(xy^3)]$$

$$g(x)\left(4y^3\right) = -g'(x)(xy^3) - g(x)y^3$$

$$\frac{g'(x)}{g(x)} = \frac{-5}{x}$$

$$\ln g(x) = -5\ln x + C_1$$

We assume $g(x) > 0$, without loss of generality.

$$g(x) = Cx^{-5}$$

for any constant $C \neq 0$.

49.   a. $\mathbf{F} = -KmM\left\langle \dfrac{x}{r^3}, \dfrac{y}{r^3}, \dfrac{z}{r^3} \right\rangle$

where $r = \sqrt{x^2 + y^2 + z^2}$

$\dfrac{\partial f}{\partial x} = \dfrac{-KmMx}{(x^2 + y^2 + z^2)^{3/2}}$

$f = \dfrac{KmM}{\sqrt{x^2 + y^2 + z^2}} + c_1(y,z)$

$\dfrac{\partial f}{\partial y} = \dfrac{-KmMy}{(x^2 + y^2 + z^2)^{3/2}} + \dfrac{\partial c_1}{\partial y}$

$\qquad = \dfrac{-KmMy}{(x^2 + y^2 + z^2)^{3/2}}$

$\dfrac{\partial c_1}{\partial y} = 0$, use $c_1(y,z) = c_2(z)$

$f = \dfrac{KmM}{\sqrt{x^2 + y^2 + z^2}} + c_2(z)$

$\dfrac{\partial f}{\partial z} = \dfrac{-KmMz}{(x^2 + y^2 + z^2)^{3/2}} + c_2'(z)$

$\qquad = \dfrac{-KmMz}{(x^2 + y^2 + z^2)^{3/2}}$

$c_2'(z) = 0$, use $c_2(z) = 0$;

$f(x,y,z) = \dfrac{kmM}{\sqrt{x^2 + y^2 + z^2}}$

$\qquad = kmM\left(\dfrac{1}{r}\right)$

Note: Physicists sometimes use

$\phi = -\dfrac{kmM}{r}$ as the gravitational

potential.

b. $W = \displaystyle\int_P^Q \mathbf{F} \cdot d\mathbf{R}$

$= f(a_2, b_2, c_2) - f(a_1, b_1, c_1)$

$= \dfrac{kmM}{\sqrt{a_2^2 + b_2^2 + c_2^2}} - \dfrac{kmM}{\sqrt{a_1^2 + b_1^2 + c_1^2}}$

51. a. Let $x = \cos t$ and $y = \sin t;\ 0 \le t \le \pi$

$$\int_{C_1} \frac{-y\,dx + x\,dy}{x^2 + y^2} = \int_0^\pi dt = \pi$$

on the upper semicircle. For $\pi \le t \le 2\pi$, we have the lower semicircle:

$$\int_{C_2} \frac{-y\,dx + x\,dy}{x^2 + y^2} = \int_{\pi}^{2\pi} dt = \pi$$

b. $M = -\dfrac{y}{x^2 + y^2}$, $N = \dfrac{x}{x^2 + y^2}$

$M_y = N_x$, but $\mathbf{F}$ is not conservative because it is not continuous at $(0,0)$.

53. The force $\mathbf{F}_1 = \langle -a, -ae^{-y}\rangle$ is also conservative since

$$\frac{\partial}{\partial y}(-a) = 0 = \frac{\partial}{\partial x}(-ae^{-y})$$

The work, computed along the line segment C_1 between $A(0,0)$ and $B(2,1)$ can be parameterized as $C_1: x = 2t,\ y = t$, $0 \le t \le 1$ is

$$W_1 = \int_{C_1} \mathbf{F}_1 \cdot d\mathbf{R}$$
$$= \int_0^1 \left[-a(2) - ae^{-t}(1)\right] dt$$
$$= a\left(e^{-1} - 3\right)$$

55. These observations suggest "bending" the path above the straight line path, to take advantage of the e^{-y} term.

57. For example, using parabolas of the form

$$y = x(b - ax)$$

the condition $y(2) = 1$ leads to $b = (1 + 4a)/2$; and a study of work as a function of a shows that the larger the a, the smaller the amount of work. But large a sends the path way "north" which is possibly unrealistic.

59. $\dfrac{d}{dt}(\mathbf{V} \cdot \mathbf{V}) = 2\mathbf{V} \cdot \dfrac{d\mathbf{V}}{dt} = 22\mathbf{V} \cdot \mathbf{A} = 2\mathbf{A} \cdot \dfrac{d\mathbf{R}}{dt}$

a. Since $F = mA$, the work done is

$$W = \int_C \mathbf{F} \cdot d\mathbf{R}$$
$$= \int_{t_0}^{t_2} m\mathbf{A} \cdot \frac{d\mathbf{R}}{dt}\,dt$$
$$= \int_{t_0}^{t_2} \frac{m}{2} \frac{d}{dt}(\mathbf{V} \cdot \mathbf{V})\,dt$$
$$= \frac{m}{2} \|\mathbf{V}\|^2 \Big|_{t_0}^{t_2}$$
$$= \frac{m}{2} \|\mathbf{V}(t_1)\|^2 - \frac{m}{2}\|\mathbf{V}(t_0)\|^2$$
$$= K(t_1) - K(t_0)$$

b. If $-f$ is a scalar potential for $\mathbf{F}$, then

$$W = \int_C \mathbf{F} \cdot d\mathbf{R}$$
$$= f(t_1) - f(t_0)$$
$$= -P(t_1) + P(t_0)$$
$$= K(t_1) - K(t_0) \qquad \textit{From part a.}$$

Thus, $P(t_1) + K(t_1) = P(t_0) + K(t_0)$.

13.4 Green's Theorem, page 1059

> **SURVIVAL HINT:** *Green's theorem expresses an important relationship between a line integral around a simple closed curve. Remember, that a curve is **simple** is there are no self-intersections and **closed** if there are no endpoints. This region, D, must have a positive orientation.*

1. $\displaystyle\oint_C (y^2\,dx + x^2\,dy) = \int_0^1 \int_0^1 (2x - 2y)\,dy\,dx$
$$= \int_0^1 (2x - 1)\,dx$$
$$= 0$$

Check:

$C_1\colon y = 0;\ C_2\colon x = 1;\ C_3\colon y = 1;\ C_4\colon x = 0$

$$\int_0^1 0\,dx + \int_0^1 dy + \int_1^0 dx + \int_1^0 0\,dy = 0$$

3. $\mathbf{F}(x,y) = (2x^2 + 3y)\mathbf{i} - 3y^2\mathbf{j}$

    where $M(x,y) = 2x^2 + 3y$, $N(x,y) = 3y^2$

    Since the orientation is clockwise,

    $$\oint_C \mathbf{F}\cdot d\mathbf{R} = -\int_0^1 \int_y^{2-y} (0 - 3)\,dx\,dy$$

    $$= 3\int_0^1 (2 - 2y)\,dy$$

    $$= 3$$

    Check:

    $C_1\colon x = t,\ y = t;\ 0 \le t \le 1$

    $C_2\colon x = 1+t,\ y = 1-t;\ 0 \le t \le 1$

    $C_3\colon x = 2-t,\ y = 0;\ 0 \le t \le 2$

    $$\oint_C \mathbf{F}\cdot d\mathbf{R}$$

    $$= \int_0^1 [2t^2 + 3t - 3t^2]\,dt$$

    $$+ \int_0^1 [2(1+t)^2 + 3(1-t) - 3(1-t)^2(-1)]\,dt$$

    $$+ \int_0^2 [2(2-t)^2(-1)]\,dt$$

    $$= \frac{7}{6} + \frac{43}{6} - \frac{16}{3}$$

    $$= 3$$

5. The orientation is clockwise, so

    $$\oint_C 4xy\,dx = -\int\int_D (-4x)\,dA$$

    $$= 4\int_0^{2\pi} \int_0^1 r\cos\theta(r\,dr\,d\theta)$$

    $$= 0$$

    Check:

    $C\colon x = \sin\theta,\ y = \cos\theta$

    $dx = \cos\theta\,d\theta,\ dy = -\sin\theta\,d\theta$

$$4\int_0^{2\pi} \cos^2\theta\sin\theta\,d\theta = -\frac{4}{3}\cos^3\theta\Big|_0^{2\pi}$$

$$= 0$$

7. $$\oint_C (2y\,dx - x\,dy) = \int\int_D (-1 - 2)\,dA$$

    $$= -3\left[\frac{1}{2}\pi(2)^2\right]$$

    $$= -6\pi$$

9. $$\oint_C (2y\,dx - x\,dy) = \int\int_D (-1 - 2)\,dA$$

    $$= -3\left[\frac{1}{2}(2)(6)\right]$$

    $$= -18$$

11. $$\oint_C (e^x\,dx - \sin x\,dy) = \int_0^1 \int_0^1 (-\cos x)\,dy\,dx$$

    $$= \int_0^1 (-\cos x)\,dx$$

    $$= -\sin 1$$

13. Using Example 3 for two ellipses (one with $a = b = R$, the other with $a = b = r$). That is, circles traveled counterclockwise for the outer circle and clockwise for the inner circle. The desired area (assuming $R > r$) is the area enclosed by the outer circle minus the area enclosed by the inner circle so

    $$\pi R^2 - \pi r^2$$

15. $$\oint_C (x\sin x\,dx - e^{y^2}\,dy) = \int\int_D 0\,dA$$

    $$= 0$$

17. $$\oint_C (x\sin x\,dx - e^{y^2}\,dy) = \int\int_D 0\,dA$$

    $$= 0$$

19. $\oint_C [(x+y)\,dx - (3x - 2y)\,dy]$

$$= \iint_D (-3 - 1)\,dA$$

$$= -4\left[\frac{1}{2}\pi(2)^2\right]$$

$$= -8\pi$$

21. $\oint_C [(x+y)\,dx - (3x - 2y)\,dy]$

$$= \iint_D (-3 - 1)\,dA$$

$$= -4\left[\frac{1}{2}(2)(6)\right]$$

$$= -24$$

23. $\oint_C [(x - y^2)\,dx + 2xy\,dy]$

$$= \iint_D (2y + 2y)\,dA$$

$$= \int_0^2 \int_0^2 4y\,dy\,dx$$

$$= \int_0^2 8\,dx$$

$$= 16$$

25. $\oint_C (\sin x \cos y\,dx + \cos x \sin y\,dy)$

$$= \iint_D \left[\frac{\partial}{\partial x}(\cos x \sin y) - \frac{\partial}{\partial y}(\sin x \cos y)\right] dA$$

$$= \int_0^2 \int_0^2 [-\sin x \sin y + \sin x \sin y]\,dy\,dx$$

$$= \int_0^2 0\,dx$$

$$= 0$$

27. $\oint_C [2x\,dx + 2y\,dy]$

$$= \iint_D \left[\frac{\partial}{\partial x}(2y) - \frac{\partial}{\partial y}(2x)\right] dA = 0$$

29. $\oint_C \sin y\,dx + x \cos y\,dy$

$$= \iint_D \left[\frac{\partial}{\partial x}(x \cos y) - \frac{\partial}{\partial y}(\sin y)\right] dA$$

$$= \iint_D (\cos y - \cos y)\,dA$$

$$= 0$$

31. $W = \oint_C \mathbf{F} \cdot d\mathbf{R}$

$$= \oint_C [(3y - 4x)]\mathbf{i}$$
$$\quad + (4x - y)\mathbf{j}] \cdot [dx\,\mathbf{i} + dy\,\mathbf{j}]$$

$$= \iint_D \left[\frac{\partial}{\partial x}(4x - y) - \frac{\partial}{\partial y}(3y - 4x)\right] dA$$

$$= \iint_D (4 - 3)\,dA$$

$$= 2(1)\pi \qquad \text{The semimajor and semiminor axis of the}$$
$$\text{ellipse have lengths 2 and 1, respectively.}$$

$$= 2\pi$$

33. Parameterize the circle by $x = 2\cos t$, $y = 2\sin t, 0 \le t \le 2\pi$

$$A = \frac{1}{2}\oint_C (-y\,dx + x\,dy)$$

$$= \frac{1}{2}\int_0^{2\pi} [(-2\sin t)(-2\sin t)$$
$$\quad + (2\cos t)(2\cos t)]\,dt$$

$$= \frac{1}{2}\int_0^{2\pi} 4\,dt$$

$$= 4\pi$$

Page 538

Check: $A = \pi r^2 = \pi(2)^2 = 4\pi$.

35. C_1: $x = t$, $y = 0$; $0 \le t \le 4$
C_2: $x = 4 - t$, $y = t$; $0 \le t \le 3$
C_3: $x = 1 - t$, $y = 3$; $0 \le t \le 1$
C_4: $x = 0$, $y = 3 - t$; $0 \le t \le 3$

$$A = \frac{1}{2}\oint_C (-y\,dx + x\,dy)$$

$$= \frac{1}{2}\int_0^4 0\,dt + \frac{1}{2}\int_0^3 [-t(-1) + (4 - t)]\,dt$$

$$+ \frac{1}{2}\int_0^1 (-3)(+1)\,dt + \frac{1}{2}\int_0^3 0\,dt$$

$$= \frac{15}{2}$$

Check:
$A = \frac{1}{2}(b_1 + b_2)h = \frac{1}{2}(1 + 4)(3) = \frac{15}{2}$.

37. $\displaystyle\oint_C \left(x^2 y\,dx - y^2 x\,dy\right) = \int\!\!\int_D \left(-y^2 - x^2\right)dA$

$$= -\int\!\!\int_D r^2\,dA$$

$$= -\int_0^\pi \int_0^a r^3\,dr\,d\theta$$

$$= -\frac{\pi a^4}{4}$$

39. $\displaystyle I = \oint_C \left[(5 - xy - y^2)\,dx - (2xy - x^2)\,dy\right]$

$$= \int\!\!\int_D (-2y + 2x + x + 2y)\,dA$$

$$= 3\int\!\!\int_D x\,dA$$

$$= 3M_y$$

$M_y = \dfrac{1}{3}I$

The square has area $A = 1$, so

$$\bar{x} = \frac{M_y}{A} = \frac{\frac{1}{3}I}{1}$$

We see $I = 3\bar{x}$.

41. $\displaystyle\oint_C x^2\,dy = \int\!\!\int_D (2x - 0)\,dA$

$$= 2\int\!\!\int_D x\,dA$$

$$= 2\bar{x}A$$

$$\oint_C y^2\,dy = \int\!\!\int_D (0 - 2y)\,dA$$

$$= -2\int\!\!\int_D y\,dA$$

$$= 2\bar{y}A$$

Thus, $\bar{x} = \dfrac{1}{2A}\oint_C x^2\,dy$ and

$\bar{y} = \dfrac{1}{2A}\oint_C y^2\,dy$.

43. $\displaystyle\oint_C \left[\left(\frac{-y}{x^2} + \frac{1}{x}\right)dx + \frac{1}{x}\,dy\right]$

$$= \int\!\!\int_D \left[-\frac{1}{x^2} + \frac{1}{x^2}\right]dA$$

$$= 0$$

45.

Let C_1 be a circle centered at the origin with radius R so small that all of C_1 is contained within the given curve C. Assume C_1 is oriented clockwise, and let D be the region between C_1 and C. Then, according to Green's theorem for doubly-connected regions,

$$\oint_C \frac{x\,dx + y\,dy}{x^2 + y^2} + \oint_{C_1} \frac{x\,dx + y\,dy}{x^2 + y^2}$$

$$= \int\int_D \left[\frac{\partial}{\partial x}\left(\frac{y}{x^2+y^2} \right) - \frac{\partial}{\partial y}\left(\frac{x}{x^2+y^2} \right) \right] dA$$

$$= \int\int_D \left[\frac{-2xy}{(x^2+y^2)^2} - \frac{-2xy}{(x^2+y^2)^2} \right] dA$$

$$= 0$$

To evaluate the line integral about C_1, use the parameterization C_1: $x = \sin\theta$, $y = \cos\theta$; $0 \le \theta \le 2\pi$. (Remember, C_1 is oriented clockwise.) Thus,

$$\oint_C \frac{x\,dx + y\,dy}{x^2+y^2} = -\oint_{C_1} \frac{x\,dx + y\,dy}{x^2+y^2}$$

$$= -\int_0^{2\pi} \frac{(\sin\theta)(\cos\theta) + (\cos\theta)(-\sin\theta)}{\sin^2\theta + \cos^2\theta}\,d\theta$$

$$= 0$$

47.

$$\oint_C \frac{-(y+2)\,dx + (x-1)\,dy}{(x-1)^2 + (y+2)^2}$$

$$= \int\int_D \frac{-[(x-1)^2 - (y+2)^2] - [(y+2)^2 - (x-1)^2]}{[(x-1)^2 + (y+2^2]^2}\,dA$$

$$= 0$$

49. $\nabla f = (2xy - 2y)\mathbf{i} + (x^2 - 2x + 2y)\mathbf{j}$

Since $\mathbf{N} = \dfrac{dy}{ds}\mathbf{i} - \dfrac{dx}{ds}\mathbf{j}$ is a unit normal to the curve.

$$\oint_C \frac{\partial f}{\partial n}\,dx = \oint_C (\nabla f \cdot \mathbf{N})\,ds$$

$$= \oint_C \left[(2xy - 2y)\,dy - (x^2 - 2x + 2y)\,dx \right]$$

$$= \int_0^1 \int_0^1 (2y + 2)\,dy\,dx$$

$$= \int_0^1 3\,dx$$

$$= 3$$

51. $\oint_C [(x - 3y)\,dx + (2x - y^2)\,dy]$

$$= \int\int_D (2 + 3)\,dA$$

$$= 5A$$

53. If $\dfrac{\partial u}{\partial y} = \dfrac{\partial v}{\partial x}$, then for any closed curve C, we have

$$\oint_C \mathbf{F} \cdot d\mathbf{R} = \oint_C [u\,dx + v\,dy]$$

$$= \int\int_D \left(\frac{\partial v}{\partial x} - \frac{\partial u}{\partial y} \right) dA$$

$$= 0$$

where D is the interior of C. Thus, $\oint_C \mathbf{F} \cdot d\mathbf{R}$ is independent of path and $\mathbf{F}$ is conservative.

Conversely, if $\mathbf{F}$ is conservative, then $\oint_C \mathbf{F} \cdot d\mathbf{R} = 0$ for any closed curve C in D. Thus,

$$0 = \oint_C [u\,dx + v\,dy]$$

$$= \int\int_D \left(\frac{\partial v}{\partial x} - \frac{\partial u}{\partial y} \right) dA$$

which is true for every closed curve C only if

$$\frac{\partial v}{\partial x} = \frac{\partial u}{\partial y}$$

55. a. $\displaystyle\oint_C f\frac{\partial g}{\partial n}\,ds = \oint_C f\nabla g\cdot\mathbf{N}\,ds$

$\displaystyle = \oint_C f(g_x\,\mathbf{i} + g_y\,\mathbf{j})\cdot(dy\,\mathbf{i} - dx\,\mathbf{j})$

$\displaystyle = \oint_C [fg_x\,dy - fg_y\,dx]$

$\displaystyle = \iint_D \left[\frac{\partial}{\partial x}(fg_x) + \frac{\partial}{\partial y}(fg_y)\right]dA$

$\displaystyle = \iint_D [f_x g_x + fg_{xx} + f_y g_y + fg_{yy}]\,dA$

$\displaystyle = \iint_D [f(g_{xx} + g_{yy}) + f_x g_x + f_y g_y]\,dA$

$\displaystyle = \iint_D [f\nabla^2 g + \nabla f\cdot\nabla g]\,dA$

b. $\displaystyle\oint_C \left(f\frac{\partial g}{\partial n} - g\frac{\partial f}{\partial n}\right)ds$

$\displaystyle = \oint_C (f\nabla g\cdot\mathbf{N} - g\nabla f\cdot\mathbf{N})\,ds$

$\displaystyle = \oint_C [f(g_x\mathbf{i} + g_y\mathbf{j}) - g(f_x\mathbf{i} + f_y\mathbf{j})]\cdot(dy\,\mathbf{i} - dx\,\mathbf{j})$

$\displaystyle = \oint_C [(fg_x - gf_x)\,dy + (fg_y - gf_y)(-dx)]$

$\displaystyle = \iint_D \left[\frac{\partial}{\partial x}((fg_x - gf_x)) + \frac{\partial}{\partial y}(fg_y - gf_y)\right]dA$

$\displaystyle = \iint_D [fg_{xx} + g_x f_x - gf_{xx} - f_x g_x + fg_{yy}$

$\displaystyle \qquad + f_y g_y - gf_{yy} - g_y f_y]\,dA$

$\displaystyle = \iint_D [f(g_{xx} + g_{yy}) - g(f_{xx} + f_{yy})]\,dA$

$\displaystyle = \iint_D [f\nabla^2 g - g\nabla^2 f]\,dA$

57. Introduce a "cut" line C_4 joining C_1 and C_2 and another C_5 joining C_2 to C_3. Then if

$$\mathbf{F} = M(x,y)\,\mathbf{i} + N(x,y)\,\mathbf{j}$$

is continuously differentiable in the shaded region D, the conditions of Green's theorem

are satisfied by the curve C formed by traveling counterclockwise around C_1, to the left along C_4, clockwise around the bottom of C_2, to the left along C_5, clockwise around C_3, to the right along C_5, clockwise around the top of C_2, the back to C_1 along C_4. Then,

$\displaystyle\oint_C (M\,dx + N\,dy)$

$\displaystyle = \iint_D \left[\frac{\partial N}{\partial x} - \frac{\partial M}{\partial y}\right]dA$

$\displaystyle = \int_{C_1}(M\,dx + N\,dy) + \int_{C_2}(M\,dx + N\,dy)$

$\displaystyle \qquad + \int_{C_3}(M\,dx + N\,dy)$

If F is conservative in D and C_2 and C_3 are traversed counterclockwise instead of clockwise, then

$$0 = \int_{C_1} - \int_{C_2} - \int_{C_3}$$

so that

$$\int_{C_1} = \int_{C_2} + \int_{C_3}$$

59. If R has two holes, the same kind of analysis as in Problem 58 yields the following possibilities for $\oint_C \mathbf{F}\cdot d\mathbf{R}$:

Case I: C surrounds neither hole;

$$\oint_C \mathbf{F}\cdot d\mathbf{R} = 0$$

Case II: C surrounds just one hole. There

are four possibilities depending on which hole is surrounded and whether the orientation of C is clockwise or counterclockwise.

Case III: C surrounds both holes. Again, there are two possibilities. Either C circles both holes with the same orientation or it circles one hole with one orientation and the other with the opposite orientation (a "figure- eight"). Thus, there are a total of seven possible values for

$$\oint_C \mathbf{F} \cdot d\mathbf{R}$$

13.5 Surface Integrals, page 1072

1. $z = 2 - y$; $z_x = 0$, $z_y = -1$

$$dS = \sqrt{0^2 + (-1)^2 + 1}\, dA$$

$$= \sqrt{2}\, dy\, dx$$

$$\iint_S xy\, dS = \int_0^2 \int_0^2 xy\sqrt{2}\, dy\, dx$$

$$= 2\sqrt{2} \int_0^2 x\, dx$$

$$= 4\sqrt{2}$$

3. $z = x + 3$; $z_x = 1$, $z_y = 0$

$$dS = \sqrt{(1)^2 + (0)^2 + 1}\, dA$$

$$= \sqrt{2}\, dy\, dx$$

$$\iint_S xy\, dS = \int_0^5 \int_0^5 xy\sqrt{2}\, dy\, dx$$

$$= \frac{25\sqrt{2}}{2} \int_0^5 x\, dx$$

$$= \frac{625\sqrt{2}}{4}$$

5. $z = 5$; $dS = dA$

$$\iint_S xy\, dS = \iint_R xy\, dA$$

$$= \int_0^{2\pi} \int_0^1 (r\cos\theta)(r\sin\theta)\, r\, dr\, d\theta$$

$$= \frac{1}{4} \int_0^{2\pi} \sin\theta \cos\theta\, d\theta$$

$$= 0$$

7. $z = 4 - x - 2y$; $z_x = -1$, $z_y = -2$

$$dS = \sqrt{(-1)^2 + (-2)^2 + 1}\, dA$$

$$= \sqrt{6}\, dy\, dx$$

$$\iint_S (x^2 + y^2)\, dS$$

$$= \int_0^4 \int_0^2 (x^2 + y^2)\sqrt{6}\, dy\, dx$$

$$= \sqrt{6} \int_0^4 \left(2x^2 + \frac{8}{3}\right) dx$$

$$= \frac{160\sqrt{6}}{3}$$

9. $z = 4$; $z_x = 0$, $z_y = 0$

$$dS = \sqrt{(0)^2 + (0)^2 + 1}\, dA$$

$$\iint_S (x^2 + y^2)\, dS$$

$$= \int_0^{2\pi} \int_0^1 r^2(r\, dr\, d\theta)$$

$$= \frac{1}{4} \int_0^{2\pi} d\theta$$

$$= \frac{\pi}{2}$$

11. $x = 4$; $x_y = 0$, $x_z = 0$

$$dS = \sqrt{(0)^2 + (0)^2 + 1}\, dA$$

$$= dy\, dz$$

$$\iint_S 7\,dS = \int_0^{2\pi}\int_0^1 7r\,dr\,d\theta$$

$$= 14\pi \int_0^1 r\,dr$$

$$= 7\pi$$

For Problems 13-18, $z = \sqrt{4 - x^2 - y^2}$;

$$z_x^2 + z_y^2 + 1 = \left(\frac{-x}{z}\right)^2 + \left(\frac{-y}{z}\right)^2 + 1$$

$$= \frac{4}{z^2}$$

$$ds = \sqrt{z_x^2 + z_y^2 + 1}\,dA = \frac{2}{z}\,dA$$

R is the circular disk $x^2 + y^2 \le 4$.

13. $\displaystyle\iint_S z\,dS = \iint_R z\left(\frac{2}{z}\right)dA$

$$= \int_0^{2\pi}\int_0^2 2\,r\,dr\,d\theta$$

$$= 8\pi$$

15. $\displaystyle\iint_S (x - 2y)\,dS$

$$= \iint_R (x - 2y)\left(\frac{2}{z}\right)dA$$

$$= \int_0^{2\pi}\int_0^2 (r\cos\theta - 2r\sin\theta)\frac{2}{\sqrt{4 - r^2}}\,r\,dr\,d\theta$$

$$= 2\pi \int_0^{2\pi} (\cos\theta - 2\sin\theta)\,d\theta$$

$$= 0$$

17. $\displaystyle\iint_S (x^2 + y^2)z\,dS$

$$= \iint_R (x^2 + y^2)z\left(\frac{2}{z}\right)dA$$

$$= 2\int_0^{2\pi}\int_0^2 r^2(r\,dr\,d\theta)$$

$$= 8\int_0^{2\pi} d\theta$$

$$= 16\pi$$

For Problems 19-24, $z = x^2 + y^2 = r^2$;
$z_x = 2x$, $z_y = 2y$

$$ds = \sqrt{(2x)^2 + (2y)^2 + 1}\,dA = \sqrt{4r^2 + 1}\,dA$$

The projected region R is the disk
$x^2 + y^2 \le 4$.

19. $\displaystyle\iint_S z\,dS = \int_0^{2\pi}\int_0^2 r^2\sqrt{4r^2 + 1}\,r\,dr\,d\theta$

$$= 2\int_0^{2\pi}\int_0^2 r^3\sqrt{r^2 + \left(\frac{1}{2}\right)^2}\,dr\,d\theta \quad \boxed{\text{Formula 88}}$$

$$= 2\int_0^{2\pi}\left[\frac{(r^2 + \frac{1}{4})^{5/2}}{5} - \frac{\frac{1}{4}(r^2 + \frac{1}{4})^{3/2}}{3}\right]\Bigg|_0^2 d\theta$$

$$= \int_0^{2\pi}\left(\frac{391\sqrt{17}}{120} + \frac{1}{120}\right)$$

$$= \frac{\pi}{60}\left(391\sqrt{17} + 1\right)$$

21. $\displaystyle\iint_S (4z + 1)\,dS$

$$= \int_0^{2\pi}\int_0^2 (4r^2 + 1)\sqrt{4r^2 + 1}\,r\,dr\,d\theta$$

$$= \int_0^{2\pi}\left[\frac{289\sqrt{17}}{20} - \frac{1}{20}\right]d\theta$$

$$= \frac{\pi}{10}\left[289\sqrt{17} - 1\right]$$

23. $\displaystyle\iint_D \sqrt{1 + 4z}\,dS$

$$= \iint\limits_D \sqrt{1+4z}\sqrt{1+4r^2}\,dA$$

$$= \iint\limits_D \left(1+4r^2\right) r\,dr\,d\theta$$

$$= \int_0^{2\pi}\int_0^2 \left(r + 4r^3\right) dr\,d\theta$$

$$= \int_0^{2\pi} 18\,d\theta$$

$$= 36\pi$$

25. $z = \sqrt{1-x^2-y^2}$

$$z_x = \frac{-x}{z},\ z_y = \frac{-y}{z}$$

$$z_x^2 + z_y^2 + 1 = \frac{x^2}{z^2} + \frac{y^2}{z^2} + 1$$

$$= \frac{x^2+y^2+z^2}{z^2}$$

$$= \frac{1}{z^2}$$

$$dS = \frac{1}{z}\,dA$$

The projected region R is $x^2 + y^2 \le 1$.

$$\iint\limits_S \left(x^2+y^2\right) dS$$

$$= \int_0^{2\pi}\int_0^1 \frac{r^2}{\sqrt{1-r^2}}\,r\,dr\,d\theta \quad \boxed{\text{Formula 113}}$$

$$= \int_0^{2\pi} \left[\frac{(1-r^2)^{3/2}}{3} - \left(1-r^2\right)^{1/2}\right]_0^1 d\theta$$

$$= \int_0^{2\pi} \frac{2}{3}\,d\theta$$

$$= \frac{4\pi}{3}$$

27. $z = 1 - x - y;\ z_x = -1,\ z_y = -1$

$$dS = \sqrt{(-1)^2 + (-1)^2 + 1}\,dA$$

$$= \sqrt{3}\,dy\,dx$$

The projected region R on the xy-plane is
$x + y \le 1;\ x \ge 0,\ y \ge 0$.

$$\iint\limits_S dS = \int_0^1\int_0^{1-x} \sqrt{3}\,dy\,dx$$

$$= \int_0^1 \left[\sqrt{3}(1-x)\right] dx$$

$$= \frac{\sqrt{3}}{2}$$

29. $z = (x^2 + y^2)^{1/2}$;

$$z_x = \frac{x}{\sqrt{x^2+y^2}},\ z_y = \frac{y}{\sqrt{x^2+y^2}}$$

$$dS = \sqrt{\frac{x^2}{x^2+y^2} + \frac{y^2}{x^2+y^2} + 1}\,dA$$

$$= \sqrt{2}\,dy\,dx$$

The projected region R is $x^2 + y^2 \le 4$.

$$\iint\limits_S 2\,dS = \int_0^{2\pi}\int_0^4 2\sqrt{2}\,r\,dr\,d\theta$$

$$= \int_0^{2\pi} 16\sqrt{2}\,d\theta$$

$$= 32\pi\sqrt{2}$$

31. $z = x + 1;\ dS = \sqrt{1^2 + 0^2 + 1}\,dA$

$$= \sqrt{2}\,r\,dr\,d\theta$$

The projected region R is $x^2 + y^2 \le 1$.

$$\iint\limits_S \left(x^2 + y^2 + z^2\right) dS$$

$$= \iint\limits_R \left[\left(r^2 + z^2\right)\sqrt{2}\right] r\,dr\,d\theta$$

$$= \int_0^{2\pi}\int_0^1 \left[r^2 + (r\cos\theta + 1)^2\right]\sqrt{2}\,r\,dr\,d\theta$$

$$= \int_0^{2\pi} \int_0^1 [r^3 + r^3\cos^2\theta + 2r^2\cos\theta + r]\sqrt{2}\,dr\,d\theta$$

$$= \int_0^{2\pi} \frac{\sqrt{2}}{12}\left(3\cos^2\theta + 8\cos\theta + 9\right) d\theta$$

$$= \frac{7\pi\sqrt{2}}{4}$$

33. $z = 2 - 5x + 4y$; $z_x = -5$, $z_y = 4$

$$dS = \sqrt{(-5)^2 + (4)^2 + 1}\,dA$$

$$= \sqrt{42}\,dy\,dx$$

$$\mathbf{N} = \frac{1}{\sqrt{42}}(5\mathbf{i} - 4\mathbf{j} + \mathbf{k})$$

$$\mathbf{F} = x\mathbf{i} + 2y\mathbf{j} - 3z\mathbf{k}$$

$$\mathbf{F} \cdot \mathbf{N} = \frac{1}{\sqrt{42}}(5x - 8y - 3z)$$

The projected region R is $0 \le x \le 1$, $0 \le y \le 1$.

$$\iint_S \mathbf{F} \cdot \mathbf{N}\,dS$$

$$= \iint_R \frac{1}{\sqrt{42}}(5x - 8y - 3z)\sqrt{42}\,dy\,dx$$

$$= \int_0^1 \int_0^1 [5x - 8y - 3(2 - 5x + 4y)]\,dy\,dx$$

$$= \int_0^1 (20x - 16)\,dx$$

$$= -6$$

35. $f(x, y, z) = x^2 + y^2 + z^2 = 1$

$$\nabla f = 2x\mathbf{i} + 2y\mathbf{j} + 2z\mathbf{k};$$

$$dx = \frac{1}{z}\,dA_{xy}\;;\; \mathbf{N} = x\mathbf{i} + y\mathbf{j} + z\mathbf{k}$$

$$\mathbf{F} \cdot \mathbf{N} = x^2 + y^2$$

$$\iint_S \mathbf{F} \cdot \mathbf{N}\,dS = \int_0^{2\pi} \int_0^1 \frac{r^3\,dr\,d\theta}{\sqrt{1 - r^2}}$$

$$= 2\pi \int_0^1 r^2 \frac{r\,dr}{\sqrt{1 - r^2}}$$

$$= \frac{4\pi}{3}$$

37. $z = y + 1$; $z_x = 0$, $z_y = 1$

$$dS = \sqrt{(0)^2 + (1)^2 + 1}; \; dA = \sqrt{2}\,dy\,dx$$

$$f(x, y, z) = z - y - 1$$

$$\mathbf{N} = \frac{\nabla f}{\|\nabla f\|} = \frac{-\mathbf{j} + \mathbf{k}}{\sqrt{2}}$$

$$\mathbf{F} \cdot \mathbf{N} = \frac{-y^2 + z^2}{\sqrt{2}} = \frac{2y + 1}{\sqrt{2}}$$

The projected region R is the disk $x^2 + y^2 \le 1$.

$$\iint_S \mathbf{F} \cdot \mathbf{N}\,dS$$

$$= \iint_R \frac{1}{\sqrt{2}}(2y + 1)\left(\sqrt{2}\,dy\,dx\right)$$

$$= \int_0^{2\pi} \int_0^1 (2r\sin\theta + 1)\,r\,dr\,d\theta$$

$$= \int_0^{2\pi} \left(\frac{2}{3}\sin\theta + \frac{1}{2}\right) d\theta$$

$$= \pi$$

39. $\mathbf{N} = \dfrac{\langle f_x, f_y, 1\rangle}{\sqrt{4x^2 + 1}}$

$$\iint_S \mathbf{F} \cdot \mathbf{N}\,dS$$

$$= \iint_S \langle y, z^2, -2z\rangle \cdot \frac{\langle f_x, f_y, 1\rangle}{\sqrt{4x^2 + 1}}\sqrt{4x^2 + 1}\,dx\,dy$$

$$= \iint_R \langle y, z^2, -2z\rangle \cdot \langle 2x, 0, 1\rangle\,dx\,dy$$

$$= \int_{-\sqrt{3}}^{\sqrt{3}} \int_0^1 [2xy - 2(3 - x^2)]\,dy\,dx$$

$$= \int_{-\sqrt{3}}^{\sqrt{3}} \left(2x^2 + x - 6\right) dx$$

$$= \left[\frac{2}{3}x^3 + \frac{1}{2}x - 6x\right]_{-\sqrt{3}}^{\sqrt{3}}$$

$$= -8\sqrt{3}$$

41. Let S be the (cylindrical) surface defined by $\mathbf{R}(u, v) = \cos v\mathbf{i} + \sin v\mathbf{j} - u\mathbf{k}$, $0 \le u \le 1$, $0 \le v \le \pi/2$. The area element of this cylindrical patch is $dS = \|\mathbf{R}_u \times \mathbf{R}_v\| \, du \, dv$, where $\|\mathbf{R}_u \times \mathbf{R}_v\| = 1$. Thus, $dS = du \, dv$. The given integral can be rewritten as

$$\iint_S uv \, dS = \int_0^{\pi/2} \int_0^1 u \, v \, du \, dv$$

$$= \left(\int_0^{\pi/2} v \, dv\right)\left(\int_0^1 u \, du\right)$$

$$= \frac{\pi^2}{16}$$

43. $\mathbf{R} = u^2\mathbf{i} + v\mathbf{j} + u\mathbf{k}$; $\mathbf{R}_u = 2u\mathbf{i} + \mathbf{k}$; $\mathbf{R}_v = \mathbf{j}$

$$\mathbf{R}_u \times \mathbf{R}_v = \begin{bmatrix} \mathbf{i} & \mathbf{j} & \mathbf{k} \\ 2u & 0 & 1 \\ 0 & 1 & 0 \end{bmatrix} = -\mathbf{i} + 2u\mathbf{k}$$

$$\|\mathbf{R}_u \times \mathbf{R}_v\| = \sqrt{1 + 4u^2}$$

$$\iint_S \left(x - y^2 + z\right) dS$$

$$= \int_0^1 \int_0^1 \left(u^2 - v^2 + u\right)\sqrt{1 + 4u^2} \, dv \, du$$

$$= \int_0^1 \left(u^2 - \frac{1}{3} + u\right)\sqrt{1 + 4u^2} \, du$$

$$= -\frac{19\ln\left(\sqrt{5} + 2\right)}{192} + \frac{17\sqrt{5}}{32} - \frac{1}{12}$$

45. $\mathbf{R} = u\mathbf{i} - u^2\mathbf{j} + v\mathbf{k}$; $\mathbf{R}_u = \mathbf{i} - 2u\mathbf{j}$; $\mathbf{R}_v = \mathbf{k}$

$$\mathbf{R}_u \times \mathbf{R}_v = \begin{bmatrix} \mathbf{i} & \mathbf{j} & \mathbf{k} \\ 1 & -2u & 0 \\ 0 & 0 & 1 \end{bmatrix} = -2u\mathbf{i} - \mathbf{j}$$

$$\|\mathbf{R}_u \times \mathbf{R}_v\| = \sqrt{1 + 4u^2}$$

$$\iint_S \left(x^2 + y - z\right) dS$$

$$= \int_0^2 \int_0^1 \left(u^2 - u^2 - v\right)\sqrt{1 + 4u^2} \, dv \, du$$

$$= -\frac{1}{2}\int_0^2 \sqrt{1 + 4u^2} \, du$$

$$= -\frac{1}{8}\left[4\sqrt{17} + \ln\left(4 + \sqrt{17}\right)\right]$$

47. $\mathbf{R} = \langle \sin u \cos v, \sin u \sin v, \cos u \rangle$

$\mathbf{R}_u = \langle \cos u \cos v, \cos u \sin v, -\sin u \rangle$

$\mathbf{R}_v = \langle -\sin u \sin v, \sin u \cos v, 0 \rangle$

$$\mathbf{R}_u \times \mathbf{R}_v = \begin{bmatrix} \mathbf{i} & \mathbf{j} & \mathbf{k} \\ \cos u \cos v & \cos u \sin v & -\sin u \\ -\sin u \sin v & \sin u \cos v & 0 \end{bmatrix}$$

$$= (\sin^2 u \cos v)\mathbf{i} - (0 - \sin^2 u \sin v)\mathbf{j}$$
$$+ (\sin u \cos u)\mathbf{k}$$

$$\mathbf{F} = \langle x^2, 0, z \rangle = \langle \sin^2 u \cos^2 v, 0, \cos u \rangle$$

$$\mathbf{F} \cdot (\mathbf{R}_u \times \mathbf{R}_v) = (\sin^2 u \cos^2 v)(\sin^2 u \cos v)$$
$$+ \cos u(\sin u \cos u)$$

$$= \sin^4 u \cos^3 v + \sin u \cos^2 u$$

$$\iint_S \mathbf{F} \cdot \mathbf{N} \, dS = \iint_D \mathbf{F} \cdot (\mathbf{R}_u \times \mathbf{R}_v) \, du \, dv$$

$$= \int_0^\pi \int_0^{2\pi} \left(\sin^4 u \cos^3 v + \sin u \cos^2 u\right) dv \, du$$

$$= \int_0^\pi 2\pi \sin u \cos^2 u \, du$$

$$= \frac{4\pi}{3}$$

49. $z = 4 - x - 2y$; $z_x = -1$, $z_y = -2$

$$dS = \sqrt{(-1)^2 + (-2)^2 + 1}\, dA$$
$$= \sqrt{6}\, dy\, dx$$

The projected region R is $x + 2y \le 4$; $x \ge 0$, $y \ge 0$.

$$m = \iint_R x\left(\sqrt{6}\, dy\, dx\right)$$
$$= \sqrt{6} \int_0^4 \int_0^{(4-x)/2} x\, dy\, dx$$
$$= \sqrt{6} \int_0^4 \frac{x(4-x)}{2}\, dx$$
$$= \frac{16\sqrt{6}}{3}$$

51. $z = x^2 + y^2$; $z_x = 2x$, $z_y = 2y$
$$dS = \sqrt{(2x)^2 + (2y)^2 + 1}\, dA$$
$$= \sqrt{1 + 4r^2}\, r\, dr\, d\theta$$

The projected region R is $x^2 + y^2 \le 1$; $\rho = z = r^2$.

$$m = \iint_R z\, dS$$
$$= \int_0^{2\pi} \int_0^1 r^2 \left(4r^2 + 1\right)^{1/2} r\, dr\, d\theta$$
$$= \int_0^{2\pi} \left(\frac{5\sqrt{5}}{24} + \frac{1}{120}\right) d\theta$$
$$= \frac{\pi}{60}\left(25\sqrt{5} + 1\right)$$

53. $z^2 = 5 - x^2 - y^2$
$$z = \sqrt{5 - r^2}$$
$$dS = \sqrt{\frac{r^2}{z^2} + 1}\, dA$$
$$= \frac{\sqrt{5}}{z}\, r\, dr\, d\theta$$

The plane $z = 1$ intersects $r^2 + z^2 = 5$ where $r^2 = 4$, so the projected region R is the disk $r^2 \le 4$;

$$m = \iint_R \theta^2\, dS$$
$$= \int_0^{2\pi} \int_0^2 \theta^2 \frac{\sqrt{5}}{\sqrt{5 - r^2}}\, r\, dr\, d\theta$$
$$= \int_0^{2\pi} \theta^2 \left(5 - \sqrt{5}\right) d\theta$$
$$= \frac{8}{3}\pi^3 \left(5 - \sqrt{5}\right)$$

55. $\rho\mathbf{V} = \rho\langle xy, yz, xz\rangle$
Let $G = x^2 + y^2 + z - 9$;
$\nabla G = \langle 2x, 2y, 1\rangle$
The rate of fluid flow is:

$$\iint_S \rho\mathbf{V} \cdot \mathbf{N}\, dS = \iint_S \rho\mathbf{V} \cdot \nabla G\, dA$$
$$= \iint_{x^2+y^2\le 9} \rho\langle xy, yz, xz\rangle \cdot \langle 2x, 2y, 1\rangle\, dA$$
$$= \iint_{x^2+y^2\le 9} \rho\left[2x^2 y + 2y^2 z + xz\right] dA$$
$$= \rho \iint_{x^2+y^2\le 9} \left[2x^2 y + (2y^2 + x)(9 - x^2 - y^2)\right] dA$$

Since $z = 9 - x^2 - y^2$

$$= \rho \iint_{x^2+y^2\le 9} \left[2r^2\cos^2\theta(r\sin\theta) \right.$$
$$\left. + \left(2r^2\sin^2\theta + r\cos\theta\right)(9 - r^2)\right] r\, dr\, d\theta$$
$$= \rho \int_0^{2\pi} \frac{81}{10}\left[12\cos^2\theta\sin\theta + 4\cos\theta + 15\sin^2\theta\right] d\theta$$
$$= \frac{243\pi\rho}{2}$$

57. a. $z = \sqrt{a^2 - x^2 - y^2}$;

$$z_x = \frac{-x}{z}, \; z_y = \frac{-y}{z}$$

$$dS = \sqrt{\left(\frac{-x}{z}\right)^2 + \left(\frac{-y}{z}\right)^2 + 1}\, dA$$

$$= \frac{a}{\sqrt{a^2 - r^2}}\, r\, dr\, d\theta$$

The sphere and cone intersect where $2(x^2 + y^2) = a^2$, so the projected region R is the disk $r \le \dfrac{a}{\sqrt{2}}$;

$$\rho = x^2 y^2 z = (r\cos\theta)^2 (r\sin\theta)^2 \sqrt{a^2 - r^2}$$

$$m = \iint\limits_{R} x^2 y^2 z\, dS$$

$$= \int_0^{2\pi}\!\int_0^{a/\sqrt{2}} r^4 \cos^2\theta \sin^2\theta \sqrt{a^2 - r^2}\left(\frac{a}{\sqrt{a^2 - r^2}}\right) r\, dr\, d\theta$$

$$= \int_0^{2\pi}\!\int_0^{a/\sqrt{2}} a r^5 \cos^2\theta \sin^2\theta\, dr\, d\theta$$

$$= \frac{a^7}{48}\int_0^{2\pi} \cos^2\theta \sin^2\theta\, d\theta$$

$$= \frac{\pi a^7}{192}$$

b. $z = \sqrt{a^2 - x^2 - y^2}$; $\quad z_x = \dfrac{-x}{z}$,

$$z_y = \frac{-y}{z}; \quad dS = \frac{ar\, dr\, d\theta}{\sqrt{a^2 - r^2}}$$

The cone with vertex angle ϕ has the equation $z = a\cot\phi$ and intersects the sphere where $r^2 + r^2\cot^2\phi = a^2$, so the projected region R is $r^2(1 + \cot^2\phi_0) \le a^2$ or $r \le a\sin\phi_0$.

$$m = \int_0^{2\pi}\!\int_0^{a\sin\phi_0} \frac{ar\, dr\, d\theta}{\sqrt{a^2 - r^2}}$$

$$= \int_0^{2\pi} a\left(a - a\sqrt{1 - \sin^2\phi_0}\right) d\theta$$

$$= 2\pi a^2 (1 - \cos\phi_0)$$

59. $z = \sqrt{a^2 - x^2 - y^2}$; $\; z_x = \dfrac{-x}{z}, \; z_y = \dfrac{-y}{z}$

$$dS = \sqrt{\left(\frac{x}{z}\right)^2 + \left(\frac{y}{z}\right)^2 + 1}\, dA$$

$$= \frac{ar\, dr\, d\theta}{\sqrt{a^2 - r^2}}$$

$$m = 2\int_0^{2\pi}\!\int_0^{a} a\left(a^2 - r^2\right)^{-1/2} r\, dr\, d\theta$$

$$= 2\int_0^{2\pi} a^2\, d\theta$$

$$= 4\pi a^2$$

Furthermore,

$$I_z = 2a\int_0^{2\pi}\!\int_0^{a} r^3\left(a^2 - r^2\right)^{-1/2} dr\, d\theta$$

$$= \frac{16a^4}{3}\int_0^{\pi/2} d\theta$$

$$= \frac{8\pi a^4}{3}$$

$$= \frac{2}{3} m a^2$$

13.6 Stokes' Theorem and Applications, page 1081

1. Green's theorem Let D be a simply connected region that is bounded by the positively oriented piecewise smooth Jordan curve C. Then if the vector field $\mathbf{F}(x, y) = M(x, y)\mathbf{i} + N(x, y)\mathbf{j}$ is continuously differentiable on D, we have

$$\oint_C (M\, dx + N\, dy) = \iint\limits_{D}\left(\frac{\partial N}{\partial x} - \frac{\partial M}{\partial y}\right) dA$$

Stokes' theorem Let S be an oriented surface with unit normal vector field $\mathbf{N}$, and assume that S is bounded by a piecewise

Page 548

smooth Jordan curve C whose orientation is compatible with that of S. If $\mathbf{F}$ is a vector field that is continuously differentiable on S, then

$$\oint_C \mathbf{F} \cdot d\mathbf{R} = \iint_S (\operatorname{curl} \mathbf{F} \cdot \mathbf{N}) \, dS$$

Green's theorem is valid on a simply connected region and Stokes' theorem is defined on an oriented surface. Both the region in Green's theorem and the surface in Stokes' theorem are piecewise smooth Jordan curves.

SURVIVAL HINT: *Problem 1 summarizes the important ideas in this section. After you have answered the question, go back and read your answer again.... and read it slowly and thoughtfully.*

3. Evaluating the line integral $\oint_C \mathbf{F} \cdot d\mathbf{R}$ where C is the curve $x = 3\cos\theta$, $y = 3\sin\theta$, $z = 0$; $0 \le \theta \le 2\pi$.

$$\oint_C \mathbf{F} \cdot d\mathbf{R} = \int_C (z\,dx + 2x\,dy + 3y\,dz)$$
$$= \int_0^{2\pi} 0 \, d\theta + 2\int_0^{2\pi} 9\cos^2\theta \, d\theta + 3\int_0^{2\pi} 0 \, d\theta$$
$$= 18\pi$$

Evaluating the integral $\iint_S (\operatorname{curl} \mathbf{F} \cdot \mathbf{N}) \, dS$:

$$\operatorname{curl} \mathbf{F} = \begin{vmatrix} \mathbf{i} & \mathbf{j} & \mathbf{k} \\ \frac{\partial}{\partial x} & \frac{\partial}{\partial y} & \frac{\partial}{\partial z} \\ z & 2x & 3y \end{vmatrix} = 3\mathbf{i} + \mathbf{j} + 2\mathbf{k}$$

$$f(x, y, z) = x^2 + y^2 + z^2 = 9;$$
$$\nabla f = \langle 2x, 2y, 2z \rangle$$

$$\mathbf{N} = \frac{1}{3}\langle x, y, z \rangle; \; z = \sqrt{9 - x^2 - y^2}$$

$$dS = \sqrt{\left(\frac{-x}{z}\right)^2 + \left(\frac{-y}{z}\right)^2 + 1} \, dA$$
$$= \frac{3}{\sqrt{9 - r^2}} r \, dr \, d\theta$$

The projected region D is the disk $x^2 + y^2 \le 9$.

$$\iint_S (\operatorname{curl} \mathbf{F} \cdot \mathbf{N}) \, dS$$
$$= \iint_S \frac{1}{3}(3x + y + 2z) \, dS$$
$$= \frac{1}{3}\int_0^{2\pi}\int_0^3 \left(3r\cos\theta + r\sin\theta + 2\sqrt{9 - r^2}\right)\frac{3r\,dr}{\sqrt{9 - r^2}} \, d\theta$$
$$= \int_0^{2\pi}\int_0^3 \frac{3r^2\cos\theta \, dr \, d\theta}{\sqrt{9 - r^2}} + \int_0^{2\pi}\int_0^3 \frac{r^2\sin\theta \, dr \, d\theta}{\sqrt{9 - r^2}}$$
$$\quad + \int_0^{2\pi}\int_0^3 2r \, dr \, d\theta$$
$$= 0 + 0 + 18\pi$$
$$= 18\pi$$

5. Evaluating the line integral $\oint_C \mathbf{F} \cdot d\mathbf{R}$ where

C_1: $\quad x + 2y = 3$
C_2: $\quad 2y + z = 3$
C_3: $\quad x + z = 3$

Parameterizing all three with $0 \le t \le \frac{3}{2}$:

C_1: $\quad x = 3 - 2t, \, y = t, \, z = 0$
$\mathbf{R} = (3 - 2t)\mathbf{i} + t\mathbf{j},$
$d\mathbf{R} = (-2\mathbf{i} + \mathbf{j}) \, dt,$
$\qquad \mathbf{F} \cdot d\mathbf{R} = [-2(x + 2z) + (y - x)] \, dt$
$\qquad\qquad = [-3x + y - 4z] \, dt$
$\qquad\qquad = (7t - 9) \, dt$
$$I_1 = \int_0^{3/2} (7t - 9) \, dt = -\frac{45}{8}$$

$$C_2: \quad x = 0, \; y = \frac{3}{2} - t, \; z = 2t$$

$$\mathbf{R} = \left(\tfrac{3}{2} - t\right)\mathbf{j} + 2t\mathbf{k},$$

$$d\mathbf{R} = (-\mathbf{j} + 2\mathbf{k})\, dt,$$

$$\mathbf{F} \cdot d\mathbf{R} = [-(y - x) + 2(z - y)]\, dt$$
$$= [x - 3y + 2z]\, dt$$
$$= \left(7t - \frac{9}{2}\right) dt$$

$$I_2 = \int_0^{3/2} \left(7t - \frac{9}{2}\right) dt = \frac{9}{8}$$

$$C_3: \quad x = 2t, \; y = 0, \; z = 3 - 2t$$

$$\mathbf{R} = 2t\mathbf{j} + (3 - 2t)\mathbf{k},$$

$$d\mathbf{R} = (2\mathbf{i} - 2\mathbf{k})\, dt,$$

$$\mathbf{F} \cdot d\mathbf{R} = [2(x + 2z) - 2(z - y)]\, dt$$
$$= [2x + 2y + 2z]\, dt$$
$$= 6\, dt$$

$$I_3 = \int_0^{3/2} 6\, dt = 9$$

$$\oint_C \mathbf{F} \cdot d\mathbf{R} = I_1 + I_2 + I_3$$
$$= -\frac{45}{8} + \frac{9}{8} + 9$$
$$= \frac{9}{2}$$

Evaluating the integral $\int\int_S (\operatorname{curl} \mathbf{F} \cdot \mathbf{N})\, dS$:

$$\operatorname{curl} \mathbf{F} = \begin{vmatrix} \mathbf{i} & \mathbf{j} & \mathbf{k} \\ \frac{\partial}{\partial x} & \frac{\partial}{\partial y} & \frac{\partial}{\partial z} \\ x + 2z & y - x & z - y \end{vmatrix}$$
$$= -\mathbf{i} + 2\mathbf{j} - \mathbf{k}$$

$$\mathbf{N} = \frac{1}{\sqrt{6}}(\mathbf{i} + 2\mathbf{j} + \mathbf{k})\, ; \quad dS = \sqrt{6}\, dA;$$

$$\int\int_S (\operatorname{curl} \mathbf{F} \cdot \mathbf{N})\, dS$$
$$= \int\int_R \frac{-1 + 4 - 1}{\sqrt{6}} \sqrt{6}\, dA$$
$$= 2(\text{AREA OF THE TRIANGLE})$$
$$= 2\left(\frac{9}{4}\right)$$
$$= \frac{9}{2}$$

7. Evaluating the line integral $\oint_C \mathbf{F} \cdot d\mathbf{R}$ where
$C: x = 2\cos\theta, \; y = 2\sin\theta, \; z = 0;$
$0 \le \theta \le 2\pi$

$$\oint_C \mathbf{F} \cdot d\mathbf{R} = \oint_C (2y\, dx - 6z\, dy + 3x\, dz)$$
$$= \int_0^{2\pi} 2(2\sin\theta)(-2\sin\theta)\, d\theta$$
$$= -8\pi$$

Evaluating the integral $\int\int_S (\operatorname{curl} \mathbf{F} \cdot \mathbf{N})\, dS$:

$$\operatorname{curl} \mathbf{F} = \begin{vmatrix} \mathbf{i} & \mathbf{j} & \mathbf{k} \\ \frac{\partial}{\partial x} & \frac{\partial}{\partial y} & \frac{\partial}{\partial z} \\ 2y & -6z & 3x \end{vmatrix} = 6\mathbf{i} - 3\mathbf{j} - 2\mathbf{k}$$

$$f(x, y, z) = x^2 + y^2 + z = 4$$

$$\nabla f = 2x\mathbf{i} + 2y\mathbf{j} + \mathbf{k};$$

$$\mathbf{N} = \frac{2x\mathbf{i} + 2y\mathbf{j} + \mathbf{k}}{\sqrt{4(x^2 + y^2) + 1}}$$

$$z = 4 - x^2 - y^2$$

$$z_x = -2x, \; z_y = -2y$$

$$dS = \sqrt{4x^2 + 4y^2 + 1}\, dA$$

$$\iint\limits_{S} (\text{curl } \mathbf{F} \cdot \mathbf{N})\, dS$$

$$= \iint\limits_{R} \frac{12x - 6y - 2}{\sqrt{4(x^2 + y^2) + 1}} \sqrt{4(x^2 + y^2) + 1}\, dA$$

$$= \int_{0}^{2\pi} \int_{0}^{2} (12r\cos\theta - 6r\sin\theta - 2)\, r\, dr\, d\theta$$

$$= \int_{0}^{2\pi} (32\cos\theta - 16\sin\theta - 4)\, d\theta$$

$$= -8\pi$$

9. The physical meaning of the integral $\oint_{C} \mathbf{F} \cdot d\mathbf{R}$ is that of work performed by the force field $\mathbf{F}$ along the curvilinear path C.

11. Since this problem involves individual student research, answers may vary. Some scientists use the name of Stokes' Theorem for a generalized version of this result (by generalizing it to the n-dimensional space $\mathbb{R}^n$). Many physicists, as well as some mathematicians, sometimes refer to the 3-dimensional version of Stokes' Theorem as the Kelvin-Stokes' Theorem, in order to distinguish it from its generalized $\mathbb{R}^n$ version.

13. $\mathbf{F} = y\mathbf{i} + x\mathbf{j} + z\mathbf{k}$

$$\text{curl } \mathbf{F} = \begin{vmatrix} \mathbf{i} & \mathbf{j} & \mathbf{k} \\ \frac{\partial}{\partial x} & \frac{\partial}{\partial y} & \frac{\partial}{\partial z} \\ y & x & z \end{vmatrix} = \mathbf{0}$$

$\mathbf{N} = -\mathbf{k}$ so curl $\mathbf{F} \cdot \mathbf{N} = 0$ which implies the integral is 0.

15. $\mathbf{F} = z\mathbf{i} + x\mathbf{j} + y\mathbf{k}$

$$\text{curl } \mathbf{F} = \begin{vmatrix} \mathbf{i} & \mathbf{j} & \mathbf{k} \\ \frac{\partial}{\partial x} & \frac{\partial}{\partial y} & \frac{\partial}{\partial z} \\ z & x & y \end{vmatrix} = \mathbf{i} + \mathbf{j} + \mathbf{k}$$

The triangle is the portion of the plane $2x + y + 3z = 6$ in the first octant, and since the orientation is clockwise, the normal is

$$\mathbf{N} = \frac{-2\mathbf{i} - \mathbf{j} - 3\mathbf{k}}{\sqrt{14}}, \text{ so curl } \mathbf{F} \cdot \mathbf{N} = \frac{-6}{\sqrt{14}}$$

Since $z = \frac{1}{3}(6 - 2x - y)$, we have $z_x = -\frac{2}{3}$, $z_y = -\frac{1}{3}$, and

$$dS = \sqrt{\left(-\frac{2}{3}\right)^2 + \left(-\frac{1}{3}\right)^2 + 1}\, dA$$

$$= \frac{\sqrt{14}}{3}\, dy\, dx$$

The projected region D is $2x + y \le 6$ for $x \ge 0$, $y \ge 0$. By Stokes' theorem

$$\oint_{C} \mathbf{F} \cdot d\mathbf{R} = \iint\limits_{S} (\text{curl } \mathbf{F} \cdot \mathbf{N})\, dS$$

$$= \int_{0}^{3} \int_{0}^{6-2x} \left(-\frac{6}{\sqrt{14}}\right)\left(\frac{\sqrt{14}}{3}\right) dy\, dx$$

$$= \int_{0}^{3} -2(6 - 2x)\, dx$$

$$= -18$$

17. $\mathbf{F} = 2xy^2 z\mathbf{i} + 2x^2 yz\mathbf{j} + (x^2 y^2 - 2z)\mathbf{k}$;

$$\text{curl } \mathbf{F} = \begin{vmatrix} \mathbf{i} & \mathbf{j} & \mathbf{k} \\ \frac{\partial}{\partial x} & \frac{\partial}{\partial y} & \frac{\partial}{\partial z} \\ 2xy^2 z & 2x^2 yz & x^2 y^2 - 2z \end{vmatrix}$$

$$= \mathbf{0}$$

$$\oint_{C} \left[2xy^2 z\, dx + 2x^2 yz\, dy + \left(x^2 y^2 - 2z\right) dz\right]$$

$$= \iint\limits_{S} (\text{curl } \mathbf{F} \cdot \mathbf{N})\, dS \qquad \textit{Stokes' theorem}$$

$$= 0 \qquad \textit{Since curl } \mathbf{F} = 0$$

19. $\mathbf{F} = y\mathbf{i} + z\mathbf{j} + x\mathbf{k}$

$$\text{curl } \mathbf{F} = \begin{vmatrix} \mathbf{i} & \mathbf{j} & \mathbf{k} \\ \frac{\partial}{\partial x} & \frac{\partial}{\partial y} & \frac{\partial}{\partial z} \\ y & z & x \end{vmatrix} = -(\mathbf{i} + \mathbf{j} + \mathbf{k})$$

Take S to be the boundary plane $x + y = 2$. Since the curve is traversed clockwise, as viewed from above,

$$\mathbf{N} = \frac{-(\mathbf{i}+\mathbf{j})}{\sqrt{2}} \text{ and curl } \mathbf{F}\cdot\mathbf{N} = \frac{2}{\sqrt{2}} = \sqrt{2}$$

Projecting onto the xz-plane, we have $y = 2 - x$, so $y_x = -1$, $y_z = 0$, and $dS = \sqrt{2}\,dA$. The plane and the sphere intersect when

$$x^2 + (2-x^2) + z^2 = 4$$

so the projected region D is the interior of the ellipse

$$(x-1)^2 + \frac{1}{2}z^2 = 1$$

By Stokes' theorem,

$$\oint_C \mathbf{F}\cdot d\mathbf{R} = \iint_S (\text{curl }\mathbf{F}\cdot\mathbf{N})\,dS$$
$$= \iint_D \sqrt{2}\left(\sqrt{2}\,dA\right)$$
$$= 2(\text{AREA OF ELLIPSE})$$
$$= 2\left[\pi(1)\left(\sqrt{2}\right)\right]$$
$$= 2\sqrt{2}\pi$$

Note: see Example 3 of 13.4 for the formula for the area of an ellipse with semimajor axis of length $a = \sqrt{2}$ and semiminor axis of length $b = 1$: $\pi ab = \pi(1)\sqrt{2}$.

21. Consider the integral

$$\oint_C (x\,dx + y\,dy + z\,dz)$$

where C is the intersection of the plane $z = 4$ and the surface $x^2 + y^2 = z$, traversed counterclockwise as viewed

from the origin. $x^2 + y^2 = z$ represents a circular paraboloid, whose symmetry axis is the z-axis, and the intersection curve C with the horizontal plane $z = 4$ is the circle $x^2 + y^2 = 4$.
$\mathbf{F} = x\mathbf{i} + y\mathbf{j} + z\mathbf{k}$ implies curl $\mathbf{F} = \mathbf{0}$. Consequently, for any surface S bounded by the curve C, we obtain

$$\oint_C \mathbf{F}\cdot d\mathbf{R} = \iint_S (\text{curl }\mathbf{F}\cdot\mathbf{N})\,dS$$
$$= 0$$

23. $\mathbf{F} = 3y\mathbf{i} + 2z\mathbf{j} - 5x\mathbf{k}$;

$$\text{curl }\mathbf{F} = \begin{vmatrix} \mathbf{i} & \mathbf{j} & \mathbf{k} \\ \frac{\partial}{\partial x} & \frac{\partial}{\partial y} & \frac{\partial}{\partial z} \\ 3y & 2z & -5x \end{vmatrix}$$
$$= -2\mathbf{i} + 5\mathbf{j} - 3\mathbf{k}$$

Take S to be the xy-plane, so $\mathbf{N} = \mathbf{k}$ and curl $\mathbf{F}\cdot\mathbf{N} = -3$. We have $dS = dA$ and the projected region D is the disk $x^2 + y^2 \le 1$. By Stokes' theorem

$$\oint_C \mathbf{F}\cdot d\mathbf{R} = \iint_S (\text{curl }\mathbf{F}\cdot\mathbf{N})\,dS$$
$$= \iint_D (-3)\,dA$$
$$= -3(\text{AREA OF }D)$$
$$= -3\pi(1)^2$$
$$= -3\pi$$

25.
$$\mathbf{F} = (y^2+z^2)\mathbf{i} + (x^2+y^2) + (x^2+y^2)\mathbf{k};$$
$$\text{curl }\mathbf{F} = \begin{vmatrix} \mathbf{i} & \mathbf{j} & \mathbf{k} \\ \frac{\partial}{\partial x} & \frac{\partial}{\partial y} & \frac{\partial}{\partial z} \\ y^2+z^2 & x^2+y^2 & x^2+y^2 \end{vmatrix}$$
$$= (2y)\mathbf{i} - (2x-2z)\mathbf{j} + (2x-2y)\mathbf{k}$$

The plane containing the triangle is
$x + y + z = 1$, and the projected region in
the xy-plane is the triangle D bounded by
the x-axis, the y-axis, and the line
$x + y = 1$. If $G = x + y + z - 1$, we have
$\nabla G = \mathbf{i} + \mathbf{j} + \mathbf{k}$, and by Stokes' theorem.

$$\oint_C \mathbf{F} \cdot d\mathbf{R}$$

$$= \int\int_S (\text{curl } \mathbf{F} \cdot \mathbf{N})\, dS$$

$$= \int\int_D (\text{curl } \mathbf{F} \cdot \nabla G)\, dA$$

$$= \int\int_D \langle 2y, 2z - 2x, 2x - 2y \rangle \cdot \langle 1, 1, 1 \rangle\, dA$$

$$= \int_0^1 \int_0^{1-x} [2y + 2z - 2x + 2x - 2y]\, dy\, dx$$

$$= \int_0^1 \int_0^{1-x} 2(1 - x - y)\, dy\, dx$$

$$= \int_0^1 (x - 1)^2\, dx$$

$$= \frac{1}{3}$$

27. $\text{curl } \mathbf{F} = \begin{vmatrix} \mathbf{i} & \mathbf{j} & \mathbf{k} \\ \frac{\partial}{\partial x} & \frac{\partial}{\partial y} & \frac{\partial}{\partial z} \\ 6x^2 e^{yz} & 2x^3 z e^{yz} & 2x^3 y e^{yz} \end{vmatrix}$

$= (2x^3 e^{yz} + 2x^3 yz e^{yz} - 2x^3 e^{yz} - 2x^3 zy e^{yz})\mathbf{i}$
$\quad - (6x^2 y e^{yz} - 6x^2 y e^{yz})\mathbf{j}$
$\quad + (6x^2 z e^{yz} - 6x^2 z e^{yz})\mathbf{k}$

$= \mathbf{0}$

Thus,

$$\int\int_S (\text{curl } \mathbf{F} \cdot \mathbf{N})\, dS = 0$$

29. The boundary curve C is a square in the xy-plane ($z = 0$):
C_1: $x = t, y = 0$ for $0 \le t \le 1$
C_2: $x = 1, y = t$ for $0 \le t \le 1$
C_3: $x = 1 - t, y = 1$ for $0 \le t \le 1$
C_4: $x = 0, y = 1 - t$ for $0 \le t \le 1$

$$\int\int_S (\text{curl } \mathbf{F} \cdot \mathbf{N})\, dS$$

$$= \oint_C (xy\, dx - z\, dy)$$

$$= \int_0^1 0\, dt + \int_0^1 0\, dt$$

$$\quad + \int_0^1 (1 - t)(-1)\, dt + \int_0^1 0\, dt$$

$$= -\frac{1}{2}$$

31. The paraboloid and the plane intersect where
$x^2 + y^2 = y$ or $r = \sin\theta$ (in polar form).
Thus, the curve C can be parameterized as:

$$x = (\sin\theta)\cos\theta = \frac{1}{2}\sin 2\theta$$

$$y = z = \sin\theta \sin\theta = \frac{1}{2}(1 - \cos 2\theta)$$

$$0 \le \theta \le \pi.$$

$$\int\int_S (\text{curl } \mathbf{F} \cdot \mathbf{N})\, dS$$

$$= \oint (xy\, dx + x^2\, dy + z^2\, dz)$$

$$= \int_0^\pi \left[\frac{1}{4}\sin 2\theta(1 - \cos 2\theta)(\cos 2\theta) \right.$$

$$\left. + \frac{1}{4}\sin^2 2\theta(\sin 2\theta) \right.$$

$$+ \frac{1}{4}(1 - \cos 2\theta)^2 (\sin 2\theta) \Bigg] d\theta$$

$$= \int_0^\pi [1 - \cos 2\theta + \sin^2 2\theta] \left(\frac{1}{4} \sin 2\theta \right) d\theta$$

$$= 0$$

33. The curve C is the circle $x^2 + y^2 = 4$ in the xy-plane: $x = 2\cos t$, $y = 2\sin t$, $z = 0$; $0 \le t \le 2\pi$

$$\iint_S (\text{curl } \mathbf{F} \cdot \mathbf{N}) \, dS$$

$$= \oint (4y \, dx + z \, dy + 2y \, dz)$$

$$= \int_0^{2\pi} [4(2\sin t)(-2\sin t) + 0 + 0] \, dt$$

$$= -16 \int_0^{2\pi} \sin^2 t \, dt$$

$$= -16\pi$$

35. Let C be the boundary of the standard unit circle in the $x\,y$-plane. Hence, $\mathbf{N} = \mathbf{k}$ and the projected region D is the disk

$$x^2 + y^2 \le 1$$

$$\mathbf{F} = (1+y)z \, \mathbf{i} + (1+z)x \, \mathbf{j} + (1+x)y \, \mathbf{k}$$

$$\text{curl } \mathbf{F} = \begin{vmatrix} \mathbf{i} & \mathbf{j} & \mathbf{k} \\ \frac{\partial}{\partial x} & \frac{\partial}{\partial y} & \frac{\partial}{\partial z} \\ (1+y)z & (1+z)x & (1+x)y \end{vmatrix}$$

$$= \mathbf{i} + \mathbf{j} + \mathbf{k}$$

$$(\text{curl } \mathbf{F} \cdot \mathbf{N}) \, dS = (1) \, dS = dA, \text{ so}$$

$$\oint_C \mathbf{F} \cdot d\mathbf{R} = \iint_S (\text{curl } \mathbf{F} \cdot \mathbf{N}) \, dS$$

$$= \iint_D 1 \, dA$$

$$= (\text{AREA OF UNIT CIRCLE})$$

$$= \pi$$

37. Let S be the part of the plane $x + y + z = 1$ that lies in the first octant.

$$\mathbf{F} = (1+y)z\mathbf{i} + (1+z)x\mathbf{j} + (1+x)y\mathbf{k}$$
$$f(x, y, z) = x + y + z - 1$$

$$\mathbf{N} = \frac{1}{\sqrt{3}}(\mathbf{i} + \mathbf{j} + \mathbf{k}) \, ; \, dS = \sqrt{3} \, dA$$

$$\text{curl } \mathbf{F} = \begin{vmatrix} \mathbf{i} & \mathbf{j} & \mathbf{k} \\ \frac{\partial}{\partial x} & \frac{\partial}{\partial y} & \frac{\partial}{\partial z} \\ (1+y)z & (1+z)x & (1+x)y \end{vmatrix}$$

$$= \mathbf{i} + \mathbf{j} + \mathbf{k}$$

$$(\text{curl } \mathbf{F} \cdot \mathbf{N}) \, dS = 3 \, dA$$

$$\oint_C \mathbf{F} \cdot d\mathbf{R} = \iint_S (\text{curl } \mathbf{F} \cdot \mathbf{N}) \, dS$$

$$= \iint_R 3 \, dA$$

$$= \frac{3}{2}$$

since the projection in the xy-plane is a triangle with vertices $(0, 0)$, $(0, 1)$, $(1, 0)$.

39. Students are encouraged to perform independent research. Stokes' theorem has several applications in thermodynamics, including in the second law. The following article is suggested as a reading: *Stokes' theorem and the geometric basis for the second law of thermodynamics, by Lawrence H. Bowen, J. Chem. Educ., 1988, 65 (1), p 50 DOI: 10.1021/ed065p50, Publication Date: January 1988.* Any other relevant reading is strongly recommended. Instructors are encouraged to request a brief abstract or essay on the topic.

41. We review the classical statement of Green's theorem. Let D be a simply connected region that is bounded by the positively oriented piecewise smooth Jordan curve C. Then, if the vector field

$$\mathbf{F}(x, y) = M(x, y)\,\mathbf{i} + N(x, y)\,\mathbf{j}$$

is continuously differentiable on D, we have

$$\oint_C (M\,dx + N\,dy) = \iint_D \left(\frac{\partial N}{\partial x} - \frac{\partial M}{\partial y} \right) dA$$

In the statement of Green's theorem, $M\,dx + N\,dy$ represents a 1-*form* in its highest generality, while

$$\left(\frac{\partial N}{\partial x} - \frac{\partial M}{\partial y} \right) dA = \left(\frac{\partial N}{\partial x} - \frac{\partial M}{\partial y} \right) dx\,dy$$

represents a specific 2-*form*. Functions are referred to as 0-*forms*. For a rigorous definition of 1-*forms* or 2-*forms*, one may consult web resources or an advanced calculus textbook.

43. $\mathbf{V} = x\mathbf{i} + (z - x)\mathbf{j} + y\mathbf{k}$

$$\text{curl } \mathbf{V} = \begin{vmatrix} \mathbf{i} & \mathbf{j} & \mathbf{k} \\ \frac{\partial}{\partial x} & \frac{\partial}{\partial y} & \frac{\partial}{\partial z} \\ x & z - x & y \end{vmatrix} = -\mathbf{k}$$

Take S to be the hemisphere $z = \sqrt{1 - x^2 - y^2}$, so $\mathbf{N} = x\mathbf{i} + y\mathbf{j} + z\mathbf{k}$
curl $\mathbf{V} \cdot \mathbf{N} = -z$, and

$$dS = \sqrt{\left(\frac{-x}{z} \right)^2 + \left(\frac{-y}{z} \right)^2 + 1}\,dA = \frac{1}{z}\,dA.$$

The projected region D is the disk $x^2 + y^2 \le y$ or $r \le \sin \theta$ (in polar form), for $0 \le \theta \le \pi$. By Stokes' theorem

$$\oint_C \mathbf{V} \cdot d\mathbf{R} = \iint_S (\text{curl } \mathbf{V} \cdot \mathbf{N})\,dS$$

$$= \iint_D (-z)\left(\frac{1}{z}\,dA \right)$$

$$= -(\text{AREA OF } D)$$

$$= -\pi \left(\frac{1}{2} \right)^2$$

$$= -\frac{\pi}{4}$$

45. $\mathbf{V} = (e^{x^2} + z)\mathbf{i} + (x + \sin y^3)\mathbf{j}$
$\qquad + [y + \ln(\tan^{-1} z)]\mathbf{k}$
curl $\mathbf{V}$

$$= \begin{vmatrix} \mathbf{i} & \mathbf{j} & \mathbf{k} \\ \frac{\partial}{\partial x} & \frac{\partial}{\partial y} & \frac{\partial}{\partial z} \\ e^{x^2} + z & x + \sin y^3 & y + \ln(\tan^{-1} z) \end{vmatrix}$$

$$= \mathbf{i} + \mathbf{j} + \mathbf{k}$$

The sphere and the cone intersect where

$$x^2 + y^2 + \left(x^2 + y^2 \right) = 1$$

or, equivalently, $z = \frac{\sqrt{2}}{2}$. Take S to be the plane $z = \frac{\sqrt{2}}{2}$, so $\mathbf{N} = \mathbf{k}$, curl $\mathbf{V} \cdot \mathbf{N} = 1$, and $dS = dA$. The projected region D is the disk $x^2 + y^2 \le \frac{1}{2}$. By Stokes' theorem

$$\oint_C \mathbf{V} \cdot d\mathbf{R} = \iint_S (\text{curl } \mathbf{V} \cdot \mathbf{N})\,dS$$

$$= \iint_D 1\,dA$$

$$= (\text{AREA OF DISK})$$

$$= \pi \left(\frac{\sqrt{2}}{2} \right)^2$$

$$= \frac{\pi}{2}$$

47. The plane can be rewritten $z = -\frac{1}{2}x - \frac{1}{2}y + 1$. Now,

$$\mathbf{F} = (x + y)^2\,\mathbf{i} - (x - y)^2\,\mathbf{j} + z^2\,\mathbf{k}$$

and $\mathbf{N} = \dfrac{\frac{1}{2}\mathbf{i} + \frac{1}{2}\mathbf{j} + \mathbf{k}}{\sqrt{\frac{3}{2}}}$

Let C be the ellipse of intersection between the plane $z = \frac{-1}{2}x + \frac{-1}{2}y + 1$ and the solid cylinder provided. The projection of C on the xy-plane is the ellipse $\dfrac{x^2}{9} + \dfrac{y^2}{4} = 1$. Let the domain D represent the interior of the ellipse $\dfrac{x^2}{9} + \dfrac{y^2}{4} = 1$, which also represents the projection of the surface S on the xy-plane. Consider the given integral

$$\oint_C [(x+y)^2\,dx - (x-y)^2\,dy + z^2\,dz]$$

A simple computation provides

$$\text{curl } \mathbf{F} = \begin{vmatrix} \mathbf{i} & \mathbf{j} & \mathbf{k} \\ \frac{\partial}{\partial x} & \frac{\partial}{\partial y} & \frac{\partial}{\partial z} \\ (x+y)^2 & -(x-y)^2 & z^2 \end{vmatrix}$$

$$= -4x\,\mathbf{k}.$$

We have that $dS = \sqrt{z_x^2 + z_y^2 + 1}\,dA$ and

$$\sqrt{z_x^2 + z_y^2 + 1} = \sqrt{\left(-\tfrac{1}{2}\right)^2 + \left(-\tfrac{1}{2}\right)^2 + 1}$$

so

$$\oint_C \mathbf{F} \cdot d\mathbf{R} = \iint_S (\text{curl } \mathbf{F} \cdot \mathbf{N})\,dS$$

$$= \iint_D (-4x)\,dA$$

Since D represents the interior of the ellipse $\dfrac{x^2}{9} + \dfrac{y^2}{4} = 1$, we need to perform the following transformation of

coordinates: $x = 3X$, $y = 2Y$. The Jacobian of this transformation is $J = 6$, that is, $dA = dx\,dy = 6\,dX\,dY$. Thus, we obtain a new domain R, in the plane, that corresponds to the elliptic domain D. R represents the interior of the unit disk

$$X^2 + Y^2 \le 1$$

Changing to polar coordinates, we obtain the integral

$$\oint_C \mathbf{F} \cdot d\mathbf{R} = \iint_D (-4x)\,dA$$

$$= \iint_R (-72\,X)\,dX\,dY$$

$$= -72 \int_0^{2\pi} \int_0^1 (r\cos\theta)\,r\ dr\,d\theta$$

$$= 0$$

49. $\oint_C (y^2\,dx + x^2\,dy + z^2\,dz)$ where C represents the intersection of the paraboloid $z = 2x^2 + 3y^2$ with the plane $z = 3y + 5x + 2$. Let S be the interior of this intersection C. The projection of the ellipse C on the xy-plane can be rewritten as

$$\dfrac{(x - \frac{5}{4})^2}{\frac{47}{16}} + \dfrac{(y - \frac{1}{2})^2}{\frac{47}{24}} = 1$$

and is obtained by eliminating the z component between the two equations. Next,
$\mathbf{F} = y^2\,\mathbf{i} + x^2\,\mathbf{j} + z^2\,\mathbf{k}$

$$\text{curl } \mathbf{F} = \begin{vmatrix} \mathbf{i} & \mathbf{j} & \mathbf{k} \\ \frac{\partial}{\partial x} & \frac{\partial}{\partial y} & \frac{\partial}{\partial z} \\ y^2 & x^2 & z^2 \end{vmatrix}$$

$$= 2(x - y)\mathbf{k}$$

Applying Stokes' theorem and passing from the cross-sectional surface S to its projection D on the xy-plane, we obtain

$$\oint_C \mathbf{F} \cdot d\mathbf{R} = \int\int_S (\operatorname{curl} \mathbf{F} \cdot \mathbf{N})\, dS$$

$$= \int\int_D 2(x - y)\, dA$$

$$= \frac{3}{2}(\text{AREA OF ELLIPSE})$$

$$= \frac{47\pi}{16}\sqrt{\frac{3}{2}}$$

51. $\mathbf{F} = y^2\mathbf{i} + xy\mathbf{j} + xz\mathbf{k}$
C: $z = 0$, $x = \cos\theta$, $y = \sin\theta$; $0 \le \theta \le 2\pi$

$$\int\int_S (\operatorname{curl} \mathbf{F} \cdot \mathbf{N})\, dS$$

$$= \oint_C (y^2\, dx + xy\, dy + xz\, dz)$$

$$= \int_0^{2\pi} \left[\sin^2\theta(-\sin\theta) + (\cos\theta\sin\theta)\cos\theta\right] d\theta$$

$$= 0$$

53. $\operatorname{curl} \mathbf{F} = \begin{vmatrix} \mathbf{i} & \mathbf{j} & \mathbf{k} \\ \frac{\partial}{\partial x} & \frac{\partial}{\partial y} & \frac{\partial}{\partial z} \\ 4z & -3x & 2y \end{vmatrix}$

$$= 2\mathbf{i} + 4\mathbf{j} - 3\mathbf{k}$$

Let C be a positively-oriented curve bounding the region (surface) D in the plane $5x + 3y + 2z = 4$. If $G = 5x + 3y + 2z - 4$, then $\nabla G = \langle 5, 3, 2 \rangle$ and by Stokes' theorem

$$\oint_C \mathbf{F} \cdot d\mathbf{R}$$

$$= \int\int_S (\operatorname{curl} \mathbf{F} \cdot \mathbf{N})\, dS$$

$$= \int\int_S (\operatorname{curl} \mathbf{F} \cdot \nabla G)\,\sqrt{38}\, dA$$

$$= \int\int_D \langle 2, 4, -3 \rangle \cdot \langle 5, 3, 2 \rangle \sqrt{38}\, dA$$

$$= \int\int_D 16\sqrt{38}\, dA$$

$$= 16\sqrt{38}(\text{AREA OF } D)$$

Thus, $\oint_C \mathbf{F} \cdot d\mathbf{R}$ depends on only the area bounded by C, and not on the curve of C itself, so

$$\oint_{C_1} = \oint_{C_2}$$

as claimed.

55. Let C be the curve of intersection of the top half of the ellipsoid with the xy-plane, oriented counterclockwise and let $-C$ denote the same curve oriented clockwise. However, in relation to the outer normal of the bottom half of the ellipsoid, $-C$ has positive orientation. Thus, by Stokes' theorem

$$\int\int_S (\operatorname{curl} \mathbf{F} \cdot \mathbf{N})\, dS$$

$$= \oint_C \mathbf{F} \cdot d\mathbf{R} + \oint_{-C} \mathbf{F} \cdot d\mathbf{R}$$

$$= \oint_C \mathbf{F} \cdot d\mathbf{R} - \oint_C \mathbf{F} \cdot d\mathbf{R}$$

$$= 0$$

The same procedure applies whenever S is a closed surface.

57. By virtue of the fourth Maxwell's Law, curl $\mathbf{B} = \mu_0 \mathbf{J}$. Using Stokes' theorem, we obtain

$$\oint_C \mathbf{B} \cdot d\mathbf{R} = \int\int_S \text{curl } \mathbf{B} \cdot \mathbf{N} \, dS$$

$$= \mu_0 \int\int_S \mathbf{J} \cdot \mathbf{N} \, dS$$

This is **Ampère's circuital law** (1826, André-Marie Ampère).

59. $\text{curl}(f \nabla g) = \begin{vmatrix} \mathbf{i} & \mathbf{j} & \mathbf{k} \\ \frac{\partial}{\partial x} & \frac{\partial}{\partial y} & \frac{\partial}{\partial z} \\ fg_x & fg_y & fg_z \end{vmatrix}$

$= \langle f_y g_z - f_z g_y, f_z g_x - f_x g_z, f_x g_y - f_y g_x \rangle$

$= \nabla f \times \nabla g \qquad$ *since* $g_{zy} = g_{yz}, g_{xz} = g_{zx}, g_{xy} = g_{yx}$

Thus, by Stokes' theorem

$$\int_C (f \nabla g) \cdot d\mathbf{R} = \int\int_S \text{curl}(f \nabla g) \cdot \mathbf{N} \, dS$$

$$= \int\int_S (\nabla f \times \nabla g) \cdot \mathbf{N} \, dS$$

13.7 Divergence Theorem and Applications, page 1093

1. The divergence theorem establishes a certain relationship between a surface integral over a closed surface and a volume (triple) integral over the solid domain enclosed by it.

3. Gauss discovered the law that is commonly known as Gauss's flux theorem, which is a law relating the distribution of electric charge to the resulting electric field. The law was formulated by Karl Friedrich Gauss in

1835, but was not published until 1867. It is one of the four Maxwell's equations which form the basis of classical electrodynamics, and a **special case** of the divergence theorem. On the other hand, it is estimated that Gauss had discovered the general case of the divergence theorem as early as 1813.

5. Evaluating the surface integral $\int\int_S \mathbf{F} \cdot \mathbf{N} \, dS$

For the top portion:
$\mathbf{F} = xz\mathbf{i} + y^2\mathbf{j} + 2z\mathbf{k};$
$z = \sqrt{4 - x^2 - y^2}$
$z_x = -\frac{x}{z}, z_y = -\frac{y}{z}; dS = \frac{2}{z} dA$
(Upward normal): $\mathbf{N} = \frac{x}{2}\mathbf{i} + \frac{y}{2}\mathbf{j} + \frac{z}{2}\mathbf{k}$

$$\int\int_{S_T} \mathbf{F} \cdot \mathbf{N} \, dS = \int\int_R \left(x^2 + \frac{y^3}{z} + 2z \right) dA$$

$$= \int\int_R \left(x^2 + \frac{y^3}{\sqrt{4 - x^2 - y^2}} \right.$$

$$\left. + 2\sqrt{4 - x^2 - y^2} \right) dy \, dx$$

$$= \int_0^{2\pi} \int_0^2 \left(r^3 \cos^2\theta + \frac{r^4 \sin^3\theta}{\sqrt{4 - r^2}} + 2r\sqrt{4 - r^2} \right) dr \, d\theta$$

$$= \int_0^{2\pi} \left(4\cos^2\theta + 3\pi \sin^3\theta + \frac{16}{3} \right) d\theta$$

$$= \frac{44\pi}{3}$$

For the bottom portion:
$\mathbf{F} = xz\mathbf{i} + y^2\mathbf{j} + 2z\mathbf{k};$
$z = -\sqrt{4 - x^2 - y^2}$
$z_x = \frac{x}{\sqrt{4 - x^2 - y^2}}; z_y = \frac{y}{\sqrt{4 - x^2 - y^2}}$

$$dS = -\frac{2}{z}dA$$

(Downward normal): $\mathbf{N} = \frac{x}{2}\mathbf{i} + \frac{y}{2}\mathbf{j} + \frac{z}{2}\mathbf{k}$

$$\iint\limits_{S_B} \mathbf{F} \cdot \mathbf{N}\, dS$$

$$= \iint\limits_{R} \left(-x^2 + \frac{y^3}{\sqrt{4 - x^2 - y^2}} \right.$$

$$\left. + 2\sqrt{4 - x^2 - y^2} \right) dy\, dx$$

$$= \int_0^{2\pi} \int_0^2 \left(-r^3\cos^2\theta + \frac{r^4\sin^3\theta}{\sqrt{4 - r^2}} + 2r\sqrt{4 - r^2} \right) dr\, d\theta$$

$$= \int_0^{2\pi} \left(3\pi\sin^3\theta + 4\sin^2\theta + \frac{4}{3} \right) d\theta$$

$$= \frac{20\pi}{3}$$

Thus,

$$\iint\limits_{S} \mathbf{F} \cdot \mathbf{N}\, dS = \frac{44\pi}{3} + \frac{20\pi}{3} = \frac{64\pi}{3}$$

Evaluating the integral $\iiint\limits_{D} \operatorname{div} \mathbf{F}\, dV$

$$\iiint\limits_{D} \operatorname{div} \mathbf{F}\, dV = \iiint\limits_{D} (z + 2y + 2)\, dV$$

$$= \int_0^{\pi} \int_0^{2\pi} \int_0^2 (\rho\cos\phi$$

$$+ 2\rho\sin\phi\sin\theta + 2)\rho^2\sin\phi\, d\rho\, d\theta\, d\phi$$

$$= \int_0^{\pi} \int_0^{2\pi} \left(4\sin\phi\cos\phi + 8\sin^2\phi\sin\theta \right.$$

$$\left. + \frac{16}{3}\sin\phi \right) d\theta\, d\phi$$

$$= \int_0^{\pi} \left(8\pi\sin\phi\cos\phi + \frac{32\pi}{3}\sin\phi \right) d\phi$$

$$= \frac{64\pi}{3}$$

7. $\mathbf{F} = x\mathbf{i} + y\mathbf{j} + z\mathbf{k}$ implies that div $\mathbf{F} = 3$. Since D is the interior of the ellipsoid

$$\frac{x^2}{a^2} + \frac{y^2}{b^2} + \frac{z^2}{c^2} \le 1$$

with a, b, c constants, the volume of this domain is $\frac{4}{3}\pi abc$ (see index for a reference where you can find this derivation). It is then immediate to verify that

$$\iint\limits_{S} \mathbf{F} \cdot \mathbf{N}\, dS = \iiint\limits_{D} \operatorname{div} \mathbf{F}\, dV$$

$$= 4\pi abc$$

9. Evaluating the surface integral $\iint\limits_{S} \mathbf{F} \cdot \mathbf{N}\, dS$

$\mathbf{F} = 2y^2\mathbf{j}$;

$f(x, y, z) = x + 4y + z - 8 = 0$

$\nabla f = \mathbf{i} + 4\mathbf{j} + \mathbf{k}$

$\mathbf{N} = \dfrac{\mathbf{i} + 4\mathbf{j} + \mathbf{k}}{3\sqrt{2}}$;

$\mathbf{F} \cdot \mathbf{N} = \dfrac{8y^2}{3\sqrt{2}}$

$dS = 3\sqrt{2}\, dA$

$$\iint\limits_{S_B} \mathbf{F} \cdot \mathbf{N}\, dS$$

$$= \int_0^2 \int_0^{8-4y} \frac{8y^2}{3\sqrt{2}} 3\sqrt{2}\, dx\, dy$$

$$= 8 \int_0^2 y^2(8 - 4y)\, dy$$

$$= \frac{128}{3}$$

Evaluating the integral $\iiint_D \text{div } \mathbf{F} \, dV$

Since div $\mathbf{F} = 4y$, we have

$$\iiint_D \text{div } \mathbf{F} \, dV = \iiint_D 4y \, dV$$

$$= 4 \int_0^2 \int_0^{8-4y} \int_0^{8-x-4y} y \, dz \, dx \, dy$$

$$= 4 \int_0^2 \int_0^{8-4y} \left(8y - xy - 4y^2 \right) dx \, dy$$

$$= 4 \int_0^2 \left(32y - 32y^2 + 8y^3 \right) dy$$

$$= \frac{128}{3}$$

11. R: source; S: source; T: sink; U: source
13. R: source; S: source; T: source; U: source
15. R: source; S: source; T: sink; U: neither
17. $\mathbf{F}(x, y) = x^2 \mathbf{i} + y^2 \mathbf{j}$
    div $\mathbf{F} = 2x + 2y$
    R, source; S, source; T, sink; U, source.
    We verify this by visually evaluating the sign of div $\mathbf{F}(x, y) = 2x + 2y$. Points where $y > -x$ and the sign of the divergence is positive are classified as sources. Points where $y < -x$ and the sign of the divergence is negative are classified as sinks. Since no coordinates are given, a rough estimate can be made with naked eye.
19. $\mathbf{F}(x, y) = (\sin x)\mathbf{i} + (\sin y)\mathbf{j}$
    div $\mathbf{F} = \cos x + \cos y$
    R, source; S, source; T, source; U, source.
    We verify this by visually evaluating the sign of div $\mathbf{F}(x, y) = \cos x + \cos y$. Points where $\cos x > -\cos y$ and the sign of the divergence is positive are classified as sources. Points where $y < -x$ and the sign

of the divergence is negative are classified as sinks. Since no coordinates are given, a rough estimate can be made with naked eye.
21. $\mathbf{F}(x, y) = 2x^2 \mathbf{i} + 3y^2 \mathbf{j}$
    div $\mathbf{F} = 4x + 6y$
    R, source; S, source; T, sink; U, neither.
    We verify this by visually evaluating the sign of div $\mathbf{F}(x, y) = 4x + 6y$. Points where $x > -\frac{3}{2}y$ and the sign of the divergence is positive are classified as sources. Points where $y < -x$ and the sign of the divergence is negative are classified as sinks. Since no coordinates are given, a rough estimate can be made with naked eye.
23. div $\mathbf{F} = 3$;

$$\iint_S \mathbf{F} \cdot \mathbf{N} \, dS = \iiint_D \text{div } \mathbf{F} \, dV$$

$$= 3V$$

$$= 3\pi$$

25. S is the cube $0 \leq x \leq 1, 0 \leq y \leq 1$, $0 \leq z \leq 1$, which encloses the solid cube D of volume 1. $\mathbf{F} = x\mathbf{i} + y\mathbf{j} + z\mathbf{k}$ implies that div $\mathbf{F} = 3$.

$$\iint_S \mathbf{F} \cdot \mathbf{N} \, dS = \iiint_D \text{div } \mathbf{F} \, dV = 3$$

27. $\mathbf{F} = (\cos yz)\mathbf{i} + e^{xz}\mathbf{j} + 3z^2 \mathbf{k}$; div $\mathbf{F} = 6z$;

$$\iint_S \mathbf{F} \cdot \mathbf{N} \, dS$$

$$= \iiint_D \text{div } \mathbf{F} \, dV$$

$$= 6 \int_0^{2\pi} \int_0^2 \int_0^{\sqrt{4-r^2}} rz \, dz \, dr \, d\theta$$

$$= 3 \int_0^{2\pi} \int_0^2 \left(4 - r^2\right) r \, dr \, d\theta$$

$$= 3 \int_0^{2\pi} 8(8 - 4) \, d\theta$$

$$= 24\pi$$

29. Let S_C denote the surface of the closed cube. Then $S_C = S \cup S_m$, where S_m is the surface of the missing face. For S_C,

$$\mathbf{F} = (x^2 + y^2 - z^2)\mathbf{i} + (x^2 y)\mathbf{j} + (3z)\mathbf{k}$$

$$\text{div } \mathbf{F} = 2x + x^2 + 3$$

$$\iint_S \mathbf{F} \cdot \mathbf{N} \, dS$$

$$= \iiint_D \text{div } \mathbf{F} \, dV$$

$$= \int_0^1 \int_0^1 \int_0^1 \left(2x + x^2 + 3\right) dz \, dy \, dx$$

$$= \int_0^1 \left(x^2 + 2x + 3\right) dx$$

$$= \frac{13}{3}$$

On S_m, $z = 0$ and $\mathbf{N} = -\mathbf{k}$, so $\mathbf{F} \cdot \mathbf{N} = 0$ and

$$\iint_S \mathbf{F} \cdot \mathbf{N} \, dS = \iint_{S_C} \mathbf{F} \cdot \mathbf{N} \, dS - \iint_{S_m} \mathbf{F} \cdot \mathbf{N} \, dS$$

$$= \frac{13}{3} - 0$$

$$= \frac{13}{3}$$

31. Let S_C denote the closed surface consisting of the paraboloid S and the top disk S_d. Then $\mathbf{F} = x\mathbf{i} + y\mathbf{j} + z\mathbf{k}$; div $\mathbf{F} = 1 + 1 + 1 = 3$,

$$\iint_S \mathbf{F} \cdot \mathbf{N} \, dS = \iiint_D \text{div } \mathbf{F} \, dV$$

$$= \int_0^{2\pi} \int_0^3 \int_{r^2}^9 3 r \, dz \, dr \, d\theta$$

$$= 3 \int_0^{2\pi} \int_0^3 \left(9 - r^2\right) r \, dr \, d\theta$$

$$= 6\pi \int_0^3 \left(9r - r^3\right) dr$$

$$= \frac{243\pi}{2}$$

33. $\mathbf{F} = x^2\mathbf{i} + y^2\mathbf{j} + z^2\mathbf{k}$;
div $\mathbf{F} = 2(x + y + z)$;

$$\iint_S \mathbf{F} \cdot \mathbf{N} \, dS$$

$$= \iiint_D \text{div } \mathbf{F} \, dV$$

$$= \iiint_D 2(x + y + z) \, dV$$

$$= \int_0^{2\pi} \int_0^2 \int_{-\sqrt{4-r^2}}^{\sqrt{4-r^2}} 2[r \cos \theta + r \sin \theta + z] \, r \, dz \, dr \, d\theta$$

$$= \int_0^{2\pi} \int_0^2 4(\cos \theta + \sin \theta) r^2 \sqrt{4 - r^2} \, dr \, d\theta$$

$$= \int_0^{2\pi} 4\pi (\cos \theta + \sin \theta) \, d\theta$$

$$= 0$$

35. div $\mathbf{F} = 1 + 1 + 2z = 2(z + 1)$

$$\iint_S \mathbf{F} \cdot \mathbf{N} \, dS = \iiint_D \text{div } \mathbf{F} \, dV$$

$$= 2 \int_0^{2\pi} \int_0^2 \int_0^1 (z + 1) \, dz \, r \, dr \, d\theta$$

$$= 3 \int_0^{2\pi} \int_0^2 r \, dr \, d\theta$$

$$= 6 \int_0^{2\pi} d\theta$$

$$= 12\pi$$

37. $\mathbf{F} = \langle xy^2, yz^2, x^2y \rangle$

Even though the region R is described in spherical coordinates, it is easier to use cylindrical coordinates because $y^2 + z^2$ is complicated when expanded in spherical form. In cylindrical coordinates, the sphere $\rho = 2$ becomes $r^2 + z^2 = 4$, the cone $\phi = \frac{\pi}{4}$ becomes $z = r$, and the projected region on the xy-plane is the disk $r \le \sqrt{2}$. Thus, we have the (somewhat easier) calculation:

$$\iint_S \mathbf{F} \cdot \mathbf{N} \, dS$$

$$= \iiint_R \operatorname{div} \mathbf{F} \, dV$$

$$= \int_0^{2\pi} \int_0^{\sqrt{2}|} \int_r^{\sqrt{4-r^2}} (4^2 \sin^2\theta + z^2) r \, dz \, dr \, d\theta$$

$$= \int_0^{2\pi} \int_0^{\sqrt{2}} \left[r^3 \sin^2\theta z + \frac{r^3}{3} z^3 \right] \Big|_r^{\sqrt{4-r^2}} dr \, dt$$

$$= \int_0^{2\pi} \int_0^{\sqrt{2}} \left[r^3 \sin^2\theta \left(\sqrt{4-r^2} - r \right) \right. $$
$$ \left. + \frac{r}{3} (4-r^2)^{3/2} - r^3 \right] d\theta$$

$$= \int_0^{2\pi} -\frac{8}{15} \left[(5\sqrt{2} - 8)\sin^2\theta + \sqrt{2} - 4 \right] d\theta$$

$$= \frac{8(16 - 7\sqrt{2})}{15} \pi$$

39. $\operatorname{div} \mathbf{F} = 3x^2 + 3y^2 + 3a^2 = 3(r^2 + a^2)$

$$\iint_S \mathbf{F} \cdot \mathbf{N} \, dS$$

$$= \iiint_R \operatorname{div} \mathbf{F} \, dV$$

$$= \int_0^{2\pi} \int_0^a \int_0^1 3(r^2 + a^2) \, r \, dz \, dr \, d\theta$$

$$= 3 \int_0^{2\pi} \int_0^a (r^3 + a^2 r) \, dr \, d\theta$$

$$= 3 \int_0^{2\pi} \frac{3a^4}{4} \, d\theta$$

$$= \frac{9\pi a^4}{2}$$

41. $\mathbf{F} = x\mathbf{i} + y\mathbf{j} + z\mathbf{k}$; $\operatorname{div} \mathbf{F} = 3$

$$V(D) = \iiint_D dV$$

$$= \frac{1}{3} \iiint_R \operatorname{div} \mathbf{F} \, dV$$

$$= \frac{1}{3} \iint_S \mathbf{F} \cdot \mathbf{N} \, dS$$

$$= \frac{1}{3} \iint_S (x\mathbf{i} + y\mathbf{j} + z\mathbf{k}) \cdot \mathbf{N} \, dS$$

43. $\|\mathbf{R}\| \mathbf{R} = a(x\mathbf{i} + y\mathbf{j} + z\mathbf{k})$;
$\operatorname{div}(\|\mathbf{R}\| \mathbf{R}) = a + a + a = 3a$

$$\iint_S \|\mathbf{R}\| \mathbf{R} \cdot \mathbf{N} \, dS = \iiint_D 3a \, dV$$

$$= 3a \left(\frac{4}{3}\pi a^3 \right)$$

$$= 4\pi a^4$$

45. Applying the divergence theorem with $\mathbf{F} = f \nabla f$, we obtain

$$\iint_S \mathbf{F} \cdot \mathbf{N}\, dS = \iiint_G \text{div}(f\nabla f)\, dV$$

Therefore, since

$$\text{div}(f\nabla f) = f\, \text{div}\, \nabla f + \nabla f \cdot \nabla f$$
$$= f\, \text{div}\, \nabla f + \|\nabla f\|^2$$

it follows that

$$\iint_S (f\nabla f) \cdot \mathbf{N}\, dS = \iiint_G \|\nabla f\|^2\, dV$$

if $f\, \text{div}\, \nabla f = 0$; that is, $\text{div}\, \nabla f = 0$.

47. $\displaystyle\iint_S \frac{\partial u}{\partial n}\, dS = \iint_S \nabla u \cdot \mathbf{N}\, dS$

$$= \iiint_D \text{div}(\nabla u)\, dV$$

$$= \iiint_D \nabla \cdot \nabla u\, dV$$

$$= \iiint_D \nabla^2 u\, dV$$

49. This identity was proved in Problem 55b, Section 13.4, but here we desire a proof using the divergence theorem.

$$\iint_S \left(f\frac{\partial g}{\partial n} - g\frac{\partial f}{\partial n} \right) dS$$

$$= \iint_S (f\nabla g - g\nabla f) \cdot \mathbf{N}\, dS \quad \begin{array}{l}\textit{Definition of}\\ \textit{normal derivative}\end{array}$$

$$= \iiint_D \text{div}(f\nabla g - g\nabla f)\, dV \;\; \textit{Divergence theorem}$$

$$= \iiint_D (\nabla f \cdot \nabla g + f\nabla^2 g - \nabla g \cdot \nabla f - g\nabla^2 f)\, dV$$

$$= \iiint_D (f\nabla^2 g - g\nabla^2 f)\, dV$$

51. Let $\mathbf{U} = u_1\mathbf{i} + u_2\mathbf{j} + u_3\mathbf{k}$; and suppose $\mathbf{F} = \text{curl}\, \mathbf{U}$; then

$$\text{curl}\, \mathbf{U} = \nabla \times \mathbf{U}$$

$$= \left(\frac{\partial u_3}{\partial y} - \frac{\partial u_2}{\partial z} \right)\mathbf{i} + \left(\frac{\partial u_1}{\partial z} - \frac{\partial u_3}{\partial x} \right)\mathbf{j}$$

$$+ \left(\frac{\partial u_2}{\partial x} - \frac{\partial u_1}{\partial y} \right)\mathbf{k}$$

$$\text{div}(\text{curl}\, \mathbf{U}) = \frac{\partial^2 u_3}{\partial x \partial y} - \frac{\partial^2 u_2}{\partial x \partial z} + \frac{\partial^2 u_1}{\partial y \partial z} - \frac{\partial^2 u_3}{\partial y \partial x}$$

$$+ \frac{\partial^2 u_2}{\partial z \partial x} - \frac{\partial^2 u_1}{\partial z \partial y}$$

Thus, $\text{div}\, \mathbf{F} = 0$; and

$$\iint_S \mathbf{F} \cdot \mathbf{N}\, dS = \iiint_V \text{div}\, \mathbf{F}\, dV = 0$$

53. $\mathbf{E} = \dfrac{q}{4\pi\epsilon} \left[\dfrac{x\mathbf{i} + y\mathbf{j} + z\mathbf{k}}{(x^2 + y^2 + z^2)^{3/2}} \right]$

$$\text{div}\, \mathbf{E} = \frac{q}{4\pi\epsilon}(x^2 + y^2 + z^2)^{-5/2}[-(2x^2 - y^2 - z^2)$$
$$- (2y^2 - x^2 - z^2) - (2z^2 - x^2 - y^2)]$$
$$= 0$$

Thus, by the divergence theorem,

$$\iint_S \mathbf{E} \cdot \mathbf{N}\, dS = \iiint_D \text{div}\, \mathbf{E}\, dV = 0$$

55. Suppose the surface S encloses the origin. Then by Gauss' law (Problem 54), we have

Page 563

$$\iint_S \mathbf{D} \cdot \mathbf{N}\, dS = \iint_S (\epsilon \mathbf{E}) \cdot \mathbf{N}\, dS$$

$$= \frac{q}{\epsilon}(\epsilon)$$

$$= q$$

$$= \iint_V \int Q\, dV$$

Let S' be a sphere centered at the origin that is entirely contained within S, and let S'' be the surface of the region inside S but outside S' (as in the proof of Problem 54). Then the divergence theorem applies to S'', and we have

$$\iint_V \int Q\, dV = \iint_{S''} \mathbf{D} \cdot \mathbf{N}\, dS$$

$$= \iint_{\text{Interior of S}} \int \text{div}\, \mathbf{D}\, dV$$

If we take S' to be smaller and smaller ($\rho \to 0$, where ρ is the radius of S'), then in the limit, we have

$$\iint_V \int Q\, dV = \iint_V \int \text{div}\, \mathbf{D}\, dV$$

Finally, since this equation holds for *every*

Chapter 13 Review

🛑 Studying for a chapter examination is a personal process, one which nobody else can do for you. Simply take the time to review what you have done.

region contained within a surface that encloses the origin, it follows that the integrands must be equal; that is,
$Q = \text{div}\, \mathbf{D}$.

57. Since $\nabla \times \mathbf{E} = -\dfrac{\partial \mathbf{B}}{\partial t}$, we have

$$\text{curl}(\text{curl}\, \mathbf{E}) = \nabla \times (\nabla \times \mathbf{E})$$

$$= \nabla \times \left(-\frac{\partial \mathbf{B}}{\partial t} \right)$$

$$= -\frac{\partial}{\partial t}(\nabla \times \mathbf{B})$$

$$= -\frac{\partial}{\partial t}[\nabla \times (\mu \mathbf{H})]$$

$$= -\mu \frac{\partial}{\partial t}(\text{curl}\, \mathbf{H})$$

59. From Problem 58, we consider the special case where $\mathbf{F} = \mathbf{E}$,

$$\nabla(\text{div}\, \mathbf{E}) - (\nabla \cdot \nabla)\mathbf{E} = \text{curl}(\text{curl}\, \mathbf{E})$$

$$= -\mu \frac{\partial}{\partial t}(\text{curl}\, \mathbf{H})$$

From Problem 57

$$= -\mu \frac{\partial}{\partial t}\left(\sigma \mathbf{E} + \epsilon \frac{\partial \mathbf{E}}{\partial t} \right)$$

Given formula for curl H

SURVIVAL HINT: Work all of Chapter 13 problems in the Proficiency Examination (whether they are assigned or not). Work through all of the problems before looking at the answers, and *then* correct each of the problems. The answers to all these problems are given in the answer section at the back of the text. If you worked the problem correctly, move on to the next problem, but if you did not work it correctly (or you did not know what to do), then look at the solutions below, look back in the chapter to study the procedure, or ask your instructor.

Finally, go back over the homework problems you have been assigned. If you worked a problem correctly, move on to the next problem, but if you missed it on your homework, then you should look back in the book or talk to your instructor about how to work the problem.

If you follow these steps, you should be successful with your review of this chapter.

Proficiency Examination, page 1097

1. A vector field is a collection S of points in space together with a rule that assigns to each point (x, y, z) in S exactly one vector $\mathbf{V}(x, y, z)$.

2. The divergence of a differentiable vector field
$$\mathbf{V}(x, y, z) = u(x, y, z)\mathbf{i} + v(x, y, z)\mathbf{j} + w(x, y, z)\mathbf{k}$$
is denoted by div $\mathbf{V}$ and is defined by
$$\text{div } \mathbf{V} = \frac{\partial u}{\partial x}(x, y, z) + \frac{\partial v}{\partial y}(x, y, z) + \frac{\partial w}{\partial z}(x, y, z)$$

3. The curl of a differentiable vector field
$$\mathbf{V}(x, y, z) = u(x, y, z)\mathbf{i} + v(x, y, z)\mathbf{j} + w(x, y, z)\mathbf{k}$$
is denoted by curl $\mathbf{V}$ and is defined by

$$\text{curl } \mathbf{V} = \left(\frac{\partial w}{\partial y} - \frac{\partial v}{\partial z} \right)\mathbf{i} + \left(\frac{\partial u}{\partial z} - \frac{\partial w}{\partial x} \right)\mathbf{j} + \left(\frac{\partial v}{\partial x} - \frac{\partial u}{\partial y} \right)\mathbf{k}$$

4. The del operator is defined by

$$\nabla = \frac{\partial}{\partial x}\mathbf{i} + \frac{\partial}{\partial y}\mathbf{j} + \frac{\partial}{\partial z}\mathbf{k}$$

5. The Laplacian of f is

$$\nabla^2 f = \nabla \cdot \nabla f$$
$$= \frac{\partial^2 f}{\partial x^2} + \frac{\partial^2 f}{\partial y^2} + \frac{\partial^2 f}{\partial z^2}$$
$$= f_{xx} + f_{yy} + f_{zz}$$

The equation $\nabla^2 f = 0$ is called Laplace's equation.

6. A line integral involves taking a limit of a Riemann sum formed by parameterizing with respect to a curve in space. An "ordinary" Riemann integral is formed by parameterizing with respect to the x-axis.

7. Let $\mathbf{F}(x, y, z) = u(x, y, z)\mathbf{i} + v(x, y, z)\mathbf{j} + w(x, y, z)\mathbf{k}$ be a vector field, and let C be the curve with parametric representation

$$\mathbf{R}(t) = x(t)\mathbf{i} + y(t)\mathbf{j} + z(t)\mathbf{k}$$

for $a \leq t \leq b$. Using

$$d\mathbf{R} = dx\,\mathbf{i} + dy\,\mathbf{j} + dz\,\mathbf{k}$$

we denote the line integral of $\mathbf{F}$ over C by $\int_C \mathbf{F} \cdot d\mathbf{R}$ and define it by

$$\int_C \mathbf{F} \cdot d\mathbf{R} = \int_C (u\,dx + v\,dy + w\,dz)$$

$$= \int_a^b \left\{ u[x(t), y(t), z(t)] \frac{dx}{dt} \right.$$

$$+ v[x(t), y(t), z(t)] \frac{dy}{dt}$$

$$\left. + w[x(t), y(t), z(t)] \frac{dz}{dt} \right\}$$

8. Let **F** be a continuous force field over a domain D. Then the **work** W done by **F** as an object moves along a smooth curve C in D is given by the line integral

$$W = \int_C \mathbf{F} \cdot d\mathbf{R}$$

9. Let f be continuous on a smooth curve C. If C is defined by

$$\mathbf{R}(t) = x(t)\mathbf{i} + y(t)\mathbf{j} + z(t)\mathbf{k}$$

where $a \le t \le b$, then

$$\int_C f(x, y, z)\, ds$$

$$= \int_a^b f[x(t), y(t), z(t)] \sqrt{[x'(t)]^2 + [y'(t)]^2 + [z'(t)]^2}\, dt$$

10. Let **F** be a conservative vector field on the region D and let f be a scalar potential function for **F**; that is, $\nabla f = \mathbf{F}$. Then, if C is any piecewise smooth curve lying entirely within D, with initial point P and terminal point Q, we have

$$\int_C \mathbf{F} \cdot d\mathbf{R} = f(Q) - f(P)$$

Thus, the line integral $\int_C \mathbf{F} \cdot d\mathbf{R}$ is independent of path in D.

11. A vector field **F** is said to be conservative in a region D if it can be represented in D as the gradient of a continuously differentiable function f, which is then called a scalar

potential of **F**. That is, $\mathbf{F} = \nabla f$ for (x, y) in D.

12. If **F** is a conservative vector field and $\mathbf{F} = \nabla f$, then f is a scalar potential for **F**.

13. A Jordan curve is a closed curve with no self intersections.

14. Let D be a simply connected region with a positively oriented piecewise smooth boundary C. Then if the vector field

$$\mathbf{F}(x, y) = M(x, y)\mathbf{i} + N(x, y)\mathbf{j}$$

is continuously differentiable on D, we have

$$\oint_C (M\, dx + N\, dy) = \iint_D \left(\frac{\partial N}{\partial x} - \frac{\partial M}{\partial y} \right) dA$$

15. $A = \dfrac{1}{2} \oint_C (-y\, dx + x\, dy)$

$$= -\oint_C y\, dx$$

$$= \oint_C x\, dy$$

16. A normal derivative of f is denoted by $\partial f/\partial n$ and is the directional derivative of f in the direction of the normal **N** pointing to the exterior of the domain of f.

$$\frac{\partial f}{\partial n} = \nabla f \cdot \mathbf{N}$$

where **N** is the outer unit normal.

17. Let S be a surface defined by $z = f(x, y)$ and R its projection on the xy-plane. If f, f_x, and f_y are continuous in R and g is continuous on S, then the surface integral of g over S is

$$\iint\limits_{S} g(x, y, z) \, dS$$

$$= \iint\limits_{R} g(x, y, f(x, y)) \sqrt{[f_x(x, y)]^2 + [f_y(x, y)]^2 + 1} \, dA$$

Similar definitions hold for projections on yz and xz planes.

18. If a surface S is defined parametrically by the vector function

$$\mathbf{R}(u, v) = x(u, v)\mathbf{i} + y(u, v)\mathbf{j} + z(u, v)\mathbf{k}$$

and $f(x, y, z)$ is continuous on D, the surface integral of f over D is given by

$$\iint\limits_{S} f(x, y, z) \, dS = \iint\limits_{D} f(\mathbf{R}) \, \|\mathbf{R}_u \times \mathbf{R}_v\|$$

19. The flux integral of a vector field $\mathbf{F}$ across a surface S is given by

$$\iint\limits_{S} \mathbf{F} \cdot \mathbf{N} \, dS$$

20. Let S be an oriented surface with unit normal vector $\mathbf{N}$, and assume that S is bounded by a closed, piecewise smooth curve C whose orientation is compatible with that of S. If $\mathbf{F}$ is a vector field that is continuously differentiable on S, then

$$\oint_{C} \mathbf{F} \cdot d\mathbf{R} = \iint\limits_{S} (\text{curl } \mathbf{F} \cdot \mathbf{N}) \, dS$$

21. If $\mathbf{F}$ and curl $\mathbf{F}$ are continuous in a simply connected region D, then $\mathbf{F}$ is conservative in D if and only if curl $\mathbf{F} = \mathbf{0}$.

22. Let D be a region in space bounded by a smooth, orientable closed surface S. If $\mathbf{F}$ is a continuous vector field whose components have continuous partial derivatives in D, then

$$\iint\limits_{S} \mathbf{F} \cdot \mathbf{N} \, dS = \iiint\limits_{D} \text{div } \mathbf{F} \, dV$$

where $\mathbf{N}$ is the outward unit normal to the surface S.

23. $\mathbf{F} = yz\mathbf{i} + xz\mathbf{j} + xy\mathbf{k} = M\mathbf{i} + N\mathbf{j} + P\mathbf{k}$

$$\text{curl } \mathbf{F} = \begin{vmatrix} \mathbf{i} & \mathbf{j} & \mathbf{k} \\ \frac{\partial}{\partial x} & \frac{\partial}{\partial y} & \frac{\partial}{\partial z} \\ yz & xz & xy \end{vmatrix}$$

$$= (x - x)\mathbf{i} - (y - y)\mathbf{j} + (z - z)\mathbf{k}$$

$$= \mathbf{0}$$

so $\mathbf{F}$ is conservative.

$$\frac{\partial f}{\partial x} = yz, \text{ so } f = xyz + a(y, z)$$

and

$$\frac{\partial f}{\partial y} = xz + \frac{\partial a}{\partial y} = xz, \text{ so } \frac{\partial a}{\partial y} = 0$$

and,

$$a(y, z) = b(z); \quad f = xyz + b(z)$$

but

$$\frac{\partial f}{\partial z} = xy + \frac{\partial b}{\partial z} = xy$$

so

$$\frac{\partial b}{\partial z} = 0, b(z) = C \text{ and } f = xyz$$

24. $\mathbf{F} = x^2 y\mathbf{i} - e^{yz}\mathbf{j} + \dfrac{x}{2}\mathbf{k}$

$$\text{div } \mathbf{F} = 2xy - ze^{yz}$$

$$\text{curl } \mathbf{F} = \begin{vmatrix} \mathbf{i} & \mathbf{j} & \mathbf{k} \\ \frac{\partial}{\partial x} & \frac{\partial}{\partial y} & \frac{\partial}{\partial z} \\ x^2 y & -e^{yz} & \frac{x}{2} \end{vmatrix}$$

$$= ye^{yz}\mathbf{i} - \frac{1}{2}\mathbf{j} - x^2\mathbf{k}$$

25. $\oint_C \mathbf{F} \cdot d\mathbf{R} = \int\int_T \left[\dfrac{\partial}{\partial x}(3y^2) - \dfrac{\partial}{\partial y}(2x + y) \right] dA$

$= \int\int_T (-1)\, dA$

$= -(\text{AREA OF } T)$

$= -\dfrac{1}{2}(2)(2)$

$= -2$

26. By Stokes' theorem,

$\oint_C \mathbf{F} \cdot d\mathbf{R} = \int\int_S (\text{curl } \mathbf{F} \cdot \mathbf{N})\, dS$

$\text{curl } \mathbf{F} = \begin{vmatrix} \mathbf{i} & \mathbf{j} & \mathbf{k} \\ \dfrac{\partial}{\partial x} & \dfrac{\partial}{\partial y} & \dfrac{\partial}{\partial z} \\ 2y & z & y \end{vmatrix}$

$= (1 - 1)\mathbf{i} - 2\mathbf{k}$

$= -2\mathbf{k}$

The intersection of the sphere and the plane:

$x^2 + y^2 + (x + 2)^2 = 4(x + 2)$

$2x^2 + y^2 = 4$

$\dfrac{x^2}{2} + \dfrac{y^2}{4} = 1$

$g(x, y, z) = z - x - 2 = 0$

$\mathbf{N} = \dfrac{-\mathbf{i} + \mathbf{k}}{\sqrt{2}}; \text{ curl } \mathbf{F} \cdot \mathbf{N} = -\dfrac{2}{\sqrt{2}};$

$dS = \sqrt{2}\, dy\, dx$

$\int\int_S (\text{curl } \mathbf{F} \cdot \mathbf{N})\, dS = -\int\int_S \sqrt{2}\sqrt{2}\, dy\, dx$

$= -2(\text{AREA OF ELLIPSE})$

$= -2(\pi ab)$

$= -2\pi \left(\sqrt{2} \right)(2)$

$= -4\pi\sqrt{2}$

27. By the divergence theorem

$\int\int_S \mathbf{F} \cdot \mathbf{N}\, dS = \int\int\int_D \text{div } \mathbf{F}\, dV$

$\mathbf{F} = x^2\mathbf{i} + (y + z)\mathbf{j} - 2z\mathbf{k};$

$\text{div } \mathbf{F} = 2x + 1 - 2$

$= 2x - 1$

$\int\int\int_D \text{div } \mathbf{F}\, dV = \int\int\int_D (2x - 1)\, dV$

$= \int_0^1 \int_0^1 \int_0^1 (2x - 1)\, dz\, dy\, dx$

$= \int_0^1 (2x - 1)\, dx$

$= 0$

28. $\mathbf{F} \cdot d\mathbf{R} = \dfrac{x\, dx}{(x^2 + y^2)^2} + \dfrac{y\, dy}{(x^2 + y^2)^2}$

$= M\, dx + N\, dy$

$\dfrac{\partial N}{\partial x} = \dfrac{-2y(2x)}{(x^2 + y^2)^3} = \dfrac{-4xy}{(x^2 + y^2)^3}$

$\dfrac{\partial M}{\partial y} = \dfrac{-2x(2y)}{(x^2 + y^2)^3} = \dfrac{-4xy}{(x^2 + y^2)^3}$

$\dfrac{\partial N}{\partial x} = \dfrac{\partial M}{\partial y}$ so $\mathbf{F}$ is conservative and independent of path. Since C is a closed path the value of the line integral is 0.

29. Since

$\text{curl } \mathbf{R} = \begin{vmatrix} \mathbf{i} & \mathbf{j} & \mathbf{k} \\ \dfrac{\partial}{\partial x} & \dfrac{\partial}{\partial y} & \dfrac{\partial}{\partial z} \\ x & y & z \end{vmatrix} = \mathbf{0}$

it follows that $\mathbf{R}$ and hence $\mathbf{F} = m\omega^2 \mathbf{R}$ are conservative. Ignore for the moment the factor $m\omega^2$.

$\dfrac{\partial f}{\partial x} = x, \text{ so } f = \dfrac{x^2}{2} + c(y, z)$

and

$$\frac{\partial f}{\partial y} = \frac{\partial c}{\partial y} = y; \, c = \frac{y^2}{2} + c_1(z)$$

$$f = \frac{x^2}{2} + \frac{y^2}{2} + c_1(z)\,; \text{but } \frac{\partial f}{\partial z} = \frac{\partial c_1}{\partial z} = z\,,$$

so

$$c_1 = \frac{z^2}{2} + C; \, f = \frac{x^2}{2} + \frac{y^2}{2} + \frac{z^2}{2}$$

Inserting the scalar factor (and taking $C = 0$):

$$\phi = \frac{m\omega^2}{2} f = \frac{m\omega^2}{2}\left(x^2 + y^2 + z^2\right)$$

is a scalar potential function for $\mathbf{F}$.

30. Since $\mathbf{F}$ is conservative with scalar potential

$$\phi = \frac{m\omega^2}{2}(x^2 + y^2 + z^2)\,, \text{ we have}$$

$$\begin{aligned}
W &= \int_C \mathbf{F} \cdot d\mathbf{R} \\
&= \phi(-3, 0, 2) - \phi(3, 0, 2) \\
&= \frac{m\omega^2}{2}\left[(-3)^2 + 0^2 + 2^2\right] \\
&\quad - \frac{m\omega^2}{2}\left[3^2 + 0^2 + 2^2\right] \\
&= 0
\end{aligned}$$

This result could have been anticipated as z is constant and $\mathbf{R}$ is symmetric about the y-axis.

Supplementary Problems, page 1098

Note: In Problems 1-6, since M and N only involve x and y, you can use $\dfrac{\partial M}{\partial y} = \dfrac{\partial N}{\partial x}$ for the criterion of conservativity. We show the more general method using curl $\mathbf{F} = \mathbf{0}$ *in our solutions to 1-5.*

1. curl $\mathbf{F} = \mathbf{0}$, so $\mathbf{F}$ is conservative and $f = 2x - 3y$.

3. curl $\mathbf{F} = \begin{vmatrix} \mathbf{i} & \mathbf{j} & \mathbf{k} \\ \frac{\partial}{\partial x} & \frac{\partial}{\partial y} & \frac{\partial}{\partial z} \\ y^{-3} & -3xy^{-4} + \cos y & 0 \end{vmatrix}$

$= \mathbf{0}$

$\mathbf{F}$ is conservative.
$f_x = y^{-3}$
$f = xy^{-3} + c(y)$
$c'(y) = \cos y$, so $c(y) = \sin y$
Thus,

$$f = xy^{-3} + \sin y$$

5. curl $\mathbf{F} = \begin{vmatrix} \mathbf{i} & \mathbf{j} & \mathbf{k} \\ \frac{\partial}{\partial x} & \frac{\partial}{\partial y} & \frac{\partial}{\partial z} \\ \frac{1}{y} + \frac{y}{x^2} & -\frac{x}{y^2} + \frac{1}{x} & 0 \end{vmatrix}$

$$= \left(-\frac{1}{y^2} - \frac{1}{x^2} + \frac{1}{y^2} - \frac{1}{x^2}\right)\mathbf{k}$$

$\neq \mathbf{0}$

$\mathbf{F}$ is not conservative.

For Problems 7-12, $x = t$, $y = t^2$ and the endpoints of the curve are $P(1, 1)$ and $Q(2, 4)$; $d\mathbf{R} = (\mathbf{i} + 2t\mathbf{j})\,dt$.

7. scalar potential: $f(x, y) = 2x - 3y$

$$\int_C \mathbf{F} \cdot d\mathbf{R} = f(2, 4) - f(1, 1) = -7$$

9. scalar potential: $f(x, y) = xy^{-3} + \sin y$

$$\begin{aligned}
\int_C \mathbf{F} \cdot d\mathbf{R} &= f(2, 4) - f(1, 1) \\
&= \left(\sin 4 + \frac{1}{32}\right) - (1 + \sin 1) \\
&= \sin 4 - \sin 1 - \frac{31}{32}
\end{aligned}$$

11. $\mathbf{F} = \left(1 + \dfrac{1}{t^2}\right)\mathbf{i} - \left(\dfrac{1}{t^3} - \dfrac{1}{t}\right)\mathbf{j}$

$$\int_C \mathbf{F} \cdot d\mathbf{R} = \int_1^2 \left(\dfrac{1}{t^2} + 1 - \dfrac{2}{t^2} + 2\right)dt$$

$$= \dfrac{5}{2}$$

13. div $\mathbf{F} = 1 + 1 + 1 = 3$

curl $\mathbf{F} = \begin{vmatrix} \mathbf{i} & \mathbf{j} & \mathbf{k} \\ \frac{\partial}{\partial x} & \frac{\partial}{\partial y} & \frac{\partial}{\partial z} \\ x & y & z \end{vmatrix} = \mathbf{0}$

15. div $\mathbf{F} = \nabla \cdot \mathbf{F}$

$$= \left(\dfrac{\partial}{\partial x}\mathbf{i} + \dfrac{\partial}{\partial y}\mathbf{j} + \dfrac{\partial}{\partial z}\mathbf{k}\right) \cdot (xy\mathbf{i} + yz\mathbf{j} + xz\mathbf{k})$$

$$= x + y + z$$

curl $\mathbf{F} = \begin{vmatrix} \mathbf{i} & \mathbf{j} & \mathbf{k} \\ \frac{\partial}{\partial x} & \frac{\partial}{\partial y} & \frac{\partial}{\partial z} \\ xy & yz & xz \end{vmatrix}$

$$= -y\mathbf{i} - z\mathbf{j} - x\mathbf{k}$$

17. div $\mathbf{F} = \nabla \cdot \mathbf{F}$

$$= \left(\dfrac{\partial}{\partial x}\mathbf{i} + \dfrac{\partial}{\partial y}\mathbf{j} + \dfrac{\partial}{\partial z}\mathbf{k}\right) \cdot (ax\mathbf{i} + by\mathbf{j} + \mathbf{k})$$

$$= a + b$$

curl $\mathbf{F} = \nabla \times \mathbf{F}$

$$= \begin{vmatrix} \mathbf{i} & \mathbf{j} & \mathbf{k} \\ \frac{\partial}{\partial x} & \frac{\partial}{\partial y} & \frac{\partial}{\partial z} \\ ax & by & c \end{vmatrix}$$

$$= \mathbf{0}$$

19. div $\mathbf{F} = \nabla \cdot \mathbf{F}$

$$= \left(\dfrac{\partial}{\partial x}\mathbf{i} + \dfrac{\partial}{\partial y}\mathbf{j} + \dfrac{\partial}{\partial z}\mathbf{k}\right)$$

$$\cdot (ax\mathbf{i} + by\mathbf{j} + cz\mathbf{k})$$

$$= a + b + c$$

curl $\mathbf{F} = \nabla \times \mathbf{F}$

$$= \begin{vmatrix} \mathbf{i} & \mathbf{j} & \mathbf{k} \\ \frac{\partial}{\partial x} & \frac{\partial}{\partial y} & \frac{\partial}{\partial z} \\ ax & by & cz \end{vmatrix}$$

$$= \mathbf{0}$$

21. $\mathbf{F} = (x^2 + y^2 + z^2)^{-1/2}(x\mathbf{i} + y\mathbf{j} + z\mathbf{k})$

div $\mathbf{F} = (x^2 + y^2 + z^2)^{-3/2}[(y^2 + z^2)$

$$+ (x^2 + z^2) + (x^2 + y^2)]$$

$$= \dfrac{2(x^2 + y^2 + z^2)}{(x^2 + y^2 + z^2)^{3/2}}$$

$$= \dfrac{2}{\sqrt{x^2 + y^2 + z^2}} = \dfrac{2}{r}$$

curl $\mathbf{F} = \begin{vmatrix} \mathbf{i} & \mathbf{j} & \mathbf{k} \\ \frac{\partial}{\partial x} & \frac{\partial}{\partial y} & \frac{\partial}{\partial z} \\ \frac{x}{r} & \frac{y}{r} & \frac{z}{r} \end{vmatrix}$

$$= \left[-\dfrac{2yz}{2r^3} + \dfrac{2yz}{2r^3}\right]\mathbf{i}$$

$$- \left[-\dfrac{2xz}{2r^3} + \dfrac{2xz}{2r^3}\right]\mathbf{j}$$

$$+ \left[-\dfrac{2xy}{2r^3} + \dfrac{2xy}{2r^3}\right]\mathbf{k}$$

$$= \mathbf{0}$$

For Problems 23-26,

$$\int_C [(3x + 2y)\, dx - (2x + 3y)\, dy]$$

$$\dfrac{\partial}{\partial y}(3x + 2y) = 2;\ \dfrac{\partial}{\partial x}(-2x - 3y) = -2$$

These are not the same, so by the cross-partials test, the field is not conservative.

23. $x = \cos t,\ y = \sin t;\ 0 \le t \le \pi$

$$\int_0^\pi [(3\cos t + 2\sin t)(-\sin t)$$

$$- (2\cos t + 3\sin t)(\cos t)]\, dt$$

$$= \int_0^\pi (-6\cos t \sin t - 2)\, dt$$
$$= -2\pi$$

25. The circular part is the same as Problem 23 and the line segment is
C_3: $x = t - 1$, $y = 0$, $0 \le t \le 2$
$$\oint_C \mathbf{F} \cdot d\mathbf{R} = -2\pi + \int_0^2 3(t-1)\, dt$$
$$= -2\pi + 0$$
$$= -2\pi$$

27. C_1: $y = 0$, $x = t$; $1 \le t \le \pi$
C_2: $x = \pi$, $y = t$; $0 \le t \le \pi$
$$\int_C [(\sin \pi y)dx + (\cos \pi x)dy]$$
$$= \int_1^\pi [0+0]dt + \int_0^\pi [0 + \cos \pi^2]\, dt$$
$$= \pi \cos \pi^2$$

29. We need to evaluate
$$\int_C (y\, dx + x\, dy + dz)$$

using the parameterization of the arc C of the helix $x = t$, $y = 3\sin t$, $z = 2\cos t$, obtained for $0 \le t \le \frac{\pi}{2}$.
$$\int_C (y\, dx + x\, dy + dz)$$
$$= \int_0^{\frac{\pi}{2}} [(3\sin t)(1) + (t)(3\cos t) + (-2\sin t)]\, dt$$
$$= \int_0^{\frac{\pi}{2}} (\sin t + 3t\cos t)\, dt$$
$$= \frac{3\pi}{2} - 2$$

31. $\mathbf{F} = yz\mathbf{i} + xz\mathbf{j} + (xy+2)\mathbf{k}$
C: $x = \tan^{-1} t$, $y = t^2$, $z = -3t$; $0 \le t \le 1$

$$\text{curl } \mathbf{F} = \begin{vmatrix} \mathbf{i} & \mathbf{j} & \mathbf{k} \\ \frac{\partial}{\partial x} & \frac{\partial}{\partial y} & \frac{\partial}{\partial z} \\ yz & xz & xy+2 \end{vmatrix} = \mathbf{0}$$

The line integral is path independent.
$$\frac{\partial f}{\partial x} = yz; \ f = xyz + c(y,z)$$
$$\frac{\partial f}{\partial y} = xz + \frac{\partial c}{\partial y} = xz$$
$$f = xyz + c_1(z)$$
$$\frac{\partial f}{\partial z} = xy + c_1'(z)$$
$$= xy + 2$$
$$c'(z) = 2, \ c_1 = 2z$$
$$f = xyz + 2z$$
The endpoints of the curve are $P(0,0,0)$ and $Q(\frac{\pi}{4}, 1, -3)$.
$$\int_C \mathbf{F} \cdot d\mathbf{R} = f\left(\frac{\pi}{4}, 1, -3\right) - f(0,0,0)$$
$$= -\frac{3\pi}{4} - 6$$

33. $\mathbf{R}_u = \mathbf{i} + \mathbf{j} + \mathbf{k}$; $\mathbf{R}_v = \mathbf{j}$
$$\mathbf{R}_u \times \mathbf{R}_v = \begin{vmatrix} \mathbf{i} & \mathbf{j} & \mathbf{k} \\ 1 & 1 & 1 \\ 0 & 1 & 0 \end{vmatrix} = -\mathbf{i} + \mathbf{k}$$
$$dS = \|\mathbf{R}_u \times \mathbf{R}_v\|\, du\, dv = \sqrt{2}\, du\, dv$$
S: $x = u$, $y = u + v$, $z = u$;
$0 \le u \le 1$, $0 \le v \le 1$
$$\iint_S (3x^2 + y - 2z)\, dS$$
$$= \int_0^1 \int_0^1 (3u^2 + u + v - 2u)\sqrt{2}\, dv\, du$$
$$= \sqrt{2} \int_0^2 \left(3u^2 - u + \frac{1}{2}\right) du$$
$$= \sqrt{2}$$

35. $\mathbf{F} = x\mathbf{i} + x\mathbf{j} - y\mathbf{k}$;
$C: x = t, y = t^2, z = t, 0 \le t \le 1$

$$\int_C (x\,dx + x\,dy - y\,dz) = \int_0^1 \left[t + t(2t) - t^2 \right] dt$$

$$= \int_0^1 \left(t + t^2 \right) dt$$

$$= \frac{5}{6}$$

37. $\mathbf{F} = x^2\mathbf{i} + y\mathbf{j}$
$$\frac{\partial N}{\partial x} = 0 = \frac{\partial M}{\partial y};$$
the integral is path independent
$$f_x = x^2; \quad f = \frac{x^3}{3} + a(y)$$
$$\frac{\partial f}{\partial y} = a'(y) = y; \quad f(x,y) = \frac{x^3}{3} + \frac{y^2}{2}$$
The endpoints of C are $P(0,1)$ and
$Q(0, 1 - 2\pi)$.

$$\int_C (x^2 dx + y\,dy) = f(0, 1 - 2\pi) - f(0, 1)$$

$$= 2\pi(\pi - 1)$$

39. $\mathbf{F} = y\mathbf{i} + x\mathbf{j} - 2\mathbf{k}$

$$\text{curl } \mathbf{F} = \begin{vmatrix} \mathbf{i} & \mathbf{j} & \mathbf{k} \\ \frac{\partial}{\partial x} & \frac{\partial}{\partial y} & \frac{\partial}{\partial z} \\ y & x & -2 \end{vmatrix} = 0$$

The line integral is path independent. Since
the curve C is closed, we have

$$\oint_C (y\,dx + x\,dy - 2\,dz) = 0$$

41. $\mathbf{F} = \left\langle \dfrac{x}{\sqrt{x^2 + y^2}}, \dfrac{-y}{\sqrt{x^2 + y^2}} \right\rangle$
$= \langle \cos t, -\sin t \rangle$
$\mathbf{R} = (a\cos t)\mathbf{i} + (a\sin t)\mathbf{j}$, for $0 \le t \le \frac{\pi}{2}$
$d\mathbf{R} = (-a\sin t\,dt)\mathbf{i} + (a\cos t\,dt)\mathbf{j}$,

$\mathbf{F} \cdot d\mathbf{R} = -a\sin t \cos t - a\sin t \cos t$
$= -2a\sin t \cos t$

$$\int_C \mathbf{F} \cdot d\mathbf{R} = -2a \int_0^{\pi/2} \cos t \sin t\,dt = -a$$

43. $\mathbf{F} = -2y\mathbf{i} + 2x\mathbf{j} + \mathbf{k}$; $\mathbf{N} = \mathbf{k}$; $dS = dA_{xy}$

$$\text{curl } \mathbf{F} = \begin{vmatrix} \mathbf{i} & \mathbf{j} & \mathbf{k} \\ \frac{\partial}{\partial x} & \frac{\partial}{\partial y} & \frac{\partial}{\partial z} \\ -2y & 2x & 1 \end{vmatrix} = 4\mathbf{k}$$

For the disk $A: x^2 + y^2 \le 1$, we use Stokes'
theorem,

$$\oint_C (-2y\,dx + 2x\,dy + dz) = \iint_S (\text{curl } \mathbf{F} \cdot \mathbf{N})\,dS$$

$$= \iint_S 4\,dA$$

$$= 4A$$

$$= 4\pi(1)^2$$

$$= 4\pi$$

45. $I = \displaystyle\iint_S (2x^3\mathbf{i} + y^3\mathbf{j} + z^3\mathbf{k}) \cdot \mathbf{N}\,dS$

$$= \iiint_D \text{div}(2x^3\mathbf{i} + y^3\mathbf{j} + z^3\mathbf{k})\,dV$$

$$= \iiint_D (6x^2 + 3y^2 + 3z^2)\,dV$$

Let $\sqrt{2}x = u, y = v, z = t$, then
$u^2 + v^2 + t^2 = 1$ is a sphere.

$$\frac{\partial(x,y,z)}{\partial(u,v,t)} = \begin{vmatrix} \frac{\partial x}{\partial u} & \frac{\partial x}{\partial v} & \frac{\partial x}{\partial t} \\ \frac{\partial y}{\partial u} & \frac{\partial y}{\partial v} & \frac{\partial y}{\partial t} \\ \frac{\partial z}{\partial u} & \frac{\partial z}{\partial v} & \frac{\partial z}{\partial t} \end{vmatrix}$$

$$= \begin{vmatrix} \frac{1}{\sqrt{2}} & 0 & 0 \\ 0 & 1 & 0 \\ 0 & 0 & 1 \end{vmatrix} = \frac{1}{\sqrt{2}}$$

$$I = \frac{3}{\sqrt{2}} \int\int\int_D (u^2 + v^2 + t^2)\, dV$$

$$= \frac{3}{\sqrt{2}} \int_0^{2\pi} \int_0^{\pi} \int_0^1 \rho^2(\rho^2 \sin\phi)\, d\rho\, d\phi\, d\theta$$

$$= \frac{3}{\sqrt{2}} \int_0^{2\pi} \int_0^{\pi} \frac{\sin\phi}{5}\, d\phi\, d\theta$$

$$= \frac{3}{\sqrt{2}} \int_0^{2\pi} \frac{2}{5}\, d\theta$$

$$= \frac{12\pi}{5\sqrt{2}}$$

$$= \frac{6\sqrt{2}\pi}{5}$$

47. $\nabla\phi = \nabla(2x + 3y) = 2\mathbf{i} + 3\mathbf{j}$;
$f(x, y, z) = ax + by + cz - 1 = 0$

$$\mathbf{N} = \frac{a\mathbf{i} + b\mathbf{j} + c\mathbf{k}}{\sqrt{a^2 + b^2 + c^2}}$$

$$dS = \frac{1}{c}\sqrt{a^2 + b^2 + c^2}\, dA$$

$$\int\int_S \nabla\phi \cdot \mathbf{N}\, dS$$

$$= \int\int_S \frac{(2a + 3b)\sqrt{a^2 + b^2 + c^2}}{c\sqrt{a^2 + b^2 + c^2}}\, dA$$

$$= \frac{2a + 3b}{c}\, (\text{AREA OF TRIANGLE})$$

$$= \frac{2a + 3b}{c}\left(\frac{1}{2}\right)\left(\frac{1}{a}\right)\left(\frac{1}{b}\right)$$

$$= \frac{2a + 3b}{2abc}$$

49. Let $\mathbf{F} = x\mathbf{i} + y\mathbf{j} + z\mathbf{k}$ and let S be the surface of the unit cube $0 \le x \le 1$, $0 \le y \le 1$, $0 \le z \le 1$. Since the conditions of the Divergence theorem are satisfied, we

will use it, because it is the simplest method for this problem.

$$\int\int_S \mathbf{F} \cdot \mathbf{N}\, dS = \int\int\int_R \text{div}\, \mathbf{F}\, dV$$

$$= \int\int\int_R 3\, dV$$

$$= 3$$

since the volume of the cube is equal to 1.

51. $\text{div}\, \mathbf{F} = 0$, so

$$\int\int_S \mathbf{F} \cdot \mathbf{N}\, dS = \int\int\int_D \text{div}\, \mathbf{F}\, dV = 0$$

53. $\text{div}\, \mathbf{F} = yz + xz + xy$

$$\int\int_S \mathbf{F} \cdot \mathbf{N}\, dS$$

$$= \int\int\int_D \text{div}\, \mathbf{F}\, dV - \int\int_{S_1} \mathbf{F} \cdot (-\mathbf{k})\, dS$$

where S_1 is the missing face

$$= \int_0^1 \int_0^1 \int_0^1 (yz + xz + xy)\, dz\, dy\, dx + \int\int_{S_1} 0\, dA$$

$$= \int_0^1 \int_0^1 \left(\frac{y}{2} + \frac{x}{2} + xy\right) dy\, dx + 0$$

$$= \int_0^1 \left(\frac{1}{4} + \frac{x}{2} + \frac{x}{2}\right) dx$$

$$= \frac{3}{4}$$

55. $\dfrac{\partial}{\partial x}(cx^2 + 4y) = \dfrac{\partial}{\partial y}(\sqrt{x} + 3xy)$

$$2cx = 3x$$

$$c = \frac{3}{2} \qquad (x \ne 0)$$

57. $\mathbf{F} = e^{yz/x}\left[\left(\dfrac{cyz}{x^2}\right)\mathbf{i} + \left(\dfrac{z}{x}\right)\mathbf{j} + \left(\dfrac{y}{x}\right)\mathbf{k}\right]$

$\text{curl } \mathbf{F} = \left[e^{yz/x} \left(\dfrac{1}{x} \right) + \dfrac{zy}{x^2} e^{yz/x} - \dfrac{1}{x} e^{yz/x} - \dfrac{yz}{x^2} e^{yz/x} \right] \mathbf{i}$
$+ \left[\dfrac{cy}{x^2} e^{yz/x} + \dfrac{cyz}{x^2} \left(\dfrac{y}{x} \right) e^{yz/x} - \left(\dfrac{-y}{x^2} \right) e^{yz/x} - \dfrac{y}{x} e^{yz/x} \left(\dfrac{-yz}{x^2} \right) \right] \mathbf{j}$
$+ \left[e^{yz/x} \left(\dfrac{-z}{x^2} \right) + \dfrac{z}{x} e^{yz/x} \left(\dfrac{-yz}{x^2} \right) - e^{yz/x} \left(\dfrac{cz}{x^2} \right) - \dfrac{cyz}{x^2} e^{yz/x} \left(\dfrac{z}{x} \right) \right] \mathbf{k}$
$= \mathbf{0} \text{ if } c = -1$

59. $\text{curl } \mathbf{F} = (1 - 1)\mathbf{i} + (0 - 0)\mathbf{j}$
$$+ \left[\left(2y - \dfrac{e^{x/y}}{xy} + \dfrac{e^{x/y}}{x^2} \right) \right.$$
$$\left. - \left(\dfrac{-e^{x/y}}{xy} + \dfrac{e^{x/y}}{x^2} + 2y \right) \right] \mathbf{k}$$
$$= \mathbf{0}$$
$\mathbf{F}$ is conservative.

61. If $\mathbf{F}$ is conservative, then
$$\mathbf{F} = \nabla \phi = \langle \phi_x, \phi_y, \phi_z \rangle$$

so
$$M = \phi_x, \, N = \phi_y, \, P = \phi_z$$

Thus, using the equality of mixed partials, we have
$$\dfrac{\partial M}{\partial y} = \phi_{xy} = \phi_{yx} = \dfrac{\partial N}{\partial x}$$
$$\dfrac{\partial N}{\partial z} = \phi_{yz} = \phi_{zy} = \dfrac{\partial P}{\partial y}$$
$$\dfrac{\partial P}{\partial x} = \phi_{zx} = \phi_{xz} = \dfrac{\partial M}{\partial z}$$

Thus,
$$\dfrac{\partial P}{\partial y} = \dfrac{\partial N}{\partial z}; \dfrac{\partial M}{\partial z} = \dfrac{\partial P}{\partial x}; \dfrac{\partial N}{\partial x} = \dfrac{\partial M}{\partial y}$$

63. $\mathbf{F}$ is conservative since

$$\dfrac{\partial}{\partial y} [f(x) + y] = 1 = \dfrac{\partial}{\partial x} [g(y) + x]$$
If h is a scalar potential for $\mathbf{F}$,
$$\dfrac{\partial h}{\partial x} = f(x) + y$$
$$h = xy + \int f(x)\, dx + c(y)$$
$$\dfrac{\partial h}{\partial y} = x + c'(y)$$
$$= g(y) + x$$
$$c'(y) = g(y); \, c(y) = \int g(y)\, dy$$
$$h(x, y) = xy + \int f(x)\, dx + \int g(y)\, dy$$

65. $\text{curl } \mathbf{F} = \begin{vmatrix} \mathbf{i} & \mathbf{j} & \mathbf{k} \\ \dfrac{\partial}{\partial x} & \dfrac{\partial}{\partial y} & \dfrac{\partial}{\partial z} \\ 2x & -x - z & y - x \end{vmatrix} \neq \mathbf{0}$
The integral depends on the path.
$x = t^2, y = t^2 - t, z = 3$
$$W = \int_C [2x\, dx - (x + z)\, dy + (y - x)\, dz]$$
$$= \int_0^1 \left[(2t^2)(2t) - (t^2 + 3)(2t - 1) \right] dt$$
$$= \dfrac{5}{6}$$

67. $\dfrac{\partial}{\partial y} \left(\dfrac{1}{x + y} \right) = \dfrac{-1}{(x + y)^2}$
$$= \dfrac{\partial}{\partial x} \left(\dfrac{1}{x + y} \right)$$
so $\mathbf{F}$ is conservative in any region of the plane where $x + y \neq 0$.
$$\dfrac{\partial f}{\partial x} = \dfrac{1}{x + y}$$
$$f = \ln|x + y| + g(y)$$

$$\frac{\partial f}{\partial y} = \frac{1}{x+y} + g'(y)$$

$$= \frac{1}{x+y}$$

$g'(y) = 0$, so $g = 0$

$f(x, y) = \ln|x + y|$

$$W = \int_C \mathbf{F} \cdot d\mathbf{R}$$

$$= f(c, d) - f(a, b)$$

$$= \ln\left|\frac{c+d}{a+b}\right|$$

69. Let $\mathbf{F} = f_1\mathbf{i} + f_2\mathbf{j} + f_3\mathbf{k}$

$\mathbf{G} = g_1\mathbf{i} + g_2\mathbf{j} + g_3\mathbf{k}$

$$\mathbf{F} \times \mathbf{G} = \begin{vmatrix} \mathbf{i} & \mathbf{j} & \mathbf{k} \\ f_1 & f_2 & f_3 \\ g_1 & g_2 & g_3 \end{vmatrix}$$

$$= [f_2 g_3 - f_3 g_2]\mathbf{i} - [f_1 g_3 - f_3 g_1]\mathbf{j} + [f_1 g_2 - f_2 g_1]\mathbf{k}$$

$\text{div}(\mathbf{F} \times \mathbf{G}) = f_2(g_3)_x - f_3(g_2)_x - f_1(g_3)_y$
$+ f_3(g_1)_y + f_1(g_2)_z - f_2(g_1)_z + (f_2)_x g_3$
$- (f_3)_x g_2 - (f_1)_y g_3 + (f_3)_y g_1 + (f_1)_z g_2 - (f_2)_z g_1$

$\mathbf{G} \cdot \text{curl } \mathbf{F} - \mathbf{F} \cdot \text{curl } \mathbf{G}$

$$= \mathbf{G} \cdot \begin{vmatrix} \mathbf{i} & \mathbf{j} & \mathbf{k} \\ \frac{\partial}{\partial x} & \frac{\partial}{\partial y} & \frac{\partial}{\partial z} \\ f_1 & f_2 & f_3 \end{vmatrix} - \mathbf{F} \cdot \begin{vmatrix} \mathbf{i} & \mathbf{j} & \mathbf{k} \\ \frac{\partial}{\partial x} & \frac{\partial}{\partial y} & \frac{\partial}{\partial z} \\ g_1 & g_2 & g_3 \end{vmatrix}$$

$= \mathbf{G} \cdot \{[(f_3)_y - (f_2)_z]\mathbf{i} - [(f_3)_x - (f_1)_z]\mathbf{j}$
$+ [(f_2)_x - (f_1)_y]\mathbf{k}\} - \mathbf{F} \cdot \{[(g_3)_y - (g_2)_z]\mathbf{i}$
$- [(g_3)_x - (g_1)_z]\mathbf{j} + [(g_2)_x - (g_1)_y]\mathbf{k}\}$

$= (f_3)_y g_1 - (f_2)_z g_1 - (f_3)_x g_2 + (f_1)_z (g_2)$
$+ (f_2)_x g_3 - (f_1)_y g_3 - f_1(g_3)_y + f_1(g_2)_z$
$+ f_2(g_3)_x - f_2(g_1)_z - f_3(g_2)_x + f_3(g_1)_y$

Thus,

$$\text{div}(\mathbf{F} \times \mathbf{G}) = \mathbf{G} \cdot \text{curl } \mathbf{F} - \mathbf{F} \cdot \text{curl } \mathbf{G}$$

71. Since $\mathbf{F}$ and $\mathbf{G}$ are conservative, curl $\mathbf{F} = \mathbf{0}$ and curl $\mathbf{G} = \mathbf{0}$. By Problem 69,

$$\text{div}(\mathbf{F} \times \mathbf{G}) = \mathbf{G} \cdot \text{curl } \mathbf{F} - \mathbf{F} \cdot \text{curl } \mathbf{G}$$
$$= 0 - 0$$
$$= 0$$

so that $\mathbf{F} \times \mathbf{G}$ is incompressible. Thus, we see

$$\text{div } \mathbf{F} = \text{div}(\text{curl } \mathbf{G}) = 0$$

so $\mathbf{F}$ is incompressible (See Problem 53 of Section 13.1.)

73. Let $\mathbf{F} = \langle f_1, f_2, f_3 \rangle$ and $\mathbf{A} = \langle a_1, a_2, a_3 \rangle$. Then,

$$\mathbf{A} \times \mathbf{F} = \begin{vmatrix} \mathbf{i} & \mathbf{j} & \mathbf{k} \\ a_1 & a_2 & a_3 \\ f_1 & f_2 & f_3 \end{vmatrix}$$

$$= (a_2 f_3 - a_3 f_2)\mathbf{i} + (a_3 f_1 - a_1 f_3)\mathbf{j}$$
$$+ (a_1 f_2 - a_2 f_1)\mathbf{k}$$

and

$\text{div}(\mathbf{A} \times \mathbf{F})$

$= a_2(f_3)_x - a_3(f_2)_x + a_3(f_1)_y - a_1(f_3)_y$
$\quad + a_1(f_2)_z - a_2(f_1)_z$

$= -a_1[-(f_2)_z + (f_3)_y] - a_2[-(f_3)_x + (f_1)_z]$
$\quad - a_3[-(f_1)_y + (f_2)_x]$

$= -\mathbf{A} \cdot \text{curl } \mathbf{F}$

75. $\displaystyle\oint_C \left(\frac{-y}{x^2} dx + \frac{1}{x} dy\right)$

$$= \iint_D \left[\frac{\partial(x^{-1})}{\partial x} - \frac{\partial\left(-\frac{y}{x^2}\right)}{\partial y}\right] dy\, dx$$

$$= \iint_D \left[-\frac{1}{x^2} + \frac{1}{x^2}\right] dy\, dx$$

$$= 0$$

$$\oint_C \left(\frac{-y}{x^2} dx + \frac{1}{x} dy\right)$$

$$= \iint_D \left[\frac{\partial(x^{-1})}{\partial x} - \frac{\partial\left(-\frac{y}{x^2}\right)}{\partial y} \right] dy\, dx$$

$$= \iint_D \left[-\frac{1}{x^2} + \frac{1}{x^2} \right] dy\, dx$$

$$= 0$$

77. D is any region that does not intersect the y-axis since

$$\frac{\partial}{\partial x}\left(\frac{-y - x^2 y}{x^2} \right) = \frac{2y}{x^3} = \frac{\partial}{\partial y}\left(\frac{1 + y^2}{x^3} \right)$$

$$f_x = (1 + y^2)x^{-3}$$

$$f = -\frac{1 + y^2}{2x^2} + a(y)$$

$$f_y = -\frac{2y}{2x^2} + a'(y)$$

$$= -\frac{y}{x^2} - y$$

$$a(y) = -\frac{y^2}{2}$$

$$f = -\frac{1 + y^2}{2x^2} - \frac{y^2}{2}$$

We have path independence and we note that C should not intersect the y-axis.

$$\int_C \mathbf{F} \cdot d\mathbf{R} = f(3, 4) - f(1, 1)$$

$$= \left(-\frac{161}{18} \right) - \left(\frac{-3}{2} \right)$$

$$= -\frac{67}{9}$$

79. The vertices of the triangle are $(1, 2)$, $\left(2, \frac{3}{2}\right)$, and $(2, 4)$. Evaluating the line integral directly:

$$C_1: x = t, \ y = \frac{5 - t}{2}; \ 1 \le t \le 2$$

$$I_1 = \int_1^2 \left[\frac{2}{5 - t} + \frac{1}{t}\left(-\frac{1}{2} \right) \right] dt$$

$$= \ln\frac{16}{9} - \frac{1}{2}\ln 2$$

$$C_2: x = 2, \ y = t; \ \frac{3}{2} \le t \le 4$$

$$I_2 = \int_{3/2}^4 \frac{dt}{2} = \frac{5}{4}$$

$$C_3: x = 2 - t, \ y = 2(2 - t); \ 0 \le t \le 1$$

$$I_3 = \int_0^1 \left[\frac{-1}{4 - 2t} + \frac{-2}{2 - t} \right] dt$$

$$= -\frac{5}{2}\ln 2$$

$$\int_C \left(\frac{dx}{y} + \frac{dy}{x} \right) = I_1 + I_2 + I_3$$

$$= \frac{5}{4} + \ln\frac{2}{9}$$

Using Green's theorem:

$$\int_C \left(\frac{dx}{y} + \frac{dy}{x} \right)$$

$$= \iint_D \left(-\frac{1}{x^2} + \frac{1}{y^2} \right) dA$$

$$= \int_1^2 \int_{(5-x)/2}^{2x} \left(-\frac{1}{x^2} + \frac{1}{y^2} \right) dy\, dx$$

$$= \int_1^2 \left(-\frac{2}{x - 5} + \frac{5}{2x^2} - \frac{3}{x} \right) dx$$

$$= \frac{5}{4} + \ln\frac{2}{9}$$

81. $\nabla^2 w = 0$;

$$\nabla \cdot (w \nabla w)$$

$$= \left(\frac{\partial}{\partial x}\mathbf{i} + \frac{\partial}{\partial y}\mathbf{j} + \frac{\partial}{\partial z}\mathbf{k} \right) \cdot [ww_x\mathbf{i} + ww_y\mathbf{j} + ww_z\mathbf{k}]$$

$$= (ww_x)_x + (ww_y)_y + (ww_z)_z$$

$$= [ww_{xx} + ww_{yy} + ww_{zz}]$$
$$+ [(w_x)^2 + (w_y)^2 + (w_z)^2]$$
$$= 0 + \|\nabla w\|^2$$
$$= \|\nabla w\|^2$$

83. $\mathbf{R}(t) = x(t)\mathbf{i} + y(t)\mathbf{j} + z(t)\mathbf{k}$;
$$\frac{d\mathbf{R}}{dt} = \frac{dx}{dt}\mathbf{i} + \frac{dy}{dt}\mathbf{j} + \frac{dz}{dt}\mathbf{k}$$
Since $\mathbf{F}$ is always perpendicular to path C,
we must have $\mathbf{F} \cdot \dfrac{d\mathbf{R}}{dt} = 0$. Thus,
$$W = \int_C \mathbf{F} \cdot d\mathbf{R} = 0$$

85. $\mathbf{F} = -m\omega^2 \mathbf{R}$ where $m = 10{,}000/9.8$ and
$\mathbf{R} = 7{,}000(x\mathbf{i} + y\mathbf{j} + z\mathbf{k})$. The semicircular
path C may be parameterized by
$x = 7{,}000\cos t$, $y = 7{,}000\sin t$, $z = 0$ for
$0 \le t \le \pi$.

$$W = \int_C \mathbf{F} \cdot d\mathbf{R}$$
$$= -\frac{10{,}000}{9.8}(7{,}000)\omega^2 \int_C (x\,dx + y\,dy + z\,dz)$$
$$= \frac{-10{,}000(7{,}000)\omega^2}{9.8} \int_0^\pi 7{,}000[\cos t(-\sin t)$$
$$+ \sin t(\cos t)]\,dt$$
$$= 0$$

87. We show $\mathbf{F}$ is conservative by finding f so
that $\mathbf{F} = \nabla f$:
$$\frac{\partial f}{\partial x} = 2xyz$$
$$f = x^2 yz + c(y, z)$$
$$\frac{\partial f}{\partial y} = x^2 z + \frac{\partial c}{\partial y} = x^2 z - \frac{1}{z}\tan^{-1}\frac{y}{z}$$
$$\frac{\partial c}{\partial y} = -\frac{1}{z}\tan^{-1}\frac{y}{z}$$
$$c(y, z) = \frac{1}{2}\ln(y^2 + z^2) - \frac{y}{z}\tan^{-1}\frac{y}{z} + c_1(z)$$

$$f = x^2 yz + \frac{1}{2}\ln(y^2 + z^2) - \frac{y}{z}\tan^{-1}\frac{y}{z} + c_1(z)$$
$$\frac{\partial f}{\partial z} = x^2 y + \frac{y}{z^2}\tan^{-1}\frac{y}{z} + \frac{1}{z} + c_1'(z)$$
$$= x^2 y + \frac{y}{z^2}\tan^{-1}\frac{y}{z}$$
$$c_1'(z) = -\frac{1}{z}; \; c_1(z) = -\ln|z|$$
$$f(x, y, z) = x^2 yz + \frac{1}{2}\ln(y^2 + z^2)$$
$$- \frac{y}{z}\tan^{-1}\frac{y}{z} - \ln|z|$$
The endpoints of the curve C are $P(0, 1, 1)$
and $Q(1, 0, 2)$.

$$W = \int_C \mathbf{F} \cdot d\mathbf{R}$$
$$= f(1, 0, 2) - f(0, 1, 1)$$
$$= \left[0 + \frac{1}{2}\ln 4 - 0 - \ln 2 \right]$$
$$- \left[0 + \frac{1}{2}\ln 2 - \frac{\pi}{4} - 0 \right]$$
$$= \frac{\pi}{4} - \frac{1}{2}\ln 2$$

89. Since for $x \ne 0$, $y \ne 0$,
$$\frac{\partial}{\partial x}\left(\frac{x}{x^2 + y^2} \right) = \frac{-x^2 + y^2}{(x^2 + y^2)^2}$$
$$\frac{\partial}{\partial y}\left(\frac{-y}{x^2 + y^2} \right) = \frac{-x^2 + y^2}{(x^2 + y^2)^2}$$
Thus, the vector field

$$\mathbf{F} = \frac{1}{x^2 + y^2}(-y\mathbf{i} + x\mathbf{j})$$

is conservative in any region that does not
contain the origin. Since the limaçon C
does contain $(0, 0)$, the fundamental theorem
on line integrals does not apply. Let C_1 be a
circle centered at the origin with radius a so

small that C_1 is contained entirely within C. Then $\mathbf{F}$ is conservative in the region between C and C_1, and Green's theorem for multiply-connected regions tells us that

$$\oint_C \mathbf{F} \cdot d\mathbf{R} = \oint_{C_1} \mathbf{F} \cdot d\mathbf{R}$$

if C_1 is oriented counterclockwise. Using the parameterization $C_1: x = a\cos\theta$, $y = a\sin\theta$ for $0 \le \theta \le 2\pi$, we obtain

$$\oint_C \frac{-y\,dx + x\,dy}{x^2 + y^2} = \oint_{C_1} \frac{-y\,dx + x\,dy}{x^2 + y^2}$$
$$= \int_0^{2\pi} \frac{(-a\sin\theta)(-a\sin\theta) + (a\cos\theta)(a\cos\theta)}{a^2\cos^2\theta + a^2\sin^2\theta}\,d\theta$$
$$= 2\pi$$

91. The normal derivative $\partial g/\partial n$ is defined as the directional derivative in the normal direction, that is

$$\frac{\partial g}{\partial n} = \nabla g \cdot \mathbf{N}$$

Applying the divergence theorem to $\mathbf{F} = f\nabla g$, and using properties of the gradient, we obtain

$$\iint_S f\frac{\partial g}{\partial n}\,dS = \iint_S f\,(\nabla g \cdot \mathbf{N})\,dS$$
$$= \iiint_D \operatorname{div}(f\nabla g)\,dV$$
$$= \iiint_D \nabla \cdot (f\nabla g)\,dV$$
$$= \iiint_D [f\nabla^2 g + (\nabla f)\cdot(\nabla g)]\,dV$$
$$= \iiint_D \nabla f \cdot \nabla g\,dV$$

Since g is harmonic $\nabla^2 g = 0$.

Similarly, we have

$$\iint_S g\frac{\partial f}{\partial n}\,dS = \iiint_D \nabla f \cdot \nabla g\,dV$$

so

$$\iint_S f\frac{\partial g}{\partial n}\,dS = \iiint_D (\nabla f)\cdot(\nabla g)\,dV$$
$$= \iint_S g\frac{\partial f}{\partial n}\,dS$$

Therefore,

$$\iint_S f\frac{\partial f}{\partial n}\,dS = \iiint_D (\nabla f)\cdot(\nabla f)\,dV$$
$$= \iiint_D \|\nabla f\|^2\,dV$$

93. $\mathbf{R}_u = (a + b\cos v)(-\sin u)\mathbf{i}$
$\qquad + (a + b\cos v)\cos u\mathbf{j}$
$\mathbf{R}_v = -b\sin v\cos u\mathbf{i} - b\sin v\sin u\mathbf{j}$
$\qquad + b\cos v\mathbf{k}$
$\mathbf{R}_u \times \mathbf{R}_v = b(a + b\cos v)\cos u\cos v\mathbf{i}$
$\qquad + b(a + b\cos v)\sin u\cos v\mathbf{j}$
$\qquad + b(a + b\cos v)\sin v\mathbf{k}$
$\|\mathbf{R}_u \times \mathbf{R}_v\| = b(a + b\cos v)$
Thus,

$$\iint_S dS = \iint \|\mathbf{R}_u \times \mathbf{R}_v\|\,dv\,du$$
$$= \int_0^{2\pi}\int_0^{2\pi} b(a + b\cos v)\,dv\,du$$
$$= \int_0^{2\pi} 2ab\pi\,du$$
$$= 4\pi^2 ab$$

95. Only the volume of the body is important, so it does not matter which corner is removed.

$$\iint\limits_{S} \mathbf{F} \cdot \mathbf{N} \, dS = \iiint\limits_{D} \operatorname{div}(x\mathbf{i} + y\mathbf{j} + z\mathbf{k}) \, dV$$

$$= 3[(\text{VOLUME OF CUBE}) - (\text{VOLUME OF CORNER})]$$
$$= 3(2)^3 - 3(1)^3$$
$$= 21$$

97. By symmetry, $\bar{x} = \bar{y} = 0$. To find $\bar{z}$, we first compute the surface area of the cone.

$$z = \sqrt{x^2 + y^2}, z_x = \frac{x}{\sqrt{x^2 + y^2}}, \, z_y = \frac{y}{\sqrt{x^2 + y^2}}$$

$$dS = \sqrt{z_x^2 + z_y^2 + 1} \, dA$$

$$= \sqrt{\left(\frac{x}{\sqrt{x^2 + y^2}}\right)^2 + \left(\frac{y}{\sqrt{x^2 + y^2}}\right)^2 + 1} \, dA$$

$$= \sqrt{2} \, dA$$

The projected region on the xy-plane is the disk $x^2 + y^2 \le 9$. Thus,

$$S = \int_0^{2\pi} \int_0^3 \sqrt{2} \, r \, dr \, d\theta = 9\sqrt{2}\,\pi$$

$$\bar{z} = \frac{1}{S} \iint\limits_{S} z \, dS$$

$$= \frac{1}{9\sqrt{2}\pi} \iint\limits_{\text{disk } x^2 + y^2 \le 9} \sqrt{x^2 + y^2} \, \sqrt{2} \, dy \, dx$$

$$= \frac{1}{9\sqrt{2}\pi} \int_0^{2\pi} \int_0^3 r\sqrt{2}(r \, dr \, d\theta)$$

$$= \frac{1}{9\sqrt{2}\pi} \int_0^{2\pi} 9\sqrt{2} \, d\theta$$

$$= 2$$

The centroid of the conical shell is $(0, 0, 2)$.

99. This is Putnam Problem 6i in the morning session of 1948. Let $\rho = \rho(s)$ be the parametric vector equation of the given curve, where s is the arc length. We assume that ρ is a periodic function of class C^2 (that is, f is twice continuously differentiable) and period L, the length of the curve, so that ρ describes the curve once as s varies from 0 to L. Then $d\rho/ds = \mathbf{T} = \mathbf{T}(s)$ is the tangent vector to the curve at $\rho(s)$, and

$$\frac{d\mathbf{T}}{ds} = \kappa \mathbf{N}$$

where the concave side of the curve and $\kappa = \kappa(s)$ is the curvature. The radius of curvature is $r = 1/\kappa$. It is given that the force on an element ds of the curve has magnitude ds/r and direction $-\mathbf{N}$; that is,

$$d\mathbf{F} = -\frac{\mathbf{N} \, ds}{r}$$

so that $d\mathbf{F} = -d\mathbf{T}$. The condition for equilibrium is that the total force should be zero and that the total moment of the force with respect to some point should be zero. Thus, we must show

$$\oint_C d\mathbf{F} = \mathbf{0} \text{ and } \oint_C \rho \times d\mathbf{F} = \mathbf{0}$$

Now

$$\oint_C d\mathbf{F} = -\oint_C d\mathbf{T} = -\mathbf{T}(s)\big|_0^L = \mathbf{0}$$

For the second requirement, note that

$$\frac{d}{ds}(\rho \times \mathbf{T}) = \mathbf{T} \times \mathbf{T} + \rho \times \frac{d\mathbf{T}}{ds}$$

Hence

$$\oint_C \rho \times d\mathbf{F} = -\oint_C \rho \times d\mathbf{T}$$
$$= -\rho \times \mathbf{T}\big|_0^L$$
$$= \mathbf{0}$$

Cumulative Review
Chapters 11-13, page 1107

1. *Answers vary.* Calculus is the study of dynamic processes (rather than static). It is the study of infinitesimals, the behavior of functions at or near a point. There are three fundamental ideas of calculus; the notions of limits, derivatives (the limit of difference quotients), and integrals (the sum of infinitesimal quantities). Calculus is also the study of transformations of reference frames (coordinate systems), and motion with respect to reference frames (vector analysis).

3. A **vector-valued function** (or, simply, a **vector function**) $\mathbf{F}$ of a real variable with *domain D* assigns to each number t in the set D a unique vector $\mathbf{F}(t)$. The set of all vectors $\mathbf{v}$ of the form $\mathbf{v} = \mathbf{F}(t)$ for t in D is the *range* of $\mathbf{F}$. That is,

$\mathbf{F}(t) = f_1(t)\mathbf{i} + f_2(t)\mathbf{j}$ *Plane* $(\mathbb{R}^2)$
$\mathbf{F}(t) = f_1(t)\mathbf{i} + f_2(t)\mathbf{j} + f_3(t)\mathbf{k}$ *Three-Space* $(\mathbb{R}^3)$

where f_1, f_2, and f_3 are real-valued (**scalar-valued**) functions of the real number t defined on the domain set D. In this context, f_1, f_2, and f_3 are called the **components** of $\mathbf{F}$. A vector function may also be denoted by $\mathbf{F}(t) = \langle f_1(t), f_2(t)\rangle$ or $\mathbf{F}(t) = \langle f_1(t), f_2(t), f_3(t)\rangle$.

5. Answers vary; the first integral is a notation representing the following limit: If f is defined on the closed interval $[a, b]$ we say f is integrable on $[a, b]$ if

$$I = \lim_{\|P\|\to 0} \sum_{k=1}^{n} f(x_k^*)\Delta x_k$$

exists. This limit, if it exists, is called the definite integral of f from a to b. The definite integral is denoted by

$$I = \int_a^b f(x)\, dx$$

The second integral is a notation for the following limit: If $f(x, y, z)$ is defined on the smooth curve C with parametric equations $x = x(t)$, $y = y(t)$, $z = z(t)$, then the **line integral** of f over C is given by

$$\int_C f(x, y, z)\, ds = \lim_{\|\Delta s\|\to 0} \sum_{k=1}^{n} f(x_k^*, y_k^*, z_k^*)\Delta s_k$$

provided that this limit exists. If C is a closed curve, we sometimes indicate the line integral of f around C by $\oint_C f\, ds$.

The first integral is a two-dimensional measure whereas the second is a three-dimensional measure.

Integral	Line integral
$\int_a^b f(x)\, dx$	$\int_C f(x, y, z)\, ds$
Subdivision on the x-axis	Subdivisions on a curve C in space

7. a. Del operator: $\nabla = \dfrac{\partial}{\partial x}\mathbf{i} + \dfrac{\partial}{\partial y}\mathbf{j} + \dfrac{\partial}{\partial z}\mathbf{k}$

   b. Gradient:
   $$\nabla f = \dfrac{\partial f}{\partial x}\mathbf{i} + \dfrac{\partial f}{\partial y}\mathbf{j} + \dfrac{\partial f}{\partial z}\mathbf{k} = f_x\mathbf{i} + f_y\mathbf{j} + f_z\mathbf{k}$$

   c. Laplacian:
   $$\nabla^2 f = \dfrac{\partial^2 f}{\partial x^2} + \dfrac{\partial^2 f}{\partial y^2} + \dfrac{\partial^2 f}{\partial z^2} = f_{xx} + f_{yy} + f_{zz}$$

9. a. $\text{div } \mathbf{F} = \dfrac{\partial u}{\partial x} + \dfrac{\partial v}{\partial y} + \dfrac{\partial w}{\partial z}$
   $$= \nabla \cdot \mathbf{F}$$

   b. $\text{curl } \mathbf{F} = \begin{vmatrix} \mathbf{i} & \mathbf{j} & \mathbf{k} \\ \dfrac{\partial}{\partial x} & \dfrac{\partial}{\partial y} & \dfrac{\partial}{\partial z} \\ u & v & w \end{vmatrix}$
   $$= \nabla \times \mathbf{F}$$

11. $f(x, y) = 2x^2 + xy - 5y^3$
   $f_x = 4x + y$
   $f_y = x - 15y^2$
   $f_{xy} = 1$

13. $f(x, y) = \dfrac{x^2 - y^2}{x - y}$
   $f_x = 1;\ f_y = 1;\ f_{xy} = 0$

15. $f(x, y) = e^{x+y} = e^x e^y$
   $f_x = e^x e^y;\ f_y = e^x e^y;\ f_{xy} = e^x e^y;$
   Notice $f_x = f_y = f_{xy} = f = e^{x+y}$.

17. $f(x, y) = x^2 y^3 + x^3 y^2$
   $f_x = 2xy^3 + 3x^2 y^2 \qquad f_y = 3x^2 y^2 + 2x^3 y$
   $f_{xx} = 2y^3 + 6xy^2 \qquad f_{yy} = 6x^2 y + 2x^3$
   $f_{xy} = 6xy^2 + 6x^2 y$
   $f_{xx} - f_{xy} + f_{yy} = 2x^3 + 2y^3$
   $f_x - f_{xy} + f_{yy} = 2x^3 + 2y^3$

19. $\displaystyle \int_0^1 \int_x^{3x} e^{y-2x} dy\, dx = \int_0^1 (e^x - e^{-x})\, dx$
   $$= e + e^{-1} - 2$$

21. $\displaystyle \int_0^1 \int_0^z \int_y^{y-z} (x + y + z)\, dx\, dy\, dz$
   $$= \int_0^1 \int_0^z \left[-\frac{1}{2}z(4y + z)\, dy\, dz \right]$$
   $$= \int_0^1 \left(-\frac{3}{2}z^3 \right) dz$$
   $$= -\frac{3}{8}$$

23. $\displaystyle \iint_R e^{x+y}\, dA = \int_0^1 \int_0^1 e^y e^x\, dx\, dy$
   $$= (e - 1) \int_0^1 e^y\, dy$$
   $$= (e - 1)^2$$

25. $\displaystyle \iint_R \sin(x + y)\, dA$
   $$= \int_0^{\pi/2} \int_0^{\pi/4} \sin(x + y)\, dy\, dx$$
   $$= -\int_0^{\pi/2} \left[\cos\left(x + \frac{\pi}{4}\right) - \cos x \right] dx$$
   $$= 1$$

27. $\displaystyle \int_0^2 \int_0^{\sqrt{4-y^2}} \frac{1}{\sqrt{9 - x^2 - y^2}}\, dx\, dy$
   $$= \int_0^{\pi/2} \int_0^2 \frac{1}{\sqrt{9 - r^2}} r\, dr\, d\theta$$
   $$= \int_0^{\pi/2} \left[-(9 - r^2)^{1/2} \right]_0^2 d\theta$$
   $$= \int_0^{\pi/2} \left[-\sqrt{9 - 4} + \sqrt{9} \right] d\theta$$
   $$= \left(3 - \sqrt{5} \right)\frac{\pi}{2}$$

Page 581

29.

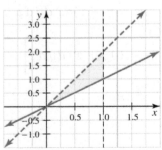

a. $\displaystyle\int_0^1 \int_x^{2x} e^{y-x}\, dy\, dx = \int_0^1 (e^x - 1)\, dx$

$$= e - 2$$

b. $\displaystyle\int_0^1 \int_{y/2}^{y} e^{y-x}\, dx\, dy + \int_1^2 \int_{y/2}^{y} e^{y-x}\, dx\, dy$

$$= -\int_0^1 (1 - e^{y/2})\, dy - \int_1^2 (e^{y-1} - e^{y/2})\, dy$$

$$= 2e^{1/2} - 3 + e - 2e^{1/2} + 1$$

$$= e - 2$$

31. $\mathbf{F} = \langle y^2 z, 2xyz, xy^2 \rangle$

$$\operatorname{curl} \mathbf{F} = \begin{vmatrix} \mathbf{i} & \mathbf{j} & \mathbf{k} \\ \frac{\partial}{\partial x} & \frac{\partial}{\partial y} & \frac{\partial}{\partial z} \\ y^2 z & 2xyz & xy^2 \end{vmatrix} = \mathbf{0}$$

Line integral is independent of path.

Choose $x = t, y = t, z = t$ for $0 \le t \le 1$.

$$\int_C (y^2 z\, dx + 2xyz\, dy + xy^2\, dz)$$

$$= \int_0^1 (t^3 + 2t^3 + t^3)\, dt$$

$$= 4 \int_0^1 t^3\, dt$$

$$= 1$$

33. $\displaystyle\oint_C \mathbf{F} \cdot d\mathbf{R}$, where $\mathbf{F} = yz\mathbf{i} - x\mathbf{k}$,

$\operatorname{curl} \mathbf{F} \neq \mathbf{0}$ so the line integral is path dependent. Parameterizing the three edges of C, we obtain

C_1: $z = 1$, $x = y = t$; $0 \le t \le 1$

C_2: $z = 1, z = 1 - t, x = 1$; $0 \le t \le 1$

C_3: $z = 1, y = 0, x = 1 - t$; $0 \le t \le 1$

Thus,

$$\oint_C \mathbf{F} \cdot d\mathbf{R} = \int_0^1 [t(1)(1) - t(0)]\, dt$$

$$+ \int_0^1 [(0)(1)(-1) - (1 - t)(0)]\, dt$$

$$= \int_0^1 t\, dt = \frac{1}{2}$$

You can check this result by using Stokes' theorem.

35. The integral is not independent of path since

$$\frac{\partial \left(\frac{x}{y} + \frac{3y}{x^2 + y^2} \right)}{\partial y} \neq \frac{\partial (e^{-xy})}{\partial x}$$

so the integral depends on path. Let C be any smooth curve from $(0, 1)$ to $(1, 1)$. We will consider the straight-line segment between $(0, 1)$ to $(1, 1)$, that is, $y = 1$, $0 \le x \le 1$.

$$\int_C \left\{ \left[\frac{x}{y} + \frac{3y}{x^2 + y^2} \right] dx + e^{-xy}\, dy \right\}$$

$$= \int_0^1 \left\{ \left[x + \frac{3}{x^2 + 1} \right] dx \right\}$$

$$= \frac{1}{2} + \frac{3\pi}{4}$$

Integral is not independent of path, so answers vary with your choice of path.

37. $\mathbf{F}$ is conservative if $\mathbf{F} = \nabla f$ where f is continuously differentiable.

39. $f(x, y) = x + 2y$, subject to
$g(x, y) = x^2 + y^2$.
$f_x = 1$, $f_y = 2$, $g_x = 2x$, $g_y = 2y$
Solve the system

$$\begin{cases} 1 = \lambda(2x) \\ 2 = \lambda(2y) \\ x^2 + y^2 = 1 \end{cases}$$

Solving this system, we find

$(x, y) = \left(\pm \frac{1}{\sqrt{5}}, \pm \frac{2}{\sqrt{5}} \right).$

$$f\left(\frac{1}{\sqrt{5}}, \frac{2}{\sqrt{5}} \right) = \frac{1}{\sqrt{5}} + \frac{4}{\sqrt{5}}$$
$$= \sqrt{5}$$

$$f\left(\frac{-1}{\sqrt{5}}, \frac{2}{\sqrt{5}} \right) = -\frac{1}{\sqrt{5}} + \frac{4}{\sqrt{5}}$$
$$= \frac{3}{\sqrt{5}}$$

$$f\left(\frac{1}{\sqrt{5}}, \frac{-2}{\sqrt{5}} \right) = \frac{1}{\sqrt{5}} - \frac{4}{\sqrt{5}}$$
$$= -\frac{3}{\sqrt{5}}$$

$$f\left(\frac{-1}{\sqrt{5}}, \frac{-2}{\sqrt{5}} \right) = -\frac{1}{\sqrt{5}} - \frac{4}{\sqrt{5}}$$
$$= -\sqrt{5}$$

Thus, the maximum is $\sqrt{5}$ and the minimum is $-\sqrt{5}$.

41. When the plane $z = C$ intersects the surface $z = f(x, y)$, the result is the space curve with the equation $f(x, y) = C$. Such an intersection is called the trace of the graph of f in the plane $z = C$. The set of points (x, y) in the xy-plane that satisfy $f(x, y) = C$ is called the level curve of f at C, and an entire family of level curves (or contour curves) is generated as C varies over the range of f.

43. A: sink; B: source; C: source

45. Write $z = x^2 + y^2$ as
$F(x, y, z) = x^2 + y^2 - z$ and
$2x^2 + 2y^2 + z^2 = 8$ as
$G(x, y, z) = 2x^2 + 2y^2 + z^2 - 8$
Then, $\nabla F = 2x\mathbf{i} + 2y\mathbf{j} - \mathbf{k}$ and
$\nabla G = 4x\mathbf{i} + 4y\mathbf{j} + 2z\mathbf{k}$ are normals to the surfaces $F = 0$ and $G = 0$ so
$$\nabla F_0 \times \nabla G_0$$
is parallel to the required tangent line. At $P_0(-1, 1, 2)$, we have
$$\nabla F_0 = -2\mathbf{i} + 2\mathbf{j} - \mathbf{k}$$
$$\nabla G = -4\mathbf{i} + 4\mathbf{j} + 4\mathbf{k}$$

$$\nabla F_0 \times \nabla G_0 = 12\mathbf{i} + 12\mathbf{j}$$
$$= 12(\mathbf{i} + \mathbf{j})$$

Thus, the tangent line has parametric equations
$$z = -1 + t, y = 1 + t, z = 2$$

47. Minimize $S = 2(xy + xz + yz)$ subject to $V_0 = xyz$;
$$\begin{cases} 2y + 2z = \lambda(yz) \\ 2x + 2z = \lambda(xz) \\ 2x + 2y = \lambda(xy) \\ xyz = V_0 \end{cases}$$
Solving this system simultaneously, we find

$$x = y = z = \sqrt[3]{V_0}$$

49. a. Let r and h be the radius and height of the cylinder.
$$V = \pi r^2 h$$
$$dV = \pi(2rh)\, dr + \pi r^2\, dh$$
so when $r = 2, h = 8$ and $|dh| \leq 0.5, |dr| \leq 0.25$

$$|dV| = \pi(4)(8)\,|0.25| + \pi(4)|0.5|$$
$$= 10\pi$$

b. $S = 2\pi rh + 2\pi r^2$

$dS = 2\pi(h + 2r)\,dr + (2\pi r)\,dh$

$|dS| = 2\pi[8 + 2(2)]|0.25| + 2\pi|0.5|$

$= 8\pi$

51. $V = \displaystyle\int_0^{2\pi}\int_0^1\int_0^{1+r^2} r\,dz\,dr\,d\theta$

$= \displaystyle\int_0^{2\pi}\int_0^1 r(1+r^2)\,dr\,d\theta$

$= \displaystyle\int_0^{2\pi}\frac{3}{4}\,d\theta$

$= \dfrac{3\pi}{2}$

53. The mass of the lamina is

$m = \displaystyle\iint_R (x+y)\,dA$

$= \displaystyle\int_0^{\pi/2}\int_0^2 (r\cos\theta + r\sin\theta)\,r\,dr\,d\theta$

$= \dfrac{16}{3}$

We have:

$\bar{x} = \dfrac{1}{m}\displaystyle\iint_R x(x+y)\,dA$

$= \dfrac{1}{m}\displaystyle\int_0^{\pi/2}\int_0^2 r\cos\theta\,(r\cos\theta + r\sin\theta)\,r\,dr\,d\theta$

$= \dfrac{1}{m}(\pi+2)$

$= \dfrac{3}{16}(\pi+2)$

$\bar{y} = \dfrac{1}{m}\displaystyle\iint_R y(x+y)\,dA$

$= \dfrac{1}{m}\displaystyle\int_0^{\pi/2}\int_0^2 r\sin\theta\,(r\cos\theta + r\sin\theta)\,r\,dr\,d\theta$

$= \dfrac{3}{16}(\pi+2)$

The center of mass is

$\left(\frac{3}{16}(\pi+2), \frac{3}{16}(\pi+2)\right)$.

55. $x = \cos t,\ y = \sin t,\ z = t$

$W = \displaystyle\int_C \mathbf{F}\cdot d\mathbf{R}$

$= \displaystyle\int_0^{2\pi} (-w)\,dt$

$= -2\pi w$

57. $V = \displaystyle\int_0^{2\pi}\int_{\pi/6}^{\pi}\int_0^4 \rho^2\sin\phi\,d\rho\,d\phi\,d\theta$

$= \displaystyle\int_0^{2\pi}\int_{\pi/6}^{\pi}\frac{64}{3}\sin\phi\,d\phi\,d\theta$

$= \dfrac{64}{3}\displaystyle\int_0^{2\pi}\left(\frac{\sqrt{3}}{2}+1\right)d\theta$

$= \dfrac{64}{3}\left(\sqrt{3}+2\right)\pi$

59. Following the same procedure as in Problem 49, Section 13.3, $\mathbf{F} = \mathbf{R}/r^3$ is conservative in D with scalar potential function $f = -1/r$. The work done in moving the object from distance r_1 to r_2 along a curve C is given by

$W = \displaystyle\int_C \mathbf{F}\cdot d\mathbf{R}$

$= \displaystyle\int_C \nabla\left(\frac{-1}{r}\right)\cdot d\mathbf{R}$

$= \left(\frac{-1}{r_2}\right) - \left(\frac{-1}{r_1}\right)$

$= \dfrac{1}{r_1} - \dfrac{1}{r_2}$

CHAPTER 14 Introduction to Differential Equations

Chapter Overview
A *differential equation* is an equation that contains a derivative or a differential. The problem of solving differential equations is the topic of one of the first post calculus courses. In this chapter, we introduce some of the useful techniques for solving differential, which allows us to example a few important applications.

14.1 First-Order Differential Equations, page 1120

1.

$$xy\,dx = (x - 5)\,dy$$

$$\int \frac{x}{x-5}dx = \int y^{-1}dy$$

$$\int \left(1 + \frac{5}{x-5}\right)dx = \int y^{-1}dy$$

$$x + 5\ln|x - 5| = \ln|y| + C_1$$

$$\ln|y| = x + \ln|x - 5|^5 + C$$

$$y = Be^{x+\ln|x-5|^5}$$

SURVIVAL HINT: *In this problem we left the answer as a function of x. However, many differential equation problems have a form which is not easily solvable for y. For this reason, it is often acceptable to leave your answer in a form which is not technically simplified. For this problem, it would be acceptable to leave your answer as*

$$\ln|y| = x + \ln|x - 5|^5 + C$$

3. $(e^{2x} + 9)\dfrac{dy}{dx} = y$

$$\int y^{-1}dy = \int \frac{dx}{e^{2x} + 9}$$

$$\ln|y| = -\frac{1}{18}\left[\ln(e^{2x} + 9) - 2x\right] + C$$

5. $9\,dx - x\sqrt{x^2 - 9}\,dy = 0$

$$\int dy = \int \frac{9\,dx}{x\sqrt{x^2 - 9}}$$

$$y = 3\sec^{-1}\left|\frac{x}{3}\right| + C$$

7. $p(x) = 2x,\ q(x) = 4x,$

$$I(x) = e^{\int 2x\,dx} = e^{x^2}$$

$$y = \frac{1}{e^{x^2}}\left[\int e^{x^2}(4x)dx + C\right]$$

$$= e^{-x^2}\left[2e^{x^2} + C\right]$$

$$= 2 + Ce^{-x^2}$$

Particular value: $0 = 2 + Ce^0;\ C = -2$
Particular solution: $y = 2(1 - e^{-x^2})$

9. $p(x) = 1,\ q(x) = \cos x,$

$$I(x) = e^{\int dx} = e^x$$

Page 585

$$y = \frac{1}{e^x}\left[\int e^x \cos x\, dx + C\right]$$
$$= \frac{1}{2}\cos x + \frac{1}{2}\sin x + Ce^{-x}$$

Particular value: $0 = \frac{1}{2}(1+0) + C$
$$C = -\frac{1}{2}$$

Particular solution:
$$y = \frac{1}{2}(\cos x + \sin x) - \frac{1}{2}e^{-x}$$

11. $x\dfrac{dy}{dx} - 2y = x^3$
$$\frac{dy}{dx} - \frac{2}{x}y = x^2$$
$$p(x) = -\frac{2}{x}, \quad q(x) = x^2,$$
$$I(x) = e^{\int(-2/x)dx} = x^{-2}$$
$$y = x^2\left[\int x^{-2}x^2\, dx + C\right]$$
$$= x^3 + Cx^2$$

Particular value: $-1 = 8 + 4C$
$$C = -\frac{9}{4}$$

Particular solution: $y = x^3 - \dfrac{9}{4}x^2$

13. $(3x-y)\,dx + (x+3y)\,dy = 0$
Show the equation is homogeneous.
$$\frac{dy}{dx} = \frac{-(3x-y)}{x+3y}\cdot\frac{\frac{1}{x}}{\frac{1}{x}}$$
$$= \frac{-\left(3-\frac{y}{x}\right)}{1+3\left(\frac{y}{x}\right)}$$
Let $f(v) = \dfrac{-3+v}{1+3v}$ where $v = \dfrac{y}{x}$

$$\frac{dv}{\frac{-3+v}{1+3v} - v} = \frac{dx}{x}$$
$$\int \frac{1+3v}{-3v^2-3}dv = \int\frac{dx}{x}$$
$$-\frac{1}{2}\ln(v^2+1) - \frac{1}{3}\tan^{-1}v = \ln|x| + C$$
$$-\frac{1}{2}\ln\left(\frac{y^2}{x^2}+1\right) - \ln|x| - \frac{1}{3}\tan^{-1}\frac{y}{x} = C$$
$$-\ln\sqrt{x^2+y^2} - \frac{1}{3}\tan^{-1}\frac{y}{x} = C$$

15. $(3x-y)\,dx + (x-3y)\,dy = 0$
Show the equation is homogeneous.
$$\frac{dy}{dx} = \frac{-(3x-y)}{x-3y}$$
$$= \frac{-\left(3-\frac{y}{x}\right)}{1-3\left(\frac{y}{x}\right)}$$
Let $v = \dfrac{y}{x}$, $f(v) = \dfrac{-(3-v)}{1-3v}$

$$\int\frac{1-3v}{3v^2-3}dv = \int\frac{dx}{x}$$
$$-\frac{1}{3}\ln\left[(v+1)^2(v-1)\right] = \ln|x| + C_1$$
$$(v+1)^2\left(\frac{y}{x}-1\right) = Cx^{-3}$$
$$\left(\frac{y}{x}+1\right)^2\left(\frac{y}{x}-1\right) = Cx^{-3}$$
$$(y+x)^2(y-x) = C$$

17. $(-6y^2 + 3xy + 2x^2)\,dx + x^2\,dy = 0$
Show the equation is homogeneous.
$$\frac{dy}{dx} = \frac{6y^2 - 3xy - 2x^2}{x^2}$$
$$= 6\left(\frac{y}{x}\right)^2 - 3\left(\frac{y}{x}\right) - 2$$

Let $f(v) = 6v^2 - 3v - 2$ where $v = \dfrac{y}{x}$.

$$\frac{dv}{6v^2 - 3v - 2 - v} = \frac{dx}{x}$$

$$\int \frac{dv}{6v^2 - 4v - 2} = \int \frac{dx}{x}$$

$$-\frac{1}{8}\ln\left|\frac{3v+1}{v-1}\right| = \ln|x| + C_1$$

$$\frac{1}{8}\ln\left|\frac{v-1}{3v+1}\right| = \ln|x| + C_1$$

$$\frac{v-1}{3v+1} = Cx^8$$

$$\frac{y-x}{3y+x} = Cx^8$$

19. $(3x^2y + \tan y)\,dx + (x^3 + x\sec^2 y)\,dy = 0$

$M(x,y) = 3x^2y + \tan y \qquad \dfrac{\partial M}{\partial y} = 3x^2 + \sec^2 y$

$N(x,y) = x^3 + x\sec^2 y \qquad \dfrac{\partial N}{\partial x} = 3x^2 + \sec^2 y$

Since $\dfrac{\partial M}{\partial y} = \dfrac{\partial N}{\partial x}$, the D. E. is exact.

$$f(x,y) = \int (3x^2y + \tan y)\,dx$$

$$= x^3 y + x\tan y + u(y)$$

$$\frac{\partial f}{\partial y} = x^3 + x\sec^2 y + u'(y)$$

Compare this with $N(x,y)$ to see
$u'(y) = 0$, so $u(y) = C_1$. Thus,

$$x^3 y + x\tan y = C$$

21. $\left[\dfrac{1}{1+x^2} + \dfrac{2x}{x^2+y^2}\right]dx + \left[\dfrac{2y}{x^2+y^2} - e^{-y}\right]dy = 0$

$M(x,y) = \dfrac{1}{1+x^2} + \dfrac{2x}{x^2+y^2} \qquad \dfrac{\partial M}{\partial y} = \dfrac{-4xy}{(x^2+y^2)^2}$

$N(x,y) = \dfrac{2y}{x^2+y^2} - e^{-y} \qquad \dfrac{\partial N}{\partial x} = \dfrac{-4xy}{(x^2+y^2)^2}$

Since $\dfrac{\partial M}{\partial y} = \dfrac{\partial N}{\partial x}$, the D. E. is exact.

$$f(x,y) = \int \left(\frac{1}{1+x^2} + \frac{2x}{x^2+y^2}\right)dx$$

$$= \tan^{-1}x + \ln(x^2 + y^2) + u(y)$$

$$\frac{\partial f}{\partial y} = \frac{2y}{x^2+y^2} + u'(y)$$

Compare this with $N(x,y)$ to see
$u'(y) = -e^{-y}$.
Integrate to find $u(y) = e^{-y} + C_1$ so

$$\tan^{-1}x + \ln(x^2 + y^2) + e^{-y} = C$$

23. $[2x\cos 2y - 3y(1 - 2x)]\,dx$
$\qquad - [2x^2\sin 2y + 3(2 + x - x^2)]\,dy = 0$

$M(x,y) = 2x\cos 2y - 3y(1 - 2x)$

$N(x,y) = -2x^2\sin 2y - 3(2 + x - x^2)$

$$\frac{\partial M}{\partial y} = -4x\sin 2y - 3(1 - 2x)$$

$$\frac{\partial N}{\partial x} = -4x\sin 2y - 3(1 - 2x)$$

Since $\dfrac{\partial M}{\partial y} = \dfrac{\partial N}{\partial x}$, the D. E. is exact.

$$f(x,y) = \int [2x\cos 2y - 3y(1 - 2x)]\,dx$$

$$= x^2\cos 2y - 3xy + 3x^2y + u(y)$$

$$\frac{\partial f}{\partial y} = -2x^2\sin 2y - 3x + 3x^2 + u'(y)$$

Compare this with $N(x,y)$ to see that
$u'(y) = -6$.

Integrate to find $u(y) = -6y + C_1$. Thus,

$$x^2\cos 2y - 3xy + 3x^2y - 6y = C$$

25. a. $\dfrac{dy}{dx} - y = x$ is a first order D. E.

$\qquad I(x) = e^{\int(-dx)} = e^{-x}$

$$y = e^x \left[\int x e^{-x} dx + C \right]$$
$$= e^x [-x e^{-x} - e^{-x} + C]$$
$$= -x - 1 + Ce^x$$

At $(1, 2)$:

$$2 = -1 - 1 + Ce$$
$$C = 4e^{-1}$$

Particular solution: $y = -x - 1 + 4e^{x-1}$.

b.

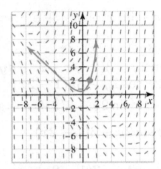

c.

n	x_n	y_n	$f(x_n, y_n)$	y_{n+1} $= y_n + 0.2f(x_n, y_n)$
0	1	2	3	2.60
1	1.2	2.60	3.80	3.36
2	1.4	3.36	4.76	4.31
3	1.6	4.31	5.91	5.49
4	1.8	5.49	7.29	6.95
5	2.0	6.95	8.95	8.74

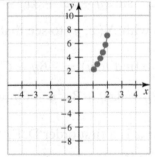

27. $\dfrac{dy}{dx} = \dfrac{x+y}{y-x} = f(x,y); \ h = 0.1$

n	x_n	y_n	$f(x_n, y_n)$	y_{n+1} $= y_n + 0.1f(x_n, y_n)$
0	0	1	1	1.10
1	0.1	1.10	1.20	1.22
2	0.2	1.22	1.39	1.36
3	0.3	1.36	1.57	1.52
4	0.4	1.52	1.72	1.69
5	0.5	1.69	1.84	1.87

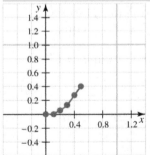

29. $\dfrac{dy}{dx} = \dfrac{5x - 3xy}{1 + x^2} = f(x,y); \ h = 0.1$

n	x_n	y_n	$f(x_n, y_n)$	y_{n+1} $= y_n + 0.1f(x_n, y_n)$
0	0	0	0	0
1	0.1	0	0.50	0.05
2	0.2	0.05	0.93	0.14
3	0.3	0.14	1.26	0.27
4	0.4	0.27	1.45	0.41
5	0.5	0.41	1.50	0.56

31. $M = y$; $N = y - x$

$g(x, y) = y^n$

We want

$$\frac{\partial}{\partial y}[y^n y] = \frac{\partial}{\partial x}[y^n(y - x)]$$

$$(n + 1)y^n = -y^n$$

$$n + 1 = -1$$

$$n = -2$$

Thus, the equation is

$$y^{-2}y\, dx + y^{-2}(y - x)dy = 0$$

$$\frac{\partial f}{\partial x} = y^{-1}; \ f = \frac{x}{y} + u(y)$$

$$\frac{\partial f}{\partial y} = \frac{-x}{y^2} + u'(y) = \frac{-x}{y^2} + y^{-1}$$

$$u'(y) = y^{-1}; \ u(y) = \ln|y| + C_1$$

Solution: $\dfrac{x}{y} + \ln|y| = C$

33. $M(x, y) = 2xy^2 + 3x^2y - y^3$

$N(x, y) = 2x^2y + x^3 - 3xy^2$

$$\frac{\partial M}{\partial y} = 4xy + 3x^2 - 3y^2$$

$$\frac{\partial N}{\partial x} = 4xy + 3x^2 - 3y^2$$

Since $\dfrac{\partial M}{\partial y} = \dfrac{\partial N}{\partial x}$, the D. E. is exact.

$$f(x, y) = \int \left(2xy^2 + 3x^2y - y^3\right) dx$$

$$= x^2y^2 + x^3y - xy^3 + u(y)$$

$$\frac{\partial f}{\partial y} = 2x^2y + x^3 - 3xy^2 + u'(y)$$

Compare with $N(x, y)$ to see that

$u'(y) = 0$ so $u(y) = C_1$. Thus,

$$x^2y^2 + x^3y - xy^3 = C$$

35. $M(x, y) = \dfrac{2x}{y} - \dfrac{y^2}{x^2}$ $\qquad \dfrac{\partial M}{\partial y} = -\dfrac{2x}{y^2} - \dfrac{2y}{x^2}$

$N(x, y) = \dfrac{2y}{x} - \dfrac{x^2}{y^2} + 3$ $\qquad \dfrac{\partial N}{\partial x} = -\dfrac{2y}{x^2} - \dfrac{2x}{y^2}$

Since $\dfrac{\partial M}{\partial y} = \dfrac{\partial N}{\partial x}$, the D. E. is exact.

$$f(x, y) = \int \left(\frac{2x}{y} - \frac{y^2}{x^2}\right) dx$$

$$= \frac{x^2}{y} + \frac{y^2}{x} + u(y)$$

$$\frac{\partial f}{\partial y} = -\frac{x^2}{y^2} + \frac{2y}{x} + u'(y)$$

Compare this with $N(x, y)$ to see that

$u'(y) = 3$ so that $u(y) = 3y + C_1$. Thus,

$$\frac{x^2}{y} + \frac{y^2}{x} + 3y = C$$

37. $e^{y-x}\sin x\, dx - \csc x\, dy = 0$ is separable.

$$\int e^{-y}dy = \int e^{-x}\frac{\sin x}{\csc x}dx$$

$$= \int e^{-x}\sin^2 x\, dx$$

$$-e^{-y} = \frac{1}{2}\int e^{-x}(1 - \cos 2x)dx$$

$$= \frac{1}{2}\int e^{-x}dx - \frac{1}{2}\int e^{-x}\cos 2x\, dx$$

$$= -\frac{1}{2}e^{-x} - \frac{1}{2}\frac{e^{-x}(-\cos 2x + 2\sin 2x)}{5} + C$$

$$e^{-y} = \frac{1}{2}e^{-x} + \frac{e^{-x}(2\sin 2x - \cos 2x)}{10} + C$$

39. $M(x, y) = 3x^2 - y\sin xy$

$N(x, y) = -x\sin xy$

$$\frac{\partial M}{\partial y} = -xy\cos xy - \sin xy$$

$$\frac{\partial N}{\partial x} = -\sin xy - xy\cos xy$$

Since $\dfrac{\partial M}{\partial y} = \dfrac{\partial N}{\partial x}$, the D. E. is exact.

$$f(x,y) = \int \left(3x^2 - y\sin xy\right)dx$$

$$= x^3 + \cos xy + u(y)$$

$$\frac{\partial f}{\partial y} = -x\sin xy + u'(y)$$

Compare with $N(x,y)$ to see that $u'(y) = 0$, so $u(y) = C_1$; thus,

$$x^3 + \cos xy = C$$

41. This is a first-order linear D. E.

$$\left(y - \sin^2 x\right)dx + (\sin x)dy = 0$$

$$\frac{dy}{dx} + \frac{y}{\sin x} = \sin x$$

$$p(x) = \csc x, \; q(x) = \sin x$$

$$I(x) = e^{\int \csc x\, dx}$$

$$= \exp\left[\ln\left(\frac{1}{\csc x + \cot x}\right)\right]$$

$$= \frac{1}{\csc x + \cot x}$$

$$y = (\csc x + \cot x)\left[\int \frac{\sin x}{\csc x + \cot x}dx + C_1\right]$$

$$\int \frac{\sin x}{\csc x + \cot x}dx = \int \frac{\sin^2 x}{1 + \cos x}dx$$

$$= \int \frac{(1 - \cos^2 x)}{1 + \cos x}dx$$

$$= \int (1 - \cos x)\, dx$$

$$= x - \sin x + C_2$$

Thus,

$$y = (\csc x + \cot x)(x - \sin x + C)$$

43. The given D. E. is homogeneous.

$$\frac{dy}{dx} = \frac{y}{x} - \sqrt{1 + \left(\frac{y}{x}\right)^2}$$

Let $f(v) = v - \sqrt{1 + v^2}$ where $v = \dfrac{y}{x}$

$$\int \frac{dv}{v - \sqrt{1 - v^2} - v} = \int \frac{dx}{x}$$

$$-\int \frac{dv}{\sqrt{1 + v^2}} = \int \frac{dx}{x}$$

$$-\ln\left|\sqrt{v^2 + 1} + v\right| = \ln|x| + C_1$$

$$-\ln\left|\sqrt{\left(\frac{y}{x}\right)^2 + 1} + \frac{y}{x}\right| = \ln|x| + C$$

$$-\ln\left|\sqrt{x^2 + y^2} + y\right| + \ln|x| - \ln|x| = C_1$$

$$\ln\left|\sqrt{x^2 + y^2} + y\right| = C$$

> **SURVIVAL HINT:** *Here is an alternate form:*
> $\sqrt{x^2 + y^2} + y = B$ (where $B = e^C$).

45. The given D. E. is homogeneous.

$$\left(x\sin^2\frac{y}{x} - y\right)dx + x\, dy = 0;$$

$$\frac{dy}{dx} = \frac{y}{x} - \sin^2\frac{y}{x}$$

Let $f(v) = v - \sin^2 v$ where $v = \dfrac{y}{x}$.

$$\frac{dv}{v - \sin^2 v - v} = \frac{dx}{x}$$

$$-\int \frac{dv}{\sin^2 v} = \int \frac{dx}{x}$$

$$\cot v = \ln|x| + C$$

$$\cot \frac{y}{x} = \ln|x| + C$$

If $x = \dfrac{4}{\pi}$ when $y = 1$, we have

$$\cot\frac{\pi}{4} = \ln\frac{4}{\pi} + C$$

$$C = 1 - \ln\frac{4}{\pi}$$

so that

$$\cot\frac{y}{x} = \ln|x| + 1 - \ln\frac{4}{\pi}$$

$$\cot\frac{y}{x} = \ln\frac{\pi|x|}{4} + 1$$

47. The D. E. is first-order linear.

$$x\frac{dy}{dx} - 3y = x^3$$

$$\frac{dy}{dx} - \frac{3}{x}y = x^2$$

$$p(x) = -\frac{3}{x}, \; q(x) = x^2$$

$$I(x) = e^{\int(-3/x)dx} = e^{-3\ln|x|} = x^{-3}$$

$$y = x^3\left[\int x^{-3}x^2\,dx + C\right]$$

$$= x^3\ln|x| + Cx^3$$

If $x = 1$, then $y = 1$, so $C = 1$, so

$$y = x^3\ln x + x^3$$

49. $M(x, y) = \sin(x^2 + y) + 2x^2\cos(x^2 + y)$

$N(x, y) = x\cos(x^2 + y)$

$\dfrac{\partial M}{\partial y} = \cos(x^2 + y) - 2x^2\sin(x^2 + y)$

$\dfrac{\partial N}{\partial x} = \cos(x^2 + y) - 2x^2\sin(x^2 + y)$

Since $\dfrac{\partial M}{\partial y} = \dfrac{\partial N}{\partial x}$, the D. E. is exact.

$$f(x, y) = \int x\cos(x^2 + y)\,dy$$

$$= x\sin(x^2 + y) + u(x)$$

$$\frac{\partial f}{\partial x} = \sin(x^2 + y) + 2x^2\cos(x^2 + y) + u'(x)$$

Compare with $M(x, y)$ to see that $u'(x) = 0$ so $u(x) = C$. Also, since $x = y = 0$, $C = 0$. Thus,

$$x\sin(x^2 + y) = 0$$

51. $ye^x\,dy = (y^2 + 2y + 2)\,dx$ is separable.

$$\int\frac{y\,dy}{y^2 + 2y + 2} = \int e^{-x}dx$$

$$\frac{1}{2}\ln|y^2 + 2y + 2| - \tan^{-1}(y + 1) = -e^{-x} + C$$

If $x = 0$, $y = -1$, so $0 - 0 = -1 + C$ or $C = 1$. Thus,

$$\tfrac{1}{2}\ln(y^2 + 2y + 2) - \tan^{-1}(y + 1) + e^{-x} = 1$$

53. If $\alpha \neq \beta$, Example 2 shows that

$$Q(t) = \frac{\alpha\beta[e^{(\alpha-\beta)kt} - 1]}{\alpha e^{(\alpha-\beta)kt} - \beta}$$

If $\alpha > \beta$, then (by l'Hôpital's rule)

$$\lim_{t\to\infty} Q(t) = \lim_{t\to\infty}\frac{\alpha\beta(\alpha - \beta)ke^{(\alpha-\beta)kt}}{\alpha(\alpha - \beta)ke^{(\alpha-\beta)kt}}$$

$$= \beta$$

If $\alpha < \beta$, then the exponent of e is negative and

$$\lim_{t \to \infty} \frac{\alpha\beta(0-1)}{\alpha(0)-\beta} = \alpha$$

If $\alpha = \beta$,

$$\frac{dQ}{dt} = k(\alpha - Q)^2$$

$$\int \frac{dQ}{(\alpha - Q)^2} = k$$

$$\frac{1}{\alpha - Q} = kt + C$$

$$\alpha - Q = \frac{1}{kt + C}$$

$$Q = \frac{-1}{kt + C} + \alpha$$

$Q(0) = 0$, so $C = \dfrac{1}{\alpha}$, and we see that

$$Q(t) = \frac{-1}{kt + \frac{1}{\alpha}} + \alpha = \frac{\alpha^2 kt}{\alpha kt + 1}$$

In "the long run,"

$$\lim_{t \to \infty} Q(t) = \lim_{t \to \infty} \left(\frac{\alpha^2 k}{\alpha k} \right) = \alpha$$

55. a.
$$\frac{dP}{dt} = P(k - \ell P) - h$$

$$\int \frac{dP}{P^2 - \frac{k}{\ell}P + \frac{h}{\ell}} = \int (-\ell) \, dt$$

Since $h < \dfrac{k^2}{4\ell}$, we have

$$P^2 - \frac{k}{\ell}P + \frac{h}{\ell}(P - r_1)(P - r_2)$$

where $r_1 = \dfrac{k + \sqrt{k^2 - 4h\ell}}{2\ell}$ and

$$r_2 = \frac{k - \sqrt{k^2 - 4h\ell}}{2\ell}. \text{ Let}$$

$D = \sqrt{k^2 - 4h\ell}$, so that $r_1 - r_2 = \dfrac{D}{\ell}$.

$$\int \frac{dP}{P^2 + \frac{k}{\ell}P + \frac{h}{\ell}} = \int \frac{dP}{(P - r_1)(P - r_2)}$$

$$= \int \frac{\ell}{D}\left[\frac{1}{P - r_1} - \frac{1}{P - r_2}\right] dP$$

$$= \frac{\ell}{D} \ln \left| \frac{P - r_1}{P - r_2} \right| + C_1$$

Since $\int(-\ell) \, dt = -\ell t + C_2$, we now have a solution to the given D. E.

$$\frac{\ell}{D} \ln \left| \frac{P - r_1}{P - r_2} \right| = -\ell t + C$$

$$P(t) = \frac{r_1 - r_2 B e^{-Dt}}{1 - B e^{-Dt}}$$

Since $P = P_0$ when $t = 0$, we have

$$P_0 = \frac{r_1 - r_2 B}{1 - B}$$

$$B = \frac{P_0 - r_1}{P_0 - r_2}$$

so

$$P(t) = \frac{r_1(P_0 - r_2) - r_2(P_0 - r_1)e^{-Dt}}{(P_0 - r_2) - (P_0 - r_1)e^{-Dt}}$$

b. $\lim\limits_{t \to \infty} P(t) = r_1$

57. a. Reverse the roles of x and y: The integrating factor is

$$I(y) = e^{\int R(y)dy}$$

and the general solution has the form

$$x = \frac{1}{I(y)}\left[\int I(y)S(y)\,dy + C\right]$$

b. $y\,dx - 2x\,dy = y^4 e^{-y}\,dy$

$$\frac{dx}{dy} - \frac{2x}{y} = y^3 e^{-y}$$

$$I(y) = e^{\int(-2y)dy} = y^{-2}$$

$$x = y^2\left[\int y^{-2}(y^3 e^{-y})\,dy\right] + C$$

$$= -y^2[(y+1)e^{-y} + C]$$

59. a. $z = \dfrac{1}{y-u}$ so that $y = u + \dfrac{1}{z}$

$$\frac{dy}{dx} = \frac{du}{dx} - \frac{1}{z^2}\frac{dz}{dx}$$

and by substituting into the Riccati equation, we obtain

$$\frac{du}{dx} - \frac{1}{z^2}\frac{dz}{dx}$$

$$= P(x)\left(u + \frac{1}{z}\right)^2 + Q(x)\left(u + \frac{1}{z}\right) + R(x)$$

$$= P(x)u^2 + Q(x)u + R(x)$$

$$+ \left[P(x)\frac{1}{z^2} + 2P(x)u\frac{1}{z} + Q(x)\frac{1}{z}\right]$$

But u is assumed to satisfy the Riccati equation, so

$$\frac{du}{dx} = P(x)u^2 + Q(x)u + R(x)$$

and

$$\frac{du}{dx} - \frac{1}{z^2}\frac{dz}{dx}$$

$$= \frac{du}{dx} + \left[P(x)\frac{1}{z^2} + 2P(x)u\frac{1}{z} + Q(x)\frac{1}{z}\right]$$

Multiply by $-z^2$:

$$\frac{dz}{dx} = -P(x) - 2P(x)uz - Q(x)z$$

$$\frac{dz}{dx} + [2P(x)u + Q(x)]z = -P(x)$$

which is a first-order linear equation, as required.

b. The first-order linear equation in part **a** can be solved to obtain

$$z = \frac{C}{w(x)} + \frac{1}{w(x)}\int[-P(x)w(x)]\,dx$$

where

$$w(x) = \exp\left(\int[2P(x)u(x) + Q(x)]\,dx\right)$$

and then

$$y = u(x) + \frac{1}{z(x)}$$

is a general solution of the Riccati equation.

c. $\dfrac{dy}{dx} = \dfrac{1}{x^2}y^2 + \dfrac{2}{x}y - 2$ implies

$P(x) = \dfrac{1}{x^2}$, $Q(x) = \dfrac{2}{x}$, and

$R(x) = -2$. To find a particular solution of the form $u = Ax$, note that $du/dx = A$ and by substituting, we obtain

$$A = \frac{1}{x^2}\left(A^2 x^2\right) + \frac{2}{x}(Ax) - 2$$

$$A = A^2 + 2A - 2$$

$$A^2 + A - 2 = 0$$

$$A = 1, -2$$

For $A = 1$, the particular solution is

$u = x$ and $y = u + \dfrac{1}{z} = x + \dfrac{1}{z}$.

Substituting into the equation

$$\frac{dy}{dx} = \frac{1}{x^2}y^2 + \frac{2}{x}y - 2$$

we obtain

$$1 - \frac{1}{z^2}\frac{dz}{dx} = \frac{1}{x^2}\left(x + \frac{1}{z}\right)^2 + \frac{2}{z}\left(x + \frac{1}{z}\right) - 2$$

$$1 - \frac{1}{z^2}\frac{dz}{dx} = 1 + \frac{4}{xz} + \frac{1}{x^2 z^2}$$

$$\frac{dz}{dx} + \frac{4z}{x} = -\frac{1}{x^2}$$

This is a first-order linear equation with the integrating factor

$$I(x) = e^{\int (4/x)dx} = e^{4\ln x} = x^4$$

so

$$z = \frac{1}{x^4}\left[\int x^4\left(-\frac{1}{x^2}\right)dx + C\right]$$

$$= \frac{1}{x^4}\left[-\frac{x^3}{3} + C\right]$$

$$= -\frac{1}{3x} + \frac{C}{x^4}$$

Thus, the general solution to the given Riccati equation is

$$y = x + \frac{1}{z}$$

$$= x + \frac{1}{\frac{C}{x^4} - \frac{1}{3x}}$$

$$= x + \frac{1}{\frac{3C - x^3}{3x^4}}$$

$$= \frac{3Cx + 2x^4}{3C - x^3}$$

14.2 Second-Order Homogeneous Linear Differential Equations, page 1132

1. $y'' + y' = 0$

$$r^2 + r = 0$$

$$r(r + 1) = 0$$

$$r = 0, -1$$

Particular solutions: $y_1 = 1$, $y_2 = e^{-x}$
General solution: $y = C_1 + C_2 e^{-x}$

3. $y'' + 6y' + 5y = 0$

$$r^2 + 6r + 5 = 0$$

$$(r + 5)(r + 1) = 0$$

$$r = -5, -1$$

Particular solutions: $y_1 = e^{-5x}$, $y_2 = e^{-x}$
General solution: $y = C_1 e^{-5x} + C_2 e^{-x}$

5. $y'' - y' - 6y = 0$

$$r^2 - r - 6 = 0$$

$$(r - 3)(r + 2) = 0$$

$$r = 3, -2$$

Particular solutions: $y_1 = e^{3x}$, $y_2 = e^{-2x}$
General solution: $y = C_1 e^{3x} + C_2 e^{-2x}$

7. $2y'' - 5y' - 3y = 0$

$$2r^2 - 5r - 3 = 0$$

$$(2r + 1)(r - 3) = 0$$

$$r = -\frac{1}{2}, 3$$

Particular solutions: $y_1 = e^{(-1/2)x}$, $y_2 = e^{3x}$
General solution: $y = C_1 e^{(-1/2)x} + C_2 e^{3x}$

9. $y'' - y = 0$

$$r^2 - 1 = 0$$
$$(r-1)(r+1) = 0$$
$$r = 1, -1$$

Particular solutions: $y_1 = e^x$, $y_2 = e^{-x}$.
General solution: $y = C_1 e^x + C_2 e^{-x}$

11. $y'' + 11y = 0$

$$r^2 + 11 = 0$$
$$r = \pm \sqrt{11}\,i$$

$\alpha = 0, \beta = \sqrt{11}$
General solution:
$$y = C_1 \cos \sqrt{11}x + C_2 \sin \sqrt{11}x$$

13. $7y'' + 3y' + 5y = 0$

$$7r^2 + 3r + 5 = 0$$
$$r = \frac{-3 \pm \sqrt{9 - 4(7)(5)}}{2(7)}$$
$$r = -\frac{3}{14} \pm \frac{\sqrt{131}}{14}i$$

$\alpha = -\dfrac{3}{14}, \beta = \dfrac{\sqrt{131}}{14}$
General solution:
$$y = e^{(-3/14)x}\left[C_1\cos\left(\tfrac{\sqrt{131}}{14}x\right) + C_2\sin\left(\tfrac{\sqrt{131}}{14}x\right)\right]$$

15. $y''' + y'' = 0$
$$r^3 + r^2 = 0$$
$$r^2(r+1) = 0$$
$$r = 0 \text{ (mult. 2)}, -1$$
$$y = C_1 + C_2 x + C_3 e^{-x}$$

17. $y^{(4)} + y''' + 2y'' = 0$

$$r^4 + r^3 + 2r^2 = 0$$
$$r^2(r^2 + r + 2) = 0$$
$$r = 0 \text{ (mult. 2)}, -\frac{1}{2} \pm \frac{\sqrt{7}}{2}i$$
$$y = C_1 + C_2 x$$
$$+ e^{-(1/2)x}\left[C_3\cos\left(\frac{\sqrt{7}}{2}x\right) + C_4\sin\left(\frac{\sqrt{7}}{2}x\right)\right]$$

19. $y''' + 2y'' - 5y' - 6y = 0$
$$r^3 + 2r^2 - 5r - 6 = 0$$
$$(r+1)(r-2)(r+3) = 0$$
$$r = 2, -3, -1$$
$$y = C_1 e^{2x} + C_2 e^{-3x} + C_3 e^{-x}$$

21. $y'' - 10y' + 25y = 0$
$$r^2 - 10r + 25 = 0$$
$$(r-5)^2 = 0$$
$$r = 5 \text{ (mult. 2)}$$
$$y = C_1 e^{5x} + C_2 x e^{5x}$$
$y(0) = 1$, so
$$1 = C_1(1) + 0$$
$$C_1 = 1$$
$$y' = 5e^{5x} + C_2(e^{5x} + 5xe^{5x})$$
$y'(0) = -1$, so
$$-1 = 5(1) + C_2(1+0)$$
$$C_2 = -6$$
$$y = e^{5x} - 6xe^{5x} \text{ or } e^{5x}(1 - 6x)$$

23. $y'' - 12y' + 11y = 0$
$$r^2 - 12r + 11 = 0$$
$$(r-11)(r-1) = 0$$
$$r = 11, 1$$
$$y = C_1 e^x + C_2 e^{11x}$$
$y(0) = 3$, so
$$3 = C_1 + C_2$$

$y' = C_1 e^x + 11 C_2 e^{11x}$
$y'(0) = 11$, so

$$11 = C_1 + 11 C_2$$

Solve $\begin{cases} C_1 + C_2 = 3 \\ C_1 + 11 C_2 = 11 \end{cases}$

Solve this system to find $C_1 = \frac{11}{5}, C_2 = \frac{4}{5}$.
$y = \frac{11}{5} e^x + \frac{4}{5} e^{11x}$

25. $y''' + 10 y'' + 25 y' = 0$
$r^3 + 10 r^2 + 25 r = 0$
$$r = 0, -5 \text{ (mult. 2)}$$
$y = C_1 + C_2 e^{-5x} + C_3 x e^{-5x}$
$y(0) = 3$, so

$$3 = C_1 + C_2 \cdot 1$$

$y'(0) = 2$, so

$$2 = -5 C_2 \cdot 1 + C_3 (1 - 0)$$

$y''(0) = -1$, so

$$-1 = 25 C_2 \cdot 1 + C_3 (-10 \cdot 1 + 0)$$

Solve $\begin{cases} C_1 + C_2 = 3 \\ -5 C_2 + C_3 = 2 \\ 25 C_2 - 10 C_3 = -1 \end{cases}$

Solve this system to find
$C_1 = \frac{94}{25}, C_2 = -\frac{19}{25}, C_3 = -\frac{9}{5}$.

$y = \frac{94}{25} - \frac{19}{25} e^{-5x} - \frac{9}{5} x e^{-5x}$

27. $W\left(e^{-2x}, e^{3x}\right) = \begin{vmatrix} e^{-2x} & e^{3x} \\ -2e^{-2x} & 3e^{3x} \end{vmatrix}$
$$= 3e^x + 2e^x$$
$$= 5e^x$$
$$\neq 0$$

29. $W(e^{-x}, xe^{-x}) = \begin{vmatrix} e^{-x} & xe^{-x} \\ -e^{-x} & (1-x)e^{-x} \end{vmatrix}$
$$= (1-x)e^{-2x} + xe^{-2x}$$
$$= e^{-2x}$$
$$\neq 0$$

31. $W(e^{-x}\cos x, e^{-x}\sin x)$
$$= \begin{vmatrix} e^{-x}\cos x & e^{-x}\sin x \\ e^{-x}(-\cos x - \sin x) & e^{-x}(\cos x - \sin x) \end{vmatrix}$$
$$= e^{-x}\cos x \cdot e^{-x}(\cos x - \sin x)$$
$$\quad - e^{-x}(-e^{-x})\sin x(\sin x + \cos x)$$
$$= e^{-2x}$$
$$\neq 0$$

33. $y'' + 6y' + 9y = 0$
$y_1 = e^{-3x}$
$y_2 = ve^{-3x}$

$$y_2' = v'e^{-3x} - 3ve^{-3x}$$
$$y_2'' = v''e^{-3x} - 6v'e^{-3x} + 9ve^{-3x}$$

$$(v'' - 6v' + 9v)e^{-3x} + 6(v' - 3v)e^{(-3x)}$$
$$+ 9ve^{-3x} = 0$$
$v'' = 0$; $v = x$ is one solution;
$y_2 = xe^{-3x}$
$W\left(e^{-3x}, xe^{-3x}\right)$
$$= \begin{vmatrix} e^{-3x} & xe^{-3x} \\ -3e^{-3x} & e^{-3x} - 3xe^{-3x} \end{vmatrix}$$
$$= e^{-6x}$$
$$\neq 0$$
General solution:

$$y = C_1 e^{-3x} + C_2 x e^{-3x}$$
$$= e^{-3x}(C_1 + C_2 x)$$

35. $xy'' + 4y' = 0$
$y_1 = 1$
$y_2 = vy_1 = v$

$$y_2' = v'$$
$$y_2'' = v''$$
$$xv'' + 4v' = 0$$
$$\frac{v''}{v'} = -\frac{4}{x}$$
$$\ln|v'| = -4\ln|x|$$
$$v' = x^{-4}$$
$$v = -\frac{1}{3}x^{-3}$$

SURVIVAL HINT: *For the consecutive integrations, constants are omitted because we are looking for particular solutions of y_1 and y_2 and their combination will remain unchanged no matter what particular independent solutions we pick.*

$$y_2 = v = -\frac{1}{3}x^{-3}$$
$$W\left(1, -\frac{1}{3}x^{-3}\right) = \begin{vmatrix} 1 & -\frac{1}{3}x^{-3} \\ 0 & x^{-4} \end{vmatrix}$$
$$= x^{-4}$$
$$\neq 0$$

General solution:
$$y = C_1 + C_2 x^{-4}$$

37. $x^2 y'' + 2xy' - 12y = 0$
$$y_1 = x^3$$
$$y_2 = vy_1 = vx^3$$
$$y_2' = v'x^3 + 3vx^2$$
$$y_2'' = v''x^3 + 6v'x^2 + 6xv$$
$$x^2\left(v''x^3 + 6v'x^2 + 6xv\right)$$
$$+ 2x\left(v'x^3 + 3vx^2\right) - 12vx^3 = 0$$
$$x^5 v'' + 8x^4 v' = 0$$
$$\frac{v''}{v'} = -\frac{8x^4}{x^5}$$
$$\frac{v''}{v'} = \frac{-8}{x}$$
$$\ln|v'| = -8\ln|x|$$

$$v' = x^{-8}$$
$$v = -\frac{1}{7}x^{-7}$$
$$y_2 = vx^3 = \left(-\frac{1}{7}x^{-7}\right)x^3 = -\frac{1}{7}x^{-4}$$
$$W\left(x^3, -\frac{1}{7}x^{-4}\right) = \begin{vmatrix} x^3 & -\frac{1}{7}x^{-4} \\ 3x^2 & \frac{4}{7}x^{-5} \end{vmatrix}$$
$$= x^{-2}$$
$$\neq 0$$

General solution:
$$y = C_1 x^3 + C_2 x^{-4}$$

For Problems 39-44, the spring constant is

$$k = \frac{16 \text{ lb}}{\frac{8}{12} \text{ ft}} = 24 \text{ lb/ft}$$

and the mass is

$$m = \frac{16 \text{ lb}}{32 \text{ ft/sec}^2} = \frac{1}{2} \text{ slug}$$

The governing equation has the form

$$\frac{1}{2}y'' + cy' + 24y = 0$$

39. With $y_0 = \frac{1}{2}$, $y_0' = v_0 = -8$, $c = 0$
(negative because velocity is upward)
$$y'' + 48y = 0$$
$$r^2 + 48 = 0$$
$$r = \pm 4\sqrt{3}i$$
General solution:
$$y = C_1\cos\left(4\sqrt{3}t\right) + C_2\sin\left(4\sqrt{3}t\right)$$
$$y' = v(t)$$
$$= -4\sqrt{3}C_1\sin\left(4\sqrt{3}t\right) + 4\sqrt{3}C_2\cos\left(4\sqrt{3}t\right)$$

Now, $y(0) = \frac{1}{2}$, so

$$\frac{1}{2} = C_1 \cdot 1 + C_2 \cdot 0$$

$$C_1 = \frac{1}{2}$$

$y'(0) = -8$, so

$$-8 = 4\sqrt{3}C_2 \cdot 1$$

$$C_2 = \frac{-2}{\sqrt{3}}$$

Solution:

$$y = \tfrac{1}{2}\cos\left(4\sqrt{3}t\right) - \tfrac{2}{\sqrt{3}}\sin\left(4\sqrt{3}t\right)$$

41. $\frac{1}{2}y'' = -24y$ with $y_0 = -\frac{8}{12} = -\frac{2}{3}$ ft, $y_0' = v_0 = 0$. Using y and y' in Problem 39,

$$y(0) = -\frac{2}{3}, \text{ so } C_1 = -\frac{2}{3}$$

$y'(0) = 0$, so

$$0 = 4\sqrt{3}C_2$$

$$C_2 = 0$$

Solution:

$$y = -\tfrac{2}{3}\cos(4\sqrt{3}t)$$

43. In this problem we have $c = 0.4$, so the governing equation is

$$0.5y'' + 0.4y' + 24y = 0$$

The characteristic equation is:

$$0.5r^2 + 0.4r + 24 = 0$$

$$r \approx -0.4 \pm 6.91665i$$

$$y \approx e^{-0.4t}[C_1\cos 6.9t + C_2\sin 6.9t]$$
$y_0 = 0.5$ so $C_1 = 0.5$

$$y' = -0.4e^{-0.4t}[C_1\cos 6.9t + C_2\sin 6.9t]$$
$$+ e^{-0.4t}[-6.9C_1\sin 6.9t + 6.9C_2\cos 6.9t]$$
Solution:

$$y = e^{-0.4t}[0.5\cos 6.9t + 0.03\sin 6.9t]$$

45. $y_1 = e^{ax}$; assume $y_2 = ve^{ax}$
$y_2' = v'e^{ax} + ave^{ax}$
$y_2'' = v''e^{ax} + 2av'e^{ax} + a^2ve^{ax}$
Substitute into the given D. E.:

$$\left(v'' + 2av' + a^2v\right)e^{ax} - 2a(v' + av)e^{ax} + bve^{ax} = 0$$

Hence,

$v'' = 0$ since $a^2 = b$
$v = x$ is one solution to $v'' = 0$ so $y_2 = xe^{ax}$

$$W(a^{ax}, xe^{ax}) = \begin{vmatrix} e^{ax} & xe^{ax} \\ ae^{ax} & (ax+1)e^{ax} \end{vmatrix}$$

$$= e^{2ax}$$

$$\neq 0$$

Thus, e^{ax} and xe^{ax} are linearly independent.

47. $y_1 = x^{-3/2}$, $y_1' = -\frac{3}{2}x^{-5/2}$,
$y_1'' = \frac{15}{4}x^{-7/2}$
$x^{-(3/2)}(15 - 18 + 3) = 0$, so we see that y_1 is a solution.

49. a. The mass is $m = \frac{100}{32}$; $v_0 = 150$, $s_0 = 0$

$$m\frac{dv}{dt} = -mg$$

$$\frac{ds}{dt} = v = -32t + 150 \qquad \textit{Since } v_0 = 150.$$

$$s = -16t^2 + 150t \qquad \textit{Since } s_0 = 0.$$
The object return to earth when
$s = 0$ or $t = \frac{150}{16} = \frac{75}{8}$ seconds

b.
$$m\frac{dv}{dt} = -\frac{1}{2}v - mg$$
$$\frac{dv}{dt} = -0.16v - 32$$
$$\int \frac{dv}{0.16v + 32} = -\int dt$$
$$v = -200 + Ce^{-0.16t}$$
$$v(0) = 150 = -200t + C, \text{ so } C = 350.$$

$$\frac{ds}{dt} = v = 200 + 350e^{-0.16t}$$
$$s(t) = -200t - \frac{350}{0.16t}e^{-0.16t} + C_1$$

$$0 = s(0) = -2{,}187.5 + C_1$$
$$C_1 = 2{,}187.5$$

$$s = -200t + 2{,}187.5\left(1 - e^{-0.16t}\right)$$

$s = 0$ when $t \approx 7.80$ seconds. It takes less time with air resistance.

51. a. $x = X + A$, $y = Y + B$ where
$$\begin{cases} aA + bB = -c \\ rA + sB = -t \end{cases}$$

Solving, we find
$$A = \frac{bt - cs}{as - br}, \quad B = \frac{at - cr}{br - as}$$

With this change the D. E. becomes
$$\frac{dY}{dX} = f\left(\frac{a(X + A) + b(Y + B) + c}{r(X + A) + s(Y + B) + t}\right)$$
$$= f\left(\frac{aX + bY + (aA + bB + c)}{rX + sY + (rA + sB + t)}\right)$$

$$= f\left(\frac{aX + bY}{rX + sY}\right)$$
$$= f\left(\frac{a + b\left(\frac{Y}{X}\right)}{r + s\left(\frac{Y}{X}\right)}\right)$$

This transformed D. E. is homogeneous.

b. $\dfrac{dy}{dx} = \dfrac{-3x + y + 2}{x + 3y + 5}$

Substitute $x = X + A$, $y = Y + B$.
Solve
$$\begin{cases} -3A + B + 2 = 0 \\ A + 3B - 5 = 0 \end{cases}$$

to find $A = \dfrac{11}{10}$, $B = \dfrac{13}{10}$.

$$\frac{dY}{dX} = \frac{-3X + Y + 0}{X + 3Y + 0} = \frac{-3 + \frac{Y}{X}}{1 + 3\left(\frac{Y}{X}\right)}$$

Let $f(v) = \dfrac{-3 + v}{1 + 3v}$ where $v = \dfrac{Y}{X}$.

$$\int \frac{dv}{\frac{-3+v}{1+3v} - v} = \int \frac{dX}{X}$$
$$\int \frac{(1 + 3v)\,dv}{3 + 3v^2} = -\int \frac{dX}{X}$$
$$\frac{1}{2}\ln(v^2 + 1) + \frac{1}{3}\tan^{-1}v = -\ln|X| + C_1$$
$$\ln\sqrt{\frac{Y^2}{X^2} + 1} + \frac{1}{3}\tan^{-1}\left(\frac{Y}{X}\right) = -\ln|X| + C_1$$

Now substitute
$$X = x - \frac{11}{12}, \quad Y = y - \frac{13}{10} \text{ to obtain}$$

$$\ln\sqrt{\left(x-\frac{11}{10}\right)^2+\left(y-\frac{13}{10}\right)^2}$$

$$+\frac{1}{3}\tan^{-1}\left(\frac{y-\frac{13}{10}}{x-\frac{11}{10}}\right)=C_1$$

$$\ln\sqrt{(10x-11)^2+(10y-13)^2}$$

$$+\frac{1}{3}\tan^{-1}\left(\frac{10y-13}{10x-11}\right)=C$$

53. The differential equation describing the motion is

$$mx''(t)+k_2x'(t)+k_1x=0$$

The characteristic equation is:

$$mr^2+k_2r+k_1=0$$

$$r=\frac{-k_2\pm\sqrt{k_2^2-4k_1m}}{2m}$$

For critical damping, $r_1=r_2=-\frac{k_2}{2m}$ with solution

$$x(t)=(C_1+C_2t)\exp\left[\left(\frac{-k_2}{2m}\right)t\right]$$

Suppose $x'(0)=0$ and $x(0)=x_0$.
$x_0=(C_1+0)\cdot1$ so $C_1=x_0$

$$x'(t)=C_1\exp\left[\left(\frac{-k_2}{2m}\right)t\right]\left(-\frac{k_2}{2m}\right)$$

$$+C_2\exp\left[\left(\frac{-k_2}{2m}\right)t\right]$$

$$+C_2t\exp\left[\left(\frac{-k_2}{2m}\right)t\right]\left(-\frac{k_2}{2m}\right)$$

If $x'(0)=0$, then

$$x'(0)=x_0\cdot1\left(-\frac{k_2}{2m}\right)+C_2\cdot1+0$$

$$C_2=\frac{k_2}{2m}x_0$$

$$x(t)=\left(1+\frac{k_2t}{2m}\right)x_0\exp\left[\left(\frac{-k_2}{2m}\right)t\right]$$

Maximum displacement occurs when $t=0$.
Next, suppose $x'(0)=v_0$ and $x(0)=0$.

$$x(t)=tC_2\exp\left[\left(\frac{-k_2}{2m}\right)t\right]$$

$$x'(t)=C_2\exp\left[\left(\frac{-k_2}{2m}\right)t\right]$$

$$+tC_2\exp\left[\left(\frac{-k_2}{2m}\right)t\right]\left(-\frac{k_2}{2m}\right)$$

The initial conditions imply $C_2=v_0$ so

$$x(t)=v_0+t\exp\left[\left(\frac{-k_2}{2m}\right)t\right]$$

Since

$$x'(t)=\frac{v_0}{m}\exp\left[\left(\frac{-k_2}{2m}\right)t\right]\left(m-\frac{1}{2}k_2t\right)$$

we see that initial velocity causes the maximum displacement from equilibrium to occur at $t=\frac{2m}{k_2}$.

55. $ay''+by'+cy=0$
The characteristic equation is:
$$ar^2+br+c=0$$
$$r=\frac{-b\pm\sqrt{b^2-4ac}}{2a}$$

Let $\alpha = -\dfrac{b}{2a}$, $\beta = \dfrac{\sqrt{4ac - b^2}}{2a}$;

$y_1 = e^{\alpha x} \cos \beta x$

$y_1' = e^{\alpha x}(\alpha \cos \beta x - \beta \sin \beta x)$

$y_1'' = e^{\alpha x}[(\alpha^2 - \beta^2)\cos \beta x + (-2\alpha\beta)\sin \beta x]$

Substitute into the original D. E.:

$ay_1'' + by_1' + cy_1$
$= ae^{\alpha x}[(\alpha^2 - \beta^2)\cos \beta x + (-2\alpha\beta)\sin \beta x]$
$\quad + be^{\alpha x}[\alpha \cos \beta x - \beta \sin \beta x] + ce^{\alpha x}\cos \beta x$
$= 0$

Thus, $y_1 = e^{\alpha x}\cos \beta x$ satisfies the differential equation and a similar computation shows that $y_2 = e^{\beta x}\sin \beta x$ is also a solution. Finally, to show that y_1 and y_2 are linearly independent, we compute the Wronskan

$W(e^{\alpha x}\cos \beta x, e^{\alpha x}\sin \beta x)$
$= e^{2\alpha x}[\beta(\cos^2 \beta x + \sin^2 \beta x)]$
$= \beta e^{2\alpha x}$
$\neq 0$

57. Assuming $\sin \theta \approx \theta$, using the results for Problem 56,

$$mL\theta'' + mg\theta = 0$$

$$\theta'' + \left(\frac{g}{L}\theta\right) = 0$$

Characteristic equation:

$$r^2 + \frac{g}{L} = 0$$

$$r = \pm\sqrt{\frac{g}{L}}i$$

$\theta(t) = C_1\cos\left(\sqrt{\frac{g}{L}}\right)t + C_2\sin\left(\sqrt{\frac{g}{L}}\right)t$

The equation is identical to the one in Problem 52 with

$$\frac{k}{m} = \frac{g}{L}$$

59. a. $m(t)s''(t) + m'(t)v_0 + m(t)g = 0$

The weight of the rocket and fuel at time t is $(w - rt)$ and its mass is

$$\frac{w - rt}{g} = m(t)$$

Thus,

$$s''(t) = -\frac{m'(t)}{m(t)}v_0 - g$$

$$= -\frac{\left(-\dfrac{r}{g}\right)}{\left(\dfrac{w-rt}{g}\right)}v_0 - g$$

$$= \frac{rv_0}{w - rt} - g$$

Integrating, we obtain

$$s'(t) = -v_0\ln|w - rt| - gt + C_1$$

and since $s'(0) = 0$, we have

$$s'(0) = -v_0\ln(w - 0) - g(0) + C_1$$

so $C_1 = v_0\ln w$. Thus,

$$s'(t) = -v_0\ln(w - rt) - gt + v_0\ln w$$

$$= -v_0\ln\left(\frac{w - rt}{w}\right) - gt$$

b. Integrating a second time, we find

$$s(t) = -v_0\left[t\ln\left(\frac{w-rt}{w}\right) - t\right.$$
$$\left. - \frac{w}{r}\ln(w-rt)\right] - \frac{gt^2}{2} + C_2$$

Since $s(0) = 0$,

$$0 = -v_0\left[0 - 0 - \frac{w}{r}\ln w\right] = 0 + C_2$$

so $C_2 = -\frac{v_0}{r}w\ln w$. Thus,

$$s(t) = -v_0\left[t\ln\left(\frac{w-rt}{w}\right) - t\right.$$
$$\left. - \frac{w}{r}\ln(w-rt) - \frac{gt^2}{2} - \frac{v_0}{r}w\ln w\right.$$
$$= \frac{v_0(w-rt)}{r}\ln\left(\frac{w-rt}{w}\right) - \frac{1}{2}gt^2 + v_0 t$$

c. The fuel is consumed when $rt = w_f$; that is, when $t = w_f/r$.

d. At the time when $t = w_f/r$, the height is

$$s\left(\frac{w_f}{r}\right) = \frac{v_0\left[w - r\left(\frac{w_f}{r}\right)\right]}{r}\ln\left[\frac{w - r\left(\frac{w_f}{r}\right)}{w}\right]$$
$$- \frac{1}{2}g\left(\frac{w_f}{r}\right)^2 + v_0\left(\frac{w_f}{r}\right)$$
$$= \frac{v_0(w-w_f)}{r}\ln\left(\frac{w-w_f}{w}\right)$$
$$- \frac{1}{2}\frac{gw_f^2}{r^2} + \frac{v_0 w_f}{r}$$

14.3 Second-Order Nonhomogeneous Differential Equations, page 1144

1. $y'' - 6y' = 0$
$$r^2 - 6r = 0$$
$$r = 0, 6$$
$$y_h = C_1 + C_2 e^{6x};\ F(x) = e^{2x}$$

$$\bar{y}_p = Ae^{2x}$$

3. $y'' + 6y' + 8y = 0$
$$r^2 + 6r + 8 = 0$$
$$r = -2, -4$$
$$y_h = C_1 e^{-2x} + C_2 e^{-4x};\ F(x) = 2 - e^{2x}$$
$$\bar{y}_p = A + Be^{2x}$$

5. $y'' + 2y' + 2y = 0$
$$r^2 + 2r + 2 = 0$$
$$r = -1 \pm i$$
$$y_h = e^{-x}(C_1\cos x + C_2\sin x);$$
$$F(x) = e^{-x}$$
$$\bar{y}_p = Ae^{-x}$$

7. $y'' + 2y' + 2y = e^{-x}\sin x$
See the solution to Problem 5.
$$y_h = e^{-x}(C_1\cos x + C_2\sin x);$$
$$F(x) = (\sin x)e^{-x}$$
$$\bar{y}_p = xe^{-x}(A\cos x + B\sin x)$$

9. $y'' + 4y' + 5y = 0$
$$r^2 + 4r + 5 = 0$$
$$r = -2 \pm i$$
$$y_h = e^{-2x}(C_1\cos x + C_2\sin x);$$
$$F(x) = e^{-2x}(x + \cos x)$$
$$\bar{y}_p = (A + Bx)e^{-2x}$$
$$+ xe^{-2x}(C\cos x + D\sin x)$$

For Problems 11-18, the characteristic equation is
$$r^2 + 6r + 9 = 0$$
$$(r+3)^2 = 0$$
$$r = -3\ (\text{mult. 2})$$
$$y_h = (C_1 + C_2 x)e^{-3x}.$$

11. $F(x) = 3x^3 - 5x$

$$\overline{y}_p = A_3 x^3 + A_2 x^2 + A_1 x + A_0$$

13. $F(x) = x^3 \cos x$
$$\overline{y}_p = (A_3 x^3 + A_2 x^2 + A_1 x + A_0)\cos x$$
$$+ (B_3 x^3 + B_2 x^2 + B_1 x + B_0)\sin x$$

15. $F(x) = e^{2x} + \cos 3x$
$$\overline{y}_p = A_0 e^{2x} + B_0 \cos 3x + C_0 \sin 3x$$

17. $F(x) = 4x^3 - x^2 + 5 - 3e^{-x}$
$$\overline{y}_p = A_3 x^3 + A_2 x^2 + A_1 x + A_0 + B_0 e^{-x}$$

19. $y'' + y' = 0$
Characteristic equation:

$$r^2 + r = 0$$
$$r(r + 1) = 0$$
$$r = 0, -1$$

$y_h = C_1 + C_2 e^{-x}$
$F(x) = -3x^2 + 7$
$$\overline{y}_p = A_2 x^3 + A_1 x^2 + A_0 x$$
$$\overline{y}_p' = 3A_2 x^2 + 2A_1 x + A_0$$
$$\overline{y}_p'' = 6A_2 x + 2A_1$$
Substitute into the given D. E.:

$$(6A_2 x + 2A_1) + (3A_2 x^2 + 2A_1 x + A_0)$$
$$= 3x^2 + 7$$

This gives rise to the system of equations

$$\begin{cases} 3A_2 = -3 \\ 2A_1 + 6A_2 = 0 \\ A_0 + 2A_1 = 7 \end{cases}$$

Solve to find: $A_0 = 1, A_1 = 3, A_2 = -1$.
Thus, $\overline{y}_p = -x^3 + 3x^2 + x$
Solution:

$$y = C_1 + C_2 e^{-x} - x^3 + 3x^2 + x$$

21. $y'' + 8y' + 15y = 0$
Characteristic equation:

$$r^2 + 8r + 15 = 0$$
$$(r + 3)(r + 5) = 0$$
$$r = -3, -5$$

$y_h = C_1 e^{-3x} + C_2 e^{-5x}$
$F(x) = 3e^{2x}$
$$\overline{y}_p = A e^{2x}$$
$$\overline{y}_p' = 2A e^{2x}$$
$$\overline{y}_p'' = 4A e^{2x}$$
Substitute into the given D. E.:

$$4A e^{2x} + 8(2A e^{2x}) + 15(A e^{2x}) = 3e^{2x}$$
$$A = \frac{3}{35}$$

Solution: $y = C_1 e^{-3x} + C_2 e^{-5x} + \frac{3}{35} e^{2x}$

23. $y'' + 2y' + 2y = 0$
Characteristic equation:

$$r^2 + 2r + 2 = 0$$
$$r = \frac{-2 \pm \sqrt{2^2 - 4(1)(2)}}{2}$$
$$r = -1 \pm i$$

$y_h = e^{-x}[C_1 \cos x + C_2 \sin x]$
$F(x) = \cos x$
$$\overline{y}_p = A \cos x + B \sin x$$
$$\overline{y}_p' = -A \sin x + B \cos x$$
$$\overline{y}_p'' = -A \cos x - B \sin x$$
Substitute into the given D. E.:

$$[-A \cos x + B \sin x] + 2[-A \sin x + B \cos x]$$
$$+ 2[A \cos x + B \sin x] = \cos x$$

This gives rise to the system of equations

$$\begin{cases} A + 2B = 1 \\ -2A + B = 0 \end{cases}$$

Page 603

which has solution $A = \frac{1}{5}$, $B = \frac{2}{5}$.
Solution:
$$y = e^{-x}(C_1\cos x + C_2\sin x)$$
$$+ \tfrac{1}{5}\cos x + \tfrac{2}{5}\sin x$$

25. $7y'' + 6y' - y = 0$
Characteristic equation:

$$7r^2 + 6r - 1 = 0$$
$$(r + 1)(7r - 1) = 0$$
$$r = -1, \frac{1}{7}$$

$y_h = C_1 e^{x/7} + C_2 e^{-x}$
$F(x) = e^{-x}(Ax + B)x$
$\bar{y}_p = e^{-x}(Ax^2 + Bx)$
$\bar{y}_p' = e^{-x}[-Ax^2 + (2A - B)x + B]$
$\bar{y}_p'' = e^{-x}[Ax^2 + (B - 4A)x + (2A - 2B)]$
Substitute into the given D. E.:

$$7e^{-x}[Ax^2 + (B - 4A)x + (2A - 2B)]$$
$$+ 6e^{-x}[-Ax^2 + (2A - B)x + B]$$
$$- e^{-x}[Ax^2 + Bx]$$
$$= e^{-x}(x + 1)$$

This gives rise to the system of equations
$$\begin{cases} -16A = 1 \\ 14A - 8B = 1 \end{cases}$$

which has solution $A = -\frac{1}{16}$, $B = -\frac{15}{64}$.
Solution:
$$y = C_1 e^{x/7} + C_2 e^{-x} + e^{-x}\left(-\tfrac{1}{16}x^2 - \tfrac{15}{64}x\right)$$
27. $\quad y'' - y' = 0$
Characteristic equation:

$$r^2 - r = 0$$
$$r(r - 1) = 0$$
$$r = 0, 1$$

$y_h = C_1 + C_2 e^x$
$F(x) = x^3 - x + 5$
$\quad \bar{y}_p = A_3 x^3 + A_2 x^2 + A_1 x + A_0$
Since we have $r = 0$, we multiply by x:
$\quad \bar{y}_p = A_3 x^4 + A_2 x^3 + A_1 x^2 + A_0 x$
$\quad \bar{y}_p' = 4A_3 x^3 + 3A_2 x^2 + 2A_1 x + A_0$
$\quad \bar{y}_p'' = 12A_3 x^2 + 6A_2 x + 2A_1$
Substitute into the given D. E.:

$$(12A_3 x^2 + 6A_2 x + 2A_1) - (4A_3 x^3 + 3A_2 x^2$$
$$+ 2A_1 x + A_0) = x^3 - x + 5$$

This gives rise to the system of equations

$$\begin{cases} -4A_3 = 1 \\ 12A_3 - 3A_2 = 0 \\ 6A_2 - 2A_1 = -1 \\ 2A_1 - A_0 = 5 \end{cases}$$

which has solution $A_0 = -10$, $A_1 = -\frac{5}{2}$,
$A_2 = -1$, $A_3 = -\frac{1}{4}$.
Solution:
$$y = C_1 + C_2 e^x - \left(\tfrac{1}{4}x^4 + x^3 + \tfrac{5}{2}x^2 + 10x\right)$$
29. $y'' + 2y' + y = 0$
Characteristic equation:

$$r^2 + 2r + 1 = 0$$
$$(r + 1)^2 = 0$$
$$r = -1 \quad \text{(mult. 2)}$$

$y_h = C_1 e^{-x} + C_2 x e^{-x}$
$F(x) = (4 + x)e^{-x}$
$\bar{y}_p = e^{-x}(Ax + B)x^2$

$\overline{y}'_p = e^{-x}[-Ax^3 + (3A - B)x^2 + 2Bx]$
$\overline{y}''_p = e^{-x}[Ax^3 - (6A - B)x^2 + (6A - 4B)x$
$\qquad\qquad + (6A - 4B)x + 2B]$

Substitute into the given D. E.:

$e^{-x}\left[Ax^3 - (6A - B)x^2 + (6A - 4B)x + 2B\right]$
$+ 2e^{-x}\left[-Ax^3 + (3A - B)x^2 + 2Bx\right]$
$+ e^{-x}\left(Ax^3 + Bx^2\right)$
$= (4 + x)e^{-x}$

This gives rise to the system

$$\begin{cases} 6A = 1 \\ 2B = 4 \end{cases}$$

which has solution: $A = \frac{1}{6}, B = 2$
Solution:

$y = C_1 e^{-x} + C_2 x e^{-x} + e^{-x}\left(\frac{1}{6}x^3 + 2x^2\right)$

$\quad = e^{-x}\left(C_1 + C_2 x + 2x^2 + \frac{1}{6}x^3\right)$

31. $y'' + y = 0;\ F(x) = \tan x$
Characteristic equation:

$$r^2 + 1 = 0$$
$$r = \pm i$$

$y_1 = \cos x,\ y_2 = \sin x$

$u' = \dfrac{-\sin x(\tan x)}{1} = -\dfrac{\sin^2 x}{\cos x}$

$u = \displaystyle\int \dfrac{-\sin^2 x}{\cos x}\,dx$

$\quad = -\ln|\sec x + \tan x| + \sin x$

$v' = \dfrac{\cos x(\tan x)}{1} = \sin x$

$v = \displaystyle\int \sin x\,dx$

$\quad = -\cos x$

$\overline{y}_p = uy_1 + vy_2$
$\quad = u\cos x + v\sin x$
$\quad = (-\ln|\sec x + \tan x| + \sin x)\cos x$
$\qquad + (-\cos x)\sin x$
$\quad = -\cos x \ln|\sec x + \tan x|$
Solution:
$y = C_1\cos x + C_2\sec x - \cos x \ln|\sec x + \tan x|$

33. $y'' - y' - 6y = 0;\ F(x) = x^2 e^{2x}$
Characteristic equation:

$$r^2 - r - 6 = 0$$
$$(r - 3)(r + 2) = 0$$
$$r = 3, -2$$

$y_1 = e^{3x},\ y_2 = e^{-2x}$
Let $D = y_1 y'_2 - y_2 y'_1 = -5e^x$.

$u' = \dfrac{-y_2 F(x)}{D} = \dfrac{-e^{-2x}(x^2 e^{2x})}{-5e^x} = \dfrac{1}{5}x^2 e^{-x}$

$u = \displaystyle\int \dfrac{1}{5}x^2 e^{-x}\,dx$

$\quad = -\dfrac{1}{5}\left(x^2 + 2x + 2\right)e^{-x}$

$v' = \dfrac{y_1 F(x)}{D} = \dfrac{e^{3x}(x^2 e^{2x})}{-5e^x} = -\dfrac{1}{5}x^2 e^{4x}$

$v = \displaystyle\int \left(-\dfrac{1}{5}\right)x^2 e^{4x}\,dx$

$\quad = -\dfrac{1}{160}\left(8x^2 - 4x + 1\right)e^{4x}$

$\bar{y}_p = uy_1 + vy_2$

$= \left[-\dfrac{1}{5}(x^2 + 2x + 2)e^{-x}\right]e^{3x}$

$\quad + \left[-\dfrac{1}{160}(8x^2 - 4x + 1)e^{4x}\right]e^{-2x}$

$= -\dfrac{1}{32}(8x^2 + 12x + 13)e^{2x}$

Solution:

$y = C_1 e^{3x} + C_2 e^{-2x}$
$\quad - \frac{1}{32}(8x^2 + 12x + 13)e^{2x}$

35. $y'' + 4y = 0$; $F(x) = \sec 2x \tan 2x$

Characteristic equation:

$r^2 + 4 = 0$

$r = \pm 2i$

$y_1 = \cos 2x$, $y_2 = \sin 2x$

Look for a particular solution of the form

$\bar{y}_p = uy_1 + vy_2 = u \cos 2x + v \sin 2x$

$u' = \dfrac{(-\sin 2x)(\sec 2x \tan 2x)}{2} = -\dfrac{1}{2}\tan^2 2x$

$u = \displaystyle\int \left(-\dfrac{1}{2}\tan^2 2x\right)dx$

$\quad = \dfrac{x}{2} - \dfrac{1}{4}\tan 2x$

$v' = \dfrac{\cos 2x (\sec 2x \tan 2x)}{2} = \dfrac{1}{2}\tan 2x$

$v = \displaystyle\int \dfrac{1}{2}\tan 2x\, dx$

$\quad = -\dfrac{1}{4}\ln|\cos 2x|$

Solution:

$y = C_1 \cos 2x + C_2 \sin 2x + \frac{1}{2}x \cos 2x$
$\quad - \frac{1}{4}\sin 2x(\ln|\cos 2x|)$

37. $y'' + 2y' + y = 0$; $F(x) = e^{-x}\ln x$

Characteristic equation:

$r^2 + 2r + 1 = 0$

$(r - 1)^2 = 0$

$r = 1 \quad \text{(mult. 2)}$

$y_1 = e^{-x}$, $y_2 = xe^{-x}$

Look for a particular solution of the form

$\bar{y}_p = uy_1 + vy_2 = ue^{-x} + vxe^{-x}$

$u' = \dfrac{-xe^{-x}e^{-x}(\ln x)}{e^{-x}(-x+1)e^{-x} - xe^{-x}(-e^{-x})}$

$\quad = -x \ln x$

$u = \displaystyle\int -x \ln x\, dx$

$\quad = \dfrac{x^2}{4} - \dfrac{x^2}{2}\ln x$

$v' = \dfrac{e^{-x}e^{-x}(\ln x)}{e^{-2x}}$

$\quad = \ln x$

$v = \displaystyle\int \ln x\, dx$

$\quad = x \ln x - x$

$\bar{y}_p = \left(\dfrac{x^2}{4} - \dfrac{x^2}{2}\ln x\right)e^{-x}$
$\quad + (x \ln x - x)xe^{-x}$

$\quad = \dfrac{1}{4}x^2(2\ln x - 3)e^{-x}$

Solution:

$y = C_1 e^{-x} + C_2 xe^{-x}$
$\quad + \frac{1}{4}x^2(2\ln x - 3)e^{-x}$

39. $y'' - y' = 0$; $F(x) = \cos^2 x$

Characteristic equation:

$r^2 - r = 0$

$r(r - 1) = 0$

$r = 0,\ 1$

$y_1 = 1$, $y_2 = e^x$

Look for a particular solution of the form
$$\overline{y}_p = uy_1 + vy_2 = u + ve^x.$$
$$u' = \frac{-e^x\cos^2 x}{(1)e^x - e^x(0)} = -\cos^2 x$$
$$u = \int \left(-\cos^2 x\right) dx$$
$$= -\frac{1}{4}\sin 2x - \frac{1}{2}x$$
$$v' = \frac{(1)\cos^2 x}{e^x} = e^{-x}\cos^2 x$$
$$v = \int e^{-x}\cos^2 x\, dx$$
$$= -\frac{1}{5}e^{-x}\left(\cos^2 x - \sin 2x + 2\right)$$
$$\overline{y}_p = (1)\left(-\frac{1}{4}\sin 2x - \frac{1}{2}x\right)$$
$$+ e^x\left[-\frac{1}{5}e^{-x}\left(\cos^2 x - \sin 2x + 2\right)\right]$$
$$= -\frac{1}{20}\sin 2x - \frac{1}{2}x - \frac{1}{5}\cos^2 x - \frac{2}{5}$$
Solution:
$$y = C_1 + C_2e^x - \tfrac{1}{20}\sin 2x - \tfrac{1}{2}x - \tfrac{1}{5}\cos^2 x$$

SURVIVAL HINT: *The term* $-\frac{2}{5}$ *in* y_p *is "absorbed" by the arbitrary constant* C_1.

---

41. From Problem 39,
$$y = C_1 + C_2e^x + 2\left(-\tfrac{1}{20}\sin 2x - \tfrac{1}{2}x - \tfrac{1}{5}\cos^2 x\right)$$

$$y(0) = C_1 + C_2 \cdot 1 + 0 + 0 - \tfrac{2}{5}$$
$$y'(x) = C_2e^x - \tfrac{1}{5}\cos 2x + \tfrac{2}{5}\sin 2x - 1$$
$$y'(0) = C_2 \cdot 1 - \tfrac{1}{5} \cdot 1 + 0 - 1$$
Solve the system:

$$\begin{cases} C_1 + C_2 = \dfrac{2}{5} \\[2mm] C_2 = \dfrac{6}{5} \end{cases}$$

which has solution $C_1 = -\frac{4}{5}$, $C_2 = \frac{6}{5}$.
Solution:
$$y = -\tfrac{4}{5} + \tfrac{6}{5}e^x - \tfrac{1}{10}\sin 2x - x - \tfrac{2}{5}\cos^2 x$$

43. $y'' + 9y = 0$; $F(x) = 4e^{3x}$;
Characteristic equation:

$$r^2 + 9 = 0$$
$$r = \pm 3i$$

$$y_h = C_1\cos 3x + C_2\sin 3x$$
$$\overline{y}_p = Ae^{3x}$$
$$\overline{y}_p' = 3Ae^{3x}$$
$$\overline{y}_p'' = 9Ae^{3x}$$
Substitute into the given D. E.:

$$9Ae^{3x} + 9Ae^{3x} = 4e^{3x}$$
$$A = \frac{2}{9}$$

General solution:
$$y = C_1\cos 3x + C_2\sin 3x + \tfrac{2}{9}e^{3x}$$
$$y(0) = C_1 \cdot 1 + C_2 \cdot 0 + \frac{2}{9} = 0$$
$$C_1 = -\frac{2}{9}$$

$$y' = \frac{2}{3}\sin 3x + 3C_2\cos 3x + \frac{2}{3}e^{3x}$$
$$y'(0) = 0 + 3C_2 \cdot 1 + \frac{2}{3} = 2$$
$$C_2 = \frac{4}{9}$$
Solution:

$$y = -\frac{2}{9}\cos 3x + \frac{4}{9}\sin 3x + \frac{2}{9}e^{3x}$$

45. $y'' + 9y = 0$; $F(x) = x$;
Characteristic equation:
$$r^2 + 9 = 0$$
$$r = \pm 3i$$

$y_h = C_1\cos 3x + C_2\sin 3x$
$\overline{y}_p = Ax + B$
$\overline{y}_p' = A$
$\overline{y}_p'' = 0$
Substitute into the given D. E.:
$$0 + 9(Ax + B) = 0$$

This gives rise to the system of equations
$$\begin{cases} 9A = 1 \\ 9B = 0 \end{cases}$$

which has solution $A = \frac{1}{9}$, $B = 0$.
Thus,
$y = C_1\cos 3x + C_2\sin 3x + \frac{1}{9}x$
$y' = -3C_1\sin 3x + 3C_2\cos 3x + \frac{1}{9}$
$y'(0) = 4 = 3C_2 + \frac{1}{9}$
$C_1 = 0$, $C_2 = \frac{35}{27}$
Solution: $y = \frac{35}{27}\sin 3x + \frac{1}{9}x$

47. $y'' - 4y' - 12y = 0$; $F(x) = 3e^{5x}$
Characteristic equation:
$$r^2 - 4r - 12 = 0$$
$$(r - 6)(r + 2) = 0$$
$$r = 6, -2$$
$y_h = C_1e^{-2x} + C_2e^{6x}$
$F(x) = 3e^{5x}$; $\overline{y}_p = Ae^{5x}$
$\overline{y}_p' = 5Ae^{5x}$; $\overline{y}_p'' = 25Ae^{5x}$

Substitute into the given D. E.:
$$25A - 20A - 12A = 3$$

which has solution $A = -\frac{3}{7}$.
General solution:
$y = C_1e^{-2x} + C_2e^{6x} - \frac{3}{7}e^{5x}$
$y' = -2C_1e^{-2x} + 6C_2e^{6x} - \frac{15}{7}e^{5x}$
$y(0) = C_1 + C_2 - \frac{3}{7} = \frac{18}{7}$
$y'(0) = -2C_1 + 6C_2 - \frac{15}{7} = -\frac{1}{7}$
which has solution $C_1 = 2$, $C_2 = 1$
Solution: $y = 2e^{-2x} + e^{6x} - \frac{3}{7}e^{5x}$

49. $F(x) = \begin{cases} 1 & \text{for } 0 \leq x < 1 \\ -\dfrac{1}{2}x + \dfrac{3}{2} & \text{for } 1 \leq x \leq 3 \end{cases}$

Characteristic equation:
$$r^2 + r - 6 = 0$$
$$(r - 2)(r + 3) = 0$$
$$r = 2, -3$$
$y_h = C_1e^{2x} + C_2e^{-3x}$

If $0 \leq x < 1$,
$\overline{y}_1 = A$; $-6A = 1$
$$A = -\frac{1}{6}$$
Solution: $\overline{y}_1 = -\frac{1}{6}$

If $1 \leq x \leq 3$,
$\overline{y}_2 = A_1x + A_2$
$\overline{y}_2' = A_1$
$\overline{y}_2'' = 0$

Substitute into the given D. E.:
$$A_1 - 6(A_1x + A_2) = -\frac{1}{2}x + \frac{3}{2}$$

This gives rise to the system of equations:

$$\begin{cases} A_1 - 6A_2 = \dfrac{3}{2} \\ -6A_1 = -\dfrac{1}{2} \end{cases}$$

which has solution: $A_1 = \frac{1}{12}$, $A_2 = -\frac{17}{72}$

Solution: $\bar{y}_2 = \frac{1}{12}x - \frac{17}{72}$

General solution:
$y = C_1 e^{2x} + C_2 e^{-3x} + G(x)$ where

$$G(x) = \begin{cases} -\dfrac{1}{6} & \text{for } 0 \le x < 1 \\ \dfrac{1}{12}x - \dfrac{17}{72} & \text{for } 1 \le x \le 3 \end{cases}$$

51.

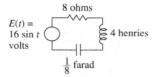

$I(0) = 0$; $I'(0) = 0$
The governing equation is:

$$L\frac{dI}{dt} + RI + \frac{Q}{C} = E$$

$$L\frac{d^2I}{dt^2} + R\frac{dI}{dt} + \frac{I}{C} = \frac{dE}{dt} \qquad \textit{Since } I = \frac{dQ}{dt}.$$

$$4\frac{d^2I}{dt^2} + 8\frac{dI}{dt} + \frac{I}{\frac{1}{8}} = 16\cos t$$

$$4I'' + 8I' + 8I = 16\cos t$$

Characteristic equation:

$$4r^2 + 8r + 8 = 0$$
$$r^2 + 2r + 2 = 0$$
$$r = \frac{-2 \pm \sqrt{4 - 4(1)(2)}}{2}$$
$$= -1 \pm i$$

$I_h(t) = e^{-t}[C_1\cos t + C_2\sin t]$
$\bar{y}_p = A_1\cos t + A_2\sin t$
$\bar{y}_p' = -A_1\sin t + A_2\cos t$
$\bar{y}_p'' = -A_1\cos t - A_2\sin t$

Substitute into the given D. E.:

$4(-A_1\cos t - A_2\sin t) + 8(-A_1\sin t$
$+ A_2\cos t) + 8(A_1\cos t + A_2\sin t)$
$= 16\cos t$

This gives rise to the system of equations

$$\begin{cases} 4A_1 + 8A_2 = 16 \\ -8A_1 + 4A_2 = 0 \end{cases}$$

which has solution: $A_1 = \frac{4}{5}$, $A_2 = \frac{8}{5}$

General solution:
$$I(t) = e^{-t}[C_1\cos t + C_2\sin t]$$
$$+ \frac{4}{5}\cos t + \frac{8}{5}\sin t$$

$$I(0) = 1 \cdot [C_1 \cdot 1 + 0] + \frac{4}{5} \cdot 1 + 0 = 0$$

$$C_1 = -\frac{4}{5}$$

$$I'(t) = -e^{-t}[C_1\cos t + C_2\sin t]$$
$$+ e^{-t}[-C_1\sin t + C_2\cos t]$$
$$- \frac{4}{5}\sin t + \frac{8}{5}\cos t$$

$$I'(0) = -1 \cdot [C_1 + 0] + 1 \cdot [0 + C_2]$$
$$-0 + \frac{8}{5} \cdot 1 = 0$$

This gives rise to the system of equations

$$\begin{cases} -C_1 + C_2 = -\dfrac{8}{5} \\ C_1 = -\dfrac{4}{5} \end{cases}$$

which has solution: $C_1 = -\frac{4}{5}$, $C_2 = -\frac{12}{5}$.

Solution:
$$I(t) = e^{-t}\left[-\frac{4}{5}\cos t - \frac{12}{5}\sin t\right]$$
$$+ \frac{4}{5}\cos t + \frac{8}{5}\sin t$$

53. The D. E. is $I'' + 3I' + 2I = 16$, since
$$\frac{dE}{dt} = 16.$$
Characteristic equation:
$$r^2 + 3r + 2 = 0$$
$$r = -1, -2$$

$I_h = C_1 e^{-t} + C_2 e^{-2t}$
$\bar{y}_p = A$. Substitute into the D. E. to obtain

$$0 + 0 + 2A = 16$$
$$A = 8$$

The general solution is
$I(t) = C_1 e^{-t} + C_2 e^{-2t} + 8$
$I(0) = C_1 + C_2 + 8 = 0$
$I'(t) = -C_1 e^{-t} - 2C_2 e^{-2t}$
$I'(0) = -C_1 - 2C_2 = 0$ which has
solution $C_1 = -16$, $C_2 = 8$.

Solution: $I(t) = -16e^{-t} + 8e^{-2t} + 8$

55. From Problem 54, $L = 1$ henry, $R = 10$ ohms, $C = \frac{1}{9}$ farad. For this problem, $E(t) = 5t\sin t$ so that
$$\frac{dE}{dt} = 5\sin t + 5t\cos t$$
We use the governing equation from the solution to Problem 51:

$$I'' + 10I' + 9I = 5\sin t + 5t\cos t$$

Characteristic equation:
$$r^2 + 10r + 9 = 0$$
$$(r+1)(r+9) = -1, 9$$

$I_h(t) = C_1 e^{-t} + C_2 e^{-9t}$
$\bar{y}_p = A_1\cos t + A_2\sin t$
$\quad + t(B_1\cos t + B_2\sin t)$

$\bar{y}_p' = -A_1\sin t + A_2\cos t + B_1\cos t$
$\quad + B_2\sin t + t(-B_1\sin t + B_2\cos t)$
$\quad = (A_2 + B_1)\cos t + (B_2 - A_1)\sin t$
$\quad + t(B_2\cos t - B_1\sin t)$

$\bar{y}_p'' = (2B_2 - A_1)\cos t + (-A_2 - 2B_1)\sin t$
$\quad + t(-B_2\sin t - B_1\cos t)$
Substitute into the given D. E.:

$[(2B_2 - A_1)\cos t + (-A_2 - 2B_1)\sin t]$
$+ t(-B_2\sin t - B_1\cos t)]$
$+ 10[(A_2 + B_1)\cos t + (B_2 - A_1)\sin t$
$+ t(B_2\cos t - B_1\sin t)$
$+ 9[A_1\cos t + A_2\sin t + t(B_1\cos t + B_2\sin t)]$
$= 5\sin t + 5t\cos t$
Equating the coefficients, we find

$$A_1 = \frac{-500}{1{,}681} \approx -0.297,$$

$$A_2 = \frac{-225}{3{,}362} \approx -0.067$$

$$B_1 = \frac{10}{41} \approx 0.244$$

$$B_2 = \frac{25}{82} \approx 0.305$$

$$I(t) \approx C_1 e^{-t} + C_2 e^{-9t}$$
$$- 0.297 \cos t - 0.067 \sin t$$
$$+ t(0.244 \cos t + 0.305 \sin t)$$

$$I(0) = C_1 + C_2 - 0.297 = 0$$

$$I'(t) = -C_1 e^{-t} - 9C_2 e^{-9t} + 0.177 \cos t$$
$$+ 0.602 \sin t + t(0.305 \cos t$$
$$- 0.244 \sin t)$$

$$I'(0) = 0 = -C_1 - 9C_2 + 0.177$$

This gives rise to the system of equations

$$\begin{cases} C_1 + C_2 = 0.297 \\ C_1 + 9C_2 = 0.177 \end{cases}$$

which has solution: $C_1 \approx 0.312$,
$C_2 \approx -0.015$
Solution:
$$I(t) \approx 0.312 e^{-t} - 0.015 e^{-9t}$$
$$- 0.297 \cos t - 0.067 \sin t$$
$$+ t(0.244 \cos t + 0.305 \sin t)$$

57. a. $x^2 y'' - 3xy' + 4y = 0$
$$y_1 = x^n$$
$$y_1' = n x^{n-1}$$
$$y_1'' = n(n-1) x^{n-2}$$

$$x^2 \left[n(n-1) x^{n-2} \right] - 3x \left[n x^{n-1} \right] + 4x^n = 0$$

Characteristic equation:

$$n(n-1) - 3n + 4 = 0$$
$$n^2 - 4n + 4 = 0$$
$$(n-2)^2 = 0$$
$$n = 2 \quad \text{(mult. 2)}$$

$y_1 = x^2$ is one solution; assume
$y_2 = vx^2$.
$$y_2' = v' x^2 + 2xv$$
$$y_2'' = v'' x^2 + 4xv' + 2v$$
$$x^2 \left[v'' x^2 + 4xv' + 2v \right]$$
$$- 3x \left[v' x^2 + 2xv \right] + 4v x^2 = 0$$
$$v'' x^2 + xv' = 0$$
$$\frac{v''}{v'} = -\frac{1}{x}$$
$$\ln|v'| = -\ln|x|$$
$$v' = x^{-1}$$
$$v = \ln|x|$$

$y_2 = x^2 \ln|x|$
The solutions $y_1 = x^2$ and $y_2 = x^2 \ln x$
are linearly independent since

$$W = y_1 y_2' - y_2 y_1' = x^3 \neq 0$$

b. We want a particular solution of

$$y'' - \frac{3}{x} y' + \frac{4}{x^2} y = \frac{\ln x}{x}$$

of the form $\overline{y}_p = u y_1 + v y_2$. The
division by x^2 is necessary because our
derivation of the formulas for variation
of parameters assumed an equation of
the form $y'' + a(x)y' + b(x)y = F(x)$.

$$u' = \frac{-(x^2\ln x)\left(\frac{\ln x}{x}\right)}{x^3} = \frac{-(\ln x)^2}{x^2}$$

$$u = \frac{1}{x}\left[(\ln x)^2 + 2\ln x + 2\right]$$

$$v' = \frac{x^2\left(\frac{\ln x}{x}\right)}{x^2} = \frac{\ln x}{x^2}$$

$$v = -\frac{1}{x}(\ln x + 1)$$

and

$$\overline{y}_p = \frac{1}{x}\left[(\ln x)^2 + 2\ln x + 2\right]x^2$$
$$+ \left[-\frac{1}{x}(\ln x + 1)\right](x^2\ln x)$$
$$= x(\ln x + 2)$$

The general solution is

$$y = C_1 x^2 + C^2 x^2 \ln x + x(\ln x + 2)$$

59. It is a first order separable differential equation. It is homogenous since

$$\frac{dy}{dx} = f\left(\frac{y}{x}\right)$$

where $f(u) = -u^{-1}$.

$$y' = -\frac{x}{y}$$

$$y\,dy = -x\,dx$$

$$\int y\,dy = \int (-x)\,dx$$

$$\frac{1}{2}y^2 + C_1 = -\frac{1}{2}x^2 + C_2$$

$$x^2 + y^2 = C$$

The orthogonal trajectories are $Y = CX$.

Chapter 14 Review

 Studying for a chapter examination is a personal process, one which nobody else can do for you. Simply take the time to review what you have done.

SURVIVAL HINT: Work all of Chapter 14 problems in the Proficiency Examination (whether they are assigned or not). Work through all of the problems before looking at the answers, and *then* correct each of the problems. The answers to all these problems are given in the answer section at the back of the text. If you worked the problem correctly, move on to the next problem, but if you did not work it correctly (or you did not know what to do), then look at the solutions below, look back in the chapter to study the procedure, or ask your instructor.

  Finally, go back over the homework problems you have been assigned. If you worked a problem correctly, move on to the next problem, but if you missed it on your homework, then you should look back in the book or talk to your instructor about how to work the problem.

  If you follow these steps, you should be successful with your review of this chapter.

Proficiency Examination, page 1146

1. A separable differential equation is one that can be written in the form

$$\frac{dy}{dx} = \frac{g(x)}{f(y)}$$

and then be solved by separating the

variables and integrating each side.

2. A differential equation of the form

$$M(x, y)\, dx + N(x, y)\, dy = 0$$

said to be a homogeneous differential equation if it can be written in the form

$$\frac{dy}{dx} = f\left(\frac{y}{x}\right)$$

In other words, dy/dx is isolated on one side of the equation and the other side can be expressed as a function of y/x.

3. A first-order linear differential equation is one of the form

$$\frac{dy}{dx} + p(x)y = q(x)$$

Its general solution is given by

$$y = \frac{1}{I(x)}\left[\int I(x)q(x)\, dx + C\right]$$

where $I(x)$ is the integrating factor

$$I(x) = e^{\int p(x)\, dx}$$

4. An exact differential equation is one that can be written in the general form

$$M(x, y)\, dx + N(x, y)\, dy = 0$$

where M and N are functions of x and y that satisfy the cross-derivative test

$$\frac{\partial M}{\partial y} = \frac{\partial N}{\partial x}$$

5. Euler's method is a procedure for approximating a solution of the initial value problem

$$\frac{dy}{dx} = f(x, y),\ y(x_0) = y_0$$

It depends on the fact that the portion of the solution curve near (x_n, y_n) is close to the line

$$y = y_n + f(x_n, y_n)(x - x_n)$$

6. The functions $y_1, y_2, \cdots, y_n$ are said to be *linearly independent* if the equation

$$C_1 y_1 + C_2 y_2 + \cdots + C_n y_n = 0$$

has only the trivial solution $C_1 = C_2 = \cdots = C_n = 0$. Otherwise the y_k's are *linearly dependent*.

7. The *Wronskian* $W(y_1, y_2, \cdots, y_n)$ of n functions $y_1, y_2, \cdots, y_n$ having $n - 1$ derivatives on an interval I is defined to be the determinant function

$$W(y_1, y_2, \cdots, y_n)$$
$$= \begin{vmatrix} y_1 & y_2 & \cdots & y_n \\ y_1' & y_2' & \cdots & y_n' \\ \vdots & \vdots & & \vdots \\ y_1^{(n-1)} & y_2^{(n-1)} & \cdots & y_n^{(n-1)} \end{vmatrix}$$

The functions $y_1, \cdots, y_n$ are linearly independent if and only if $W \neq 0$ on I.

8. **a.** The *characteristic equation* of

$$ay'' + by' + cy = 0$$

is the equation

$$ar^2 + br + c = 0$$

b. If r_1 and r_2 are the roots of the characteristic equation $ar^2 + br + c = 0$, then the general solution of the homogeneous linear equation $ay'' + by' + cy = 0$ can be expressed in one of these forms:

$b^2 - 4ac > 0$: The general solution is

$$y = C_1 e^{r_1 x} + C_2 e^{r_2 x}$$

$b^2 - 4ac = 0$: The general solution is

$$y = C_1 e^{-bx/2} + C_2 x e^{-bx/2}$$
$$= (C_1 + C_2 x)e^{-bx/2}$$

$b^2 - 4ac < 0$: The general solution is

$$y = e^{-bx/2}\left[C_1 \cos\left(\frac{\sqrt{4ac - b^2}}{2a}x\right)\right.$$
$$\left. + C_2 \sin\left(\frac{\sqrt{4ac - b^2}}{2a}x\right)\right]$$

9. Let y_p be a particular solution of the nonhomogeneous second-order linear equation $y'' + ay' + by = F(x)$. Let y_h be the general solution of the related homogeneous equation

$$y'' + ay' + by = 0$$

Then the general solution of

$$y'' + ay' + by = F(x)$$

is given by the sum $y = y_h + y_p$.

10. To solve $y'' + ay' + by = F(x)$ when $F(x)$ is one of the following forms:
   (1) $F(x) = P_n(x)$, a polynomial of degree n

(2) $F(x) = P_n(x)e^{kx}$
(3) $F(x) = e^{kx}[P_n(x)\cos \alpha x + Q_n(x)\sin \alpha x]$, where $Q_n(x)$ is another polynomial of degree n

Outline of the procedure:
1. The solution is of the form

$$y = y_h + y_p$$

where y_h is the general solution and y_p is a particular solution.

2. Find y_h by solving the homogeneous equation

$$y'' + ay' + by = 0$$

3. Find y_p by picking an appropriate trial solution $\overline{y}_p$:

   a. Form:
   $$P_n(x) = c_n x^n + \cdots + c_1 x + c_0$$
   Corresponding trial expression:
   $$A_n x^n + \cdots + A_1 x + A_0$$

   b. Form: $P_n(x)e^{kx}$
   Corresponding trial expression:
   $$[A_n x^n + \cdots + A_1 x + A_0]e^x$$

   c. Form:
   $$e^{kx}[P_n(x)\cos \alpha x + Q_n(x)\sin \alpha x]$$
   Corresponding trial expression:
   $$e^{kx}[(A_n x^n + \cdots + A_0)\cos \alpha x$$
   $$+ ([B_n x^n + \cdots + B_0])\sin \alpha x]$$

4. If no term in the trial expression $\overline{y}_p$ appears in the general homogeneous solution y_h, the particular solution

Page 614

can be found by substituting $\overline{y}_p$ into the equation

$$y'' + ay' + by = F(x)$$

and solving for the undetermined coefficients.

5. If any term in the trial expression $\overline{y}_p$, appears in y_h, multiply $\overline{y}_p$ by x^k, where k is the smallest integer such that no term in $x^k\overline{y}_p$ is in y_k. Then proceed as in Step 4, using $x^k\overline{y}_p$ as the trial solution.

11. To find the general solution of

$$y'' + P(x)y' + Q(x)y = F(x)$$

1. Find the general solution,

$$y_h = C_1y_1 + C_2y_2$$

to the related homogeneous equation

$$y'' + Py' + Qy = 0$$

2. Set $y_p = uy_1 + vy_2$ and substitute into the formulas:

$$u' = \frac{-y_2F(x)}{y_1y_2' - y_2y_1'} \qquad v' = \frac{y_1F(x)}{y_1y_2' - y_2y_1'}$$

3. Integrate u' and v' to find u and v.
4. A particular solution is

$$y_p = uy_1 + vy_2$$

and the general solution is

$$y = y_h + y_p$$

12. $$\frac{dy}{dx} = \sqrt{\frac{1 - y^2}{1 + x^2}}$$
$$\int \frac{dy}{\sqrt{1 - y^2}} = \int \frac{dx}{\sqrt{1 + x^2}}$$
$$\sin^{-1}y = \sinh^{-1}x + C$$

13. $$\frac{x}{y^2}\,dx - \frac{x^2}{y^3}\,dy = 0$$
$$\int x^{-1}\,dx = \int y^{-1}\,dy$$
$$\ln|x| = \ln|y| + C$$
$$y = Bx$$

14. $$\frac{dy}{dx} = \frac{2x + y}{3x} = \frac{2}{3} + \frac{y}{3x}$$
$$\frac{dy}{dx} - \frac{1}{3x}y = \frac{2}{3}$$
$$p(x) = -\frac{1}{3x}, \; q(x) = \frac{2}{3}$$
$$I(x) = e^{\int[-1/(3x)]\,dx} = x^{-1/3}$$
$$y = x^{1/3}\left[\int x^{-1/3}\left(\frac{2}{3}\right)dx + C\right]$$
$$= x^{1/3}\left[x^{2/3} + C\right]$$
$$= x + C\sqrt[3]{x}$$

15. $xy\,dy = (x^2 - y^2)\,dx$
$$\frac{dy}{dx} = \frac{x^2 - y^2}{xy} = \frac{x}{y} - \frac{y}{x}$$
Let $f(v) = v^{-1} - v$ where $v = \frac{y}{x}$.
$$\int \frac{dv}{\frac{1}{v} - v - v} = \int \frac{dx}{x}$$
$$\int \frac{v\,dv}{1 - 2v^2} = \int \frac{dx}{x}$$
$$-\frac{1}{4}\ln|1 - 2v^2| = \ln|x| + C_1$$
$$1 - 2v^2 = Cx^{-4}$$

$$y = C_1 x + C_2 \left[\frac{x}{2} \ln \left| \frac{x-1}{x+1} \right| + 1 \right]$$

45. Let $p = y'$. Substitute into $xy'' + 2y' = x$ to obtain

$$xp' + 2p = x$$

$$p' + \frac{2}{x} p = 1$$

$I(x) = e^{\int 2/x \, dx} = e^{2 \ln x} = x^2$. Thus,

$$y' = p = x^{-2} \left[\int x^2 (1) \, dx + C_1 \right]$$

$$= x^{-2} \left[\frac{x^3}{3} + C_1 \right]$$

$$= \frac{1}{3} x + \frac{C_1}{x^2}$$

and

$$y = \int \left(\frac{1}{3} x + \frac{C_1}{x^2} \right) dx$$

$$= \frac{1}{6} x^2 - \frac{C_1}{x} + C_2$$

47. $\dfrac{dx}{dy} - \dfrac{x}{y} = ye^y$

$$I(x) = e^{\int (-1/y) \, dy} = e^{-\ln y} = \frac{1}{y}$$

$$x = y \left[\int \frac{1}{y} (ye^y) \, dy + C \right]$$

$$= y[e^y + C]$$

49. Let the mass of the chain be m and let $x(t)$ be the length of the chain (in feet) which has moved over the peg at time t (in sec). At time t there are $(3 - x)$ ft on one side and $(7 + x)$ ft on the other. The excess

$(4 + 2x)$ ft on the "long" side produces an unbalanced force of

$$(4 + 2x) \frac{mg}{10} \text{ lb}$$

The governing equation is

$$m \frac{d^2 x}{dt^2} = (4 + 2x) \frac{mg}{10}$$

$$x'' - \frac{g}{5} x = \frac{2g}{5}$$

The characteristic equation:

$$r^2 - \frac{g}{5} = 0$$

$$r = \pm \sqrt{\frac{g}{5}} \approx \pm 2.53$$

$$x_h = C_1 e^{2.53t} + C_2 e^{-2.53t}$$
$$\bar{x}_p = A \text{ and}$$

$$-\frac{32}{5} A = 2 \left(\frac{32}{5} \right)$$

$$A = -2$$

General solution:

$$x(t) = C_1 e^{2.53t} + C_2 e^{-2.53t} - 2$$

The velocity of the chain is

$$x'(t) \approx 2.53 \left[C_1 e^{2.53t} + C_2 e^{-2.53t} \right]$$

When $t = 0$, $x = 0$, and $v = 0$, so

$$\begin{cases} C_1 + C_2 - 2 = 0 \\ 2.53(C_1 - C_2) = 0 \end{cases}$$

which has solution $C_1 = C_2 = 1$. Thus,

$$x(t) = e^{2.53t} + e^{-2.53t} - 2$$

$$t = \frac{1}{2.53} \ln \left[\frac{(x+2) + \sqrt{x^2 + 4x}}{2} \right]$$

The chain slides off the peg when $x = 3$;
that is

$$t = \frac{1}{2.55} \ln \left[\frac{(3+2) + \sqrt{3^2 + 4(3)}}{2} \right]$$

$$\approx 0.62 \text{ sec}$$

51. a. $EI \dfrac{d^2 y}{dx^2} = WLx - \dfrac{1}{2} Wx^2$

Integrate both sides, to obtain

$$EI \frac{dy}{dx} = \frac{WL}{2} x^2 - \frac{W}{6} x^3 + C_1$$

$$\int EI \, dy = \int \left(\frac{WL}{2} x^2 - \frac{W}{6} x^3 + C_1 \right) dx$$

$$EI y = \frac{WL}{2} \frac{x^3}{3} - \frac{W}{6} \frac{x^4}{4} + C_1 x + C_2$$

$$y(0) = 0, \; C_2 = 0; \; y(L) = 0$$

$$\frac{WL^4}{6} - \frac{W}{24} L^4 + C_1 L = 0$$

$$C_1 L = -\frac{WL^4}{8}$$

$$C_1 = -\frac{WL^3}{8}$$

Thus,

$$y(t) = \frac{WLx^3}{6EI} - \frac{Wx^4}{24EI} - \frac{WL^3 x}{8EI}$$

$$= \frac{Wx}{24EI} \left(4Lx^2 - x^3 - 3L^3 \right)$$

b. $\dfrac{dy}{dx} \dfrac{WL}{2EI} x^2 - \dfrac{W}{6EI} x^3 - \dfrac{WL^3}{8EI} = 0$

$$\frac{L}{2} x^2 - \frac{1}{6} x^3 - \frac{L^8}{8} = 0$$

Assume $x = mL$, then

$$\frac{m^2}{2} L^3 - \frac{m^3}{6} L^3 - \frac{L^3}{8} = 0$$

$$\frac{m^2}{2} - \frac{m^3}{6} - \frac{1}{8} = 0$$

$$12m^2 - 4m^3 - 3 = 0$$

Use technology to find $m \approx 0.554$. The
maximum deflection occurs at
$x \approx 0.554L$. Thus,

$$y_{\text{max}} = y(0.544L)$$

$$= \frac{WL^4}{EI} \left[\frac{(0.554)^3}{6} - \frac{(0.554)^4}{24} - \frac{0.554}{8} \right]$$

$$\approx \frac{WL^4}{EI} (-0.045)$$

53. a. $y = x^m; \; y' = mx^{m-1};$
$y'' = m(m-1)x^{m-2}$
Substitute into the homogeneous part of
the given D. E.

$$x^2 \left[m(m-1)x^{m-2} \right] + Ax \left[mx^{m-1} \right] + Bx^m = 0$$

$$[m(m-1) + Am + B]x^m = 0$$

$$m^2 + (A-1)m + B = 0$$

The discriminant for this equation is

$$D = (A-1)^2 - 4(1)B$$

b. If $D > 0$, there are 2 real roots. This
occurs when

$$(A-1)^2 > 4B$$

In this case, the characteristic equation
has two distinct real roots m_1, m_2, and
$y_1 = x^{m_1}$, $y_2 = x^{m_2}$ are linearly
independent solutions of the Euler
equation. The general equation is

$$y = C_1 x^{m_1} + C_2 x^{m_2}$$

c. If $D = 0$, the real root is repeated. This occurs when

$$(A-1)^2 = 4B$$

In this case, $m_0 = \frac{1}{2}(1-A)$ is a repeated root. One solution of the Euler equation is $y_1 = x^{m_0}$. A second solution y_2 can be found by reduction of order.

$$y_2 = vx^{m_0}$$
$$y_2' = v'x^{m_0} + m_0 vx^{m_0-1}$$
$$y_2'' = v''x^{m_0} + 2m_0 v'x^{m_0-2}$$

Substituting into the Euler equation, we obtain

$$x^2[v''x^{m_0} + 2m_0 v'x^{m-1}$$
$$+ m_0(m_0-1)vx^{m_0-1}] + Bvx^{m_0} = 0$$
$$x^{m_0+2}v'' + [2m_0 + A]x^{m_0+1}v'$$
$$+ [m_0(m_0-1) + Am_0 + B]x^{m_0}v = 0$$

Since $m_0(m_0-1) + Am_0 + B = 0$, *we have*

$$xv'' + [2m_0 + A]v' = 0$$
$$\frac{v''}{v'} = -\frac{(2m_0+A)}{x}$$
$$= -\frac{1}{x} \quad \text{since } m_0 = \frac{1}{2}(1-A)$$
$$\ln|v'| = -\ln|x|$$
$$v' = \frac{1}{x}$$
$$v = \ln|x|$$

Thus, $y_2 = x^{m_0}\ln x$ so the general solution is

$$y = C_1 x^{m_0} + C_2 x^{m_0}\ln|x|$$

d. If $D < 0$, there are complex conjugate roots. This occurs when
$$(A-1)^2 < 4B y_1 = x^\alpha \cos(\beta \ln|x|)$$

where $\alpha = \dfrac{1-A}{2}$ and
$\beta = \frac{1}{2}\sqrt{4B - (A-1)^2}$. Find
$$y_1' = \alpha x^{\alpha-1}\cos(\beta \ln|x|) - \beta x^{\alpha-1}\sin(\beta \ln|x|)$$
$$y_1'' = x^{\alpha-2}[\alpha^2 - \alpha - \beta^2]\cos(\beta \ln|x|)$$
$$- x^{\alpha-2}\beta(2\alpha - 1)\sin(\beta \ln|x|)$$

Substitute these values into the Euler equation. Reduction of order yields a second solution of the form
$$y_2 = vx^\alpha \cos(\beta \ln|x|)$$
where
$$v = \tan(\beta \ln|x|)$$

Thus, the general solution is
$$y = x^\alpha[C_1\cos(\beta \ln|x|) + C_2\sin(\beta \ln|x|)]$$

55.
$$\frac{dx}{dt} = a_{11}x - a_{12}y$$
$$\frac{dy}{dt} = a_{21}x - a_{22}y$$

Differentiate the first equation with respect to t, then substitute for dy/dt from the second equation:
$$\frac{d^2x}{dt^2} = a_{11}\frac{dx}{dt} - a_{12}\frac{dy}{dt}$$
$$= a_{11}\frac{dx}{dt} - a_{12}\underbrace{[a_{21}x - a_{22}y]}_{\text{Second equation}}$$

Now substitute for y using the first equation:
$$\frac{d^2x}{dt^2} - a_{11}\frac{dx}{dt} + a_{12}a_{21}x$$
$$= a_{12}a_{22}y$$
$$= a_{22}\underbrace{\left[a_{11}x - \frac{dx}{dt}\right]}_{\text{First equation}}$$

$$\frac{d^2x}{dt^2} + (a_{22} - a_{11})\frac{dx}{dt}$$
$$+ (a_{12}a_{21} - a_{11}a_{22})x = 0$$

Solve this second order linear differential equation to obtain $x(t)$. Then (from the first equation),

$$y(t) = \frac{1}{a_{12}}\left[a_{11}x(t) - \frac{dx}{dt}\right]$$

57. This is Problem 9 of the afternoon session of the 1938 Putnam examination. Let $v = y'$, $y'' = v\dfrac{dv}{dy}$. Then,

$$yy'' - 2(y')^2 = 0$$

$$vy\frac{dv}{dy} - 2v^2 = 0$$

$$\int \frac{dv}{v} = \int \frac{2\,dy}{y}$$

$$\ln|v| = 2\ln|y| + C_1$$

$$v = Cy^2$$

$$\frac{dy}{dx} = Cy^2 \qquad \text{\textit{since } } v = \frac{dy}{dx}$$

$$\int \frac{dy}{y^2} = \int C\,dx$$

$$-\frac{1}{y} = Cx + D$$

Since $y = 1$ when $x = 1$, we have

$$-1 = C + D$$
$$D = -1 - C$$

and

$$y = \frac{1}{1 + C(1 - x)}$$

59. This is Problem 6ii from the morning session of the 1948 Putnam examination. Let

$$f(x) = x + \tfrac{2}{3}x^3 + \tfrac{2\cdot4}{3\cdot5}x^5 + \tfrac{2\cdot4\cdot6}{3\cdot5\cdot7}x^7 + \cdots$$

Then,

$$f'(x) = 1 + 2x^2 + \frac{2}{3}(4)x^4 + \frac{2\cdot4}{3\cdot5}(6)x^6 + \cdots$$

$$= 1 + x\frac{d}{dx}\left[x^2 + \frac{2}{3}x^4 + \frac{2\cdot4}{3\cdot5}x^6 + \cdots\right]$$

$$= 1 + x\frac{d}{dx}[x\,f(x)]$$

$$= 1 + xf(x) + x^2f'(x)$$

Thus, we have

$$(1 - x^2)f'(x) = 1 + xf(x)$$

$$f'(x) - \frac{x}{1 - x^2}f(x) = \frac{1}{1 - x^2}$$

The integrating factor is

$$I = e^{\int[-x/(1-x^2)]dx}$$

$$= e^{\ln(1-x^2)/2}$$

$$= \sqrt{1 - x^2}$$

so

$$f(x) = \frac{1}{\sqrt{1 - x^2}}\left[\int \sqrt{1 - x^2}\,dx + C\right]$$

$$= \frac{1}{\sqrt{1 - x^2}}\left[\frac{x}{2}\sqrt{1 - x^2} + \frac{1}{2}\sin^{-1}x\right] + C$$

Since $f(0) = 0$, we have

$$f(0) = \frac{1}{2}|0 + C| = 0$$

so $C = 0$ and

$$f(x) = \frac{1}{\sqrt{1 - x^2}}\left[\frac{x}{2}\sqrt{1 - x^2} + \frac{1}{2}\sin^{-1}x\right]$$

$$= \frac{x}{2} + \frac{\sin^{-1}x}{\sqrt{1 - x^2}}$$